FOUNDATIONS OF MODERN GLOBAL
SEISMOLOGY

FOUNDATIONS OF MODERN GLOBAL SEISMOLOGY

SECOND EDITION

CHARLES J. AMMON
Department of Geosciences
The Pennsylvania State University
University Park, PA, United States

AARON A. VELASCO
Department of Geological Sciences
University of Texas at El Paso
El Paso, TX, United States

THORNE LAY
Earth & Planetary Sciences
University of California Santa Cruz
Santa Cruz, CA, United States

TERRY C. WALLACE
Los Alamos National Laboratory
Los Alamos, NM, United States

ACADEMIC PRESS

An imprint of Elsevier

ELSEVIER

Academic Press is an imprint of Elsevier
125 London Wall, London EC2Y 5AS, United Kingdom
525 B Street, Suite 1650, San Diego, CA 92101, United States
50 Hampshire Street, 5th Floor, Cambridge, MA 02139, United States
The Boulevard, Langford Lane, Kidlington, Oxford OX5 1GB, United Kingdom

Notices

Knowledge and best practice in this field are constantly changing. As new research and experience broaden our understanding, changes in research methods, professional practices, or medical treatment may become necessary.

Practitioners and researchers must always rely on their own experience and knowledge in evaluating and using any information, methods, compounds, or experiments described herein. In using such information or methods they should be mindful of their own safety and the safety of others, including parties for whom they have a professional responsibility.

To the fullest extent of the law, neither the Publisher nor the authors, contributors, or editors, assume any liability for any injury and/or damage to persons or property as a matter of products liability, negligence or otherwise, or from any use or operation of any methods, products, instructions, or ideas contained in the material herein.

Library of Congress Cataloging-in-Publication Data
A catalog record for this book is available from the Library of Congress

British Library Cataloguing-in-Publication Data
A catalogue record for this book is available from the British Library

ISBN: 978-0-12-815679-7

For information on all Academic Press publications
visit our website at https://www.elsevier.com/books-and-journals

Publisher: Candice Janco
Acquisitions Editor: Amy Shapiro
Editorial Project Manager: Lindsay Lawrence
Production Project Manager: Sruthi Satheesh
Designer: Miles Hitchen

Typeset by VTeX

We Dedicate This Book to

Our Families

For Their Love, Patience, and Support

Contents

9. Tsunami and tsunami warning

10. Earth structure

II
THEORETICAL FOUNDATIONS

11. Elasticity and seismic waves

12. Body waves and ray theory – travel times

13. Body-waves and ray theory – amplitudes

20. Imaging Earth's interior

Preface

This textbook is intended for upper-division undergraduate and first-year graduate survey courses in seismology. Students using the book should be familiar with basic calculus, complex numbers, and differential equations and have some elementary knowledge of geology. The focus is on the fundamental physics of seismic waves and the application of this understanding to extract information about Earth's internal structure and dynamic processes that is contained in seismograms, instrumental recordings of the planet's vibrations. Most of the text is developed in the context of a global seismology perspective, meaning large-scale Earth structure and earthquake source processes. However, the principles underlying elastic-wave propagation, seismic instrumentation, and techniques for extracting Earth structure and source information from seismograms are common to applications in exploration, mining, and explosion seismology. These disciplines use seismic waves to develop high-resolution images of the upper crust for oil and mineral resource exploration, and to monitor seismic events associate with mining, and industrial as well as underground nuclear explosions. The basic principles are in no way restricted to the Earth, and in the future they will, we hope, be applied to many other celestial bodies (exciting work has already been performed on the Sun, Moon, and Mars).

This text is a substantial modification of the first edition, which was derived from class notes for introductory seismology courses taught by coauthors at the University of Michigan, the University of California at Santa Cruz (T.L.), and the University of Arizona (T.C.W.). Those class notes, in turn, have a complex legacy, in large part tracing back to lecture notes of Professor Hiroo Kanamori at the Seismological Laboratory of the California Institute of Technology, who taught an inspirational introductory seismology course to the original coauthors. Other material was drawn from numerous introductory geophysics textbooks and current research publications. The revision includes a substantial reorganization reflecting more recent teaching experience of the new coauthors, who for the last few decades have taught similar courses at Saint Louis University and Penn State (C.J.A.), and the University of Texas El Paso (A.V.). Material has been freely shifted between chapters, and chapters have been shortened to provide more convenient topical choices to instructors. The revision also includes some broadening of background material and topical updates that reflect the vigorous research environment that has pervaded earthquake seismology for the last 25 years.

Even in revision, the effort to distill a thorough, yet accessible, introductory survey of global seismology involves many compromises, particularly in abbreviated treatment of such topics as transient wave solutions, synthetic seismogram calculation, and rock mechanics. We have added a limited amount of material related to the societal aspects of earthquakes, mostly because for obvious reasons, students are interested and engaged by such discussions. Fortunately, the discipline is fully spanned by several advanced theoretical seismology texts (some classic, some revised classics, and some new). These include Quantitative Seismology (Aki and Richards, 2009), Seismic Waves and Sources (Ben-Menahem and Singh, 1981), Theoretical Global Seismology (Dahlen and Tromp, 1998), Seismic Wave Propagation in Stratified Media (Kennett, 2009), Fundamentals of Seismic Wave Propagation (Chapman, 2004). Classic and updated elementary earthquake overviews are also readily accessible (these days, often electronically). These include Earthquakes (Bolt 2006), Inside the Earth (Bolt, 1982), Elementary Seismology (Richter, 1958), Nuclear Explosions and Earthquakes (Bolt, 1976); and fracture mechanics textbooks: Earthquake Mechanics (Kasahara, 1981), The Mechanics of Earthquakes and Faulting (Scholz, 2019), Principles of Earthquake Source Mechanics (Kostrov and Das, 1988). Seismology survey texts that offer alternate presentations of the material in this text are also available. These include Introduction to Seismology (Båth, 1979), and Plate Tectonics (Gubbins, 1990), An Introduction to the Theory of Seismology (Bullen and Bolt, 1985), Introduction to Seismology, Earthquakes and Earth Structure (Stein and Wysession, 2003), Principles of Seismology (Udías and Buforn, 2018), and Introduction to Seismology (Shearer, 2019). Many additional texts reviewing the various fields of solid Earth geophysics provide additional resources.

Perhaps the most important resource is no longer books – online information is generally less curated, but the immediate access to information about earthquakes and the Earth is an incredible resource for instructors and students. The international community of earthquake scientists has increased the amount of high-quality seismic instruments on as much of the planet as possible for as long as possible to gather data central to observational seismology earthquake science. Others using physical and numerical laboratories have labored creatively to provide measurements and calculations that facilitate the understanding of those observations. These same communities have worked very hard for decades to establish relatively easy-to-use, responsive, and reliable data archives and data access methods. Much of the seismological investment was facilitated through the Incorporated Research Institutions for Seismology (IRIS) community, corporation, and facility – which with a talented dedicated staff, helped by a community governance, has supported dynamic research within the community. Modern seismologists owe much to the previous generation of seismologists who conceived and established IRIS.

Like most modern scientific disciplines, global seismology advances quickly, and any text can at best give an instantaneous and limited version of our knowledge of Earth structure and earthquakes. Students and other readers must strive independently to stay current as new discoveries are made. We have tried to provide a book that diligent instructors can use to help inspire students to invest the hard work to learning the fundamentals and to staying current. We are indebted to the students who have endured preliminary versions of this material and invite them to discard their early versions and replace them with this updated and improved presentation. In this regard we particularly thank Chengping Chai, Jonas Kintner, Chanel Deane, Azucena Zamora, Richard Alfaro-Diaz, Mariana Benitez, Frankie Enriquez, Sandra Harding, and Mohan Pant, along with John Ebel, Paul Richards, and Susan Schwartz, who provided helpful comments and corrections on the first edition. Finally, we especially thank our families for their endless patience as we worked on this revision.

The overall structure of the revised text

In our years of teaching, we have accepted several trends that motivated a substantial restructuring of the material. When facing a geoscience class of varying quantitative skills (many programs no longer even require multi-variable calculus) starting with an arrangement that begins with a rigorous introduction of continuum mechanics is problematic. Our pragmatic solution has been to separate the material into two parts, the first focused more on observations, the second on foundations. In particular, we moved the most derivation-laden topics to later in the book. After an initial overview chapter, we provide a semi-quantitative introduction to the mechanics of earthquakes and seismic waves, introducing topics that interest students but are difficult to reach if one waits until all the background material is covered in depth (magnitude scales were detailed in Chapter 9 of the first edition, tectonic patterns were described in Chapter 11). We stress the importance of continuum mechanics as the foundation used to understand earthquakes and deformation and the roles of elastic and non-elastic models in seismology and earthquake science, but we do not start with the details. Thus, **Part I** is more introductory in nature, focused on observational seismology, and can be taught in one semester of study. **Part II** provides students a theoretical foundation in seismology, from first principles to modeling sources and estimating earth structure. To make the book a more flexible teaching tool, we have re-organized material into shorter chapters that may be ignored or included at the comfort of the instructor. Although some cross-referencing is inevitable, we believe that the design allows more educational elasticity and the potential for a variety of paths through the topic.

Part I – Observational foundations of global seismology

Part I begins with **Chapter 1**, an overview defining some of the history and the scope of global seismology. **Chapter 2** is an introduction to the concepts of continuum mechanics including stress, strain, elasticity, and the mechanical foundations underlying the study of earthquakes and seismic waves. We introduce the conceptual physics underlying fault rupture (friction) and the observational framework used to characterize earthquake ruptures. We discuss how seismic waves are conceptually discussed as elastic waves, but in reality are nearly-elastic waves. **Chapter 3** is a review of *seismotectonics* and a description of how seismology provides information about active, ongoing deformation processes in the Earth. Quantification of earthquake faulting characteristics such as fault orientation, sense of slip, and cumulative displacement played a major role in the evolution

of the theory of plate tectonics and together with satellite-based geodesy, continue to provide valuable information on the first-order tectonic force systems. In parallel with the rapid advances in our knowledge of Earth structure has come a comparable expansion of our understanding of earthquake faulting and its role in global dynamics, including plate tectonics and lithospheric deformation.

Chapter 4 is a description of how we measure ground motion as a function of position and time, or *seismograms*, which are the basic data used to study seismic-wave excitation and propagation. A major challenge posed by the substantial frequency range (bandwidth) and amplitude range (dynamic range) of seismic observations has been to design and to build sensors and recording systems capable of registering all useful signals against a continuous background of ambient motion produced by earth, atmospheric, and oceanic coupling and human activity. No single instrument records the full spectrum of motions with a linear response, so a suite of instruments that record portions of the seismic spectrum have been developed and deployed. However, technological advances have enabled the development of seismic recording systems that provide remarkable bandwidth and dynamic range for global seismological investigations. The chapter is a review of the remarkable instrument technology involved in the field of *seismometry*, or recording of ground motion.

Chapter 5 is an introduction to seismogram interpretation. We use seismogram-based observations to quantify seismic sources (location, size, rupture history), one of the most important tasks in observational seismology. More practically, seismology is easier to learn once you "know your way around a seismogram". Two of the earliest fundamental problems in seismogram analysis were how to use them to locate and compare the size of earthquakes. An earthquake location is specified using the latitude (or colatitude), longitude, and depth, and the time of an earthquake's *initiation*. Accurately estimating earthquake locations requires adequate observations of first and secondary arrivals that azimuthally surround a source, adequate signal-to-noise for accurate timing, and an appropriate velocity model. The size of an earthquake can be estimated by measuring amplitude of seismic waves, and which again, should be calibrated to a region. We review location and magnitude estimation in **Chapters 6 and 7**.

Although we still cannot predict the time, location, or size of impending earthquakes, early warning systems take advantage of the nature of the seismic wavefield to estimate earthquake source parameters using the first-arriving seismic arrivals (P waves) in order to warn of impending ground shaking produced by secondary arrivals (S and surface waves). These topics and earthquake forecasting are reviewed in **Chapter 8**. Tsunami, large waves created by large subduction zone earthquakes can have great societal impact. The 2004 Andaman Island (M_W 9.2) earthquake killed over 225,000 people. We review characteristics of tsunami in **Chapter 9**.

Seismological investigations provide the highest resolution of Earth's internal structure of any geophysical observation because of all geophysical fields (gravitational, magnetic, or thermal), seismic waves have the shortest wavelengths that propagate to significant distance. Seismic ground motions are also relatively easy to observe. The fundamental simplicity of elastic waves, which transmit disturbances over great distances through the Earth with little, or mostly predictable distortion, allows useful information to be extracted from seismograms. Most of the ground motions can be interpreted quantitatively given our current knowledge of Earth's interior, which is reviewed in **Chapter 10**.

Part II: Theoretical foundations of seismology

To understand seismological observations deeply, we must develop a theoretical foundation based on continuum physics, which is the focus of **Part II**. Seismic waves are a consequence of the physics of (nearly) elastic solids, which is fully described by the theory of elastodynamics. The theoretical foundation of seismic wave propagation is founded on continuum mechanics, linear elasticity, and applied mathematics, introduced in **Chapter 11**. Application of these ideas to body wave propagation in the Earth (P- and S-waves) is reviewed in **Chapters 12 and 13**. Application of the ideas to surface waves and free oscillations are discussed in **Chapters 14** and **15**.

We describe quantitative representations of seismic sources, focusing on shear-faulting, in **Chapters 16, 17 and 18**. The chapters start with point sources and point forces and expand to the kinematic and dynamic characteristics of earthquakes, providing a physical basis for understanding fundamental seismological concepts such as magnitude, moment, radiation patterns, and rupture directivity. Combining our knowledge of Earth structure and our models of seismic sources allows us to compute predicted ground motions to compare with observations. This comparison serves as a basis for seismic estimation of earthquake faulting parameters. This capability has led to an appreciation that faulting is a heterogeneous process with nonuniform stress release over the fault surface. **Chapter 19** is a review of classic source analysis procedures common in global earthquake seismology.

Since we are able to record seismic-wave motions only at, or very near, Earth's surface, seismology draws heavily upon mathematical methodologies for solving systems of equations that are collectively described as *geophysical inverse theory*. Many seismological applications and results of inverse theory are described in this text. The essence of all seismic inverse problems is that inferences about the source or the Earth are made by applying mathematical operations derived from elastodynamic theory to the observed ground motions. The ground motions can be viewed as the output of a sequence of linear filters with properties that we wish to determine. We can treat instrument, propagation, and source excitation effects as separate filters, and we have concentrated parts of the text on each factor that shapes an observed seismogram. Inversions may suffer from uniqueness problems, and strong, difficult-to-resolve trade-offs often exist between source and propagation effects. The history of seismological advances is one of alternating progress in characterizing source properties or in improving models of Earth structure, and clever strategies have been employed to overcome the intrinsic trade-offs in the signal analysis. Impressive resolution of Earth structure is now achieved using analysis methods and data, some classic approaches are introduced in **Chapter 20**.

When we began the revision, we planned to include problems at the end of every chapter. Two realities changed that plan. First, the extra space required to include *interesting* problems based on real data was extensive and would have added to the cost of what is already a significant financial investment (consolidation of references was driven by the same issue). Second, it is impractical to include a significant number of lists of measurements or plots of seismograms, that soon become stale. Modern seismological analysis is executed on computers, where the data storage is not an issue and displays are flexible and dynamic. Designing engaging observation-based problems is an enterprise for professors. So we leave the exercises for the educators, who also know their students better than textbook writers and who can include observations from current earthquakes and seismic stations that better engage the students.

PART I

Observational foundations

1

An overview of global seismology

Chapter goals

- Introduce global seismology as a field of study.
- Introduce seismograms as the observational foundation of global seismology.
- Provide a historical context for seismology.
- Define classical and current research objectives.
- Introduce earthquake source and Earth structure studies.

The Earth is composed of silicate and iron-alloy materials that remarkably, over the wide range of pressure and temperature conditions within the planet, respond nearly elastically to small-amplitude transient forces and viscously to slowly-changing convective forces. The Earth "rings like a bell" in response to the sudden slip of a large earthquake or the detonation of a large explosion, while at the same time convection-driven flow reshapes the planet's surface and cools the interior over longer, geological, time scales. The planet's quasi-elastic vibrations result from the excitation and propagation of elastic (seismic) waves in the interior. *Seismic* waves induce ground motions that *seismographs* record and preserve for scientific analysis. In this text, we describe the excitation and propagation of seismic waves and explore methods used to extract information from their recordings. We review how the waves illuminate the planet's interior and how we can use that to help constrain the processes at work in the global dynamic system. We hope that the text also provides insight into the processes that produce destructive earthquakes, such as the January 17, 1994, Northridge, California event, which killed more than 50 people and caused more than $40 billion in damage to Los Angeles, and the catastrophic 12 January, 2010 Haiti earthquake that killed more than 90,000 and perhaps injured more than 300,000 people. Seismology has always had a role on the world stage whenever civilization has experienced a catastrophic earthquake.

The precise definition of seismology has evolved over time; it is often defined as the study of earthquakes. We adopt a simple, but broader definition: *seismology* is the quantitative study of the excitation, propagation, and recording of the vibrations of the Earth and other celestial bodies. Our definition avoids implying that only seismologists study earthquakes – earthquake science is a broad interdisciplinary field that ranges across historical investigations, laboratory experiments, satellite geodesy, and of course, seismological analyses. Seismology does not solely focus on earthquakes, but also on natural processes and human activities that can excite *seismic waves*, which transmit energy and information outward from their source as a result of transient

3

stress imbalances. This book provides a review of both observational and theoretical developments in seismology.

A deep understanding of Earth's tectonic processes, including earthquakes, requires investigation of processes occurring throughout the lithospheric deformation "cycle", but such cycles last many years. For example, to accumulate observations at different stages in the earthquake process efficiently and to observe the diversity of behavior possible during earthquake rupture requires that we investigate earthquake phenomena globally, across all active faults and fault ruptures. Thus, the distribution of earthquakes, along with a requirement for extensive surface coverage with seismometers needed to unravel complex seismic signals, has made *global seismology* a planet-wide discipline. Unprecedented international collaboration and data exchange throughout seismology's history remains an ongoing accomplishment of the field. Thousands of seismological observatories operate around the world today, with nearly every nation participating in the effort to record seismic waves continuously. Recognition of the need for international cooperation began early and continues through principles of open but fair arrangements for data exchange, including well-defined standardized data formats and software tools that enable individuals across the globe to access and to analyze seismic data of high quality and substantial quantity.

Seismological analyses span a tremendous range of scales dictated by both the sources and seismic-wave characteristics. The smallest detectable micro-earthquake has a *seismic moment* (an important measure of earthquake size equal to the product of the fault-rupture area, the rock rigidity, and the average displacement over the rupture) on the order of 10^5 Nm; great earthquakes have seismic moments as large as 10^{23} Nm. Because seismic-wave amplitudes are directly proportional to seismic moment, seismic displacements also span an enormous amplitude range. Ground motions of interest also cover a broad frequency range. Seismic waves commonly used to explore for hydrocarbon resources include frequencies as high as 200 Hz and some environmental imaging extends to the kilohertz ranges. The longest-period standing waves excited by great earthquakes have frequencies as low as 3×10^{-4} Hz and solid Earth tide frequencies are around 2×10^{-5} Hz. These observations span a frequency range of 10^7–10^8 Hz. Studies of seismic sources extend the frequency range of interest to zero (effectively-static deformations) measured near faults and explosions using ground- and satellite-based geodesy. Finally, the range of propagation distances of interest in seismological investigation also spans many orders of magnitude. Local crustal investigations may use waves that travel only $10's$ of meters, while analysis of global structure may involve waves that transit the globe many times, traveling more than 10^8 m over Earth's surface.

Finally, seismology is a science with a substantial societal impact, both in assessing and reducing the danger from natural hazards and in revealing present Earth structure and buried resources. The relative sluggishness of mantle convective flow, or thermal inertia of the system, ensures that knowledge of the present-day structure reflects processes that have been occurring in the Earth over the past several hundred million years and, to a certain extent, over the entire evolution of the planet.

1.1 The foundation of seismology: seismograms

Seismograms serve as the foundation for global seismology. An example seismic recording is shown in Fig. 1.1. Three orthogonal components of motion (up–down, north–south, and east–west) record the total (vector) ground displacement as a function of time. The example data were recorded at seismic station SSPA

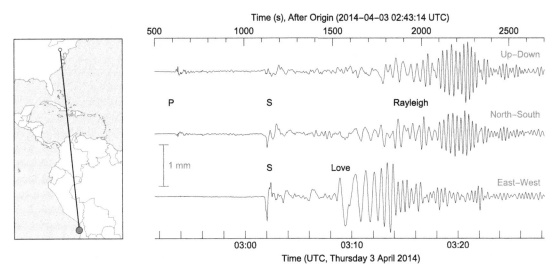

FIGURE 1.1 Recordings of the ground displacement history at station SSPA (Standing Stone, Pennsylvania, USA) produced by seismic waves from an M_W 7.7 earthquake along the coast of northern Chile on 3 April, 2014. (left) Map showing the earthquake and station locations and the great-circle arc connecting the two. The earthquake is an epicentral distance of about 62° (angular distance measured from center of the Earth) from SSPA. (right) The three seismograms correspond to vertical (U–D), north–south (N–S), and east–west (E–W) displacements. The direction to the source is almost due south, so horizontal displacements transverse to the path appear on the east–west component. The first arrival is a P wave that produces ground motion along the direction of wave propagation. The S motion is large on the horizontal components. The Love wave includes only transverse motions recorded predominantly on the E–W component, and the Rayleigh wave includes predominantly vertical and north-south motions.

(Standing Stone, Pennsylvania, USA). The source of the motions was a distant M_W 7.7 (M_W is called a magnitude, a quantity we use to represent the size of the earthquake) earthquake that struck the Chilean coast about 50 km southwest of Iquique, Chile in 2014 (a large aftershock of an M_W 8.2 event that occurred a few days earlier). The earthquake-induced ground motions at SSPA commenced about 10 minutes after the fault rupture initiated, the length of time it took for the fastest seismic waves to travel through the Earth from the source region to the station. The primary body waves (waves that travel through the body of the Earth) and surface waves (waves that travel along the surface of the Earth) are labeled in the figure. Ground displacements in Pennsylvania were as large as 1 mm, but the waves were much stronger near the source in Chile. A complex sequence of wave arrivals produced ground motions at the station that continue for hours following the earthquake. The later recorded motions are very small, with ground displacements of only a few microns; such motions are imperceptible without sensitive instrumentation. However, every wiggle on the seismogram carries information about the source and the part of Earth's interior through which the waves have traveled. The fault that hosted the 2014 Chile earthquake shown in Fig. 1.1 ruptured for about 30–40 km, and sliding across that rupture lasted for roughly 40 s, yet the earthquake induced motions detectable on the Pennsylvania seismometer for hours. The prolonged duration of the vibrations on the seismograms is a result of the seismic-wave interactions with Earth's interior. Later arriving energy includes waves that repeatedly circle the globe. The fundamental simplicity of elastic waves, which transmit disturbances over great distances through the Earth with little, or mostly predictable distortion, allows useful information to be gleaned from seismograms.

1.2 The historical development of global seismology

Seismology is a relatively young science, having awaited both the evolution of the theory of elasticity and the development of instrumental data. Although the Chinese had the first operational seismic-wave detector around 132 AD, the theoretical side of the science was considerably ahead of the observational side until the late 1800s. From the introduction in 1660 of Hooke's law, indicating a proportionality between stress and strain, to the development of equations for elasticity theory by Navier and Cauchy in 1821–1822, our understanding of the behavior of solid materials evolved rapidly. In the early 1800's, the laws of conservation of energy and mass were combined to develop the equations of motion for solids. In 1830, Poisson used the equations of motion and elastic constitutive laws to show that two (and only two) fundamental types of waves propagate through the interior of homogeneous solids: *P* waves (compressional waves involving volumetric disturbances, and directly analogous to sound waves in fluids) and *S* waves (shear waves with only shearing deformation and no volume change, which can therefore not propagate in fluids). The sense of ground deformation, or particle motions, relative to the direction of propagation for *P*- and *S*-wave disturbances is shown in Fig. 1.2. *P* (primary) waves travel faster than *S* (secondary) waves and are thus the first motion (preceded only by very small-amplitude gravitational perturbations) to be detected from any source in an elastic solid. These types of disturbance are called *body waves*, because they can traverse the interior of the medium, that is, the body of the Earth.

In 1887, contemporaneous with the recording of the first seismograms, Lord Rayleigh demonstrated the existence of additional solutions to equations of motion for bodies with free surfaces (such as Earth's surface). His solutions, called Rayleigh waves, involve motions confined near and propagating along the surface. By 1911, a second type of surface-wave motion, produced in a bounded body with depth-dependent material properties was characterized by A.E. Love; now called Love waves. Rayleigh and Love waves are collectively called *surface waves* and result from the interaction of *P* and *S* waves with a boundary at which the shear stresses vanish (a boundary condition). The sense of particle motion for the surface waves is indicated in Fig. 1.2. Love waves cause horizontal ground motion transverse to the propagation direction; Rayleigh waves cause retrograde elliptical motion in the vertical plane containing the source and observing position. Body and surface waves are influenced by both smooth and abrupt changes in material properties in the Earth with depth. Abrupt changes (e.g. the crust-mantle and core-mantle boundaries) in the Earth can reflect energy; these interactions can be analyzed by solving boundary-value problems, and they are often expressed in terms of reflection and transmission coefficients.

The association of earthquakes and propagating waves in western science originated with John Michell's 1760 report on the 1755 Lisbon Earthquake. He went so far as to estimate seismic wave speeds using imprecise human observations and recollections, but it was obvious more precise observations were needed. Theoretical solutions to elastodynamics problems for a solid medium also helped motivate the development of instruments capable of recording ground-motion time histories at a fixed location. International efforts led to the invention of the first seismometer by Filippo Cecchi in Italy in 1875. The sensitivity of early seismometers im-

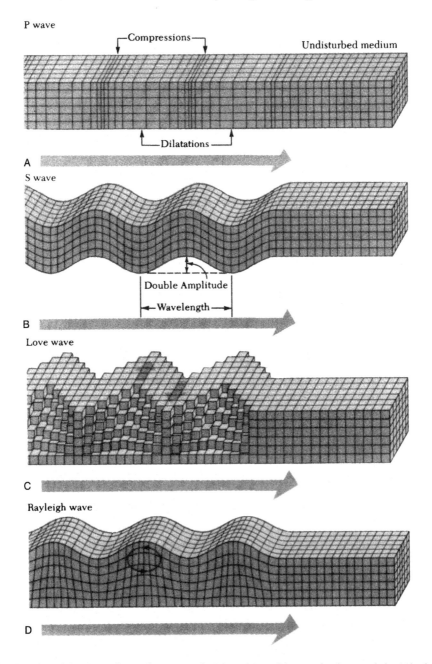

FIGURE 1.2 Schematic of the sense of particle motions during passage of the two fundamental elastic body waves, (top) *P* and (middle top) *S* waves, as well as the two surface waves in the Earth, (middle bottom) *Love*, and (bottom) *Rayleigh* waves. The waves are all propagating from left to right, with the surface of initial particle motion corresponding to the wavefront. The relative velocity of each wave type decreases from top to bottom. Observations of all four wave types at a single seismic station is shown in Fig. 1.1 (from Bolt, 1976).

proved rapidly, and by 1889 the first accurate recording of waves from a distant earthquake were recognized on a recording from Potsdam, Germany – the waves arrived 15 min after an earthquake struck Japan. More than 100 years of quantitative ground motion observations confirm the existence of P, S, Rayleigh, and Love waves in the Earth, as well as many other (mostly) understood seismic-wave arrivals. The observations and analyses demonstrate that the Earth behaves as a (nearly) elastic body in the frequency band of most seismic observations. The seismograms in Fig. 1.1 clearly exhibit distinct P-, S-, Love-, and Rayleigh-wave arrivals, and their particle motions are a close match to the theoretical expectations in Fig. 1.2.

In 1892, while working in Tokyo, John Milne developed a seismometer that was sufficiently compact that it could be installed in about 40 observatories around the world. Milne's network began the systematic collection of global seismic data. Around the turn of the century, seismometer technology improved significantly and body-wave data accumulated rapidly. The observations revealed systematic patterns relating body-wave travel time and distance from their sources. The data transformed the field and fueled a period of rapid first-order discovery of the main features of Earth's interior and earthquakes. Oldham discovered the Earth's core in 1906, and in 1913 Gutenberg determined an accurate depth to the core of about 2900 km (the current preferred value is 2889 km). In 1909, Mohorovičić discovered an abrupt velocity transition that we now refer to as the *Moho*, which is interpreted as the base of the crust (in most instances). In 1936, Inge Lehmann discovered the Earth's inner core (and later discovered a boundary in the uppermost mantle). Sir Harold Jeffreys compiled the travel times of thousands of seismic arrivals and developed the first detailed cross section of the Earth from surface to center by 1939 and what we call the J–B tables. The J–B tables predict the arrival times of P waves to any point on the Earth's surface

to within a remarkable 0.2% accuracy, limited primarily by the existence of three-dimensional variations in structure not allowed for in the tables. These travel-time tables were used routinely for over 50 years to locate global earthquakes.

In parallel with the advances in Earth structure, seismology and field observations were revealing the nature of earthquakes. In the late 1800's, Gilbert in the United States and Koto in Japan identified faulting as an intrinsic part of the earthquake process and not simply a consequence of the shaking. In 1905, Anderson developed an idealized theory of faulting that related the style of fault to the orientation of stresses near Earth's surface and Coulomb's failure criteria. In 1910, Reid enunciated the *elastic rebound theory* of earthquake-related faulting that remains a powerful conceptual model for understanding earthquakes. The year 1928 brought the recognition of the existence of deep-focus earthquakes by Japan's Wadati. In the mid-1930's, Richter developed the first quantitative measure of relative earthquake size, the local magnitude scale (M_L), referred to as the "Richter magnitude". By 1940 the global distribution of earthquakes was accurately mapped out, clearly defining major belts of activity that we now associate with boundaries between lithospheric plates.

Although the first half of the twentieth century revolutionized our knowledge of the Earth and earthquakes, seismology remained a rather obscure science, with only a small number of active seismology research programs. A significant problem was the limited number of worldwide seismic stations. Furthermore, the instrument characteristics of these stations were not standardized, which made it difficult to analyze the details of the ground motion when the data were compared and combined. Not until the advent of underground nuclear testing, with geopolitical implications, did seismology become a truly modern science. Seismology allows for monitoring underground nuclear tests because underground explosions excite seismic waves that can be detected at great distances. In fact, data from a seismic station at Tucson, AZ were used by Gutenberg to estimate the detonation time of the first nuclear explosion (Trinity) on July 16, 1945, when timing equipment at the test site failed (the Trinity device was suspended above ground, but sufficient energy coupled into the ground from the blast to excite seismic waves.) The first underground nuclear explosion (designated Rainier) was detonated in 1957 by the United States, and the 1963 Limited Test Ban Treaty banned atmospheric, oceanic, and deep space testing of nuclear devices by all of its 116 signatory nations. At that point, the U.S. government recognized the need to support research to understand seismic-wave propagation in complex structures in order to monitor foreign underground tests, and so it created the VELA UNIFORM program. One of the first accomplishments of this program was the deployment of the World-Wide Standardized Seismograph Network in the late 1950's and early 1960's. This 120-station global network of high-quality, well-calibrated, well-timed stations caused observational seismology to leap ahead of theoretical developments, bringing about major investments in university research programs. Around the same time, advances in computer technology enabled sophisticated analysis of increasing volumes of seismic data (a trend that continues). Although many first-order discoveries regarding Earth's interior had been made in the pioneering time prior to 1960, the field of global seismology flourished thereafter, and we will concentrate primarily on developments since that transformational decade. Today, global seismology continues to evolve with thousands of globally distributed, high-quality seismometers that can record small to large events on-scale over a wide range of frequencies.

1.3 The topics of global seismology

Having introduced the basic nature of seismology, we now undertake an overview of the topics and contributions of global seismology. Subsequent chapters provide the background required to understand how we obtain quantitative results from seismic recordings. The fundamental nature of linear elasticity allows us to treat the process of excitation, propagation, and recording of seismic waves as a sequence of linear filters that combine to produce observed seismograms. In other words, an observed ground displacement history, $u(t)$, is the result of a source function, $s(t)$, operated on by a propagation function, $p(t)$, and combined with an instrument recording function, $i(t)$. The filter operations are time-domain convolutions, which is an ubiquitous operation in linear, time-invariant systems. Convolution is a blending of two functions to produce a composite, let $x(t)$, $y(t)$ and $z(t)$ be functions of time, and further, let $x(t)$ represent the convolution of $y(t)$ and $z(t)$. Then

$$x(t) = \int_{-\infty}^{\infty} y(\tau) z(t - \tau) \, d\tau . \qquad (1.1)$$

If we denote convolution as $x(t) = y(t) * z(t)$, then we can express the ground motion or seismogram as

$$u(t) = s(t) * p(t) * i(t) . \qquad (1.2)$$

Modern seismology strives to describe mathematically each of the filters contributing to the observed displacements, and seismological research efforts classically bifurcate into two major categories: (1) study of the seismic source terms and their associated phenomena, and (2) study of the propagation terms and the associated Earth structure. The instrument transfer function is always the best-known filter but involves an interesting body of theory in its own right. Much of the organization of this textbook (as

TABLE 1.1   Major topics of global seismology.

Seismic-source topics	Earth-structure topics
Classical objectives	
A. Source location (position and time)	A. Basic layering (crust, mantle, core)
B. Energy release (magnitude, seismic moment)	B. Continent-ocean differences
C. Source type (earthquake, explosion, other)	C. Subduction zone geometry
D. Faulting geometry, area, slip	D. Crustal layering, structure
E. Earthquake spatial & temporal distribution	E. Physical state of layers (fluid, solid)
Current research objectives	
A. Slip distribution of fault rupture	A. Lateral variations in properties
B. Stresses on faults and in Earth	B. Boundary topography/sharpness
C. Faulting initiation/propagation/termination	C. Anelastic properties
D. Ground-motion and earthquake prediction	D. Thermal/compositional inferences
E. Analysis of landslides, eruptions, etc.	E. Anisotropy, mineralogy, flow

well as almost every other seismology book) tends to focus sequentially on these filters. However, the convolutional nature of seismological observations makes clear that any analysis of $u(t)$ must consider the combined source, $s(t)$, and propagation characteristics, $p(t)$. Table 1.1 is a list of some of the many topics of classical and current interest in the two major categories. Below, we survey some basic results of seismological analysis in each category before developing the theory and procedures used in global seismology.

1.3.1 Seismic sources

Elastic waves are generated whenever a transient stress imbalance is produced within or on the surface of an elastic medium. A great variety of physical phenomena in the Earth involve rapid motions that excite detectable seismic waves. These sources can be grouped into those that are external to the solid Earth and those that are internal. Table 1.2 lists some common seismic sources, all of which involve processes of interest to Earth scientists. Induced seismicity from oil and gas exploration are not included in the table, as they can be classified as earthquake faulting that have been advanced in their earthquake cycle, and new studies continue to explore this phenomena.

Mathematical descriptions and physical models for all of these source types have been developed, although most are kinematic descriptions rather than first-principle theories. In order to represent these complex physical phenomena mathematically, we usually determine dynamically equivalent, idealized force systems that mathematically replace the actual physical process. "Dynamically equivalent" means that the elastic motions produced by the idealized equivalent forces are the same as those of the actual physical process. We can then use these force representations in the equations of motion (derived from the conservation of energy and momentum) to predict the resulting waves accurately.

External sources are usually easier to represent mathematically than internal sources. In most cases, external sources can be treated as time-varying tractions applied to the Earth's surface (a traction is a vector stress created by application of a force to a surface). The source's variation with time creates a stress imbalance that is balanced by deformation of the medium, which propagates outward as seismic waves. The mathematics of this process are given in Part II. Internal force systems may be relatively simple, as in the three-dipole force system needed to represent an ideal, isotropic explosion, or quite complex, as in the spatial distri-

TABLE 1.2 Primary sources of seismic waves.

Internal	External	Mixed
Earthquake faulting	Wind, atmospheric phenomena	Volcanic eruptions
Buried explosions	Ocean waves and tides	Landslides
Hydrological processes	Rapid sediment transport	
Magma movements	Sudden glacial changes	
Abrupt phase changes	Explosions, meteorites, bollides	
Mining and induced activity	Human activity (cultural noise, energy production)	

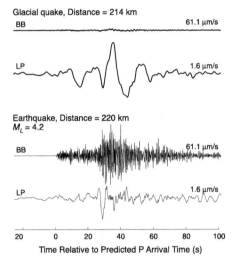

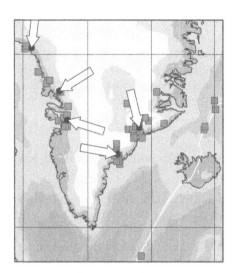

FIGURE 1.3 (left) Seismograms from an Alaska glacial event and a nearby, tectonic, M_L 4.2 earthquake recorded at comparable distances. Top trace shows broad-band (BB) (0.00833 to 20 Hz) ground velocity record, and bottom shows filtered long-period (LP) (0.01 to 0.2 Hz) ground velocities. (right) Map of southern Greenland and vicinity. The white line shows the location of the northern mid-Atlantic ridge. Squares show the locations of detected glacial events and/or earthquakes in the region. The black hexagon at the tip of each arrow indicates the estimated source centroid location (modified from Ekström et al., 2003).

bution of double-couple forces needed to represent a large earthquake. All sources produce body and surface waves, but the relative excitation and the frequency and amplitude characteristics of these waves depend strongly on the source type and force-time history. For example, the seismic recordings of nuclear explosions can usually be discriminated from natural earthquakes by their very strong excitation of high-frequency P waves relative to lower-frequency surface waves. Such dependencies are extremely valuable because they lead to inverse problems

that allow us to estimate source properties from observed seismograms.

Although the sources of primary interest in global seismology are shear-faulting earthquakes, many of the sources listed in Table 1.2 can produce globally observable seismic signals. The top two seismograms on the left of Fig. 1.3 are plots of the ground motions produced by glacial ice fracturing events in two different frequency ranges. Typical earthquake signals in the same frequency ranges and observed at a comparable distance are shown in the bottom two seismograms. The signal characteristics for the two sources are quite different, but the underlying physics is the same. Strain release during the glacial process is much slower than the typical earthquake and very weak short-period motions are excited – they go unnoticed in analysis focused on short-period signals. Global investigations using lower frequency surface-wave detections have detected numerous glacial events in areas such as Alaska, Greenland, and Antarctica. The frequency of glacial events has provided valuable information on the response of ice sheets and glaciers to global warming. As exotic as the glacial source may be, the resulting seismic motions produced by the glacial events behave predictably according to the theory of elastic waves.

1.3.2 Earthquake sources involving shear faulting

Historically, ground disruption and surface faulting were commonly associated with earthquakes, but often these effects were absent (mostly likely from deeper earthquakes), confusing observers as to which was cause and which was effect. Several notable geologists including Gilbert and Koto, had described the relationship between earthquakes, faulting, and surface uplift in the late 1800's, but the development of *equivalent force systems* for natural earthquakes required an understanding of the earthquake-rupture process, which was not available before the 20th century. After analysis of observations from the 1906 San Francisco earthquake, H.F. Reid enunciated a model that related shaking and faulting in a simple, but quantitative way. Reid had carefully studied the well-exposed permanent ground motions that occurred at the time of the 1906 earthquake. The horizontal deformations in the vicinity of the San Andreas fault exhibited a simple symmetry that led him to formulate the *elastic rebound theory* of earthquakes. This partly empirical, partly intuitive theory outlines how an earthquake is a sudden offset of the rocks sep-

arated by a *fault*. A fault is a fracture in the rock across which some previous displacement has occurred and compared with the surrounding, competent rock, the fault is a zone of relative weakness. Before an earthquake, crustal stresses, generally resulting from lithospheric motion, cause strain to accumulate in the immediate vicinity of the fault. The strained rocks exert a stress on the fault, but fault motion is resisted by friction. Motion continues and strain increases. At some point the strain exceeds a threshold related to the fault's frictional properties, abrupt sliding occurs, and rapidly releases the accumulated strain energy. Much of the energy is consumed by heating and fracturing of the rocks near the fault, but a portion is converted into seismic waves that propagate outward from the rebounding volume surrounding the rupture. The lithospheric motions continue, which leads to cycles of strain accumulation and release during the active lifetime of the fault.

Examples of the historical *geodetic* (measured permanent ground deformation) observations that provided early evidence for the elastic rebound model, collected largely in Japan where there are numerous shallow crustal faults and frequent earthquakes, are listed in Table 1.3. These examples indicate that the crust cannot accumulate strains much larger than about 10^{-4} without failure, where strain is calculated as slip on the fault divided by the distance perpendicular to the fault over which there are significant coseismic displacements. Most events involve strains from 10^{-5} to 10^{-4}, at least in typical continental situations, a fundamental result that we return to in Chapter 11. A large number of such observations of faulting and ground displacement have given rise to the hypothesis that most *shallow* (less than 70 km deep) earthquakes result from shear dislocations on faults, even though most such events occur below the depth of direct observation.

Systematic analysis of seismic waves from thousands of earthquakes and several decades of satellite-based GPS and Interferometric Syn-

TABLE 1.3   Classic observations of faulting strain.

Earthquake	Fault length (km)	Average offset (m)	Decay distance (km)	Strain
1906 San Francisco, CA $M_S = 7.8$	200	5	20	2.5×10^{-4}
1927 Tango, Japan $M_S = 7.8$	30	3	30	1.0×10^{-4}
1943 Tottori, Japan $M_S = 7.4$	40	2	15	1.3×10^{-4}
1946 Nankaido, Japan $M_S = 8.2$	80	0.7	100	1.0×10^{-5}
1971 San Fernando, CA $M_S = 6.6$	30	2	20	1.0×10^{-4}

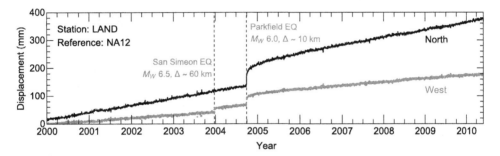

FIGURE 1.4 Observed position of a GPS station located on the west of the San Andreas Fault near Parkfield, California, USA in the NA-12, North-America-Fixed, geodetic reference frame. Most of the time, the station moves to the NNW at a steady rate of about 28 mm/yr as strain accumulates near the San Andreas Fault. Sudden station displacements associated with two earthquakes are identified by the dashed lines. The first, more clear on the west component, was cause by the 2003 M_W 6.5 San Simeon, CA earthquake, about 60 km away. The second offset was associated with the M_W 6.0 2004 Parkfield earthquake on the segment of the San Andreas Fault adjacent the GPS observatory. During the Parkfield earthquake, the station moved about 50 mm north and about 20 mm west. Data from the UNR Geodetic Laboratory, Blewitt et al. (2016).

thetic Aperture Radar (InSAR, a radar technique that measures ground deformation) observations provide strong empirical evidence supporting this hypothesis. Fig. 1.4 is a plot of surface offsets measured at a high-sample-rate GPS station (LAND) located about 10 km from the San Andreas Fault in central California. The observations illustrate the elastic-rebound process. For most of the time shown, steady plate motions cause the station on the west side of the San Andreas Fault to transit in a north-northwest direction as the Pacific Plate moves relative to North America. During this time, no motion oc-

curs at the fault, which is locked and held in check by friction. Two sudden offsets associated with earthquakes, identified by the dashed lines, first occurred in late 2003 60 km away, while the second is from the nearby segment of the San Andreas. The station is within the zone that accumulates strain during the interval of steady motion at a rate lower than the relative plate motion, and then rebounds suddenly during the earthquake to catch up with the far-field plate motions. The fact that an event 60 km away causes some rebound at this site is an indication of the complex deformation field associated with the myriad of faults that are active in the San Andreas Fault system. The pattern of surface displacement produced by a vertical fault that slips horizontally is particularly simple. Shallow faults with other orientations produce more complicated, but still easily predictable patterns of horizontal and vertical motion. The governing equations for these static deformations are derived in Chapter 11.

The 1906 San Francisco earthquake was also scientifically important because it was widely recorded on the early generation of seismometers available near the turn of the century. Fig. 1.5 shows a horizontal component of ground motion recorded on an Omori seismometer that was operated in Tokyo. Classic helicorder seismograms are read like a book, line by line, starting in the upper left. The time at the end of the top line corresponds to the beginning of the second line and so on. The 1906 recording shows an initial P-wave arrival followed by a much larger S-wave arrival (first line) and then a complex sequence of surface waves (second line). The shearing nature of earthquake fault offsets is partly responsible for the greater amplitude of the S wave relative to the P wave. The combination of a conceptual model for faulting and the constraints on source force systems provided by observed amplitudes and polarities of P and S waves enabled the development of simple equivalent force systems to model earthquakes. The **double couple** and the **moment tensor** are force models routinely used in seismology to model and to characterize shear-faulting sources (Chapter 16).

These simple equivalent-force models can be used to model some of the first-order features in seismic and geodetic displacements caused by earthquakes. But observations from numerous continuously operating seismic and high-sample-rate GPS stations (such as those shown in Fig. 1.4) have led to the recognition that earthquakes are part of a broad spectrum of strain release processes that operate along plate boundaries and major faults. For many years,

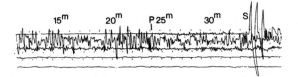

FIGURE 1.5 A classic seismic recording of the 1906 earthquake made by an Omori seismometer located in Tokyo, Japan. The ground motion is horizontal, east–west. The station was 75° distant from the source (1° = 111 km). Time increases toward the right on each line in the recording, and the tick marks indicate 60-s intervals. The P-wave arrives first, the S-wave arrives about 10 min later. The record was recorded on a translating drum; time wraps from one line to the next.

we have known that certain *aseismic* "creeping" segments of major faults released strain nearly continuously (often accompanied by frequent micro-earthquakes). Recent observations near subduction margins have revealed that other deformation processes are also at work. One way to classify the range of processes is using the characteristic time scale of the deformation, which we'll represent using τ. Slip during a "standard" earthquake generally lasts less than 10 s at any point on the fault ($\tau < 10\,\text{s}$). Other seismic strain transients include large creep events ($10^3\,\text{s} < \tau < 10^6\,\text{s}$), earthquake afterslip ($10^6\,\text{s} < \tau < 10^8\,\text{s}$), low- and very-low frequency earthquakes ($10\,\text{s} < \tau < 100\,\text{s}$), slow-slip events ($10^5\,\text{s} < \tau < 10^7\,\text{s}$), and episodic tremor and slip ($10^6\,\text{s} < \tau < 10^7\,\text{s}$). Aseismic creep typically occurs at rates of 1–100 mm yr^{-1}. The nonstandard earthquakes and other strain transients are an important research focus because of their similarities to, and differences from, standard earthquakes. They provide a new perspective and new insights to earthquake processes. Also important is the relationship between strain release on aseismic fault segments to strain changes on more hazardous *seismogenic* fault segments and the potential role of aseismic creep in standard earthquake nucleation. Although our primary focus in this book is on the standard earthquake processes, when appropriate,

TABLE 1.4 Characteristic seismic wave periods.

Wave type	Period (s)
Body waves	0.01–50
Surface waves	10–350
Free oscillations	350–3600

we will connect our discussion with other strain transients that transfer strain in or out of seismogenic systems.

1.3.3 Seismic waves and seismograms

In general, earthquake body waves (P and S waves) have shorter characteristic periods of vibration than surface waves (Rayleigh and Love waves), which in turn have shorter periods than *free oscillations* of the Earth (standing modes of vibration of the entire planet, which are detectable only for large earthquakes) (Table 1.4). Furthermore, the ground displacements for body waves generated during a large earthquake may be only 10^{-3} cm after traveling 1000 km, but long-period surface waves may have amplitudes of several meters after traveling the same distance (such distant, long-period displacements correspond to small accelerations, so are not generally a hazard far from the source). These differences result from source excitation and propagation effects that depend on wave type and the Earth's structure.

Each seismic wave transmits energy spanning a range of frequencies, and the ground motion produced by the same wave may have a different appearance depending on the characteristics of the recording system. A seismic signal in one frequency band may not accurately represent the behavior in other frequency bands. A very broad-band seismometer records many frequencies of ground motion, as shown in Fig. 1.6, which are for the October 1989 M_W 6.9 Loma Prieta, CA earthquake that ruptured a fault in the Santa Cruz Mountains. The top panel shows a time window of 12 hrs. The Loma Prieta earthquake shaking is a large transient on the long-period sinusoidal signal with a period of 12 hrs. The long-period signal is the solid Earth tide; the seismometer sensitivity at such long periods is low, the Earth rises and falls about 40 cm at station ANMO every day in response to solid-earth tides caused by the gravitational attraction of the Sun and Moon. The seismogram only shows a fraction of the total signal because it is filtered to exclude periods shorter than 50 s. Rayleigh-wave arrivals that travel east along the short arc between the source and station (R_1) and west along the long arc (R_2) are labeled. R_3 is a second pass of the R_1 arrivals after circling the Earth once more. The middle panel shows a time window of 20 min containing the main signal from the Loma Prieta earthquake. The largest signal is the Rayleigh wave (R_1), which has an amplitude of about 1–2 cm and shows a clear *dispersion* (different frequencies arrive at different times because they travel at different speeds). The P and S waves are much smaller in amplitude. The bottom panel shows the first 240 s of the P arrival and its coda, which has many higher-frequency oscillations. The theoretical arrival time for the S waves is about 320 s, but the vertical S amplitude is quite low at ANMO. The higher-frequency P energy is complex, reflecting that propagation and source effects have a strong frequency dependence.

The spherical shape of the planet has a large influence on seismic observations, seismic waves can be expected to approach a seismometer from two different directions, the short (minor) arc and the long (major) arc connecting the source and the station. Fig. 1.7 is a plot of long-period Rayleigh waves produced by the 2004 Sumatra – Andaman Islands earthquake (M_W 9.2) recorded at globally distributed seismometers. R_1 waves travel along the short arc of the great circle from the source to the receiver and then continue to circle the Earth, reappearing as R_3 at the same station 3 hrs later. R_2 travels along the long arc of the great circle and arrives at the station again as R_4 and then as R_6, etc. in 3-hr shifts.

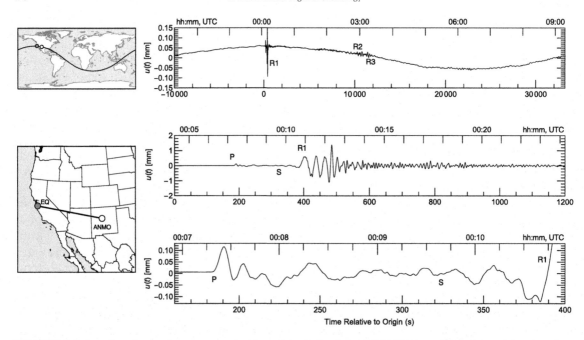

FIGURE 1.6 Broadband vertical-component recording of the 18 October, 1989 (UTC) Loma Prieta (M_W 6.9) earthquake at station ANMO (Albuquerque, New Mexico). Maps on the left show the great-circle path followed by surface waves and the regional great-circle segment connecting the source (gray circle) with the seismic station. The top panel is 20 hrs in duration (the earthquake is the rider on the long-period signal); tidal effects dominate. The middle panel is for a 30-min interval, and the bottom panel is for a 240-s interval. The signal in the top, long-period panel, was filtered to exclude periods shorter than about 50 s to highlight longer-period vibrations.

The multi-orbit surface waves slowly decrease in amplitude as they circle the Earth because of energy losses due to *attenuation* (anelastic losses) and increasing dispersion of the energy. Longer-period oscillations dominate later in the traces because both attenuation and dispersion have a strong frequency dependence. We must account for these effects when studying source processes, but they reveal information about Earth structure when directly studied. At the *antipode*, the location on the opposite side of the planet from the source (where R_1 and R_2 arrive together in Fig. 1.7), seismic waves can interfere and create large amplitude arrivals. Of course, focussing at such great distance poses no hazard, but the sharpness of focus provides information on the planet's symmetry.

1.3.4 Quantification of earthquakes

Prior to instrumental recording, comparisons of earthquakes were based mainly on shaking damage and *seismic intensity* scales were developed based on variations in damage. Intensity levels can be contoured, defining *isoseismals*, or regions of common shaking damage, and typically the highest intensities are close to the rupture. Although such earthquake measures are strongly influenced by proximity of the event to population centers, construction practices, and local site effects, seismic intensities are often all that we know about pre-instrumental earthquakes, and they play a major role in regions where most known large events occurred before routine seismogram collection began. To be able to compare recent with older events, we con-

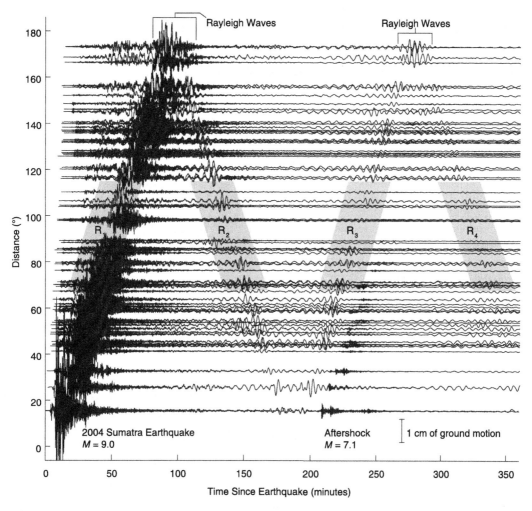

FIGURE 1.7 Long-period Rayleigh waves produced by the 2004 Great Sumatra – Andaman Islands Earthquake ($M_W =$ 9.2) as recorded by the United States Geological Survey (USGS)/Incorporated Research Institutions for Seismology (IRIS)-National Science Foundation (NSF) Global Seismographic Network. The vertical axis is the angular distance along the surface from the source, and time is from the earthquake origin time. R_1 and R_2, are Rayleigh waves traveling along the minor and major arcs of the great circle from source to station, respectively; R_3 is the next passage of the R_1 and R_4 is the next passage of the R_2 wave after circling the entire globe (modified from an IRIS poster).

tinue to track seismic intensity measurements. The modern version is the U.S. Geological Survey's "Did You Feel It?" (DYFI) project that uses online reports from internet users to construct modern intensity maps. The digital nature of the DYFI analysis also enables comparison of intensity estimates with instrumental ground shaking measured using seismic sensors. Crowd sourcing has also been used to detect earthquakes using social media reports, and several interesting projects that engage the public in "citizen seismology" to collect damage photos and reports or even operate inexpensive seismic sensors have produced helpful information on earthquake activity. However, most quantitative measures of earthquake size and characteristics are based on recorded ground motions.

Until the 1980's, it was common to use different seismometers, sensitive to distinct, limited frequency bands and with varying amplitude sensitivity, to record different seismic waves based on their specific frequency content (see Table 1.4). Narrow-band sensors were easier to build and more stable. But narrow-band instruments recorded a small component of the total ground motion. The diversity of instruments recording different segments of the signals led to the development of many scales to compare the relative size of earthquakes (using different seismic waves and frequency ranges), termed seismic *magnitudes*. Even with their complexities, we use seismic waves to compare earthquake size because it can be done systematically and quantitatively and because it does not rely on damage or other macroscopic phenomena that are strongly affected by factors other than the source (such as variable construction standards and surface topography).

Most seismic magnitude scales are based on the logarithm of the amplitude of a particular seismic wave on a particular seismometer (which limits the frequency range), with an adjustment for the distance to the source. Examples of the commonly used seismic magnitude scales are listed in Table 1.5 and are compared with their period of measurement in Fig. 1.8. An important consequence of seismology's observational history is that an earthquake may have several different seismic magnitudes, because the measurements are made using different waves at different frequencies. This fact often leads to confusion because most people (reasonably) tend to expect a given earthquake to have only a single magnitude. Changes in measurement practice and sensor characteristics, including the distribution and number of measurements available to average, can impact measurements that have accumulated over many decades. Working with seismic catalog magni-

TABLE 1.5 Examples of seismic magnitude scales.

Symbol	Name	Period of measurement (s)
M_L	Richter magnitude	0.1–1.0
m_b	Body-wave magnitude	1.0–5.0
M_S	Surface-wave magnitude	20
M_w	Moment magnitude	> total source duration

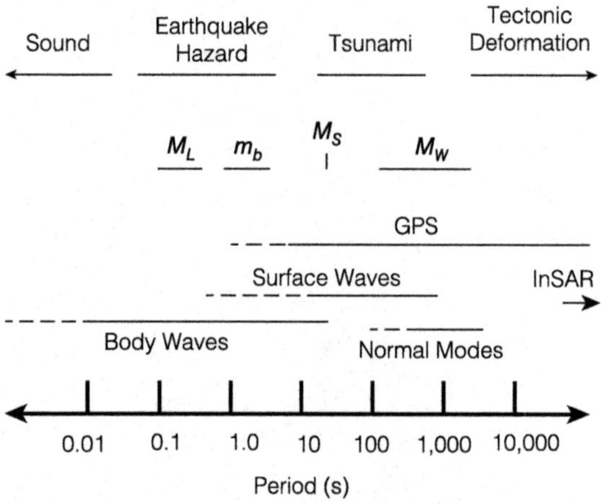

FIGURE 1.8 The range in period of deformation phenomena in the Earth along with the characteristic periods of body waves, surface waves, and different seismic magnitude scales. Each magnitude scale tends to be associated with a particular wave type and frequency range and often a particular instrument. For example, the Richter magnitude, M_L, is associated with the Wood-Anderson instrument that was widely used in southern California in the 1930's when Richter developed his local magnitude scale.

tudes often requires an appreciation of the history of magnitude estimation practices.

Magnitude is an empirical estimate of event size, with only a limited theoretical basis (described in Chapter 7). Nonetheless, seismic magnitudes are widely used in earthquake and explosion analyses. The essential measurements needed to estimate a seismic magnitude are the amplitude and period of the peak ground motion and the distance from the source to the seismometer. Measurements are made at as many stations as possible and averaged to produce a

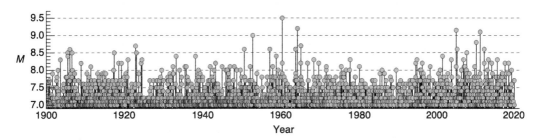

FIGURE 1.9 Major ($M \geq 7.0$) and great ($M \geq 8.0$) earthquakes from the USGS PAGER catalog (events through mid-2008) and the USGS catalog (for more recent events). The magnitude used for modern earthquakes post-1977 is a moment magnitude, earlier estimates are a mix of M_W and M_S.

single value. Since wave amplitudes decrease systematically with distance, an adjustment, or correction to, a reference distance is included in the definition of magnitude. The precise amplitude-distance relation varies from region-to-region and for different wave types and frequency ranges, so in a strict sense Richter's Local Magnitude (M_L) scale is restricted to Southern California, where it was developed. Also, since many factors in addition to distance influence seismic-wave amplitudes, the spread of magnitudes can be relatively large, and magnitudes are seldom used to a precision better than 0.1 magnitude units. Most formulas include the logarithm of the period-normalized amplitude, so at a fixed distance, an increase of one magnitude unit corresponds to an increase in shaking amplitude by a factor of ten.

Magnitude is not the only way to characterize earthquake size. Earthquakes can be quantified by determining physical characteristics of the fault rupture, such as the rupture length, area, average displacement, sliding (particle) velocity or acceleration, duration of faulting, radiated energy, heterogeneity of slip distribution, or combinations of these or similar quantities. Although we can determine many of these characteristics through detailed seismic-wave analysis, any single magnitude scale can at best qualitatively describe the complex process that occurs at the source. We shall see that from a tectonic viewpoint, the most suitable quantity

with which to represent an earthquake size is the *seismic moment*, which is a measure of the static characteristics of the total fault motion. The units of seismic moment are force × distance (in Physics, known as work), and the numbers can be large and less intuitive to work with than magnitude. For this reason, a *moment magnitude scale*, M_W, is based on logarithmic scaling of seismic moment to give numerical magnitudes that are roughly comparable with traditional magnitude scales. However, structural damage from earthquakes is often controlled by high-frequency waves, so short-period magnitudes remain useful.

Earthquakes with $7.0 \leq M < 8.0$ are called *major* earthquakes (here we use a generic symbol, M to represent surface-wave, M_S, or moment magnitude, M_W). Major earthquakes roughly correspond to events having more than 1 m of average displacement (or slip) along ruptures that are more than 30 km long. *Great earthquakes* are events with $M \geq 8.0$ and involve larger ruptures and greater slip. We have a complete record of major and great earthquakes since about 1900 (Fig. 1.9). The largest instrumentally recorded event is the 1960 Chilean earthquake ($M_W = 9.5$), which involved an average of 20 m of displacement during a rupture that took a few-minutes to complete and that extended along a roughly 150-km-wide×1000-km-long fault segment. The 2011 Tohoku earthquake (M_W 9.0) produced slip up to approxi-

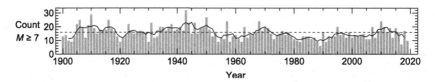

FIGURE 1.10 The annual number of major ($M \geq 7.0$) and great ($M \geq 8.0$) earthquakes from the USGS PAGER catalog (events through mid-2008) and the USGS common catalog (for more recent events). The mean number of large events is about 16. The solid curve indicates a five-year running mean through the values and the dashed line identifies the mean number of events per year, which lies between 15 and 16. The magnitude used for modern earthquakes post-1977 is a moment magnitude, earlier estimates are a mix of M_W and M_S.

mately 60 m in the shallow regions of the rupture, which generated a catastrophic tsunami that devastated the northeast coast of Honshu. The largest earthquakes are capable of producing tsunami that impact coastal areas throughout entire ocean basins. Since 1900 we know of six earthquakes with $M_W \geq 8.8$, all have occurred in subduction systems, which are sometimes referred to as *giant earthquakes*. The 2004 Sumatra–Andaman Islands earthquake killed more than 230,000; most of those deaths were a result of large tsunami that impacted the entire northern Indian Ocean region.

The annual average number of major and great ($M \geq 7.0$) earthquakes is about 16. Fig. 1.10 is a plot of the observed number of major and great earthquakes each year from 1900 to 2018. The record suggests a more or less steady annual rate of large earthquakes, but the pattern is not uniform. More uncertainty exists in magnitude estimates before the advent of standardized global seismic networks since the 1960's. On average, one but sometimes no great event may occur in a single year; however, more frequent smaller events can also produce a catastrophic loss of life and damage. In the first decade of the current century, there was a surge of great earthquakes (Fig. 1.9).

The largest earthquakes have not caused the greatest amount of human suffering. An example of the destructive potential of earthquakes is the 1976 (M_W 7.7) Tangshan, China earthquake, which took approximately 250,000 lives (some estimates put the toll as high as 700,000).

More recently, the 2010 (M_W 7.0) Haiti earthquake struck close to the densely populated Haitian capital of Port Au Prince and killed between 90,000 and 300,000 people. The data in Fig. 1.11 suggest that an equivalent average annual number of earthquake fatalities is just under 20,000 (earthquakes of course are discrete events). Although historically, the largest number of earthquake fatalities have been concentrated along the Alpine-Himalayan collision belt, modern earthquakes have devastated regions geographically more broad. Earthquake hazard varies dramatically with location around the world, but the less shaking-resistant construction practices found commonly in developing nations often accentuates earthquake damage. Circum-Pacific countries tend to experience more frequent large events. The 1985 Mexico City (M_W 8.0) earthquake struck a city with moderate construction standards, located near a zone of frequent earthquakes. Although Mexico City was 250 km from the earthquake, at least 7000 people lost their lives, mainly due to building collapse. Soil conditions under the building foundations, construction practices in the city, and an unusually long rupture duration have been blamed for that catastrophe.

1.3.5 Earthquake geographic distributions

One of the classical problems in seismology has been the systematic mapping of earthquake

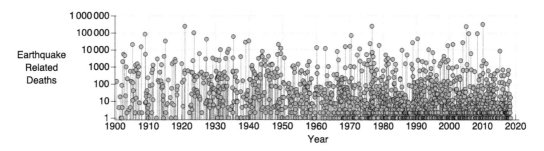

FIGURE 1.11 Earthquake fatalities from the USGS PAGER and National Geophysical Data Center significant earthquake catalogs since 1900. Fewer low-fatality counts early in the record reflects less common reporting of events with small death tolls. The number of fatalities correlates little with magnitude. Even small earthquakes can cause extensive loss of life in regions with poor building construction, or if secondary hazards such as fires or landslides enhance the damage. We continue to have catastrophic earthquakes in modern times, but world population has increased roughly five-fold during the time interval shown.

distributions on a variety of spatial scales. Efforts began in the 1800's using historical data, and later instrumental-based results played a key role in the evolution of the theory of *plate tectonics*, which describes the large-scale relative motions of a mosaic of lithospheric plates on the Earth's surface. Observations of seismic-wave arrival times and a model of the Earth's velocity structure are the basis of classic seismic-source location methods. Historically, the development of velocity models and improved source locations has developed in a seesaw fashion, with occasional, independently known source locations and origin times providing first-order models of the structure, which could be improved over time as they were used to locate more earthquakes. Earthquake location methods are described in Chapter 6.

By 1941, through work by Beno Gutenberg and Charles Richter, the global distribution of major earthquake belts was quite well determined, and the enhanced location capabilities of the modern global network now allow routine location of all events greater than about magnitude 4.5 (Fig. 1.12). The spatial distribution and time history of seismic events, or *seismicity*, is clearly nonuniform. Most events occur around the Pacific margin, but mid-ocean ridge and fracture zone structures are also quite ac-

tive. Continental seismicity tends to be diffuse except where it is concentrated in seismic belts along the Pacific margins. Studies by Turner and Wadati in the early 1920's revealed the occurrence of earthquakes at depths greater than 70 km. We call events with depths less than 70 km *shallow*; events with depths between 70 and 300 km are called *intermediate-depth*; and events with depths between 300 and 700 km depth are called *deep*. Intermediate and deep events are found primarily in linear belts around the Pacific, under Europe, and under parts of Asia (Fig. 1.12). These events presumably occur within downwelling portions of oceanic lithosphere that is sinking into the mantle. Shallow events are much more frequent and release much more energy than intermediate and deep earthquakes. The exact nature of deep sources remains somewhat puzzling because frictional sliding should be difficult for the enormous pressures at such great depths. Yet, the seismic radiation from deep events appears indistinguishable from shallow shear-faulting earthquakes, discussed in Chapter 3.

The distribution of lower-magnitude seismicity is also a focus of seismological investigation, particularly in densely inhabited areas where the earthquake hazard is an issue. Epicenters of earthquakes located in southern California

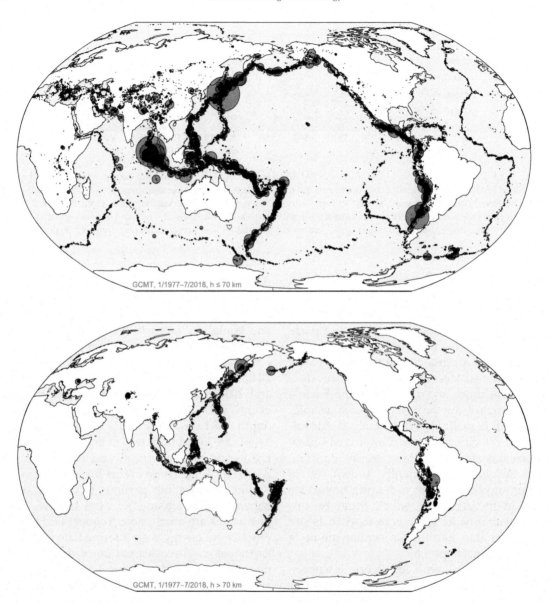

FIGURE 1.12 Shallow (top) post-1977 earthquake epicenters showing events with depths less than or equal to 70 km (top) and events with depths greater than 70 km (bottom). Most events shown have moment magnitudes larger than 5.0 and symbol area is proportional to seismic moment. Locations and size estimates are from the Global Centroid Moment Tensor Catalog (GCMT).

are shown in Fig. 1.13. The seismic distribution is exceedingly complex and does not strictly adhere to the mapped faults observed at the surface. Dense arrays of seismometers are installed in areas of intense seismicity, or *seismogenic* zones, in order to obtain precise earth-

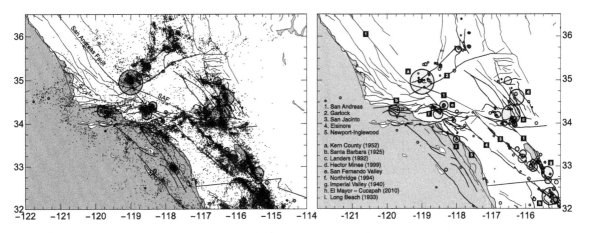

FIGURE 1.13 (left) Map of roughly 60,000 southern California earthquakes that occurred during the years 1925–2019 and have $M \geq 2.5$. Most of the events are very small and occurred more recently (when monitoring networks were more complete); a dense network of seismometers is deployed in the region to locate all of the earthquakes accurately. Small earthquakes are shown using small points, larger events are shown with gray circles. The circle area is scaled proportional to the events estimated seismic moment. (right) Selectively labeled faults and large or significant earthquakes (data are from the Southern California Earthquake Center and the U.S. Geological Survey).

quake locations and to study the faulting motions that must be taking place in the region. Note that if we use only a short-term seismicity pattern to identify faults in a region, we may fail to identify the major fault that produces the largest earthquakes. For example, the San Andreas fault, during this time interval (and most of the time), hosts few small events. The last large earthquakes on the San Andreas fault in this region were in 1857 (the northern part of the map) and earlier on the southernmost fault segment (over three hundred years ago). Activity drops off rather quickly to the east, indicating the distribution of the major part of the plate boundary system.

The need to assess large-earthquake hazard leads global seismologists to look at the historic record of large earthquakes around the world over longer periods of time. The global distribution of great earthquakes since 1900 is shown in Fig. 1.14; surface-wave magnitude, M_S, and/or moment magnitude, M_W, values are listed for each event (when known) in the Appendix (Tables 1.6 and 1.7). Estimating the

magnitude in the early part of a 20th century remains a challenge, so the uncertainty is larger for older events. The great event distribution mirrors the overall seismicity pattern. The largest events occur around the Pacific margins, but numerous events, many of them devastating, occur within the Middle East and Asia. For great earthquakes, even a record of $\tilde{1}20$ years is short – given the list, one would not identify the Cascadia subduction zone or the southern San Andreas fault as capable of producing major earthquakes. To estimate large earthquake hazards, earthquake scientists must push the record back to times preceding instrumental recording by using descriptive reports of historical events, geological records of landslides and uplift, and by digging into near-surface faults to examine the history of motions preserved in the soil and rock disruptions. The efforts reveal that the southern San Andreas fault has indeed had great earthquakes, the most recent in 1857, with many previous events recurring about every 130 years. Historic and geologic evidence indicate that the Cascadia Subduction

Great Earthquakes, M ≥ 8.0, 1900–2020 (June)

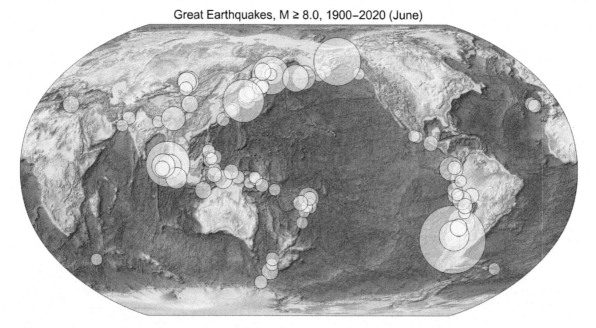

FIGURE 1.14 The global distribution of great earthquakes since 1900. Symbol area is proportional to the earthquake's seismic moment. The events are listed in Table 1.7.

Zone hosted a $M \sim 9$ earthquake in January of 1700 and was struck by other large earthquakes in the previous few thousand years. Other significant U.S. events revealed by historical accounts occurred in regions such as southeastern Missouri, where a sequence struck in 1811–1812, and South Carolina in 1886. In some places, such as Missouri, Arkansas, and Tennessee, current small-magnitude seismicity alerts us to the local earthquake potential, whereas others, like South Carolina, remain relatively inactive. Events in stable continental regions with low strain accumulation remain difficult to forecast. Some have suggested large earthquakes in plate interiors produce extended aftershock sequences (such as modern New Madrid seismicity). Even if that speculation is true, a moderate aftershock could be consequential in the region and more research is needed. Earthquake hazards issues are discussed further in Chapter 8.

1.3.6 Global faulting patterns and earthquake models

To understand the distribution and fundamental causes of earthquakes, we must observe the fault-related motions that are involved. But only a few earthquakes rupture the surface and allow direct observation. Again, analysis of seismic waves assists us greatly because the seismic waves traveling outward from the earthquake carry information on the faulting-related deformation. Remarkably, after accounting for propagation effects, we can use the seismograms to constrain the fault orientation and direction of slip, even though the seismometer may be thousands of kilometers from the source (Fig. 1.15). Faulting geometry affects the amplitudes and polarities of all seismic waves, so we can routinely use entire seismograms to infer the faulting geometry (in the signal band where we know enough about Earth's structure to correct for source effects) of moderate and large

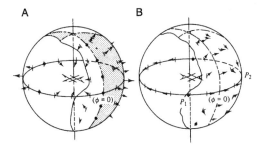

FIGURE 1.15 As elastic waves propagate away from an earthquake they preserve the sense of ground motion near the source. The directions of *P*- and *S*-wave particle motions on the expanding wavefront are shown above. (A) *P*-wave motions are perpendicular to the wavefront and reflect the initial motion either toward or away from the source. (B) *S*-wave motions are parallel to the wavefront and their shearing direction is controlled by the orientation of the shearing in the source. By observing seismic waves, it becomes possible to relate distant motions to near-source motions and determine the source geometry (from Kasahara, 1981).

earthquakes all over the world, even at inaccessible depths and under water. The routine determination of earthquake faulting geometry for the global distribution of earthquakes by seismic-wave analysis is one of the remarkable accomplishments of modern global seismology. The Global Centroid Moment Tensor Project (GCMT) begun in the early 1980's by A. Dziewonski and J. Woodhouse and continued presently by Dziewonski students and colleagues G. Ekström and M. Nettles ranks as a profound contribution. The project database now includes estimates of the faulting geometry and seismic moment for more than 50,000 globally distributed earthquakes more or less complete in the M 5.0–5.5 range since 1977. Fig. 1.16 includes maps that show the GCMT faulting geometry estimates for earthquakes with $M_w \geq$ 5.0. The circular symbols that identify each event are stereographic projections that show the direction of *P*-wave motion near the source called *focal mechanisms*. Focal mechanisms communicate fault-plane orientation, strain axes orientations, and reveal the sense of fault slip at the

source (introduced in Chapter 3 and described fully in Chapter 16). We separated the GCMT results into four panels corresponding to reverse and normal dip-slip faults, strike slip faulting, and oblique faulting, which is a mix of strike-slip and reverse, or strike-slip and normal. The GCMT data base is a tremendous resource for earthquake and global and regional tectonics analyses, and extensively used to account for source effects in investigations of Earth's interior using seismic waveforms.

Characterizing the average fault orientation and seismic moment of the events is only the first step in earthquake analysis. More detailed seismic analysis are routinely used to explore the rupture sequence for large events, from onset to termination of faulting. An example composite point-source rupture model is shown in Fig. 1.17 for a large earthquake in 1976 in Guatemala that resulted from the rupture of a nearly vertical fault called the Motagua Fault. The earthquake model parameterizes the large event as a *superposition* of spatially localized point sources. The rupture started at location 1 and spread along the fault, offsetting the two sides as it propagated. Detailed analysis of *P* waveforms recorded around the world for this event indicate that the radiation of energy was heterogeneous during the rupture and that the faulting geometry varied along the rupture. The complex energy release time history is a common attribute of large earthquakes. The effects can be seen in seismograms and confirmed in earthquake-produced surface displacement along ruptures that breach the surface. Chapter 14 is a description of the procedures that are used in such studies.

For more recent earthquakes, the quality of global seismic data is more plentiful and greatly improved over earlier decades. Even great earthquakes seismograms are recorded on-scale (earlier analog systems "clipped" large-amplitude records from large events) with greater bandwidth than previously possible. Improved data allows more detailed analysis of seismic rup-

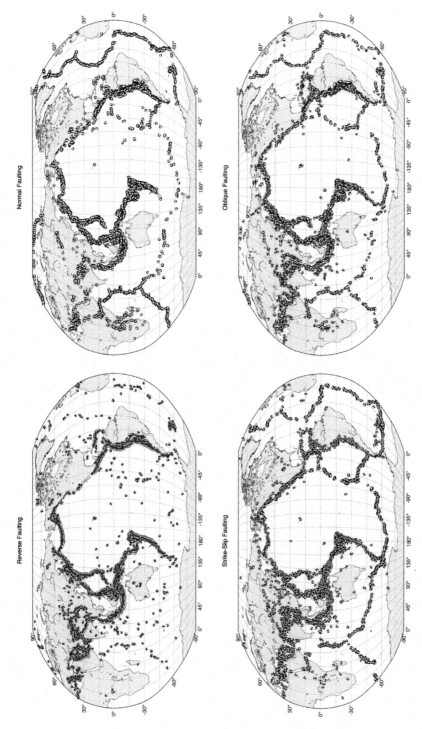

FIGURE 1.16 Source mechanisms for all earthquakes in the GCMT Catalog from 1976-July, 2018. The events are grouped into faulting styles and are identified by the seismic focal mechanism. In each mechanism, the unshaded areas indicate dilatational (toward the source) motions. Source mechanisms are detailed in Chapter 16. The GCMT Catalog is a major resource for seismological and tectonic investigation and is one the principal contributions in the history of instrumental seismology.

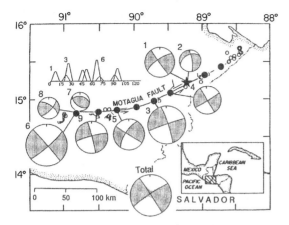

FIGURE 1.17 A map of the Motagua fault, which ruptured during the 1976 Guatemala earthquake. The seismologically determined history of strain energy release is shown in the upper left. Each pulse corresponds to radiation from different sections of the fault as the rupture spread away from the initiation point (star). Each subevent has a source orientation determined in the analysis, with the projections of the *P*-wave nodal radiation planes being shown in stereographic projections. Darkened areas represent compressional *P*-wave motions. The fault orientation changed during rupture, and the strength of radiation was not uniform along the fault (from Kikuchi and Kanamori, 1991).

tures, including the estimation of displacement as a function of time and position along the rupture. An example is shown in Fig. 1.18, which was constructed by combining large distance (teleseismic) *P*, *S*, and Rayleigh-Wave observations. The data suggest a rupture size of 200–300 km in length and about 100 km in width (down-dip). The slip is heterogeneous with local slip maxima 100 and 150 km from the *hypocenter* (the origin of the coordinate system). The rupture lasted over two minutes. For all events larger that $M \geq 7$ several groups, including the United States Geological Survey (USGS), produce kinematic "finite-fault" models. For the largest events these models provide important information for rapid estimation of the event impact. The rapid estimates are often followed by models produced by the academic community. Faulting analyses are important for doc-

umenting the diversity of earthquake rupture and accumulate the observations needed to test models of dynamic earthquake processes. The broad range of seismic source investigations is described further in Chapter 19.

1.3.7 Earth's interior: radial Earth layering

A second and equal emphasis of global seismology is the exploration of Earth's interior. Most of what we know about Earth's deep interior, including constraints on composition, layering, dynamics, physical state, and temperature has been based on the combination of seismic observations of the otherwise inaccessible interior regions and laboratory measurements of mineral properties. In addition, just as it was for seismic sources, pioneers working in the early 1900's, such as Oldham, Mohorovičić, Gutenberg, Lehmann, Jeffreys, Bullen, solved many of the first-order Earth structure problems. They discovered the outer core, deduced that it must be fluid because it does not transmit *S*-wave energy, then discovered the solid inner core. They identified the crust-mantle boundary, explored the mantle and mapped the transition zone. But, as is the case for understanding earthquakes, resolving second-order details remains critical to developing a deep understanding of the Earth's dynamic processes. For example, the presence of several-hundred-degree lateral temperature differences deep in the mantle may produce only a 1% change in seismic velocity but is sufficient to drive convective flow of the interior on long time scales. Similarly, grossly different models of the chemistry of the interior differ in their elasticity parameters by a few percent or less. Important deep earth dynamic problems require an intense effort to constrain the internal structure with high precision.

A single seismogram can be complicated, but patterns caused by first-order features in Earth's interior are clear in a collection of seismograms covering range of source-to-station distances.

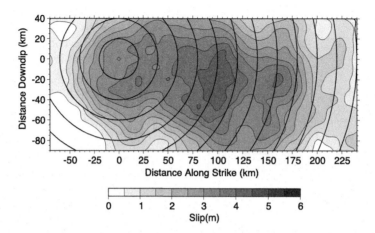

FIGURE 1.18 Contoured coseismic slip estimates for the 15 November 2006 (average slip 4.6 m) Kuril Islands earthquake. The rupture began at the grid origin and the propagation isochrons (10 s spacing) correspond to a 2 km/s uniform rupture speed. This model was obtained by analysis of distant (teleseismic) *P*, *S*, and Rayleigh waves (from Ammon et al., 2008).

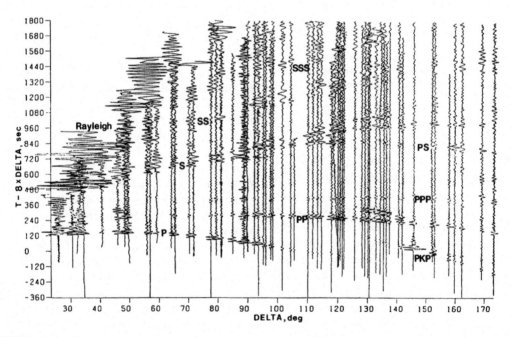

FIGURE 1.19 A collection of vertical-component seismograms for a single event that occurred near Sumatra, plotted at the angular distance to each station. The records are from the World-Wide Standardized and Canadian Seismic Networks. Upward motion on each trace is toward the left. Note that coherent arrivals can be tracked from trace to trace. These define the travel-time behavior for different paths through the Earth. The start time of each trace has been reduced by a value of 8Δ s, where Δ is the angular distance. Thus, traces on the right begin much later than traces on the left (modified from Müller and Kind, 1976).

A display of seismograms as a function of distance, often called a *seismic profile*, reveals much about seismic wave propagation within the Earth's interior. An example is shown in Fig. 1.19. The records show the distinct, but systematic travel-time variations of seismic waves as a function of distance. For example, at epicentral distances (measured in angular degrees along the Earth's circumference between source and station) of less than 100°, a clear *P*-wave arrives at the onset of ground motions. The disruption of the *P* arrival branch near 100° is due to the low-velocity core of the planet. The systematic timing as a function of distance, or *travel-time curve*, for each seismic phase can be analyzed using geophysical inverse methods to estimate the internal structure of the Earth. Fitting the travel times of seismic arrivals constrains the variation in seismic wave speed as a function of planetary radius. Complicated wave interactions can also constrain abrupt velocity changes that may accompany significant changes in material properties. Abrupt changes produce reflections, including *P*-to-*S* and *S*-to-*P* converted waves that can corroborate information from direct wave travel times and seismic-wave amplitudes. Many complex low-amplitude wave interactions can be observed only after stacking (averaging) many observations at a given distance, which reduces background noise. Record sections and waveform stacks for all large earthquakes are now routinely computed and shared online by the International Research Institutions for Seismology (IRIS) Consortium.

The observed seismic wavefield is complicated by the existence of both body and surface waves, by conversions and reflections of body waves off the core and other internal discontinuities, by the spherical geometry of the Earth and multiple reflections of body waves off the surface, as well as by relatively smaller lateral variations in structure. But millions of seismic travel-time observations have accumulated in the routine process of locating earthquakes throughout the history of instrumental seismology. Modern observations are available online, in catalogs, which also list hypocentral information on all detected earthquakes. The USGS and the International Seismological Center (ISC) compile earthquake bulletins with travel-time reports from stations around the world. From the observations, master travel-time curves such as those in Fig. 1.20 can be constructed for different source depths and earth models can be built to allow the calculation of interactions so subtle that they are unobserved in the data (unless you have some guidance on where to look).

Once seismic-wave travel-time curves are available, it is possible to invert for *P*- and *S*-wave velocity as a function of depth using the methods described in Chapter 20. Through analysis of body waves, long-period surface waves, and free oscillations, global seismologists have developed one-dimensional models of the elastic velocities and density of the entire Earth. One of the most frequently used models is shown in Fig. 1.21. This model, different from the first generation of global models developed in the 1930's in subtle but important ways, indicates the major subdivisions of the interior: the solid inner core, the fluid outer core, the lower mantle, and the upper mantle. The crust is a very thin veneer on the surface. Chapter 20 includes a discussion of the seismological constraints on each region. Radial models of the Earth's elastic structure are used in many applications (including earthquake rupture modeling) and are critical for efforts to determine the composition and state of the interior. However, radial models fail to express the complexity of what we know to be a dynamic, evolving system, so seismology is now striving to develop fully three-dimensional models for the interior at all scale lengths.

1.3.8 Heterogeneous Earth models

That simple layered models are a poor approximation of the Earth's crust has been known for many years. The exposed surface geology

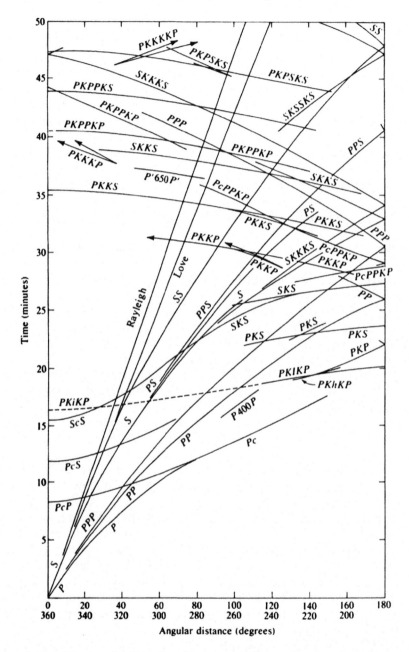

FIGURE 1.20 Average travel time as a function of angular epicentral distance from a surface seismic source. Curve labels identify a seismic phase code, which depends on the wave type P or S, and path. For example, c represents upward reflection off the core-mantle boundary and PcP is a P-to-P refection. To represent propagation in the core, we use the characters I (inner core), and K (outer core) and the character i represents upward reflection off the inner core. P' is sometimes used to abbreviate PKP. Reflections of Earth's surface are represented concatenating the symbol for incident and reflected wave types, such as PP or PS (from Bolt, 1982).

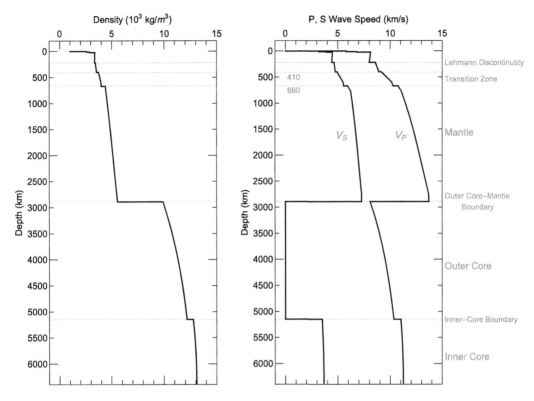

FIGURE 1.21 The Preliminary Reference Earth Model (PREM) of density (left) *P* velocity (V_P), *S* velocity (V_S) (right) as a function of depth in the Earth. After Dziewonski and Anderson (1981). Despite the "preliminary" in its name, PREM has served the seismological community as a reasonable and reliable reference model for almost 40 years.

provides clear evidence of complexity, and the difference between oceanic and continental regions illustrates broad-scale, first-order differences. Gravitationally driven geological processes that produce layering, such as sediment deposition, chemical precipitation, and lava flows, all do so on limited spatial scales, and subsequent crustal motions deform locally stratified rocks. Efforts to study crustal structure, driven on the one hand by resource exploration and on the other by Earth science efforts to understand how the crust evolved, have led to many attempts to develop two- and three-dimensional models for crustal regions. Dense distributions of seismometers and very high frequency recordings are required to see the

complex, laterally discontinuous arrivals reflected from deep structure, which generally show deformed layering offset by faults. In many crustal locations two-dimensional models are inadequate to describe the subsurface, particularly for complex formations that may trap oil; thus three-dimensional images are developed in numerous crustal investigations. Development of three-dimensional imaging for both exploration and global analyses requires fast computers with large data storage capabilities. Although we have made great advances during the rapid technological development of the last decades, high-resolution seismology efforts still challenge our computational resources.

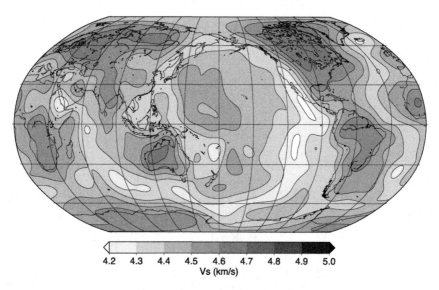

FIGURE 1.22 Map of estimated shear-velocity variations at a depth of 100 km. Darker regions correspond to higher-velocity regions of the upper mantle. This model was constructed by analysis of body and surface waves and normal-mode measurements. Fluctuations with scale lengths of less than roughly 500–1000 km are not resolved, so small features like the Yellowstone hot spot in Fig. 1.24 are not resolved. The model is from Moulik and Ekström (2014).

During the 1950's and 1960's, the first computer-assisted analyses of long-period surface waves began to reveal systematic lateral variations in deeper Earth structure below the crust. By the 1980's many global seismologists were actively analyzing different types of seismic data to constrain three-dimensional structure at depth by methods collectively identified as *seismic tomography* (based on mathematical similarities to medical imaging tomography, which is used to image internal structure of the human body without surgery). These early researchers found that every region of the interior, with the possible exception of the outer core, appears to have detectable aspherical heterogeneity. The ability to resolve perturbations from average one-dimensional radially-symmetric Earth models, and the recognition of its importance for internal dynamics, prompted a revolution in geophysical investigations of the deep interior. Original approaches used body-wave travel-time and surface-wave dispersion observations,

modern approaches integrate these with normal-mode and direct waveform modeling. Coupled with the increasing quantity of global seismic stations, earth models have improved steadily since the 1980's. An example is shown in Fig. 1.22. The image shows the variation in speed at a depth of 100 km, where the main features of surface plate tectonics are visible. Low speeds along the young oceanic ridge-system correspond to warm mantle passively upwelling in response to the plate separation and high speeds beneath old continental cratons correspond to thermal and compositional properties in the mantle lithospheric beneath these relatively long-undisturbed regions.

The steady improvement in global models as data become more abundant is not the only advance in large-scale earth structure investigation. The U.S. National Science Foundation EarthScope/USArray funded project provided data to image much of North America at resolutions hard to achieve without a major effort.

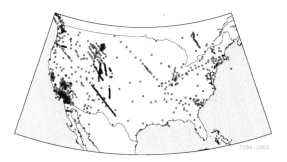 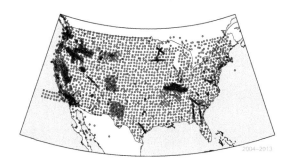

FIGURE 1.23 Broadband seismic station coverage in the ten years from 1994 to 2003 and the ten years from 2004 to 2013. The USArray-related deployments provided a large, high-quality, spatially dense sampling of the conterminous United States (similar coverage in Alaska is not shown). Not all stations shown operated for the whole decade, most of the Transportable Array stations operated in a location for just over one year and then were moved in a leap-frog manner from west to east to complete the coverage. These data have led to major advances in subsurface imaging that benefits from the relatively regular spacing of the stations.

The seismology component of EarthScope, US-Array, was managed by the community through IRIS. The ambitious project included a Transportable Array (TA) striving to complete a uniform sampling of the conterminous US and Alaska, and an instrument base to construct temporary dense flexible arrays to investigate smaller-scale tectonic features. Fig. 1.23 shows the project's substantial impact in increasing coverage of broadband stations across the U.S. The map on the left shows the broad-band stations operated at any time during the ten years prior to EarthScope (including temporary and permanent stations), while the map on the right shows the stations deployed during the ten years of the EarthScope project. In both cases, linear seismometer patterns represent temporary portable deployments, and the regional networks are apparent as broad, relatively dense sensors clusters. The relatively uniformly spaced TA dominates the EarthScope station map. Not every site was operated simultaneously, as only 400–500 stations recorded simultaneously at any given time. After over one year of recording, stations were migrated from west to east over the ten-year program. Flexible array deployments were embedded within the TA in a series of university-led field projects that focussed on specific geologic targets. The availability of spatially broad and relatively dense seismic networks drove innovations in regional seismic analysis and seismic imaging, particularly for the analysis of surface-waves crossing the sensors. The results are widely available in the seismological community. The sensors were also used to analyze large, distant earthquake ruptures using array-processing approaches, small earthquake occurrence and characteristics, including human triggered events in the central and eastern U.S., local and regional wave propagation, scattering, and attenuation across the continent, temporal and spatial patterns in seismic background motions, and imaging the Earth deep beneath central North America.

An example deep image of the three-dimensional velocity variation beneath the western U.S., including the Yellowstone Hot Spot (YHS), is shown in Fig. 1.24. The map on the left shows the location of the cross-section and the dots correspond to the dots on the cross section. The left end of the image is in the Pacific Ocean. The image on the right shows the perturbations in S-wave speed from a radially symmetric reference model constructed using core-phase travel times observed across USArray seismic stations.

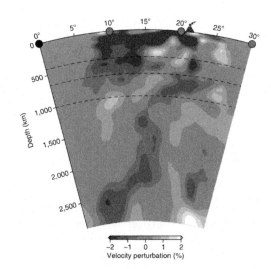

FIGURE 1.24 Depth cross-section through the plume structure showing its connection with the Yellowstone hotspot. The cross section sliced north-northeast through the western Mexico, United States and Canada. The left edge is southwest of Baja and the right edge is in southwest-central Saskatchewan. The surface position of the Yellowstone hotspot is shown by the cartoon volcano. Dashed lines indicate 410, 660 and 1000 km depth. The shading was chosen to highlight lower-mantle features. Anomalies in the upper mantle mainly reflect the choice of starting model (modified from Nelson and Grand, 2018).

Darker regions identify regions of lower seismic wave speed. The substantially low speeds in the upper mantle beneath the tectonically active western United States are clear (they are clipped). The palette was chosen to highlight a channel of relatively low speeds that connects the region near the base of the mantle to the YHS. Imaging subtle speed variations, especially low-speed features remains a challenge but matters greatly to dynamic models of mantle dynamics. The likelihood that the variations in velocity are at least in part due to thermal variations (higher-velocity material being colder and lower-velocity material being hotter at a given pressure), combined with the fact that any thermal variations cause density variations, suggests that the three-dimensional seismic models reveal density heterogeneity. Density heterogeneity results in long-term stresses (due to gravitational pull) that cause earth materials to flow, with upwellings and downwellings being driven by gravity as the Earth system transports heat to the surface. Thus, remarkably, imaging the Earth with elastic waves provides a means for determining the ongoing dynamic convection of the mantle. Chapter 10 includes details on more earth-structure investigations.

1.3.9 Summary

We hope that this introduction has made it clear that modern global seismology is a rapidly advancing, quantitative discipline essential to the investigation of a vast array of important physical phenomena in the Earth System. There is beauty and elegance in the mathematical procedures used in the discipline and in the richness and complexity of seismological data. The challenge of extracting information from seismic signals continues to draw researchers into the field, with both applied and basic science emphases. In this text, we develop much of the basic theory and touch upon many of the major observations and results of modern global seismology.

1.4 Appendix: Great earthquakes, 1900–mid2020

A well-trained earthquake seismologist knows some earthquake history. This appendix includes a list of the largest, ($M \geq 8.5$), earthquakes and a list of great ($M \geq 8.0$) earthquakes from the USGS PAGER and Common earthquake catalogs. The $M\,8$ cutoff is nothing spe-cial, the difference between a great earthquake and an $M\,7.9$ is not that significant. For many old events, the data are sparse and some of the events included in the list may not have been great events, just close, and some events are undoubtedly left off the list for similar reasons. Some of the tabulated events have depths of zero, which is an indication of an uncertain depth, but most of the events are most likely shallow.

TABLE 1.6 Great earthquakes from 1900–mid2020 with $M \geq 8.5$.

Origin Time (UTC)	Latitude (°)	Longitude (°)	Depth (km)	M	Region
1960-05-22T19:11:17	−38.294	−73.054	35	9.5	Near Coast Of Central Chile
1964-03-28T03:36:12	61.019	−147.630	6	9.2	Southern Alaska
2004-12-26T00:58:53	3.270	95.860	22	9.2 (9.1)	Off W Coast Of Northern Sumatra
2011-03-11T05:46:24	38.297	142.370	29	9.1	Near East Coast Of Honshu, Japan
1952-11-04T16:58:26	52.750	159.500	–	9.0	Off East Coast Of Kamchatka
2010-02-27T06:34:11	−36.122	−72.898	23	8.8	Near Coast Of Central Chile
1965-02-04T05:01:21	51.210	178.500	29	8.7	Rat Islands, Aleutian Islands
1922-11-11T04:32:45	−28.553	−70.755	35	8.7	Central Chile
2012-04-11T08:38:36	2.327	93.063	20	8.6	Off W Coast Of Northern Sumatra
2005-03-28T16:09:36	2.050	97.060	34	8.6	Northern Sumatra, Indonesia
1963-10-13T05:17:55	44.763	149.800	26	8.6	Kuril Islands
1957-03-09T14:22:33	51.587	−175.420	35	8.6	Andreanof Islands, Aleutian Is.
1950-08-15T14:09:30	28.500	96.500	–	8.6	Eastern Xizang-India Border Reg.
1906-01-31T15:36:00	1.000	−81.500	–	8.6 (8.8)	Off Coast Of Ecuador
2007-09-12T11:10:26	−4.440	101.370	34	8.5	Southern Sumatra, Indonesia
1923-02-03T16:01:48	53.853	160.760	35	8.5	Near East Coast Of Kamchatka
1917-06-26T05:49:42	−15.500	−173.000	–	8.5	Tonga Islands
1906-08-17T00:40:00	−33.000	−72.000	–	8.5 (8.2)	Off Coast Of Central Chile
1905-07-09T09:40:24	49.000	99.000	–	8.5 (8.4)	Mongolia

TABLE 1.7 Great earthquakes from 1900–mid2020 with $M \geq 8.0$ from the USGS PAGER and Common earthquake catalogs.

Origin Time (UTC)	Latitude (°)	Longitude (°)	Depth (km)	M	Region
1903-01-04T05:07:00	−20.000	−175.000	400	8.0	Tonga Islands
1903-08-11T04:32:54	36.360	22.970	80	8.3	Southern Greece
1905-07-09T09:40:24	49.000	99.000	0	8.5 (8.4)	Mongolia
1905-07-23T02:46:12	49.000	98.000	0	8.4	Russia-Mongolia Border Region
1906-01-31T15:36:00	1.000	−81.500	0	8.6 (8.8)	Off Coast of Ecuador
1906-08-17T00:40:00	−33.000	−72.000	0	8.5 (8.2)	Off Coast of Central Chile

continued on next page

TABLE 1.7 (*continued*)

Origin Time (UTC)	Latitude (°)	Longitude (°)	Depth (km)	M	Region
1906-09-14T16:04:18	−7.000	149.000	0	8.0	New Britain Region, P.N.G.
1908-12-12T12:08:00	−14.000	−78.000	60	8.2	Off Coast of Peru
1911-06-15T14:26:00	28.000	130.000	90	8.1	Ryukyu Islands, Japan
1917-05-01T18:26:30	−29.000	−177.000	0	8.0	Kermadec Islands, New Zealand
1917-06-26T05:49:42	−15.500	−173.000	0	8.5	Tonga Islands
1918-08-15T12:18:16	5.653	123.560	35	8.2	Mindanao, Philippines
1919-04-30T07:17:12	−19.823	−172.220	35	8.2	Tonga Islands Region
1920-12-16T12:05:54	36.601	105.320	25	8.3	Western Nei Mongol, China
1922-11-11T04:32:45	−28.553	−70.755	35	8.7	Central Chile
1923-02-03T16:01:48	53.853	160.760	35	8.5	Near East Coast of Kamchatka
1924-04-14T16:20:41	7.023	125.950	35	8.2	Mindanao, Philippines
1924-06-26T01:37:32	−56.407	158.490	15	8.3	Macquarie Island Region
1929-06-27T12:47:14	−54.714	−29.549	35	8.0	South Sandwich Islands Region
1932-05-14T13:11:05	0.258	126.170	35	8.1	Northern Molucca Sea
1933-03-02T17:31:00	39.224	144.620	35	8.4 (8.4)	Off East Coast of Honshu, Japan
1934-01-15T08:43:25	26.773	86.762	35	8.0 (8.1)	Nepal-India Border Region
1935-05-30T21:32:56	28.894	66.176	35	8.1	Pakistan
1935-09-20T01:46:42	−3.920	141.330	35	8.1	New Guinea, Papua New Guinea
1938-02-01T19:04:21	−5.050	131.620	35	8.4	Banda Sea
1938-11-10T20:18:46	55.328	−158.370	35	8.0 (8.2)	Alaska Peninsula
1941-11-25T18:03:58	37.171	−18.960	25	8.1	Azores-Cape St. Vincent Ridge
1942-11-10T11:41:28	−49.726	29.947	15	8.0	South of Africa
1943-04-06T16:07:15	−30.750	−72.000	0	8.2	Off Coast of Central Chile
1944-12-07T04:35:42	33.750	136.000	0	8.1 (8.1)	Near S. Coast of Western Honshu
1945-11-27T21:56:50	24.500	63.000	0	8.0	Off Coast of Pakistan
1946-04-01T12:28:54	52.750	−163.500	0	8.0 (8.1)	South of Alaska
1946-12-20T19:19:05	32.500	134.500	0	8.1 (8.1)	Shikoku, Japan
1948-01-24T17:46:40	10.500	122.000	0	8.1	Panay, Philippines
1948-09-08T15:09:11	−21.000	−174.000	0	8.0	Tonga Islands
1949-08-22T04:01:11	53.750	−133.250	0	8.0 (8.1)	Queen Charlotte Islands Region
1950-08-15T14:09:30	28.500	96.500	0	8.6 (8.6)	Eastern Xizang-India Border Reg.
1952-03-04T01:22:43	42.500	143.000	0	8.1	Hokkaido, Japan Region
1952-11-04T16:58:26	52.750	159.500	0	9.0 (9.0)	Off East Coast of Kamchatka
1957-03-09T14:22:33	51.587	−175.420	35	8.6 (8.6)	Andreanof Islands, Aleutian Is.
1957-12-04T03:37:51	45.182	99.219	25	8.1 (8.1)	Mongolia
1958-11-06T22:58:09	44.311	148.650	35	8.4	Kuril Islands
1959-05-04T07:15:46	53.370	159.660	35	8.0	Near East Coast of Kamchatka
1960-05-21T10:02:54	−37.825	−73.379	12	8.2	Near Coast of Central Chile
1960-05-22T19:11:17	−38.294	−73.054	35	9.5 (9.5)	Near Coast of Central Chile
1963-10-13T05:17:55	44.763	149.800	26	8.6	Kuril Islands
1964-03-28T03:36:12	61.019	−147.630	6	9.2 (9.2)	Southern Alaska
1965-01-24T00:11:17	−2.453	125.960	29	8.2	Ceram Sea
1965-02-04T05:01:21	51.210	178.500	29	8.7 (8.7)	Rat Islands, Aleutian Islands
1966-10-17T21:41:57	−10.799	−78.680	34	8.2	Near Coast of Peru
1968-05-16T00:49:00	40.901	143.350	26	8.3	Off East Coast of Honshu, Japan
1969-08-11T21:27:37	43.477	147.820	46	8.2	Kuril Islands
1971-07-14T06:11:30	−5.519	153.900	44	8.0	New Ireland Region, P.N.G.
1971-07-26T01:23:22	−4.889	153.180	37	8.1	New Ireland Region, P.N.G.

continued on next page

TABLE 1.7 (*continued*)

Origin Time (UTC)	Latitude (°)	Longitude (°)	Depth (km)	M	Region
1975-05-26T09:11:51	35.971	−17.646	4	8.1	Azores-Cape St. Vincent Ridge
1976-08-16T16:11:07	6.292	124.090	58	8.0	Mindanao, Philippines
1977-06-22T12:08:33	−22.908	−175.750	64	8.0	Tonga Islands Region
1977-08-19T06:08:55	−11.125	118.380	21	8.3	South of Sumbawa, Indonesia
1979-12-12T07:59:03	1.602	−79.363	24	8.1	Near Coast of Ecuador
1985-09-19T13:17:47	18.455	−102.370	20	8.0 (8.0)	Michoacan, Mexico
1989-05-23T10:54:46	−52.507	160.600	2	8.0	Macquarie Island Region
1994-06-09T00:33:16	−13.877	−67.532	635	8.2 (8.2)	Northern Bolivia
1994-10-04T13:22:55	43.832	147.330	33	8.3	Kuril Islands
1995-07-30T05:11:23	−23.336	−70.265	40	8.0	Near Coast of Northern Chile
1995-10-09T15:35:53	19.052	−104.210	26	8.0	Near Coast of Jalisco, Mexico
1996-02-17T05:59:30	−0.919	136.980	36	8.2	Irian Jaya Region, Indonesia
1998-03-25T03:12:25	−62.901	149.610	20	8.1	Balleny Islands Region
2000-11-16T04:54:56	−3.990	152.260	28	8.0 (8.0)	New Ireland Region, P.N.G.
2001-06-23T20:33:14	−16.380	−73.500	32	8.4 (8.4)	Near Coast of Peru
2003-09-25T19:50:06	41.860	143.870	27	8.3 (8.3)	Hokkaido, Japan Region
2004-12-23T14:59:04	−49.330	161.420	4	8.1 (8.1)	North of Macquarie Island
2004-12-26T00:58:53	3.270	95.860	22	9.2 (9.1)	Off Coast of Northern Sumatra
2005-03-28T16:09:36	2.050	97.060	34	8.6 (8.6)	Northern Sumatra, Indonesia
2006-05-03T15:26:40	−20.150	−174.100	55	8.0	Tonga Islands
2006-11-15T11:14:13	46.580	153.270	10	8.3 (8.3)	Kuril Islands
2007-01-13T04:23:21	46.230	154.550	10	8.1	East of Kuril Islands
2007-04-01T20:39:58	−8.430	157.060	10	8.1	Solomon Islands
2007-08-15T23:40:57	−13.380	−76.610	39	8.0	Near Coast of Peru
2007-09-12T11:10:26	−4.440	101.370	34	8.5 (8.5)	Southern Sumatra, Indonesia
2009-09-29T17:48:10	−15.489	−172.100	18	8.1 (8.1)	Samoa Islands Region
2010-02-27T06:34:11	−36.122	−72.898	23	8.8	Near Coast of Central Chile
2011-03-11T05:46:24	38.297	142.370	29	9.1	Near East Coast of Honshu, Japan
2012-04-11T08:38:36	2.327	93.063	20	8.6	Off W Coast of Northern Sumatra
2012-04-11T10:43:10	0.802	92.463	25	8.2	Off W Coast of Northern Sumatra
2013-02-06T01:12:25	−10.799	165.110	24	8.0	Santa Cruz Islands
2013-05-24T05:44:48	54.892	153.220	598	8.3	Sea of Okhotsk
2014-04-01T23:46:47	−19.610	−70.769	25	8.2	Near Coast of Northern Chile
2015-09-16T22:54:32	−31.573	−71.674	22	8.3	Near Coast of Central Chile
2017-09-08T04:49:19	15.022	−93.899	47	8.2	Near Coast of Chiapas Mexico
2019-05-26T07:41:15	−5.812	−75.269	123	8.0	Peru

2

An overview of earthquake and seismic-wave mechanics

Chapter goals

- Introduce the concepts of continuum mechanics including stress, strain, elasticity.
- Introduce the mechanical foundations underlying earthquakes and seismic waves.
- Introduce the conceptual physics underlying fault rupture (friction).
- Introduce the observational framework used to characterize earthquake ruptures.

The seismological approach to earthquake study is to match observed earthquake-related patterns with continuum-mechanics based models of faulting and seismic-wave propagation. Our models simplify a complicated process, and we use observations to guide development of more sophisticated models. Shallow earthquakes occur on *faults*, fractures in rocks across which the two sides have moved relative to each other. Although observations clearly support faults as the primary structure involved in shallow earthquakes, the exact mechanics of deep earthquakes remains somewhat obscure. But, the radiation of seismic waves from deep sources very closely resembles the radiation of seismic waves from shallow sources and we can match observations from both deep and shallow earthquakes using the same simplified mechanical models. In this introductory overview, we

focus on shallow earthquakes, but key results have value for all earthquakes.

In elementary physics, the relationship between the forces acting on particles and the particle motions that apply to individual particles also apply to a continuum of particles that we use to represent the atmosphere, hydrosphere, cryosphere, solid Earth and fluid outer core of the planet. Our physical models of earthquakes and seismic-wave propagation are based in an extension of particle-mechanics that is called ***continuum mechanics***. We use continuum mechanics in seismology as a model, not as a literal representation of the planet. In particle mechanics, we were concerned with point masses and forces, in continuum mechanics, we are concerned with surfaces, volumes, and *stresses* (forces per unit area). The use of a continuous model may seem strange considering that the central structure in the earthquake process is a discontinuous fracture – a fault rupture. But decades of experience have demonstrated that treating Earth as a continuum with boundaries (including the fault) is an effective way to understand earthquakes and seismic-wave excitation and propagation.

To model earthquakes, the obvious surface of interest is a fault, along which rocks spend most of the time locked together by *friction* (the force resisting relative motion of the rock on either side of the fault). An earthquake is the sudden slip of the rocks adjacent to the fault past one other. The abrupt movement results in a reduction in shear stress over the rupture surface, which we call the *stress drop*. There is a corresponding drop in strain energy in the rock volume surrounding the fault. To model seismic-wave propagation, we focus on the interaction of surfaces bounding parts of the material that exert stresses upon one another as seismic waves propagate deformation and transmit energy outward from the source. The initial onset of motion is called a *wavefront*, and is a surface across which the deforming volume of the material (initially the source volume) exerts an elastic stress on material at rest until the seismic waves travel reached the region. Equally important are boundaries between volumes with different geologic properties, including Earth's surface, the boundaries between the crust and mantle and the mantle and core, as well as any smooth or abrupt change in any region. Seismic waves interact with these boundaries, changing direction and creating reflected and converted (*P*-to-*S*, *S*-to-*P*) waves that travel in different, but predictable, directions.

Continuum mechanics leads to new mathematical tools and to more general physical laws that will be developed and explored with more rigor later (see Part II). Here, we provide a simpler overview of stress, strain, and elasticity, which we will use to explore the conceptual physics of earthquakes and elastic wave propagation. We also describe modest extensions of the simple model that include some of the effects of anelasticity. We introduce a number of concepts and terms that are routinely involved in seismogram analysis and that will be revisited often during subsequent chapters.

2.1 Stress

A mechanical *stress* is an exertion of one surface in a material on another. The word exertion may seem vague, but the more familiar term from elementary mechanics, force, is commonly used to represent an exertion applied at a point (not a whole surface). A surface-to-surface exertion, or stress, represents a spatial distribution of force with a specific direction and a magnitude quantified as a force per unit area. Imagine standing and holding a chair – you exert a force per unit area on the floor beneath your feet. If you stand on the chair, the same weight will exert a different force per unit area on the floor because the area beneath the feet of the chair differs from the area beneath your feet.

In seismology we are concerned with exertions across fault surfaces, where the most important stresses are the driving stresses produced by plate tectonics and the resistance produced by fault friction. Earthquakes are ultimately a result of plate motions (part of Earth's cooling process) modulated by fault frictional resistance. When fault slip is sudden, the rapid reduction of strain in rocks adjacent to the fault initiates accelerations that propagate outward as seismic waves. Seismic waves propagate by transmitting stress changes originating in the sudden near-fault accelerations, from the source region outward into the planet. Seismic waves propagate by exerting stresses across the surface (the wavefront) separating a deforming region and the as-yet undeformed volume surrounding the wave source. We quantify the stress as a function of position and time using a tensor that describes stress independently of the surface upon which it acts. A *tensor* is a coordinate-system independent mathematical tool used to represent physical quantities that may have directional characteristics. The use of tensors allows us to consider the stress at a particular position acting on any surface of interest passing through that position (e.g., faults with different orientations).

All the materials in the Earth system experience some stress. Perhaps the most familiar stress is atmospheric pressure, which near Earth's surface is roughly 101 kPa (the Pascal

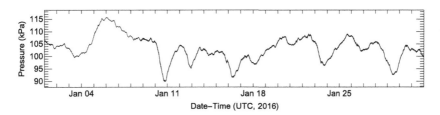

FIGURE 2.1 Atmospheric pressure variations recorded during January, 2016 at seismic station TA-N59A, located in eastern Pennsylvania. The atmospheric pressure averaged roughly 104 kPa during the month as several winter storms passed to the east (low pressures dips on the diagram). One Pa is 1.02×10^{-5} atm, so the numbers average roughly 1.06 atm.

(Pa) is the SI unit for pressure, which represents a Newton per meter-squared – a force per unit area). Weather-related variations of atmospheric pressure are typically on the order of a few kPa (kilo-pascal) (Fig. 2.1). The pressure in a typical car tire is roughly about 220 kPa. *Hydrostatic pressure* within a body of water or within a continuously connected set of fractures in the shallow Earth is larger than atmospheric pressure and measured in mega-pascals (MPa). *Lithostatic pressure* within Earth is much larger and is generally measured in giga-pascals (GPa).

To a first approximation, hydrostatic and lithostatic pressure can be computed from the density, $\rho(z)$, and gravitational acceleration, $g(z)$, within the water or rock columns,

$$P(z) = \rho(z)\, g(z)\, z \,, \qquad (2.1)$$

where z represents the depth within the planet. For first-order calculations, we ignore the atmospheric pressure, which is only significant very close to the surface. The hydrostatic pressure gradient is roughly 10 MPa/km. The lithostatic pressure gradient within the continental crust is roughly 0.03 GPa/km; the pressure near the crust-mantle boundary is roughly 1 GPa. More accurate models of the lithostatic pressure account for variations in density resulting from compositional changes as well as changes induced in materials because of compression and mineralogical changes. Pressure is called an isotropic stress because it has the same value regardless of the orientation of the surface on

which you choose to examine it. At the same location, a pressure is the same on a vertical surface, a horizontal surface, or any other surface. Within Earth, pressure ranges from atmospheric pressure at the surface, ~ 100 kPa, to a value over 350 GPa at the center (Fig. 2.2).

For earthquakes and seismic waves, the most important stresses are the deviations from the lithostatic pressure field, which are often called *deviatoric stresses*. At a particular location within Earth and at a particular time, the deviatoric stresses depend on the orientation of the surfaces involved in the deformation process. For any complete analysis, we must specify the orientation of the surface and the value of the stresses. A *traction* is the stress associated with a particular surface, described as a vector. We are often interested in how one surface pushes and/or shears another, so we often resolve the traction acting on a surface into two components, a *normal stress* that operates perpendicular to the surface and a *shear stress* that operates parallel to the surface. Friction is a shear stress since it resists motion across the fault – but as we will see, the frictional resistance also depends on the normal stress acting on the surface.

Specifying a different value of stress for each surface at each location would be tedious, so fortunately it's not necessary. At a particular location, the stress on any surface can be calculated if we know the value of stress on three orthogonal planes (at the same position). The nine components specifying the stress on the three planes

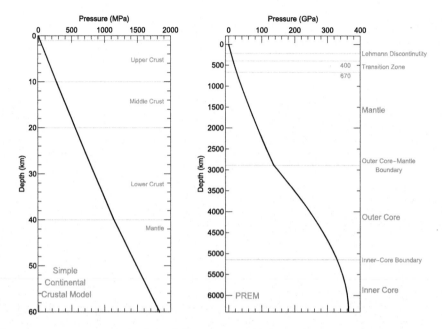

FIGURE 2.2 Pressure variations with depth for (left) a generic crustal model (constructed for analyses of seismic waves in the central United States by Robert Herrmann) and (right) the PREM, data from Peter Bormann, in the New Manual of Seismological Observatory Practice (NMSOP).

can be ordered to form the stress tensor,

$$\sigma_{ij} = \begin{pmatrix} \sigma_{11} & \sigma_{12} & \sigma_{13} \\ \sigma_{21} & \sigma_{22} & \sigma_{23} \\ \sigma_{31} & \sigma_{32} & \sigma_{33} \end{pmatrix} , \qquad (2.2)$$

where the σ_{ij} represent the components of the stress acting on planes parallel to the coordinate axes (three orthogonal spatial directions). The first index, i, identifies a plane that is perpendicular to a vector in the i-direction. The second index, j, identifies the component of the force acting in the j-direction. In three dimensions, we have nine stresses. We adopt a positive sign convention for stress components that are positively directed forces acting on positive faces (faces with normals in the positive component directions) and for negatively directed forces on negative faces (faces with normals in the negative component directions). See Box 2.1 for a more general description of index notation, which we will use throughout the text.

The conservation of momentum requires that the stress tensor is symmetric, so $\sigma_{ij} = \sigma_{ji}$ and only six of the nine tensor components are unique. Given the stresses acting on the three orthogonal planes, Cauchy worked out how to calculate the stress acting on any particular plane with an orientation described using a unit-magnitude surface normal vector, n_i. If we represent the stress or traction, as t_j, then

$$t_j = \sigma_{ij} n_i . \qquad (2.3)$$

For simplicity, we often show the traction acting on a surface using a single vector (Fig. 2.3). But the entire surface is experiencing force, not just the location illustrated. The strength and orientation of the force may change with position along the surface, but in many physical situations, the stress changes slowly across the region of interest and a single vector suffices.

Box 2.1 Index notation

The large number of elements in the stress and strain tensors and the numerous terms involving their spatial derivatives make it useful to adopt index (or indicial) notation to represent quantities. In general, the stress and strain components are prescribed with respect to some convenient reference system (e.g., the Cartesian system, x_1, x_2, x_3), and we use subscripts (indexes) to indicate surfaces and directions in the reference system. A surface (e.g., the x_2x_3 plane) can be identified by the direction of the normal to the surface. The components of a vector are identified using subscripts, for example, u_i is used to represent the vector

$$\mathbf{u} = u_1\hat{\mathbf{x}}_1 + u_2\hat{\mathbf{x}}_2 + u_3\hat{\mathbf{x}}_3 , \tag{B2.1.1}$$

where $\hat{\mathbf{x}}_i$ are unit vectors in the coordinate directions. The term u_i is understood to take on values $i = 1, 2, 3$, as appropriate in a given equation. The nine terms of the displacement gradient can be represented by a single term

$$\frac{\partial u_i}{\partial x_j} = \begin{pmatrix} \frac{\partial u_1}{\partial x_1} & \frac{\partial u_1}{\partial x_2} & \frac{\partial u_1}{\partial x_3} \\ \frac{\partial u_2}{\partial x_1} & \frac{\partial u_2}{\partial x_2} & \frac{\partial u_2}{\partial x_3} \\ \frac{\partial u_3}{\partial x_1} & \frac{\partial u_3}{\partial x_2} & \frac{\partial u_3}{\partial x_3} \end{pmatrix} \tag{B2.1.2}$$

where the index representation denotes the appropriate component for given values of i and j. This quantity can be written even more compactly using a comma to represent differentiation,

$$u_{i,j} \equiv \frac{\partial u_i}{\partial x_j} . \tag{B2.1.3}$$

Special quantities such as the *Kronecker Delta* can be written compactly with index notation,

$$\delta_{ij} = \left\{ \begin{array}{ll} 0 & \text{for } i \neq j \\ 1 & \text{for } i = j \end{array} \right. \quad i, j = 1, 2, 3 . \tag{B2.1.4}$$

Throughout this text we also assume the *Einstein Summation Notation*, in which repetition of indices within a term explicitly requires summation over that index. For example,

$$\varepsilon_{nn} = \varepsilon_{11} + \varepsilon_{22} + \varepsilon_{33} \tag{B2.1.5}$$

the repeated index n implies summation. The summation rule holds for repeated indices within any single term,

$$x_i y_i = x_1 y_1 + x_2 y_2 + x_3 y_3$$

$$y_{i,i} = \frac{\partial y_1}{\partial x_1} + \frac{\partial y_2}{\partial x_2} + \frac{\partial y_3}{\partial x_3} . \tag{B2.1.6}$$

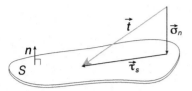

FIGURE 2.3 A stress vector, or traction, t acting on a surface, S, can be resolved into components normal and parallel to the surface upon which it is acting ($t = \sigma_n + \tau_s$). We call the normal component the normal stress (σ_n), and the parallel component shear stress (τ_s).

2.2 Strain and rotation

In elementary mechanics, if you apply a force to a rigid object it translates and/or rotates. In continuum mechanics, if you stress a material, it translates, rotates, and/or deforms. For small deformations (infinitesimal), we can represent the deformation as a combination of linear strains and rotations. For larger deformations (finite), translation, such as that across a fault also becomes important. An analysis of long-term faulting requires that we track and account for large deformations. For a single earthquake, we can treat the isolated rupture using a model of an elastic material containing a crack. Seismic-wave analyses, for the most part, involve small deformations and a linearized theory is sufficient. A more rigorous development of the mechanics is presented later (see Part II), our goal here is to describe the key relations between displacement gradients, strains, and rotations.

A *strain* is a change in length per unit length, a quantity with no physical units. Strain is a deformation that transforms an initial into a final shape. Strains are categorized into two types, *longitudinal (normal) strains* and *shear strains*. To be precise, we need to introduce a position vector x_i, and a displacement vector u_i. For small deformations, we can represent the strain field as a tensor, ε_{ij} defined using displacement gradients,

$$\varepsilon_{ij} = \frac{1}{2}\left(\frac{\partial u_i}{\partial x_j} + \frac{\partial u_j}{\partial x_i}\right). \qquad (2.7)$$

Longitudinal strains correspond to $i = j$ and the shear strains correspond to $i \neq j$. Longitudinal strains represent elongations or contractions along the coordinate axes directions; shear strains represent the change in angles between parts of the material aligned with coordinate directions before and after the deformation. The strain tensor is symmetric by definition, $\varepsilon_{ij} = \varepsilon_{ji}$.

Not all deformations can be described by strains, to be complete, we must account for small-scale rotations within the material. Focusing again on small deformations, the rotation field, ω_{ij}, can be described by

$$\omega_{ij} = \frac{1}{2}\left(\frac{\partial u_i}{\partial x_j} - \frac{\partial u_j}{\partial x_i}\right). \qquad (2.8)$$

The rotation tensor is anti-symmetric since by definition $\omega_{ij} = -\omega_{ji}$. The strain and rotation tensors are the even and odd parts of the displacement gradient tensor,

$$\frac{\partial u_i}{\partial x_j} = \varepsilon_{ij} + \omega_{ij}, \qquad (2.9)$$

and together, they quantify small deformations of a continuum completely. Modern seismology is mostly based on translational motions, which are observed using seismometers. Seismic-arrays (sets of seismic stations configured in a small geographic region) allow us to measure displacement spatial gradients that can be used to estimate the strains and rotations. Work continues on developing rotational sensors to measure the rotations directly. Single-station rotations combined with more traditional translations can constrain wave direction and local phase velocities (similar to an array) and accurate near-field rotation measurements could provide complementary constraints on the slip distribution along an earthquake rupture. Perhaps most important, rotations may contribute to damaging motions near the source, so observing them is an important step towards deepening our understanding of them. Regardless, the seismogram-based analyses on which we focus illustrate the rich store of information on earthquakes and Earth's interior that seismic waves carry.

2.3 Hooke's law

Robert Hooke's famous law, originally developed in 1676 for a spring, is a statement that the restoring force in a deformed elastic material is in direct proportion to the deformation. In an ideal elastic system, energy is converted between kinetic (motion) to elastic potential (strain) energy without loss (all energy is recoverable). In a general sense, Hooke's Law encapsulates a pattern observed for elastic materials – stress varies linearly in proportion to strain

$$Stress \propto Strain \, . \qquad (2.10)$$

This is an example of a **constitutive relationship** – an expression that relates a kinematic quantity (position, velocity, etc.) and dynamic quantity (force, stress, etc.). His idealization is generally applicable to rocks as long as the strains are small and the time-scale of the deformation is short. These two conditions are reasonable first-order approximations for simple models of faulting and seismic-wave propagation. The mechanical behavior of rocks depends on the time scale but most of the Earth responds nearly elastically for short-term deformations produced by seismic waves. Even the most important part of the mechanical system of earthquakes (the part storing energy) behaves in a nearly manner elastic manner on times scales of earthquake rupture and slip.

As a result, we can derive much insight into both earthquakes and seismic waves by studying elastic models. For example, Hooke also re-

alized that the vibration of a linear elastic material is described by what we now call **simple harmonic motion**. Harmonic (frequency-based) approaches are central tools in seismogram analysis. However, not all the deformation associated with faulting or with seismic-wave propagation is elastic. Some deformation during the earthquake cycle may produce permanent rock deformation, and nonelastic processes attenuate seismic waves as they travel through the planet. Elasticity is a good first-order approximation such that we formulate many of the conceptual models of earthquake faulting and seismic wave propagation with elastic assumptions and then modify them to account for non-elastic effects.

Hooke's Law has been generalized to allow for more abstract analyses using tensors:

$$\sigma_{ij} = C_{ijkl}\, \varepsilon_{kl} \, , \qquad (2.11)$$

where the C_{ijkl} represent elastic moduli (with units of stress) that depend on the material properties and the summation convention for repeated indices is assumed. Different earth materials possess different elastic moduli, and these play a role in both the excitation and propagation of waves. A general, **anisotropic** elastic solid responds differently depending on the orientation of the stresses and strains. In the most general case, symmetries and thermodynamic principles reduce the number of unique moduli (C_{ijkl}) a material may possess from a theoretical maximum of 81 to 21. Observation suggests that rocks are anisotropic and the analysis of anisotropic materials is an important part of modern seismology. The topic will be revisited later and is explored in more detail in more advanced texts. However, experience suggests that we can derive valuable insight into seismic sources and seismic waves from analysis of the more simple **isotropic** elastic materials, for which the deformation properties includes no directional dependence.

For an isotropic material, only two of the C_{ijkl} values are unique. We can work with Lamè's parameters, λ and μ (also called the shear modulus), or the bulk modulus, K, and μ. For an isotropic material, Hooke's Law reduces to:

$$\sigma_{ij} = \lambda \delta_{ij} \varepsilon_{kk} + 2\mu \varepsilon_{ij}$$

$$= 3K \frac{\varepsilon_{kk}}{3} \delta_{ij} + 2\mu \left(\varepsilon_{ij} - \frac{\varepsilon_{kk}}{3} \delta_{ij}\right) \qquad (2.12)$$

where δ_{ij} is a Kronecker Delta, which has a value of one for $i = j$ and zero for $i \neq j$, and the summation convention of index notation for repeated indices is assumed (i.e., $\varepsilon_{kk} = \varepsilon_{11} + \varepsilon_{22} + \varepsilon_{33}$). The index-notion expression of Hooke's Law is exceptionally compact. Hooke's Law includes nine equations – using the integers 1, 2, 3 to represent the coordinate directions, the expanded form of Eq. ((2.12), Lamè form) is

$$\sigma_{11} = \lambda \left(\varepsilon_{11} + \varepsilon_{22} + \varepsilon_{33}\right) + 2\mu\varepsilon_{33}$$
$$\sigma_{12} = 2\mu\varepsilon_{12}$$
$$\sigma_{13} = 2\mu\varepsilon_{13}$$
$$\sigma_{21} = 2\mu\varepsilon_{21}$$
$$\sigma_{22} = \lambda \left(\varepsilon_{11} + \varepsilon_{22} + \varepsilon_{33}\right) + 2\mu\varepsilon_{22}$$
$$\sigma_{23} = 2\mu\varepsilon_{23}$$
$$\sigma_{31} = 2\mu\varepsilon_{31}$$
$$\sigma_{32} = 2\mu\varepsilon_{32}$$
$$\sigma_{33} = \lambda \left(\varepsilon_{11} + \varepsilon_{22} + \varepsilon_{33}\right) + 2\mu\varepsilon_{33} . \qquad (2.13)$$

Not only does index notation save space, it often clarifies the relationships by reducing the number of terms to a manageable number.

2.3.1 Elastic potential energy

The most important property of an elastic material is the ability for the material to store energy in a recoverable manner. For both seismic-wave propagation and earthquake-related deformation adjacent to faults, rocks behave nearly elastically (much of the energy is recoverable). Decades of experience suggest that considering earth materials to be elastic and then adjusting our models for anelastic behavior (which absorbs energy) is a fruitful approach to understanding the key processes involved in earthquake rupture and in seismic-wave propagation. For example, energy stored in shear deformation adjacent to a friction-locked fault can be related to the strain using

$$E_\varepsilon \sim \int_V \frac{1}{2} \mu \varepsilon^2 \, dV , \qquad (2.14)$$

where ε represents the strain and V the strained volume of rock. The largest deviatoric strains typical in crustal rocks are roughly about 10^{-3}, those involved in earthquake processes are often on the order of 10^{-4}–10^{-5}. Using the maximal value and a shear modulus of about 30 GPa leads to an upper bound for the maximum energy density of about $15\,\mathrm{kJ\,m^{-3}}$ and a typical rough estimate for that involved in crustal faulting are on the order of 1–$100\,\mathrm{J/m^3}$. A joule is a small quantity of energy, but each cubic kilometer includes 10^9 cubic meters, so the energy released in larger earthquakes is substantial.

In the remainder of this chapter, we consider earthquake and seismic-wave mechanics from an ideal linear elastic viewpoint, and then follow with a discussion of allowances that are made to relax the approximation of elasticity. We start with earthquakes, which are in fact, a sudden release of elastic energy. But longer-term views of earthquake processes reveal important and easily observable non-elastic phenomena. After discussing earthquakes, we discuss seismic waves, which are well understood from a linearly elastic perspective, but which also are observably affected by the anelastic behavior of earth materials. Our approach is the same in both instances, we start with simpler elastic models that provide clear insight, then adapt our elasticity-based models to account for anelastic processes.

2.4 Earthquakes: conceptual models

The large range and diversity of physical environments experiencing tectonic deformation results in a spectrum of fault offset processes and a wide range of slip rates varying from slow steady creep to the sudden motions that occur during earthquakes. Continuous motion along plate boundary faults occurs at rates of a few tens of millimeters per year ($\sim 10^{-8}$ m/s). Average intraplate long-term fault slip rates are typically much lower. Our estimates of the slip rates during earthquakes are ~ 1 m/s, so fault slip rates of interest span at least nine orders of magnitude. Our focus is strain release that occurs rapidly enough to generate seismic waves. These are the processes most people refer to when they use the word earthquake. But such "quick" earthquakes are part of a spectrum of slip rates that occur along different fault segments and different regions. Further, earthquake slip rates are not uniform, and the variation is an important observation for unraveling the physics of earthquakes.

We can use the concepts of stress and strain to explore simple models of fault slip and fault rupture, i.e. earthquakes. We'll start with a tectonic/geodetic view, elastic rebound, based mostly on kinematic quantities including plate-motions, deformation, and strain (but with an essential role for friction). The view provides a solid conceptual understanding of the cause of earthquakes. To understand the shaking produced by earthquakes (particularly the strong shaking close to the source) we must understand the details of the fault-rupture process – a process more conveniently examined using dynamic quantities such as friction, absolute stress, and stress drop. But since the excitation of seismic waves occurs during the sudden stress reduction along the fault, seismic observations are more easily related to stress drop rather than absolute stress or strain.

2.4.1 Elastic rebound

The elastic rebound model (Chapter 1; Fig. 2.4) is a summary of the elastic stress and strain changes of earthquake-related processes. Earthquakes occur because the friction on faults resists plate motion and prevents offset of the rocks on either side of the fault for long periods of time. While the rocks along the fault are locked and not offsetting, away from the fault, the plates move steadily. This differential motion strains the rocks adjacent to the fault, storing recoverable elastic energy in the system. As strain accumulates, it exerts a stress on the fault that is matched by friction up to a level that depends on the material and physical properties within the fault (fluid content, frictional characteristics, stiffness, temperature, surface roughness, etc.). When the stress on the fault nears the capacity of the fault to resist motion, an instability can form that releases the strain energy by offsetting the rocks across the rupture and dropping the stress on the ruptured fault segment. Most of the strain energy release is dissipated within and near the fault as heat and some is used to fracture areas of the fault that had annealed since the last earthquake. Some energy fractures and damages the rock adjacent to the fault. When the slip is sudden, seismic waves are excited, and they carry a fraction of the energy away from rupture and produce seismic shaking.

Although greatly simplified, elastic rebound is a good conceptual model of the tectonic earthquake process. The blocks represent plates and the fault-locking-strain-building nicely explains the accumulation of elastic-potential (strain) energy in the rocks adjacent to the fault. The model was developed in 1910 by H.F. Reid to explain geodetic observations from the 1906 San Francisco earthquake. Understanding why the blocks moved (plate tectonics) was developed in the 1950's and 1960's. Thus, to first order, we understand why earthquake occur – they are part of Earth's plate motions, driven by thermal pro-

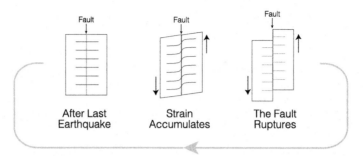

FIGURE 2.4 Cartoon illustration of the elastic rebound conceptual model of earthquakes. The left panel shows a fault a short time after the previous earthquake (start of the next earthquake cycle). The horizontal lines are reference markers to illustrate the deformation pattern before and during the next earthquake. The middle panel shows the fault with the fault-adjacent rocks held together by friction as motion continues at distance. The result is strain accumulation in the fault-adjacent rocks. As the strain increases, it exerts an increasingly larger stress on the fault surface, eventually overcoming the frictional resistance and resulting in a sudden offset – an earthquake. The offset relieves the strain and the stress on the fault drops leaving a permanent offset across the fault, shown in the right-most panel. The gray arrow indicates that in the model, the process repeats as long as the motions away from the fault continue.

cesses related to Earth formation and cooling history. However, this simplest elastic-rebound based model omits important details. Earthquakes do not occur between isolated blocks but within the lithosphere that on the time scales of earthquake processes includes regions of both elastic and ductile behavior. Dynamic rock friction plays an essential role in the release of strain energy once sliding starts. Seismic wave excitation depends on the slip and slip-rate distribution during sliding. Aftershocks are affected by the radiation of mainshock waves, the stress and strain changes during faulting, and interactions between elastic and ductile regions of the source volume. We briefly discuss some of what we have learned about these processes next, but many of the details remain a focus of current earthquake-science research.

2.4.2 Rock friction and frictional sliding

To understand the earthquake rupture process, we must understand a little about rock friction. Similar to simple models from introductory physics, we can think of rock friction in terms of *static* (no-motion) and *dynamic* (in motion) friction. In both cases, frictional resistance is pro-

portional to normal stress acting on the surface (Fig. 2.5, left). The *coefficient of friction* (COF) is defined as the ratio of the shear-to-normal stress required to cause sliding. At low normal stresses rock COF varies substantially. At high normal stresses associated with all but the shallowest part of a large fault, the friction required to cause sliding is remarkably relatively independent of rock type and corresponds to a static COF in the range of 0.60 to 0.85. However, while a fault is locked, static friction increases in proportion to the logarithm of fault "hold" time (how long the fault has been locked). This is the result of microscopic processes that slowly increase the contact areas across the juxtaposed, rough surfaces. The temporal changes could cause differences between faults with short earthquake recurrence intervals relative to long earthquake recurrence intervals, but other factors also contribute to overall fault behavior.

When sliding begins, friction transitions from static to dynamic, and two primary responses are possible. In some materials and environments, sliding can induce a consistent increase in resistance – and the system is called *stable* because it resists further sliding. In other environments, sliding causes an initial increase in resis-

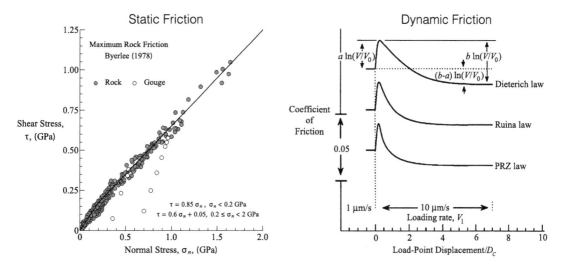

FIGURE 2.5 (left) After (Byerlee, 1978) – at relatively high normal stresses, the static friction that must be overcome to cause sliding is relatively independent of rock type and corresponds to a coefficient of static friction in the range of 0.6 to 0.85. (right) From Marone (1998) – calculated variations in the coefficient of dynamic friction versus normalized displacement for the three rate and state friction laws. Constitutive parameters defined at the top apply to each curve, $a = 0.01$, $b = 0.015$, $D_c = 20\,\mu m$, and $k = 0.01\,\mu m^{-1}$. The effect of a step increase sliding speed is shown for each law. The initial sliding velocity was $V_0 = 1\,\mu m/s$ and a near-instantaneous, order-of-magnitude increase (but still much lower than fault sliding speeds) was applied.

tance, but if slip extends beyond a critical distance, D_c, a significant decrease in resistance occurs – and the system is called *unstable* (Fig. 2.5). Several quantities influence the frictional state, including the temperature, effective pressure, composition, rate of sliding, and system stiffness (a ratio of the rock shear modulus to the sliding area's spatial dimension). The dependence of dynamic resistance on sliding velocity is often used to describe the friction as *velocity strengthening* (stable) and *velocity weakening* (unstable). In quantitative rock-friction empirical models, earthquakes only nucleate in unstable regions, but ruptures can propagate into regions of *conditional slip weakening* (a region that is stable for slower sliding but unstable for faster sliding).

The frictionally unstable region of a fault is called the *seismogenic zone*, where earthquakes occur. Earthquake nucleation is restricted to this region primarily as a result of two effects. Tem-

perature limits the depth at which fault-adjacent rocks can both store elastic strain (without ductile flow slowly releasing it) and the depth at which the fault remains velocity weakening. Regions of conditional stability occur at any depth along a large fault, but are most common at the shallowest depths and just beneath the base of the seismogenic zone. Existing rate-state dependent friction laws predict this basic behavior well. The shallow transition is a result of mineralogy, fluids, and low normal stresses; the deeper transition results from increased temperature. The deeper transition is often located in the mid-to-lower crustal, precisely where depends on lithospheric thermal properties. We also observe differences between continental and oceanic systems because the transition occurs $\sim 300°C$ in quartz-rich continental crust and near $\sim 600°C$ in the olivine-rich upper mantle. The seismogenic zone width (depth extent) influences the maximum earthquake size in a tectonic system because it limits the along-dip extent of co-seismic slip during large earthquakes. In vertical strike-slip fault systems, the down-dip width of the seismogenic zone may be a few 10's of kilometers. In shallow-dipping subduction fault systems, the down-dip width may be 100–200 km. The result is much larger seismogenic zones in subduction systems, which is why they host the largest earthquakes. In seismically active regions, the zone can be mapped reasonably accurately using small-magnitude earthquake locations (e.g., Fig. 3.15).

Also of great interest in earthquake science are the subtleties of earthquake nucleation, which depend on the details of the transition from static to dynamic friction (Fig. 2.5, right). The value of D_c is important in earthquake nucleation processes and may indicate a minimum size of earthquakes, if one exists. But D_c is not well known outside the laboratory, and accurately scaling the laboratory value to earthquake dimensions is a challenge. Efforts to unravel the details have important implications for earthquake triggering and earthquake early warning efforts (see Chapter 8), which could benefit from any indication of a nascent earthquake's ultimate capacity to grow into a large hazardous event. D_c is certainly much smaller than offsets produced during larger earthquakes, so most of the slip in a large earthquake occurs at the lower dynamic friction value. Thus, most of the heat produced and most of the strain is released under the dynamic resistance. Frictionally controlled spatial variations in the rate and roughness of sliding during an earthquake affect the intensity and frequency content of seismic shaking. Near the source, this translates into vibrations that may or may not resonate human-made constructions causing damage. Understanding these effects is an important goal of **strong-motion** seismology and essential for accurate estimates of seismic hazards. Eventually, when slip sufficient to lower the strain significantly has occurred, the dynamic friction is enough to resist further motion. Sliding stops.

2.4.3 Anelastic processes and postseismic relaxation

The elastic view of strain accumulation and release, even with the addition of friction does not explain all the first-order observations of the earthquake process. Fig. 2.6 shows the surface deformation observed just before, during, and after the 2004 Parkfield earthquake. The pattern we expect for an ideal elastic system includes steadily (linearly) accumulating surface deformation, a sudden offset during the earthquake, and a return to linearly accumulating motion after the earthquake. The observations are different. The preseismic motion is roughly linear, and the earthquake (indicated by the vertical line) produces a sudden offset. However, the motion following the earthquake is not a simple return to the pre-earthquake rates. The motion is nonlinear, initially fast but steadily slowing to a rate comparable to the pre-event motion. The nonlinear post-seismic motions are caused by a combination of aftershocks, slow afterslip in the earthquake rupture zone, and the response of the ductile material near the rupture and depending on the earthquake size, can take days, months, or years to return to pre-event rates.

The elastic rebound model is an ideal approximation of the tectonic earthquake process. A slightly more representative model is illustrated in Fig. 2.7. This still-simplified physical model includes an elastic layer overlying a ductile layer that includes a time-dependent response to deformation. The model resembles a very simple view of a large, vertical strike-slip fault such as the San Andreas in California. Strain accumulates in the overlying elastic layer, which except during an earthquake, is locked by friction. As they strain accumulates, the lower portions of the boundary continue to move past one another, producing a pattern of surface displacement that can be well-observed at the surface adjacent to large strike slip faults (Fig. 2.6). Throughout the relatively long build-up of strain, a small region of the ductile material coupled to the elastic layer lags behind the

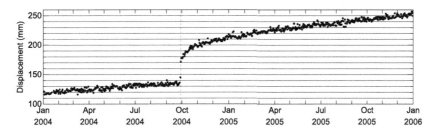

FIGURE 2.6 GPS observations of the 2004 Parkfield Earthquake showing the effect of anelastic deformation on the surface motions. The nonlinear motion immediately after the earthquake is a result of aftershocks, afterslip, and ductile deformation in the region surrounding the rupture (dominated by the sudden elastic recovery). These same data were shown in Fig. 1.4 of Chapter 1.

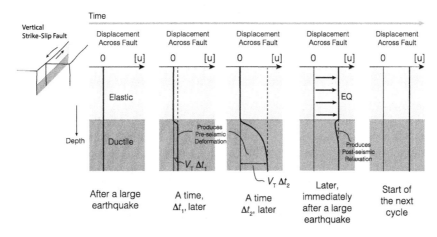

FIGURE 2.7 Summary of deformation processes operating during an idealized earthquake cycle.

deeper ductile deformation. The elastic strain is released by the earthquake suddenly, but the ductile material lags behind, deforming in the days, months, and even years following the rupture – postseismic relaxation. The return to background motions may take centuries in intraplate regions where the strain rates are low and the adjustments are slow.

The simple elastic and slightly modified anelastic rebound models provide a good conceptual understanding of first-order origin of tectonic earthquakes. The models originated to explain geodetic observations collected near faults, starting early last century and extending to the present. Elastic fault models also help

us understand important patterns in seismic observations, which demonstrate that at least to first order, large and small earthquakes are *self-similar* – that is, several key quantities that characterize the faulting and rupture processes maintain a roughly constant ratio across a large range of earthquake sizes.

2.4.4 Earthquake scaling relations & stress drop

Two parameters that are relatively easy to measure following large shallow earthquakes are the seismic moment (M_0), and the rupture area, A. Seismic moment is directly estimated

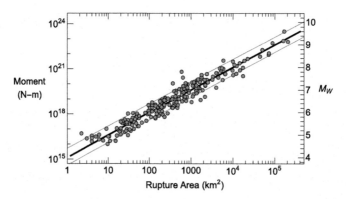

FIGURE 2.8 Rupture area (estimated from early aftershock distribution) versus seismic moment (estimated from seismograms), one of the fundamental patterns in observational earthquake seismology. Data from Wells and Coppersmith (1994) and Henry and Das (2002) with some more recent events added. The line has a slope of 3/2 with the optimal intercept, $\log_{10} M_0 = (3/2)\log_{10} A + 15.1$. The gray lines bounding most of the observations have identical slopes and correspond to intercepts equal to 14.6 and 15.6.

from seismograms and rupture area can be estimated from the spatial extent of the early aftershocks (first day for large earthquakes because the region of aftershocks often expands slightly with time following the mainshock). Rupture area can also be estimated from the frequency content of seismic waves, if we adopt a simple model of earthquake rupture. Such spectral modeling is the most common method used to estimate the size of small earthquakes. Neither moment nor area can be measured perfectly, yet they can be measured routinely and their uncertainties can be estimated.

Fig. 2.8 is a plot of the logarithms of seismic moment versus rupture area and is an important earthquake pattern. Larger earthquakes correspond to larger ruptures and the logarithm of rupture area varies linearly with the logarithm of the seismic moment. A least-square-optimal fit to the observations produces an estimated slope of 1.47 ± 0.03, or roughly 3/2. This is the same slope chosen earlier by Gutenberg and Richter to relate seismic-energy and surface-wave magnitude (Chapter 7). Tsuboi (1959) suggested that a slope of (3/2) in the energy relationship is expected if the strain energy driving earthquakes is roughly constant, the 3/2 repre-

sents a ratio of the source volume (the volume that contributed to strain energy release) and the projection of that volume onto the fault (the area that ruptured during the earthquake). Tusboi's idea is not a detailed mechanical explanation, but observed earthquake patterns can be related to simplified theoretical earthquake models to provide insight into the physics of earthquake processes.

For example, we can combine the empirical moment-area relationship (Fig. 2.8) with the definition of seismic moment ($M_0 = \mu A \bar{D}$), which originates in an elastic model of fault slip, to explore how earthquake slip, $\bar{D}$, varies with rupture area. Rewriting the definitions in terms of logarithms, we have

$$\log_{10} M_0 = \log_{10} \mu + \log_{10} A + \log \bar{D}, \quad (2.15)$$

where $\mu \sim 30\,\mathrm{GPa}$ is a typical value for the shear modulus. For convenience, convert the empirical relationship from Fig. 2.8 such that the area is expressed in m^2 (then everything is in SI units),

$$\log_{10} M_0 = \frac{3}{2} \log_{10} A + 6.1, \quad (2.16)$$

where M_0 is in N-m and A is in m^2. Equating the definition and the empirical relationship we have,

$$\log_{10} A + \log \bar{D} = \frac{3}{2} \log_{10} A + 6.1$$

$$\log \bar{D} = \frac{1}{2} \log_{10} A + 6.1 - \log_{10} \mu \qquad (2.17)$$

$$\log \bar{D} \sim \frac{1}{2} \log_{10} A - 4.4 .$$

The pattern in Fig. 2.8 combined with the elastic model that physically defines seismic moment suggests that the average slip during an earthquake scales with the square root of area. We made a significant, but reasonable assumption in our derivation ($\mu \sim 30\,\mathrm{GPa}$). Our result is identical to the relationship developed by Leonard (2010), using many of the same data.

The observed pattern provides more information, if we combine it with simple elastic earthquake models. Stress drop, $\Delta\sigma$, the difference between the stress at a location on the fault before and after an earthquake is an important quantity in the rupture process and the excitation of seismic waves. In general, the stress drop varies temporally and spatially during an earthquake. An earthquake's **static stress drop** is the final stress drop averaged over the rupture area and a quantity that can be computed for ideal, simple elastic earthquake models (Table 2.1). Consider the earthquake ruptures shown conceptually in Fig. 2.9. We represent the rupture length with L, the rupture width with w, and the average displacement with $\bar{D}$. Small and moderate-size events have rupture dimensions less than the minimum dimension of the seismogenic zone (usually w_L in the figure) – they are often reasonably well approximated by ruptures with an aspect ratio of roughly one. Large events typically rupture much of the down-dip extent of the seismogenic zone and grow in size by expanding along strike – they sometimes have a length several to many times greater than their width. Rupture aspect ratio plays a role in important simplifications of the model formulas.

TABLE 2.1 Stress drop and seismic moment relations for three fault types with shallow, near-surface rupture.

	Circular (radius = a)	Strike slip (L » w)	Dip slip
$\Delta\sigma$	$\frac{7\pi}{16}\mu\left(\frac{D}{a}\right)$	$\frac{2}{\pi}\mu\left(\frac{D}{w}\right)$	$\frac{4(\lambda+\mu)}{\pi(\lambda+2\mu)}\mu\left(\frac{D}{w}\right)$
M_0	$\frac{16}{7}\Delta\sigma a^3$	$\frac{\pi}{2}\Delta\sigma w^2 L$	$\frac{\pi(\lambda+2\mu)}{4(\lambda+\mu)}\Delta\sigma w^2 L$

Depending on the model, the strain change associated with an average fault offset of $\bar{D}$ is proportional to $\bar{D}/w$, $\bar{D}/L$, or $\bar{D}/\sqrt{A}$.

We can relate an earthquake's strain change to the static stress drop, $\Delta\sigma$, using Hooke's Law:

$$\Delta\sigma = C\mu\left(\frac{\bar{D}}{\tilde{L}}\right), \qquad (2.18)$$

where $\tilde{L}$ is a **characteristic rupture dimension** (equal to either $\sqrt{A}$, L, or w, often assumed to be the smaller of L or w) and C is a non-dimensional constant that depends on the fault geometry (roughly equal to one). Table 2.1 includes the constant C for several static earthquake models with different geometries. If, for simplicity, we assume that the earthquakes used to construct Fig. 2.8 produced circular ruptures, then from Table 2.1,

$$M_0 = \frac{16}{7}\Delta\sigma a^3$$

$$= \left(\frac{16}{7}\Delta\sigma\right)\frac{A^{3/2}}{\pi^{3/2}} . \qquad (2.19)$$

Taking the logarithm of both sides of Eq. (2.19), and simplifying, we find,

$$\log M_0 = \frac{3}{2}\log A + \log\left(\frac{16}{7\pi^{3/2}}\Delta\sigma\right). \qquad (2.20)$$

A static, elastic, circular earthquake model (in fact, any model with a uniform aspect ratio) predicts a linear trend between the logarithms of moment and rupture area matching the observed slope of $3/2$ (Fig. 2.8). We retain the linear trend only if the stress drop is constant, and in

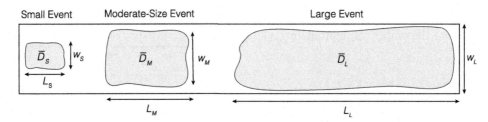

FIGURE 2.9 Dimensional analysis of a fault. The length is nearly always equal to or larger than the fault width. L scales in proportion to w up to the dimensions of the seismogenic zone (for small and moderate-size events). For large events, L increases while w, which equals the width of the seismogenic zone, remains constant.

that case, stress drop can only shift the trend up or down on the chart. The line through the data in Fig. 2.8 corresponds to a stress drop of 3 MPa. The bounding trend lines in the figure correspond to stress drops of 1 and 10 MPa. Most earthquake stress drops are between these values, although order-of-magnitude variations higher or lower have been observed.

A relatively uniform stress drop implies a relatively uniform strain drop,

$$\frac{\bar{D}}{\tilde{L}} = C^{-1}\frac{\Delta\sigma}{\mu} \sim \frac{3\,\text{MPa}}{30,000\,\text{MPa}} \sim 10^{-4}\,. \quad (2.21)$$

We noted this in Chapter 1 and it is the basis of Tsuboi's argument for roughly uniform strain energy that drives earthquake faulting. Regardless of specific numbers, an important implication is that the slip scales in proportion with the characteristic fault dimension $\bar{D} \propto \tilde{L}$. A complication is that the precise relationship changes with fault type and rupture aspect ratio. Work continues to better define the ranges and fault types over which $\tilde{L}$ represents w, L, or $\sqrt{A}$. Eq. (2.17) suggests that

$$\bar{D} \approx 10^{-4.4}\,A^{1/2} \approx 4 \times 10^{-5}\,\sqrt{A},$$
$$\bar{D} \text{ in m, } A \text{ in m}^2. \quad (2.22)$$

Other observations lead to slightly different relationships for slip scaling versus length – these can be found in the seismological research literature. The relationships are important because they are used in seismic hazard estimation.

Scaling relationships suggest that small and large earthquakes (at least over the range of moderate-to-large events) are *self similar*, the ratios of some first-order characteristics (slip, area, length) remain relatively constant over a significant range of earthquake sizes and faulting geometry types. Some recent observations hint at variations in some of the similarity ratios. Thus, the topic remains a focus of study. In the next section, we use another simple, elastic model and the relative uniformity of stress drop to explore characteristics of ground motions close to an earthquake.

2.4.5 Stress drop, particle velocity, and rupture velocity

Static models ignore the temporal history of slip and stress drop, but these are important in the excitation of seismic waves. If we rearrange Eq. (2.18) to isolate the average slip, then $\bar{D} = (\tilde{L}\Delta\sigma)/\mu C$, which implies that $D(t)$, slip as a function of time, and $\Delta\sigma$ must have an identical time history. Brune (1970) developed a simple model to explore the temporal history of stress drop and its relation to near-rupture ground motion. He considered a crack with surface area A in a homogeneous material (Fig. 2.10). Initially, a traction, σ_0, is applied to the surface of A and at time $t = 0^+$ this traction starts to relax to a new value, σ_1. The stress drop, $\Delta\sigma = \sigma_0 - \sigma_1$, takes a time τ_r to reach the final stress state. Fig. 2.10 shows one side of A and a point P away from

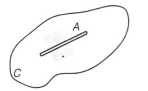
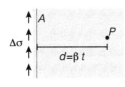

FIGURE 2.10 Geometry of a fault, A, which cuts a continuous medium. One surface of A is subjected to a stress that produces a SH wave that travels to a point P.

the fault. The relaxation of σ_0 at time $t = 0^+$ is equivalent to applying a negative shear stress to the fault surface. This shear stress creates a stress imbalance that propagates through the material at the shear-wave speed, β. At any later time, a shear-wave pulse will have propagated a distance $\beta \cdot t$ into the medium towards P. We can calculate an instantaneous strain from the displacement, $u(t)$, at P,

$$\varepsilon = \frac{u(t)}{\beta t} = \frac{\Delta\sigma}{\mu} , \qquad (2.23)$$

or,

$$u(t) = \frac{\Delta\sigma \, \beta t}{\mu} . \qquad (2.24)$$

This simple model predicts that the particle displacement close to the rupture should have a ramp-like character, and if $\Delta\sigma$ is nearly constant, $\dot{u}(t)$ should also be nearly earthquake independent. The model also suggests that the initial ground velocity, $\dot{u}(t)$, is directly proportional to the stress drop: $\dot{u}(t) = \Delta\sigma \cdot (\beta/\mu)$. Using typical values for the stress drop, shear-wave speed and shear modulus, the result suggests that $\dot{u}(t)$ is on the order of $1\,m/s$, which is consistent with seismogram modeling.

This simple model neglects the effects of rupture propagation and finiteness that are important properties of an earthquake. Eq. (2.24) is based on the assumption that the stress relaxes along the entire rupture simultaneously. In actuality, a rupture propagates across the fault with a **rupture velocity**, v_r. As long as $v_r < \beta$ (the usual

case), point P will sense a stress change from the ruptured segment of the fault before the rupture front reaches the fault adjacent to P. Depending on its position, the point may also sense the effects of the finite size of the rupture. The net effect is a reduced particle velocity,

$$\dot{u}(t) = \frac{\Delta\sigma}{2\mu} \beta e^{-(t/\tau)} , \qquad (2.25)$$

where τ is a time similar to $\tilde{L}/v_r$. The modification produces a ground velocity that approaches zero following the event, which of course also makes more sense physically.

Stress drop appears to have no role in controlling the rupture velocity, v_r. Fracture energy conditions usually require that v_r starts small and then increases to a final value when the fault rupture front exceeds some critical dimension. Most earthquakes rupture with speeds ranging from 0.5-0.8 × β. Ruptures that exceed the shear-wave speed ($v_r > \beta$) are called **super-shear**, and have been most commonly associated with strike-slip faults (but some exceptions and deep events have included similar characteristics). Super-shear segments of strike-slip ruptures appear to produce fewer aftershocks and may be associated with relatively smooth fault segments. Rapid rupture propagation can produce strongly focused strong motions, and so understanding the physics of the process is an important goal for seismic hazard estimation.

2.5 Seismic-waves: the elastic equations of motion

If we consider the stresses acting on a small volume of an elastic material in light of Newton's Second Law (force imbalances produce accelerations), we can derive the seismological *equations of motion*,

$$\rho \ddot{u}_i = \sigma_{ij,j} + f_i , \qquad (2.26)$$

Box 2.2 What is the stress on a fault?

In our discussions of fault dynamics, we found that most processes scale with stress drop, which averages less than 10 MPa. We have not talked about the absolute level of stress of a fault, which was once a fundamental controversy in earthquake mechanics. Two basic data types provide conflicting expectations for fault stress: (1) laboratory measurements of rock friction and measurements of static strength suggest that faults are inherently strong, while (2) dynamic measurements of seismic energy and frictional heat production suggest faults are inherently weak. The dynamics arguments are based on the earthquake energy budget; the work done in sliding the sides of the fault past each other must equal the sum of the energy released by generation of seismic waves (E_S) or energy converted to heat (E_h) while overcoming frictional resistance to fault motion. The energy per unit fault area can be written as

$$\frac{E}{A} = \frac{E_s}{A} + \frac{E_h}{A} = \bar{\sigma}\eta D + \sigma_r D, \tag{B2.2.1}$$

where $\bar{\sigma}$ is the average of the stress on the fault before and after the earthquake, σ_r is the frictional resistance stress, η is the seismic efficiency, and D is the final displacement on the rupture. Seismic efficiency is defined as the fraction of the total energy that is partitioned into seismic radiation:

$$\eta = \frac{E_s}{E} = \frac{\sigma_s}{\bar{\sigma}}, \tag{B2.2.2}$$

where σ_s is the stress that caused the seismic radiation (presumably stress drop). σ_s is usually estimated using

$$\sigma_s = \mu\frac{E_s}{M_0}, \tag{B2.2.3}$$

where $\log E_s = 11.8 + 1.5\,M$. For any reasonable value of μ, σ_s is less than 5 MPa, which implies that $\bar{\sigma}$ must be small unless the seismic efficiency is less than 0.1. σ_r can be estimated by measurements of the heat flow around active faults. The San Andreas fault in California has been extensively sampled for heat flow, and no significant heat-flow signature associated with the fault, suggesting that the frictional resistance is less than 10 MPa.

The *normal* stress on the fault at a depth of 8–10 km (the middle of a large rupture surface) will be related to the lithostatic load, $\rho g h$. If the crust behaves hydrostatically, the normal stress will equal the lithostatic stress; if the crust behaves as a *Poisson solid*, the normal stress will be ~ 0.3 times the lithostatic load. Byerlee's laboratory measurements (Fig. 2.5, left) indicate that most faults will have internal coefficients of friction ~ 0.6. In either case, the average shear stress on the fault will exceed 50 MPa, a value five times that suggested by the heat-flow measurements. Herein lies the controversy. We can reconcile these observations in several possible ways: (1) significant earthquake energy may go into neglected processes such as chemical or phase changes along the fault surface, (2) the heat-flow anomaly is "washed" away by ground-water transport, or (3) stick-slip sliding mechanisms must be modified to include rupture phenomena that reduce the effective normal stress on the fault and reduce frictional heat generation. The exact nature of the process remains a research focus.

where ρ is the material density, $\ddot{u}_i$ is the vector acceleration, $\sigma_{ij,j}$ is the spatial stress gradients (the comma represents differentiation), and the term, $\sigma_{ij,j}$, also includes a summation (the index j is repeated, see Box 2.3). The last term, f_i, represents body forces (such as gravity, or seismic-source related quantities). Gravity is generally only important for the longest period seismic waves that are sensitive to the changes in the gravity field associated with large-scale vibrations of the planet. To represent seismic sources as a vector f_i, we use continuum based relationships known as **representation theorems** that allow us to represent elastic models of slip on a fault surface as a set of equivalent body forces. The equations of motion and representation theorems are the foundation of quantitative models of seismic-wave excitation and propagation (see Part II). They are also central to the modeling of co-seismic geodetic deformation (elastic rebound), and even provide some useful approximate solutions for fault strain accumulation (which as discussed above also involves non-elastic, time-varying deformation processes).

For practical applications, we often work with relationships that are derived from the equations of motion under assumptions about the geologic model and/or the characteristics of the seismic source. Many of those relationships are the focus of much of this text. For example, P and S wave equations can be derived from (2.26) directly with little effort – for P waves in a uniform material and one spatial dimension (u_x), we have

$$\frac{\partial^2 u_x(x,t)}{\partial t^2} = \frac{1}{\alpha^2}\frac{\partial^2 u_x(x,t)}{\partial x^2}, \qquad (2.27)$$

where u_x is displacement, x and t are position and time, and α is the P-wave speed. The square of the P-wave speed is the ratio of elastic to inertial quantities,

$$\alpha^2 = \frac{\lambda + 2\mu}{\rho} \quad \text{or} \quad \alpha^2 = \frac{K + 4/3\mu}{\rho}. \qquad (2.28)$$

An analogous wave equation can be derived for shear waves, which travel at a speed satisfying an analogous relation,

$$\beta^2 = \frac{\mu}{\rho}. \qquad (2.29)$$

Fig. 1.21 is a plot of the average radial variations in P and S waves within the planet. Seismic waves travel at speeds of thousands of kilometers per hour, so we usually express the values in units of km/s.

For earth models composed of layers of spherically-symmetric shells, we can derive differential equations that together with appropriate initial and boundary conditions on stress and displacement, quantify the propagation of seismic waves in those models. Although few of the differential equations have analytic solutions, they are routinely solved using approximate and numerical methods. **Ray-theory** is often used to analyze body waves and **normal-mode theory** is often used to analyze surface waves. Ray approaches construct approximate solutions to the equations of motion by tracking the propagation of energy through the material; normal-mode approaches satisfy the equations of motion by combining solutions for a system's preferred frequencies of vibration. Laplace and Fourier Transform based solutions are also directly numerically evaluated. For earth models that lack structural symmetry (3D models of geologic heterogeneity), we compute numerical solutions of the equations of motion directly using finite-difference and finite-element based methods).

Seismic sources are approximated using force systems, f_i, that radiate seismic waves in a manner equivalent to the physical models of shear-faulting, explosions (chemical, nuclear, and volcanic), mine-collapses, human-, atmosphere- and ocean-induced background motions, etc. **Double-couple** and **seismic moment-tensor** models are equivalent body-force representations of shear-faulting that are commonly used to model seismic-wave excitation from earthquakes and explosions. Both are also routinely estimated

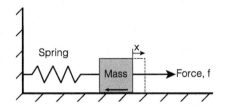

FIGURE 2.11 Phenomenological model for seismic attenuation. The spring represents elastic processes in the Earth described by spring constant, or stiffness, k. The force, $\mathbf{f}$, represents a driving force on the mass. The force at the base of the mass represents friction opposing the motion of the mass.

for all moderate and large earthquakes. Again, many of the underlying details are described in the chapters that follow. In this overview, we start with a simple view of an ideal, oscillating system – the mass and spring. The model allows us to introduce the concept of simple harmonic motion – a characteristic of elastic systems.

2.5.1 Harmonic motion

Harmonic motion is a fundamental characteristic of elastic and nearly-elastic deformation. The underlying mathematics can be developed for an oscillating mass connected to a spring. Consider Fig. 2.11, where a mass, m, attached to a spring with spring constant, k (k is a measure of the spring's elastic properties), slides across a surface displaced from its equilibrium by an imposed force impulse, with friction acting opposite to the direction of motion on the base of the mass. We first consider a frictionless case and examine the homogeneous equation (no explicit driving force after initial deflection from equilibrium) – solutions to problems with specific forces can be derived in terms of these homogeneous solutions. The equation of motion for this system relates the elastic restoring force of the spring to the inertial force imparted by the moving mass,

$$m\,\ddot{x}(t) + k\,x(t) = 0 \,. \tag{2.30}$$

The equation of motion quantifies how the mass vibrates in the x-direction. Solutions of this equation include the familiar sine and cosine functions,

$$x(t) = A' \cos \omega_0 t + B' \sin \omega_0 t \tag{2.31}$$

where

$$\omega_0 = \sqrt{k/m} \,, \tag{2.32}$$

and A' and B' are constants that depend on initial conditions and the force driving the system. The natural angular frequency of the system, ω_0, is equal to $2\pi f_0$. The time it takes for the system to complete a single oscillation is the period, $T_0 = 1/f_0$. A stiffer spring produces a higher natural frequency, and larger mass produces a lower natural frequency.

For wave propagation problems it is often more convenient to convert the sinusoids to complex exponential form. The general solution to this equation of motion is a harmonic oscillation that can be represented in complex form (Box 2.3) as

$$x(t) = A\,e^{i\,\omega_0 t} + B\,e^{-i\,\omega_0 t} \,, \tag{2.33}$$

where A and B are constants that depend on the initial conditions and the nature of the force exciting the vibration. The use of complex quantities simplifies algebraic analysis, the actual position of the mass must be a real-valued quantity, the real part of $x(t)$.

Once the motion starts in our frictionless system, it will continue forever, oscillating at the natural frequency of the system ω_0. Of course, for motion to start, we need a force to perturb the system from equilibrium. In the continuum case, we need a stress or stress change.

Although the physical system described above appears elementary, the same equation arises from elastic behavior for all forms of the equations of motion that include a time dependence. Not only are harmonic oscillators a component of the solutions used to model seismic waves, they also provide insight into seismometers,

Box 2.3 Complex numbers

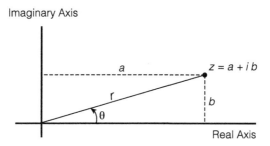

Imaginary Axis

Real Axis

$z = a + ib$

a

b

r

θ

FIGURE B2.3.1 An Argand Diagram of a complex quantity, $z = a + ib$, shown in the complex plane. The modulus or magnitude r and phase, θ are defined geometrically as shown.

Solutions of differential equations such as (2.34) often involve complex numbers of the form $c = a + ib$, where $i = \sqrt{-1}$. Complex numbers are a mathematical tool for simplifying algebraic analyses. Observations are of course always real-valued. We refer to a as the real part and b as the imaginary part of the complex number c. An Argand diagram is a plot of complex numbers with the real part along the horizontal axis and the imaginary part along the vertical axis. We often refer to the diagram as the complex plane. As suggested by Fig. B2.3.1, a complex number, c, can be represented in a *polar* coordinate form as

$$c = a + ib = re^{i\theta} = r\cos\theta + ir\sin\theta , \tag{B2.3.1}$$

where the **magnitude** of c is $|c| = r = (a^2 + b^2)^{1/2}$, and the **phase** of c is the angle $\theta = \tan^{-1}(b/a)$. Addition of two complex numbers, c and d, is computed by summing the real and imaginary parts,

$$\begin{aligned} c + d &= (a_1 + ib_1) + (a_2 + ib_2) \\ &= (a_1 + a_2) + i(b_1 + b_2) . \end{aligned} \tag{B2.3.2}$$

Complex number multiplication (note: $i \cdot i = -1$) is given by

$$\begin{aligned} c \cdot d &= (a_1 + ib_1) + (a_2 + ib_2) \\ &= (a_1 a_2 - b_1 b_2) + i(a_1 b_2 + b_1 a_2) \end{aligned} \tag{B2.3.3}$$

or in *polar form*

$$c \cdot d = r_1 e^{i\theta_1} r_2 e^{i\theta_2} = r_1 r_2 e^{i(\theta_1 + \theta_2)} . \tag{B2.3.4}$$

The **complex conjugate** of a number is denoted by c^* and is computed as

$$c^* = re^{-i\theta} . \tag{B2.3.5}$$

The product $c \cdot c^* = r e^{i\theta} r e^{-i\theta} = r^2$ equals the square of the magnitude of c. Complex numbers lying on the unit circle in the complex plane, $r = 1$ satisfy $e^{i\theta} = \cos\theta + i\sin\theta$, and $e^{-i\theta} = \cos\theta - i\sin\theta$. The value of the function repeats periodically with phase changes of 2π, just as do the cosine and sine functions. By adding and subtracting the exponentials, we obtain useful relations:

$$\cos\theta = \frac{e^{i\theta} + e^{-i\theta}}{2}$$

$$\sin\theta = \frac{e^{i\theta} - e^{-i\theta}}{2i}.$$

(B2.3.6)

These expressions are two of Euler's relations and are essential for working with and interpreting complex quantities.

resonating structures and equipment, tsunami, normal modes, etc. In an elastic continuum, a more general model would include a mass connected in all directions and allowing motion in three spatial dimensions. The more general cases also require that we consider oscillations with a range of vibration frequencies and the solution to general problems is a sum over frequency that will appear many times throughout the text. You will observe different ways we use or combine those solutions to derive insight into seismological phenomena. In the next section we examine seismic wave attenuation using a slight modification to the spring-mass model to include friction. That same mathematical model opens a door to insight into the movement of rocks during fault slip and many other physical systems.

2.5.2 Seismic-wave attenuation

Elasticity is an approximation – a useful one we that will adopt throughout much of the course to understand many aspects of seismic sources and seismic-wave propagation. In a purely elastic Earth model, seismic-wave amplitudes decrease with time as the energy is scattered throughout the planet. But once excited, those seismic waves persist indefinitely. The real Earth is not perfectly elastic, and prop-

agating waves **attenuate** with time due to various anelastic energy-loss mechanisms. The successive conversion of potential energy (particle position) to kinetic energy (particle velocity) as a wave propagates is not perfectly reversible, and other work, such as microscopic movements along mineral dislocations or shear heating at grain boundaries, taps the wave energy. We often describe these processes collectively as *internal friction*, and we "model" internal-friction processes using simpler phenomenological descriptions because our macroscopic observations blur the small-scale details.

Damped harmonic motion

Once harmonic motion starts in a perfectly elastic system, it will continue forever, oscillating at the natural frequency, ω_0. We can model *attenuation* by adding a damping force, such as friction between the moving mass and the underlying surface (Fig. 2.11). As before, the equation of motion for this system relates the restoring force of the spring to the inertial force imparted by the moving mass, but now includes a force that resists motion and is proportional to the velocity of the mass,

$$m\,\ddot{x}(t) + \gamma\,\dot{x}(t) + k\,x(t) = 0 \qquad (2.34)$$

or rewriting,

$$\ddot{x}(t) + \epsilon\,\omega_0\,\dot{x}(t) + \omega_0^2\,x(t) = 0, \qquad (2.35)$$

where $\epsilon = (\gamma\,m^{-1}\,\omega_0^{-1})$, and $\omega_0 = (k/m)^{1/2}$. The quantity γ is a *coefficient of friction* and ϵ is a ratio of frictional and inertial terms. The solution of (2.35) is of the form

$$x(t) = A_0\,e^{-\epsilon\,\omega_0\,t}\,\sin\left(\omega_0\,t\,\sqrt{1-\epsilon^2}\right). \qquad (2.36)$$

This is a harmonic oscillation that decays exponentially with time. If $\epsilon = 0$ (no attenuation), (2.36) reverts to simple harmonic motion, Eq. (2.33). Before exploring the solution details, it is helpful to translate some of the quantities into definitions commonly used in discussions of seismic waves.

The quality factor, Q

Seismic wave attenuation is often quantified using a quantity called the quality factor, Q. We can express ϵ in Eq. (2.36) in terms of a quality factor by defining

$$\epsilon = \frac{\gamma}{m\,\omega_0} = \frac{1}{2Q}. \qquad (2.37)$$

Q is the ratio of the mass- and spring-related terms to the coefficient of friction, γ. Q has an inverse relationship with attenuation, the smaller Q, the larger the attenuation. Higher Q indicates that friction has less influence on the mass' motion. Using (2.37), we can write the amplitude, $A(t)$, of the vibration as a function of time

$$A(t) = A_0\,e^{-\omega_0 t/2Q}. \qquad (2.38)$$

Now consider the amplitude in two successive cycles of deformation. The change in peak amplitudes measured one period ($T_0 = 2\pi/\omega_0$) apart is

$$A(t + T_0) = A(t)\,e^{-\pi/Q}. \qquad (2.39)$$

The energy in a harmonic wave, $E(t)$ is proportional to the square of displacement amplitude,

$x(t)$, so the change in energy associated with propagation one period longer decreases the energy as

$$E(t + T_0) = E(t)\,e^{-2\pi/Q}. \qquad (2.40)$$

If we define $\Delta E = E(t + T_0) - E(t)$, to first order, we can relate Q to the fractional change in energy over one deformation cycle using

$$2\pi\,Q^{-1} = \frac{-\Delta E}{E}. \qquad (2.41)$$

The quality factor, Q has no physical units and is related the fractional loss of energy per cycle of oscillation. Although we have focused on Q effects over successive cycles measured temporally, a seismic wave propagates, and we can also develop an expression for amplitude as a function of distance traveled. Let c represent the wave speed, and r represent the distance, then $t = r/c$ and the amplitude as a function of distance traveled is

$$x(r) = A_0\,\exp\left(-\frac{f\,\pi\,r}{c\,Q}\right). \qquad (2.42)$$

A few simple examples will help make the ideas more clear. The amplitudes for a monochromatic wave traveling from left-to-right in a rock with granite-like properties are shown in Fig. 2.12. The gray curve identifies a signal with no attenuation (no amplitude change). The black curve identifies an attenuation of a two-second period wave with $Q = 250$. The amplitude decreases systematically with time (or distance traveled, assumed speed $= 6\,\text{km/s}$). For a Q of 250, the amplitude decreases by about a factor of two in about 600 km, or 50 wavelengths. Thinking of attenuation in terms of the number of wavelengths traveled is a useful habit. Since a wave travels a distance of one wavelength, λ, in one period ($\lambda = c\,T_0$), Eq. (2.39) corresponds to propagation one wavelength. Thus for a propagation time equal T_0 or distance equal to λ, a 1% decrease in amplitude occurs for a $Q \sim 300$,

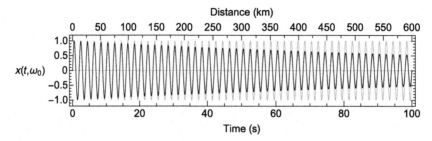

FIGURE 2.12 A monochromatic solution to damped harmonic motion (black line) for a sinusoidal displacement driven steadily at the left (origin) and propagating to the right. The amplitude decreases steadily with time (or distance). The gray curve is an undamped solution shown for reference.

a 5% amplitude decrease occurs for a $Q \sim 60$, and a 10% amplitude decrease for a $Q \sim 30$.

We have considered attenuation in terms of a very simple physical spring-mass-friction model because the ideas carry over into more sophisticated models of seismic wave propagation. We discuss details later, but the amplitude (2.38) and energy (2.41) relations in terms of Q for the simple spring-mass-friction model can be used in the interpretation of seismic-wave amplitude attenuation, as long as we integrate the effects over a range of frequencies.

2.5.3 Seismic wave attenuation in Earth

Two processes work to decrease the amplitudes of seismic waves in Earth. To the effects of intrinsic (anelastic processes) that we just described, we must add the effects of scattering. We will discussion scattering in more detail later, our focus here is on more on elementary mechanical processes, not details of wave propagation. But we acknowledge that at all frequencies, and particularly at frequencies above 1 Hz, the amplitudes of seismic waves are strongly influenced by the heterogeneity in the planet. The effect is a first-order feature of seismograms and gives rise the ever-present *coda* signals that follow all major seismic arrivals.

The spring-slider attenuation model provides insight into attenuated seismic signals but corresponds to a single vibration frequency. Vibra-

tions of interest in seismology span a broad frequency range. The corresponding value of Q depends on material properties, temperature, pressure, fluid content, etc. For lower frequencies ($< 0.5\,\text{Hz}$) Q appears to have only a weak frequency dependence. At higher frequencies, the bulk Q generally is a composite of anelastic and scattering effects and may have a strong frequency dependence. Although we routinely account for attenuation in seismogram modeling, attenuation is difficult to measure precisely because other factors from the source, gradients in seismic wave speed, and scattering also affect seismic-wave amplitudes. But the attenuation properties of rocks are quite illuminating of the material properties and hence accurate measures are quite valuable. The relative low attenuation in subducting slabs compared with more typical upper mantle played a role in the confirmation of lithospheric subduction in the early investigations surrounding plate tectonics.

In general, Q increases with material density and seismic wave speed. Q for P waves in the Earth is systematically larger than Q for S waves, and we refer to the corresponding quantities as Q_α and Q_β, respectively. Current anelastic attenuation models (described in more detail later) assume that intrinsic attenuation occurs almost entirely in shear deformation (even for P-waves), associated with lateral movements of lattice effects and grain boundaries. For a material with all losses due to only shearing mecha-

TABLE 2.2 Approximate Q values.

Rock type	Q_α	Q_β	$\sim Q_\beta/Q_\alpha$
Shale	30	10	3/9
Sandstone	58	31	5/9
Granite	250	110	4/9
Peridotite	650	280	4/9
Upper Mantle (LVZ)	195	80	4/9
Mantle Transition Zone	360	140	4/9
Lower mantle	800	312	4/9
Outer Core	> 4000	0	0
Inner Core	100–800	–	–

nisms,

$$Q_\beta = \frac{4}{3}(\alpha/\beta)^{-2}Q_\alpha ,\qquad (2.43)$$

and for a Poisson solid (a reasonable approximation throughout much of the solid earth), $(\alpha/\beta)^2 = 3$, and

$$Q_\beta \approx \frac{4}{9}Q_\alpha .\qquad (2.44)$$

Table 2.2 is a list of Q-values for several rock types and regions of the Earth. The ratio of Q_β to Q_α is often constrained to be near 4/9. The data are consistent with that ratio, but more work is needed to resolve the ratio independently to confirm the underlying ideas.

We will return to attenuation later when we discuss seismic waves, normal modes, and Earth's interior in more detail. Our point here was to introduce the concept of attenuation as a slight modification to the elastic model that underlies much of quantitative seismology. The quantities t^* (Box 13.2) and Q are frequently referred to in seismology because they provide a valuable phenomenological description of seismic-wave attenuation. A key insight worth remembering is the increasing importance of attenuation on waves with increasing frequency and with increasing propagation distance. More generally, seismic wave attenuation is complicated by both frequency dependence

for periods of less than $1\,s$ and strong spatial variations at all periods (see Chapter 10).

2.6 Summary

The mathematical models that underlie global seismology are based on continuum mechanics – a generalization from the simple point-mass and rigid-body mechanics of elementary physics. Forces are replaced with stresses and positions are replaced by strains. The deformations are mostly small and the theories used are appropriate for small deviatoric stresses and strains. Continuum-mechanics based analyses have led to great success in the investigation of earthquakes and Earth's interior using a simple assumption of linear elasticity based on Hooke's Law. Adjustments can be made for the frictional and ductile processes that are an essential part of earthquake process, and for the microscopic loss of energy experienced by seismic waves that travel through Earth's imperfectly elastic interior. The models rarely explain the observations exactly because models describe a macroscopic view of complex processes and because all observations contain components inexplicable by simple models. We seldom hope to explain a seismogram wiggle for wiggle, nor do we know enough of the distribution of geologic features throughout the planet to do so. We do better with long-wavelength observations, where work with simpler models is easier because of the averaging performed by the longer-wavelength observations. Short-wavelength observations are often still interpreted at least in part statistically. Seismological analysis often involves a comparison of observations with a suite of simplified models of complex processes. Indeed, learning what is and what isn't a good match to the observations is one of the challenges of becoming a seismologist and one of the skills we hope to communicate through the text.

3

Earthquakes and plate tectonics

Chapter goals

- Review basic elements of plate tectonics.
- Review plate boundary types.
- Review the relationship of earthquake locations and type of earthquakes to plate boundary type.

Modern geoscience is conducted in the context of *plate tectonics*, an overarching theory that describes the dynamics of the Earth's outer shell. Earthquake seismology has played a major role in developing the concept of plate tectonics, and the relationship between earthquake occurrence and tectonic processes is known as *seismotectonics*. The spatial distribution of earthquakes can be used to determine the location of plate boundaries, faulting orientation can be used to infer the directions of relative motion between plates, and the *tectonic rate of motion* and cumulative displacement of earthquake occurrence can be used to infer the relative velocity between plates. Because tectonic motions cause earthquakes, it is critical to understand plate tectonics when trying to reduce the societal hazard associated with faulting. In this chapter, we will develop a foundation for plate tectonics and earthquakes. Only a few of the many existing aspects of *seismotectonics* can be discussed here, and we restrict our attention to the most basic elements of *plate tectonics*, as current journal arti-

cles are filled with many additional applications of seismology to the study of plate tectonics.

The basic concept in modern plate tectonics, introduced in the mid-1960's, is that of a mobile *lithosphere.* The lithosphere is a high-viscosity region that translates coherently on the Earth's surface. Relative motions between a mosaic of surface lithospheric plates are accommodated at plate boundaries. Fig. 3.1, from a classical paper by Isacks et al. (1968), shows a cartoon of a mantle stratified on the basis of rheological behavior. The essence of lithospheric motions is large horizontal transport, from creation to destruction of lithospheric plates. The lithosphere behaves rigidly on geologic time scales of millions of years. This requires that the material within the lithosphere be at low enough temperatures to inhibit solid-state creep. The lower boundary of the lithosphere is sometimes defined by a temperature, or *isotherm*, usually about 1300°C, although above 600°C, the material will begin to experience ductile deformation, which defines a "thermal" lithosphere. In general, the thermal lithosphere is about 100 km thick, although in some regions its thickness is essentially zero and in others it may be several hundred kilometers. Below the lithosphere is the *asthenosphere*, a region of relatively low strength that may be nearly decoupled from the lithosphere. The asthenosphere behaves like a viscous fluid on relatively short time scales (10^4 yr). The classic example of this viscous behavior is the present-

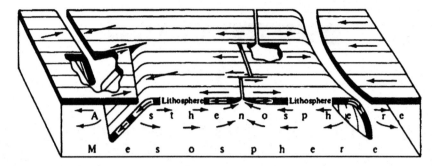

FIGURE 3.1 The basic elements of plate tectonics. The lithosphere is a region of high viscosity and strength that forms a coherently translating plate. Relative motions among plates cause them to interact along convergent, transcurrent, and divergent boundaries. The lithosphere rides on the asthenosphere, a region of low strength (from Isacks et al., 1968).

day uplift of Fennoscandia postdating the removal of the ice sheet associated with the last ice age. Beneath the asthenosphere, Isacks et al. (1968) defined the *mesosphere*, a region of relatively high strength that they believed to play only a passive role in tectonic processes. The concept of the mesosphere is dated, but below a depth of approximately 350 km the viscosity of the upper mantle does appear to increase. Many geophysicists also believe that the viscosity of the lower mantle is significantly higher than that of the upper mantle. In a modern context, the lithosphere is associated with the stiff upper portion of the surface thermal boundary layer of a mantle convection system.

Movements of lithospheric plates involve shearing motions at the plate boundaries. Much of this motion occurs by aseismic creep (movement without earthquakes), but shear-faulting earthquakes are also produced. The release of strain energy by seismic events is restricted to regions where there is an inhomogeneous stress environment and where material is sufficiently strong for brittle failure. For the most part, this means that earthquakes are confined to plate boundaries. Fig. 3.2 shows the earthquake locations from the Bulletin of the International Seismic Center (ISC) from 1960 to 2015, shaded as a function of depth. Distinct zones of shallow seismicity mark divergent and transform boundaries, while deeper seismicity correspond

to convergent boundaries (see Fig. 3.2). The seismicity is largely concentrated in narrow bands that represent the plate boundaries; 12 major plates are fairly obvious, but on this scale it is difficult to recognize the dozen or so additional smaller plates. Three basic types of plate boundary are characterized by different modes of plate interaction:

1. *Divergent boundaries*, where two plates are moving apart and new lithosphere is produced or old lithosphere is thinned. *Midoceanic ridges* and *continental rifts* are examples of divergent boundaries.
2. *Convergent boundaries*, where lithosphere is thickened or consumed by sinking into the mantle. *Subduction zones* and *alpine belts* are examples of convergent plate boundaries.
3. *Transcurrent boundaries*, where plates move past one another without either convergence or divergence. *Transform faults* and other strike-slip faults are examples of transcurrent boundaries.

In general, divergent and transcurrent plate boundaries are characterized by shallow seismicity (focal depths less than 30 km). In Fig. 3.2, the mid-oceanic ridge system is the best manifestation of this shallow extensional faulting. Subduction zones and regions of continental collision can have much deeper seismicity. Fig. 3.2 shows epicenters as a function of focal depth,

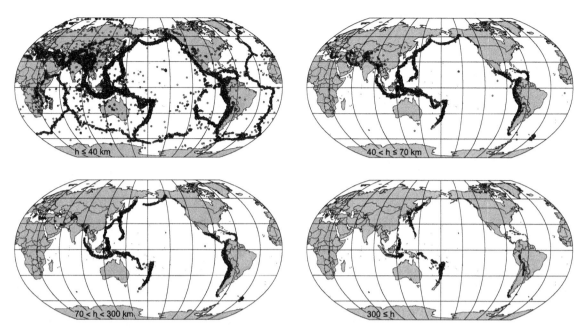

FIGURE 3.2 Earthquake locations separated into shallow (upper left and upper right), intermediate (lower left), and deep (lower right) depth categories (Chapter 1). Each symbol represents an earthquake of magnitude 5.0 or larger contained in the USGS global earthquake catalog (event origin times range from 1970-mid2020). The hypocentral depth (*h*) range for each plot is shown at the bottom of each map. The shallow depth category was separated to show the relatively shallow nature of most continental and mid-ocean ridge earthquakes. Deep events occur within subducted oceanic lithosphere, most beneath the western Pacific and South America, but also in several more spatially localized regions, notably beneath southern Spain, the Tyrrhenian Sea west of Italy, and the Hindu-Kush Region of central Asia.

which isolates the subduction zones in Fig. 3.2 (light to dark gray). Fully 80% of the world's seismicity is along the circum-Pacific margin and occurs mostly in subduction zones. Fig. 3.2 also shows clusters of deep Eurasian earthquakes, which are the result of the collision of India and Eurasia.

Before we develop the concept of seismotectonics further, it is best to review some of the basics of plate motion. The fundamental concept that underlies motion of rigid plates on a sphere is *Euler's theorem*, which states that the relative motion of two plates can be described by a rotation about a pole (called an *Euler pole*). Fig. 3.3 shows the relative motion between two plates. If a plate boundary is *perpendicular* to a small circle about the pole of rotation, convergence or divergence must be occurring between the plates. If

the boundary is *parallel* to a small circle, the relative plate motion is transcurrent. The relative velocity between two plates depends on the proximity of a plate boundary to the pole of rotation. The rotation is described by an angular velocity ω. As the distance to the boundary increases, the relative motion on a boundary also increases. Several investigators have used the orientation of plate boundaries and relative velocities derived from the analysis of magnetic anomalies to construct relative plate motions for the entire world. *Absolute plate motions* are more difficult to determine because the whole surface is in motion, but it is possible if one assumes a fixed hotspot or other reference frame. Fig. 3.4 shows the absolute plate velocities determined from the NNR-MORVEL56 model (Argus et al., 2011). This model included 56 distinct plates and

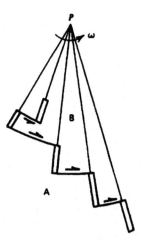

FIGURE 3.3 Motion of plate boundaries is determined by rotation about an Euler pole. Transcurrent boundaries are oriented along the arcs of small circles about the pole. Divergent and convergent boundaries are oriented along radial spokes through the pole (from Wilson, 1965).

shows that rates vary slightly along each plate boundary. Plate motion may be accommodated both seismically and aseismically.

In this chapter, we discuss the various categories of plate boundary and the associated types of seismic activity. We also develop the relationship between plate tectonic processes and earthquake recurrence. The concept that the processes driving plate tectonics are steady and long-term implies that the rate of strain accumulation and release at a plate boundary must also be relatively steady. We also relate faulting orientation (see Box 3.1) to stresses that generate different fault types. In general, we can identify fault orientation from the determination of focal mechanisms (see Box 3.2 for an introduction). If the mechanical response of a fault along a plate boundary has regular stick-slip behavior, a characteristic earthquake cycle may exist. This idea is of fundamental importance to earthquake hazard assessment and has given rise to the concepts of *recurrence intervals* and *seismic gaps* in earthquake forecasting and prediction (see Chapter 8). One major societal goal of seismology is to predict earthquakes to minimize loss of life. Although models have been developed that can generally describe earthquake recurrence, a predictive capability is still elusive and may intrinsically be unattainable. However, earthquake damage can be reduced by modifying construction standards in regions with an established long-term seismic hazard.

3.1 Divergent boundaries

Oceanic ridges and continental rift zones are regions of the Earth in which the relative motion across the ridge or rift is *divergent* or *extensional*. In divergent settings the minimum principal stress, σ_3, is horizontal and directed perpendicular to the ridge or rift. The maximum principal stress, σ_1, is vertical. Most of the earthquakes in this environment involve normal faulting. Oceanic ridges are by far the most important divergent boundaries, for it is at oceanic ridges that new oceanic crust and lithosphere are actually created and partitioned to the diverging plates on either side of the ridge. For this reason, ridges are also known as *spreading centers*.

The creation and evolution of oceanic lithosphere strongly influence the seismicity at oceanic ridges. Fig. 3.5 shows a schematic cross section through a ridge. At the plate boundary the lithospheric thickness is essentially *zero*. As the material added to the plates at the spreading center cools below a critical isotherm, it acquires the mechanical property, high viscosity, of "lithosphere." As the lithospheric plate moves away from the plate boundary, it continues to cool, which has two principal consequences: (1) the lithosphere beneath a point on the plate thickens, and (2) the depth to the ocean floor increases. The lithospheric thickness can be calculated by assuming that an isotherm for a cooling boundary layer is the lithospheric boundary:

$$th = 2c\sqrt{kt}, \tag{3.1}$$

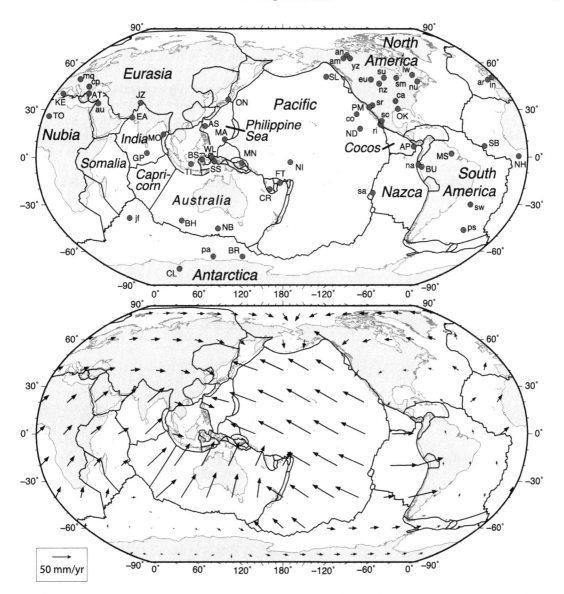

FIGURE 3.4 (Top) Plate boundaries and rotation poles for plate-motion model MORVEL56 (Argus et al., 2011). Large plates are labeled, plate rotation poles are identified as following: am (Amur), an (Antarctica), ar (Arabia), au (Australia), cp (Capricorn), ca (Caribbean), co (Cocos), eu (Eurasia), in (India), jf (Juan de Fuca), lw (Lwandle), mq (Macquarie), nz (Nazca), na (North America), nb (Nubia), pa (Pacific), ps (Philippine Sea), ri (Rivera), sw (Sandwich), sc (Scotia), sm (Somalia), sa (South America), su (Sunda), sr (Sur), yz (Yangize), AS (Aegean Sea), AP (Altiplano), AT (Anatolia), BR (Balmoral Reef), BS (Banda Sea), BH (Birds Head), BU (Burma), CL (Caroline), CR (Conway Reef), EA (Easter), FT (Futuna), GP (Galapagos), JZ (Juan Fernandez), KE (Kermadec), MN (Manus), MO (Maoke), MA (Mariana), MS (Molucca Sea), NH (New Hebrides), NI (Nivafo'ou), ND (North Andes), NB (North Bismarck), OK (Okhotsk), ON (Okinawa), PM (Panama), SL (Shetland), SS (Solomon Sea), SB (South Bismarck), TI (Timor), TO (Tonga), WL (Woodlark). (Bottom) Sample of plate motion vectors computed for the model (showing data only on major plates). Figure courtesy of Matthew Herman.

Box 3.1 Quantifying faulting geometry

For abstract analyses such as the introduction of stress and strain, specific coordinate directions are unimportant (as long as they are orthogonal). But in any observational analysis, we must use well-defined coordinate directions. For example, we represent the orientation of a surface in geographic coordinates using a strike and dip (Fig. B3.1.1). The *strike* is the direction measured clockwise from north to the intersection of the surface with a horizontal reference plane (such as Earth's surface). We use a convention to choose one of the two possible strike directions (ϕ and $\phi + 180°$). In our convention, when you look in the strike direction, the fault dips to your right (the right-hand rule). In this convention, the dip direction is $\phi + 90°$. The *dip* is the angle measured downward from horizontal to the surface in the direction of fault dip (Fig. B3.1.1). To quantify the direction of slip during the earthquake, we measure the angle between a horizontal line on the fault surface (parallel to Earth's surface) and the direction that the *hanging wall* moved relative to the *foot wall*. The angle, called the *rake*, is measured counterclockwise from the horizontal reference in the plane of the fault (Fig. B3.1.1).

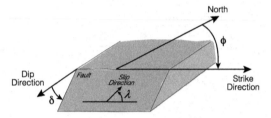

FIGURE B3.1.1 Fault block diagram showing the footwall of a dipping fault. The hanging wall is the rock block above the fault that is not shown in the cartoon (for clarity). Geographically referenced angles used to describe a fault orientation are strike, ϕ and dip, δ, and the angle used to describe the slip direction is rake, λ.

Earthquakes with rakes near $0°$ or $180°$ are called left-lateral and right-lateral *strike-slip* earthquakes respectively. Earthquakes with rakes near $-90°$ and $90°$ are called *normal* and *reverse* dip-slip earthquakes respectively. Earthquakes with rakes significantly different from these angles are called oblique-slip events. An exact difference threshold is not defined for terminology, but a rule of thumb would be $\pm 20°$. Far more convenient for theoretical analysis is a representation of a surface orientation using a unit-length vector perpendicular to the surface. The relationship between the normal vector and the strike, ϕ, the dip, δ, and slip angle or rake, λ is

$$\hat{\mathbf{n}} = (-\sin\delta\,\sin\phi,\ \sin\delta\,\cos\phi,\ -\cos\delta)^T\ , \tag{B3.1.1}$$

where *the vector components are north, east, and down, respectively*. The slip vector, the direction of hanging wall movement, $\hat{\mathbf{s}}$ in geographic coordinates is

$$\hat{\mathbf{s}} = \begin{pmatrix} \cos\delta\,\sin\lambda\,\sin\phi + \cos\lambda\,\cos\phi \\ \cos\lambda\,\sin\phi - \cos\delta\,\sin\lambda\,\cos\phi \\ -\sin\delta\,\sin\lambda \end{pmatrix} . \tag{B3.1.2}$$

The slip vector, $\hat{\mathbf{s}}$, lies on the fault surface and is orthogonal to the normal vector. The geometry of faulting has important applications in tectonic investigations of the origin of a particular style of faulting and deformation.

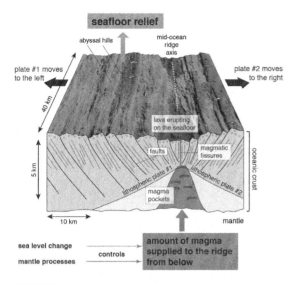

FIGURE 3.5 Bathymetry and schematic cross section of a segment of the intermediate-spreading Chile Ridge located at 39° 12′S, 91° 30′W (white star) looking south, showing the major tectono-magmatic processes that shape the seafloor. The lithosphere is defined by an isotherm, so as the plate cools and moves away from the ridge, its thickness increases. This associates the oceanic lithosphere with a thermal boundary layer. The schematic also shows faulting and the process in which melt supplies and is emplaced as new oceanic crust (from Olive et al., 2015).

where c and k are heat diffusion properties of the material. This implies that th is proportional to $\sqrt{t}$, the square root of the age of the plate at a given position. If the plate moves away from the spreading center at a constant rate, called the *spreading rate, th* is proportional to $\sqrt{d}$, where d is the distance from the ridge. The map of the age of ocean plate exhibits the systematic age increase away from ridges (Fig. 3.6).

The second thermal effect, the increase of oceanic depth with age, requires the concept of isostasy. In the simplest case, isostasy requires two columns of rock at equilibrium on the Earth to have the same total mass. If one column is less dense, the column will have a higher surface elevation. The density of most Earth materials decreases as temperature increases. In other words, as temperature rises, most ma-

terials expand. Conversely, as temperature decreases, density increases, and isostasy requires a *smaller* column. At oceanic ridges the temperature of the shallow mantle is very high, which is manifested as high elevation, or shallow ocean depths. As lithosphere moves away from the ridge and cools, it becomes more dense and "sinks"; thus the ocean depth increases. Fig. 3.7 shows the ocean depth as a function of age for two plates and a $\sqrt{t}$ model. The correlation of th with ocean depth for this model is excellent, except for oceanic regions older than 100 million years.

The thermal structure of young oceanic lithosphere has a profound influence on the observed seismicity. Oceanic ridges that are spreading rapidly have a broad, smooth topographic signature. The lithospheric thickness is very small near the ridge, which means that the available "width" of a fault plane is also small. This condition results in shallow, moderate- to small-sized earthquakes. Shallow earthquakes can occur along the mid ocean ridges defining divergent boundaries (Fig. 3.2), and also along the faults connecting the offset ridges (called *Transform Faults or Fracture Zones* – see next section). Although large earthquakes can occur along these faults, oceanic plates can have large **intraplate earthquakes** (earthquakes that occur within a plate and not along a plate boundary.) For example, outer rise earthquakes occur within a subducting plate in the region of plate bending due to subduction (Craig et al., 2014). Other earthquakes appear to have broken within the oceanic plate. For example, the 2017 Tehuantepec, Mexico (M8.2) ruptured through the oceanic plate below the locked section of the Mexican subduction zone (Ye et al., 2017). In 2012, one of the largest strike-slip earthquakes ever recorded (M8.7) occurred within the oceanic plate about 100–200 kilometers southwest of the Sumatra subduction zone. This earthquake was shortly followed within two hours by a great (M8.2) aftershock that occurred along an orthogonal fault (Yue et al., 2012). The 1998 Antarctic plate earth-

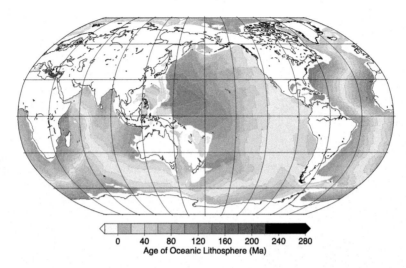

FIGURE 3.6 Age (in millions of years, *Ma*) of the oceanic lithosphere from Müller et al. (2008). Ocean floor age is estimated using geological and marine geophysical observations. The width of the youngest regions (0–20 Ma) provides an indication of geologically recent plate spreading rates. Continents, continental margins, and flooded transitional crust (modified oceanic or continental) are unshaded.

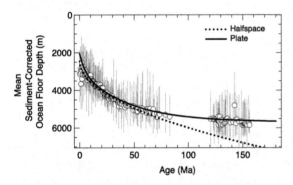

FIGURE 3.7 Estimated mean sediment-corrected ocean floor depth as a function of estimated ocean floor age Rowley (2019). The oceanic lithosphere is diverse and its properties depend on the lithospheric history. Simplifying the observations requires decisions about which areas to exclude because they were modified by other tectonic processes, how to weight different areas to account for sampling, how to correct for sediment load, etc. Each mean value shown was estimated using at least 40 observations; the vertical lines represent the spread in those observations. The curves correspond to predicted patterns computed for two simple oceanic lithosphere cooling models. The halfspace (thermal boundary layer) model is an optimal fit to the observations for ages less than 84 Ma. The plate model includes an optimal plate thickness of 117 km (see Rowley (2019) for details).

quake appears to have ruptured a series of faults 300-km long in the middle of the plate (Nettles et al., 1999).

Strictly speaking, most normal-faulting earthquakes associated with oceanic ridges do not occur exactly at the plate boundary but are on faults associated with a feature known as the *axial valley*. To illustrate this, Fig. 3.5 shows the bathymetry along a segment of the intermediate-spreading Chile Ridge located at $39°12'S$, $91°30'W$ looking south, showing the major tectono-magmatic processes that shape the seafloor. Normal faults flank the axial valley, and the lithospheric thickness increases as the plate cools and moves away from the ridge. The normal faults are probably produced in response to isostatic adjustment.

Continental rifts are also divergent plate boundaries, although we usually do not recognize the opposite sides of a rift as distinct, separate plates until rifting produces new oceanic plate. The faulting in continental rifts is usually much more complicated than that at oceanic ridges, probably reflecting the nature of conti-

TABLE 3.1   Continental rifts.

	Location	Length (km)	Maximum earthquake size
Baikal Rift	54 °N, 110 °E	2000	7.5
East African Rift	15 °N, 38 °E	3000	7.5
Rheingraben	49 °N, 8 °E	300	5.0–5.5
Rio Grande Rift	35 °N, 107 °W	700	6.5

nental crust and the difference between active and passive rifting. In general, continental rifts are much wider, and the seismicity is more diffuse than at oceanic ridges. The lithosphere may be thinned at a rift, but it does not go to zero until continental breakup occurs. Table 3.1 lists the major continental rifts; the "spreading" rates for these rifts are much smaller than for most oceanic ridges.

The most active continental rift is the East African Rift System (EARS), which is part of a spreading system that includes the Red Sea and the Gulf of Aden, which are floored by oceanic material. This rift system broke apart the African and Arabian plates, and it is further breaking apart the African plate. Fig. 3.8 shows the complex East African Rift zone. Rifting within East Africa began at the Afar triple junction approximately 25–40 million years ago and has since propagated southward at an estimated rate of 2–5 cm/yr (Oxburgh and Turcotte, 1974). The rift splits into a western and an eastern branch just north of the equator and reconnects at about 8° south (Fig. 3.9). The largest events along the EARS are $M_s > 7.5$; this can be generalized to state that the maximum magnitude expected in a continental divergent setting will be significantly larger than that along an oceanic ridge. This reflects the thermomechanical properties of the continental crust. Although the heat flow in the EARS is much higher than in surrounding regions, the seismogenic zone still extends to at least 10 km.

3.2 Transcurrent boundaries

Transcurrent boundaries, involving movement between horizontally shearing plates, are of two types: (1) *transform faults*, which offset ridge segments, and (2) *strike-slip faults* that connect various combinations of divergent and convergent plate boundaries. One of the interesting features of midoceanic ridges is that they are *offset* by lineaments known as *fracture zones* (see Fig. 3.10). The portion of the fracture zone between the ridge crests is actively slipping as plate A shears past plate B. On the basis of the apparent left-lateral offset of the ridges, one might expect the ridge segments to be separated by a left-lateral strike-slip fault. However, this is not the case, and the relative position of the ridges does not change with time. Wilson (1965) proposed the concept of *transform faults*, shear faults that accommodate offset ridge spreading, which was subsequently confirmed primarily with seismic focal mechanisms. The ridge segments initiated in their offset position largely as a result of preexisting zones of weakness in continental crust that has rifted apart. For the transform fault in Fig. 3.10, the mechanisms are thus right-lateral strike-slip. The fracture zone beyond the ridge-ridge segment is a pronounced topographic scarp, but it is wholly contained within a single plate, so no horizontal relative motion and little or no seismicity occur on the "scar." Fig. 3.10 also shows a schematic cross section along the transform fault. As discussed in the previous section, cooling effects

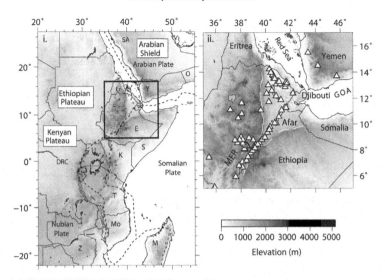

FIGURE 3.8 (left) The east African Rift system (EARS). The bold box identifies the triple-junction region (right). Dashed black lines represent the plate boundaries (Stamps et al., 2014); O–Oman; SA–Saudi Arabia; Y–Yemen; Er–Eritrea; D–Djibouti; E–Ethiopia; S–Somalia; K–Kenya; U–Uganda; DRC–Democratic Republic of Congo; Mo–Mozambique; Z–Zambia; T–Tanzania; and M–Madagascar. (right) Enlargement of the Northern East-Africa Rift (Box in (left)). White lines delineate the major border faults bounding the rift zone. White triangles are the major active volcanoes. MER – Main Ethiopian Rift. GOA – Gulf of Aden (modified from Civiero et al., 2015).

control the ocean floor bathymetry. In general, the fracture zone separates material of different ages and thus of different bathymetry and lithospheric thickness. This scarp is most pronounced at a ridge crest, where the fracture zone juxtaposes material of zero age with material that has been transported from the adjoining ridge crest.

The size of earthquakes on a transform fault largely depends on two factors: (1) the length of offset between ridges and (2) the rate of spreading. The earthquake size is limited by the lithospheric thickness available for brittle failure. If a transform fault is long, connecting slowly spreading ridges, a substantial seismogenic source region is available. In most cases, the largest expected magnitude for transform events is 7.0–7.5. Fig. 3.11 shows the total seismic moments and length of the ridge for earthquakes on many transform faults, and in general, the maximum moments are smallest for fast spreading systems. The total moment scales

to $\sum M \approx L^{3/2}V^{1/2}$ (Boettcher and Jordan, 2004). Fig. 3.12 shows seismicity and focal mechanisms along the Romanche and Chain transform faults along the Mid-Atlantic Ridge (Abercrombie and Ekström, 2001), where the Romanche transform offsets 50-Myr lithosphere while the neighboring Chain transform offsets younger lithosphere (17 Myr) and is shorter (Abercrombie and Ekström, 2001). In all cases, the east-west plane is believed to be the fault plane, and vertical, right-lateral, strike-slip motion occurs on the faults, although there appears to be an opposite sense of dip between the faults of unknown origin.

Transform faults can also offset convergent boundaries, but for simplicity, we reserve the name transform fault for ridge-ridge offsets. We use the term strike-slip faults for transcurrent boundaries that juxtapose continental material. These types of transcurrent boundaries are usually complicated, and their formation is somewhat problematic–clearly they cannot form as a simple ridge-ridge offset. There are

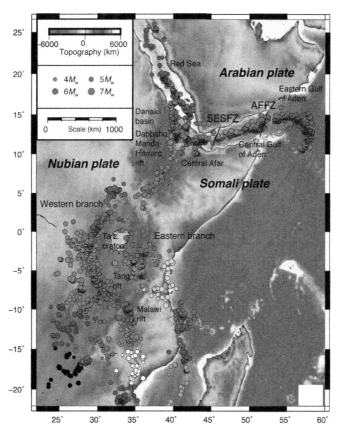

FIGURE 3.9 Distribution of National Earthquake Information Center (NEIC) earthquakes during 1973–2013 in Africa and Arabia plotted and categorized as 32 clusters. Earthquakes are scaled to magnitude and shaded according to which cluster they are in. The clusters correlate well to morphological basins along the East African rift (EAR). Tang. rift, Tanganyika rift; Tanz. craton, Tanzanian craton; SESFZ, Socotra Hadbeen fracture zone; AFFZ, Alula-Fartak fracture zone (modified from Hall et al., 2018).

many theories on the formation of these features, such as plate boundary "jumps," subduction of ridges, highly oblique subduction, or continent-continent collision.

Perhaps the most famous continental strike-slip fault is the San Andreas fault in California, which separates the North American and Pacific plates. Atwater (1970) described a plausible scheme for the development of the San Andreas fault. Thirty million years ago, the Farallon plate was subducting beneath western North America. The midocean ridge separating the Pacific plate and the Farallon plate was west of North

America (Fig. 3.13). The relative motion between the North American and Farallon plates was convergent. Because the Farallon-North American convergence rate was greater than the spreading rate between the Farallon and Pacific plates, eventually the ridge impinged on the Farallon-North America subduction boundary. Farallon plate production (along those ridge segments) ceased and the Pacific-North American plate boundary was created. Faulting along the new boundary was dictated by the Pacific-North America relative motion, which produced

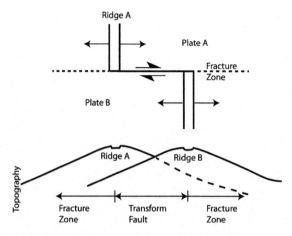

FIGURE 3.10 (Top) Schematic diagram showing the offset of two ridge segments by a transform fault. (Bottom) Schematic cross section showing topographic variations along the transform fault.

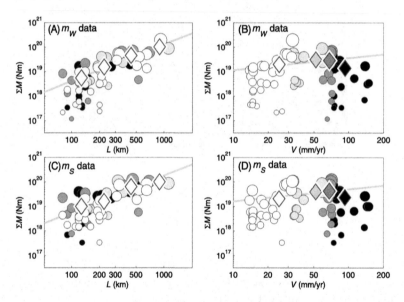

FIGURE 3.11 (left) Total seismic moment $\sum M$ versus fault length L and (right) slip rate V for (top) the Harvard CMT catalog and (bottom) recalibrated ISC M_S catalog. Points show cumulative moments for individual Ridge Transform Faults (RTFs) (circles) and maximum likelihood estimates for the binned data (diamonds). The abscissa values for the diamonds are the averages of L and V in each bin weighted by the plate tectonic moment release rate $\mu A_T V$. Solid lines correspond to the model scaling relation, $\sum M \approx L^{3/2} V^{1/2}$ (modified from Boettcher and Jordan, 2004).

a transcurrent boundary that includes the San Andreas fault (SAF).

An implication of this tectonic model is that the plate boundary is *growing*, with its northern end near Cape Mendocino (Fig. 3.14) propagating northward. Marine geophysical and geological evidence support the model and a boundary age of 30 million years. An interesting aspect

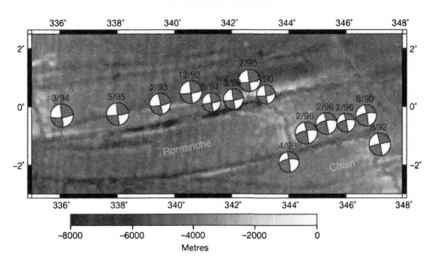

FIGURE 3.12 Map showing focal mechanisms along the Romanche transform. Focal mechanisms are joined to their National Earthquake Information Center hypocenter locations (circles), with radius proportional to magnitude. Earthquakes on the Romanche transform dip to the south at approximately 80°, while those on the Chain transform dip to the north at approximately 73°. The active faults thus appear planar, but the reason for their opposite sense of dip is unknown (modified from Abercrombie and Ekström, 2001).

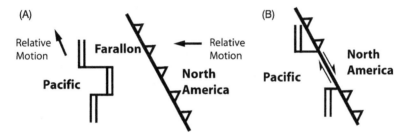

FIGURE 3.13 Plausible tectonic evolution of the San Andreas fault, which separates the North American plate from the Pacific plate. (A) Plate motions prior to subducting Farallon Ridge. (B) San Andreas fault forms to accommodate transcurrent motion along the stretch of annihilated ridge; a gap forms in the underthrust Farallon plate called a slabless window (modeled after Atwater, 1970).

of the boundary is that the SAF is more than 100 km *inboard* of the coast in central California. The basic Atwater model predicts that the transcurrent boundary should juxtapose continental and oceanic material, but the actual situation is more complex. Fig. 3.14 shows a fault map of present-day California. Although the San Andreas is presently the dominant fault in the system, the Pacific-North American boundary is complex. The boundary is actually a *system* of faults that accommodate the plate motion and plate boundary evolution. The present-day San Andreas fault may be positioned along a lineament of preexisting weakness within what used to be the upper plate on a convergent margin. The SAF actually accommodates about 2/3 of the relative Pacific-North American plate motion, with distributed deformation accounting for the rest.

Box 3.2 An introduction to focal mechanisms

Seismologists use a set of symbols called *seismic focal mechanisms* to represent earthquake faulting geometry on diagrams and maps. The symbolism originates from a seismic method to estimate the fault orientation and slip direction using the initial, far-field ground-motion (up or down) of *P*-waves. Up or down motion at the receiver can be projected back to an initial push-away or pull-towards the source. We discuss the details later, here we introduce the symbolism because it is widely used to represent the faulting geometry for earthquakes. Imagine shrinking an earthquake rupture to a point located at the center of a *focal sphere*. The pattern of *P*-wave motions (up or down) mapped onto that sphere separates the sphere into four quadrants. The spatial orientation of the pattern is related to the faulting geometry (Box 3.1). The examples shown in Fig. B3.2.1 identify regions with upward-directed *P*-motion (at a seismic station) as shaded, and those with downward-directed *P*-motion (at a seismic station) as unshaded.

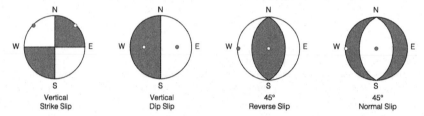

FIGURE B3.2.1 Sample seismic focal mechanisms. P-wave polarity is indicated by shading, dark regions have P-motions that are directed away from the source, unshaded regions correspond to P-wave motions toward the source. A vector is represented by a point in the circle (the dots in diagram). In three dimensions the pattern always has four quadrants. A hemispheric projection includes between two and four quadrants.

The focal sphere is represented on a chart or map using a *stereographic projection*. Specifically, half of the sphere (an upper or lower hemisphere) is projected onto a 2D surface. The projection is equivalent to looking at the hemisphere from directly above or directly below the sphere's center. Upper hemispheres are generally used for local-distance studies, for which *P*-waves leave the source in an upward direction. Lower hemispheres are used for regional- and teleseismic-distance analyses, for which the *P* waves leave the source region with a downward trajectory. Understanding focal mechanisms requires that you understand how vectors (such as *P*-wave take-off directions or stress axes) and surfaces (such as faults) appear in the projection.

The hemisphere projects to a circle on the diagram; the circle represents the rim of the hemisphere, the center of the sphere projects to the center of the circle. Imagine a vector extending from the focal sphere's center and piercing the sphere. The intersection of a vector and the sphere is a point, so vectors are displayed on the focal sphere as points (Fig. B3.2.2). The azimuth of the vector is measured clockwise from north to a line extending from the mechanism center through the point and to the circle's edge. The takeoff angle (in 3D, the angle measured from the downward direction up to the vector (this is 90° minus the vector's *plunge*) is roughly proportional to the distance from the focal mechanism center. An exact relationship (for an equal-area projection) is $r \propto \sin i/2$ where i is the takeoff angle and r is the distance from the center to the vector (the circle radius is the proportionality constant).

Fig. B3.2.2 presents three examples and uses one to define the azimuth ϕ and the vector's projected distance, r. The vector **a** is northeast with a takeoff angle near $\sim 50°$; vector **b** is just east of south with a takeoff angle near $\sim 90°$ (near horizontal); and vector **c** leaves the source region due west with a takeoff angle near ~ 60–$70°$.

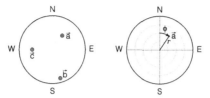

FIGURE B3.2.2 Points on a focal sphere represent vectors. A vector has a an azimuth and a takeoff angle (angle measured from the downward vertical up to the vector). The azimuth is $0°$ for a northward directed vector, $90°$ for an eastward directed vector and so on. The take-off angle is $0°$ for a vertically downward directed vector, $90°$ for a horizontal vector, and $180°$ for a vector directed vertically upward.

Now imagine a plane passing through the center of the focal sphere – slicing the sphere into two halves. If the plane is horizontal, the sphere is split into a lower and an upper hemisphere. The intersection of the plane with the sphere is a great circle that bisects the sphere. On the projection, a great circle is a curve. Fig. B3.2.3 is a plot of four north-striking planes with a range of dips intersecting a lower-hemisphere projection. The smaller the dip, the larger the curvature of the curve representing the plane. The dip is roughly proportional to the distance from the projection edge to the curve measured along the plane's dip direction.

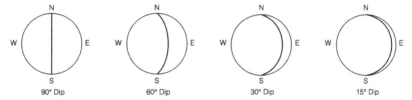

FIGURE B3.2.3 Curves on the sphere plane the intersection of the sphere and a plane. The larger the curvature, the shallower the dip of the plane. The thick lines represent north-striking planes that dip to the right (east). A vertical plane is a line with no curvature, a horizontal plane intersects at the edge of the projection, and has maximum curvature.

Examination of the focal mechanisms shown in Fig. B3.2.1 indicates that each mechanism include two vectors (small circles) and two planes (the two curves). One of the curves represents the fault surface, the other represents a surface called the *auxiliary plane*. The auxiliary plane is not physical, it represents an alternative surface that could produce the exact same P-wave polarity pattern as that produced by the actual faulting geometry. Since an earthquake on the auxiliary plane could have generated the same P-wave pattern – we can't tell the difference between the two planes using only P-wave polarities. So, we always show both planes. The fault and auxiliary planes are orthogonal and separate the focal mechanism into regions of distinct P-wave polarities. The planes are sometimes referred to as *nodes* because they represent directions corresponding to small (zero) P-wave amplitudes. To transition from toward- to away-from-the-source motion, you must cross a node of no motion. The quadrants are referred to as *compressional* (motion away the source) and *dilatational* (motion toward the source) – these terms refer to how motions at the source affect the region surrounding the source.

Interestingly, the auxiliary plane's normal vector is equal to the fault-plane's slip vector (see Box 3.1) and to produce the same P waves with the auxiliary plane, the auxiliary-slip vector must equal the fault plane's normal vector. Of great interest in tectonic studies are the directions associated with the stress patterns that produce the strain accumulation and release. We often approximate the directions of tectonic forces using three stress or strain axes, the *pressure* (**p**), *tension* (**t**), and *intermediate* (**b**) axes. Unit vectors for each axis can be computed from the fault normal, $\hat{\mathbf{n}}$ and fault slip vector $\hat{\mathbf{s}}$,

$$\hat{\mathbf{p}} = \frac{\hat{\mathbf{s}} - \hat{\mathbf{n}}}{\sqrt{2}}$$

$$\hat{\mathbf{t}} = \frac{\hat{\mathbf{s}} + \hat{\mathbf{n}}}{\sqrt{2}} \qquad (B3.2.1)$$

$$\hat{\mathbf{b}} = \hat{\mathbf{n}} \times \hat{\mathbf{s}} = \hat{\mathbf{p}} \times \hat{\mathbf{t}}.$$

The stress axes are often of more interest than the specific fault geometry since the axes provide an idealized indication of the directions of the stress causing tectonic deformation and they directly indicate the direction of principal strain axes. On a focal mechanism, the compressional direction, **p** locates exactly in the center of the dilatational quadrant and the tensional direction, **t** locates exactly in the center of the compressional quadrant. The reason is that we are now considering the forces exerted on the fault from regions outside the source region. The intermediate axis, **b** is located at the intersection of the fault and auxiliary planes. The **p** and **t** vectors are shown on the focal spheres by the shaded and unshaded small circles on the focal mechanisms in Fig. B3.2.1.

The examples below, in Fig. B3.2.4 show a set of north-striking faults that dip to the east. The auxiliary planes dip to the west (it's horizontal for the vertical fault).

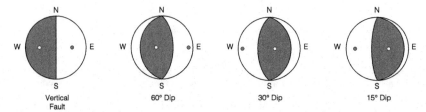

FIGURE B3.2.4 Sample seismic focal mechanisms for a suite of north-striking faults with decreasing dips. Shaded region is called the compressional quadrant, shaded vector identifies the direction of the compressional stress. The unshaded regions identify the dilatational quadrants, the unshaded vector identifies the direction of the (relatively) tensional stress.

Note that as the fault plane rotates, so do the stress axes and the auxiliary plane. Up to this point our examples have been either perfectly strike-slip or dip-slip. The examples included in Fig. B3.2.5 show the effect of *oblique faulting* – where the rake is not equal to 0° or 180° or ±90°. From left to right, the rake varies from 90° (reverse dip slip) to 15° (almost strike slip). The fault plane is identical in each of the diagrams, but rakes associated with oblique faulting move the **b** axis away from the edge (or center) of the focal sphere and substantially change the orientation of the auxiliary plane and alter the pattern of P-wave polarities.

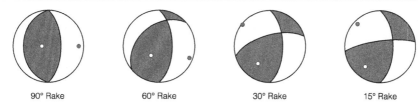

FIGURE B3.2.5 Sample mechanisms for a north-strike fault with a suite of rake values illustrating the character of focal mechanisms for earthquakes with oblique slip. From left to right the mechanism shows the transition of a dip-slip reverse faulting geometry to a slightly oblique left-lateral faulting geometry.

Typically, in the upper 10's of kilometers, the stresses that produce earthquakes often orient such that one of the stress directions is roughly perpendicular to Earth's surface. This orientation favors slip roughly in the strike or dip directions. Although not always the case, the obliquity in a focal mechanism may result from uncertainty and instability in the modeling used to estimate the faulting geometry. Finally, note that we did not include the compass directions (N,S,E,W) in Fig. B3.2.5. The directions are usually not shown – assume north is at the top of the diagram (or along the meridian on a map). The only times this is not the case is when we are looking at a side-perspective in a vertical cross section – then the top of the focal sphere represents the upward direction.

Finally, the examples we have considered correspond to shear-faulting. Other seismic sources include ideal explosions (the entire mechanism is shaded because all motion is away from the source), implosions (the entire mechanism is unshaded), or combinations of volume change (e.g. dike injection, mine collapse) and shear faulting, or a combination different faults failing simultaneously. In this case, the compression and dilatational quadrants do not split into regions divided by great circles. We refer to the difference from an ideal faulting geometry as *non-double-couple components* (we discuss the details later in the text). For events with substantial non-double-couple characteristics, the focal mechanism takes on a smoother character as shown in Fig. B3.2.6.

FIGURE B3.2.6 Sample mechanisms with non-double-couple components and their best-matching double couple planes (the stress axes are the same for both the non-double-couple and shear-faulting models).

From left-to-right in the figure, the focal mechanisms represent faulting with a mostly reverse, normal, and strike-slip character. Non-double couple components in a focal mechanism may be an indication of a complex source (multiple involved faults), but often is a result of inadequacy of the earth model used in the analysis to estimate the faulting geometry. We often work with the ideal shear-faulting geometry closest to the estimated focal mechanism. These "best double couples" are shown by the curves on each solution in Fig. B3.2.6. The stress axes for the original estimate and best approximation are identical.

The 3D visualization required to understand and interpret focal mechanisms takes time to develop. Most students become comfortable with focal mechanisms through practice interpreting them. Tectonics and seismological literature contains many examples that can be explored. The global centroid moment tensor catalog (GCMT) has more than 40,000 examples online (http://globalcmt.org/) that you can use to practice.

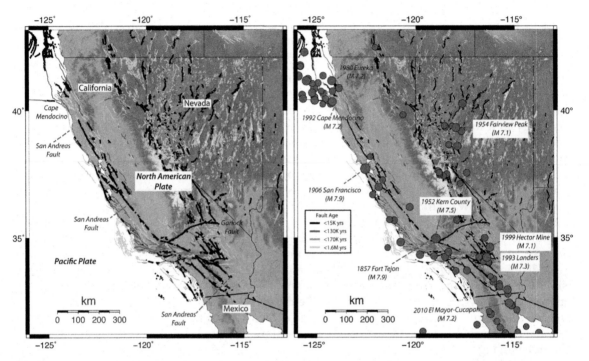

FIGURE 3.14 (left) Major faults and topography in California and Nevada using the USGS classification for faults (USGS website). The San Andreas fault marks the boundary between the North American and Pacific Plates. Thick black lines show Quaternary faults (<15,000 years) with well constrained location; Dark gray show late Quaternary faults (<130,000 years) with moderately constrained location; Medium gray lines show middle and late Quaternary faults (<750,000 years); Light gray show undifferentiated Quaternary faults (<1.6 million years) that are well constrained in location. (right) Earthquakes with $M \geq 6.0$ in same region since 1900, with the events with $M \geq 7.0$ labeled (USGS website). Note the diffuse nature of seismicity along the plate boundary.

The San Andreas fault has numerous bends and appears to have numerous *strands*, especially at the southern end. This complicates predicting the seismic behavior of the SAF. Some sections of the fault slip only as great earthquakes; others slip continuously through *creep*. Fig. 3.14 also shows the historic seismic record for California. This record contains two great SAF earthquakes: (1) the 1906 San Francisco

earthquake, which ruptured the northern third of the SAF, and (2) the 1857 Fort Tejon earthquake, which ruptured the south central SAF. Both events were about $M_s = 7.9–8.0$. Since 1900, only 9 earthquakes occurred with $M \geq 7.0$ in California, Nevada, and northern Baja California, Mexico. Many other significant and damaging earthquakes have occurred with $M \geq 6.0$ (Fig. 3.14) but only one, the 1989 Loma Prieta, CA (M_W 6.9) earthquake, appears to be directly related to the SAF. Even the Loma Prieta earthquake is problematic in that it did not rupture to the surface, and the fault-plane solution indicates oblique thrust motion, inconsistent with the SAF orientation. The Loma Prieta earthquake occurred in a region with a small bend in the SAF, which produces compression across the fault accommodated on several splay faults.

Transcurrent faults in continental settings are extremely important from a seismic hazard point of view. These faults typically traverse regions of high population density, and in this century earthquakes on these faults have been responsible for the largest number of fatalities, with the most recent in Haiti in 2010, which was responsible for over 90,000 deaths. Other major plate-boundary continental transcurrent faults are the Anatolian fault in Turkey, the Alpine fault in New Zealand, and the Sigang fault in Burma. Some continental transcurrent faults occur far from plate boundaries, and we will discuss these further in the section on intraplate earthquakes.

3.3 Convergent boundaries

3.3.1 Subduction zones

There are two types of convergent boundaries, *subduction zones* and *continental-continental collision zones*, where the lithospheric plate type (oceanic or continental) determines the type of interaction. Continental collision zones, although rare, produce major mountain ranges and plateaus, such as the Himalaya and the Tibetan Plateau. Currently more than 90% of the world's convergent boundaries are subduction zones, which form at zones of oceanic-oceanic or oceanic-continental convergence. Convergence at these boundaries is accommodated by underthrusting of one lithospheric plate beneath another. The underthrust

plate descends through the asthenosphere and represents "consumed" lithosphere, which balances the surface area of new lithosphere created at mid-oceanic ridges. Subduction of the lithosphere is one of the most important phenomena in global tectonics; no oceanic crust older than Jurassic ($\sim$ 200 million years) exists, yet we find continental crust as old as 3.5 Bya. Summing the area of ocean floor that has opened since Jurassic time, we find that 20 billion km^3 of material has been subducted. At the present rate of subduction, an area equal to the entire surface of the Earth will be cycled into the interior in 160 million years.

The initiation of subduction is a topic of considerable controversy. Although no generally accepted model exists to explain how subduction initiates, the thermal buoyancy of the plate explains why it *continues*. If a slab of lithosphere is displaced into the asthenosphere, it will be "colder" and hence more dense than the surrounding asthenosphere. This results in a negative buoyancy force that causes the lithosphere to sink. The sinking of a subducting plate through the asthenosphere is a complex function of the age of the lithosphere, the rate of convergence, and the age of the subduction zone. As the name *thermal buoyancy* implies, the effect is a manifestation of the temperature differences between the slab and surrounding mantle. This temperature difference also has a signature on the seismic velocities, which will be discussed later in Chapter 10 along with tomography, which can be used to image the subducting slab. As will be discussed in Chapter 10, the ultimate fate of the subducted plate is very controversial; models range from continuous subduction to great depths–perhaps to the core-mantle boundary–before thermal equilibration and assimilation take place, to no subduction below 660 km depth. The seismic activity along the slab can be used to address these hypotheses.

Three categories of seismic activity occur in subduction zones. The first is the interaction be-

Box 3.3 Creeping plate boundaries

When a fault slips continuously, we refer to this as *aseismic slip* or *creep.* The best known example of a creeping fault is a 100-km-long segment of the San Andreas between San Juan Bautista and Parkfield, California. Fig. B3.3.1 shows the rate of slip along this creeping section of the San Andreas. The fault creep was discovered when the Cienega Winery was built on the San Andreas in 1948, and the walls of the winery were progressively offset at a steady rate of approximately 11 mm/yr.

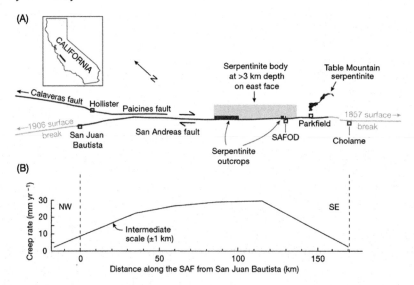

FIGURE B3.3.1 (A) The creeping section lies between areas of the fault that ruptured during great earthquakes in 1857 and 1906. Serpentinite occurs in rare surface exposures of the fault and in the probable active trace of the fault encountered at approximately 3 km vertical depth in the San Andreas Fault Observatory at Depth (SAFOD) drill hole. (B) Recently updated creep rates measured from 10 m to 1 km from the fault. Total offset rates along the San Andreas system in the creeping section are considered to be between 20 and 34 mm/yr (modified from Moore and Rymer, 2007).

Faults that creep require special frictional behavior; either the fault has unusually low normal stresses, or the fault is lined with very ductile or weak material. The creeping segment of the San Andreas is characterized by a wide zone (more than 2 km) that has a pronounced low seismic velocity. This low-velocity zone is absent in the adjacent locked sections of the San Andreas, suggesting that the fault zone is filled with "fault gouge," a poorly consolidated, clay-rich material. The San Andreas Fault Observatory at Depth (SAFOD), drilled in 2005, provides direct measurements of fault material at seismogenic depths (Zoback et al., 2010). Analysis of this material shows very low frictional strength of fault material at depth (Lockner et al., 2011), thus the majority of the fault's slip is accommodated aseismically.

tween the two converging lithospheric plates. Typically, an oceanic plate is flexed at a 10° to 30° angle and dives below the overriding plate. This results in a large contact zone be-

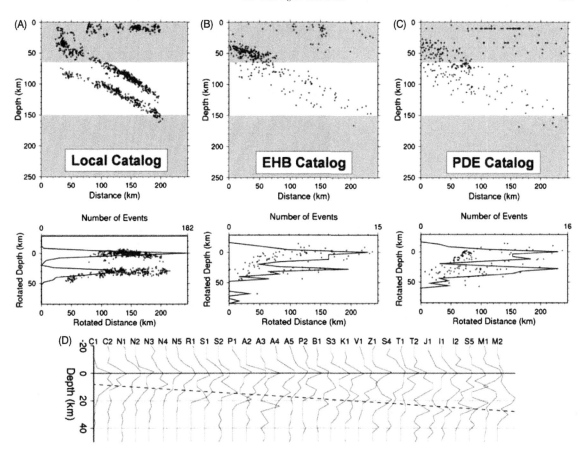

FIGURE 3.15 Analysis of intraplate double Benioff zone (DBZ) separation using slab normal distributions. (A) Results for northeastern Japan using relocated, local-network hypocenters. Top panel shows events (crosses) in typical crosssectionview (gray areas not analyzed), and bottom panel shows events after rotation into down-dip and slab-normal locations. This provides the benchmark for comparison with results using global hypocentral catalogs of (B) EHB and (C) PDE. Similar estimates of DBZ separation among the three data sets at several other subduction zones confirm that global catalogs are sufficient to characterize DBZs. (D) Histograms showing slab-normal distribution of EHB events for all segments analyzed, sorted by subducting plate age. DBZ separation is estimated by multiple Gaussian fits shown in green, and dashed linear best fit highlights a significant increase with age (modified from Brudzinski et al., 2007).

tween two plates on which frictional sliding can take place, producing interplate seismicity. The largest earthquakes occur along these contacts and involve thrust faulting. The 1960 Chilean earthquake had a fault length of more than 1000 km, a fault width of 200 km, and an *average* displacement of 24 m! The M_w of the Chilean event was 9.5. The second category of seismicity is related to the internal deformation of the over-riding plate commonly associated with back-arc extension or upper-plate compression. The final category of seismic activity has to do with internal deformation within the subducting plate that results from the slab's interaction with surrounding mantle. It is this **intraplate seismicity** that allows the seismologist to map the subducting plate and to infer the mechanical state of the mantle. In the early 1930's, K. Wadati

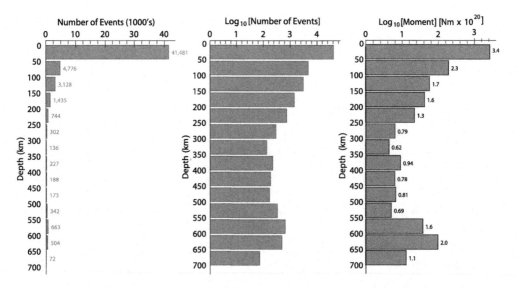

FIGURE 3.16 Depth distribution of the total number of earthquakes and accumulated seismic moment from global earthquakes that occurred between 1976–2017, derived from the Global Centroid Moment Tensor catalog (www.globalcmt.org). Note the minimum in energy release between 250 and 550 km depth.

first observed deep zones of seismicity beneath Japan. With the advent of plate tectonics, these zones were recognized as an expression of the Pacific plate subducting beneath the Eurasian plate. H. Benioff detailed the occurrence of deep seismicity zones in many regions of the world in the 1940's, and we now refer to these deep seismic belts as Wadati-Benioff zones. Fig. 3.15 shows a depth distribution of earthquakes for neartheastern Japan, highlighting a double Benioff zone (DBZ) that results from old oceanic lithosphere subducting and unflexing (Brudzinski et al., 2007). Of note is that global catalogs have enough uncertainty in location that the DBZ is less evident, while the local catalogs clearly show the DBZ.

Internal Slab Deformation. The state of stress in a subducting plate depends on the balance of two forces: (1) the negative buoyancy of the descending slab and (2) the resistance force of mantle that is being displaced by the subducting plate. These forces depend strongly on the viscosity structure, phase transformations in the slab, the rate of subduction, the age of the

subduction zone, and the depth of slab penetration. Fig. 3.16 includes plots of the number of earthquakes between 1976 and early 2020 observed globally versus depth, alongside the corresponding accumulated seismic moment as a function of depth. Most earthquakes and the greatest energy release occur in the upper 200–250 km. This is the region of interplate interaction and slab bending. Normal frictional sliding processes dominate at these depths. Below about 50 km all events are intraplate rather than on the plate interface. The earthquake energy release is at a minimum between 250 and 550 km depth, where the subducting lithosphere is interacting with *weak* asthenosphere. Frictional sliding may occur at these depths only if hydrous phases destabilize and release water or other fluids to allow high pore pressures to exist. Below 400 km, the number of earthquakes increases with depth and some slabs strongly distort. Increasing resistance to slab penetration is often inferred, but frictional sliding mechanisms are generally not expected at these depths (unless further hydrous phases exist at these depths and release fluids as they destabilize), so other mechanisms such as phase changes may be operating (Box 3.4). All earthquake activity ceases by a depth of 700 km. Some of the largest deep events are found near the maximum depth of seismicity in different slabs, so there is not a simple tapering off of activity. This maximum depth is conspicuously similar to the velocity discontinuity near 660 km depth (see Chapter 10). A phase change in the slab may occur that suppresses earthquake failure. The termination of seismicity is a first-order observation of the fate of the subducted slab, but its implication is controversial, because an aseismic slab extension may exist in the lower mantle.

The focal mechanisms of the earthquakes along the Wadati-Benioff zone can be used to map the stress orientation in the slab. The stress orientation is controlled by the slab geometry and the balance of thermal, resistive, and negative buoyancy forces. If negative buoyancy dominates, the slab will be in downdip *extension*. As the resistive force becomes more important, the slab experiences downdip *compression*. However, subduction zones exhibit a wide variety of behavior. For example, Fig. 3.17 shows focal mechanisms for earthquakes along the central Betics in the Gibraltar Arc slow convergence region. In this case, the internal arc is in compression at depth while the external section of the slab is in extension. Isacks and Molnar (1971) recognized a corresponding range of behavior in subducting slabs by examining numerous subduction zones. Fig. 3.18 shows the stress patterns for various subduction zones and Isacks and Molnar's model for accounting for the stress conditions. *Strength* is a time-dependent concept, and most geophysicists would replace strength with viscosity in Isacks and Molnar's model. In addition, given the uncertain mechanism associated with deep earthquakes, it is not clear what stress variations are actually required.

Although the general model for the state of stress in a slab outlined in the previous paragraphs explains the gross features of Wadati-Benioff zone seismicity, some subduction zones show very interesting variations at depths of 50 to 200 km. In the Japan, Kuril, and Tonga trenches, the Wadati-Benioff zone is made up of two distinct planes. Each plane is defined by a thin, well-defined cluster of epicenters; the planes are separated by 30–40 km. As shown previously, Fig. 3.15 shows an example of the double Wadati-Benioff zones in northern Japan. There are several proposed models to explain double Wadati-Benioff zones that involve the unbending/bending of the plate, thermal stresses, and the breakdown of hydrous minerals. In one model, bending a thin plate causes extension in the outer arc of the bend, while the underside of the plate is in compression. The extensional zone is separated from the compressional zone by a neutral surface. Shallow tensional events and deep compressional events are, in fact, observed in the outer rise of subduction zones. If the plate is suddenly released from the torquing force, it will "unbend" and experience forces opposite to those imposed during bending. This could explain the double Benioff zone stresses at intermediate depth. In another model, lower plane seismicity relates to the subduction of large amounts of water due to hydration from outer rise faulting, shown in Fig. 3.19. Water carried in the lithospheric mantle may be transported to the transition zone, which are serpentinized. Release of fluids essentially increases pore pressure on faults, reducing normal stress, thus inducing earthquakes.

Even today, the ultimate fate of subducted slabs penetrating as deep as 660 km remains controversial (Fig. 3.20), although many results show slabs penetrating into the lower mantle and others that flatten out at the upper-lower mantle boundary. The observation that seismicity ceases near this depth indicates that something significant happens associated with the 660-km boundary. Two basic theories explain the maximum depth of seismicity. In the first, the 660-km discontinuity is viewed as impenetrable to subduction. When the slab encounters the boundary, it must flatten out, although it may depress the discontinuity. What could cause such a "strong" boundary? There are two choices: (1) the 660-km discontinuity may be a boundary between chemically distinct lower and upper mantles, or (2) the viscosity across the boundary may increase by more than several orders of magnitude, enough to prevent penetration. In the case of the chemical boundary, the lower mantle must have a composition with a high enough density to exceed the thermally induced density anomaly and inertial effects of the slab. The existence of strong viscous or compositional stratification would cause mantle convection to be separated into upper- and lower-mantle convective regimes. The second theory is that the 660-km discontinuity is a phase change, with conversions to high-pressure Bridgmanite being expected at this depth. Although viscosity may increase at this transition, the slab can usually penetrate the phase boundary, with seismicity terminating as the phase transformation occurs. Most geophysicists agree that the seismic boundary is such a phase change.

The most direct way for seismologists to determine the ultimate fate of the slab is to search for a seismic signature. Since all seismicity stops at ~ 660 km depth, only the seismic velocity difference between the slab and surrounding mantle can provide this signature. Unfortunately, accurately measuring this velocity contrast is difficult, as discussed in Chapter 10. High-resolution tomographic results suggest that penetration may depend on the individual slab, as some slabs appear to deflect and broaden at the 660-km discontinuity, while others appear to penetrate at least a few hundred kilometers (Fig. 3.20). Resolving this issue will contribute to solving the problem of whole-mantle versus layered-mantle convection and will elucidate the evolution of the Earth.

The variations between subduction zones in terms of maximum earthquake size and rupture length is remarkable. Here, we introduce the concept of *seismic coupling*, which is a measure of the seismogenic mechanical interaction between the subducting and overriding lithospheric plates. In some subduction zones, such as the Marianas, the largest thrusting earthquakes are relatively small, and a significant component of aseismic slip must occur. In other regions, such as southern Chile, the subduction slip occurs primarily in very large earthquakes. Furthermore, recent work has identified *slow slip events (SSE)* and *episodic tremor and slip*

Box 3.4 Mechanisms for deep earthquakes

Deep earthquakes have long posed a problem for seismologists. Laboratory experiments indicate that the pressures at a few hundred kilometers depth should prohibit brittle fracture and frictional sliding processes. Yet earthquakes as large as $M_W = 8.3$ have occurred at 650 km. The deep seismicity has many characteristics that are similar to those of shallow earthquakes. Most important, the deep earthquakes have radiation patterns consistent with double couples, which implies shear faulting. In 1945, P. Bridgman proposed that phase transformations, the reordering of solid phases to higher-density structures, could produce nonhydrostatic stresses that may cause the deep earthquakes. Other work has shown that phase transformations and fault growth may be a self-feedback system, where 'anti-cracks' of densified material in the presence of deviatoric stress initiates shear failure localization. The localization of the transformation and shear strain that develops rapidly after fault growth can, in turn, cause the fault to grow.

This mechanism for deep earthquakes depends strongly on the thermal structure of the subducting plate. Two major phase changes are postulated to occur in the mantle (see Chapter 10): one at 400 km depth and one at ∼ 660 km depth. Since the slab is *colder* than the surrounding mantle, the 400-km phase transition (olivine → β-spinel) should occur at shallower depths in the slab. However, experiments show that this equilibrium reaction may be kinetically suppressed, meaning that olivine can actually persist to greater depth. The olivine → β-spinel transformation can occur rapidly on planes of maximum shear stress, giving rise to double-couple-like energy release. The net effect is a thin metastable wedge of olivine-like material in the slab. Recent work on the Great 2013 Sea of Okhotsk earthquake (M8.3, depth 607 km) and the Great 1994 Bolivia earthquake (M8.3, depth 637 km) suggest two fundamentally different faulting mechanisms: Phase transformational faulting occurs inside the metastable olivine wedge while shear melting occurs inside warm slabs (see Fig. B3.4.1).

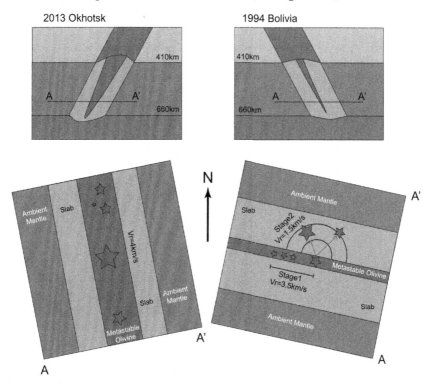

FIGURE B3.4.1 Conceptual models of the Okhotsk and Bolivia earthquakes in cross section (top panels) and map view (bottom panels). Due to differences in the thermal states of the subducting slabs in which the two earthquakes occurred, the widths of the metastable olivine wedges in the slab cores are also different. This causes different dominant faulting mechanisms for the two largest deep earthquakes. The Okhotsk earthquake is inferred to have ruptured mostly inside the relatively thick metastable olivine wedge, whereas the Bolivia earthquake's major rupture was outside the relatively thin metastable olivine wedge (modified from Zhan et al., 2014).

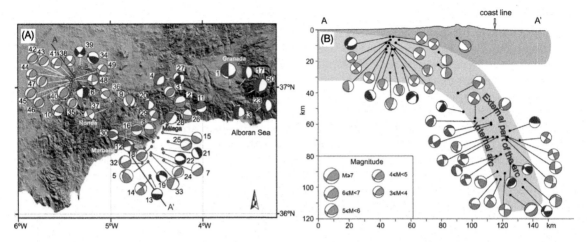

FIGURE 3.17 (A) Earthquake focal mechanism solutions in the western Betic Cordillera where numbers refer to specific events (see Ruiz-Constán et al. (2011)) and shading to stress tensor groups (shallow earthquakes in dark grey; intermediate events in lighter and light gray for the external and internal arc of the slab, respectively; deep earthquakes in very dark grey; non-explained solutions are in black). (B) Cross section along line A–A' showing vertical distribution of focal mechanism solutions in equal area and vertical projection. Earthquake focal mechanism solutions are scaled by magnitude and compressional quadrants are shaded; the topographic profile is shown with a vertical exaggeration of a factor of 4 for legibility (modified from Ruiz-Constán et al., 2011).

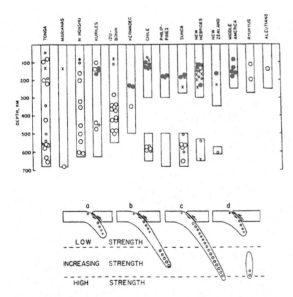

FIGURE 3.18 Distribution of downdip extension (solid hypocenters) and downdip compression (open hypocenters) in various subduction zones. The mechanical interpretation is shown below (from Isacks and Molnar, 1971).

(ETS) beneath coupled regions of subduction zones (see Box 3.5). The concept of coupling is illustrated in Fig. 3.21, which shows fault segments that fail in large ruptures and the size of high stress drop and relatively large displacement patches (called asperities) within those segments (Lay and Kanamori, 1981). (Note that this definition of asperity need not correspond to protrusions on the fault, as associated with asperities on the small scale discussed in Chapter 7.) A strongly coupled subduction zone will have a greater portion of its interface covered by asperities compared with less coupled zones. The motivation for this model of asperity coupling comes from the complexity of source time functions for large thrust events and the size of the fault zone inferred from aftershocks and surface-wave models of the source finiteness.

What causes the variability in coupling? These ideas about seismic coupling are very qualitative, and the actual frictional properties on the thrust faults are undoubtedly further complicated by the history of prior slip, hydrological

Box 3.5 Episodic tremor and slip

The deployment of Global Positioning System (GPS) networks and analysis of the data they produce led to the discovery of unexpected episodic slip, deep along the Cascadia Subduction Zone (CSZ) off the coast of western North America. The locked seismogenic zone of the CSZ exhibits little seismicity (see Box 9.2). Contemporaneous analyses of high-frequency seismic ground motions in other subduction zones (e.g. Obara, 2002) and the CSZ revealed emergent seismic motions that lasted for hours to days and correlated with the occurrence of the slow, GPS-detected episodic movement (Fig. B3.5.1). The high-frequency motions are called *tremor*, or *non-volcanic tremor*. The current explanation for both phenomena is movement along the deep subduction interface, referred to as *Episodic Tremor and Slip (ETS)*, that occurs down-dip (and in some places up-dip) of the locked portion of the plate boundary (Fig. B3.5.1. An important implication is that ETS may episodically load the deep regions of the seismogenic zone at rates much higher than typically plate motions.

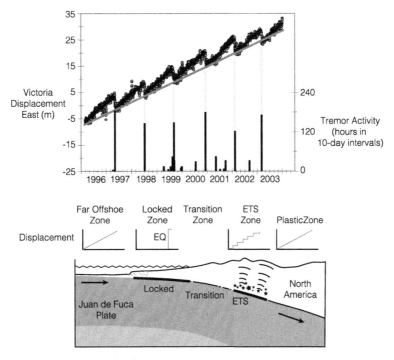

FIGURE B3.5.1 (Left) Observed surface motion (circles) at GPS station ALBH (Victoria) with respect to DRAO (Penticton), which is assumed fixed on the North America plate. Continuous line shows the long-term (interseismic) eastward motion of the site. Saw-tooth line segments show the mean elevated eastward trends between the slip events which are marked by the reversals of motion every 13 to 16 months. The central peaked curve is a plot of the normalized cross-correlation of the de-trended GPS time series and a 160-point zero-mean sawtooth segment used to determine the dates for the slip events. The curve along the bottom is the number of hours of tremor activity observed at southern Vancouver Island within a sliding 10-day interval (complete annual records examined from 1999 onward). Ten days corresponds to the nominal duration of a slip event. In each case, pronounced tremor activity coincides with slow slip events. (Right) Conceptual distribution of slip deficit at CSZ across the northern Cascadia margin (modeled after Dragert et al., 2004).

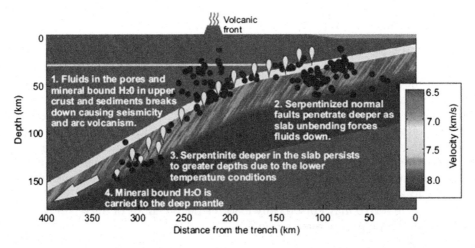

FIGURE 3.19 Summary of subduction zone structure where hydration of slab mantle is the dominant mechanism for seismicity (black dots) in a double Wadati-Benioff zone in northern Japan. Fault depth increase with distance from outer rise is due to downward forcing of fluids. Approximate depth at which mineral-bound water is released is shown by white balloon shapes. Some mineral-bound water is delivered to mantle transition zone, indicated by white arrow. Depth of penetration of these hydrated fault zone structures is controlled by temperature at which serpentinite is stable. As slab is heated at depth, stability field of serpentinite becomes restricted to cool core of slab (modified from Garth and Rietbrock, 2014).

variations, and thermal regime of the plate interface. The notion of asperities is useful primarily as a qualitative characterization of the stress heterogeneity, not as a model for dynamic slip processes. A number of factors, such as age of lithosphere and convergence rate, contribute to coupling. Furthermore, recent work has identified slow slip events (SSE) beneath coupled regions of subduction zones. The coupling force has been modeled as a function of slab length and age of the subducting slab and the sea anchor force as a function of upper plate velocity and slab length (Scholz and Campos, 2012),

$$\Delta F_n = F_{SA} \sin(\phi) + F_{SU} \cos(\phi), \qquad (3.2)$$

where ϕ is the dip of the subduction interface. The trench suction force (ΔF_{SU}) is a function of slab length and age of the subducting plate and the sea anchor force ΔF_{SA} is a function of upper plate velocity and slab length. Fig. 3.22 shows the coupling coefficients plotted versus ΔF_n. The various estimates of seismic coupling coefficient plotted versus ΔF_n confirms the idea

that the normal force correlates with earthquake size for most subduction zones, with a few notable exceptions (Scholz and Campos, 2012).

Within the subducting plate relatively few earthquakes occur that are associated with the bending of the plate, oceanward of the trench, in what is called the **outer rise**. For trenches in which the subducting and overriding plates are weakly coupled, these intraplate events can be very large. Tensional stress from the negative buoyancy of the slab concentrates in the outer rise. This tensional stress results in large normal-faulting earthquakes, which may actually "break" the subducting lithosphere. The 1933 Sanriku, 1977 Sumbawa, and the 2017 Tehuantepec, Mexico (M8.2) earthquakes are examples of this type of earthquake.

3.3.2 Continental collisions

A second type of convergent boundary is a continent-continent collision. Unlike oceanic lithosphere, continental lithosphere is too buoy-

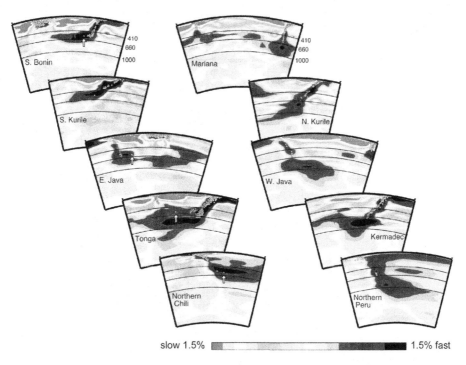

slow 1.5%   1.5% fast

FIGURE 3.20 Tomographic images showing a variety of subduction zones and the nature of the subducting plate. Some plates show a stagnant slab in the transition zone (left) and of a trapped slab in the uppermost lower mantle (right) in the same subduction zone. Subhorizontal distributions of deepest shocks as marked by arrows are associated with stagnant slabs in the transition zone. Steeply dipping distributions are associated with penetrating slabs across the 660 or with trapped slabs below it (modified from Fukao and Obayashi, 2013).

ant to subduct. When two continental masses come into contact along a convergent boundary, the relative motion between the plates is taken up in lithospheric shortening and thickening or, in extreme cases, lateral expulsion of lithosphere away from the collision along strike-slip faults. Although continent-continent collisions account for only a small fraction of the present convergent boundaries, they profoundly affect topography. The continent-continent collision between India and Eurasia produced the Himalayas and the Tibetan Plateau, which are the most conspicuous topographic features in the world. The two major drainages from the Himalayas (the Ganges and Indus) carry more than 50% of the world's river-delivered sediment load. Former continent-continent collisions are recognized as suture zones. The Appalachian and Ural mountains are examples of ancient collisions.

The seismotectonics of a continent-continent collision are complicated. Usually, faulting is dominated by thrusting, which is a manifestation of the lithospheric shortening. In many of the collisions, well-developed, low-angle thrusts occur beneath the suture zone. These low-angle faults can generate very large earthquakes ($M_w > 8.5$). The effects of collision can also be propagated large distances inboard of the suture. Strike-slip faulting in China and normal faulting at Lake Baikal are the result of the Indian-Eurasian collision 2000 km away.

The Indian-Eurasian collision remains the quintessential continent-continent collision, and

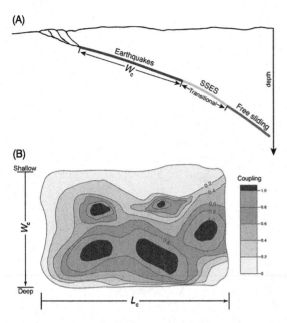

FIGURE 3.21 Schematic diagram for stress heterogeneity on the fault plane that results in different orders of coupling. (A) Schematic cross-section of a subduction zone showing the coupled (seismogenic) width Wc. Below that is a transitional region of partial coupling in which the slip occurs in slow slip events (SSEs). (B) A cartoon of seismic coupling on the coupled area, where filled regions represent high stress drop and relatively large displacements during major earthquakes, called asperities. Some regions have more heterogeneous stress conditions, which give rise to more complex earthquake ruptures (modified from Scholz and Campos, 2012).

began fifty million years ago where oceanic lithosphere north of the Indian shield was subducting beneath Eurasia. By 40 million years before present, the oceanic material had been completely consumed, and the Indian shield began to collide with the Eurasian continental mass. Because the continental lithosphere of India is too buoyant to subduct, the collision created uplift of the Himalayan front and the Tibetan Plateau (with a doubling of the crustal thickness), plus "ejection" of Southeast Asia away from the collision along a series of strike-slip faults. India continues to move northward relative to Eurasia at a rate of 5 cm/yr, making the

India-Eurasian collision zone one of the most seismically active regions in the world.

Fig. 3.23 shows a tectonic map of the India-Eurasia collision zone. The boundary between India and Eurasia used to be the Indus-Tsangpo suture. The boundary is now marked by an arcuate system of thrust faults over 2000 km long known as the Main Boundary Thrust (MBT) and the Main Central Thrust (MCT). At either end of the collisional arc are sharp "kinks" known as syntaxes. In the northeast, the Indian plate is terminated by the Indo-Burma syntaxis. In the west, it is terminated by the Pamir and Chaman faults in Pakistan. The Tibetan Plateau is a region of extreme elevation (in excess of 5 km) and thick crust (in places, in excess of 70 km). Although the exact mechanism of the uplift of Tibet is still a source of controversy, it represents significant lithospheric shortening. A series of strike-slip faults occur north and east of Tibet that "move" Eurasia away from the collision zone and elevate Tibet. These are some of the most active strike-slip faults in the world, and the long written history in China allows for the investigation of the earthquake cycle. The incremental strain field based on earthquake focal mechanisms of earthquakes shows the difference stress regimes in the collision zone. Tibet and its Eastern margin deform by transtension, as does a narrow region in Myanmar, while the rest of the collision zone is under transpression or contraction, with transpression being the most dominant mode of deformation in the collision zone.

The area affected by the India-Eurasia collision is extremely active seismically. The collision belt accounts for approximately 15% of yearly global seismic energy release. The majority of the energy is released along the MBT and MCT during major earthquakes. The Indian shield is being thrust beneath the Himalaya along a plane dipping 10°–20°N. There have been four great earthquakes along the MBT in the past 100 yr (1897, 1905, 1934, 1950). The four great earthquakes have rupture zones up to 300 km long,

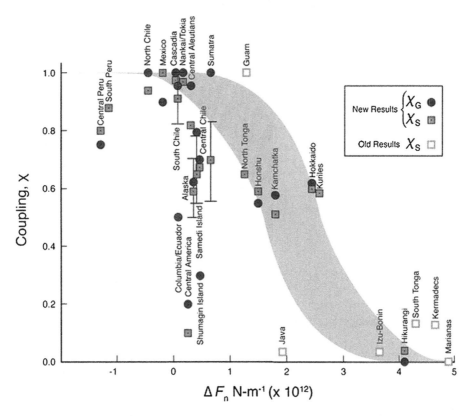

FIGURE 3.22 Plot showing the coupling coefficients plotted versus the normal force acting on the subducting plate. A typical coupling width (Wc) value of 100 km results in a change in ΔF_n of $1 \times 10^{12}\,\mathrm{Nm}^{-1}$, which is equivalent to a mean change of normal stress of 10 MPa, so that the entire scale range is 70 MPa. Since the absolute value of normal stress is not known, these changes are relative to a reference state. The curves delineating gray areas were obtained, independently of the data shown, from calibration with the Izu-Bonin/Marianas system (modified from Scholz and Campos, 2012).

and this area has a very serious seismic hazard (Fig. 3.24). The 1897 event, located near the Indo-Burma syntaxis, is one of the largest earthquakes in recorded history. It caused about 1542 fatalities and damaged masonry structures within a region of northeastern India comparable to the size of England. It may not have been on the shallow dipping boundary as there was large uplift around Shillong. If we assume that most of these great thrust faults relieve nearly all the strain accumulation due to the movement of the Indian plate, events on each segment should repeat every 200–500 year.

3.4 Intraplate earthquakes

Although the vast majority of seismic moment release occurs at plate boundaries, some regions of seismicity are far removed from plate boundaries. This seismicity is referred to as *intraplate*, and it represents internal deformation of a plate not related to plate boundaries. The nature of intraplate seismicity is often quite complicated, and the tectonic driving mechanisms are poorly understood. Natural intraplate earthquakes can create extensive damage and fatalities, such as the 2008 *M*7.9 Wenchuan, China

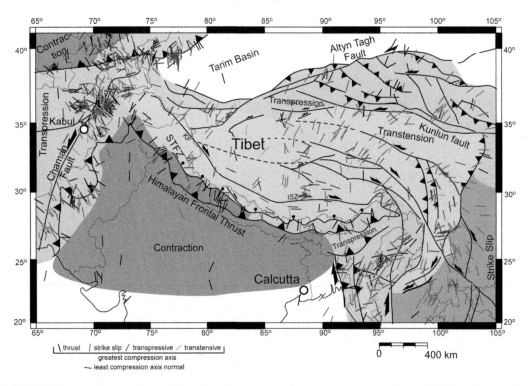

FIGURE 3.23 Tectonic map of the India-Eurasia collision zone showing the incremental strain field based on earthquake focal mechanisms of earthquakes. Tibet and its Eastern margin deform by transtension, as does a narrow region in Myanmar. The rest of the collision zone is under transpression or contraction, with transpression being the most dominant mode of deformation in the collision zone (modified from Andronicos et al., 2007).

earthquake (Fig. 3.25). Furthermore, an increase of intraplate seismicity since 2009 in the U.S. likely relates to anthropogenic activity, and has generated $M5.9$ earthquakes in Oklahoma. (See Fig. 3.29.)

Natural (non-induced) intraplate and inter-plate earthquakes differ in two important ways. First, the recurrence interval of natural intraplate events is generally much longer than that of in-terplate events, and second, intraplate events typically have much higher stress drops. These two observations may be coupled; since in-traplate faults fail infrequently, they appear to be "stronger" than interplate faults. There are several possible explanations for this. Faults that move frequently with a high slip rate will pro-duce a well-defined gouge zone, which weakens

the fault. Furthermore, **_interseismic healing_**, in-volving chemical processes that progressively cement a fault, will be more effective for faults with long recurrence times. Natural intraplate seismicity is difficult to characterize seismically because the recurrence times are long and we have few examples of repeat events. Often, earthquakes occur on old zones of crustal weak-ness such as sutures or rifts that are _reactivated_ by the present stress field (Sykes, 1978). In many cases, the origin of the stress field that causes in-traplate earthquakes is somewhat problematic. It appears to stem mainly from the driving forces of plate tectonics, such as slab pull or ridge push, far from where faulting takes place. The stress is probably "localized" or concentrated by weak structures or by shear stress from flow around

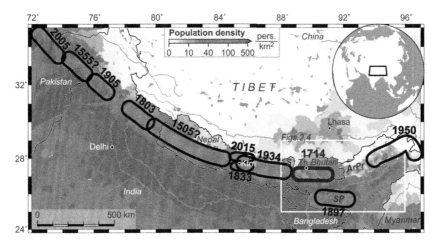

FIGURE 3.24 Great earthquakes in and near the Himalayas since 1500 A.D. The rupture areas of magnitude 7.5 and larger events are schematic and represent the published along-arc extent estimates from many authors. Possible hypocenter loci of the 1714 A.D. Bhutan earthquake. ArPr, Arunachal Pradesh; Ktm, Kathmandu; S, Sikkim; SP, Shillong Plateau; and Th, Thimphu. Inset locates main map on the globe (modified from Hetényi et al., 2016).

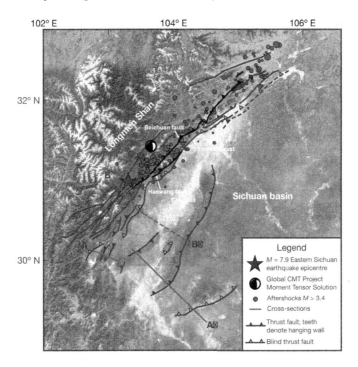

FIGURE 3.25 Map of the Longmen Shan and western Sichuan basin, showing the epicenter and focal mechanism of the 2008 Wenchuan (M7.9) earthquake, aftershocks, and major faults. Emergent and blind thrusts are distinguished; gray faults are the Beichuan and Hanwang faults, which ruptured in the 2008 Wenchuan earthquake (modified from Hubbard and Shaw, 2009).

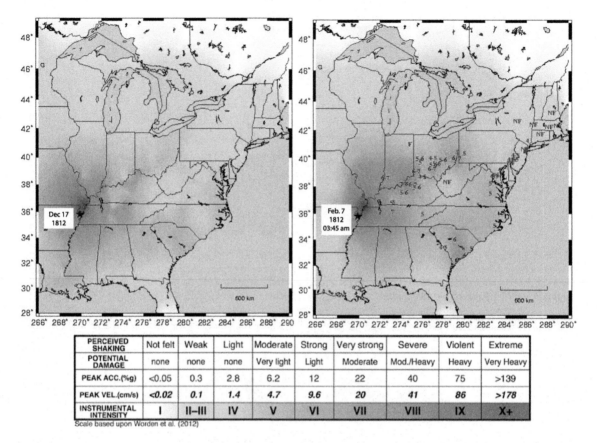

PERCEIVED SHAKING	Not felt	Weak	Light	Moderate	Strong	Very strong	Severe	Violent	Extreme
POTENTIAL DAMAGE	none	none	none	Very light	Light	Moderate	Mod./Heavy	Heavy	Very Heavy
PEAK ACC.(%g)	<0.05	0.3	2.8	6.2	12	22	40	75	>139
PEAK VEL.(cm/s)	<0.02	0.1	1.4	4.7	9.6	20	41	86	>178
INSTRUMENTAL INTENSITY	I	II–III	IV	V	VI	VII	VIII	IX	X+

Scale based upon Worden et al. (2012)

FIGURE 3.26 Maps of shaking intensity interpolated from historic accounts of the December 16, 1811 (M 7.2–7.3) and February 7, 1812 (M 7.3–7.4) mainshocks of the New Madrid sequence (Hough et al., 2000). Maps courtesy of Susan Hough, USGS.

continental lithospheric roots. The majority of intraplate events have P and T axes that are consistent with regional stress provinces. A few localized driving forces, such as removal of a glacial load, stresses induced by surface topography, loading of reservoirs, ascent of mantle plumes, or delamination of the deep crust, may also cause intraplate earthquakes.

The faults that produce intraplate earthquakes are not easily recognized at the surface. This is because the faulting is usually several kilometers deep, and little cumulative offset occurs because of the long recurrence intervals. The eastern and midcontinent regions of the United States are usually thought of as "stable" and seismically inactive, but in fact, some very large earthquakes have occurred in these regions. In 1886, an earthquake devastated the Charleston, South Carolina, area; it was felt as far away as Chicago. Furthermore, the fault that created the 2008 M7.9 Wenchuan, China earthquake and caused more than 87,000 fatalities produced little shortening across the range front, although the topographical relief is greater than anywhere else on the Tibetan plateau (Hubbard and Shaw, 2009). Thus, the earthquake was not expected, illustrating the difficulty for estimating seismic hazard (covered in Chapter 8.)

New Madrid Seismic Zone - Quaternary Fault Localities

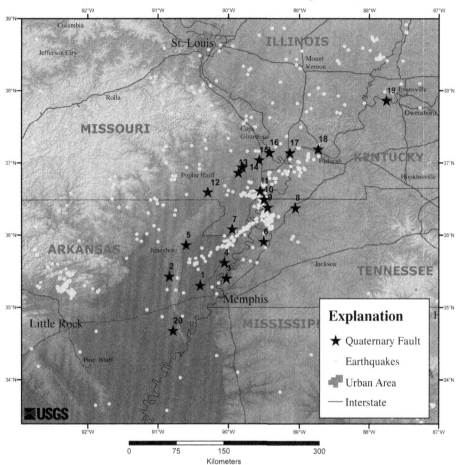

FIGURE 3.27 This map of the New Madrid seismic zone shows earthquakes with magnitudes larger than 2.5 as light circles (University of Memphis and USGS Professional Paper 1527). The dark stars represent localities where Quaternary faulting, sites that are generally less than about 75,000 years old, has been detected in the subsurface. Earthquake data is from USGS 1974 to June 2014 (modified from https://earthquake.usgs.gov/learn/topics/nmsz/images/nmszseis.pdf.)

A sequence of earthquakes near New Madrid, Missouri, during the winter of 1811–1812 represents the largest seismic energy release episode in the historical record of the eastern United States. Fig. 3.26 shows the shakemap (contours of equal shaking damage) for two of the three large New Madrid events on December 16, 1811 ($M7.2$–7.3), January 23, 1812 ($M7.0$), and February 7, 1812 ($M7.3$–7.4) (Hough et al., 2000). The earthquakes were felt as far away as New York City. These events caused a major change in the course of the Mississippi River, and the land along the banks of the Mississippi sank up to 5 m in several places. Reelfoot Lake in western Tennessee was formed by the subsidence of swamp land (the lake is 30 km long, 10 km wide, and 2 to 4 m deep). Fig. 3.27 shows the recent (1974–2014) seismicity in the New Madrid seis-

TABLE 3.2 Large basin and range earthquakes.

Name	Date	Latitude (°N)	Longitude (°W)	Magnitude (M_s)
Owens Valley	March 26, 1872	36.5	118.3	8.3
Pitaycachi	May 3, 1887	31.0	109.2	7.2
Pleasant Valley	October 10, 1915	40.5	117.5	7.5
Clarkson	June 27, 1925	46.0	111.2	6.7
Cedar Mountain	December 21, 1932	38.8	118.0	7.2
Excelsior Mountain	January 30, 1934	38.3	118.5	6.3
Harsel Valley	March 13, 1934	41.8	112.9	6.6
Virginia City	October 18, 1935	46.6	112.0	6.2
Fallon	July 6, 1954	39.3	118.4	6.6
Fallon	August 24, 1954	39.4	118.4	6.8
Fairview Peak	December 16, 1954	39.2	118.0	7.0
Dixie Valley	December 16, 1954	39.7	117.9	6.8
Dixie Valley	March 23, 1959	39.4	118.0	6.1
Hebgen Lake	August 17, 1959	44.8	111.1	7.1
Pocatello Valley	March 28, 1975	42.2	112.5	6.0
Mammoth Lakes	May 25, 1980	37.6	118.6	6.4, 6.3
Mammoth Lakes	May 27, 1980	37.6	118.6	6.3
Borah Peak	October 28, 1983	44.0	113.9	7.0
Big Pine	May 17, 1993	37.2	117.8	6.1
Wells Nevada	February 21, 2008	41.1	114.9	6.0
Challis Idaho	March 31, 2020	44.5	115.1	6.5
Monte Cristo Range Nevada	May 15, 2020	48.2	117.9	6.5

mic zone. There are two prominent trends in seismicity: (1) a 120-km-long lineament trending from Arkansas to the tip of Kentucky and (2) a nearly orthogonal segment in southeastern Missouri. The first feature parallels the Mississippi Embayment, a trough that increases in depth to the south. The Mississippi Embayment has been postulated to be a failed arm of a triple junction, which would indicate that it is a rift-like structure. Focal mechanisms of microearthquakes within the New Madrid seismic zone indicate that the fault slip is a mixture of thrusting and right-lateral strike-slip. The usual explanation for the earthquakes is reactivation of an ancient zone of weakness under the action of the current regional stress system, which is completely different from the one that formed the Mississippi Embayment. The large size of the 1811–1812 earthquakes makes the New Madrid

seismic zone an important seismic hazard. It is very difficult to determine the recurrence interval for the 1811–1812 sequence, since there is no surface expression of the faulting. Using the b values determined from the instrumental record in the last 30 years, the recurrence is thought to be on the order of 1000 years.

The western United States is one of the most seismically active intraplate environments. The seismicity is located primarily in the Basin and Range Province, and nearly all the earthquakes are associated with normal faulting or a mixture of normal and strike-slip faulting. The Basin and Range is a greatly extended terrain. Some investigators have suggested that it is not a true intraplate setting but a complex "back arc" to a convergent boundary. On the other hand, the earthquake recurrence rate on most of the faults is longer than 1000 years, and the stress

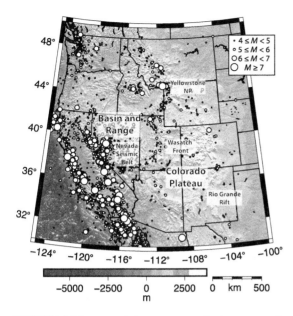

FIGURE 3.28 Map of the western U.S showing earthquakes from the USGS catalog from 1850–2020 with $M \geq 4.0$. The Nevada Seismic Belt in west central Nevada had significant activity from 1887–1954. The Wasatch fault zone and the Rio Grande Rift are regions with little seismic activity, significant fault scarps, creating a seismic hazard that can significant impact on urban populations.

drops of most recent events are consistent with those of intraplate earthquakes. Table 3.2 is a list of the largest earthquakes to occur in the Basin and Range since 1872. The earthquakes appear to be clustered around the edges of the province. Although the San Andreas is the most active feature, the Basin and Range events account for nearly the same total energy release. Of course, a magnitude 9.0 thrust event along the Oregon-Washington coast, which is not implausible, would dwarf all of these events.

The Basin and Range marks a unique tectonic province in the western U.S. that is a result of extension within the North American plate, yielding basins (valleys) and ranges (mountains) through normal faulting. Fig. 3.28 shows the seismicity in the western U.S. for $M \geq 4$ earthquakes from 1850–2020. The apparent cluster-ing of Basin and Range seismicity around the margins of the province may be an artifact of a short period of observation, but extension of the Basin and Range probably takes place at its edges. There are several prominent "seismic belts," with one defined as the Nevada Seismic Belt in west central Nevada, where there was a "burst" of activity along the Nevada Seismic Belt in the last century (1887, 1915, 1932, 1934, and three events in 1954). Other regions with significant topography and little to no recent seismic activity can have significant seismic hazard, potentially impacting cities that have little appreciation for the potential for earthquakes (see Chapter 8). The Wasatch fault zone in central Utah does not have large, historic events but is associated with many fault scarps, and the fault zone lies along the major population centers in Utah, such as Salt Lake City. The Rio Grande Rift (RGR), a continent rift, lies at the southeastern edge of the Basin and Range province and has fault scarps from north of Albuquerque, NM to El Paso, TX; thus the RGR has a potential for moderate to large earthquakes to occur. Although individual faults in the Basin and Range have long recurrence intervals (> 1000 years), it appears that large seismic events may cluster temporally within a restricted geographic region, such as what occurred along the Nevada Seismic Belt. This region may have been quiet for the previous millennia, and thus may not be active for thousands of years. Other regions that have not been active may indeed become active. Temporal clustering has profound implications for seismic hazard evaluation. During a temporal burst a region will have an elevated hazard; once the burst ceases, the hazard is greatly reduced. There is no good explanation for the apparent cyclic seismic energy release, but apparently the strain release must be related to regional deformation in the lowermost crust or the upper mantle.

Some intraplate earthquakes are *induced*; this means that human activity triggers the earthquakes. This phenomenon was first observed

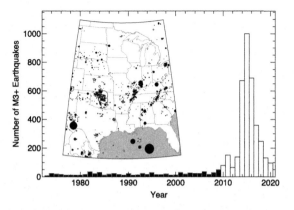

FIGURE 3.29 Annual number of earthquakes with a magnitude of 3.0 or larger in the central and east-central United States, 1973–2020. The long-term rate of approximately 25 earthquakes per year increased sharply starting around 2009, likely the results of activity related to oil and gas extraction. The decrease in the annual event number starting in 2016 is a result of mitigation efforts instituted in regions with ongoing fluid-injection related seismicity (data from the USGS).

with the filling of Lake Mead behind Hoover Dam beginning in 1935. Before the reservoir was filled, the background seismicity along the Nevada-Arizona border was low; beginning in 1936 the seismicity began to rise, culminating in a magnitude 5.0 earthquake in 1940. After this largest event, the seismicity has declined. The epicenters of the earthquakes near Lake Mead are all shallow (most less than 6 km depth) and appear to cluster on steep faults on the east side of the lake. Since the Lake Mead experience, more than 30 examples of reservoir-induced seismicity (RIS) have been documented. Furthermore, a dramatic increase, starting around 2009, in the annual number of earthquakes with a magnitude of 3.0 or larger in the central and eastern United States, 1970–2016 has been attributed to activity related to oil and gas exploration and extraction (Fig. 3.29). The increase in seismicity in Oklahoma has been attributed to salt water disposal in a unit above basement rock, where ancient faults have been reactivated.

What causes induced seismicity? No consensus exists at this time. For RIS, it is unlikely that the weight of the water alone would cause fault failure, since the weight of the water would only add a tiny fraction of the total stress 2 or 3 km below the surface. For both RIS and induced seismicity related to deep injection waste water disposal, an increase in subsurface *pore pressure* reduces normal stress such that existing shear stresses allow for movement along preexisting fault planes. Injection of high pressure fluid to fracture geological formations, called hydraulic fracturing ("fracking"), releases trapped oil and gas, but can also create clusters of earthquakes in space and time. Furthermore, new research is suggesting that poroelastic effects, that is the elastic response to the injected volume, can also trigger earthquakes.

3.5 Summary

This chapter reviewed the basic elements of plate tectonics, where the brittle lithosphere moves over time and overlies the ductile asthenosphere. The asthenosphere convects over time, driven by heat provided by the cooling planet. At the boundaries of the lithosphere, known as tectonic plates, stresses build up are then released during earthquakes. There are three fundamental boundary types: transform (plates sliding by one another), convergent (plates moving towards each other), and divergent (plates moving apart). Each boundary type has distinct geological attributes and behavior. For example, subduction zones, where an oceanic plate subducts beneath a continent or another oceanic plate, can have very large earthquakes caused by wide faults, and these large earthquakes can generate tsunamis if they move the ocean floor. These boundaries also have volcanoes that can be very destructive. Mid-ocean ridges dominate the divergent boundaries, and generate the bulk of the new crustal material.

Transform boundaries connect these ridges, and also transect continent regions, such as the San Andreas Fault in California. Finally, the Chapter discussed earthquake triggering (how earthquakes interact), and new concerns over human induced earthquakes.

4

Earth motions & seismometry

Chapter goals

- Review some history of seismic observation.
- Review the basic characteristics of Earth's ground motions.
- Review the basic theory of inertial seismometers.
- Provide a basic understanding of instrument effects in modern seismic data.

4.1 Introduction

Direct, quantitative analysis of seismic disturbances requires instrumentally recorded vibrations. The instrumentation must (1) be able to detect the transient vibrations within a moving reference frame (the instrument moves with the Earth as it shakes), (2) operate continuously with a very sensitive detection capability and with absolute timing so that the ground motion can be recorded as a function of time, producing a *seismogram*, and (3) have a fully known response to ground motion, or instrument calibration, which allows the seismic recording to be accurately related to the amplitude and frequency of the ground motion. Such a recording system is called a *seismograph*, and the ground-motion sensor that converts vibrations into some form of signal is called a *seismometer*

(or *geophone* in exploration seismology). Modern seismic instruments are the product of at least 200 years effort to develop equipment to accurately record Earth's vibrations. The focus has been primarily on translational motions (displacements, velocities, and accelerations), but progress has been made and work continues on methods to measure ground strain and rotation.

The first known attempts to simply register the occurrence of ground motion were conducted as early as 132 AD. At that time, a Chinese philosopher, Zhang Heng, developed the first *seismoscope*, an instrument that documents the occurrence of motion but does not produce a recording as a function of time. His instrument presumably involved a pendulum system inside a 6-ft-diameter jar, from which eight dragon heads protruded at principal compass directions. Balls were placed in the mouths of the dragons, and the internal pendulum was designed so that ground shaking would dislodge the ball from the dragon mouth in the direction of shaking. The underlying technology for this seismoscope appears to have been lost, and significant further development of ground-motion sensors was not pursued until the 1700s.

Italian scientists developed numerous seismoscopes in the early eighteenth century, motivated by the frequent earthquakes in the Mediterranean. In 1751 Andrea Bina described a pendulum system with a pointer that etched patterns in sand. Increasingly sophisticated pendulum systems were incorporated in seismoscopes over the next 100 years and seismoscopes were widely distributed around the world during the nineteenth century. The first attempt to record the time of shaking was probably made in 1784, when A. Cavalli placed seismoscopes (involving bowls filled to the brim with mercury) above rotating platforms perforated with cavities keyed to the time of day, which would collect any mercury slopped out of the bowls. In 1851 Robert Mallet initiated the field of explosion seismology when he employed a ground-motion sensor that used optical reflection from a basin of mercury to measure the speed of elastic waves in surface rocks.

The first true seismograph, which recorded the relative motion of a pendulum and the Earth as a function of time, was built by Filippo Cecchi in Italy in 1875. A seismoscope started a clock and a recording device at the onset of shaking. The oldest known seismic record produced by this system is dated February 23, 1887. A period of rapid instrument development and improvement occurred after 1875. A group of British seismologists teaching in Japan, including John Milne, James Ewing, and Thomas Gray, led to the first relatively long-period systems (mainly sensitive to ground displacements produced by nearby events) and the first vertical-component seismographs. In these early systems, mechanical or optical mechanisms amplified the mass motion, but friction provided the only damping of the pendulum oscillations. Systems continued to evolve and in 1889 the first known seismogram of a distant earthquake was made on a photographically recording instrument located in Potsdam and designed by Ernst von Rebeur-Paschwitz to study the horizontal tilt from earth tides. By 1900 a global network of 40–50 photographically recording horizontal-component seismographs established by John Milne, along with other observatory instruments operating in Europe and Japan, provided the initial seismograms used to apply elastic-wave theory in the quest to understand Earth's vibrations.

4.1.1 Seismic stations, networks, and arrays

Individual seismograph locations are often referred to as *seismic stations* or observatories. Deployed collections of seismographs are called *seismic arrays* or *seismic networks*. Traditionally, seismic networks are deployed to monitor seismicity in a region, and the sensors may be sparse and/or irregularly placed. Sometimes the region of focus is the entire planet – Fig. 4.1 is a map of station locations in the Global Seismographic Network (GSN) operated with support from the U.S. National Science Foundation and the U.S. Geological Survey. The GSN includes the largest collection of high-quality very broadband (VBB) sensors, but each station also typically includes a secondary broadband and a strong-motion accelerometer to record motions when the other instruments are driven off scale. Details are provided later, but broadband sensors have a uniform sensitivity to ground motions from short periods to at least 30 s period, spanning the classical seismic analysis bands (e.g., center periods of 1 s short-period instruments to record body-waves and 20 s long-period instruments to record surface waves); VBB sensors include additional sensitivity to the longest period motions ($T \gtrsim 200$–300 s), suitable for analysis of Earth's long-period free oscillations. Together with France's GEOSCOPE (The French Global Network of VBB Seismic Stations), most of the accessible land surface is instrumented with stations placed about every 1000 km. For long-term monitoring, ocean coverage is limited by island locations. The large southern oceans results in relatively sparse coverage throughout the southern third of the planet. The combined networks are relatively sparse, but many countries now operate national seismic networks that provide more dense sampling of the seismic wavefield. Over the last two decades, seismology has become relatively data rich, modern global seismologists work with substantially more and higher-quality data than ever before.

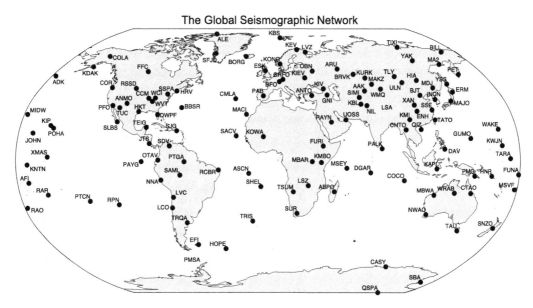

FIGURE 4.1 Locations of seismic stations (black circles) in the U.S. Global Seismographic Network (The GSN). The stations are maintained by the US Geological Survey and the IDA Program at the University of California, San Diego (funded by the US National Science Foundation, NSF). Each GSN station includes a high-quality very broadband sensor, a secondary broadband sensor with a more limited long-period response, and an accelerometer to capture nearby strong motions. Many stations also include additional scientific sensors such as atmospheric pressure transducers and some are colocated with infrasound sensors. Data from the network are openly available in near real time from the NSF-Supported IRIS Consortium, https://doi.org/10.7914/SN/IU.

Seismic arrays are traditionally spatially tighter sensor configurations deployed to record ground motions densely enough to allow analysis of the signals as a *wave-field*. A properly designed seismic array can improve signal-to-noise ratios and reduce seismic detection thresholds by "stacking" or averaging signals from nearby sensors (assuming the signal is coherent and the noise is not). In addition, arrays can be used to measure directional information and the apparent horizontal speed of seismic waves. Both observations provide valuable information on the source and nature of the signals. Seismologists deploy both permanent and temporary seismic networks and arrays and recent field deployments have blurred the distinctions between arrays and networks and temporary and permanent installations. Such work is not unique, complementary efforts are applied to

study other Earth motions such as rotation, tilting, and strain.

To understand and use seismic data quantitatively, you must understand some of the physics of seismometers and seismographic recording systems. We present a brief overview, and refer the student to more advanced treatments in, for example, Aki and Richards (2009), Scherbaum (2001), Havskov and Alguacil (2004) for more details. But before exploring seismic instrumentation details, it is worthwhile to first discuss the nature of seismic ground motions, since these are the observations we wish to collect. Historically, of course, our understanding of Earth's motions advanced simultaneously with the technology developed to measure them. If the instrument designers had known as much about Earth motions as we do, the goals of instrument design would have been clearer and development could have proceeded much faster. Still, the advances made in understanding and using the recorded ground motions in the first few decades after the deployment of John Milne's global network of horizontal seismometers were dramatic.

Earthquake and explosion signals are transient – they vibrate the Earth for a relatively short time following excitation by the source. Most of the time, seismometers are recording the continuous background motions of the planet. Background motions can be a nuisance when your interest is the transient signals with which they interfere, and when that is the case, the background motions are called *noise*. However, Earth's incessant motions also carry information on the rocks through which they travel and the processes that produce the motion. Some common sources of the background motion are atmospheric processes including winds and pressure fluctuations, ocean processes including near-shore oscillations and storm-induced pressure and ocean-wave changes, cryospheric processes including cracking and calving of glaciers, or, any other process that vibrates the ground. For example, human machinery (trains, traffic, pumps, pipelines, etc.) induce detectable ground vibrations that are commonly observed on seismic systems. All these vibrations propagate, transmit energy, and carry information on their sources and on the geology through which they travel. At one time or another, seismic analyses have been used to study each of these processes, but our primary focus is of course earthquake-related ground motion.

4.2 Earthquake-related ground motions

Much of this text is devoted to understanding the sources and propagation of seismic ground motions. This section is a brief overview of the ground-motion ranges of interest – enough to place seismograph design requirements in context. The scope of important ground motions in earthquake science ranges from slow movement of plates to rapid oscillations of signals with audible frequencies. Near a large earthquake the ground motions may include fault offsets of many meters, but at great distances shaking is as small as a few nanometers. Observing motions with amplitudes spanning a range of 10 orders of magnitude (~ 10 to 10^{-9} m) is a substantial challenge, but one that can be met with modern geophysical facilities (Box 4.1). Accurately measuring and recording signals over such a large deformation range requires deploying a suite of instruments. Global Positioning System (GPS) sensors can record the large offsets adjacent to an earthquake rupture, accelerometers can measure the strong short-period shaking within the epicentral region, and seismometers and seismometer arrays can provide sensitivity to detect small seismic events.

Many more earthquakes are observed from a distance than from close by, so most of the motions used by seismologists to investigate earthquakes and Earth's interior are relatively small, ranging roughly from microns to millimeters, and are recorded with seismographs. For this reason, we focus on seismometer physics, which must be understood to effectively use modern seismic data. High-sample rate GPS are derived in a very different manner, but can be readily combined with the seismometer-based observations in uniform analyses. In the last 10–15 years, we have been able to capture some earthquake deformation signals with bandwidths that range from effective static offsets to 10's of Hertz. A good principle to keep in mind is that resolution increases with bandwidth, so seismologists always seek the broadest band observations we can get. Integrating the broad-spectrum of information into improved earthquake and earth models remains an important challenge.

The amplitude of earthquake-induced seismic ground motions depends on the earthquake size and depth, observation distance from the earthquake, ground-motion frequency, seismic

Box 4.1 Measuring complete ground motions

This chapter is focused on seismic instruments designed to record transient ground motions, but the spectrum of deformation processes associated with earthquakes spans a temporal range from plate motions to seismic-wave accelerations (Fig. B4.1.1). Advances in high-sample-rate GPS measurement and processing have bridged the gap from very long-period motions (as long as the technique has existed) to intermediate-period seismic motions. Satellite-based interferometry (InSAR) can provide high-quality, spatially-dense measurements of surface deformation in the direction towards a satellite.

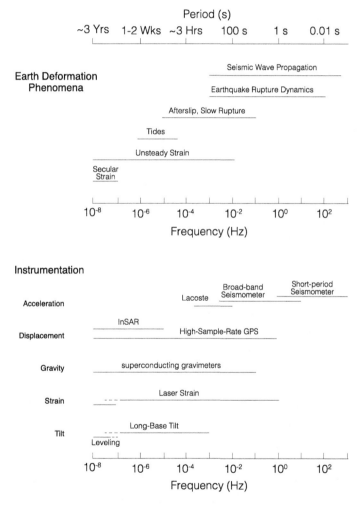

FIGURE B4.1.1 Ground-motion measurements and corresponding phenomena are studied using a variety of instrumentation – from satellites to seismometers (modified from an original by D. Agnew).

> Displacement is not our only interest. Specialized instruments like gravimeters have been used to measure gravitational changes associated with mass redistribution, and strain and tilt meters have been used to measure strain changes along faults and on or near volcanoes.

wave type, direction relative to the faulting geometry, and geological structure along the wave's path. Close to a shallow moderate-to-major earthquake, peak shear-wave ground accelerations in the 1 to 10 Hz band may range from roughly 0.1 g to as much as 2 g ($g \sim$ 9.81 m/s^2, is the average gravitational acceleration at Earth's Surface). These are the motions that damage buildings, bridges, and other human constructions. These strong, short-period oscillations are attenuated relatively quickly (by scattering and intrinsic energy absorption) during propagation, so longer-period signals dominate at larger distances from the fault. At *regional* distances ($\sim$ 1000 to 3000 km), shear-wave ground motions from strong earthquakes are roughly a few 10's of microns ($T \sim 1$ Hz). Distant *teleseismic* body-wave displacements may be a few *mm* for the largest events, but are typically in the range of a few *microns*. Intermediate- and long-period surface waves from large shallow earthquakes may move the ground a few *cm* at distances thousands of kilometers from their source, but their accelerations are small, and the waves are unfelt. Even longer-period free oscillations excited by large earthquakes are typically a fraction of a *cm* and correspond to accelerations of $\sim 10^{-9}$ g. Although small, these motions can be detected for weeks following large earthquakes.

Analysis of these earthquake ground motions is the central theme of this text, so we will return to these observations many times. For now, we focus our attention on Earth's ambient background motions, or as they were once called, seismic noise.

4.3 Earth's continuous background motion

The ground never rests. Deformation produced by tides, storms and coastal ocean interactions, atmospheric pressure variations, diurnal heating of the surface, and human-induced vibrations produce continuous background motions called *earth tremors* or *microseisms*. The terms are inherently fuzzy, we'll use earth tremor to represent all background motions of the planet, because microseism often has a particular association with oceanic-generated signals. As such, earth tremors include all the motions produced by all natural processes and vibrations from human activity, which is often referred to as *cultural noise*.

Milne (1903) notes that the earliest detection of earth tremors (both human-induced and natural) was the result of the development and use of precise instruments in astronomy (careful pendulum observations or reflections from a tray of mercury) and high-power microscopes. High precision work at Greenwich Observatory was occasionally impossible because of earth tremors produced by crowds celebrating a holiday at nearby Greenwich Park. In an 1874 expedition to make astronomical measurements of the transit of Venus from New Zealand, H.S. Palmer found it necessary to place his instruments in one-meter deep trenches to avoid vibrations from a railway roughly 400 meters away. A similar practice remains part of seismometer installation today. We often bury our sensors, although more to reduce pressure, wind, and thermal fluctuations than noise from railways, which we simply try to avoid. An analogous interaction remains part of modern astronomy. Earth tremors are carefully accounted for as part of the processing of measurements in gravitational wave detectors such as the Laser Interferometer Gravitational-Wave Observatory (LIGO).

The term microseisms, coined by Italian Jesuit scientist Timoteo Bertelli, pre-dates anything resembling a modern seismometer. Italian scientists, led by Bertelli used microscopic observations of suspended pendulum oscillations and correlated the motions with barometric fluctuations produced by storms. The association of microseisms with storms continued to be investigated theoretically and pragmatically. During World War II, Jesuit seismologist J.E. Ramirez, working at Saint Louis University, used multiple tripartite (three-station) arrays to measure the microseismic wave travel direction and to try to track storms. The method provided some success, but precision suffered from the fact that the exact relationship between the signals and the storms remained a puzzle for decades. Only fast-moving storms appear to generate microseisms near their center, which complicates storm tracking.

More modern investigations, using larger and better seismic arrays and three-component sensors that allow the measurement of the wave propagation direction, speed, and wave type have provided more detailed information on the sources of microseisms. Most of the energy appears to be arriving from near or offshore. *Primary microseisms* with periods in the range from 10–16 s appear to be caused by waves crashing on the shore (which occurs with a similar frequency). *Secondary microseisms* are larger and have periods from roughly 4–10 s, and appear to be excited by standing waves in the ocean initiated by wave interference patterns (waves traveling toward the shore constructively interfering with waves reflected off the shore). The standing waves produce pressure fluctuations on the ocean bottom, resulting in microseismic vibrations that can propagate across entire continents. Primary microseisms

are predominantly Rayleigh (with some Love) waves often arriving at the station from one direction, especially when a strong storm approaches the shore. Secondary microseisms are dominated by surface-wave motion, but also include detectable body waves because storms in the ocean can produce short period P-waves.

Although relatively slowly changing in statistical properties, Earth's continuous background motions are temporally and spatially variable and also vary with frequency. Any detection of transient wave arrivals such as those produced by earthquakes is made in the presence of these more slowly changing background vibrations. The character of microseismic motions strongly influenced the design of seismic recording systems until late last century. Fig. 4.2 is a plot of the observed background motions at a seismic station in Waverly, Tenessee, USA recorded in early 2018. The original unfiltered displacement is shown at the top, the lower traces show the displacements in small period ranges centered on the value listed in the lower left of each panel. The ground displacement is quite small, about one half of a micron, and much smaller at the high frequencies. Amplitudes of the signals with central periods near 10 s and frequencies near 3 Hz were boosted so the details would be visible. Since ground motion size varies with period (or frequency) we often analyze seismic motions as a function of frequency.

4.3.1 Ambient background motion power spectra

Transient signals, like those from earthquakes and explosions, last a finite amount of time and have finite energy. Earth's background motions continue indefinitely and formally have infinite energy, but finite power (energy per unit time). Such signals are analyzed using power spectra, a plot showing the relative signal power as a function of frequency of vibration. Fig. 4.3 is a plot of the variation of the strength of ambient ground acceleration versus frequency (Box 4.2) for sta-

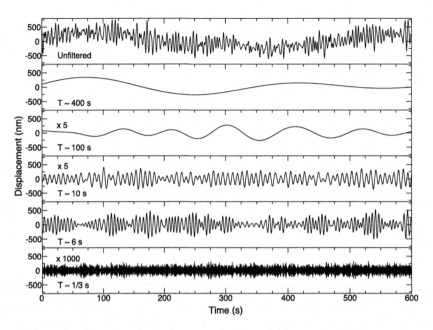

FIGURE 4.2 Ambient seismic ground displacements (top panel) recorded in early 2018 at seismic station WVT, Waverly, Tennessee, USA. Lower panels show motions in select period ranges contained in the unfiltered signals. The center period is listed in the lower left of seismogram and plot amplification factors required to boost the low-amplitude signals for 100 s, 10 s and 3 Hz are listed above and to the left of the seismogram.

tions at three different distances from the coast. Strength of the signal is shown using a *spectrum*, a plot of the relative ground acceleration *power spectral density* (PSD) as a function of the oscillation frequency in Hertz (Hz) or period in seconds. Spectra are powerful tools used extensively in the analysis of seismic signals. The PSD is used to show the relative amount of signal at each frequency in the near-stationary background motions. We can compute an estimate of the power spectral density, $\hat{P}(\omega)$, from the Fourier Transform of a signal, $F(\omega)$,

$$\hat{P}(\omega) = \frac{1}{T} |F(\omega)|^2 . \tag{4.1}$$

Values computed using this simple PSD estimate (called a periodogram) have large uncertainties, so we often average estimates from at least a few (and many more if possible) segments of a signal to produce a smoother and more statistically robust PSD spectrum. The name, power-spectra, is somewhat of a misnomer. Under certain assumptions the power in the signal is proportional to the squared amplitude of the signal, but a PSD, particularly one computed from ground accelerations, is not a direct measure of energy flow per unit time.

The ground-acceleration PSD spectra (Fig. 4.3) illustrate a number of well-established patterns in seismic background motions. Most seismic stations typically have high noise at frequencies from 0.15–0.2 Hz caused by oceanic processes (including the primary and secondary microseismic peaks). The ground-acceleration PSD varies by a factor of about 10^5 to 10^7 over the frequency range shown, and the high-frequency noise peak will tend to overwhelm other signals on any seismometer with uniform sensitivity unless it has a dynamic range that can resolve

Box 4.2 Time series, Fourier transforms, and spectra

Seismograms, and many other observations that vary with time are called *time series*. In seismology and many other fields, great mathematical flexibility arises from our ability to represent functions of time by equivalent functions of frequency. The Fourier Transform links the time and frequency "domain" representations of the information. Fourier Transforms are integral expressions that project the original signal onto harmonic functions. The Fourier Transform of a time series, $f(t)$, is represented by complex-valued quantities, $F(\omega)$, computed using the integral

$$F(\omega) = |F(\omega)| \, e^{i\varphi(\omega)} = \int_{-\infty}^{\infty} f(t) \, e^{-i\omega t} dt \,, \tag{B4.2.1}$$

where $\omega = 2\pi f$ is called the angular frequency, and $|F(\omega)|$ is the amplitude and $\phi(\omega)$ is the phase of the Fourier Transform. The functions $F(\omega)$, $|F(\omega)|$, and $\phi(\omega)$ are called the spectrum, amplitude spectrum, and phase spectrum respectively. Specifically,

$$|F(\omega)| = \sqrt{F(\omega) \times F^{\dagger}(\omega)} \,,$$

$$\phi(\omega) = \tan^{-1} \frac{Re\,\{F(\omega)\}}{Im\,\{F(\omega)\}} \,, \tag{B4.2.2}$$

where $F^{\dagger}(\omega)$ represents the complex conjugate of $F(\omega)$ and $Re\{\}$ and $Im\{\}$ represent the real and imaginary parts respectively. The transform is fully invertible using

$$f(t) = \frac{1}{2\pi} \int_{-\infty}^{\infty} F(\omega) \, e^{i\omega t} d\omega$$

$$= \frac{1}{2\pi} \int_{-\infty}^{\infty} |F(\omega)| \, e^{i\varphi(\omega)} \, e^{i\omega t} d\omega \,. \tag{B4.2.3}$$

We will refer to (B4.2.1) as the forward transform (*from time-to-frequency*) and (B4.2.3) as an inverse transform (*from frequency-to-time*). Since an integral is a sum, (B4.2.1) indicates that an arbitrary time series, even an impulsive one, can be expressed as a sum of monochromatic periodic functions. This may seem surprising, but destructive and constructive interference between the harmonics insures that they sum exactly to the original time series.

We often represent seismological observations using their spectra, which contain all of the information contained in the original seismogram, as long as both amplitude and phase are retained. Additionally, since seismograms are real-valued quantities, their Fourier Transform have a symmetry about the origin,

$$F(-\omega) = F^{\dagger}(\omega) \,. \tag{B4.2.4}$$

This relationship allows us to display spectra using only positive frequencies, but an account for both positive and negative frequencies is required during analysis. Modern seismograms are uniformly sampled with a sample interval Δt. For digital seismogram analysis, Fourier spectra are computed using discrete equivalents of the continuous transforms. The largest frequency accurately represented in a sampled seismogram is called the Nyquist frequency, $f_{Nyquist}$,

$$f_{Nyquist} = \frac{1}{2\,\Delta t}\,. \tag{B4.2.5}$$

Finally, the definitions of the Fourier Transform are not unique. Aki and Richards (2009), for example, use the opposite convention for the sign of the complex exponent (positive for the forward and negative for the inverse). As long as a particular analysis uses consistent definitions, the result will be the same. But awareness of the potential differences might save you a great deal of time.

large variations in signal amplitudes. The observations also suggest that in this frequency band, island sites (RPN is on Easter Island) will be noisier than sites well removed from the coast. The factor of 10 variations in noise levels in the 0.1–1.0 Hz passband also indicates that seismic-event detection will be nonuniform and that amplitude and arrival time measurement uncertainties vary from station to station. At longer periods ($T \gtrsim 50\,\mathrm{s}$), all three stations are high-quality and exhibit relatively low background acceleration power spectra.

4.3.2 Power spectral density and time-domain ground motions

We can relate the PSD amplitude over a specific bandwidth to the expected amplitude of the corresponding time-domain signal (Aki and Richards, 2009). The relationships are approximate, but we can estimate the Root-Mean-Square (RMS) amplitude using an integral of the PSD over a frequency range. We have to be careful to keep track of the bandwidth for the estimate, because without that reference, the numbers are meaningless. If we represent the mean PSD value in a specified bandwidth with

$\bar{P}(f)$, then the RMS ground motion, $u_{rms}(\Delta f)$ is

$$u_{rms}(\Delta f) \sim \sqrt{2\bar{P}(f)\Delta f}\,, \tag{4.2}$$

where the factor of 2 accounts for the negative frequency contribution, Δf is the bandwidth of interest, and the explicit dependence on bandwidth is included. The stationary signal peak amplitude is often estimated using $u_{peak}(\Delta f) \sim 1.25\,u_{rms}(\Delta f)$. For these calculations bandwidth is often specified using octaves. Let f_0 be the center of the bandwidth of interest and let n represent the frequency octaves spanned by the bandwidth (n may be a fraction, and in seismology is often $1/3$). Then the lower frequency bound is $f_l = 2^{-n/2}\,f_0$ and the higher frequency bound is $f_h = 2^{n/2}\,f_0$. The bandwidth is

$$\Delta f = f_0 \times (2^{n/2} - 2^{-n/2})\,. \tag{4.3}$$

Bandwidth specified this way is a fixed ratio of the central frequency (f_0). For one octave, the bandwidth is $\Delta f = f_0/\sqrt{2}$.

Once we choose a bandwidth, we can use (4.2) to estimate the RMS amplitude of ground acceleration. For example, across the frequency range from about 2 to 4 Hz (roughly one octave centered on 3 Hz) we can use the values for station WVT shown in Fig. 4.3 to estimate the RMS

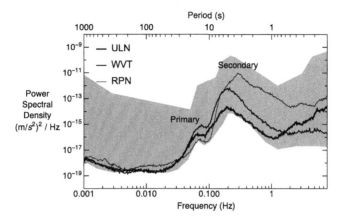

FIGURE 4.3 Amplitude spectra of average background vertical ground acceleration recorded over the first week of 2018 at each seismic station: ULN (1800 km from coast) – Ulaanbaatar, Mongolia; WVT (600 km from coast), Waverly, Tennessee, USA; RPN (2 km from coast), Rapanui, Easter Island, Chile. The gray region indicates the range of the Peterson (1993) low and high noise acceleration power spectra. Noise model values are from Bormann and Wielandt (2013). Note that the peak in noise near 0.2–0.3 Hz at all stations and the decrease in 0.1-1 Hz background motion amplitude with distance from the coast. The primary microseism peak is located near 14 s period; the secondary microseism peak is located near 3–7 s period. The long-period ($T \geq 30$ s) motions at each station are comparable. To convert the vertical scale to db, multiply the value's exponent by 10.

shaking level in that frequency band, which we will call $\langle a_{rms} \rangle_{WVT}$. The WVT PSD value near 3 Hz is roughly $10^{-16} (m/s^2)^2$ per Hz, so

$$\langle a_{rms} \rangle_{WVT} \sim \sqrt{2 \times (1 \times 10^{-16}) \times (4-2)} \ m/s^2$$
$$\sim \ 30 \times 10^{-9} \ m/s^2 \ . \qquad (4.4)$$

Using the same figure, the expected RMS acceleration value in the 2 to 4 Hz band at station RPN on Easter Island is roughly ten time larger, $\langle a_{rms} \rangle_{RPN} \sim 325 \times 10^{-9} \ m/s^2$, than that at station WVT in central Tennessee, USA. The long-period ($T \geq 50$ s) acceleration values for both stations are similar and much smaller. But that does not mean that the long-period ground motion is smaller, in fact, the ground displacements are much larger at long periods than at shorter periods.

The convention in seismology is to summarize and to compare background motions using acceleration power spectra. Acceleration spectra are chosen to level the PSD curves, easing channel-to-channel comparisons. But we

can compute velocity or displacement PSDs (Fig. 4.4) with the same approaches. Since most global seismogram analyses are performed using velocity or displacement seismograms, it makes sense to have an intuitive appreciation of how acceleration PSD spectra relate to those for ground velocity and displacement. In the time domain, we can obtain the velocity by integrating the acceleration, and displacement by integrating the velocity. In the frequency domain we can calculate the spectrum of the integrated signal using division by $i \omega = i2\pi f$. Integration of a signal increases low-frequency PSD amplitudes and decreases high-frequency PSD amplitudes. The PSD value is unchanged when $\omega = 1$ or $f = 1/2\pi$, which is indicated by a vertical grid line in Fig. 4.4. However, the affects on the lowest and highest frequencies are substantial in Fig. 4.4.

As we did earlier for accelerations, we can estimate the short-period and long-period RMS ground motion levels for station WVT using the values in Fig. 4.4. The vertical gray lines

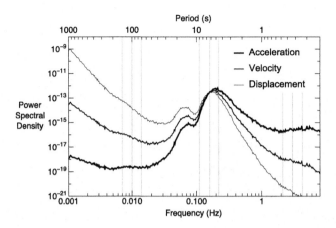

FIGURE 4.4 Acceleration, velocity, and displacement PSD spectra for vertical background ground motions recorded and averaged over the first week of 2018 at seismic station WVT (600 km from coast), Waverly, Tennessee, USA. The relative amount of signal changes dramatically depending on whether you examine acceleration $((m/s^2)^2/Hz)$, velocity $((m/s)^2/Hz)$ or displacement (m^2/Hz). Integration of the time-domain signal boosts the long-period signal, so the ground displacement is dominated by long-period background motions that are much larger than they appear based on acceleration spectra. The vertical gray lines identify three frequency bands discussed in the text.

in the figure show the central frequencies and bandwidths we will use. In the 2–4 Hz band, we have velocity and displacement PSD values of roughly $1 \times 10^{-18}(m/s)^2/Hz$ and $4 \times 10^{-21} m^2/Hz$ respectively. Using (4.2), these correspond to approximate RMS amplitudes of roughly 2 nm/s and 0.1 nm – very small motions. During a quiet hour from the same time range used to construct the PSD, the observed RMS vertical velocities and displacements in the 2-to-4 Hz band were 1.5 nm/s and 0.1 nm, quite close to the values estimated from the PSD. For a one octave-bandwidth around a frequency of 0.01 Hz (100 s period), the expected RMS background motions at WVT are roughly 1 nm/s and 26 nm respectively. Using a two-hour segment from the same time range, the measured RMS values in the 0.007 Hz to 0.014 Hz band were 0.9 nm/s and 16 nm – reasonably good agreement. Near the peak of the microseismic band, for a frequency of $1/2\pi$, the PSD of all three ground-motions (acceleration, velocity, and displacement) are equal and the corresponding RMS amplitude is roughly 260 $[nm/s^2, nm/s, nm]$. Measured RMS values

in the 0.11 to 0.22 Hz band are roughly 130 – 175 $[nm/s^2, nm/s, nm]$, a little lower than estimated using the PSD, but the PSD changes rapidly in this frequency range and we picked a value near the high end of the frequency band (Fig. 4.4).

The vertical component motions at Waverly, Tennessee, USA are quite small and examining one channel of motion at one station does not provide much intuition for the range of background motions that can expected when looking at observations across a broad selection of seismic stations. To provide some reference, we can map low and high-noise model PSDs to RMS ground displacements using a one-octave bandwidth Peterson (1993). The results are of course approximate, and are shown in Fig. 4.5. Some of the lowest intermediate- and long-period motions are observed at high-quality stations in the 30–50 s period range, which partly for this reason, is roughly the band used to model body waves in the GCMT analyses. In general, the expected ground motions at a quiet seismic station are less than 10 microns for periods shorter than 100 s. The vertical lines show the periods and

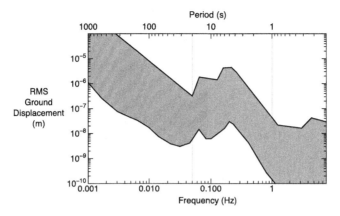

FIGURE 4.5 Mapping of low and high noise models to approximated RMS ground displacement (from Peterson, 1993). A one-octave bandwidth was assumed for the calculation. Noise model values are from Bormann and Wielandt (2013).

frequencies corresponding to the measurement of surface-wave (M_S, 0.05 Hz) and local- and teleseismic body-wave magnitudes (m_b, M_L, 1 Hz). Using measurements in relatively low-noise regions means that we will have more observations of smaller events to estimate the magnitudes.

4.3.3 Horizontal and vertical ambient ground motions

Each seismic station has unique noise characteristics, but a number of spatial and temporal characteristics are typical. Fig. 4.6 is a plot of displacement power-spectral densities for the three components of motion observed using the surface instrument on Wake Island. The sensor is located 18 m above sea level on a concrete pad resting on coral limestone. The station originally had a borehole sensor, but the current sensor is Streckheisen 2.5 instrument at the surface. The key point illustrated in the figure is the relatively large ratio of horizontal to vertical background motions. The horizontal noise begins to increase relative to the vertical at a period of about 3 s. At 10–20 seconds, the horizontals to vertical PSD ratio is about five, about 15 times at 100 s period, and increases to a factor of sixty at 200–300 s pe-

riod. The RMS values would scale as the square root of these ratios. At the long periods, the observed rate of increase in the logarithm of the PSD magnitude with the logarithm of period is roughly 2 for the vertical and 2.75 for the horizontals.

The substantial difference in vertical and horizontal seismic background motions is a result of the fact that horizontal instruments are more susceptible to ground tilt, which partitions part of the large gravitational force onto horizontal seismometers. To the seismometer, a change in the horizontal component of gravity is equivalent to a horizontal acceleration of the ground. Ground tilting can introduce large apparent displacements on horizontal seismometers and the effects are much larger at long-periods. Time-varying ground tilts can be caused by human activity, volcanic processes, atmospheric pressure fluctuations, etc. Micrometer scale deformations spread over kilometers of Earth's surface varying over periods of minutes can produce significant horizontal accelerations. Associated apparent horizontal displacements with variations in ground tilt increase roughly proportional to the period squared. Vertical seismometers are also affected, but the effect is smaller. Many seismic observatories are instrumented with barometric sensors to help monitor atmospheric changes

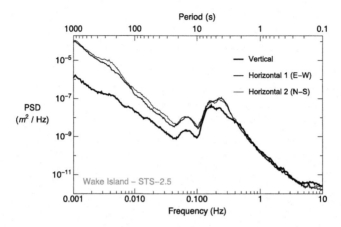

FIGURE 4.6 Comparison of the vertical and horizontal power-spectra for a sample of seismic background motions at station WAKE (Wake Island).

that may produce apparent long-period motions on the seismograms.

4.3.4 Diurnal variation in ambient ground motions

Another common variation in seismic background motions is a variation with noise between day and night. The variations are strongest in regions where human activity is prevalent, but seen in most locales. For seismometers nearby urban centers, the rhythms associated with weekdays and weekends can also be important. In Fig. 4.7 is a plot of one month of the 10 Hz vertical ambient background motions observed in downtown St. Louis, MO, USA. The 10 Hz noise produced by people, traffic, machinery, etc. increases during weekdays by about a factor of five in this band. The diurnal peaks occur in groups of five, representing activity on the five weekdays of each week. A similar pattern is observed for frequencies above one Hz at all stations near urban centers. In more remote regions, diurnal variations from the differences in temperature and wind patterns between night and day produce an observable pattern, but the variation in signal strength may be much smaller than that observed in St. Louis. Human

activity and day/night atmospheric variations are responsible for diurnal variations in seismic background motions. In most places, Earth is quieter at night. The variation can affect small earthquake detection thresholds, which could produce patterns in small events numbers that have to be accounted for in any seismic analysis.

4.3.5 Seasonal ambient ground motion variations

Ambient ground motions also exhibit seasonal variations, even at sites well removed from coastlines. Fig. 4.8 is a plot of the amplitude spectrum of earth motions observed at seismic station SSPA in Standing Stone, Pennsylvania. The observed motions were averaged weekly over the time range from 2012 to 2017. The variation in noise power is apparent and the seasonal patterns are clear and repeatable. The small triangles identify the origin times of major and great earthquakes during the period that produce significant motion and complicate the spatial patterns. The seismometer was updated in September of 2017 and an abrupt and clear reduction of the background motion power spectral density is apparent in the last quarter of 2017.

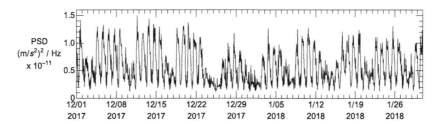

FIGURE 4.7 Temporal variation in the 10 Hz power spectral density observed at station SLM, located in midtown St. Louis, MO, USA. Hourly PSD estimates were obtained from the IRIS MUSTANG seismic-data quality data base. A few values (spikes) with PSD values larger than 1.5×10^{-11} $(m/s^2)^2/Hz$ were removed from the observations. City rhythms are clear in the 10 Hz signal.

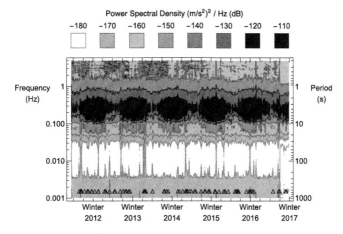

FIGURE 4.8 Seismic background motion power-spectra density observed on the borehole instruments at station SSPA (central Pennsylvania, USA) shown as a function of time and frequency. Each column in the image is the average of 168 two-hour segments of ground motion that span one week. Darker contours indicate higher levels of noise. Triangles identify the times of major and great earthquakes, which may raise the level of ground motion sufficiently to affect the averages.

Seasonal variations in ambient seismic motions reflect surface-water processes, ground-water freezing, development of polar ice caps (for nearby stations) changes in ocean and atmospheric patterns, and temperature variations of the recording sensors. The pattern in the 5–15 s microseismic band is associated with seasonal variations in oceanic processes, which lead to stronger microseisms in winter. Fig. 4.9 is a map of average power in the microseismic band from 5 to 12 s period for the month of July, 2017 at the stations of the GSN. The area of the symbols on the map represent the relative size of the PSD amplitudes in the oceanic microseismic band. A clear pattern in the figure is the relatively larger ambient motions in the southern hemisphere (winter at the time of the sample). However, the effect is not all due to winter, many of the stations in the south also are on islands, which tends to increase the noise and enhance the seasonal pattern. A comparable plot for a time sample during northern hemisphere winter, reverses the general trend at continental stations, but background motion on oceanic islands remains relatively strong.

Average Seismic Motions 5s ≤ T ≤ 12s / July, 2017

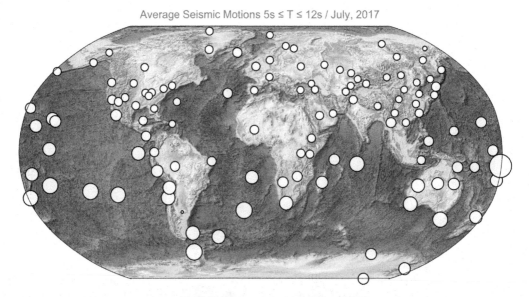

FIGURE 4.9 Relative seismic power in noise for stations in the GSN averaged throughout July, 2017. Each symbol represents a station and the symbol area is proportional to the noise power in the period range from 5–12 s. The values reflect the facts that microseism motions is larger near the coast (and thus island stations) and microseism motions is larger in the winter (southern hemisphere). Both tendencies combine to make the southern hemisphere noisier in July. Data acquired from the IRIS Mustang System.

4.3.6 Reducing ambient motions in seismic data

If your primary goal is to study transient ground motions produced by earthquakes and explosions, reducing the size of ambient background motions observed on seismograms is an important goal. In the 1960s many studies of ambient ground-motion characteristics were conducted to improve seismograph design. Observational seismologists discovered that placing instruments in mines or in boreholes could significantly reduce the background motion levels. Installing a seismometer in a deep borehole can enhance signal-to-noise ratios by an order of magnitude and enable better transient-event detection. Although more expensive to install (and repair if necessary) borehole instrumentation is particularly important for noisy island sites (although horizontal tilting cannot be so easily eluded). Many of the primary GSN sensors are located in boreholes, caves, or mines and the same practices are employed generally in the siting of seismic stations in regional networks. Even burying a seismometer in a relatively shallow hole can reduce noise from wind and thermal fluctuations significantly, and this is standard practice in temporary seismic deployments.

Microseism ground motions range from 10^{-8} to 10^{-3} cm and no seismogram is totally free of ambient background motions. Most seismological analyses should explicitly allow for ambient motion and transient wave interference. Like all waves, ambient background motions transmit information. For example, as described above, the locations of large storm centers have been inferred from noise characteristics of sets of stations. In the last few decades, seismologists have discovered that the cross correlation of continuous background motions and stacking (averaging) can be used to extract seismic propagation responses for the earth structure between two seismographs. The result is the ability to treat each seismometer as a source radiating seismic waves to other stations, nearby and distant. Such ambient noise analyses provide valuable information on Earth's interior, especially in regions of low seismicity that are not often illuminated by earthquake signals.

4.4 Seismographic systems

With some understanding of the background seismic wavefield, we now turn our focus to seismometers and seismographs, the instruments used to sense and to record transient and background earth motions. The period range of seismic waves spans from roughly an hour to small fractions of a second. The *dynamic range* (the ratio of largest to smallest amplitudes) is also large; our interest extends as small as nm (10^{-9} m) and as large as meters (10's of meters for fault offsets near large earthquakes). To record these signals requires sensors sensitive to a broad range of frequencies and capable of recording large amplitudes ranges. One way to summarize the recording goals for modern seismic instruments is to chart the amplitude and frequency range of interest (Fig. 4.10).

Construction of the diagram is tricky because we want to compare quantities measured in power (continuous background motions) with those measured in amplitude (transient earthquake-excited motions). The *equivalent* peak acceleration range is the compromise. Fig. 4.10's vertical axis is specified in units of decibels (dB). The decibel originated in acoustics and is defined in terms of a scaled logarithm of a ratio of signal power. Or if we choose a unit power reference, one dB equal to $10\log_{10}(signal\ power)$. Divide the values by 20 to get the equivalent peak acceleration as a decimal quantity. For example, $-200\,(\mathrm{m/s^2})^2$ *in* dB $= 10^{-20}\,(\mathrm{m/s^2})^2 = 10^{-10}\,\mathrm{m/s^2}$. Such diagrams are sometimes la-

beled using $20\log_{10}(signal\ amplitude)$, the square is absorbed into the factor of 20 in front of the logarithm. Decibels exist to make things easier, but for what it's worth, using a logarithm of a quantity with physical units is almost always more clear.

Fig. 4.10 also includes the dynamic ranges and bandwidths of usable signals recorded on the classic WWSSN short and long-period seismographs. These instruments, deployed to help monitor underground nuclear explosions, provided much of contemporary seismic evidence that supported the developing ideas of Plate Tectonics. Modern systems are much broader band and have much more dynamic range. The thick lines indicate the spectral amplitudes expected for body-waves and surface waves at 30° distance from earthquakes with magnitudes ranging from 5.0 to 9.0. One of the GSN instrument design goals was to capture large earthquakes signals on-scale, which it succeeds in doing at many stations when the 2004 Sumatra Andaman Islands ($M\,9.2$) and 2011 Tohoku ($M\,9.0$) earthquakes occurred. The P-wave spectral values shown at short period are subject to large ranges.

4.4.1 Inertial pendulum seismometers

The design and development of seismic recording systems is called *seismometry*, and many successful seismometers have been developed over the past 140 years. Almost all seismometers are based on damped inertial-pendulum systems of one form or another. Simple vertical and horizontal seismometer designs are illustrated in Fig. 4.11. The frame of the seismometer is rigidly attached to the ground, and the pendulum is designed so that movement of an internal proof mass, m, is delayed relative to the ground motion by the mass's inertia. Each pendulum system has an equilibrium position in which the mass is at rest and to which it will return following small transitory disturbances. The orientation of the pendulum determines the

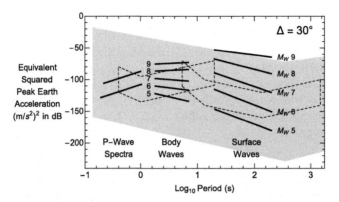

FIGURE 4.10 The range of equivalent approximate squared ground acceleration and period of ground motions spanned by IRIS/USGS Global Seismic Network (GSN) compared with capabilities of the WWSSN instrumentation (dashed lines) and expected ground accelerations from magnitude 5.0 to 9.0 earthquakes at a distance of 30° (angular distance). GSN-type instruments have become dominant for quality global seismic observatories since 1986. Modified after illustrations from IRIS and in the original IRIS proposal (by Kanamori).

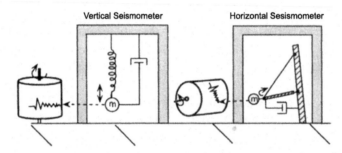

FIGURE 4.11 Schematics of inertial-pendulum vertical and horizontal seismographs. Actual ground motions displace the pendulums from their equilibrium positions, inducing relative motions of the pendulum masses. The dashpots represent a variety of possible damping mechanisms. Mechanical or optical recording systems with accurate clocks are used to produce the seismograms.

direction of ground motion that will induce pendulum motion.

Ground displacements, $u(t)$, are communicated to the mass via attached springs or lever arms, with favorably oriented motions perturbing the system from its equilibrium position, leading to periodic oscillation of the mass. Friction or viscous damping, represented by the dashpots, is generally proportional to the velocity of the mass and acts to restore the system to its equilibrium position. Small-scale fluctuations in the springs and damping elements determine an intrinsic **instrument noise** level, below which

ground motions cannot be detected. Although many early seismometers were designed empirically without mathematical analysis, the equation of motion for simple, damped harmonic oscillators provides insight into instrument characteristics.

We will assume that the mass translates in one direction only and that no rotation occurs. Let the displacement of the pendulum mass in an absolute reference frame be $z(t)$, the displacement of the mass from its equilibrium position be $y(t)$, and the ground motion be $u(t)$. The total motion of the mass is $z(t) = y(t) + u(t)$. The mass will accelerate in response to an external force operating in the direction of sensor sensitivity, $f_y(t)$, or in response to a force operating through the spring and dashpot (caused by a ground acceleration). The force from the spring, $-K\,y(t)$, is directly proportional to movement of the mass from its equilibrium position and the spring constant K. The damping force, $-D\,\dot{y}(t)$, is proportional to the velocity of the mass relative to its equilibrium position and D, the damping coefficient. Newton's Second Law requires that the forces are balanced by the mass acceleration

$$f_y(t) - D\,\dot{y}(t) - K\,y(t) = m\,\ddot{z}(t)\,. \qquad (4.5)$$

Since our primary interest is in the relationship between the ground motion and the motion of the mass relative to its equilibrium position (which we can measure), we rearrange (4.5) using $z(t) = y(t) + u(t)$, to produce

$$m\,\ddot{y}(t) + D\,\dot{y}(t) + K\,y(t) = f_y(t) - m\,\ddot{u}(t)\,. \quad (4.6)$$

The two terms on the right-hand side of (4.6) are equivalent. Either a ground acceleration or action of an external force can cause the mass to move from its equilibrium position. We'll ignore the external forces for now, assuming $f_y(t) = 0$, but note that a time-varying change in direction of the gravitational force caused by ground tilt could produce a nonzero force that will induce motions of the pendulum (and on a seismogram). Neglecting $f_y(t)$ and dividing both sides of (4.6) by the mass and introducing two new quantities, a *resonance frequency*, $\omega_0 = \sqrt{K/m}$, and a *damping factor*, $\gamma = D/2\sqrt{Km}$, we have

$$\ddot{y}(t) + 2\gamma\,\omega_0\,\dot{y}(t) + \omega_0^2\,y(t) = -\ddot{u}(t)\,. \qquad (4.7)$$

Eq. (4.7) relates the ground motion to the movement of the mass and demonstrates that if we measure the motion of the mass, we can recover the motion of the ground.

The physical significance of ω_0 can be illustrated by considering the undamped ($\gamma = 0$) and

unforced ($\ddot{u}(t) = 0$), system. Eq. (4.6) then reduces to the harmonic equation for the mass displacement,

$$\ddot{y}(t) + \omega_0^2\,y(t) = 0\,, \qquad (4.8)$$

which has purely harmonic solutions of the form $y(t) = \cos\omega_0 t$, $\sin\omega_0 t$, or $e^{\pm i\omega_0 t}$, where $\omega_0 = 2\pi f_0$ is called the *free*, *natural* or *resonant* radian frequency of the undamped system. A mechanical seismometer resonates at a period proportional to the square root of the ratio of the mass to the spring stiffness

$$T_0 = \sqrt{\frac{m}{K}}\,. \qquad (4.9)$$

Longer natural periods require a larger mass and/or a lower spring constant (softer spring).

Seismographs translate the pendulum motion into a seismogram, $x(t)$, and generally include filters to modify the output. We'll ignore these for simplicity, but we will explicitly include a ground-motion magnification factor, G, such that

$$\ddot{x}(t) + 2\gamma\omega_0\dot{x}(t) + \omega_0^2 x(t) = -G\,\ddot{u}(t). \qquad (4.10)$$

This differential equation relates the seismogram and its derivatives to ground motions and forms the basis of converting seismograph output to ground motion. We can solve the equation for $x(t)$ to characterize the instrument and then use the instrument characteristics to compute $u(t)$ from seismograms created with the instrument.

Eq. (4.10) is readily solved using Laplace transforms (for transient motions) or Fourier transforms (for stationary ground oscillations). We will explore the solutions by assuming an input displacement of the form $u_I(t) = 1 \cdot e^{-i\omega t}$ which has unit amplitude and zero phase. Of course, ground motion and the seismograms are real-valued quantities, but it is easier to analyze a complex-valued form of $u(t)$ and then select the real-valued part of the output. For an input of the form $u_I(t)$, the corresponding seismogram will have the form $x_I(t) = X_I(\omega)\,e^{-i\omega t}$, where $X_I(\omega)$ is generally a complex quantity. Direct substitution of these forms into (4.10) leads to

$$X_I(\omega) = \frac{-G\,\omega^2}{\omega^2 - \omega_0^2 + 2\,i\,\omega\,\gamma\,\omega_0}. \qquad (4.11)$$

$X_I(\omega)$ is called the instrument's Fourier *transfer function* and quantifies how the seismometer modifies input (ground) motions of radian frequency ω. The complex quantity $X_I(\omega)$ can be represented in polar form $X_I(\omega) = |X_I(\omega)|\,e^{i\phi_I(\omega)}$ with

$$|X_I(\omega)| = \frac{G\omega^2}{\sqrt{\left(\omega - \omega_0^2\right)^2 + 4\omega^2\omega_0^2\gamma^2}}$$

$$\phi_I(\omega) = -\tan^{-1}\left(\frac{2\omega\omega_0\gamma}{\omega^2 - \omega_0^2}\right) + \pi, \qquad (4.12)$$

where $|X_I(\omega)|$ is the **amplitude response** and $\phi_I(\omega)$ the **phase response**. A plot of $|X_I(\omega)|$ versus frequency represents the ratio of input ground motion amplitude to output motion amplitude (on the seismogram) and illustrates the frequency-dependent sensitivity of the instrument. When $|X_I(\omega)|$ is large, the instrument responds strongly to signals of that frequency; when $|X_I(\omega)|$ is small, the instrument responds weakly to signals of that frequency.

Sample pendulum responses are shown in Fig. 4.12. Peak seismograph sensitivity depends on the natural radian frequency, ω_0, and the damping parameter, γ. For nonzero damping, a period slightly larger than the natural period ($T_0 = 2\pi/\omega_0 = 2\pi\sqrt{m/K}$) has the maximum amplitude response. For $\gamma \to 0$ (which we call undamped), the amplitude response increases rapidly as $\omega \to \omega_0$, which is called **resonance.** If $\gamma \ll 1$ (which we call underdamped), the mass responds primarily to periods near the pendulum period, and the signal tends to "ring" or to oscillate at that period for many cycles. For

$\gamma = 1$ the response is called critically damped and oscillation is minimized; the mass quickly returning to rest as ground motion ceases. For $\gamma > 1$ (overdamped), no oscillations occur, but the mass returns to rest more slowly. Most instruments are designed to operate with near-critical damping (slightly underdamped) as a compromise between sensitivity and ringing.

If the ground-motion frequency is much lower than the seismometer frequency ($\omega \ll \omega_0$), the amplitude response is proportional to ω^2/ω_0^2, and the seismogram records ground acceleration. Thus, accelerometers, instruments intended to record strong acceleration at frequencies in the range 5–10 Hz, are seismometers with very high resonant frequencies. If the ground-motion frequency is much higher than the natural frequency ($\omega \gg \omega_0$), the seismogram is directly proportional to ground displacement. Much of the early developmental work in seismometry sought to reduce ω_0 to yield displacement recordings for regional-distance seismic signals. As discussed in more detail later, most modern seismometers are primarily sensitive to ground velocity because motions of the pendulum mass are converted to an output voltage signal proportional to the mass velocity.

The seismogram corresponding to $X_I(\omega)$ is called an *impulse response* and corresponds to

$$x_I(t) = Re\,\frac{1}{2\pi}\int\limits_{-\infty}^{\infty} |X_I(\omega)|\,e^{i\phi(\omega)}e^{i\omega t}d\omega. \qquad (4.13)$$

The impulse response summarizes an instrument's characteristics and represents the response of the sensor to an impulsive input. Since seismograms can be thought of as a sum of impulses, they can be represented as a sum of time-delayed and scaled impulse responses. More concisely a seismogram is the convolution of the instrument response with the ground motion,

$$x(t) = \int\limits_{-\infty}^{+\infty} u(\tau)\,x_I(t-\tau)\,d\tau = x_I(t) * u(t)\,, \qquad (4.14)$$

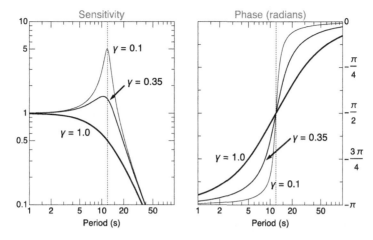

FIGURE 4.12 Pendulum amplitude and phase responses for a natural period of 12 s and three different damping factors (0.1, 0.35, and 1.0). The vertical gridline shows the pendulum's natural period and the damping factors corresponding to each response are labeled. Sensitivity is the ratio of the earth motion to the pendulum motion. At period much shorter than the natural period, the pendulum responds directly to displacement, at periods higher than the natural period, the displacement response decreases in proportion to T^{-2}, and the instrument is measuring accelerations.

where the asterisk represents the convolution process.

4.4.2 Electromagnetic seismographs

The earliest mechanical seismometers recorded the motion of the pendulum almost directly onto sand, glass, or eventually paper. Friction between the stylus and the paper provided damping, and magnification was achieved mechanically. Gradually, other damping systems were tested and introduced and magnification was increased using mirrors reflecting beams of light off the moving mass (and photographically recorded). In 1914, Russian scientist B. Galitzin, introduced an electromagnetic moving-coil transducer to convert pendulum mass motion into an electric current. Motion of a wire coil in the presence of a magnetic field generates a voltage that is proportional to the mass velocity, which Galitzin used to rotate a galvanometer coil. Light reflected from a mirror on the galvanometer was recorded on photographic paper, and a relatively long optical dis-

tance was used to produce large magnifications. This type of electromagnetic system dominated instrumentation throughout the 1900's. Optical recording eliminated friction. The coupling of a seismometer pendulum, electromagnetic transducer, and galvanometer also allowed shaping the instrument response to emphasize a particular frequency passband. The electromechanical transfer function of the galvanometer can be approximated by a solution of the form of (4.12), but with a different damping and resonant frequency corresponding to the galvanometer characteristics. The product of the pendulum, transducer, and galvanometer frequency responses controls the overall instrument response, leading to responses that are peaked at the pendulum period.

Although few seismologists work with data recorded on the oldest instruments, an appreciation of the breadth of instrumentation in the field provides important insight. Amplitude response curves for some classical mechanical and a Galitzin electromagnetic seismograph are shown in Fig. 4.13. The Galitzin re-

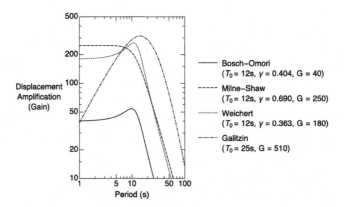

FIGURE 4.13 Magnification instrument-response curves for a suite of historic seismometers as a function of ground-motion period. The instrument resonance period is listed in the legend. All but the Galitzin are mechanical systems. The Galitzin, which has the highest gain, uses an electromagnetic sensor of mass position and a galvanometer-magnified signal (after Kanamori and Stewart, 1979).

sponse achieves higher gains due to the optical recording, but the sensitivity is more narrow-band (i.e., record a narrower frequency range) than early mechanical instruments such as the Wiechert, Bosch-Omori, and Milne-Shaw instruments. These instruments have a rapid decrease in sensitivity at long periods, proportional to $T^{-2}(\omega^2)$, where the response is proportional to ground acceleration. The seismic background motion power spectra described earlier suggest that a clear advantages of the Galitzin electromagnetic systems is that response at short periods, where the instruments respond directly to ground velocity (slope $\propto T$), is reduced in the period range of the large noise peaks (5 to 6 s). During times before digital recording, analysis, and filtering, this was a great advantage.

Early seismographs had limited long-period responses and resonated in the intermediate period range of 5–30 s. Among other things, the lack of long-period signals made estimating the size of the largest earthquakes difficult, and this difficulty impacts the accuracy of our large earthquake catalogs. The problem was solved with advances that increased recordable dy-

namic range and with the development of force feedback sensors.

4.4.3 Digital recording and force-feedback sensors

Beginning in the early 1970s seismographic systems began to forgo low-dynamic-range analog recording by ink, photographic systems, or analog tape recording in favor of digital recording. The first digital observatory stations were the High Gain Long Period (HGLP) stations deployed by Columbia University from 1969 to 1971 at sites in Alaska, Australia, Israel, Spain, and Thailand. The signal that leaves the seismometer is generally a continuous voltage. Analog-to-digital conversion of that voltage greatly extends the dynamic range (range of amplitudes that can be measured) of the system. A generous estimate of the dynamic range of old paper records is roughly just over two orders of magnitude (from the thickness of the line to the maximum amplitude recorded on the paper). A modern digitizing system can register seismometer voltages across an amplitude range of roughly 10^7. The difference is substantial – to record the modern digital range of values would

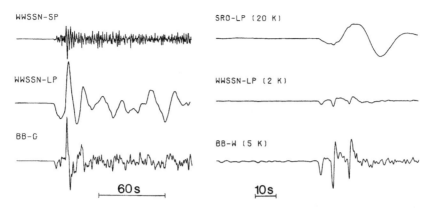

WWSSN-SP

WWSSN-LP

BB-G

SRO-LP (20 K)

WWSSN-LP (2 K)

BB-W (5 K)

60 s

10 s

FIGURE 4.14 Comparison of seismograms with varying instrument responses for the same ground motion. The records on the left compare a teleseismic *P* wave from the March 4, 1977 Bucharest event, as it would appear on WWSSN short- and long-period seismograms, with the broadband signal (proportional to ground velocity) actually recorded at station A1 of the Gräfenberg seismic array in Germany. The broadband recording contains much more information than either WWSSN recordings alone or combined. The example on the right compares GDSN (SRO-LP), WWSSN-LP, and broadband ground-displacement recordings for a *P* wave from the April 23, 1979 Fiji earthquake that has traversed the Earth's core. The broadband recording contains much more information that can reveal details of the core structure (modified from Harjes et al., 1980).

require paper with a dimension of about 10 km. Digital recording also has a tremendous advantage in that exact copies of the data can be made, compressed, and transmitted from a seismic station in near real-time.

Digital electronics developments also enabled great advances in seismic sensor design. Modern broadband and very broadband seismic sensors employ force-feedback systems that involve a negative feedback in which a force proportional to the inertial mass displacement is applied to the mass to cancel its relative motion. An electrical transducer converts the mass motion into an electrical signal to assess how much feedback force to apply. The magnitude of the force required to hold the pendulum at rest corresponds to the ground acceleration. Force-feedback greatly extends the bandwidth and linearity of a seismometer, because the mass cannot make large excursions that bend the springs or levers within the sensor. Force-feedback systems of various types have actually existed at least since 1926, when de Quervain and Piccard used one in a 21-ton seismograph in Zurich. Much

of the challenge in designing BB and VBB seismometers has been in the development of stable force-generating systems that can respond accurately over the whole range of motions of interest. Since 1973 all broadband seismic sensors have incorporated force feedback, particularly borehole sensors, which intrinsically cannot accommodate large pendulum motions due to the compact size of the sensors.

Fig. 4.14 conceptually illustrates the merits of the STS-1 very broadband seismograph relative to WWSSN stations. Broadband sensors avoid the artificial separation of signal energy into separate short- and long-period channels as was done in the WWSSN instrumentation. The dynamic range of the system is so great that using separate channels that straddle Earth noise peaks is no longer necessary. Fig. 4.14 directly illustrates the improvements of modern BB and VBB recordings compared with signals that would have been recorded on historical instruments using ground motion produced by a regional and teleseismic P waves. The broad frequency band and high dynamic range of the modern BB and VBB systems provides much more detail on the ground motions, which translates into more information on earth structure and seismic sources.

4.5 Working with modern seismograms

As described in previous sections, any measurement device distorts the observation it measures, but with good design, the distortion can be minimized. To quantitatively analyze seismic data requires that we account for distortions of the natural vibrations by the observing instruments. In the simplest case, this may be an instrument gain that is the product of a factor that converts motion to voltage and a factor that converts voltage to digital counts. A *count* is a quantity that represents the smallest division of a voltage range used to digitize an analog voltage. Digital signals in counts may resemble ground motions, but the units have no meaning until gain corrections are applied. More generally, instrument distortions are frequency dependent and include both an amplitude modification and a phase shift.

4.5.1 Digital seismic recording systems

A modern seismic recording system (a digital seismograph) produces a digital record of ground vibration. Such systems include a cascade of filters that start at the sensor (seismometer) and end with a digitized recording of a signal that is related to the ground motions (displacement, velocity, or acceleration). Most sensitive broadband and very broadband seismometers output signals proportional to ground velocity; most strong motion sensors output signals directly related to acceleration; and most GPS instruments produce signals directly related to ground displacement. Within the limits of noise, we can translate between displacement, velocity and acceleration using numerical integration and differentiation. In addition to the

seismometer, a digital seismograph may include a suite of analog and digital filters that limit the signal's frequency range in preparation for digitization, digitize the signal, and/or filter and resample the digitized signal to create separate data streams (e.g. broadband, long-period, and ultra-long-period *channels*).

The seismological community has long shared data internationally. Such cooperation is essential to study a global phenomenon such as earthquakes. To enable data sharing on the broad scale of modern seismology requires established standards for the description of seismographic responses and data formats. Seismology data standards are arranged through the International Federation of Digital Seismograph Networks (FDSN). The SEED manual (you can find The SEED Reference Manual in the publications section of the FDSN or IRIS web site). In earthquake seismology, a commonly used and efficient data storage standard is the Standard for the Exchange of Earthquake Data (SEED) format. Sharing data in SEED format requires some effort by those recording seismic data, but open-source tools for working with SEED are available online from the IRIS Consortium and other seismological organizations.

4.5.2 Removing instrument effects

Raw, digitized seismograms are related to the ground motions by convolution. Let $u(t)$ represent ground displacement, $x_I(t)$ represent the instrument response, and $x(t)$ represent the seismograph output (the raw seismogram). The relationship is

$$x(t) = u(t) * x_I(t) , \qquad (4.15)$$

where $*$ represents convolution. In the frequency domain, convolution becomes multiplication and

$$X(\omega) = U(\omega) \cdot X_I(\omega) . \qquad (4.16)$$

Quantitatively using seismic signals requires that predicted signals be convolved with the

FIGURE 4.15 A modern digital seismograph converts ground motion to a digital signal using a sequence of filtering, amplification, and digitizing stages. Diagram is from The SEED Reference Manual, published by IRIS and available online.

same instrument response, or that the effects of instrument be removed using **deconvolution**. Convolution is straightforward if the $x_I(t)$ is known accurately. We will discuss removing instrument effects using deconvolution later, after we describe instrument transfer functions.

4.5.3 Poles and zeros

We usually parameterize seismic recording instrument responses as a ratio of polynomials with real-valued coefficients. Although coefficients are the most common representation for a polynomial, another approach is to express the polynomial as a product of factors using it's roots – the values for which the polynomial vanishes. Consider a polynomial, $F(s)$, of order n, and call the n roots of the polynomial, r_k. Then,

$$F(s) = s^n + a_{n-1}s^{n-1} + \ldots + a_0$$
$$= (s - r_n) \cdot (s - r_{n-1}) \cdot \ldots \cdot (s - r_0) \quad (4.17)$$
$$= \prod_{k=1}^{n} (s - r_k) \,.$$

For simplicity, we have assumed that the coefficient of leading term in the polynomial is unity. Since the coefficients of the polynomial are real, the response is real, even though the polynomial roots may be complex.

For an instrument response, the roots of the numerator are called **zeros**, these are the frequency values for which the instrument response is zero (has no sensitivity). The roots of the denominator are called **poles**, these are the frequency values associated with an instrument resonance. By common convention in seismic instrument response expressions, the variable s is

the Laplace Transform variable and has units of radians per second. The response is then

$$X_I(s) = C_0 \frac{\prod_{k=1}^{n} (s - z_k)}{\prod_{k=1}^{m} (s - p_k)} \,, \quad (4.18)$$

where C_0 is a constant factor that includes the physical units associated with the system (conversion from displacement, velocity, or acceleration to voltage). We usually work with frequency, $\sigma = 0$, so $s = i\omega = i2\pi f$. The instrument frequency response is

$$X_I(f) = C_0 \frac{\prod_{k=1}^{n} (2i\pi f - z_k)}{\prod_{k=1}^{m} (2i\pi f - p_k)} \,. \quad (4.19)$$

Some properties are immediately apparent. A zero located at the origin corresponds to multiplication by

$$2i\pi f = i\omega \,, \quad (4.20)$$

which is the equivalent to temporal differentiation. Often an extra zero is added to the instrument response of a velocity transducer so that dividing by the response results in a signal representing ground displacement in place of ground velocity. Although the original polynomial coefficients are real, the roots of the polynomial may be complex numbers (think about the quadratic formula). However, since the polynomial is real-valued, complex-valued poles and zeros must appear in conjugate pairs.

Consider the poles and zeros for the VBB Steckheisen STS6A seismograph, a borehole sen-

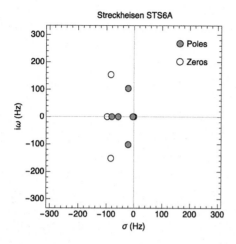

FIGURE 4.16 Poles and zeros for an STS6A borehole seismometer. The poles are shown in the complex Laplace variable plane, $s = \sigma + i\omega$, where in our case $\omega = 2i\pi$ is the radian frequency. The response is stable if all the poles are to the left of the origin.

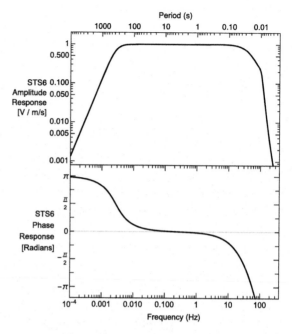

FIGURE 4.17 (top) Amplitude response corresponding to an STS6A borehole seismometer. The input is an impulse in ground velocity. The response is flat between corners of roughly 0.0033 to 3 Hz. (bottom) Phase response corresponding to an STS6A borehole seismometer. Positive phase indicates a lag (signal is shifted late). Since the size of the time shift is proportional to the period, shifts at high frequency (the period is small) are negligible. Shifts in the range between 1000 and 50 seconds are significant and must be accounted for when modeling time-domain signals of frequency-domain phase.

sor deployed at high-quality seismic observatories. Shaping the very broadband response requires 9 zeros and 13 poles. The STS6A seismograph poles and zeros are shown in Fig. 4.16. By convention, the real parts of pole/zero values are less than or equal to zero. Since real part of the poles and zeros are nonpositive, all locate on or to the left of the real-axis origin in an argand diagram. The instrument's frequency response is evaluated along the vertical axis of the diagram.

In terms of the Laplace Transform variable s, the response is real valued. But we usually work numerically with frequency, ω, not s, and then the response is generally a complex quantity that can be evaluated using (4.19). The amplitude and phase responses for positive frequencies are shown in Fig. 4.17. The STS6A is a velocity transducer, so the sensitivity shown is to ground velocity. The amplitude spectrum has *corner frequencies* near 0.0033 Hz ($T \sim 300\,\mathrm{s}$) and about 3 Hz. By corner we simply mean that a function seems to bend (like around a corner) near that value. Since the corner is not a sharp

feature, an exact corner frequency is hard to define. Between the corners, the response is relatively flat and records ground velocity in that frequency band with a near uniform sensitivity. The decrease in sensitivity beyond the corner frequencies is roughly proportional to the second power of frequency.

As do most seismic instruments, the STS6A also delays and distorts or shifts the phase of the ground motion. The phase response is shown in Fig. 4.17. STS6A phase varies smoothly from π to less than $-\pi$. Interpreting an instrument phase response is slightly more compli-

TABLE 4.1   STS6A phase lags.

Period (s)	Lag (radians)	Lag (s)
1000	2.60	414
500	1.99	159
200	0.84	27
100	0.40	6.4
50	0.20	1.56
20	0.076	0.24
10	0.034	0.054

cated than interpreting instrument amplitude response. A phase shift of π represents a polarity reversal, but other values indicate a frequency- or period-dependent time shift of the harmonic components of the signal. Causality requires that the signals are delayed relative to their arrival at the station. The time shift associated with each phase distortion can be calculated by multiplying the phase shift in radians by $T/2\pi$, where T represents the signal period. Representative phase shifts and time lags are listed in Table 4.1. For the STS6A, phase distortion is significant near and below the low-frequency corner, and must be accounted for when analyzing long-period seismic signals. The phase shifts, or time lags Table 4.1 range from a few seconds to several hundred seconds. For long-period analysis, correction for the instrument response by deconvolution of the response from the signal, or convolution of the instrument response with a predicted (synthetic) seismogram is essential.

Fig. 4.18 is a summary of common instruments employed in modern seismic analyses. The VBB STS6A borehole seismograph response is shown in each figure to provide a common reference. The sensors are categorized into observatory sensors, portable-deployment sensors, strong-motion sensors, and short-period sensors, but the categories overlap. In fact modern seismic observatories often operate a suite of instruments that may span the entire category range. The portable broadband sensors are impressive instruments given their ruggedness, but they do not match some of the ob-

servatory sensors at the longest periods. They do provide the flexibility to record shorter-period signals if configured with high sample rates. Although we no longer deploy the WWSP (Worldwide Standardized Seismographic Network (Box 4.3) short-period) and the Wood-Anderson sensors, these instrument responses played important roles in the history of seismology and remain useful filters for seismogram analysis. The WWSP is for example an excellent filter for isolating short-period body-wave signals in teleseismic observations. Other short-period sensors are often part of large station deployments to densely sample the seismic wavefield. The accelerometers have a limited bandwidth but can record strong shaking very close to large earthquakes on-scale with good fidelity.

4.5.4 Digital filters and signal decimation

The seismometer is not the only important filter in a modern seismograph response (Fig. 4.15). The seismometer's continuous output voltage is digitized in a process called analog-to-digital conversion (ADC). Two choices are important in the ADC, the sample rate and the amplitude quantization resolution. Higher sample rates produce more accurate signals but use more storage and require more computational effort (energy). Amplitude quantization is performed with as much accuracy as affordable, and is usually described by the number of bits used to divide the seismometer's output voltage range. A twenty-six bit digitizer means that we can represent $2^{26} = 67\,108\,864$ digital numbers (from all bits zero to all bits one). Let x_i represent the seismogram samples, and ΔV represent the maximum range of voltage output from the seismometer. Then the amplitude resolution, or minimum discernible difference in amplitude, is

$$\Delta x_{min} = \frac{\Delta V}{2^{26} - 1} . \qquad (4.21)$$

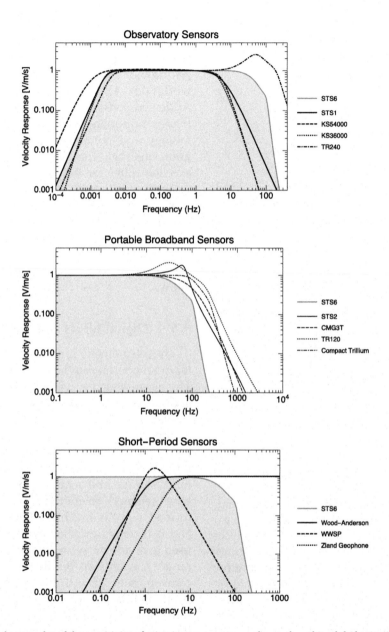

FIGURE 4.18 Reference plot of the sensitivity of seismic sensors commonly employed in global seismic analyses. Individual gains are adjusted to clarify the display. Note that the plot-axis ranges change for each plot. The VBB STS-6A response is shaded gray and used as a plot-to-plot reference, but this instrument is by no means designed to be used at the highest frequencies in accelerometer diagram.

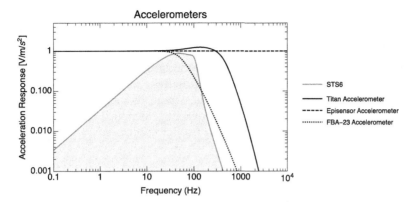

FIGURE 4.18 (*continued*)

The minus-one is to allow for a zero amplitude value. The more bits we use, the better the amplitude resolution. Any amplitude discretization introduces noise, but we can reduce this noise to some degree by over-sampling the signal (that spreads the quantization noise across a broader bandwidth) and then decimating the signal to a manageable number of samples.

Once a signal is digital, additional processing is performed using digital filters (Fig. 4.15), which can be classified as two types, finite-impulse response (FIR) or infinite-impulse response (IIR). Both construct their output from a sum of the input. The IIR uses a recursive scheme that also depends on earlier output values, which can result in a never ending filter response. We'll focus on FIR filters used as part of the digital seismogram desampling or decimation process, which is performed at least once as part of the data flow in a modern seismogram. The output of a FIR filter, y_k, is a sum of the inputs, weighted by filter coefficients, b_n.

$$y_k = \sum_{n=0}^{n=L} b_n\, x_{k-n}\,, \qquad (4.22)$$

where L is the number of coefficients in the filter.

Decimation is a procedure through which you reduce the number of digital samples in a signal. Although the name implies downsampling by a factor of ten, the process can be performed with a range of sample rate reductions. To maintain signal fidelity, you cannot just discard samples from the original signal; you must retain representative values of the original signal that include information from both the retained and discarded samples. This is possible only if we insure that the signal contains no components with frequencies above the target Nyquist frequency. Prior to discarding samples, we low-pass filter the signal, and that's when the FIR filter is applied. Ideally we would like a FIR filter with a flat response below some threshold frequency, that retains as much of the signal band as possible, and that introduces a simple, correctable phase shift to the original signal. Most of the FIR filters employed produce a linear phase shift, which corresponds to an easily accounted for constant time shift.

FIR filter coefficients used in the Seismic Analysis Code (SAC) are shown in Fig. 4.19. A larger decimation factor results in a more band-limited output. The length and breadth (as measured by the main lobe of the time-domain FIR filter coefficients) increase with the decimation reduction ratio to accommodate a narrower band low-pass filter. Each FIR filter in Fig. 4.19 has a linear phase shift equal to $\Delta t \cdot (N/2 - 1)$. The constant time shift is easy to include in the

Box 4.3 The World-Wide Standardized Seismograph Network

Short-period Benioff and long-period Sprengnether electromagnetic instruments based on the Galitzin design were deployed in the World Wide Standardized Seismic Network (WWSSN) in the 1960s. By design, the instrument responses straddled the strong oceanic microseism spectral peaks. Short-period instruments had 1-s pendulum periods and 0.7-s galvanometers; long-period instruments had either 15- or 30-s-period pendulums with 100-s-period galvanometers. In the early 1960s, as part of the VELA-Uniform project sponsored by the Department of Defense and related to treaties that moved nuclear weapons testing underground, a global array of these instruments was deployed. Each station had three short-period and three long-period instruments. The initial 30-s-period pendulums in the long-period Sprengnether instruments were excessively sensitive to barometric pressure variations. By 1965 more stable 15-s-period configurations were deployed. The distribution of the WWSSN stations (Fig. B4.3.1) was extensive, reflecting the global collaboration typical of seismology, although clear gaps exist due to both political situations and ocean basins. The WWSSN global network was more extensive than any preceding deployment and was equipped with accurate timing by crystal clocks and standardized instrumentation.

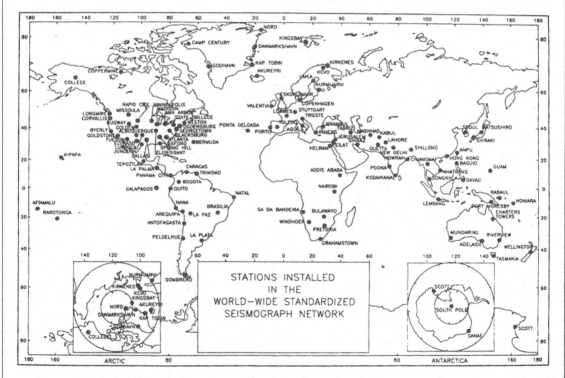

FIGURE B4.3.1 Global map indicating the locations of stations of the World-Wide Standardized Seismograph Network (courtesy of the U.S. Geological Survey).

WWSSN recordings were utilized extensively because the original photographic records were and are available on 35- or 70-mm microfiche, and copies were provided to major seismic data centers, where magnified paper copies could be printed. The WWSSN impact was tremendous, arriving at the time of the plate tectonics revolution, the accurate seismic recordings proved critical for mapping global faulting patterns. The accurate timing and response standardization facilitated many fundamental seismic studies of Earth structure and earthquakes from the 1960s through the 1980s. The successor to the WWSN, the GSN (Fig. 4.1), was completed just in time to record a surge of large earthquake activity beginning in the early 2000's.

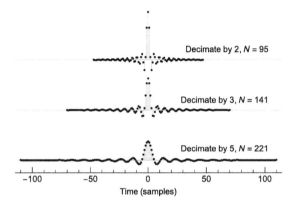

FIGURE 4.19 SAC FIR filters used to decimate by factors of 2, 3, and 5 (from top to bottom). Prior to discarding samples, the original signal is convolved with the FIR filter to reduce the spectral content of the original signal and avoid aliasing.

result – but the filters are also acausal, complicating onset picking and analysis of subtle features in seismic wave onsets. If you are picking times, always use the highest sample rate that you have for your seismograms.

Decimation can be applied efficiently since we don't need to compute the convolution at points we are going to discard, so a time-domain application of the FIR on the samples that you plan to retain can save some time. If you want to retain the seismogram start time, start sampling with the first sample in the original seismogram. Decimation factors are applied in series, so a decimation of 10 can be accomplished using filters that decimate by 10's prime factors, 2 and 5.

Consider decimation of a signal from an initial sample rate of 20 samples per second (sps) to a sample rate of 1 sps. The Nyquist frequency, which represents the highest frequency that can be represented accurately in a digital signal is

$$f_N = \frac{1}{2\,\Delta t}\,. \tag{4.23}$$

The before/after decimation ratio of Nyquist frequencies is the same and the ratio of sample rates, so transitioning from 20 sps to 1 sps corresponds to a reduction in maximum frequency content from 10 Hz to 0.5 Hz. Application of filters decimating by a factor of two, two, and five will produce the desired reduction in sample numbers by 20.

4.5.5 Removing an instrument response by deconvolution

Since the instrument effects are the result of a convolution, removing instrument effects is a deconvolution problem. Inverting equation (4.16) for the ground motion, we have

$$U(\omega) = \frac{X(\omega)}{X_I(\omega)} = \frac{X(\omega)\,X_I^*(\omega)}{X_I(\omega)\,X_I^*(\omega)} = \frac{X(\omega)\,X_I^*(\omega)}{|X_I(\omega)|^2}\,. \tag{4.24}$$

Deconvolution involves a spectral division, which can be unstable when the magnitude of the complex number in the denominator is near zero. Obviously, if the response is zero, then you can't recover the ground motion at that frequency since the instrument has no sensitivity to those motions.

In fact, whenever the denominator is small, amplification of noise in the numerator can be a problem. If you look back at the amplitude response of the STS6A seismometer (Fig. 4.17), you will notice that the places where the sensitivity is relatively small are the frequency bands beyond the instrument corner frequencies – corresponding to the longest and shortest period signals. How small is small? If the data have no noise, the answer is related to the precision of the computer arithmetic in use. In more realistic cases, the amplitude of the noise limits the range of stable division. For example, you may be able to recover 300-second period signals from a M 8 earthquake, but not from a M 6 event. The useable bandwidth is signal dependent because of the influence of seismic background motions and noise.

For seismic instruments, the problem is generally managed by limiting the bandwidth of over which the ground motion is desired. This is done by applying a filter during the deconvolution and the filter may be another seismic instrument (if you want all your data to have the same response), or by applying a band-limiting filter, $F(\omega)$, to the deconvolution result.

$$\hat{U}(\omega) = \frac{X(\omega)\,X_I^*(\omega)}{|X_I(\omega)|^2}\,F(\omega)\,. \qquad (4.25)$$

The inverted caret above U indicates that we no longer are estimating the ground motion spectrum, we are estimating a filtered version of the spectrum, $\hat{U}$.

Seismologists use a number of approaches to stabilize deconvolution (deconvolution arises often in seismology because a seismogram is the result of source, propagation, and instrument convolutions). A common approach to instrument correction is to apply a simple band-limiting filter to the deconvolution result. An often-used function is a cosine taper define by

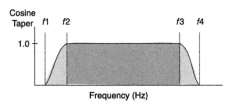

FIGURE 4.20 Definition of the frequency parameters in a simple cosine taper used to band limit an instrument deconvolution.

frequencies.

$$\hat{U}(\omega) = \frac{X(\omega)}{X_I(\omega)} \cdot C(f_1, f_2, f_3, f_4)\,. \qquad (4.26)$$

The filter is shown in Fig. 4.20 and is zero for $f \leq f_1$ and $f \geq f_4$, unity for $f_2 \leq f \leq f_3$ and has a cosine taper shape for $f_1 \leq f \leq f_2$ and $f_3 \leq f \leq f_4$.

An example application of the spectral taper is shown in Fig. 4.21. The original seismogram is shown at the top left and estimates of the ground displacement are shown beneath for a range of low-frequency spectral tapers. Only the low-frequency taper has any effect on the results since the anti-alias filter applied decimating the signal removed the high-frequency noise (the nyquist frequency is less than the instrument response corner). The tapered frequency ranges are listed in the second column, along with a plot of the spectral tapers. The first seismogram estimate is completely dominated by noise. The second begins to reveal the displacement signals, but it and the next estimate still have substantial long-period noise. The pre-signal long-period noise has a period of roughly 500 s. The fourth estimate, which is reasonably stable, was computed using a low-frequency taper between 500 and 100 s period. As shown by the next seismogram, continuing to extend the taper discards useful information from the stable frequency band. This result is fine for an analysis of signals with periods shorter than 50 s, but information is lost. The plots of the third column show the

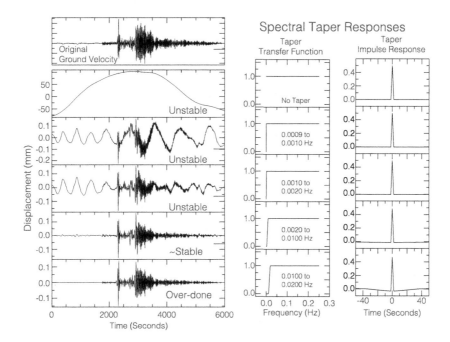

FIGURE 4.21 Sample instrument deconvolution by spectral division with a frequency-domain cosine taper. Original, raw seismogram is shown in the upper right. Estimates of the displacement seismograms using a range of cosine spectral tapers shown in the second columns are shown beneath the original seismogram. Third column shows the spectral taper response, which must be accounted for in any modeling of the signals in the spectral range beyond the taper corner.

time-domain response of the spectral taper. Multiplication by the spectral taper corresponds to convolution (in the time domain) with these signals. Most of the signals are effectively digital impulses, but as the taper narrows, the time-domain response includes low-frequency sidelobes that are visible in the "stable" estimated signal and are especially noticeable in the overcompensated displacement estimate. With the exception of the inaccurate top estimate that corresponds to no spectral taper, each cosine filter distorts the true displacement signal. The distortion is acasual and can produce long-period troughs around impulsive arrivals in the original signal. Such effects must be included in any synthetic seismograms used in waveform models if you are fitting signals near the frequency limits with which you defined the spectral taper.

4.6 Seismometry's future

We live during a time of rapid technological development. Recent improvements in seismic instrumentation, data collection, archiving, and distribution would astound the earliest seismic researchers. Manufacturing improvements lower the cost of production and lower prices lead to larger seismic networks. The number of quality broadband sensors has increased dramatically in the last 15 years. As a result, data sets are larger and complex seismic analyses are now often completed routinely, in near-real time. Since data are easier than ever to acquire and analyze, opportunities for innovative research are plentiful.

We have not presented a detailed discussion of seismic arrays, which since the 1940's have played an important role in global seismology.

Arrays are indispensable facilities for measuring not only the polarization of the waves, but also the variability and the direction and apparent speed of waves crossing the array. For the most part, their sensor requirements are similar to those of a single seismic station. The key parameters of seismic arrays are their *aperture* (spatial dimension) and their station density, which control the resolution of the direction and apparent wave speed and determine an array's utility for increasing signal-to-noise ratios. Arrays lower event and signal detection thresholds, allow directional probing of the seismic wavefield. With very dense seismic arrays, we can begin to explore spatial gradients in the wavefield that lead to deeper appreciation of the heterogeneity of the seismic wavefield including strains and rotations. At the intuitive level, broad aperture, dense seismic arrays have enabled visualizations for seismic deformation that illuminate the processes underlying seismic wave interaction near the surface. No seismologist would ever argue that an array would not be a better than a single station. But the added expense of many versus few sensors has resulted in a much larger number of isolated seismic stations than seismic arrays. Arrays are certain to be an important part of seismology's future.

We have focused on the most commonly used seismic instrumentation, inertial sensors that measure the ground's translational motions. A full description of earthquake deformation on land includes strains and rotations and pressure within the oceans. In the oceans, hydrophone pressure measurements of P-wave energy have already contributed to earthquake investigations and offer one affordable approach to instrumenting remote oceanic regions with floating seismic sensors. The development of more general earth-motion measurement tools continues and advances in high-sample rate GPS and satellite InSAR have made fundamental contributions to our understanding of Earth's deformation and promise more as data accumulates. Strain sensors have been deployed on a rela-

tively large scale near some plate boundaries. Scientists have used light scattering within fiberoptic cables to measure cable strain resulting from ground movements. Distributed Acoustic Sensing creates an opportunity to monitor small areas with dense coverage, leading to large densely-sampled seismic ground motion observations. Work also continues to design and to produce sensors to measure the ground rotation that accompanies seismic deformation, particularly in the near field as a contribution to strong motions near large earthquakes. Rotations and translations combined can among other things provide single-station information on wave direction and local phase velocities (similar to an array) and near-field rotations could provide complementary constraints on the slip distribution along an earthquake rupture surface.

4.6.1 Seismometers everywhere

A consequence of the ongoing mobile-computing revolution is the fact that anyone with a smart phone generally carries an accelerometer. Populated areas, then, are covered with hundreds of millions of seismometers. Work continues on identifying the best ways to harvest the large, but noisy (most phones are not ideally positioned to record motions) data recorded on mobile phones, tablets, and portable and home game consoles and computers. The intrinsic instrumental noise in current inexpensive sensors is significantly above the high-noise model for seismic instruments. Small vibrations on these systems are overwhelmed by internal instrument noise which limits their utility to monitor small earthquakes or distant seismic sources. However, when at rest, these devices are not bad strong-motion sensors and produce relatively accurate estimates of accelerations strong enough to be felt.

In addition to the sensor, smart phones also include relatively precise differential GPS sensors, large storage, networking capability, and have superb graphics and improving computations potential. The production of large numbers of mobile phones has also led to electronics manufacturing improvements and costs reductions that make available inexpensive microprocessors, sensors such as MEMs accelerometers, and analog to digital conversion boards, and signal processing components. The technology has been incorporated into innovative efforts to engage the public in data collection efforts, connecting an inexpensive accelerometer to a home computer, or creating a custom seismic data acquisition system with an inexpensive geophone and microprocessor-based linux computer. Such developments foretell an exciting future for data-driven earthquake science and seismology.

4.7 Summary

Earth is in continuous motion as a result of natural processes and cultural noise. Seismic background motions vary from place to place and include diurnal and seasonal variations. The largest background accelerations are produced by off-shore processes in the period range from about 5 to 15 s. Seismographs are sensitive instrument systems that convert vibrations to voltages that can be digitized, transmitted, and archived. While a few decades ago, the data rates presented a challenge for transmission and storage, modern equipment with large low-power storage make sampling signals less of a problem. A single three-component seismic station with a sampling rate of 50 sps storing 26-bit integer values would accumulate only 42 Mb per day. The actual value is even smaller, since the digital data are stored in an efficient compression algorithm that greatly reduces the file size. Effectively using seismograms, particularly to examine the subtle features of seismic signals, requires some understanding of the issues involved in data collection and processing. More detailed treatments and examples can be found in the references.

Seismogram interpretation and processing

Chapter goals

- Introduce basic observations of seismograms.
- Introduce how seismograms can be classified based on distance from a source.
- Review basic characteristics of seismograms.
- Review basic principles for interprepting a seismogram using signal processing and travel time curves.

The foundation of seismology relies on the observations of the ground motion as a function of time (*seismograms*) recorded from sensitive equipment (*seismic stations*). Our knowledge of the velocity structure of the Earth and of the various types of seismic sources is the result of *interpreting* those seismograms. The more fully we quantify all of the ground motions in a seismogram, the more fully we understand the Earth's structure and its dynamic processes. Furthermore, anything that shakes the ground (e.g., cars, machinery, glacier movement, nuclear explosions, earthquakes, etc.) can be characterized on a seismogram. Seismograms are a complicated mixture of source radiation effects (e.g., the spectral content and relative amplitude of the P- and S-wave energy that is generated at the source), $s(t)$, propagation phenomena (e.g., multiple arrivals produced by reflection and transmission at seismic impedance boundaries or at

the surface), $p(t)$, and frequency band-limiting effects of the recording instrument, $I(t)$, such that:

$$u(t) = s(t) * p(t) * I(t), \qquad (5.1)$$

where $u(t)$ represents the ground displacement at a seismic station, * represents a convolution (a filter operator) discussed later in this chapter, and $I(t)$ is the instrument response discussed in Chapter 4. The challenge for seismology is that observationally, we must assume $s(t)$ (the source of the ground motion) to study $p(t)$ (Earth structure) or we must assume $p(t)$ to study $s(t)$. Thus, any analysis of a seismogram generally requires that we assume we know the source to study the propagation, or visa versa. We discuss various approaches to minimize the impacts of these assumptions in Chapters 19 and 20.

Seismograms are fascinating to explore, as each can tell a story of the Earth or a source that can be interpreted by a seismologist. Experience and a sound foundation in elastic-wave theory helps guide a seismologist to decipher the complexities of the source or propagation on a seismogram. Expertise with seismograms can allow for the identification of subtle features, such as identifying coherent vibrations produced by reflections off deep layers from background noise or from other arrivals scat-

Foundations of Modern Global Seismology
https://doi.org/10.1016/B978-0-12-815679-7.00012-4

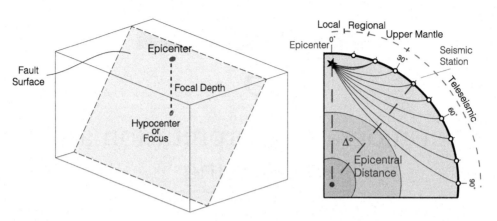

FIGURE 5.1 Cartoon showing faulting terminology used for earthquakes and seismic stations. (left) The hypocenter marks the location of rupture initiation, and the epicenter is the vertical projection of that location onto the surface of the Earth. The fault plane represents the surface on which slip occured during an earthquake. (right) The epicentral distance (Δ) to a seismic station can be measured in degrees from source to the station relative to the center of the Earth, generally used for teleseismic distances (distances from 30° to approximately 95°). Also shown are definitions for local, regional, and upper mantle distances.

tered by the Earth's three-dimensional heterogeneity. However, a book format does not allow for the full exploration of unique and interesting seismograms, and although we present many examples of seismograms throughout the book, any student of seismology should explore the large number of high-quality seismograms that currently exist in open databases, such as the databases at the Data Management Center (DMC) at the Incorporated Research Institutions of Seismology (IRIS). Because we cannot explore the digital world of seismograms in a book format, this chapter focuses on the essence of using seismograms for analysis, with examples of how simple measurements lead to important results, such as the location of the source or the identification of complex rupture processes produced by large earthquakes. In modern practice, almost all approaches that use seismogram are implemented on computers, which allows for vast quantities of data currently collected to be analyzed.

5.1 Terminology for seismograms

The coordinates of an earthquake point source (where an earthquake rupture begins) are known as the *hypocenter* (Fig. 5.1). The hypocenter is usually given in terms of latitude, longitude, and depth below the surface. The *epicenter* is the surface projection of the hypocenter (the latitude and the longitude), and the *focal depth* is the depth below the surface, shown in Fig. 5.1. *Epicentral distance* is the distance separating the epicenter and the recording seismic station, and can be measured in degrees at the center of the Earth (Fig. 5.1). Since the nature of a seismogram depends on this distance, especially for body waves, we define distance ranges to categorize seismograms (see below). For large earthquakes, the finiteness of the source volume is not negligible, and then these terms usually refer to the point at which the rupture initiates. Other terms such as the earthquake *centroid* will be introduced later to define the effective center of stress release of the source (Chapter 16).

When a packet of energy arrives at seismic stations, it results in displacement at the station at a given time (seismogram), and we refer to this as an *arrival* (arrival of energy) or a *seismic phase*. The keys to identifying these arrivals involve assessing their timing and behavior as a function of distance, measuring the type of ground motion they produce, and establishing their consistency from event to event. Additional, later arrivals are primarily reflections from velocity discontinuities at depth or from the free surface of the Earth. The timing of the various arrivals is a predictable function of the depth of the source and the distance between the seismic source and receiver. The identification of seismic phases is by no means a trivial exercise, and in fact many modern-day seismologists have little direct experience in the routine "reading" or "picking" of seismic-phase travel times and amplitudes. Systematic cataloging of the absolute and differential travel times of all phases on seismograms provides information that we can use to determine the structure of the Earth and to generate travel-time tables that can be used to locate other earthquakes. For many seismic sources, the P and S waves are radiated from a concentrated volume, which can be approximated as a point source.

Three-component seismograms record motions in three-dimensional space by recording motion in the vertical (Z) and horizontal (North: N; East: E) directions. E and N horizontal component recordings can the be rotated to Radial (R) and Transverse (T) (as long as the station and epicentral locations are known), and different seismic phases can be identified on R and T components, depending on the polarization of the seismic phase. For example, Love and horizontally polarized S-waves (SH) can be identified on the T component because of their transverse polarization, while Rayleigh, P-waves, and vertically-polarized (SV) S-waves can be identified on the vertical and radial components. The nature of arrivals can also be impulsive or emergent, depending on source mechanism and propagation. Finally, whether the first swing of a P-wave arrival is up or down represents the polarity and can be used to determine the source mechanism (see Chapter 17).

Fig. 5.2 shows an example broadband (B), high-gain (H), three-component (R, T, Z) seismic recordings from the March 11, 2011 Tohoku (M_W 9.0) earthquake recorded at PASC (Pasadena, CA) seismic station, about 85° from the epicenter. The labeled P and S waves arrivals show impulsive arrivals with some duration, while *Love waves* (G) and *Rayleigh waves* (R) show dispersion (different frequencies traveling at different speeds), which extends the duration of these arrivals. Thus, the P and S arrivals have different characteristics from those of Love wave and Rayleigh waves due to fundamental differences in propagation. The P- and S-waves travel through the body of the Earth (called *body waves*) while the *Love* and *Rayleigh* waves (*surface waves*) are trapped energy near the Earth's surface. Furthermore, because of the size of the earthquake which ruptured a fault approximately 150-km wide by 300-km long km, the P and S phases are extended in length, reflecting the long source duration of this giant earthquake (e.g., Ammon et al., 2011). The Tohoku (M_W 9.0) earthquake was one of the largest earthquakes recorded in the last 100 years, and because the earthquake involved very large slip (approximately 40–50 meters) in the shallow section of the megathrust fault off of Tohoku, Japan, it generated a very large tsunami that caused severe destruction and loss of life (see Chapter 9 for more details on this event).

Seismic stations distributed worldwide have been systematically reporting major seismic phase arrival times to the International Seismological Centre (ISC) since 1964. The Global Seismic Network (GSN) continues to record data from over 140 broadband seismic stations distributed globally, which originally developed from the World Wide Standardized Seismic Network (WWSSN) (see Chapter 4). Currently, over

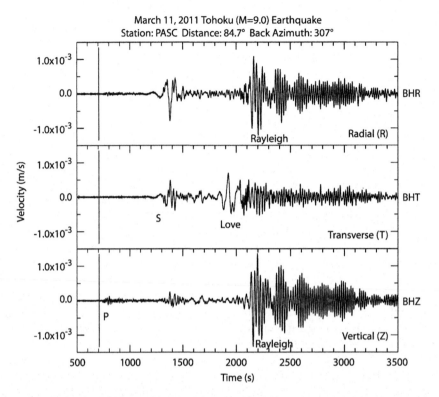

FIGURE 5.2 Broadband seismic recordings of the March 11, 2011 Tohoku, Japan (M_W 9.0) mega-thrust earthquake recorded at seismic station PASC (Pasadena City College, California). These ground motion, velocity (m/s), recordings have had the instrument (I) removed by deconvolving instrument response (see Chapter 4), and have been rotated from North-South (BHN) and East-West (BHE) to Radial (BHR) and Transverse (BHT). The rotation involves only the horizontal components and is performed using the station back-azimuth (in this case 307°). *P*, *S*, *Rayleigh* and *Love* waves are labeled.

10,000 stations are sending data to the Data Management Center (DMC) run by the Incorporated Research Institutes for Seismology (IRIS). Seismic waveform data, $u(t)$, can be easily retrieved from the DMC using a series of tools developed by IRIS. Open access to this data near-real time has transformed the field of global seismology to allow rapid analysis of any event that could cause ground motion, from large earthquakes to induce earthquakes to missiles tests to nuclear explosions.

Once direct *P* arrivals at different stations have been associated with a particular event and that event is located, one can seek to interpret additional arrivals. The ISC data base has com-

piled millions of arrival times that have been attributed to more than 25 seismic phases, each with a specific structural interaction, or path, through the Earth. Fig. 5.3 shows over 416,000 ISC travel-time picks as a function of epicentral distance for shallow (≤ 30 km) earthquakes and explosions with $M \geq 6.0$ from 1964–2016 (ISC, 2016). Clear lineaments exist that represent the travel-time branches of various phases such as direct *P* and *S*, as well as phases that have more complicated travel paths. One can view this as the Earth's "fingerprint," uniquely characterizing the complexity imparted into seismic wavefields by its structure. A seismogram at any particular distance will record the correspond-

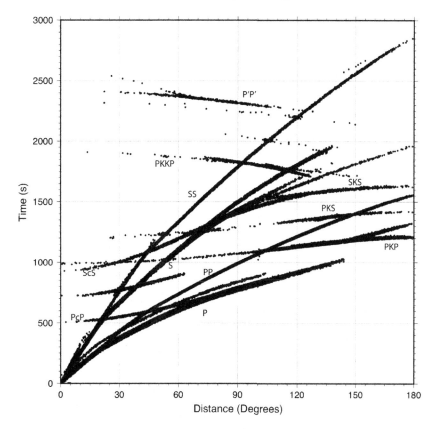

FIGURE 5.3 Plot showing over 416,000 travel times picked from phases of select shallow (≤ 30 km) earthquakes and explosions with $M \geq 6.0$ from the International Seismic Centre (ISC) catalog from 1964-2016 (ISC, ISC, 2016). The phases are named using a convention that describes the wave's path through the Earth derived from the IASPEI travel time curves (e.g., Kennett and Engdahl, 1991). For example, *PcP* is the *P* wave reflected from the Earth's core.

ing time sequence of arrivals, although source radiation and depth differences may make seismograms at the same distance appear dissimilar. We develop a nomenclature for the various arrivals and some simple guidelines for identifying seismic phases. The fact that coherent travel-time branches are so pronounced in Fig. 5.3 demonstrates the gross radial symmetry of the Earth's layered velocity structure. On the other hand, some of the unidentified arrivals as well as some of the scatter about the mean for any given branch are manifestations of three-dimensional velocity heterogeneity. Assuming a radially symmetric, layered velocity structure

enables us to predict the arrival times of most seismic phases to within a few percent, which provides the basis for most earthquake location procedures. Chapter 6 discusses several techniques for locating earthquakes, including some that can be adapted to three-dimensional structures.

5.2 Characteristics of body wave seismograms

The basic character of seismograms depends strongly on the epicentral distance. At short epi-

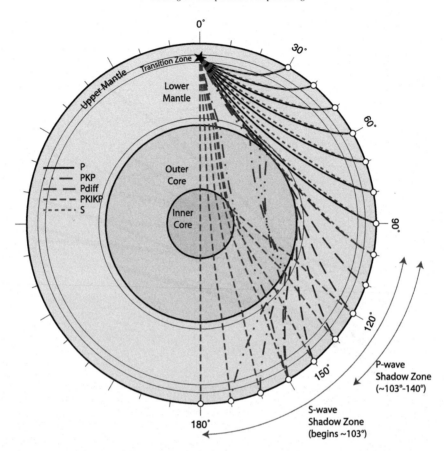

FIGURE 5.4 Schematic diagram showing body waves that travel and interact with various boundaries in the Earth, including the and lower mantle, the transition zone, and the inner and outer core. The liquid outer core creates a *shadow zone* that prevents direct *S* waves from arriving at seismic stations over 103° from a source. A *P* wave shadow zone exists from 103° to approximately 140°, where little *P* wave energy arrives, although various diffractions (*Pdiff*), refractions (*PKP*, *PKIKP*), and free surface reflections (such as *PP*, *PS*, *PPP* that are not shown) can arrive at these distances.

central distances, the character of seismograms can be complex, dominated by seismic waves that scatter within the highly heterogeneous crustal structure. At large distances, seismograms are relatively simple in nature, reflecting the relatively smoothly varying velocity structure of the deep mantle and core.

There are four general classifications of seismograms based on epicentral distance (Fig. 5.1): (1) *Local distances* are defined as travel paths of less than 100 km. (2) *Regional distances* are defined as $100 \leq X \leq 1400$ km ($1° \leq \Delta \leq 13°$), where X and Δ are the epicentral distance in kilometers and angular degrees, respectively. (3) *Upper-mantle distances* are defined as $13° \leq \Delta \leq 30°$, and seismograms recorded at these distances are dominated by seismic energy that turns in the depth range of 70 to 700 km below the surface (Fig. 5.4). This region of the Earth has a very complex velocity distribution, with a *low-velocity zone* in the upper mantle and at least two major velocity discontinuities (400 and

660 km depths) within what is called the *transition zone*. (We will discuss the details of these velocity structures in Chapter 10.) The direct *P* and *S* phases at upper-mantle distances have complex interactions with these discontinuities. (4) *Teleseismic distances* are defined as $\Delta \geq 30°$ (Fig. 5.1). The direct *P*- and *S*-wave arrivals recorded at teleseismic distances out to $\Delta \approx 95°$ are relatively simple, indicating a smooth velocity distribution below the transition zone, between 700 and 2886 km depth. The simplicity of teleseismic direct phases at distances between 30° and 95° makes them invaluable for studying earthquake sources because few closely spaced arrivals occur that would obscure the source information (Chapter 16). The overall seismogram at these distances can still be complex because of the multiplicity of arrivals that traverse the mantle, mainly involving surface and core reflections (see Fig. 5.9). Beyond 95°, the direct phases become complicated once again due to interactions with the Earth's core. Since the character of seismograms depends on the epicentral distance, the nomenclature for phases is also distance dependent.

Different recording distances and observed seismic phases all have distinct wave attributes. Many phases can be identified based on timing and amplitude, which can be estmated from travel time tables derived from standard earth models. Other attributes of seismic arrivals include frequency, where higher frequencies attenuate more rapidly as a function of distance from a source. For example, phases from local earthquakes can be identified using a high-pass filter at 5 Hz, while teleseismic *P*-waves and *S*-waves tend to have little energy for periods shorter than 1 and 4 s, respectively. Surface waves have a dominate period of 10 to 20 s at teleseismic distances and can be removed from a seismogram with a high-pass filter. Other phases, such as *T* phases, arrive only at stations near the coast or an island (see Box 5.1).

5.2.1 Local, regional, and upper mantle

Seismic recordings at local distances are strongly affected by shallow crustal structure, and relatively simple direct *P* and *S* phases are followed by complex reverberations. Regional-distance seismograms are dominated by seismic energy refracted along or reflected several times from the crust-mantle boundary. The corresponding waveforms tend to be complex because many phases arrive close in time. Fig. 5.5 shows an example profile of seismograms for a *M*4.0 earthquake in New Mexico with recordings from local to regional distances. By high-pass filtering the seismograms, different seismic arrivals can be seen, and for these distances for this event, the noise level is reduced.

At local and regional distances, a special nomenclature is used to describe the travel paths. Fig. 5.6 shows a very simplified crustal cross section with primary raypaths, plus raw and high pass filtered regional-distance seismograms from seismic station CMB from the Berkeley (BK) for the 2008 *M*6.0 Wells, NV earthquake. Note how different the ground motion appears for the different frequency bands. The higher-frequency signal allows ready identification of discrete arrivals, but there is a continuous flux of short-period energy, much of which is scattered in the crust. The direct arrivals at these short distances are usually referred to as P_g and S_g. Depending on the source depth, the velocity gradient within the shallow crust, and the distance between the source and the station, these arrivals may be either upgoing or downgoing phases. The *g* subscript is from early petrological models that divided the crust into two layers: an upper *granitic* layer over a basaltic layer. In 1910, Croatian researcher Andrija Mohorovičić identified an abrupt increase in velocity beneath the shallow rocks under Europe, which is the crust-mantle boundary termed as the *Moho*. Arrivals that travel as head waves along, or just below, the Moho are known as P_n and S_n. The frequency dependence of these

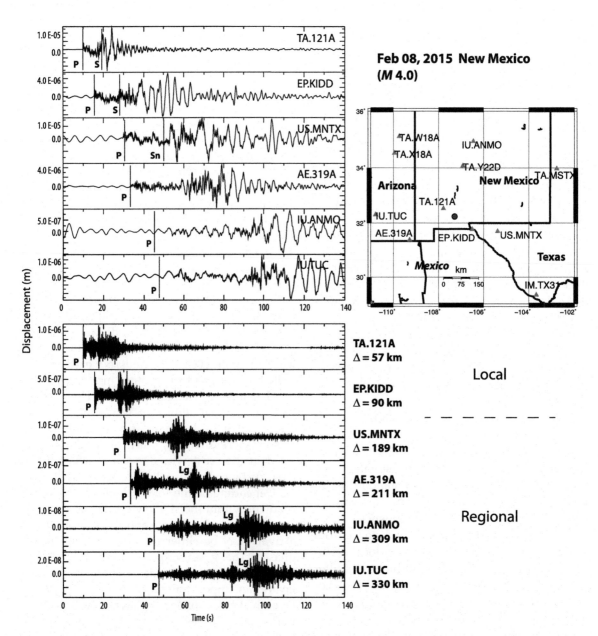

FIGURE 5.5 Profile of broadband local (top two) to regional (bottom four) seismic recordings of a M 4.0 earthquake in southern New Mexico. (Top) Unfiltered and (Bottom) 5 Hz high-pass filtered displacement seismograms. Map shows station locations labeled by network and station name (triangles) along with the epicenter (filled circle).

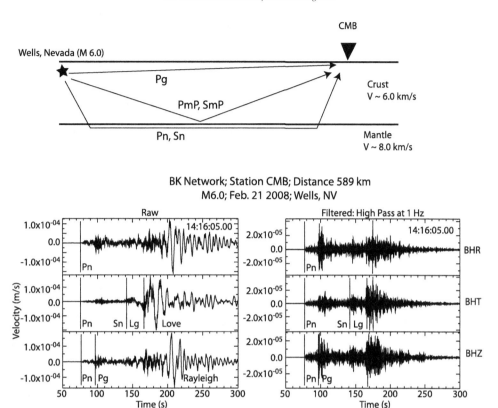

FIGURE 5.6 (top) A simplified cross section of a one-layer crust and corresponding raypaths for various phases observed at regional distances. (bottom) Raw and filtered broadband seismograms for the 2008 $M6.0$ Wells, NV earthquake recorded at regional seismic station CMB from the BK network. The high pass filter highlights the crustal phases and essentially removes the longer period surface waves. Regional phases are complex due to multiple travel paths within the crust.

head waves tends to make them longer period (Chapter 12). Moho reflections are labeled *PmP, PmS, SmP,* or *SmS*. (Note that each leg of the ray is named, and *m* denotes a reflection at the Moho.) At distances less than about 100 km, P_g is the first arrival. Beyond 100 km (depending on the crustal thickness), P_n becomes the first arrival, as in Fig. 5.6. The phase labeled L_g in Fig. 5.6 can also be labeled S_g, and is considered a higher mode surface wave, which will be described later. In many regions of the Earth, additional regional arrivals are observed that have classically been interpreted as head waves traveling along a midcrustal velocity discontinuity,

known as the *Conrad discontinuity*, called P^* and S^*, respectively. These are observed only in certain regions. In older literature, P^* is written as P_b (*b* denotes the *basaltic* layer). At distances beyond 13°, P_n, amplitudes typically become too small to identify the phase, and the first arrival is a ray that has bottomed in the upper mantle. The standard nomenclature for this arrival is now just *P* or *S*, although subscripts are used to identify different triplication branches for the transition zone arrivals.

Regional seismic recordings, although sometimes difficult to decipher because of their complexity, can be used to distinguish different

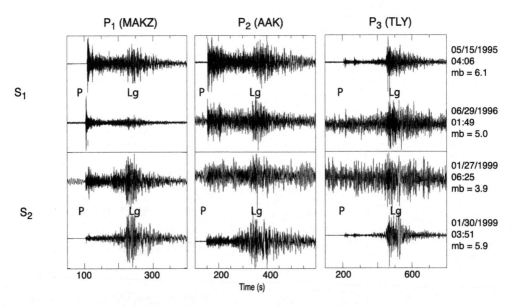

FIGURE 5.7 Broadband regional seismic recordings of two different sources: nuclear explosions S_1 and earthquakes S_2 recorded at similar distances for each station. The three seismic stations (MAKZ P_1, AAK P_2, TLY P_3) represent different propagation paths in the Earth. Note the differences in the characteristics of the sources and the paths, as regional seismograms are very complex due to the interaction of the seismic wavefield to complex regional earth structure.

source types. For example, Fig. 5.7 shows broadband regional seismic recordings for two different sources: nuclear explosions S_1 and earthquakes S_2 recorded at three seismic stations. Each column shows recordings for seismic stations (MAKZ P_1, AAK P_2, TLY P_3) and represents different propagation paths in the Earth. Each source is approximately the same distances for each station. The physics of an explosion (compressional source) is fundamentally different than an earthquake (shear faulting source). Thus, P waves are preferentially excited with nuclear explosions (or any explosion), yet at regional distances, ambiguity between earthquakes and explosions can occur as a result of the interaction of the seismic wavefield with complex regional earth structure. For example, Fig. 5.7 shows weak P wave excitation and a large Lg at station TLY for the nuclear explosion.

5.2.2 Teleseismic

The direct P- and S-wave arrivals recorded at teleseismic distances out to $\Delta \approx 95°$ show little complexity compared to regional seismic recordings, indicating a smooth velocity distribution below the transition zone, between 700 and 2886 km depth. The simplicity of teleseismic direct phases between 30° and 95° makes them invaluable for studying earthquake sources because few closely spaced arrivals occur that would obscure the source information (Chapter 16). However, the overall seismogram at these distances can still be complex because of the multiplicity of arrivals that traverse the mantle, mainly involving surface and core reflections (Fig. 5.2).

The simplest and most frequently studied body-wave phases are the direct arrivals. They travel the minimum-time path between source and receiver and are usually just labeled P or S.

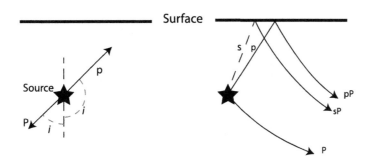

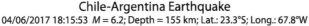

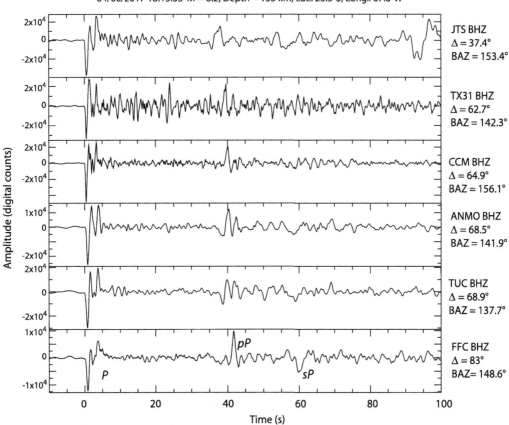

FIGURE 5.8 (Top) Schematic diagram showing the geometry of (top left) upgoing and downgoing rays along with the different take-off angles (*i*) and (top right) the geometry of depth phases where the dashed line represents the upgoing *s* that converts to *P* upon reflection from the free surface. (Bottom) Unfiltered, displacement seismograms showing *P* arrival and depth phases from a 2017, intermediate depth (155 km), *M*6.0 earthquake near the border of Chile-Argentina border recorded at selected stations in the U.S.A. Note there is little moveout of the depth (*pP*, *sP*) phases as a function of distance, which allows for depth to be determined.

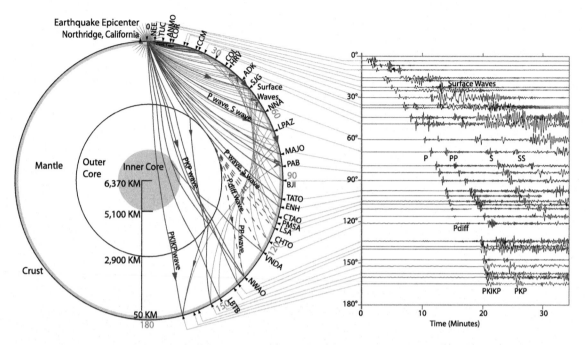

FIGURE 5.9 (left) Raypaths for seismic arrivals for the general earth structure, including inner core, outer core, mantle, and crust and (right) a seismogram profile for the 1994 Northridge, California $M6.7$ earthquake. Raypaths show curved paths due to the increase of velocity as a function of depth. Because P waves and S waves travel at different speeds, the arrival times between the P, S and Surface waves become greater with increasing distance. The sharply decreasing P and S wave amplitudes seen at stations beyond about 100 (1100 km) from the earthquake indicate the existence of the Earth's outer core. Although P waves travel through the outer core (PKP waves), S waves do not because liquids cannot support shear motion, indicating that the outer core is fluid. Seismograms show variations as a function of distance due to propagation through the Earth (modified from IRIS poster Explore Earth).

At epicentral distances greater than a few tens of kilometers in the Earth, direct arrivals usually leave the source *downward*, or away from the surface, and the increasing velocities at depth eventually refract the wave back to the surface. Fig. 5.8 illustrates two rays leaving a seismic source. The angle, i, that the ray makes with a downward vertical axis through the source is known as the *takeoff angle*. If the takeoff angle of a ray is less than 90°, the phase, or that segment of the raypath, is labeled with a capital letter: P or S. If the seismic ray has a takeoff angle greater than 90°, the ray is *upgoing*, and if it reflects from the surface or is a short upgoing segment of a composite raypath, it is signified by a lowercase

letter: p or s. Upgoing rays that travel from the source up to the free surface, reflect, and travel on to the receiver are known as *depth phases*.

The various portions of the path a ray takes, for example, between the source and the free surface, are known as *legs*. Each leg of a ray is designated with a letter indicating the mode of propagation as a P or S wave, and the phase is designated by stringing together the names of legs. Thus, there are four possible depth phases that have a single leg from the surface reflection point to the receiver: pP, sS, pS, and sP (Fig. 5.8). The relative timing between the direct arrivals and the depth phases is very sensitive to the depth of the seismic source (hence the name depth phases). Fig. 5.8 shows examples of the *pP* depth phase for two events. The *pP* arrivals must arrive later than direct *P* because they traverse a longer path through the Earth, but their relative amplitudes can vary due to the source radiation pattern. The *sP* phase, which always arrives after *pP*, is present but not impulsive in these examples.

Seismic phases that reflect at a boundary within the Earth are subscripted with a symbol representing the boundary. For example, *P*-wave energy that travels to the core and reflects is called *PcP*, the *c* indicating reflection at the core. In a spherical Earth, it is possible for a ray to travel down through the mantle, return to the surface, reflect, and then repeat the process (Fig. 5.8). Because the original ray initially traveled downward, the phase is denoted by a capital letter. The free-surface reflection is not denoted by a symbol; rather, the next leg is just written *P* or *S*. This type of phase is known as a *surface reflection*. Some common surface reflections are *PP, PS,* and *PPP,* where *PP* and *PS* each have one surface reflection (involving conversion for *PS*), and *PPP* has two surface reflections. Multiple reflections from both the core and surface occur as well, such as *PcPPcP, ScSScS* (*ScS₂*), and *ScSScSSScS* (*ScS₃*)

(see Fig. 5.9). Both reflected phases and surface reflections can be generated by depth phases. In this case, the phase notation is preceded by a lowercase *s* or *p*, for example, *pPcP* and *sPP* (Fig. 5.9). All of these phases are a natural consequence of the Earth's free surface and its internal layering, combined with the behavior of elastic waves.

The amplitude of body-wave phases varies significantly with epicentral distance. This occurs both because reflection coefficients depend on the angle of incidence on a boundary and because the velocity distribution within the Earth causes focusing or defocusing of energy, depending on the behavior of geometric spreading along different raypaths. Thus, the fact that a raypath can exist geometrically does not necessarily mean it will produce a measurable arrival. For example, the *P*-wave reflection coefficient for a vertically incident wave on the core is nearly zero (the impedance contrast is small), but at wider angles of incidence the reflection coefficient becomes larger. Thus, *PcP* can have a large amplitude in the distance range $30° < \Delta < 40°$. The surface reflections *PS* and *SP* do not appear at distances of less than 40°, but they may be the largest-amplitude body waves beyond 100°. Progressive energy losses

---

Box 5.1 Seismic waves in the ocean

In the early 1940's, D. Lineham reported a class of seismic waves that were observed only on coastal and island seismic stations. These seismic waves, denoted *T* waves (*T*ertiary waves, compared to primary and secondary waves), travel at very low phase velocities and correspond to sound waves trapped in the oceanic water layer. The normal salinity and temperature profile of the ocean conspires to decrease the compressional velocity of seawater from 1.7 km/s at the surface to about 1.5 km/s at a depth of 800–1300 m. Below this depth the velocity increases. This low-velocity channel is known as the SOFAR (sound fixing and ranging channel), and it traps sound waves very efficiently. Sound waves that enter the SOFAR channel can bounce back and forth between the top and the bottom of the channel (beyond the critical angle), and since the attenuation of seawater is very low, the energy can travel very long distances, eventually coupling back into solid rock at ocean coastlines. For some shallow volcanic events the observed *T* waves may be larger than the *P* and *S* arrivals by a factor of 5 or more.

The multiply reflected nature of T waves results in a complex wave packet. The T phase does not have a sharp onset and may produce ringing arrivals that last longer than 2 min. They are high-frequency waves (never observed at periods larger than 2 s) and are usually monochromatic. T waves are best observed on ocean-bottom seismometers (OBS), although they are occasionally observed as converted phases at island seismic stations. These converted phases are referred as TPg, TSg, or TRg. Fig. B5.1.1 shows an example T phase. Considerable research has been done on T phases for two reasons: (1) submarine noise can generate T phases that have been observed up to 1000 km away, and (2) they are a powerful tool for discriminating between underwater nuclear explosions and natural earthquakes. In the case of nuclear explosions, the sound is injected directly into the SOFAR channel and can be 30 times larger than the P or S waves.

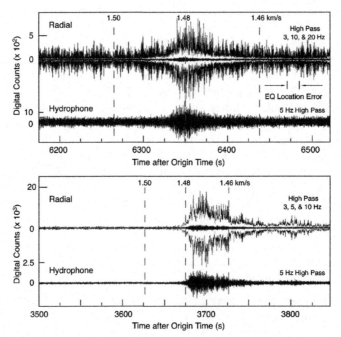

FIGURE B5.1.1 T waves recorded on the seismometer and hydrophone at H2O have an apparent velocity of about 1.48 km/ s, corresponding with the velocity near the axis of the SOFAR channel 4 km above the seafloor site. (top) Observations of an Aug. 8, 2001 (M_w 6.7) earthquake that occurred on the South Pacific Ridge, 9400 km in distance from the station. Uncertainty of apparent velocity is estimated from epicentral uncertainty. (bottom) Observations of an Oct. 8, 2001 (M_W 6.5) earthquake near the coast of Kamchatka at a distance of 5440 km. Radial components are shown in successively high-pass filtered in three stages plotted overlapping black, white, then black, respectively for low, intermediate, and highest frequency filter bands for direct comparison of amplitude and time. The hydrophone is high-pass filtered at 5 Hz (modified from Butler, 2006).

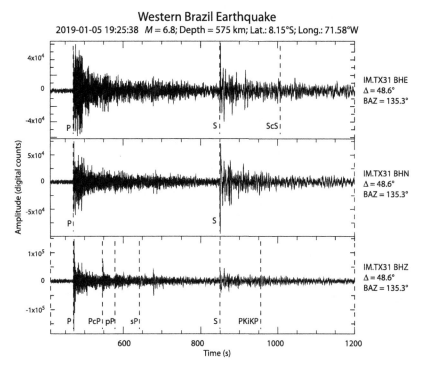

FIGURE 5.10 Filtered, band pass from 10 to 1 s, raw seismograms showing core phases from a 2019, deep (575 km), *M*6.8 earthquake in western Brazil recorded at TX31 station in the Lajitas, TX, U.S.A.

due to attenuation cause multiple reverberations to become smaller (Fig. 5.9). Amplitudes are further complicated by variability of excitation, which depends on the orientation of the seismic source. Fig. 5.10 shows a three-component recording with various phases identified, showing how the polarization of ground motion also critically influences the amplitude of individual arrivals.

Direct *P* waves that travel beyond 95° show rapidly fluctuating, regionally variable amplitudes. Beyond 100° the amplitudes decay rapidly, and short-period energy nearly disappears beyond 103°. Short-period *P* waves reappear beyond 140° but with a discontinuous travel-time branch (see Fig. 5.3). The distance range 103° < Δ < 140° is called the ***core shadow zone*** and is caused by a dramatic drop in seismic velocities that occurs going from the base of the mantle

into the core. Body waves that pass through the core have their own nomenclature. The legs of *P* waves traversing the outer core are denoted by a *K* (from Kernwellen, the German word for core). As discussed in the next chapter, the outer core is a fluid, so only *P* waves can propagate through it. Thus a *P* wave that travels to the core, traverses it, and reemerges as a *P* wave is denoted as *PKP* (or abbreviated *P'*). Similarly, it is possible to have phases *PKS*, *SKS*, and *SKP*. The leg of a *P* wave that traverses the inner core (which is solid) is denoted with an *I* (e.g., *PKIKP*); an *S* wave that traverses the inner core is written as *J* (e.g., *PKJKP*). A reflection from the inner core-outer core boundary is denoted with an *i* (e.g., *PKiKP*). Fig. 5.9 shows the raypaths for several different core phases. There is a great proliferation of phase combinations, not all of which will have significant energy.

Since the core-mantle boundary is such a strong reflector, it produces both topside (e.g., *PcP*) and bottomside (e.g., *PKKP*) reflections. *P* waves reflected once off the underside of the boundary are denoted *PKKP*, and other phases include *SKKS, SKKP,* and *PKKS*. Paths with multiple underside reflections are identified as *PmKP, SmKS*, etc., where *m* gives the number of *K* legs and *m* − 1 gives the number of underside reflections. Seismic arrays have provided observations of *P7KP*. Fig. 5.10 shows some examples of core phases. The outer core has little *P*-wave attenuation, so short-period *P* signals can be observed even for phases with long path lengths in the core. Multiple *PKP* branches can be observed at a given distance due to the spherical structure of the core and velocity gradients within it. Note the decrease in amplitude of the *P, PcP,* and *PKiKP* phases in Fig. 5.10. These results mainly from geometric spreading in the Earth and from weak reflection coefficients at different boundaries for the latter phases.

The reader should be careful not to confuse the multiplicity of seismic arrivals with complexity of the source process or with the existence of more than one initial *P* and one initial *S* spherical wavefront released from the source. First, remember that seismic rays are an artifice for tracking a three-dimensional wavefront and that wave interactions with any boundary or turning point in the Earth have frequency-dependent effects. Interactions with the Earth strongly distort the initial outgoing *P* wavefront, folding it back over on itself and begetting secondary wavefronts as energy partitions at boundaries. The body-wave nomenclature simply keeps track of the geometric complexity involved. The energy that arrives at one station as *P* may arrive at another station as *PP* with additional propagation effects. It is thus constructive to think of this as a wavefield that has been selectively sampled at different locations as a function of time rather than as discrete energy packets traveling from source to receiver. If we knew the Earth's structure exactly,

we could reverse the propagation of the entire wavefield back to the source, successfully reconstructing the initial outgoing wavefront. Of course, sources can also have significant temporal and spatial finiteness, often visualized as subevents, each giving rise to its own full set of wave arrivals that superpose to produce very complex total ground motions. Because of our imperfect knowledge of planetary structure, as described in the following chapters, there are limits to how well we can separate source and propagation effects.

5.3 Surface-waves

Surface waves generally dominate a seismogram both in amplitude and in duration, which makes them easily identifiable. However, unlike body waves with impulsive arrivals, surface wave energy is emergent in nature and dispersive, where different frequencies travel at different velocities. Even though a surface wave is energy trapped near the surface, they sample the Earth at depth because long-period (long-wavelengths) energy samples greater depths. Since the Earth's velocity structure generally increases as a function of depth, longer-period surface wave energy will usually travel faster than shorter period energy, which samples shallower structure (such as the crust) and thus arrives later on a seismogram. Because of dispersion, the duration of surface waves can last minutes to hours, and generally does not reflect the duration of an earthquake rupture.

The nomenclature for surface waves is far simpler than that for body waves. This, of course, results from the fact that all surface waves travel along the surface, and the complex interference of *P* and *S* waves that yields the surface wave is treated collectively rather than as discrete arrivals. Most of the nomenclature for surface waves is related to the frequency band of the observation. At local and regional distances, short-period (< 3 s) fundamental mode Rayleigh

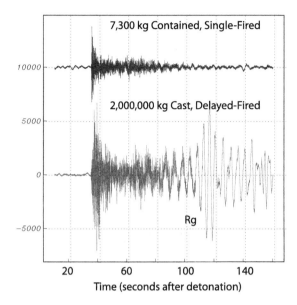

FIGURE 5.11 Comparison of vertical seismograms at near regional (200 km) distances from two mining explosions illustrating R_g excitation from different explosion configurations: 7300 kg single-fired and a 2,000,000 km cast, delayed fired explosions (modified from Stump, 2002).

waves are labeled R_g. R_g excitation is very dependent on the focal depth; if the source depth is greater than 3 km, R_g is usually absent. R_g propagation depends only on the seismic properties of the upper crust, for which most paths have an average group velocity of about 3 km/s. In most regions, R_g is rapidly attenuated, and it is rare to identify it beyond a few hundred kilometers. High-frequency overtones, or higher-mode Rayleigh waves, as well as some high-frequency Love-wave overtone energy combine to produce a phase called L_g. L_g waves have a typical group velocity of about 3.5 km/s and can be large-amplitude arrivals on all three components of motion (vertical, radial, and transverse) out to 1000 km. L_g phases are the main high-frequency arrival at regional distances in regions of thick continental crust. Fig. 5.11 shows examples of R_g, and examples of L_g can be seen in Fig. 5.5 and Fig. 5.7.

In general, Rayleigh waves with periods of 3 to 60 s are denoted R or LR, and Love waves are denoted L or LQ (the Q is for Querwellen, a German word used to describe Love waves). Very long-period surface waves are often called *mantle waves*. The periods of mantle waves exceed 60 s, with corresponding wavelengths of several hundred to about 1200 kilometers. Mantle waves from large earthquakes can reappear at a seismic station as they make a complete circuit around the globe on a great-circle path. Fig. 5.12 shows profiles of long-period ground motions recorded globally for the Jan. 23, 2018 Alaska ($M7.9$) earthquake. Love waves are polarized such that they are seen on only the horizontal transverse component, whereas Rayleigh waves are seen on both the vertical and horizontal longitudinal components. The Rayleigh waves are labeled R_1, R_2, R_3, etc., indicating wave packets traveling along the minor arc (odd numbers) or major arc (even numbers) of the great circle. R_3 is the same packet of energy as R_1, except it has traveled an additional circuit around the Earth, and R_4 is the next passage of the R_2 wave. Long-period Love waves are labeled G_1, G_2, etc. after Gutenberg. On the radial components of motion additional arrivals between R_n arrivals correspond to higher-mode Rayleigh waves, which have group velocities that differ significantly from those of the fundamental modes. These are labeled variously as O_1, O_2 or X_1, X_2, etc. The overtone wave groups are more sensitive to deeper mantle structure than are fundamental modes of comparable period. Also, a coupled Love wave can also be identified on the radial component, illustrating the complexities of surface wave propagation. In general, when using surface waves for analysis in source or propagation studies, we focus on using the vertical components for Rayleigh waves and the transverse components for Love waves. Chapter 14 focuses on the theoretical foundation of surface waves.

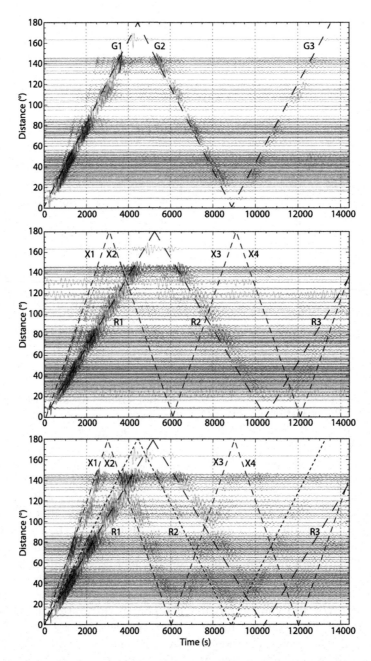

FIGURE 5.12 Profiles of (top) transverse, (middle) vertical, and (bottom) radial component low-pass filters (50 s) Global Seismic Network (GSN) seismograms for the Jan. 2018 Alaska ($M7.9$) earthquake that occurred offshore. Great-circle arrivals of Love waves (G_i; long dashed line), Rayleigh waves (R_i, long dashed line), and Rayleigh-wave overtones (X_i, short dashed line) are labeled. The Radial components show an additional arrival that is a coupled Love wave (long dashed line) (e.g., Park and Yu, 1992).

5.4 Travel-time curves

Numerous seismologists have compiled large arrival-time data sets like that shown in Fig. 5.3. Average fits to the various families of arrivals are known as *travel-time curves* or *charts*. The first widely adopted empirical travel-time curves were published by Sir Harold Jeffreys and Keith Bullen in 1940; the tabular form of these travel-time curves, called *travel-time tables*, is referred to as the *J-B* tables (Jeffreys and Bullen, 1940a). These represented painstaking data-collection efforts over the first four decades of the century, using a global array of diverse seismic stations. Careful statistical treatments were used to smooth the data so that meaningful average travel times are given by the tables. One can also use travel-time tables to calculate the ray parameter (the derivative of the travel-time curve) for a particular phase at a given distance and to calculate source depth. The J-B tables are remarkably accurate, and for teleseismic distances they can predict the travel times of principal seismic phases to within a few seconds. For a typical teleseismic *P*-wave travel time of 500 s, the tables are accurate to within a fraction of a percent of the total travel time. The J-B times are less useful at regional and upper-mantle distances, where strong heterogeneity affects times. Much of the inaccuracy in the travel-time tables comes from uncertainty in the origin time of the earthquake sources that generated the waves. In 1968 Eugene Herrin and colleagues attempted to improve the accuracy by using only well-located earthquakes and underground nuclear explosions. The resulting travel-time curves, known as the *1968 tables* (Herrin et al., 1968), improved the J-B tables slightly at teleseismic distances and more at upper-mantle distances. Kennett and Engdahl (1991) used the complete International Seismic Centre (ISC) catalogue of arrival picks to construct the most accurate, radially symmetric travel-time curves yet available, known as *iasp91*, and later updated to *ak135* (Kennett, 2005). Fig. 5.13 shows the *ak135* curve

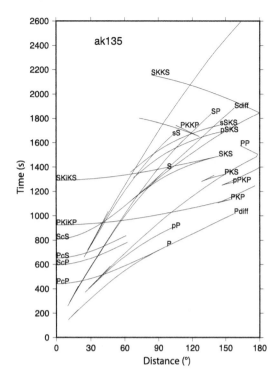

FIGURE 5.13 Travel-time curves for the empirical model ak135 for a 600-km-deep source. This model prediction indicates the arrival times of the major depth phases. The additional depth phase travel time curves add complexity relative to the surface-focus travel time curve. Phases that extend beyond 180° have travel-time curves whose times increase to the left (generated from TauP, Crotwell et al., 1999).

for a 600-km-deep seismic source. The shape of the direct *P*-wave branch in Fig. 5.13 is generally consistent with a gradual increase in velocity with depth in the mantle (see Chapter 10). On the scale of the figure, complexity of the *P*-wave branch in the distance range 15°–24° is not clear, but triplications from the transition zone are included; this complexity will be discussed in detail in the next chapter. The later branches are identified by finding paths through the Earth that are consistent with the observed times.

The details of a travel-time curve depend strongly on the depth of the source; seismic sources not at the surface have separate curves for all depth-phase branches. The depth phases are most dramatically affected, but all the travel times will change. For example, the core shadow onset is at 103° for a surface focus, but it starts at 95° for a 600-km-deep earthquake. With the advancement of instrumentation and the addition of many high-quality seismic stations globally, stacking seismograms reveals the intrica-

---

Box 5.2 Travel-time curves obtained by stacking digital seismograms

The availability of large data sets of digitally recorded seismograms makes it possible to construct "travel-time curves" without actually picking individual phase arrivals. If seismograms of many earthquakes are ordered in distance and plotted as a function of travel time, the corresponding figure is known as a *record section*. The moveout of the various phases in the record section produces coherent lineaments that correspond to travel-time branches. The coherence arises because the high-amplitude phases arrive in a systematic fashion, and therefore seismograms of similar epicentral distance will have a similar character. It is possible to sum together the seismograms of several events or event-station pairs over a small window of epicentral distances (e.g., $1° \pm 0.5°$), thus enhancing coherent signals and diminishing the amount of random noise. This is known as *stacking* a record section. Stacking seismograms directly has several problems; for example, the size of individual phases depends on the size of the event. This means that the stacked section will mostly depend on the largest events. Second, the polarity of various phases depends not only on propagation phenomena such as reflections but also on the orientation of the seismic source. In an attempt to correct for these factors, most stacked record sections actually sum seismograms that have been normalized to a reference phase amplitude, and only the *relative* amplitude of the signal is kept. When these corrected seismograms are stacked, coherent information gives a large-amplitude arrival. The stacked record section provides a travel-time curve that should be devoid of arrival-picking errors or systematic bias in picking procedures. Perhaps the biggest advantage of stacking is that some relative-amplitude information is preserved. Various phases will be strong at certain distances but very small at other distances, and this provides important information about the elastic properties of the Earth.

Shearer (1991) developed stacking procedures for global data sets and investigated the details of the upper-mantle velocity structure. Others have stacked over 33,000 traces from the IRIS FARM archive (1988–1994) to illuminate the global seismic wavefield (Astiz et al., 1996). Fig. B5.2.1 shows a stacked record section of the short-term to long-term average of long-period digital seismograms representing over 2500 earthquakes. Comparing Fig. B5.2.1 with Fig. 5.2 allows identification of the major travel-time branches (the arrival of the Rayleigh wave is marked by the strongest arrival across the section). Notice how the strength of direct *P* rapidly diminishes beyond 100°. The energy that is present is called P_{diff} and represents *P* waves diffracted along the core surface. Another advantage of using digital data to produce stacked travel-time curves is that the data contain the frequency signature of the various arrivals. If the seismograms are high-pass filtered prior to stacking, only sharp velocity boundaries are imaged. Short-period stacks typically show strong *PKKP* and *P'P'* phases; long-period stacks show *PPP* and *SSS*, which lack high frequencies due to attenuation.

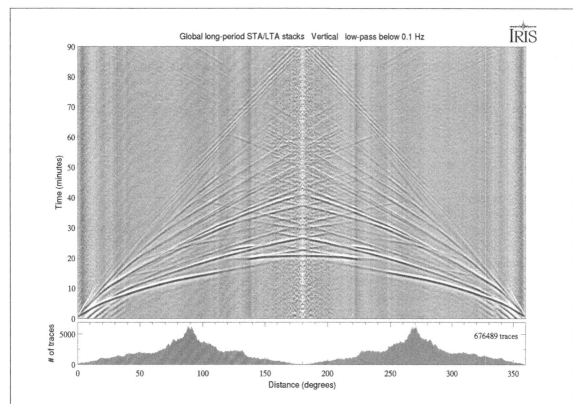

Global long-period STA/LTA stacks   Vertical   low-pass below 0.1 Hz   IRIS

FIGURE B5.2.1 Earth's travel time curve as defined by stacking global long-period (10–20 s) seismograms using the short-term to long-term average as a function of distance from the source. (From IRIS DMC (2014), Data Services Products: globalstacks Global stacks of millions of seismograms, https://doi.org/10.17611/DP/GS.1.)

cies of Earth structure through travel times (see Box 5.2.)

Many of the branches of the travel-time curve are related. For example, for a surface focus, the PP travel time can be equated to twice the travel time of P at half the distance: $t_{PP}(\Delta) = 2t_P(\Delta/2)$. Similarly, $t_{SP}(\Delta) = t_P(\Delta_1) + t_S(\Delta_2)$, where $\Delta = \Delta_1 + \Delta_2$, and the ray parameter of the P wave with $t_P(\Delta_1)$ equals that of the S wave with $t_S(\Delta_2)$. These types of simple relationships make it possible to predict the travel times of various phases and to determine a window in which to expect an arrival. These relationships also provide a tool for imaging the deviation of Earth structure from an ideal spherically symmetric velocity structure. Numerous investigators have mapped the differences between the observed and predicted times onto three-dimensional velocity models. These results are discussed more fully in later chapters.

5.5 Signal processing basics

As mentioned in the introduction, a seismogram, a function of time, comprises information about the source and propagation. Time series analysis, generally referred to as digital signal processing, deserves attention beyond the scope of this book However, to interpret seismograms, we must learn some basic concepts on time series, frequency analysis, and convolution.

5.5.1 Time representation of seismic signals

A time-invariant signal refers to signals that do not change as a function of time. Although seismograms will change as a function of distance, we can apply time-invariant, discrete (sampled), theory to signals. Digital seismograms are not *continuous* functions in the time domain, but are discretely sampled, allowing them to be analyzed on computers. In fact, they are measured values of ground displacement or velocity at regular time intervals. We refer to this as a *sampled* time series, and instead of integral transforms, we use *discrete* transforms. The principles are similar to those already developed, although there are a number of variations and limitations. It is possible to think of a digital seismic signal as a collection of N individual data points that happen to be spaced at a discrete time interval Δt.

5.5.2 Frequency-domain representation of seismic signals

In Chapter 4, we presented the concept of Fourier transforms and the equivalence of the frequency and time domains. The Fourier transform integral provides a simple procedure to convert $F(t)$ to $\hat{F}(\omega)$. If the seismic signal is an impulse ($N = 1$) with amplitude A, then the Fourier transform is given by

$$F(t) = \begin{cases} A & \text{at } t = t_1 \\ 0 & \text{elsewhere} \end{cases}$$

$$\hat{F}(\omega) = \int_{-\infty}^{\infty} F(t) e^{-i\omega t} dt = A e^{-i\omega t_1}. \quad (5.2)$$

This is the Fourier transform of a *delta function*, $\delta(t_1)$, multiplied by an amplitude A. The amplitude spectrum of this time series is $\left| A e^{-i\omega t_1} \right| = A$, and the phase spectrum is $\phi(\omega) = \omega t_1$. We can generalize this to a longer time series. If $N > 1$ and the amplitude is unity, then we can consider a discrete time series

$$S(t, \Delta t) = \begin{cases} 1 & \text{at } t = n\,\Delta t, \ n = 1, \ldots, N \\ 0 & \text{elsewhere} \end{cases}$$

$$\hat{S}(\omega) = \sum_{n=-N}^{N} e^{-i\omega n\,\Delta t}. \quad (5.3)$$

The time series $S(t, \Delta t)$ is sometimes called the *Shah* or *sampling function*. The Fourier series of $S(t, \Delta t)$ is given in terms of discrete frequencies, $\omega_m = 2\pi m/T$, by

$$S(t, \Delta t) = \sum_{m=-\infty}^{\infty} F_m e^{i\omega_m t} \quad (5.4)$$

$$F_m = \frac{1}{T} \int_{-\tau/2}^{\tau/2} S(t, \Delta t) e^{-i\omega_m t} dt. \quad (5.5)$$

For $T = \Delta t$, evaluation of (5.5) gives $F_m = 1/(\Delta t)$. Then Eq. (5.4) can be rewritten

$$S(t, \Delta t) = \sum_{n=-\infty}^{\infty} e^{i2m\pi t/\Delta t} \cdot 1/(\Delta t). \quad (5.6)$$

Eq. (5.6) is $2\pi/\Delta t$ times the inverse Fourier transform of a Shaw function in the frequency domain spaced at $(2\pi/\Delta t)$; $S(\omega; 2\pi/\Delta t)$. Thus the transform of a Shah function gives a Shaw function with sampling in time of Δt giving sampling in angular frequency of $2\pi/\Delta t$. We can use this to determine the spectrum of any sampled time signal:

$$x(t, \Delta t) = x(t) S(t, \Delta t). \quad (5.7)$$

$$\hat{x}(\omega, \Delta\omega) = \hat{x}(\omega) * (2\pi/\Delta t) S(\omega, (2\pi/\Delta t)). \quad (5.8)$$

Thus, the spectrum of the sampled time series is periodic in the frequency domain; every $2\pi/\Delta t$ the spectrum repeats itself.

The discretization of a time series introduces the concept of **bandwidth.** If $\hat{x}(\omega)$ is zero for $|\omega| > \pi/\Delta t$, then the spectra of successive frequency points, $\Delta\omega$, do not overlap. However, if $\hat{x}(\omega)$ is *nonzero* for $|\omega| > \pi/\Delta t$, the spectra of adjacent frequency points overlap, and we cannot decipher the individual contributions. This results in a phenomenon called **aliasing.** The only way we can avoid aliasing is to decrease the sampling interval, Δt, such that $\pi/\Delta t$ is a higher frequency than the highest angular frequency content of the signal. The frequency $f_n = 1/2\Delta t$ is called the **Nyquist frequency.**

When spectra are presented for digital data, the highest frequency shown is the Nyquist frequency. For IRIS broadband seismic stations, $\Delta t = 0.05$ s, so the Nyquist frequency is 10 Hz. The lowest frequency in a spectrum is given by the inverse of the length of the time window being investigated.

5.5.3 Convolution

A seismogram can be considered the output of a series of filters that represent different processes such as propagation (reflection and transmission at various boundaries), attenuation, and recording on a frequency-band-limited instrument, as discussed at the beginning of this Chapter. Each filter distorts an input signal based on prescribed rules, which can be thought of as a **transfer function**. The mathematical link between an input signal, the transfer function, and the output signal is known as a **convolution.** Mathematically the convolution is written

$$g(t) = S(t) * I(t) = \int_{-\infty}^{\infty} S(\tau) I(t-\tau) d\tau, \quad (5.9)$$

where $S(t)$ is the input signal and $I(t)$ is the filter ($*$ denotes the convolution operator). The integral is simple to understand if you think of $S(t)$

as a collection of single time point amplitudes that are passed through the filter. For the ith element in $S(t)$, the entire signal $I(t)$ is multiplied by the amplitude of the ith point. The $(i+1)$th point in $S(t)$ also serves as a multiplier of $I(t)$; this new series is shifted by dt, the spacing between the ith and $(i+1)$th point in $S(t)$. The two series are summed, and the resultant signal is the convolution. Consider an example of $S(t)$ being a delta function and $I(t)$ being a boxcar:

$S(t) = 0$ everywhere except at $t = 10$,

where $S = 1$

$I(t) = 0$ between 0 and 5, 1 between 5 and 10,

and 0 elsewhere. $\hspace{2cm}$ (5.10)

$S(t) * I(t)$ can be seen graphically in Fig. 5.14. The convolution of a time series with a delta function is just the same time series. Now consider the convolution of two boxcars (see Fig. 5.14). The resulting output function is a trapezoid.

It is far easier to perform a convolution in the frequency domain than in the time domain. It turns out that the convolution operator is just multiplication in the frequency domain:

$$F(g(t)) = F(S(t) * I(t)) = \hat{S}(\omega)\,\hat{I}(\omega). \quad (5.11)$$

This can be proved by considering the definition of the Fourier transform,

$$\int_{-x}^{x} g(t) e^{-i\omega t} dt$$

$$= \hat{g}(\omega) = \int_{-x}^{x} \left[\int_{-x}^{x} S(\tau) I(t-\tau) d\tau \right] e^{-i\omega t} dt.$$

$$(5.12)$$

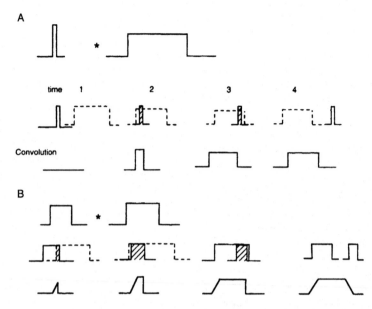

FIGURE 5.14 Graphical representation of convolution. (A) Convolution of a delta function and a boxcar gives the same boxcar. (B) Convolution of the boxcars gives a trapezoid.

Let $z = (t - \tau)$, and change the variable of integration from dt to dz; $t = (z + \tau)$, $dt = dz$:

$$\hat{g}(\omega) = \int_{-\infty}^{\infty} \left[\int_{-\infty}^{\infty} S(\tau) I(z) d\tau \right] e^{-i\omega z} e^{-i\omega \tau} dz \tag{5.13}$$

$$= \int_{-\infty}^{\infty} \left[\int_{-\infty}^{\infty} S(\tau) e^{-i\omega \tau} d\tau \right] I(z) e^{-i\omega z} dz \tag{5.14}$$

$$= \int_{-\infty}^{\infty} \hat{S}(\omega) I(z) e^{-i\omega z} dz = \hat{S}(\omega) \hat{I}(\omega). \tag{5.15}$$

5.6 Picking arrival times

One of the basic tenets of seismology is to pick the travel times and amplitudes of seismic arrivals in order to determine location, magnitude, focal mechanism, or velocity structure. Although this may sound very simple, picking a seismogram and knowing what seismic phases you are picking can be a challenge, especially in the presence of noise. Furthermore, we can different take approaches, depending if we know the source location and distance to a station. If we are trying to locate an event without knowing any information about it, we focus on picking the first arrivals as best we can (generally P waves) and use the arrival time to invert for the location, discussed in detail in the next chapter. If we know the source location, we can use travel-time curves to interpret secondary arrivals. There are other descriptions of how to pick seismograms, and in fact, many are much more descriptive than presented here. We encourage any student of seismology to seek other information and data to explore and refine seismogram interpretation skills, especially in the art of picking seismic arrivals.

The application of signal processing concepts also plays a key role in being able to analyze and interpret a seismogram. For example, a typical teleseismic P wave may have a dominant period of about 1 s, while a typical teleseismic S waves have a dominant period of around 4 s. Surface waves have higher amplitude energy from 10–20 s. Thus, to understand seismograms and to identify the characteristics needed to study both source and propagation, one must understand the basics of signal processing, Fourier Analysis, seismogram recording channels, and instrument sensitivity (covered in Chapter 4). Virtually all the figures in this book that show seismograms have applied filtering to highlight the seismic arrivals of interest. For example, Fig. 5.12 shows surface waves that have been low-passed filtered at 50 s to essentially remove body wave arrivals. High-pass filters are commonly used to highlight local to regional phases such as L_g (e.g., Fig. 5.6). One has to be cautious when applying filters, as they can distort signals in amplitude content and timing. Thus, filters can alter results for location and tomography if not taken into account.

Since surface waves are emergent, when we discuss identifying seismic arrivals, we limit this discussion to body waves. There are several basic concepts to picking a seismic arrival seismogram that involve carefully reviewing a seismogram:

- Change in amplitude
- Change in frequency
- Signal to noise

When an arrival is impulsive in nature, it can be easy to identify a change in amplitude from the background noise (Fig. 5.15). When the amplitude exceeds the background noise, we can state that the signal to noise ratio (SNR) is greater than 1, and we can pick the timing of the arrival as our arrival time. Furthermore, when working with smaller SNR signals, a change of the dominant frequency can indicate a seismic phase arrival. Filtering can also reduce the noise

levels (Fig. 5.15), making picking much easier, but recognize that there could be timing shifts due to the filter being used, as discussed above. Many times we can ignore this shift for earthquake location, as uncertainties in propagation, depth, and origin time are much greater than the error introduced by filtering. If the location of the source is not independently known, the usual procedure is first to determine an approximate epicentral distance. This usually amounts to picking the P-wave arrival time and the arrival time of either an S wave or Rayleigh wave, then comparing the measured differential time to the travel-time curve. More advanced approaches are discussed in the next chapter.

Once the earthquake location is known, travel-time curves can then be used for interpreting a seismogram and identifying phases. Computer codes exist, such as *TauP*, to compute travel times based on source depth and epicentral distance, and essentially the output serves as a guide to identifying arrivals on a seismogram. Once the distance is approximated, the travel-time curve predicts times of other arrivals such as PP and ScS, and the consistency of these predicted times with the times of the observed sequence of arrivals ensures reasonably accurate identification of the reference phases. Depth phases are sought as well, with differential times such as $pP - P$ or $sS - S$ used to determine source depth. One will not observe all phases shown on a travel-time curve on any particular seismogram because of source radiation or propagation effects. Another problem results from the extreme frequency dependence of the amplitudes of some phases. A strong short-period arrival (~ 1 s) may be absent at long periods (~ 10 s), or vice versa. Further, all three components of ground motion should be used to interpret phases. For example, SKP is much stronger than PKS on the vertical component because it emerges at the surface as a P wave with a steep incidence angle, causing the ground to move mainly vertically. Even experienced seismologists can misidentify arrivals, given the many

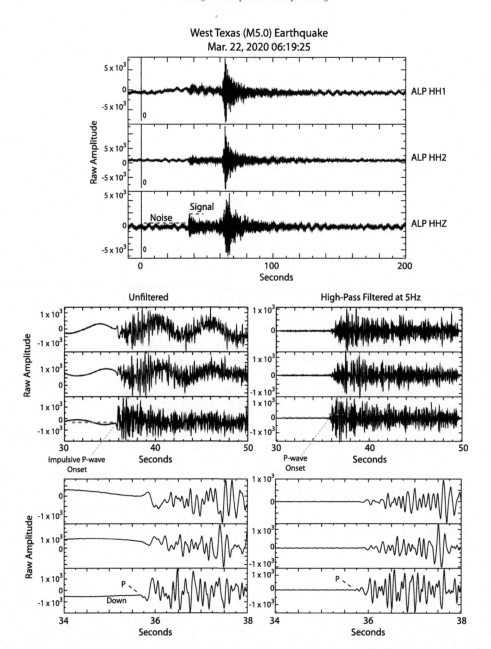

FIGURE 5.15 Three-component, regional seismograms for an earthquake (M_W 5.0) that occurred in west Texas recorded in Alpine, TX, about 217 km in distance from the epicenter. (top) Raw seismogram shows regional P and L_g seismic phases, with the signal and noise (dashed lines) also being illustrated. (left bottom) Raw and (right bottom) filtered seismograms showing the P wave in more detail. Note the reduction in noise for the filtered seismogram, along with a change in character of the onset of the P wave. To obtain a more accurate P wave arrival, it would be best practice to use even a smaller window around the P wave.

possibilities, and it is often necessary to inspect other stations that record the event to establish the travel-time moveout of the phases in question. The ISC often re-identifies phases picked by station operators who do not have accurate location estimates.

5.7 Summary

The foundation of seismology relies on the observations of the ground motion as a function of time (*seismograms*) recorded from sensitive equipment (*seismic stations*). Our knowledge of the velocity structure of the Earth and of the various types of seismic sources is the re-

sult of *interpreting* those seismograms. The more fully we quantify all of the ground motions in a seismogram, the more fully we understand the Earth's structure and its dynamic processes. Furthermore, anything that shakes the ground (e.g., cars, machinery, glacier movement, nuclear explosions, earthquakes, etc.) can be characterized on a seismogram. By interpreting seismograms, or attributes of seismograms, we can study the rupture dynamics of earthquakes and the inner workings of the Earth. The difficulty in interpreting seismograms is that they are a convolution of both the source and the propagation, and in order to study one, one has to assume the other, creating an inherent uncertainty.

6

An introduction to earthquake location

Chapter goals

- Review some of the history of earthquake location and travel time curves.
- Describe simple travel time and geometric methods used to infer earthquake location.
- Describe wave-polarization approaches used to infer earthquake location.
- Describe seismic array approaches used to infer earthquake location.
- Describe approaches to estimate earthquake location using geophysical inverse approaches.

Earthquake location has a long history. The seismoscope constructed by Zhang Heng in 132 AD (Chapter 4) was designed to provide directional information on the source of the shaking. The earliest earthquake locations were based on damage and felt reports. Since roughly the 1850's we have had reasonable global maps of earthquake-prone regions constructed from historical event descriptions. Once seismologists had access to instruments that recorded the time of shaking, wave arrival times could be used to estimate a location, provided you knew something about seismic wave speeds. This is an illustration of the fundamental coupling of earthquake and earth-structure investigations – the more you know about one, the more you can infer about the other.

Quantitatively locating seismic sources is one of the most important tasks in observational seismology. An earthquake location in space and time is specified using the latitude (or colatitude), longitude, and depth, and the time of an earthquake's *initiation*. We refer to the location as the *hypocenter* and the instant of initiation as the *origin time*, but these are not the only quantities that are used to describe an earthquake's location. The *epicenter* is the location on Earth's surface directly above the hypocenter (i.e. the latitude and longitude). For some analyses it is more convenient to use an earthquake's "geographic center" to represent its position. The *spatial centroid* is a location on the rupture surface representing the "center" of slip induced by the earthquake. For small earthquakes, which have small ruptures, the hypocenter and spatial centroid are not very far apart. But for large earthquakes, the hypocentral coordinates can be 10's to 100's of kilometers from the centroid. Knowing both quantities is essential for understanding a large earthquake.

We also commonly estimate an earthquake's *temporal centroid*, which provides information on the duration of strain release during the event (the centroid can roughly be thought of as a half duration). However, it is usually simpler to talk about duration directly. Thus, when you see the term centroid used without a spatial or temporal modifier, assume that it refers to the spatial quantity, unless the context clearly indicates otherwise. Centroid estimation requires seismogram analyses that are described else-

Foundations of Modern Global Seismology
https://doi.org/10.1016/B978-0-12-815679-7.00013-6

where (See Chapter 19). The focus of this chapter is on hypocentral and origin time estimation using seismic wave arrival times. For convenience, we will refer to a combination of a hypocenter and an origin time as a *seismic origin*.

In general, seismic origin estimation requires the identification of seismic waves and the measurement of their arrival times. To quantitatively locate an event, we must solve two problems, a forward problem and an inverse problem. The *forward problem* is the easier one – given an estimate of a seismic-source location and a seismic velocity model, calculate a travel time of wave traveling from the source to a seismic station. Adding the travel time to an estimated origin time produces an arrival-time prediction that we can compare with measurements. Earthquake location is usually posed as an *inverse problem* in which we measure the arrival times, assume an earth model, and estimate a seismic origin that is consistent with the observations and the model. Analysis of inverse problems is an essential tool for quantitative geophysical investigation. Although we explain and use key results from discrete geophysical inverse theory, an in depth treatment is beyond our scope. We encourage the reader to visit the references listed at the chapter's end for more complete and thorough descriptions.

Below, we review earthquake (and explosion) location approaches, beginning with the earliest inferences performed before data from multiple seismic stations were so readily available. Such approaches rely on understanding the simple propagation physics and polarization relationships of seismic waves. These simple methods remain worth understanding because they can provide valuable checks on more sophisticated data processing results that may fail as a result of issues with data quality. We start with methods suitable for observations from a single station and single event, then move on to methods suitable when multiple stations observe one earthquake, and conclude with methods applicable to the simultaneous location of multiple events observed at multiple stations.

6.1 Seismic arrival times

The arrival time of a seismic wave depends on the source location, x_0, and origin time, t_0, and the geology between the source and the station, which we'll represent with the seismic wave speed, c. The **travel time** is the time a wave takes to travel along its path, ℓ_0, and the **arrival time** is the sum of the travel time and the origin time. An observed arrival time, t_{obs}, which is often called a seismic wave **pick**, satisfies

$$t_{obs} = = t_0 + \int_\ell s(\ell)\,d\ell + \epsilon_{obs}, \qquad (6.1)$$

where ℓ represents the path followed by the wave from the source to the station, $s(\ell)$ is the slowness ($1/c(\ell)$) along that path, and ϵ_{obs} represents the error in the arrival-time measurement. For signals with limited bandwidth (finite-frequency signals), the wave travel path actually represents a volume of the Earth deformed by the measured wave, but we can visualize and approximate the volume with a narrow seismic ray.

The path itself depends on the source and station locations (endpoints), the physics of wave propagation, and the geologic structure the wave traverses. One tricky thing about the integral in (6.1) is that the integration path is known only approximately because we don't know the Earth's seismic wave-speed distribution with complete detail. We assume that our formulation is reasonably correct, but acknowledge that perturbations in the origin time, source location, and geologic heterogeneity (slowness variations) along the path contribute to the difference between a model-based prediction and t_{obs}. Assuming that the unknowns produce small perturbations to our initial calculation, we write

$$t_{obs} \approx t_0 + \delta t_0 + s_0\,\delta\mathbf{x_0} + \int_{\hat{\ell}} \left[\hat{s}\,(\ell) + \delta s(\ell)\right] d\ell$$

$$+ \epsilon_{obs}\,, \qquad (6.2)$$

where s_0 represents the slowness value near the source, $\hat{s}$ represents the model slowness, δs represents an unknown model slowness perturbation (the difference from the actual earth structure), δt_0 represents the origin time error, and $\delta\mathbf{x_0}$ the hypocentral location error. Splitting the integral, we have to first order,

$$t_{obs} = \underbrace{t_0 + \int_{\hat{\ell}} \hat{s}\,(\ell)\,d\ell}_{\substack{\text{Predicted} \\ \text{Time}}} + \underbrace{\int_{\hat{\ell}} \delta s(\ell)\,d\ell}_{\substack{\text{Slowness} \\ \text{Error}}} + \underbrace{\delta t_0 + s_0\,\delta\mathbf{x_0}}_{\substack{\text{Location} \\ \text{Error}}}$$

$$+ \underbrace{\epsilon_{obs}}_{\substack{\text{Pick} \\ \text{Error}}}\,. \qquad (6.3)$$

The first term is the predicted travel time and the last is the measurement uncertainty. The second term defines a seismic tomography (earth-structure estimation) problem, the third defines the earthquake location problem. The two slowness-dependent integrals are related since the path is not known perfectly. We will ignore this subtle complication under an assumption that the model is reasonably accurate. Also, since geologic heterogeneity can be large near the surface, small **station corrections** for elevation and near-surface slowness variations may be included in the travel-time prediction. Usually, station corrections are empirically estimated using the arrival times from many earthquakes and would appear in (6.3) as a station-dependent constant.

More importantly, (6.3) shows that the tomography and location problems are coupled, the more accurately location is known the better the tomographic images, the more accurately the slowness structure is known, the better the location estimates. Both problems can, and often are addressed simultaneously, but we'll focus on estimating the origin time and hypocenter of a seismic event assuming that we know the model well enough to ignore the slowness-perturbation term. Since our locations depend on our assumed earth model, it should be communicated with the seismic origins.

The mathematical goal of seismic event location is to reduce the difference between the two sides of (6.3) by modifying the origin time and hypocenter (estimating $\delta\mathbf{x_0}$ and δt_0). Our results are affected by the size of the slowness terms (the appropriateness of our velocity model) and the pick errors. Large arrival-time uncertainties or limited observations (a small number or imbalanced station distribution) can lead to imprecise locations, an inappropriate earth model can lead to inaccurate locations.

6.1.1 Seismic travel-time curves

We'll frequently make a simplifying assumption that the travel time is only a function of the source depth and the epicentral distance from the source, which corresponds to assuming a radially uniform earth model. Such models can match a body-wave travel time at great distance to within a few percent. The travel-time curves are not only useful for event location, but the character of the curve for each depth and seismic phase P, S, etc. depends seismic wave speed with depth (see Chapter 5). Modern travel-time curves are model-based. We construct an earth model consistent with observations, then compute travel times for the model. Early travel time curves were empirical and constructed by collecting, assessing, averaging, and smoothing observed travel times. The empirical curves provided important information on Earth's interior and estimating the curves was a fundamental research goal in the early to middle 20th century.

Oldham (1900) constructed some of the first global travel-time curves using earthquake recordings that included P, S, and "Long" (Rayleigh) waves observed to distances beyond 100° (leading to his discovery of the core). In the following decades, many other seismic phases were identified and studied by Dahm, Gutenberg, Lehmann, Mohorovičić, Macelwane, and others, and the first-order features of Earth's interior were mapped. To refine the models required more accurate travel times. From the 1930's to the 1960's seismologists at Saint Louis University (Father James B. Macelwane's group), Cambridge (Sir Harold Jeffreys' group) and other institutions refined travel time curves for many phases. Jeffreys, a geophysicist/astronomer, was also an accomplished mathematician who strongly defended the statistical approaches of Thomas Bayes and Pierre-Simon Laplace that are now known as Bayesian Statistics. He and his students, particularly Keith E. Bullen, developed statistical approaches for interpreting the spatially uneven and variable-quality seismic data to produce smooth travel time curves for many seismic phases. Their work is now known as the **J-B Tables** (Jeffreys and Bullen, 1940b). A revision to improve regional travel time prediction was completed in 1968 (Herrin et al., 1968), but the J-B Tables remained a mainstay of global seismology until they were replaced with model-based curves in the 1990's (Kennett and Engdahl, 1991; Kennett, 2005).

Routine, modern approaches used to estimate earthquake locations are similar to the methods developed in the early 1900's (Geiger, 1912), but we have superior data and improved earth models. Fig. 6.1 is a plot of observed P and S travel times from January 2015. Data are so plentiful that good representations of the associated travel time curves are clear from just the small data sample. More recent developments include the location of multiple events simultaneously, which can greatly improve the relative positions of the earthquakes. We will refer back to this figure several times throughout our discussion. The travel time curves, and their slopes, $dT/d\Delta$, are fundamental tools in observational seismology.

6.2 Earthquake location with information from a single station

In general, the arrival times of several seismic phases observed at multiple seismic stations are used to estimate an earthquake hypocenter and origin time. We begin with methods for using the information available from a single seismic station to infer something about an earthquake's location. Learning and practicing these approaches will help develop the skills needed for effective observational seismology. In addition, these approaches remain applicable – given the enormous cost associated with interplanetary seismometer deployment, most of our observations on other planetary bodies will likely start with solitary seismic stations.

6.2.1 Inferring seismic source properties from seismogram characteristics

The character of a seismogram communicates valuable information about the source, including its location. The frequency and duration of the shaking provide clues to the source-to-seismometer distance. Obviously, larger earthquakes can be seen at greater distance, smaller events only nearby. Small-event signals are clearer and often simpler the closer you are to the source. As we discuss seismogram characteristics for greater distances, we assume that the earthquake magnitude is large enough to be seen that far away. Precise detection thresholds at any station depend on regional geology and seismic background motion levels (Chapter 4), so we cannot specify precise event magnitude thresholds. Earthquakes as small as $M \sim 1$ are often detectable at nearby seismic stations and $M \sim 2–3$ earthquake signals are generally clear to up to 100 km. Earthquakes with $M \sim 3–4$ or larger are generally observable to 1000–3000 kilometers, those with $M \sim 4–5$ or larger are often observable beyond 3000 km. Most $M \gtrsim 5$ events are observable globally. Regular analysis of data from particular stations

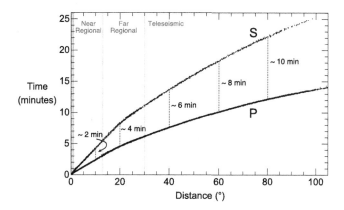

FIGURE 6.1 Observed *P* and *S* travel times extracted from the ISC catalog for the month of January 2015. Phases shown are those identified in the catalog as direct *P* or *S* arrivals. Earthquakes with depths less than 33 km were selected to produce consistent curves. The vertical gray lines divide the distance ranges into local, near-regional, far-regional or upper mantle, and teleseismic. The gray box along the left identifies the local distance range. Approximate *S*-minus-*P* travel time differences are shown at distances of $10°, 20°, 40°, 60°,$ and $80°$.

helps develop intuition about the specific observational thresholds in that region.

Local distance ($\Delta \leq 100$ km) earthquake seismograms often include clear P-wave and S-wave packages, each with a relatively short *seismic coda* – the scattered waves that follow the large direct phases and cause a gradual decay of shaking with time. Seismograms from a nearby earthquake are generally rich in short-period ($T \leq 1$ s) vibrations. The amplitude and duration of shaking produced by a larger earthquake is generally larger than that produced by a smaller event. The strongest arrivals from a nearby earthquake generally last anywhere from a few seconds to a few minutes (increasing with distance and magnitude). The time between the *S* and *P* wave arrivals, $t_S - t_P$, is less than one or two minutes.

At near-regional distances ($1° \leq \Delta \leq 13°$), the shortest period signals attenuate to levels below noise, although periods near 1 s remain clear. The signals are spread over a longer time interval because waves traveling at different speeds separate with distance. The time difference $t_S - t_P$ approaches several minutes. Longer period ($T \gtrsim 5$ s) shaking remains relatively simple and

lasts less than about 15 minutes. For shallow sources, surface waves develop within a few wavelengths distance of the source and dominate the seismograms. Deeper sources continue to produce strong and clear *P* and *S* arrivals and coda. At far-regional or upper-mantle distances ($13° \leq \Delta \leq 30°$), body-waves become complex as the waves interact strongly with mineralogical phase changes occurring in the mantle transition zone. Waves traveling through the crust continue the evolve towards lower frequency and spread their energy over longer time intervals, roughly 20–25 minutes.

At teleseismic distances ($\Delta \geq 30°$), short-period vibrations decrease in amplitude, especially the short-period *S* waves, which may not be seen for shallow sources. The duration of detectable ground motions increases beyond about 45 minutes, including an increase in $t_S - t_P$ to beyond about 5 minutes. The body-wave train expands from simple *P* and *S* wave packages to include reflections and multiple reflections off the surface and the core, some of which can be relatively large in the intermediate period band ($T \sim 5 - 50$ s). The duration of the measurable shaking can be hours following moderate-to-large magnitude earthquakes and days following large earthquakes. Deep events are notable for comparatively sharp body-wave pulses that contrast with a generally more complex character of shallow earthquakes. Large earthquakes push the dominant vibrations to longer periods (a result of longer rupture durations).

The causes for these distance varying characteristics are explored throughout this book. Experience is the best way to develop the skills needed to infer information on the source from an examination of one or many seismograms. Seismograms have never been easier to acquire. Explore them.

6.2.2 Inferring station-to-source distance & origin time using arrival times

Measurements at a single seismic station can be used to obtain a rough estimate of a seismic event location. Consider two seismic waves that travel at different speeds and assume that the temporal separation between the wave arrivals is a predictable function of distance, either from a simple theoretical approximation or empirical travel time curves (Fig. 6.1). Most often one of the waves is the first arrival (a P wave), which is easiest to measure or pick. Uncertainty is larger on secondary arrival time measurements, but this is unavoidable, and not severe enough to affect the first-order estimate that we seek. Of course, the overall signal-to-noise affects the accuracy and precision of the estimate. Using the arrival time difference also removes clock errors from the analysis, which was important early in observational seismology, when clocks were less accurate.

To construct a concrete example, consider direct P and S observations from local distance ranges and assume a simple uniform speed for each wave. Then, the measured time between the wave arrivals, Δt, is related to the distance,

D, by

$$\Delta t = t_S - t_P = \frac{D}{\beta} - \frac{D}{\alpha} \quad (6.4)$$

$$= D\left(\frac{1}{\beta} - \frac{1}{\alpha}\right) \quad (6.5)$$

where β is the S-wave speed and α is the P-wave speed. The time interval is the product of the distance and the difference in the wave slownesses. For most crustal events, using approximate values for the wave speeds, the slowness difference is roughly 1/8, and a useful rule of thumb to keep in mind is that for a nearby earthquake, $D = (t_S - t_P) \times 8$. At larger distances, travel-time tables can be used to estimate the distance from $t_S - t_P$ (Fig. 6.1). For a shallow source, the P and S time difference is roughly 2 minutes for a distance, $\Delta \sim 10°$, 4 min for $\Delta \sim 20°$, 6 min for $\Delta \sim 40°$, 8 min for $\Delta \sim 60°$, 10 min for $\Delta \sim 80°$, and just under 12 min for $\Delta \sim 100°$. Use a seismic-wave travel-time calculator or curve to be more precise, but these round numbers are useful to remember.

Everything that we just did for P and S can be applied to any two, or more than two phases on the seismogram. Comparing differential times between more than one pair of phases using travel-time curves can corroborate and improve a $t_S - t_P$ distance estimate. When paper records were the norm, travel-time charts with the same time scale as standard seismograph recordings were used to allow alignment of as many phases as possible to estimate the distance (see for example, Richter (1958)). If clearly identifiable depth phases (surface-reflections) are present, you can also estimate source depth using data from a single station.

Once we have an estimate of the distance, we can subtract the P travel time from the P arrival time to estimate the earthquake origin time. And so, with arrival-time measurements from a single seismogram, we can estimate the distance to and origin time of an earthquake. Knowing the distance means we know that the epicenter lies along a small circle with a radius equal

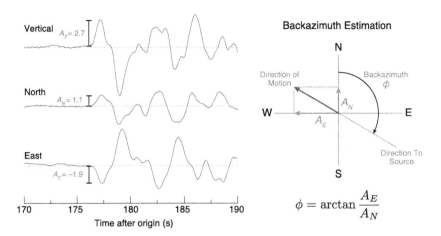

FIGURE 6.2 Example of the procedure for estimating the backazimuth, the angular direction from the station to the source, measured clockwise from north, using the three-components of P-wave ground motion. The direction is computed from the horizontal motions, but the vertical is needed to determine whether the horizontal ground motion vector points in a direction away (vertical motion up) or towards the source (vertical motion down).

to the estimated station-to-source distance centered at the station. In many instances, knowing something about where earthquakes occur commonly, the distance can give you a reasonable guess at the location (but only a guess). We can narrow the possibilities if we have seismograms recording three orthogonal components of motion.

6.2.3 Inferring station-to-source direction using ground motion polarization

If a seismic station records three-component of ground motion, we can be more specific in our single-station location. Because P waves are vertically and radially polarized, the P-wave motion can be used to infer the direction, or backazimuth, to the epicenter. In the geosciences, azimuths are directions measured by the angle clockwise from north (0° is north, 90° is east, 180° is south, 270° is west). If the vertical motion of the P wave is upward, the radial component motion of the P wave is directed away from the epicenter. If the vertical component of

the P wave is downward, the radial component is directed back toward the epicenter. Unless the event is at a back azimuth such that the horizontal P wave motion is naturally rotated onto a single component, both horizontal seismometers will record the radial component of the P wave motion. Thus, the ratio of the amplitudes on the two horizontal components can be used to find the vector projection of the P wave path to the seismic source.

An example is shown in Fig. 6.2. The data are from seismic station ANTO, Ankara, Turkey and show the ground motions produced by a M_W 5.5 earthquake that occurred on January 11, 2018 in the west-central Zagros Mountains, near the Iran-Iraq border. The station is just over 12° distance from the event and at an azimuth of −55.5° (or 304.5°). The back azimuth, the angle towards the source measured at the station and estimated using the seismograms is 150°, which is about 30° from the direction expected based on the event and station geographic coordinates ($\sim$ 117°). That level of accuracy is not great, but it provides some information on the source location.

Near surface geologic complexity at the station can make it hard to isolate P-wave motion on the seismograms – note the substantial differences in relative amplitudes between the first pair of peaks and troughs on the north and east components in the seismograms. Simple polarization analyses are less accurate at distances greater than about 20° because the P wave arrives steeply, and its horizontal component is too small to give a reliable estimate of the azimuth to the source. Similar analyses can be performed using shear waves and surface waves. For shear waves, the angles are more complicated because of the motion transverse to the direction of propagation (which is the direction from the source to the receiver). But often we can get a quick azimuth estimate from the horizontal direction of vibration of Rayleigh and Love waves. The important point is that combining an understanding wave polarization with some simple trigonometry provides a reasonable estimate of the direction towards the source.

6.3 Earthquake location with information from a seismic network

When observations from several stations are available, a location can be estimated by using P and/or S arrival times. If the event is at local distances, the two principal phases on the seismogram are P and S. Modern location methods

are based on optimizing the fit of observed P and S arrival times using predicted times computed for an earth model appropriate for the region hosting the earthquake. The same approach generalizes for regional and global applications, although the earth model may need to be adjusted. We'll begin our exploration of multiple station location using a method often described in classes for other non-science students and used in introductory geology courses.

6.3.1 Epicenter estimation with $t_S - t_P$ measurements

In our discussion of single-station location methods, we described an approach to use the S minus the P arrival time difference, $t_S - t_P$, to estimate the distance from a seismic station to the epicenter,

$$D \approx (t_S - t_P) \left(\frac{1}{\beta} - \frac{1}{\alpha} \right)^{-1}. \qquad (6.6)$$

The propagation model underlying this expression is greatly simplified (horizontal wave propagation in a uniform material), but useful for local distances, especially when the source depth is much less than the source-to-station distance. If we have multiple stations distributed around the source, each associated with a distance circle, the epicenter must be the intersection of the station-centered circles. Thus, a straight-forward generalization of the approach is to look for the

---

Box 6.1 Location using seismic array observations

A seismic array is a spatially distributed collection of sensors deployed in configuration that enables the estimation of the apparent speed and direction of approach of seismic waves. Larger aperture (the greatest distance between two stations) arrays are needed to resolve seismic waves with high apparent speeds, smaller apertures are better for waves with lower apparent speeds. Arrays and array processing approaches have been part of both exploration and earthquake seismology for many decades. But the added expense of additional stations has limited their use in academic investigations.

We use array stations to track the direction and speed with which a wave sweeps across the array. Most efficient and accurate approaches are based on computing the signal power of the sum of the array's seismograms, each shifted by an amount expected for a wave traveling across the array with a particular direction and speed. We perform the summation for a gridded range of slownesses and identify the slowness (direction and speed) that maximizes the power of the summed seismograms. Fig. B6.1.1 summarizes the signal power as a function of slowness represented by east (S_x) and north (S_y) components. Each point on the diagram corresponds to a wave sweeping across the area from a unique direction (read it like a compass) and with a specific speed (the apparent wave speed is the inverse of the distance from the slowness origin).

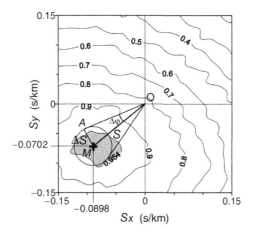

FIGURE B6.1.1 Slowness spectrum obtained by analysis of an earthquake's P waves crossing a seismic array. The region of the spectrum's largest values is shaded, and the maximum is indicated with a cross. The wave originated from the southwest. The maximum's coordinates define a P-wave slowness vector (y is north and x is east). The circle of radius ΔS defines an estimated uncertainty in the location of the spectral maximum. The associated uncertainty in the back azimuth, $\delta\phi$, is the angle subtended by $A–M$ at point O. From La Rocca et al. (2008).

The stack of the waveforms corresponding to the slowness that maximizes the signal power is called a *beam*, and the procedure is often called *beamforming*.

Directional information from arrays is generally quite good and the apparent wave speed, which is the slope of the wave's travel-time curve, can identify a wave type (P or S). For arrays at regional distances from a shallow source, the travel-time curve slope changes slowly and the apparent speed carries little information on source distance. At teleseismic distances, the travel-time curve slope changes steadily with distance, so the slowness value can be matched to the travel time curve slope to estimate the source distance. For a more complete review of seismic array methods, see Rost (2002).

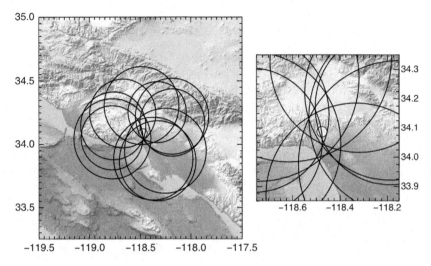

FIGURE 6.3 Station-centered $t_S - t_P$ circles for an earthquake near the coast of southern California shown at two different spatial scales. The circles intersect over a range of latitude and longitude along −118.5E and roughly between 34.0N and 34.5N. The white circles identify the optimal location estimated by the U.S. Geological Survey using P and S arrival times.

intersection of the circles (graphically or algebraically). However, because of uncertainty in the arrival-time measurements and the simplifying assumptions in the time-distance expressions used to convert $t_S - t_P$ into circles, the intersection is not always clear. Several areas of intersection may occur, or with sparse data the intersections may outline a region.

An example application of this simple approach to a 2017 M_L 3.1 earthquake that occurred near the coast of southern California is shown in Fig. 6.3. The radii of the station-centered circles were computed using $D\,[km] = (t_S - t_P) \times 8\,[km/s]$. The panel on the left shows a regional view, that indicates a location near the coast. The center of the region containing circle intersections may be taken as an approximate location, and the result is not bad for such a crude approach. But modern methods described later are used to produce a more refined estimate (and to include source depth in the problem). Finally, this method also doesn't allow us to include observations from stations that may have clear P arrival times, but not-so-clear S-wave arrival times, which is common.

We can extend the $t_S - t_P$ method to a spherical Earth by drawing geographic small (equidistant) circles around each observing station and look for the region of intersection. As it is for the local and regional distances, the global application can be imprecise, but it's not a bad back-of-the-envelope approach to check data consistency or to estimate a rough location.

6.3.2 Origin-time estimation with Wadati diagrams

The circle-based triangulation approach provides an estimate of the epicentral location, but no information on the depth or the origin time. Origin time information was removed from the data when we subtracted t_P from t_S. We can use the same information to estimate the origin time with a simple graphical technique called a *Wadati diagram*. Kiyoo Wadati (1902–1995) developed and used earthquake similar location methods to discover deep earthquakes. The seismic bands formed by these events are now called *Wadati-Benioff Zones* and later understood to be earthquakes occurring within subducting slabs. Wadati's 1928 comparison of the maximum shaking cause by shallow and deep earthquakes was also part of the inspiration that led Richter to develop his earthquake magnitude scale.

In a Wadati diagram, the time separation of the S and P phases $(t_S - t_P)$ is plotted versus the P-wave arrival time. An example constructed with the California earthquake observations (Fig. 6.3) is shown in Fig. 6.4. Since $t_S - t_P$ goes to zero at the hypocenter, a straight-line fit on the Wadati diagram gives the approximate origin time at the intercept with the P arrival time axis. The result is circled in Fig. 6.4 and the origin-time estimate from the U.S. Geological Survey obtained using more modern and sophisticated methods is shown by the vertical gray line. The Wadati diagram estimate is about 0.5 s earlier.

Wadati diagrams are not used very often these days to estimate origin times, but they can be useful to check a result. The diagrams are still used to estimate the ratio of P-wave to S-wave speeds (V_P / V_S) in a region of local seismicity. A V_P / V_S ratio can help characterize regional geology, and variations in the parameter can be related to variations in composition, temperature, and/or fluid pressures. If the Wadati-diagram data obey a linear pattern, the slope of a Wadati-diagram trend, m, is $(\alpha / \beta - 1)$. Rearranging, we have

$$V_P / V_S = \alpha / \beta = m + 1 . \tag{6.7}$$

For the example in Fig. 6.4, $\alpha / \beta \sim 1.74$, which is very close to the oft-used approximation of a Poisson solid, $\alpha / \beta = \sqrt{3} \sim 1.73$.

6.3.3 Refining locations using arrival-time residuals

Using triangulation and Wadati-diagram methods described above, we usually can produce a rough, but useful initial estimate of an earthquake's location and origin time. We can use

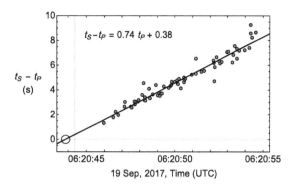

FIGURE 6.4 An example of the Wadati diagram method for estimating the origin time of a local earthquake. The origin time is given by the intercept with the P arrival time axis. The data are from an earthquake near Westwood, CA, USA a few miles northwest of Los Angeles. The thin gray lines show the zero level and the USGS estimated origin time using a larger data set and more sophisticated algorithms that include a layered earth model. The linear fit to the data is shown with the solid line and the corresponding slope and intercept are shown in the equation.

patterns in the differences in *residuals*, or between the observed and predicted (computed using the initial location estimate and an appropriate earth model) arrival times to refine the seismic origin estimate. The essential idea is that if the assumed location is incorrect, but not too far off, the residuals exhibit spatial patterns that can lead to a better location estimate. For simplicity, we'll ignore the source depth, which is the most difficult parameter to estimate precisely. We can estimate a correction to initial epicenter and origin-time by iteratively employing a four-step algorithm modified from a procedure originally described in Båth (1979).

Let the arrival time of the ith observation be represented by t_i^{obs}.

1. Using the current estimated origin, compute a distance, x_i, and a predicted arrival time, t_i^{pred}, for each station. Then compute the travel-time residuals, $t_i^{obs} - t_i^{pred}$ for each station and mean residual δt_0. Subtract δt_0, from each individual residual, and call the zero-mean values, δt_i.

2. Compute a distance correction, δx_i, for each observation by mapping δt_i to a distance using the slope of that wave's travel time curve, dT/dx_i (at the current distance),

$$\delta x_i = \frac{\delta t_i}{dT/dx_i} .$$

Different waves travel at different speeds and have different dt/dx_i values, so mapping the travel time residuals to distance corrections puts all the observations in a uniform system.

3. Plot the distance correction for each station, δx_i, as a function of the azimuth of the station (ϕ_i) from the current epicentral estimate. If the presumed epicenter is correct and the earth model used to predict the times is reasonably appropriate, then the δx_i will vary about zero. If the presumed epicenter significantly misplaced, the variation of δx_i with ϕ_i will be roughly sinusoidal.

4. An improved origin time estimate can be computed by shifting the original time by δt_0. An improved epicenter can be obtained by shifting original epicenter an amount equal to the amplitude of the distance-azimuth sinusoid, in the direction of the azimuth of the maximum in the sinusoid.

Once new epicenter and origin time estimates are computed, they can be used as a new starting location and the procedure can repeated.

The method works well when azimuthal coverage is broad. In that case, the number of observations above and below the mean is roughly the same and a reliable mean residual can be estimated. If observations are less well distributed azimuthally, it may be impossible to accurately estimate either a reliable mean residual or sinusoid amplitude. In other words, observations limited to a narrow azimuthal range introduce a possible *trade-off* between origin time and location. A trade-off is a situation where we can accommodate a change in the value of one parameter by changing the value of another.

Let's apply the approach to the 2017 southern California earthquake that we used above. We'll assume an initial location for the 2017 southern California earthquake example in the vicinity of the southernmost cluster of circle intersections in Fig. 6.3. This location is about 7.7 km south of the USGS location estimate. Following the algorithm outline above, we estimate a predicted travel time for each observed *P*-wave. The left panel of Fig. 6.5 is a map of the P-wave arrival time *residuals* (observed minus predicted time) shown centered on each station location. Symbol size is proportional to the size of the residual; gray-filled symbols correspond to stations with predicted arrival times that are too early; white-filled symbols correspond to stations with predicted arrival times that are too late. As expected, since our initial guess is south of the optimal location, the predicted *P* arrival times are too early at stations in the south, and are too late at stations in the north. The arrival time residuals range from -2.8 to 2.4 s. The unusually large residual surrounded by many small residuals to the northwest is an indication of an *outlier*, a value that deserves close examination for a picking error or timing problem at the station. Outliers are part of many data sets, and learning to spot them is a useful skill.

The right panel of Fig. 6.5 shows the distance correction pattern that arises assuming a uniform earth model and applying the procedure outline above to the 2017 California earthquake data. Each circle represents an observation at one station and the curve indicates the ideal sinusoid that we would expect if the earthquake was located at the USGS location estimate (the sinusoid amplitude is 7.7 km). Because the event is near the coast, the stations do not completely surround the earthquake, but the azimuthal coverage is sufficient to see that the suggested distance corrections form a systematic pattern. The pattern is not a perfect sinusoid because the crust in this region has some substantial heterogeneity. Positive distance corrections (initial source-to-station distances are too

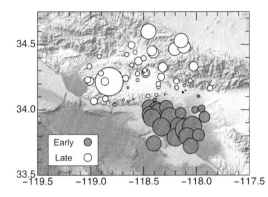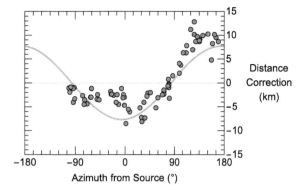

FIGURE 6.5 Spatial distribution of travel time residual values (left) and distance corrections inferred from those residuals as a function of source-to-station azimuth (right). These P-wave arrival time residual patterns are associated with a location error of about 8 km (relative to the accepted USGS event location). That's a large error for a California earthquake, but the patterns illustrate the information contained in observed seismic travel times. The sinusoidal curve shows the ideal pattern that would be observed in a uniform earth model and the erroneous location. California, of course, is not uniform, and the deviations from the curve are the fundamental data used in travel-time tomography to image the lateral variations in subsurface structure.

large) correspond to stations to the north, negative corrections correspond to stations to the south (initial source-to-station distances are too small). We can make the suggested corrections and improve our epicenter and origin time locations.

6.4 Earthquake location as an inverse problem

We finally discuss approaches that are routinely used to locate seismic events. The algorithm of the previous section includes a series of origin refinements based on a quantitative comparison of observed and predicted travel times. When we find an origin that closely matches the observations, we declare that the best location consistent with the assumed earth model or travel-time curve. In this section, we adopt a metric for closely matches, a least-squares fit of the observed and predicted times. Least-squares location methods began with the early work of Geiger (1912). Least squares methods were developed by Gauss and Laplace in the late 1700's

and early 1800's. The methods are powerful and commonly employed in quantitative analyses across many disciplines. Modern earthquake location approaches do not always rely on a least-squares, some may use an absolute-value based metric, other may use cross-correlation metrics (which are similar to least squares). Yet, the ideas are fundamental, and the concepts are worth understanding since they form the basis of many seismological analyses.

For convenience, let the seismic origin parameters be arranged in a vector,

$$\mathbf{m} = (\text{latitude, longitude, depth, origin time})^T . \tag{6.8}$$

Use the symbol F to represent a function that combines the seismic-origin, $\mathbf{m}$, with a known earth model, wave type, and station location to produce a travel-time prediction

$$t_i^{pred} = F_i(\mathbf{m}) , \tag{6.9}$$

where the subscript on F indicates the dependence of the computation on the wave type and station location corresponding to the ith observation. The location problem is to identify seismic origin parameters, $\mathbf{m}$ to satisfy the equations

$$t_i^{obs} = F_i(\mathbf{m}), \quad i = 1, \ldots, n \qquad (6.10)$$

as completely and accurately as possible. The left-hand side is a list of observations, the right-hand side is a list of forward calculations. The unknowns are the components of $\mathbf{m}$ and estimating them is an inverse problem.

A simple example can make the discussion less abstract. Consider location in a uniform half-space with wave speed, c or slowness, $s = 1/c$. Although not applicable on large scales, this simple example is a useful starting point for more sophisticated analyses. Many of the concepts it introduces generalize to problems based on depth-dependent seismic velocity models which have distance-dependent travel times. Such models remain a mainstay for routine earthquake location estimation at local to global spatial scales.

In a homogeneous half-space, the waves follow straight-line ray paths that connect the source and receiver. The travel time is the Euclidean distance multiplied by the wave slowness. Let the Cartesian coordinates of the true hypocenter and the ith seismic station be represented by (x_e, y_e, z_e) and (x_i, y_i, z_i), respectively, where x is positive north, y is positive east, and z is positive into Earth. Let t_e, t_i^{obs}, and t_i^{pred} represent the event origin time and the observed and predicted arrival times at the ith station, respectively.

For an arbitrary source location, $\mathbf{m} = (x, y, z, t)^T$, the travel time prediction function is

$$F_i(\mathbf{m}) = t + s\sqrt{(x_i - x)^2 + (y_i - y)^2 + (z_i - z)^2}. \qquad (6.11)$$

Earth model dependence is incorporated in the assumption of straight-line paths and the assumed slowness value. Station dependence is indicated by the subscript identifying the values of x_i, y_i, z_i. For the true event origin, $\mathbf{m_e} = (x_e, y_e, z_e, t_e)^T$, we have

$$t_i^{obs} \approx F_i(\mathbf{m_e})$$
$$= t_e + s\sqrt{(x_i - x_e)^2 + (y_i - y_e)^2 + (z_i - z_e)^2}. \qquad (6.12)$$

The relationship is approximate because the earth model is approximate, and the observed arrival times are subject to measurement errors. The quantities x_e, y_e, z_e, and t_e are the values that we wish to estimate. The relationship between these quantities and the observations is nonlinear. Let n_{obs} be the number of arrival times measurements. Then what we need to do is solve the system of equations

$$t_1^{obs} \approx t_e + s\sqrt{(x_1 - x_e)^2 + (y_1 - y_e)^2 + (z_1 - z_e)^2}$$

$$t_2^{obs} \approx t_e + s\sqrt{(x_2 - x_e)^2 + (y_2 - y_e)^2 + (z_2 - z_e)^2}$$

$$\vdots$$

$$t_{n_{obs}}^{obs} \approx t_e + s\sqrt{(x_n - x_e)^2 + (y_n - y_e)^2 + (z_n - z_e)^2}, \qquad (6.13)$$

to estimate the values and uncertainties of x_e, y_e, z_e, and t_e.

Nonlinear equations are generally not easy to solve, and no analytic solution to (6.13) exists. We approach the problem using an iterative method, similar to what we described to use the residual azimuthal patterns described earlier. We begin our iterative process with an initial guess of the seismic origin parameters, $\mathbf{m_0}$, and then iteratively improve our estimate using the information in the arrival-time residuals associated with each successive seismic origin estimate.

For notational convenience, organize the observed arrival times into a vector, $\mathbf{d}$,

$$\mathbf{d} = (t_1^{obs}, t_2^{obs}, \ldots, t_{n_{obs}}^{obs})^T. \qquad (6.14)$$

Then we can express ((6.10) or (6.13)) in an abstract form of a nonlinear relation between the data, $\mathbf{d}$ and model parameters, $\mathbf{m}$,

$$d_i \approx F_i(\mathbf{m}) . \tag{6.15}$$

The model *is* the relationship (6.15), the problem unknowns are the model parameters, which are contained in $\mathbf{m}$. This is an important distinction, but for brevity in the discussion below, we will refer to $\mathbf{m}$ as a model.

The next step is to linearize the nonlinear problem. We expand the nonlinear function $F_i(\mathbf{m_0})$ using a Taylor Series about the initial model, $\mathbf{m_0}$,

$$d_i \approx F_i(\mathbf{m_0}) + \frac{\partial F_i}{\partial x_0} \delta x + \frac{\partial F_i}{\partial y_0} \delta y + \frac{\partial F_i}{\partial z_0} \delta z + \frac{\partial F_i}{\partial t_0} \delta t$$
$$+ \ldots . \tag{6.16}$$

The first term on the right-hand side represents the computable predicted times corresponding to the current model estimate, $\mathbf{m_0}$. The remaining listed terms describe the sensitivity of the predicted arrival time to perturbations in the hypocentral position, δx, δy, and δz, and the origin time, δt. Specific partial derivative expressions are developed below, they are evaluated using the current estimated solution, initially $\mathbf{m_0}$. The partial derivative is essentially a measure of how sensitive the travel time is to the model parameter x_0. If the partial derivative is large, then a change in the model parameter, x, y, *etc.*, produces a large change in the predicted time.

Next we ignore the higher order terms to develop a linear approximation to $F_i(\mathbf{m_0})$ in the region surrounding $\mathbf{m_0}$. With the first-order Taylor Series expansion, we have transformed the nonlinear problem to estimate $\mathbf{m}$ directly, into a

linear problem to estimate the parameters in $\delta \mathbf{m}$.

$$d_1 \approx F_1(\mathbf{m_0}) + \frac{\partial F_1}{\partial x_0} \delta x + \frac{\partial F_1}{\partial y_0} \delta y + \frac{\partial F_1}{\partial z_0} \delta z$$
$$+ \frac{\partial F_1}{\partial t_0} \delta t$$
$$d_2 \approx F_2(\mathbf{m_0}) + \frac{\partial F_2}{\partial x_0} \delta x + \frac{\partial F_2}{\partial y_0} \delta y + \frac{\partial F_2}{\partial z_0} \delta z$$
$$+ \frac{\partial F_2}{\partial t_0} \delta t$$
$$\vdots$$
$$d_n \approx F_n(\mathbf{m_0}) + \frac{\partial F_n}{\partial x_0} \delta x + \frac{\partial F_n}{\partial y_0} \delta y + \frac{\partial F_n}{\partial z_0} \delta z$$
$$+ \frac{\partial F_n}{\partial t_0} \delta t . \tag{6.17}$$

In each equation above, the observation, d_i, is linearly proportional to each unknown, δx, δy, δz, and δt. We want to use these equations to estimate a more optimal seismic origin. Call the improved seismic origin estimate $\mathbf{m_1}$, where $\mathbf{m_1} = \mathbf{m_0} + \delta \mathbf{m}$ and $\delta \mathbf{m} = (\delta x, \delta y, \delta z, \delta t)^T$. Using only the first term of a truncated Taylor series provides a linearization, but also precludes the perturbations from immediately converging to the true $\mathbf{m}$.

If we rearrange (6.17) by moving the predicted time to the left-hand side

$$d_1 - F_1(\mathbf{m_0}) \approx \frac{\partial F_1}{\partial x_0} \delta x + \frac{\partial F_1}{\partial y_0} \delta y + \frac{\partial F_1}{\partial z_0} \delta z + \frac{\partial F_1}{\partial t_0} \delta t$$
$$d_2 - F_2(\mathbf{m_0}) \approx \frac{\partial F_2}{\partial x_0} \delta x + \frac{\partial F_2}{\partial y_0} \delta y + \frac{\partial F_2}{\partial z_0} \delta z + \frac{\partial F_2}{\partial t_0} \delta t$$
$$\vdots$$
$$d_n - F_n(\mathbf{m_0}) \approx \frac{\partial F_n}{\partial x_0} \delta x + \frac{\partial F_n}{\partial y_0} \delta y + \frac{\partial F_n}{\partial z_0} \delta z$$
$$+ \frac{\partial F_n}{\partial t_0} \delta t , \tag{6.18}$$

then the left-hand side is the travel-time residual computed using the current seismic origin estimate. Just as we used the residuals to estimate the location correction geometrically in Section 6.3.3, (6.18) allows us to use the information in the residuals to compute an origin adjustment that will improve the match of the observed and predicted arrival times.

We can compactly rewrite (6.18) using matrix-vector expressions. Let $\delta d_i = d_i - F_i(\mathbf{m_0})$ and $\delta \mathbf{d} = (\delta d_1, \delta d_2, \ldots, \delta d_n)^T$ and arrange the partial derivatives into a matrix

$$
G_{ij} = \begin{pmatrix}
\frac{\partial F_1}{\partial x_0} & \frac{\partial F_1}{\partial y_0} & \frac{\partial F_1}{\partial z_0} & \frac{\partial F_1}{\partial t_0} \\[6pt]
\frac{\partial F_2}{\partial x_0} & \frac{\partial F_2}{\partial y_0} & \frac{\partial F_2}{\partial z_0} & \frac{\partial F_2}{\partial t_0} \\[6pt]
\vdots & \vdots & \vdots & \vdots \\[6pt]
\frac{\partial F_n}{\partial x_0} & \frac{\partial F_n}{\partial y_0} & \frac{\partial F_n}{\partial z_0} & \frac{\partial F_n}{\partial t_0}
\end{pmatrix} . \tag{6.19}
$$

$\mathbf{G}$ is a summary of the problem physics: the columns of $\mathbf{G}$ quantify the sensitivity of the arrival times to each model parameter; the rows of $\mathbf{G}$ summarize the sensitivity of the *ith* observation (residual) to all the parameters. Using these matrix definitions, we can write the relationship between the residual arrival times and the origin parameters compactly,

$$
\delta \mathbf{d} = \mathbf{G} \, \delta \mathbf{m} , \tag{6.20}
$$

a form convenient to employ the powerful methods of linear algebra to estimate a model correction vector that produces an improved fit to the equations. Each row of the matrix equation corresponds to an observation.

6.4.1 A least-squares optimal location estimate

Uncertainty in the earth model, arrival-time measurement uncertainty, and approximations used to compute the predicted travel times and their partial derivatives generally make a perfect solution of (6.20) impossible. In place of an exact equality, we seek a solution that is least-squares optimal and minimizes the sum of the squared differences between both sides of the equation. Call the misfit, ϵ_2, then we seek a solution (a seismic origin) that minimizes

$$
\epsilon_2 = \sum_{i=1}^{n} |\delta d_i - (\mathbf{G} \, \delta \mathbf{m})_i|^2 . \tag{6.21}
$$

In essence we are trying to adjust the model parameters (the location and origin time) such that $\mathbf{G} \, \delta \mathbf{m}$ closely matches the residuals – in other words, just as in Section 6.3.3, we are using the information in the residuals to adjust the location parameters to produce a better overall fit between the observed and predicted arrival times. Although beyond the scope of our discussion, the solution of a system of linear algebraic equations using least-squares is described in many works on applied mathematics and geophysics. One powerful and informative approach to solve these systems of equations using the Singular Value Decomposition (SVD) is described in Press et al. (2007). We borrow some key results from the established literature on least-squares as needed, but our focus is on the form of the solution and its reliability.

To construct the least-squares optimal solution, we pre-multiply both sides of (6.20) by the transpose of $\mathbf{G}$, which we represent as $\mathbf{G}^T$. The transpose of $\mathbf{G}$ is obtained by exchanging the rows and columns of the matrix. The first column of $\mathbf{G}$ becomes the first row of $\mathbf{G}^T$, the second column becomes the second row, and so on. We then have,

$$
\mathbf{G}^T \delta \mathbf{d} = \mathbf{G}^T \mathbf{G} \, \delta \mathbf{m} . \tag{6.22}
$$

The equations represented by (6.22) are called the **normal equations** based on a geometric interpretation of least-squares problems. Ideally, we can solve the normal equations directly using the square matrix's inverse

$$
\delta \mathbf{m} = (\mathbf{G}^T \mathbf{G})^{-1} \mathbf{G}^T \delta \mathbf{d} . \tag{6.23}
$$

$\mathbf{G}^T\mathbf{G}$ is a square, symmetric, $m \times m$ matrix, which provides a number of useful properties, but does not guarantee that $(\mathbf{G}^T\mathbf{G})^{-1}$ exists. If the inverse doesn't exist, we can still solve the problem under certain conditions, but we may not be able to determine all of the model parameters. For our discussion, we'll assume that our data are sufficient to constrain the seismic origin parameters; then $(\mathbf{G}^T\mathbf{G})^{-1}$ exists and $\delta\mathbf{m}$ is least-squares optimal in the sense that it minimizes ϵ_2.

No solution is valuable unless we have at least an estimate of the model-parameter uncertainties. The model-parameter uncertainties depend on the data uncertainty and the stability of the matrix inverse, $(\mathbf{G}^T\mathbf{G})^{-1}$, which is a way of saying the uncertainties depend on the problem physics. Uncertainty can be summarized in the model-parameter *covariance matrix* that describes the model-parameter variances and potential trade-offs between parameters. For the simple least-squares solution, in the case of uniform but independent measurements with data variance is σ_d^2, the model covariance matrix, **cov m**, is

$$\mathbf{cov\,m} = \sigma_d^2 \, (\mathbf{G}^T\mathbf{G})^{-1} \,. \qquad (6.24)$$

The covariance matrix is symmetric (the inverse of a symmetric matrix is symmetric). The diagonal elements of **cov m** are the individual parameter variances (squares of the standard deviations), and the off diagonal elements are called *covariances*. In this context a covariance describes how one can maintain a good fit to the data when a change in one parameter is accommodated by a change in another parameter. The covariance between latitude and longitude is often used to draw error ellipses used to represent estimated epicentral uncertainties.

6.4.2 Halfspace arrival-time partial derivatives

To use any of the above analysis, we must compute the partial derivatives that populate the matrix $\mathbf{G}$. For a uniform halfspace, we can derive analytic expressions for the partial derivatives. For example, from (6.11), using simple calculus, we have

$$\frac{\partial F_i}{\partial x_0} = \frac{-s\,(x - x_0)}{\sqrt{(x - x_0)^2 + (y - y_0)^2 + (z - z_0)^2}} \,, \qquad (6.25)$$

and analogous expressions exist for the y_0 and z_0 derivatives. The origin time derivative is unity since a change in origin time produces a direct and equal change in the arrival time. Note that the physical units of the spatial partial derivative are those of slowness $[T/L]$, the multiplication of the derivative by a length produces a time.

We can directly use expressions such as (6.25) to construct the normal Eqs. (6.23). But since the travel time in a halfspace varies only with epicentral distance and source depth (not direction) and since the same is true for layered and spherically symmetric models, it is worth some effort to derive more intuitive expressions for the partial derivatives. For a single source depth, the travel time is only a function of epicentral distance. Then the amount an arrival time at a particular station is changed by perturbing x_0 is related to how the distance from the source to that station is changed by that x_0-perturbation. Using the symbol Δ_i to represent the epicentral distance to the ith station, we have

$$\frac{\partial F_i}{\partial x_0} = \frac{\partial T}{\partial\Delta}\bigg|_{\Delta_i} \frac{\partial\Delta_i}{\partial x_0} \,. \qquad (6.26)$$

The partial derivative is the product of the change in time with distance (the slope of the travel time curve) and the change in distance from a perturbation in the x_0.

We now rewrite our expressions in terms of two angles we can use to quantify the direction a wave leaves the source region (Fig. 6.6). The *azimuth*, ϕ_i, is the horizontal, angular direction that the wave leaves the source. By convention, azimuth is measured clockwise from

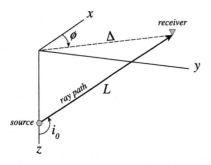

FIGURE 6.6 Simple straight-ray diagram for a halfspace earth model. The epicentral distance is Δ_i, ϕ_i is the azimuth from source to receiver, the take-off angle is i_0, and L is the path length.

north. The **take-off angle**, i_0, is the angle that the wave leaves the source measured from the vertical axis (0 is down, π is up).

Define

$$
L = \sqrt{(x_i - x_0)^2 + (y_i - y_0)^2 + (z_i - z_0)^2}
$$

$$
= \sqrt{\Delta_i{}^2 + (z_i - z_0)^2} , \tag{6.27}
$$

where

$$
\Delta_i{}^2 = (x_i - x_0)^2 + (y_i - y_0)^2 . \tag{6.28}
$$

To compute partial derivatives, we need to use the general expressions, but for convenience we'll let the epicenter be located at $(0, 0, z_0)$ and the station be located at $(x_i, y_i, 0)$, as shown in Fig. 6.6. With these definitions and the geometric relations in Fig. 6.6, the change in travel time with epicentral distance for the ith station is

$$
\left.\frac{\partial T}{\partial \Delta}\right|_{\Delta_i} = \left.\frac{\partial [L/v]}{\partial \Delta}\right|_{\Delta_i} = \left.\frac{1}{v}\frac{\partial}{\partial \Delta}\left[\sqrt{\Delta^2 + z_0^2}\right]\right|_{\Delta_i}
$$

$$
= \left.\frac{1}{v}\frac{\Delta}{\sqrt{\Delta^2 + z_0^2}}\right|_{\Delta_i}
$$

$$
= \left.\frac{\sin i_0}{v}\right|_{\Delta_i} , \tag{6.29}
$$

where we assumed that $s = 1/v$, $z_i = 0$ and $\sin i_0 = \sin(\pi - i_0)$. The quantity $dT/d\Delta$ is a hor-

izontal slowness and represents the inverse of the apparent speed of the wave observed on the halfspace surface.

Next we convert the spatial derivatives of distance to expressions involving the azimuth. From Fig. 6.6, we have $x_i = \Delta \cdot \cos \phi_i$, so

$$
\frac{\partial \Delta_i}{\partial x_0} = \frac{\partial \sqrt{(x_i - x_0)^2 + (y_i - y_0)^2}}{\partial x_0} = \frac{-(x_i - x_0)}{\Delta}
$$

$$
= \frac{-x_i}{\Delta} = -\cos \phi_i , \tag{6.30}
$$

where for the last two steps we used $x_0 = 0$. Combining (6.29) and (6.30), we have

$$
\frac{\partial F_i}{\partial x_0} = \left.\frac{\partial T}{\partial \Delta}\right|_{\Delta_i} \frac{\partial \Delta_i}{\partial x_0} = -\left.\frac{\sin i_0}{v}\right|_{\Delta_i} \cdot \cos \phi_i . \tag{6.31}
$$

Analogous relationships for the y_0 coordinate lead to

$$
\frac{\partial F_i}{\partial y_0} = \left.\frac{\partial T}{\partial \Delta}\right|_{\Delta_i} \frac{\partial \Delta_i}{\partial y_0} = -\left.\frac{\sin i_0}{v}\right|_{\Delta_i} \cdot \sin \phi_i . \tag{6.32}
$$

Changing the depth changes the ray length and travel time, but not the epicentral distance. Using a straightforward application of calculus, the vertical derivative is

$$
\frac{\partial F_i}{\partial z_0} = \frac{-s\,(z - z_0)}{\sqrt{(x - x_0)^2 + (y - y_0)^2 + (z - z_0)^2}} . \tag{6.33}
$$

Incorporating the geometry shown in Fig. 6.6 (the stations are located at $z = 0$), we have

$$
\frac{\partial F_i}{\partial z_0} = \frac{s\,z_e}{\sqrt{x^2 + y^2 + (-z_0)^2}} = \frac{s\,z_0}{\sqrt{x^2 + y^2 + z_0^2}}
$$

$$
= -\frac{\cos i_0}{v} , \tag{6.34}
$$

where we have used $\cos(\pi - i_0) = -\cos i_0$. The vertical derivative leads to a vertical slowness and represents the inverse of the apparent speed of the wave in the halfspace observed along the vertical direction. Note that for the up-going ray shown in the diagram, the quantity $\cos i_0 < 0$, so an increase in depth increases the predicted travel time, as expected.

If we introduce symbols for the horizontal and vertical slownesses, $p = \sin i_0 / v$ and $\eta = \cos i_0 / v$ then, the partial derivatives assume a compact form,

$$\frac{\partial F_i}{\partial x_0} = -p \cos \phi, \quad \frac{\partial F_i}{\partial y_0} = -p \sin \phi, \quad \frac{\partial F_i}{\partial z_0} = -\eta,$$

$$\frac{\partial F_i}{\partial t_0} = 1 . \tag{6.35}$$

Azimuth and slowness differ for each observation (slowness values depend on wave type and are distance-dependent). With these expressions, $\delta \mathbf{d} = \mathbf{G} \, \delta \mathbf{m}$ becomes

$$
\begin{pmatrix} \delta d_1 \\ \delta d_2 \\ \vdots \\ \delta d_n \end{pmatrix}
=
\begin{pmatrix}
-p_1 \cos \phi_1 & -p_1 \sin \phi_1 & -\eta_1 & 1 \\
-p_2 \cos \phi_2 & -p_2 \sin \phi_2 & -\eta_1 & 1 \\
\vdots & \vdots & \vdots & \vdots \\
-p_n \cos \phi_n & -p_n \sin \phi_n & -\eta_n & 1
\end{pmatrix}
$$

$$
\times
\begin{pmatrix} \delta x \\ \delta y \\ \delta z \\ \delta t \end{pmatrix} . \tag{6.36}
$$

A key reason to prefer these forms is that this approach generalizes to more complicated earth models that provide travel times as a function of distance (Box 6.2).

A numerical location example

We return to the 2017 southern California earthquake example to illustrate the inverse-approach. We selected 76 P-waves and 70 S-waves (the same data used earlier for the circle-based location and wadati-diagram analyses).

As noted earlier, the azimuthal distribution of stations is good, but few stations are located to the southwest in the offshore region (Fig. 6.5). The USGS source parameters for this event are origin time, 2017-09-19 06:20:44 UTC, epicentral location, 34.087N, −118.476E, and source depth, 10.5 km. The USGS solution has the benefit of a calibrated, layered earth model and the use of station corrections to account for some of the geologic heterogeneity in the region. This model and the station corrections were developed using observations from many earthquakes and improve the predicted travel times. Using a half-space, we are unlikely to produce an identical origin estimate.

A halfspace-model fit to the observed arrival times using the USGS origin is shown in Fig. 6.7. We assumed a P-wave speed of 5.7 km/s, which fits the P-observations relatively well, and a Vp/Vs ratio of 1.74, which we estimated earlier with a Wadati diagram analysis (Section 6.3.2). The corresponding shear-wave speed is 3.28 km/s. The geologic structure in the region is far from homogeneous, but the overall trends in the P and S observations are reasonably well matched (Fig. 6.7). We also need an initial set of origin parameters. Using the circle-based location analysis of Fig. 6.3, we'll assume an initial epicentral location of 34.018N and −118.468E and we'll simply guess a starting source depth of 8 km. This initial location is 7.7 km south-southeast (azimuth = 174.5°) and about 2 km above from the USGS estimate. For the initial origin time, we'll use a value that is about 0.5 s before the USGS origin time, which is what we estimated from the Wadati diagram analysis (Fig. 6.4).

We start by examining the partial derivatives that make up the $\mathbf{G}$ matrix. The partial derivatives represent the sensitivity of the observations (the arrival-time residuals) to a perturbation in each of the four origin parameters. The temporal derivative is always unity. Values of the spatial derivatives for each observed arrival time are shown in Fig. 6.8. Two important character-

---

Box 6.2 Location in a radially-symmetric sphere

For spherically symmetric earth models, we can compute the partial derivatives for a spherical location problem in an analogous approach that we used for the halfspace. We assume that we have either empirical or theoretical travel time curves as a function of distance and depth. Let λ be latitude, ϕ be longitude, and the subscript i identify a station quantity and the subscript zero identify a source quantity. Let Δ_i be the epicentral distance from the source to the ith station, ξ_i be the source-to-station azimuth, h_0 and t_0 are the source depth and origin time. The colatitude is the angle measured from the north pole so that $\theta = \pi/2 - \lambda$. Then the source-to-station epicentral distance satisfies

$$\cos \Delta_i = \cos \theta_0 \cos \theta_i + \sin \theta_0 \sin \theta_i \cos(\phi_i - \phi_0) , \qquad (B6.2.1)$$

and source-to-station azimuth satisfies

$$\sin \xi_i = \frac{\sin \theta_i \sin(\phi_i - \phi_0)}{\sin \Delta_i} . \qquad (B6.2.2)$$

The necessary epicentral partial derivatives are

$$
\begin{aligned}
\frac{\partial F_i}{\partial \theta_0} &= \frac{\partial T}{\partial \Delta}\bigg|_{\Delta_i} \frac{\partial \Delta_i}{\partial \theta_0} = \frac{\partial T}{\partial \Delta}\bigg|_{\Delta_i} \cos \xi_i \\
\frac{\partial F_i}{\partial \phi_0} &= \frac{\partial T}{\partial \Delta}\bigg|_{\Delta_i} \frac{\partial \Delta_i}{\partial \phi_0} = -\frac{\partial T}{\partial \Delta}\bigg|_{\Delta_i} \sin \theta_0 \sin \xi_i .
\end{aligned}
\qquad (B6.2.3)
$$

The quantity $\partial T/\partial \Delta$ is the slope of the travel time curve with physical units $[s/1^o]$, which we can represent using the symbol p_i. The derivative with respect to depth, $\partial T/\partial h_e$, is again the vertical slowness near the source, which we'll call η_i. The derivative with respect to origin time is again unity. The perturbation equations that form the basis of the location algorithm are

$$d_i \approx F_i(\theta_0, \phi_0, h_0, t_0) + \frac{\partial F_i}{\partial \theta_e}\delta\theta_0 + \frac{\partial F_i}{\partial \phi_0}\delta\phi_0 + \frac{\partial F_i}{\partial h_0}\delta h_0 + \delta t_0 \qquad (B6.2.4)$$

Or, in terms of residuals,

$$d_i - F_i(\theta_0, \phi_0, h_0, t_0) \approx p_i \cos \xi_i \, \delta\theta_0 - p_i \sin \theta_0 \sin \xi_i \, \delta\phi_0 + \eta_i \, \delta h_0 + \delta t_0 . \qquad (B6.2.5)$$

The procedure to locate an earthquake is the same as that described for the halfspace. Use the linearized normal equations to map the information in the residuals iteratively to construct a least-squares estimate of the seismic origin parameters. For precise work, corrections for Earth's ellipticity should be included.

---

istics of the sensitivity are immediately apparent. The location parameters are more sensitive to S-wave arrival-times than to P-wave arrival times, and distant observations are more sensitive to the epicentral coordinates than to the source depth. Enhanced S-wave sensitivity is a

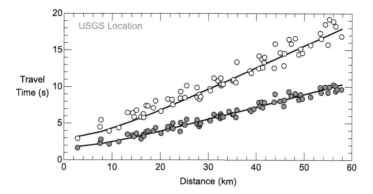

FIGURE 6.7 Observed and halfspace-predicted travel times (assuming the USGS origin time) for the 2017 southern California earthquake used in the location examples. Unshaded symbols correspond to *S*-wave arrivals, shaded symbols correspond to *P*-wave arrivals. The curves show the halfspace predicted times using the USGS location and origin time estimates.

result of the slower speed of the waves, but the increased sensitivity is balanced by the larger uncertainties associated with *S*-wave arrival time picks (see Fig. 6.8). First-arriving *P*-waves are much more easily measured and show less scatter than the corresponding *S* waves.

Depth sensitivity is largest above the source and decreases with distance. In addition, near the source, the travel time curve is less well approximated by a linear approximation, so the suggested depth correction is less reliable. At greater distances, the travel time curve is roughly a straight line and the linear approximation assumed for the Taylor Series expansion is quite good. Physically, the waves to greater distance travel roughly horizontally and the fraction of time traveling vertically is smaller and the vertical slowness, and hence sensitivity to depth, decreases. Good depth resolution generally requires a station located within 1.0–1.5 times the source depth. Without close observations, depth can be quite difficult to estimate accurately.

Our limited depth sensitivity makes estimating a good source-depth correction for this earthquake difficult. The graphical approaches we used earlier suffer the same problem, which is why we ignored the depth during our discussion of those methods. Using the normal

Eqs. (6.23) for a single iteration, the model adjustments are

$$\delta \mathbf{m} = \begin{pmatrix} 8.3 \text{ km} \\ 1.3 \text{ km} \\ 9.0 \text{ km} \\ 0.4 \text{ s} \end{pmatrix}. \qquad (6.37)$$

In words, the arrival-time residual patterns suggest that we move the source 8.3 km north, 1.3 km west, 9 km deeper, and delay the origin by 0.4 s. These epicentral adjustments move our estimate to a position less than 1 km away from the USGS estimate, and the origin-time shift results in a time 0.1 seconds difference from the USGS estimate. These are quite close to the USGS values considering our simple propagation model. However, the change in depth is problematic. Increasing the depth to 17 km would produce a rather poor match to observations at the closest stations. Extrapolating from our linearized expressions has led to an over estimate in the depth adjustment because most of the data have little sensitivity to that parameter and the linear approximation to $F(\mathbf{m_0})$ is less accurate in the region with sensitivity. Inadequacy of the linear approximation warrants that we adopt a small change in model parameters if we plan to continue the iterative estimation process.

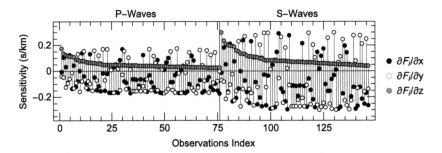

FIGURE 6.8 Partial-derivative values for the 146 arrivals used to estimate the location of the 2017 southern California earthquake. The horizontal axis identifies observation index number. The first 76 observations are *P*-wave observations, the last 70 are *S*-wave observations. A vertical line divides the two data sets. Within both data sets, the observations are sorted by distance from the initial hypocenter and all three spatial derivatives are shown using different symbol shading as indicated.

The relative difficulty estimating depth is apparent in the model-parameter covariance matrix

$$\mathbf{cov}\,\delta\mathbf{m} = \begin{pmatrix} 1.0 & 0.3 & -0.4 & 0.1 \\ 0.3 & 0.7 & 0.1 & 0.0 \\ -0.4 & 0.1 & 8.2 & -0.6 \\ 0.1 & 0.0 & -0.6 & 0.1 \end{pmatrix}.$$

(6.38)

The columns (rows) correspond to $\delta x, \delta y, \delta z$, and δt respectively. The corresponding diagonal values are the variances (square of the standard deviation). The epicentral variance estimates are roughly 1 km, that for depth is roughly eight times larger. Depth is difficult to estimate. If we perform an inversion with the depth fixed at 8 km (remove the depth dependency from the equations), the results for the epicenter are generally the same, but we end up slightly farther from the USGS location, and the origin time is shifted later than the USGS estimate. Adapting the inversion to weight the observations from closer stations more than those from the distance stations and/or using a more accurate earth model that allows us to fit the data more closely could help improve depth resolution.

The general approach described and applied above are similar to that developed more than 100 years ago by Geiger (1912). During the early decades of global seismology these computations were performed by hand – without the power of calculators or computers. Many measurements were averages to reduce the number of observations but each iteration was costly. Computational power is a less serious limitation and least-squares is no longer the only approach used to locate earthquakes. Powerful and flexible nonlinear optimization algorithms are often employed in earthquake location estimation. But many of the standard catalog locations are performed using methods similar to those we have described, and these methods illustrate a number of important concepts in data analysis and modeling that generalize to other problems in observational seismology. In addition, we located our example earthquake in isolation. Routine location of many earthquakes in a region can allow us to improve the earth model and to develop and apply station corrections that account for near-station heterogeneity. In tectonically active regions, large data sets with many thousands of arrival-time observations lead to improved locations and eventually to tomographic mapping of the variations in seismic wave speeds. These advances improve our ability to fit the data and constrain the model parameters. But there are also other ways to reduce the effects of heterogeneity on earthquake location estimates.

6.5 Relative earthquake location methods

For well-observed events, the main challenge to obtaining accurate and precise seismic origins is Earth's heterogeneity (the slowness error term in (6.3)). Models used to predict times are limited and while they produce reasonably good results, model errors affect the estimated seismic origins. Although model deviations from the actual Earth can occur along the entire source-to-station path, we can divide the errors into three types: (1) deviations near the source, (2) deviations near the station, and (3) deviations along the travel path. For a single event-station pair we cannot separate these errors. But if we focus on *clusters* of nearby events, we can use the collective information in all of the events to determine something about the errors associated with the model and to improve the relative locations of the events within the cluster. Since spatial and temporal clustering is a first-order characteristic of earthquake processes, such approaches have broad applicability.

6.5.1 Master-event methods

The simplest case, often called the "Master Event" approach is to locate the best-recorded event in the cluster as described above, and then user the residuals from that event as *path corrections* that account for heterogeneity along the path from just outside the source region to the seismic station. Our linearized equations (6.16) become

$$d_i \approx F_i(\mathbf{m_0}) + \frac{\partial F_i}{\partial x_e} \delta x + \frac{\partial F_i}{\partial y_e} \delta y + \frac{\partial F_i}{\partial z_e} \delta z$$
$$+ \frac{\partial F_i}{\partial t_e} \delta t + \delta s_i , \qquad (6.39)$$

where δs_i represents the path correction for the ith observation. For this to work the master event must have observations at all the stations for which the smaller events have data. The equations are only slightly different than what

we had before,

$$d_i - F_i(\mathbf{m_0}) - \delta s_i \approx \frac{\partial F_i}{\partial x_e} \delta x + \frac{\partial F_i}{\partial y_e} \delta y + \frac{\partial F_i}{\partial z_e} \delta z$$
$$+ \frac{\partial F_i}{\partial t_e} \delta t . \qquad (6.40)$$

All other events can be located relative to the master. The absolute location of all the events inherits the uncertainty of the master event. Uncertainty in the relative locations arises from the implicit assumption of uniform seismic structure near the sources and of course, pick errors in the measured arrival times.

6.5.2 Joint epicenter/hypocenter determination methods

We do not have to rely on a single master event to provide the path corrections, we can use the observations from all events to extract that information. We recast the original location problem to one of estimating n path corrections and m earthquake origins. Let the subscript i refer to the path (wave type / station) and j refer to the current estimate of the jth seismic origin (we are now locating more than one earthquake). Then

$$\delta d_{ij} = d_{ij} - F_i(\mathbf{m_0}_j)$$
$$\approx \frac{\partial F_i}{\partial x_j} \delta x_j + \frac{\partial F_i}{\partial y_j} \delta y_j + \frac{\partial F_i}{\partial z_j} \delta z_j + \frac{\partial F_i}{\partial t_j} \delta t_j$$
$$+ \frac{\partial F_i}{\partial s_i} \delta s_i , \qquad (6.41)$$

where δd_{ij} is the residual, or misfit, at the ith station for the jth earthquake. In matrix form, (6.41) is

$$\delta \mathbf{d} = \mathbf{G} \delta \mathbf{m} + \mathbf{S} \delta \mathbf{s}, \qquad (6.42)$$

where $\delta \mathbf{d}$ is the misfit vector with the residuals corresponding to each earthquake arranged sequentially. That is,

$$\delta \mathbf{d} = (\delta d_{1,1}, \ldots, \delta d_{1,n_1}, \ \delta d_{2,1}, \ldots, \delta d_{2,n_2},$$
$$\delta d_{k,1}, \ldots, \delta d_{k,n_k})^T , \qquad (6.43)$$

where the first index refers origin and the second to the arrival time. The jth origin includes n_j observations. The total length of $\delta \mathbf{d}$ is equal to the total number of all observed arrival times. The model correction vector includes k sets of seismic origin parameter corrections

$$\delta \mathbf{m} = (\delta m_{1,1}, \ldots, \delta m_{1,4}, \ \delta m_{2,1}, \ldots, \delta m_{2,4},$$
$$\ldots \ \delta m_{k,1}, \ \ldots, \delta m_{k,4})^T . \qquad (6.44)$$

The first index refers to the origin, the second to the four origin correction values $(\delta x_j, \delta y_j, \delta z_j, \delta t_j)$. The path correction vector, δs includes a correction for each wave type and station.

Douglas (1967) proposed the above method to estimate epicenters, (Dewey, 1972) extended the approach to the full hypocenter and origin time estimation. The approach is known as Joint Hypocentral Determination or simply JHD. Numerous authors (Herrmann et al., 1981; Pavlis and Booker, 1983; Pujol, 1988) have described efficient inversion schemes for solving (6.42). If efficiency is not an issue, the terms on the righthand side can be combined

$$\delta \mathbf{d} = (\ \mathbf{G} \ \ \mathbf{S} \ ) \begin{pmatrix} \delta \mathbf{m} \\ \delta \mathbf{s} \end{pmatrix} . \qquad (6.45)$$

In general, the submatrices $\mathbf{G}$ and $\mathbf{S}$ are relatively sparse. If the data are uncorrelated, then $\mathbf{S}$ is a diagonal identity matrix. Of course, writing the equations does not mean that we can solve them. As expressed, the equations in (6.45) are not independent and the solution is unstable. If we add a **constraint** that the sum of the adjustments to the path corrections equals zero, then we can stabilize the inversion. For example,

$$\begin{pmatrix} \delta \mathbf{d} \\ 0 \end{pmatrix} = \begin{pmatrix} \mathbf{G} & \mathbf{S} \\ \mathbf{0} & \mathbf{1} \end{pmatrix} \begin{pmatrix} \delta \mathbf{m} \\ \delta \mathbf{s} \end{pmatrix} , \qquad (6.46)$$

where $\mathbf{0}$ is a row vector of zeros of dimension equal to four times the number of events, and $\mathbf{1}$ is a row vector of ones of dimension equal to the number of path corrections. The appended equation simply expresses the requirement that the sum of the δs_i equals zero. We can formulate the normal equations for (6.46) and compute a least-squares optimal solution.

Relative locations obtained by JHD are often better than those determined by inversion using more complete and complex velocity models. The resulting locations often give a clearer picture of the seismicity. Fig. 6.9 is a comparison of hypocenters estimated by conventional inversion and by JHD. The relative locations show improved alignment and are more consistent with a reasonable geologic interpretation of the slab geometry.

6.5.3 Double-difference methods

Master event and JHD are not the only way to combine the information in arrival times from an event cluster to estimate accurate relative locations. In the double-difference method, instead of estimating station corrections, we simply remove the common path corrections by subtracting the arrival times observed at the same location for each event pair in the cluster. The result is a set of coupled equations that under the assumption of relatively uniform slowness in the source region, can constrain the relative locations of the cluster precisely.

Consider the first-order expressions for the predicted time of a single phase (e.g. P or S) observed at the ith station following two earthquakes labeled 1 and 2,

$$d_{i1} \approx F_i(\mathbf{m_1}) + \frac{\partial F_i}{\partial x_1} \delta x_1 + \frac{\partial F_i}{\partial y_1} \delta y_1 + \frac{\partial F_i}{\partial z_1} \delta z_1 + \delta t_1 ,$$
$$(6.47)$$

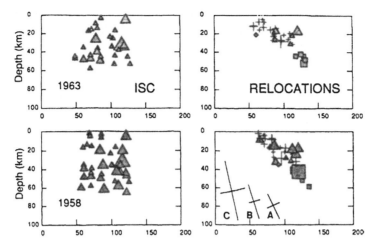

FIGURE 6.9 Comparison of earthquake locations for conventional procedures of ISC (left) and JHD relocations (right). These events are located in the Kurile subduction zone along the rupture zones of large thrust events in 1963 and 1958, and the vertical cross sections traverse the interplate thrust zone from left to right, with the slab dipping toward the right. Note that the JHD relocations reduce scatter and define a dipping plane, which is the main thrust contact. From Schwartz et al. (1989).

and

$$d_{i2} \approx F_i(\mathbf{m_2}) + \frac{\partial F_i}{\partial x_2}\delta x_2 + \frac{\partial F_i}{\partial y_2}\delta y_2 + \frac{\partial F_i}{\partial z_2}\delta z_2 + \delta t_1 .$$

(6.48)

The left-hand sides are the observations, the right-hand sides are predictions and sensitivities to seismic origin perturbations based on the current estimate of the earthquake locations. The eight perturbations, δx_1, δy_1, etc. are quantities that we would like to estimate. Now difference the equations (subtract the second from the first).

$$[d_{i1} - d_{i2}] - [F_i(\mathbf{m_1}) - F_i(\mathbf{m_2})]$$

$$\approx \frac{\partial F_i}{\partial x_1}\delta x_1 - \frac{\partial F_i}{\partial x_2}\delta x_2 +$$

$$\frac{\partial F_i}{\partial y_1}\delta y_1 - \frac{\partial F_i}{\partial y_2}\delta y_2 +$$

$$\frac{\partial F_i}{\partial z_1}\delta z_1 - \frac{\partial F_i}{\partial z_2}\delta z_2 +$$

$$\delta t_1 - \delta t_2 .$$

(6.49)

The left-hand side is a **double-difference**, the difference of the observed and predicted arrival-time differences for the two events. We can compute a double-difference for each observation common to two earthquakes, and for each pair of events in a cluster. For n_e events with n_s common seismic phases, that would lead $n_{dd} = n_e(n_e - 1)/2 \times n_s$ double-difference equations. As before, we can construct a set of linear algebraic equations

$$\delta\delta\mathbf{d} = \mathbf{G}\,\delta\mathbf{m},$$

(6.50)

where $\delta\delta\mathbf{d}$ is a n_{dd}-dimensional vector of double differences, each line of $\mathbf{G}$ represents and equation like (6.49) with appropriate indexing, and $\delta\mathbf{m}$ includes all correction terms for the n_e seismic origins. The equations contain no information on absolute locations, but provide superb relative locations as a result of the removal of unknown path effects and the interlinking of observations from multiple sources to insure internal consistency in the results. Another important benefit of this formalism is the ability to measure the double-differences by cross corre-

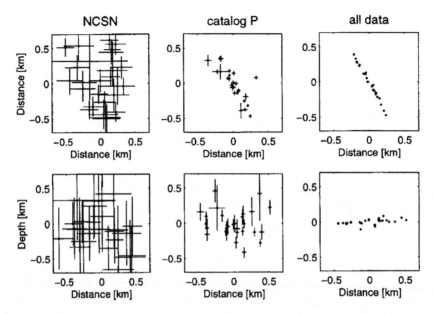

FIGURE 6.10 Relocation of 28 highly correlated events in an earthquake cluster recorded near Berkeley, CA, USA with three data sets. (left) NCSN locations, (center) Double-difference locations computed using the difference of NCSN catalog travel times for each event pair; (right) P-wave picks, P- and S-wave cross-correlation time shifts. The improvement in relative locations is dramatically improved (modified from Waldhauser, 2000).

lating waveforms for the same phase observed from two events at each station. Cross correlation can provide higher precision temporal resolution than phase picking and allows the use of signals that are impossible to pick, such as dispersed surface waves.

Illustrative inversions of 28 earthquakes observed near Berkeley, CA and relocated using double differences computed from picked times and cross-correlation measured relative time shifts are shown in Fig. 6.10. Each event is shown with a cross that represents the estimated location uncertainties. The original time-pick based locations show substantially more spread than the locations based on both absolute time picks and double differences. The narrowness of the slipping fault and the intriguing horizontal region involved in these repeating earthquakes are only revealed when information on relative origin positions is included in the inversion procedure.

6.6 Summary

Location (in time and space) of earthquakes is one of the oldest and most important problems addressed in observational seismology. Most seismic location approaches are based on arrival-time observations and require some assumptions regarding the seismic structure of Earth. Within dense seismic networks absolute location uncertainty can be reduced to a kilometer or smaller. If we sacrifice the requirement of absolute locations and instead locate clusters of events relative to one another, we can reduce the uncertainty by an order of magnitude.

We have reviewed the simplest, but still commonly used approaches to seismic origin estimation. The simple least-squares formal covariance estimates that we discussed provide a rough approximation to the true uncertainties, but often grossly underestimate the true uncertainty. Estimation of meaningful uncertainties in earthquake location remains a challenge. Important advances in direct nonlinear and probabilistic approaches that require fewer assumptions or include more realistic arrival-time and earth model uncertainties are promising. Descriptions can be found in the bibliography and suggested reading lists at the chapter's end. Understanding of the approaches described here provides a foundation for understanding more subtle and advanced approaches common in seismological research.

7

Earthquake size & descriptive earthquake statistics

Chapter goals

- Define seismic moment, a physical measure of earthquake size.
- Describe the origin of earthquake magnitude scales.
- Describe commonly used magnitude scales.
- Describe the relationship between magnitude and energy.
- Use magnitude to explore patterns in earthquake size and frequency of occurrence.
- Describe earthquake spatial and temporal aftershock patterns including magnitude.
- Discuss common earthquake catalogs.

As seismometer design and manufacture improved and standardized instruments became available, the goal of quantitatively comparing earthquakes came within reach. Much of the earthquake rupture process remained unclear (the elastodynamics of faulting were not yet fully developed) but seismogram-based measurements could be compared, combined, and cataloged. An obvious quantity to investigate was earthquake size and if possible, estimates of earthquake energy. More quantitative descriptions of earthquake geography and frequency were also important. Large events were obviously less frequent than small events, and after-

shock sequences were long noted and investigated in terms of felt-event numbers. But with an ability to measure seismic-wave amplitudes, earthquake size could be quantified and combined with estimated locations (see Chapter 6) to better explore spatial and temporal earthquake patterns.

As mentioned in Chapter 1, the modern, preferred tectonic measure of an earthquake size is the seismic moment, M_0,

$$M_0 = \mu \, A \, \bar{D} \, , \qquad (7.1)$$

where μ, the elastic shear modulus, is a measure of the stiffness of rock surrounding the ruptured portion of the fault, A is the area of the portion of the fault that ruptured, and $\bar{D}$ is the average slip over the rupture surface. Seismic moment has the intuitive interpretation that earthquake size is a product of a measure of the rupture-adjacent-rock's ability to store energy, the area of the fault that moved, and the distance the rocks moved (one side relative to the other). The units of seismic moment are *N-m* or *dyne-cm*, both of which represent a force times a distance (the lever arm). Energy has the same fundamental units ($M L^2 T^{-2}$) but we distinguish these physically different quantities by using *N-m* for the moment and *joules* for the energy. From a dimensional-analysis viewpoint, one would expect a unit-free scale factor (possibly a combina-

197

tion of physical parameters) to relate energy and moment.

Seismic moment is a static measure of an earthquake's size, a fundamental earthquake property central to any description of a seismic event, and a quantity relatively well-constrained using modern seismic observations. Although the ground motion amplitude is directly proportional to M_0, moment is not the only source characteristic that influences seismic shaking intensity. Ground-motion amplitude and duration also depend on dynamic properties that are not included in the seismic moment. The short-period spectral content of shaking, including periods that damage human-made structures, trigger liquefaction and landslides, etc., are strongly influenced by the spatial distribution and temporal rate of slip accumulation during an earthquake, as well as the speed and complexity of the earthquake's rupture propagation. Source rupture and slip characteristics are represented in the seismic source time-function, $s(t)$, or its Fourier Transform, the source spectrum, $S(\omega)$. The concepts of seismic moment and source time functions arose from simple earthquake rupture models developed after magnitudes had begun to be measured, compiled, and analyzed. The key point is that magnitude is a mixture of both the static and dynamic earthquake properties, which contributes to its value, but also makes detailed interpretations of magnitude subtle and complex.

7.1 The energy in seismic waves

For hazards and engineering, an ideal measure of earthquake size would be the energy transported by seismic waves, E_S, as well as the frequency and duration of shaking. We can quantify the energy carried by a seismic wave by considering the history of a particle as it responds to a transient deformation. As a wave passes, the particle, which has an elastic potential energy by virtue of its connection to nearby

particles, moves, and movement embodies kinetic energy. The sum of the potential and kinetic energies integrated over time equals the energy expended moving the ground.

As an example, consider a location in the ground situated directly above a monochromatic seismic energy source. The displacement of the ground, $x(t)$, at that position can be represented using

$$x(t) = A \cos\left(\frac{2\pi t}{T}\right), \tag{7.2}$$

where A is the wave amplitude and the period is T. The ground velocity is

$$\dot{x}(t) = -\left(\frac{2\pi A}{T}\right) \sin\left(\frac{2\pi t}{T}\right). \tag{7.3}$$

The kinetic energy of a unit mass at the location of interest is given by $\frac{1}{2}\rho \dot{x}^2(t)$. If we average this quantity over a complete deformation cycle, we obtain the **kinetic energy density**, $e(\rho, A, T)$,

$$e(\rho, A, T) = \frac{1}{2}\frac{\rho}{T} \int_0^T \dot{x}^2(t)\, dt$$

$$= \left(\frac{\rho}{2T}\right)\left(\frac{2\pi A}{T}\right)^2 \int_0^T \sin^2\left(\frac{2\pi t}{T}\right) dt$$

$$= \rho\, \pi^2 \frac{A^2}{T^2}. \tag{7.4}$$

The energy density is proportional to the wave amplitude squared and the frequency squared ($f = 1/T$). Thus for the same vibration amplitude, a higher-frequency signal carries more energy than a lower-frequency signal. Over the course of a cycle of deformation, the mean potential and kinetic energies are equal, so $e_s(\rho, A, T) = 2\, e(\rho, A, T)$.

Now imagine that we have a spherical wave source radiating energy in all directions about the source. The wave amplitude decreases as the wave propagates as a result of the expansion of the wavefront, and anelastic attenuation. If we integrate our expression for the seismic-wave energy density at one location over an entire spherical wavefront and include the wave attenuation effects, we obtain an expression of the form

$$E_s = F(r, \rho, c) \times \left(\frac{A}{T} \right)^2, \qquad (7.5)$$

where E_S is the energy radiated by the source as seismic waves, r is the distance the wave has traveled, ρ is the density near the wavefront, and c is the wave speed. The logarithm of the energy is proportional to the logarithm of amplitude over period, A over T:

$$\log E_s = 2 \log \left(\frac{A}{T} \right) + \log F(r, \rho, c). \qquad (7.6)$$

We represent the common, base-ten, logarithm as log, and will use ln for the natural logarithm. For a spherical wave propagating in a uniform medium, a functional form of $F(r, \rho, c)$ can be derived, but the form is more complicated for the Earth. We could relate seismic energy to seismic wave amplitude and period directly using the above equation if $F(r, \rho, c)$ were known.

As detailed below, Richter and others defined magnitude scales to include $\log(A/T)$ at least partly because the seismic-wave energy was proportional to $\log(A/T)$, and because amplitude and period could be measured from seismograms. This was a good start and the idea worked better than expected. But construction of a detailed relationship between magnitude and energy remained a challenge because the energy in the waves also depended on unknown source physics and amplitudes depended on propagation effects. Relating magnitude to energy was approached empirically. Gutenberg and Richter (1956) estimated a form for $F(r, \rho, c)$ and showed that observed values and the early empirical energy-magnitude relationships were

reasonable. But as with all things magnitude, relationships are reasonably clear for first-order comparisons, but the use of the quantities for precise work requires a deep understanding of the problem.

7.2 Earthquake magnitude scales

The concept of *earthquake magnitude*, a relative-size scale based on measurements of seismic wave amplitudes, was developed by Kiyoo Wadati and Charles Richter in the 1930s, more than 30 years before the first seismic moment was estimated in 1964. Even with paper records, ground-motion amplitude measurements could be made efficiently. However, traditional magnitudes are limited because the character of seismic waves is strongly influenced by fault rupture details. Earthquakes with similar seismic moments may have very different source-time functions, which leads to different short-period seismic-wave amplitudes. Even in the simplest cases, the duration of rupture can produce frequency-dependent effects that influence wave amplitudes in systematic ways that may be difficult to recognize without lengthy analysis. Despite these limitations, earthquake-size estimates based on wave amplitude measurements remain useful because of their simplicity and long history. Short-period magnitudes are also important because high-frequency shaking in a narrow frequency band is often responsible for earthquake-related damage.

In the 1930's, Richter was investigating earthquake processes in southern California and collecting observations from a seismic network outfitted with Wood-Anderson seismic sensors that had a standardized instrument response. A uniform seismometer sensitivity is important if you want to compare quantitatively the ground motions from earthquakes at multiple seismic stations. When comparing measurements, Richter noticed that for individual earthquakes, the logarithm of the maximum amplitude observed

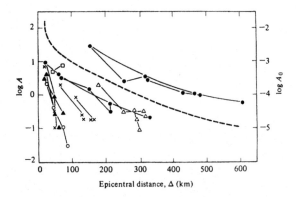

Epicentral distance, Δ (km)

FIGURE 7.1 Origin of the local magnitude scale, based on the systematic decrease of seismic amplitudes with epicentral distance. Different symbols identify amplitudes measured from different events. The data are for Southern California earthquakes in January, 1932. The dashed line shows the formula developed by Richter to match the general trend of the observations (from Richter, 1958).

on a Wood-Anderson seismogram seemed to follow a systematic pattern of decreasing amplitude with increasing epicentral (source-to-station) distance (Fig. 7.1). The amplitude patterns suggested that a single number could, to first order, characterize earthquake size, and led Richter to develop the first magnitude scale.

Magnitude scales are based on two simple assumptions: (1) given the same source-receiver geometry and two earthquakes of different size, the "larger" event will typically produce larger-amplitude arrivals; (2) the effects of source depth and geometric spreading and attenuation can be removed from the amplitude measurements. The general form of all magnitude-scale formulas is

$$M = \log(A/T) + f(\Delta, h) + C_s + C_r, \qquad (7.7)$$

where A is the absolute maximum ground displacement of the seismic wave on which the scale is based, T is the period of the signal, f is a correction for epicentral distance (Δ) and focal depth (h), C_s is a correction for the siting of a station (e.g., variability in amplification due to local rock type), and C_r is a regional correction. A log-

arithmic scale is used because the seismic-wave amplitudes of earthquakes vary enormously. A unit increase in magnitude corresponds to a 10-fold increase in ground displacement. Magnitudes are averages of estimates computed from multiple stations to overcome amplitude biases caused by source propagation effects not accounted for in the simple formula: radiation pattern, directivity, and anomalous path properties. Four primary magnitude scales are in use today: M_L, m_b, M_S, and M_W.

7.2.1 Local magnitude (M_L)

Richter's first seismic magnitude scale was motivated by his goal to issue the first catalogue of California earthquakes. This catalogue contained several hundred events, whose size ranged from barely perceptible to large, and Richter felt that an earthquake description should include a more objective size measurement to communicate the event's significance. As described above, he had observed that for measurements from uniform instruments, the logarithm of maximum ground motion for several earthquakes decayed with distance along parallel curves (Fig. 7.1). He surmised that relative size of events could be calculated by comparing event amplitudes with amplitudes corresponding to a *reference event*:

$$\log A - \log A_0 = M_L, \qquad (7.8)$$

where A and A_0 are the displacements of the earthquake and the reference event respectively, both observed at a prescribed distance. Richter chose his reference earthquake, with $M_L = 0$, such that A was $10^{-6}\,m$ at an epicentral distance of 100 km. Using Richter's reference event definition and amplitude correction curve ($\log A_0$), we can approximate Eq. (7.8) using

$$M_L = \begin{cases} \log(A) - 0.27 + 1.65 \log \Delta \, , \\ \quad 0 \le \Delta \le 200 \, \text{km} \\ \\ \log(A) - 3.55 + 3.07 \log \Delta \, , \\ \quad 200 < \Delta \le 600 \, \text{km} \end{cases} \quad (7.9)$$

where A is the amplitude in millimeters. At first glance Eq. (7.9) is not in the form of (7.7), but Richter made a number of restrictions that can be factored out of (7.7). First, all the instruments used were narrowband and identical, and thus the maximum-amplitude was always of a single dominant period, $T \sim 0.8 \, s$. Second, all the seismicity was shallow (less than 15 km deep), and the travel paths were confined to southern California. Thus the corrections for regional dependence and focal depth are approximately constant, and Eq. (7.9) is a simplified form of (7.7). The scale is not without issues however, for example, the original assumed instrument gain is now believed to have been overestimated.

Earthquakes with $M_L \le 2.5$ are called *microearthquakes* and are rarely felt. The smallest events that are recorded have magnitudes less than zero, and the largest M_L recorded is about 7, which corresponds to seven orders of magnitude in ground displacement. In practice, M_L is usually a measure of the regional-distance S wave. The magnitude for each of the horizontal components is averaged in a least-squares sense to give an M_L for a given station. M_L may vary considerably from station to station, due not only to station corrections but also to variability in the S-wave radiation pattern. The values of M_L from all stations are averaged to give the "magnitude."

M_L in its original form is rarely used today because Wood-Anderson torsion instruments are uncommon and, of course, because most earthquakes do not occur in southern California. Modern data can be analyzed by digitally simulating the response of the Wood-Anderson instrument, but doing so has led to the recognition that both the period and the gains were slightly different from those assumed. The discrepancies are not important for a first-order use of M_L values to categorize earthquakes, but subtle seismological analyses should account for details. The modern version of M_L for a region with seismic wave attenuation characteristics similar to California, recommended by the International Association of Seismology and Physics of the Earth's Interior (IASPEI), one of the eight Associations that comprise the International Union of Geodesy and Geophysics (IUGG), is

$$M_L = \log A + 1.11 \log R + 0.00189 \, R + 0.91 \, , \quad (7.10)$$

where R is the hypocentral distance ($R = \sqrt{h^2 + \Delta^2}$) where h is the hypocentral depth (both in km). A is the amplitude in microns observed on a horizontal-component seismogram filtered to replicate a Wood-Anderson standard seismograph with unit magnification. The standard Wood-Anderson short-period instrument had a magnification thought to equal 2800, but which was later estimated to equal 2080, see Borrmann and Dewey (2014). For regions with different attenuation characteristics, a similar form is recommended, but calibrated to the region. M_L remains a very important magnitude scale because it was the first widely used earthquake "size measure," and all other magnitude scales are to some extent calibrated to M_L. In addition, M_L remains a very useful scale for engineering. Many structures have natural periods close to that of a Wood-Anderson instrument (0.8 s), and the extent of earthquake damage is closely related to M_L.

7.2.2 Body-wave magnitude

Local magnitude works fine at local and near-regional distances, but the limitations imposed by instrument type and the distance correction make it impractical for global characterization of earthquake size. Gutenberg measured magnitudes (m_B) of distant or teleseismic, earthquakes using the intermediate-period (roughly 5 s to 20 s) period body-wave train that included P, PP, and S. The use of the largest observed phase helped reduce the effects of radiation pattern and the use of PP extended the distance range to distances for which direct P does not exist (beyond the P and S core shadow zones). A modern intermediate-period body-wave magnitude, $m_B(BB)$, has been defined and utilized more recently. The formula is

$$m_B(BB) = \log \frac{V_{max}}{2\pi} + Q(h, \Delta), \qquad (7.11)$$

where V_{max} is the largest amplitude in an absolute sense in a ground velocity seismogram (microns/s) observed in the P-wave train (largest amplitude before PP arrives). The distance and depth correction factor, $Q(h, \Delta)$ is described below and is the same correction used for the short-period body-wave magnitude, m_b. Although the period is not explicit in the $m_B(BB)$ expression because it is based directly on ground velocity, measurement of the period is encouraged because it contributes useful information on the spectral content of the maximum ground velocity.

In the early 1960's, with the establishment of the Worldwide-Standard Seismographic Network (WWSSN), which included standardized short- and long-period seismometers, it became convenient to define **body-wave magnitude** based on the amplitude of short-period teleseismic P waves. The short-period observations were less noisy than longer-period observations and the change led to a significant increase in the number of small events that could be quantified using teleseismic observations. The move to shorter periods also enabled incorporation of measurements from short-period seismic arrays that were established around this time. These advantages were particularly important for underground nuclear explosion monitoring, which was a primary reason the WWSSN network was established. The resulting body-wave magnitude is represented using the symbol m_b. Tradi-

tionally, this magnitude is based on amplitudes measured in the first few cycles of the P-wave arrival and is computed using

$$m_b = \log(A/T) + Q(h, \Delta), \qquad (7.12)$$

where A is the ground-motion amplitude in micrometers and T is the corresponding period in seconds. The reason for using several cycles of the P wave is that the effects of radiation pattern and wave interactions with Earth's surface for shallow events can result in a complicated waveform. In practice, the period at which m_b is usually measured is 1 s (historically, the WWSSN and many regional network instruments had a response "peaked" near 1 Hz). For a given event, scatter of the order of ± 0.3 m_b units is not unusual, so robust estimates require extensive averaging.

The correction for distance and depth, $Q(h, \Delta)$, is empirical. Fig. 7.2 includes plots of $Q(h, \Delta)$ correction surfaces from GR, Gutenberg and Richter (1956), and VC, Veith and Clawson (1972). Corrections are often shown as contour plots, but here we show them using distance curves for a set of sampled source depths. Each row of the plot represents the amount added to $\log(A/T)$ to produce a magnitude for the corresponding source depth. The height of each bar represents the size of the correction; a bar with zero height would correspond to a m_b 5.0 earthquake. The GR corrections are used by the USGS and other agencies, the VC corrections are used by the Comprehensive Test Ban Treaty Organization's (CTBTO) International Data Center (IDC). Corrections are fairly uniform in the 30° to 90° degree range, but are complicated at closer, upper-mantle distances, which reflects the complexity of the body waves in this epicentral distance range. The correction dramatically *decreases* at 20° because P-wave interactions with the mantle transition zone structure can result in very large amplitude arrivals. The corrections dramatically increase near 100° because diffraction along the core dramatically reduces

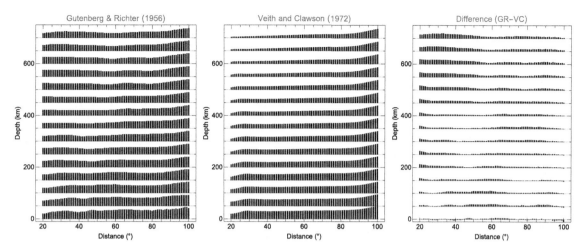

FIGURE 7.2 The correction $Q(h, \Delta)$, Gutenberg and Richter (1956), and $P(h, \Delta)$, Veith and Clawson (1972), that are applied to determine the body-wave magnitude (computed to adjust observations for an m_b 5 reference event). The third panel shows the difference between the others and all plots are on the same scale. The GR corrections remain in use by the USGS and other agencies, the VC corrections are used by the CTBTO IDC. Note that in the teleseismic range (30 to 90 degrees), the curves vary relatively smoothly. The steep increase beyond 100 degrees is an attempt to correct for the loss of short-period energy as the wave diffracts along the core-mantle boundary. The complexities in regional distances are related to upper mantle complexity with both depth and geographic location within the shallow Earth.

short-period P amplitudes. The VC corrections are smoother than the GR values. For shallow events, the corrections are similar, but include differences up to 0.25 magnitude units. For deeper sources, especially at regional distances, the differences are as great as one magnitude unit. The use of different correction surfaces can lead to systematic differences in magnitude, but correction surface issues are not the only complication of working with m_b. Differences in measurement practice, including the instrument/filter response used to isolate the short-period signals on a digital seismogram, and the length of time over which the maximum amplitude is measured also affect estimated m_b. The IDC's primary interest is in nuclear explosions, so they restrict the m_b measurement window to the first five seconds of the P-waveform. This is perfectly acceptable for explosions, which release their energy very quickly. However, large earthquake rupture durations are much longer than five seconds, and the largest amplitude

may arrive much later. The USGS and others who monitor earthquakes use a time window that extends as late as the arrival of PP or 60 seconds to capture large amplitude P waves. For large earthquakes, the time-window difference is significant and the IDC m_b will be a lower bound on the USGS measurements. Perhaps the most important take-away is that although magnitude is viewed as a simple metric, the details are rather complex and care is needed to work with magnitudes in a detailed, quantitative way. For most applications, where magnitude represents a rough indicator of earthquake size, these issues are less important.

7.2.3 Surface-wave magnitude (M_S)

Beyond about 600 km, the long-period seismograms of shallow earthquakes are dominated by surface waves, usually with a period of approximately 20 s. Earth's crustal structure is such that Rayleigh waves with periods near 20 s (roughly 18 to 22 s, sometimes called the crustal Airy phase) propagate efficiently, which makes them a good choice for measuring earthquake size. However, surface-wave amplitudes decrease with the source depth and rather than try to correct the surface-wave amplitude for depth, we use surface waves only to compare shallow earthquake size using *surface-wave magnitude*.

In modern notation, the classic M_S is represented as $M_S(20)$, that is, the magnitude determined from the amplitude measured at 20 s. Since surface-wave amplitudes depend on distance differently than those of body waves, we have a different form for the distance corrections, and in fact, a number of different formulas have been proposed. The equation for computing surface-wave magnitude is

$$M_S(20) = \log A/T + 1.66 \log \Delta + 3.3, \quad (7.13)$$

where A is the amplitude in micrometers (or microns) of Rayleigh wave observations with a period close to 20 s. In general, the amplitude of the Rayleigh wave on the vertical component is used for the distance range $20° \leq \Delta \leq 160°$, but this has not always been the case. M_S has long been used as a better metric of the size of larger earthquakes than m_b, and the energy relationship between M_S and seismic energy was used in the definition of moment magnitude, M_W. So $M_S(20)$ remains an important parameter for many historic and modern earthquakes. The existence of modern digital data offers opportunities to produce more robust and consistent metrics. The broad-band M_S recommended by IASPEI has the form

$$M_S(BB) = \log \frac{V_{max}}{2\pi} + 1.66 \log \Delta + 3.3, \quad (7.14)$$

where V_{max} is the maximum absolute ground velocity in microns/s within the vertical component Rayleigh-wave train within a period range of $3\,s \leq T \leq 60\,s$ and applicable across the distance range $2° \leq \Delta \leq 160°$.

These are not the only formulas used to estimate surface-wave magnitude. A similar, but revised, distance correction is used by groups that monitor underground nuclear explosions. As was the case with m_b, changes to magnitude scales such as M_S can provide more consistent metrics for earthquake size suitable to application with modern digital observations, but they also introduce a time-variable complexity that requires earthquake scientists to be aware of the subtle but systematic differences between modern and historic magnitude estimates.

7.2.4 Other magnitude scales

The total number of magnitudes is substantial and we cannot review them all. Small earthquakes recorded by local networks at short distances have been categorized using the duration of the coda following the S-wave. The practice allows estimates of nearby earthquakes, even when the amplitude is driven off-scale or clipped by the instrument. The more qualitative seismic intensity measurements have been calibrated with seismogram-based magnitudes so that the maximum intensity can provide an estimate of the size of large historical earthquakes. The magnitude of large historic tsunamigenic earthquakes have been measured using tsunami effects calibrated by modern measurements of magnitude and modern tsunami.

Regional magnitude, $m_b(Lg)$

In large more stable regions of the continents such as eastern North America, which for many years hosted fewer seismic stations so that the local data were less common, M_L has been replaced with $m_b(Lg)$, a measure based on the regional-distance L_g arrival. L_g is a superposition of shear-wave and higher mode energy traveling with a group speed roughly in the range from 3.2 to 3.6 km/s (regionally variable). L_g is a mixture of energy leaving the source from a range of directions that average some of the directional variations in wave amplitude. The U.S. Geological Survey uses a piecewise continuous

Box 7.1 Magnitudes and Explosion Monitoring

Seismic magnitudes are an integral part of modern approaches to monitoring large underground nuclear explosions. We can split seismic explosion monitoring analyses into two tasks, explosion identification and explosion characterization - magnitudes have been central to each. m_b, was in part developed for the analysis of explosion sources and smaller-magnitude earthquakes that were difficult to quantify using m_B or $M_S(20)$ during the era of paper records (and to some extent, even with modern digital data). After some investigation, it was discovered that the ratio of short-period P-wave to intermediate-period Rayleigh-wave energy in explosive sources differed from that typical of earthquakes. Thus examination of the ratio of $m_b/M_S(20)$ provided a way to identify candidate explosive sources from the far more numerous earthquakes. For smaller explosions, often observed only at regional distances, the use of teleseismic-based magnitudes is problematic. Regional discriminants based on the same principles but quantified using short-period spectral ratios of P-, S-, or L_g-wave amplitudes are used in place of m_b and M_S.

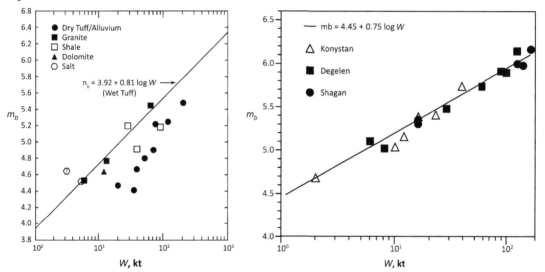

FIGURE B7.1.1 Explosion m_b-yield observations and relationships for Nevada Test Site (NTS) (left) and Semipalatinsk Test Site (STS) (right). Scatter in the NTS observations is partially from differences in explosion emplacement conditions (from Maceira et al., 2017).

In addition to their use in event discrimination, approximate relationships between magnitude and explosion yield have also been developed. P-wave-based relations are more reliable for this purpose. In Fig. B7.1.1 we show known yield versus m_b for the Nevada and Semipalatinsk (former Soviet Union) test sites. The STS relationship is relatively consistent

$$m_b = 4.45 + 0.75 \log_{10} W , \tag{B7.1.1}$$

where W is the explosion yield in kilotons of TNT equivalent. The yield relationship for the Nevada Test Site is more complicated. For hard-rock sources

$$m_b = 3.92 + 0.81 \log_{10} W \, . \tag{B7.1.2}$$

Clearly, explosions with the same m_b from each test site correspond to significantly different yields. The systematic difference in m_b for a given yield is caused by the strong attenuation beneath the U.S. test site within the Basin and Range (underlain by a hot and attenuative upper mantle) relative to the less attenuating upper mantle under the Soviet test site. In the 1970's and 80's, estimating yields from the measured magnitudes before the mantle variations were understood caused political issues between the two superpowers. In addition to geologic effects, inherent in these yield relationships are assumptions about the explosion depth and emplacement conditions, which can complicate the broad use of simple relations. Transporting a relationship from one test site to another (or even a different region of a single test site) requires careful consideration of the geologic characteristics of both sites.

formula in the region

$$m_b(L_g) = \begin{cases} 3.75 + 0.90 \log \Delta + \log(A/T) \, , \\ \quad 0.5° \leq \Delta \leq 4° \\ 3.30 + 1.66 \log \Delta + \log(A/T) \, , \\ \quad 4° < \Delta \leq 30° \end{cases} \tag{7.15}$$

where A is the peak ground motion amplitude in microns calculated using a seismogram filtered with a short-period seismometer response filter, T is the period in seconds, and Δ is the epicentral distance in degrees.

The recommended IASPEI formula for $m_b(Lg)$ is

$$m_b(L_g) = \log A + 0.833 \log R \\ + 0.04343 \gamma (R - 10) + 2.13 \, , \tag{7.16}$$

where A is the maximum absolute amplitude observed on a regional seismogram filtered with a short-period instrument response in microns, γ is a regional attenuation parameter, and R is the hypocentral distance. The Lg signals propagate well through regions with relatively uniform crustal thicknesses but do not traverse regions with significant changes in crustal thickness, including continent-ocean transitions. The value of γ has to be calibrated for each region of application. For example, attenuation in

the western conterminous U.S. is much stronger than in the east, which leads to a more substantial amplitude correction in the west relative to the east.

Seismic coda magnitude

Richter and others chose to use the amplitude of the major seismic arrivals to estimate the size of an earthquake. As described earlier in this chapter, one of the factors that reduces the size of the direct waves during propagation is the scattering of the wave's energy by heterogeneity in the Earth. Although P-waves scattered from S-waves can arrive before the S wave, most of the scattered energy arrives after the major arrival because the scattered waves travel a longer distance (from the source to the scatterer then to the receiver). Scattered waves follow all the major seismic arrivals (P, S, L_g, etc.) and form what we call a seismic wave's *coda*. The scattered waves that arrive closest in time to the major arrival have the largest amplitudes because they experience fewer scattering interactions and propagated only a slightly larger distance than the major arrival. Amplitudes decrease steadily as the time after the major arrival increases because the later-arrival scattered waves may have been scattered multiple times and/or simply followed a longer path to the station. The result is a roughly exponential decay in the amplitude of the signal behind the major arrivals. We can predict the pattern theoretically (and gain insight into the nature of seismic wave speed heterogeneity) or we can measure the coda shape empirically, since it appears to be a relatively stable function of frequency within a particular region.

The coda amplitudes carry information on earthquake size since the overall amount of scattering is proportional to the size of the waves that are being scattered. If we know the characteristic shape of the coda in a particular region, then we can use the coda level to estimate source size. The averaging properties of coda smooth out focusing or defocusing effects that can substantially increase or decrease the peak amplitude of a seismic wave, particularly at short periods. Also, since the coda includes waves that leave the source in all different directions they can help reduce the effect of earthquake radiation patterns (provided enough of the coda to insure sufficient averaging is observable). Major seismic wave amplitudes are modulated by the faulting geometry of the earthquake, which complicates earthquake size estimates that do not include a correction for the effect (radiations patterns are discussed in more detail in Chapter 17). As a result of the intrinsic averaging of the seismic wavefield by the scattering process, magnitudes based on calibrated coda-shapes can be used to estimate robust earthquake magnitudes using fewer seismic stations. In practice, a coda analysis is performed for a range of frequencies using narrowband coda signals to construct an estimate of the source spectrum (Fig. 7.3). The magnitude is then estimated from the source spectrum. The source spectrum is discussed in more detail in the next Chapter. Coda magnitudes are an important tool for constraining earthquake size and spectra, but the effects at short periods and short coda durations needed to model close, small-magnitude earthquake may require cali-

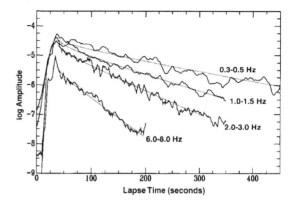

FIGURE 7.3 Example shear-wave narrowband-filtered coda-envelopes (solid lines) estimated by smoothing the logarithmic average of the codas of the two horizontal components. Dashed curves show synthetic envelope fits for the coda amplitude measurements. After a short time, the logarithm of the coda amplitude decreases linearly with distance. Note that the higher-frequency coda attenuates faster because it corresponds to signals with shorter wavelengths that make it more susceptible to intrinsic attenuation and scattering. The differences in the levels of the coda correspond to source properties or near-station geology (from Mayeda, 2003).

bration to account for lateral variations in scattering.

7.2.5 Magnitude saturation

Both $M_S(20)$ and m_b were designed to be as compatible as possible with M_L. Sometimes all three magnitudes have the same value - but, this is not always the case. For small earthquakes differences may be caused by signal-to-noise issues. For example, $M_S(20)$ is hard to measure for small or deep earthquakes. For large earthquakes, when signal-to-noise is not an issue, magnitude differences reflect earthquake physics not well understood at the time that the magnitude scales were developed. Each of M_L, m_b, and $M_S(20)$ depends on an amplitude measurement at a different period (and wave type), roughly 0.8 s (usually S), 1.0 s (usually P), and 20 s (Rayleigh), respectively. If earthquakes ra-

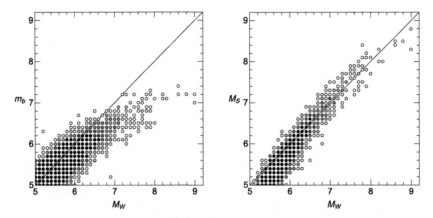

FIGURE 7.4 Illustration of magnitude saturation for the high-frequency magnitude m_b versus the lower-frequency magnitudes M_W and the start of saturation for M_S at large magnitudes. The data are from the U.S. Geological Survey measurements contained in the ISC catalog. For several of the earlier large earthquakes, the GCMT M_W is used since the values in the ISC catalog reflect initial underestimated values. The data are from the years 2004-2017 and the symbol shading indicates the number of observations at each coordinate pair. The darker symbols indicate more data, lighter symbols indicate fewer observations.

diated seismic waves with a uniform amplitude at all periods, the magnitudes would agree as long as distance and depth corrections were accurate. However, earthquakes radiate uniform amplitudes as a function of period only for periods shorter than roughly π times the earthquake duration, τ_r. A rough estimate of a typical earthquake duration in seconds can be computed using

$$\tau_r \, [s] \sim 10^{0.6\,M-2.8} \,. \tag{7.17}$$

This value is approximate, because the spread of durations for any particular magnitude is easily a factor of 2 of this "typical" value, so do not use the relation for precise work. For comparing magnitudes, we must be concerned about a decrease in seismic wave amplitude that occurs when the measurement period is shorter than roughly $\sim 3\,\tau_r$.

Physically, what happens is that longer period components of the signal, which correspond to waves with wavelengths much larger than the spatial dimensions of the rupture and periods that smooth out temporal variations in the rupture process, do not sense any rupture

complexity. Small earthquakes have short durations and M_L, m_b, and $M_S(20)$ are all measured at periods longer than $3\,\tau_r$ and the magnitudes generally agree. But for large earthquakes, the duration of the event increases first beyond the short-periods used to measure M_L and m_b and the magnitudes stop increasing with earthquake size because the short-period peak amplitudes stop increasing as rupture grows and complexity causes complex interference patterns among the shorter period components of the seismogram. We call this effect ***magnitude saturation***.

Classic M_L and m_b saturate roughly near magnitude 6.5 (which has a duration of roughly 10 s). $M_S(20)$ remains a reasonably consistent measure of earthquake size until the duration of the event is roughly a minute, and it begins to saturate for magnitudes just larger than 8.0. The effect is illustrated in Fig. 7.4, which is a comparison of USGS m_b estimates with M_W estimates on the left and USGS $M_S(20)$ with M_W estimates on the right. Ideally, the magnitudes would agree and follow the lines, which have a slope of unity. The saturation effect is quite clear for m_b and although the number of large events is limited, saturation is also apparent for $M_S(20)$. Note also that the relationship for smaller magnitudes is not simple either, neither m_b nor M_S follow the one-to-one line with M_W. Broad-band magnitude, $m_B(BB)$ and $MS(BB)$ are measured using a broader frequency ranges than the classic magnitudes. As a result, they saturate at higher magnitudes than their classic counterparts, but they too saturate. The first step towards a non-saturating magnitude scale was an empirical relation between magnitude and energy.

7.3 Seismic energy, magnitude, and moment magnitude

Gutenberg and Richter constructed empirical relationships between m_B and M_S and E_S in ergs (1 erg = 10^{-7} joules):

$$\log E_s = 5.8 + 2.4 m_B ,\qquad (7.18)$$

and

$$\log E_s = 11.8 + 1.5 M_S .\qquad (7.19)$$

Obviously, the energy calculation in Eqs. (7.18) and (7.19) suffers from all the problems of the magnitude estimation. In particular, E estimated using a saturated magnitude will be underestimated. Eq. (7.19) is fairly robust, since M_S does not saturate until large magnitudes. Eq. (7.19) provides an interesting insight into the tremendous range of earthquake size. The difference between the energy released in an $M_S = 7.0$ and an $M_S = 8.0$ earthquake is a factor of $10^{1.5}$, or ~ 32. In other words, the seismic energy released in a magnitude 8.0 earthquake is over 30 times greater than that released in a magnitude 7.0 earthquake, and it is three orders of magnitude greater than that released in a $M_S = 6.0$ earthquake.

Kanamori et al. (1993) related E_S computed from local and regional distance seismograms with M_L in southern California. Their analysis suggests that for small and moderate-magnitude earthquakes in southern California, M_L is a stable estimator of earthquake energy. The relationship, valid for $M_L \leq 6.5$, is

$$\log E_s = 9.05 + 1.96 M_L ,\qquad (7.20)$$

where E_S is in ergs. The relationship has a slope of 2, which indicates energy in the short-period ($T \sim 0.8\ s$) range used to estimate M_L, increases by a factor of 100 for a unit increase in M_L. This differs with the usual relationship between M_S and energy for large earthquakes because M_S and M_L are measured at different frequencies and the source spectrum of earthquakes is frequency dependent. The M_L expression is not valid for large earthquakes because M_L saturates because of the same source-spectrum frequency dependence. For first-order approximations, the M_S relationship is usually the one chosen, but the important point is that the complexity of earthquake rupture can complicate magnitude's interpretation. Using magnitude for other than direct event comparisons requires understanding of the history, limitations and subtleties of this important and oft-quoted quantity.

We can also relate seismic moment to seismic energy. An earthquake's energy budget can be quantified in terms of the change in stress caused by the fault rupture and offset. Kostrov (1974) showed that the radiated seismic energy is proportional to the rupture area, fault offset, and the stress decrease on the fault produced by the earthquake (usually called the stress drop):

$$E_s \approx \frac{1}{2}\Delta\sigma \bar{D} A ,\qquad (7.21)$$

or, rearranging terms using the definition of $M_0 = \mu A \bar{D}$,

$$E_s \approx \frac{\Delta\sigma}{2\mu} M_0.\qquad (7.22)$$

We can use this expression to relate M_0 to magnitude through Eq. (7.19). If we assume that earthquake stress drop is roughly constant and equal to about 3 MPa, this yields the relation

$$\log M_0 = 1.5\, M_S + 16.1\,, \qquad (7.23)$$

where the moment is in units of dyne-cm. This equation gives a simple way to relate magnitude to seismic moment and, in fact, can be used to define a new magnitude scale, M_W, called *moment magnitude*:

$$M_{\mathrm{w}} = \frac{\log M_0 - 16.1}{1.5}. \qquad (7.24)$$

The moment-magnitude scale, derived by Kanamori (1977), is calibrated to M_S but will not saturate because M_0 does not saturate. Generally, estimating M_0 is more complicated than estimating magnitude, but modern seismic analyses routinely provide M_0 estimates for all events larger than $M_W \sim 5.0$. Using the non-saturating M_W magnitude scale, the largest earthquake recorded was the 1960 Chilean earthquake, with $M_W = 9.5$.

7.4 Descriptive earthquake statistics

Once earthquake size was quantified, observations from around the world could be compared and contrasted and patterns identified. Just as earthquake locations greatly refined evidence that earthquakes were spatially concentrated in relatively narrow regions, important patterns were identified in earthquake temporal, spatial, and magnitude distributions. With the tools of instrument-based earthquake location and magnitude estimation, the patterns could be investigated more broadly and more quantitatively because analyses could be extended to smaller and more remote earthquake sequences, which greatly increased the observations of earthquake processes.

7.4.1 The Gutenberg-Richter relationship

For example, there is an important scaling relationship between earthquake size and frequency of occurrence. Richter and Gutenberg first proposed that in a given region and for a given period of time, the frequency of occurrence can be represented by

$$\log N = A - b\, M_S \qquad (7.25)$$

where N is the number of earthquakes with magnitudes in a fixed range around magnitude M_S, and A and b are constants. The numerical value of A varies with the length of time over which seismicity is examined, the area of the region studied, and the area's seismicity rate. Mathematically, A represents the logarithm of the number of $M = 0$ events. For practical reasons the lower magnitude threshold for analysis, M_a, is larger than zero, and we replace M_S with $(M_S - M_a)$ in the relation. Then a is the number of events at the magnitude threshold). Often, $N(M)$ represents the number of events equal to or larger than M. The form of the relationship remains the same, but the constant, a, must be interpreted differently. The constant b in (7.25) is called the **b value** and often, if not typically, has a value of 1.0. We can rearrange terms in (7.25) and substitute M_0 from (7.23) for M_S to obtain

$$N(M_0) = A'\, M_0^{-(b/1.5)}. \qquad (7.26)$$

This type of power-law size distribution arises from earthquake *self-similarity*. Self similarity has a number of interpretations, but generally means that small and large earthquakes are similar. That means that the ratio of fundamental physical quantities such as fault slip to fault length or fault slip to fault area scale in proportion to event size. The same ideas lead to a roughly constant earthquake stress drop (actually stress drop varies over a limited range). Fig. 7.5 is a histogram of the number

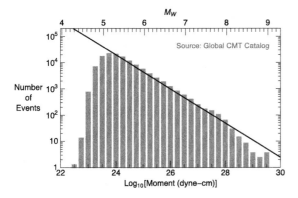

FIGURE 7.5 Number of earthquakes as a function of the logarithm of seismic moment (or magnitude). This is a global data set for all earthquakes since 1977. The solid line is a straight line fit to the central part of the distribution. The equation of the line is $\log N = 19.9 - 0.65 \log M_0$, which has a slope close to $-2/3$, and a b-value for the magnitude dependence close to 1.0.

of earthquakes as a function of seismic moment and magnitude. The data set includes over 50,000 events with magnitudes greater than about M_W 5 that have occurred since 1977. In the intermediate size range (where most of the earthquakes contribute), the b value is roughly 1.0. In general, b values are between $\frac{2}{3}$ and 1 and do not show much regional variability. If we consider Eq. (7.25) determined *per year*, then the A value gives the maximum expected earthquake (assuming $b = 1$). Globally, we expect roughly one earthquake per year for which $M_W \geq 8.0$. A b-value of unity implies that each year we should observe roughly 10 or more events for which $M_W \geq 7.0$ (see Figure in chapter 1) and 100 or more events for which $M_W \geq 6.0$. With this in mind, you can understand the seismologist's frustration with the question, "Is this the big one?" which is heard any time an earthquake does damage in California. Earthquakes of magnitude 6.0 are hardly unusual phenomena, at least globally.

The one case in which the b-value departs significantly from ~ 1.0 is **earthquake swarms**. Swarms are sequences of earthquakes that are clustered in space and time and are not associated with an identifiable mainshock. The b values for swarms can be as large as 2.5, which implies that no large earthquakes accompany the occurrence of small-magnitude events. Swarms most commonly occur in volcanic regions, and the generally accepted explanation is that faults simply are not large or continuous in this environment and that stress is substantially heterogeneous. Thus, the maximum moment expected for a given earthquake is small, and many smaller events must accommodate the strain accumulation (since the energy associated with a magnitude of 5.0 is 1000 times smaller than that for a magnitude of 7.0, we mean *many* smaller events!)

Eq. (7.25) is often used in seismic hazard analysis to determine the *maximum credible earthquake* during a specified time window. If we assume that A and b values determined for a given time range are self-similar in both size and frequency of occurrence, the Gutenberg-Richter relation can be extrapolated to larger time windows. The recurrence time of an earthquake of magnitude M is proportional to N^{-1}. For example, if we monitored a region for one year and found that $b = 1$ and $A = 3.5$, we then would expect a magnitude 4.5 earthquake in the next 10 years and a magnitude 5.5 earthquake in the next century. Obviously, this type of analysis is loaded with assumptions; a maximum magnitude will eventually be reached, and a 1-yr window may poorly represent the earthquake frequency of occurrence for all magnitudes.

We can use Eq. (7.25) to estimate the yearly energy release from earthquakes. The largest event in a given year usually accounts for approximately 10% of the total seismic energy release, and the events with magnitudes greater than 7.0 account for more than 10% of the total. Proceeding this way, the annual energy release by earthquakes averages $1.0 - 2.0 \times 10^{17}$ J/yr. The average energy release is smaller than that released in large megathrust earthquakes. The 1960 Chilean megathrust earthquake had a seismic moment of roughly 2×10^{23} N-m. Using Eq. (7.22) and assuming a 3 MPa stress drop and shear modulus of 40 GPa, the estimated energy release is about 80×10^{17} J, at least 40 times the annual average. The annual earthquake energy release is comparable to that expected from a magnitude in the range from 8.3 to 8.5. Earthquake energy release is relatively small compared to the energy release of other geophysical processes. The 1991 eruption of Mt. Pinatubo in the Philippines alone released roughly 8×10^{19} J. The heat flux out of the solid Earth is roughly 44 terawatts, or about 1.4×10^{21} J/year. For comparison, global human energy consumption is roughly 6×10^{20} J/yr.

7.4.2 Earthquake occurrence rates

Estimation of the probability of an earthquake occurrence is central to seismic hazard estimation, which is a fundamental problem for earthquake science. In many physical systems, random events are modeled as a Poisson process - events are assumed to occur randomly, but at a uniform rate. Under such assumptions probabilities of event occurrence can be estimated using a Poisson distribution, which is parameterized completely by the rate of event occurrence. For earthquakes in general, the assumption of a Poisson process is inappropriate. Earthquakes occur in sequences and many events are not independent. But a Poisson model has value when we consider the largest earthquakes that may occur in a particular region. If these events can be considered independent, their occurrence rate can be estimated using earthquake history, earthquake catalogs, and paleo-seismic investigations. The results can be used to estimate earthquake occurrence probabilities. More generally, earthquake occurrence patterns are viewed as the sum of two components, background earthquakes that occur as part of a tectonically driven Poisson process, and triggered earthquakes that are initiated by other earthquakes. The separation of

earthquakes into background and triggered activity requires separating or ***de-clustering*** events in an earthquake catalog. Many approaches for earthquake catalog de-clustering have been developed - early approaches were heuristic, more recent approaches have incorporated statistical models of earthquake occurrence.

On a global scale, we can crudely assume that most large earthquakes are mainshocks, not triggered by other events (not completely accurate, but for a simple experiment, it's a reasonable starting point). Fig. 7.6 shows the cumulative distribution of the number of large events per year observed for the last ~ 120 years. We used the USGS PAGER catalog updated with the USGS Common Catalog for more recent years for the computation. We chose to define large with a minimum magnitude of 7.0, but the difference between an $M\,6.9$ and an $M\,7.0$ earthquake is of no physical significance. The number of cataloged events from 1900 to 2018 with $M \geq 7$ is 1872. The cumulative large-event count-per-year distribution is consistent with our estimates of the mean number of earthquakes per year in Chapter 1 - roughly 15-16 (we can consider the mean value as an estimate of the mean of the distribution). The match to the distribution is good, and the pattern is reasonably consistent considering that we did not remove aftershocks, which we know do not follow the same Poisson-rate assumptions as mainshocks.

To refine the analysis, we applied a very simple, heuristic declustering algorithm to the original earthquake list. We assumed that any event occurring within four weeks and 200 km of an earlier large event is an aftershock. Applying this rule to the original 1872 large events identified 164 (roughly 10%) of the events as suspected aftershocks. The cumulative histogram of the de-clustered events is compared with a Poisson distribution parameterized by 14 events per year in Fig. 7.7. The largest visual change in the histogram is the removal of any year with more than 28 events (four of the original 32 events in 1943 were classified as aftershocks). The cumu-

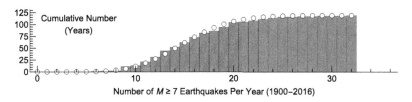

FIGURE 7.6 Cumulative distribution of the number of years (count) versus the number of large, $M \geq 7$, in each year. The dotted curve shows the predictions for a Poisson distribution with a global large earthquake rate between 15-16 events per year.

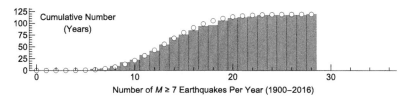

FIGURE 7.7 Cumulative distribution of the number of years (count) versus the number of large, $M \geq 7$, in each year with suspected aftershocks removed. The dotted curve shows the predictions for a Poisson distribution with a global large earthquake rate between 14 events per year.

lative Poisson distribution fit is slightly better for the de-clustered data. But we must keep in mind that we employed a simple de-clustering rule. Changing the rule to classify events in larger or smaller temporal and spatial ranges as aftershocks would affect the pattern. A decrease in the mean number of events per year by roughly one event per year (from 15-16 to 14) is more consistent with de-clustered cumulative distribution. Of course any estimate of average number of large earthquakes in a given year should include the aftershocks, so the number of 15-16 remains the appropriate answer to the general question of how many large earthquakes typically occur each year? The larger the events we want to analyze, the fewer data we have and the less certain the analysis. Using the Onur and Muir-Wood historical catalog of giant earthquakes (estimated $M_W \geq 8.8$), from about 1550 to the present, a similar analysis suggests a mean of roughly 5-6 giant earthquakes per century. Assuming a Poisson distribution, this suggests a 55% chance of experiencing five or fewer and an 70% chance of experiencing six of fewer

of these giant events in any hundred-year interval.

The global pattern of large earthquakes appears roughly compatible with a Poisson description, but this is far from a complete characterization of earthquake occurrence patterns. A substantial number of earthquakes are triggered by other earthquakes and are not independent in their timing, location, or size. For these events, the uniform-rate assumption associated with a classic Poisson distribution is inapplicable.

7.5 Patterns in earthquake sequences

Earthquakes occur in sequences. Nearly all large earthquakes are followed by a sequence of smaller earthquakes, known as *aftershocks*, which are spatially and temporally related to the large event. We assume that the same process also occurs for even the smallest events, but those aftershocks are too small for routine detection. Some events are preceded by seismicity, which after the fact are called *foreshocks*. The largest earthquake in the sequence is called the *mainshock*, but these labels are only certain after the sequence is complete. The terminology is valuable to enable discussion of events as they occur, but must remain flexible as a result of the diversity of earthquake sequence types. Much of this text can be considered an analysis of mainshock properties - in this section we discuss foreshock and aftershock processes and their associated patterns.

7.5.1 Foreshock patterns and earthquake nucleation

Many earthquakes appear to be preceded by detected precursory seismic activity called foreshocks and the percentage is even larger (over 70%) for large earthquakes (for which foreshock activity may be easier to detect). Foreshocks themselves are no different from other earthquakes, so identifying them as foreshocks is the challenge. Given the complexity of earthquake rupture, it makes some sense that fault ruptures would start with precursory activity. The 1960 Chile earthquake began with a sequence of major earthquakes that may have been triggered by precursory aseismic slip. The 33 hours prior to the great event included at least three major events. Although at a smaller scale, several recent large earthquakes also began with extended sequences of foreshocks, including the 2011 Tohoku and the 2014 Northern Chile earthquakes (Fig. 7.8). Just over three weeks before the nucleation of the M_W 9.0 Tohoku earthquake, precursory activity began near and migrated towards the large event's hypocenter. Two days before the great event, an M_W 7.3 event rupture a region close to the Tohoku earthquake's hypocenter. The 2014 Tarapacá, Chile (M_W 8.1) earthquake showed a similar pattern - it was preceded by nearly two weeks of moderate to large offshore earthquakes in a region that had not registered a large earthquake since a very large event in 1877.

However, earthquake sequences quite similar to the recent foreshock sequences have been observed in other regions believed to be large-earthquake capable, but no large earthquake followed. Thus, it remains difficult to distinguish elevated earthquake activity that occurs prior to a larger event from elevated earthquake activity that is not followed by a large event. The interpretation of seismic activity as precursory remains ambiguous. In the case of Tohoku, the earliest activity could easily have been interpreted as foreshocks of the M_W 7.0 event, not the larger M_W 9.0 event. One research avenue that might reduce the ambiguity is work to collect and study information on possible aseismic deformation that may drive foreshock activity. Since most events are offshore, that effort will require marine deformation networks, which are under development but require substantial resources. In addition, other observations, such as those obtained before the 2004 Parkfield earthquake, suggest that expecting a simple, clear pattern may not be realistic. But even an approach that works part of the time could have substantial value. Given the complexity of earthquake precesses, earthquake scientists must continue to inform nonspecialists that any notable earthquake in a region is a indication of a potential for future shaking and a reminder for those that live in earthquake-prone regions to stay prepared.

7.5.2 Aftershock patterns and rupture area

Awareness of aftershocks must be very old, since large earthquakes without aftershocks are exceptionally rare. But the term aftershock originated in descriptions of the 1755 Lisbon earthquake. The largest earthquake in a sequence, known as the *mainshock*, introduces a nearby strain change that perturbs adjacent segments of the same fault as well as the complex system of faults generally found in all fault systems. Faults and fault segments adjacent to the rup-

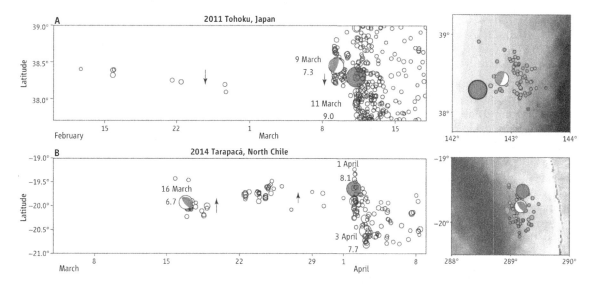

FIGURE 7.8 Plate boundary thrust sequences for the 2011 Mw 9.0 earthquake in Tohoku, Japan (A), and the 1 April 2014 Mw 8.1 earthquake in north Chile (B). Shaded dots indicate great earthquake epicenters. The fault geometries of large foreshocks are indicated by focal mechanisms. Arrows indicate migration direction. In the maps on the right of each panel, shade denotes depth of foreshocks (from Brodsky and Lay (2014)).

ture respond to stress changes induced by the strain change, and *aftershocks* are produced. Aftershocks typically begin immediately after a mainshock and are distributed throughout the volume around the rupture affected by the strain adjustment (the *source volume*). Aftershocks can be dangerous as a result of the damage to structures caused by the mainshock. But in most earthquake sequences, the sum of the aftershock seismic moments rarely exceeds 10% of the mainshock moment. For a typical shallow interplate $M_S = 7.0$ earthquake, thousands of small aftershocks may occur. In general, for shallow earthquakes, the largest aftershock magnitude is roughly a unit smaller than the mainshock (1.2 units smaller on average, this is called *Båth's Law*). These patterns vary with tectonic environment. Oceanic transform earthquakes, oceanic intraplate earthquakes, and deep earthquakes (earthquakes that rupture within oceanic lithosphere) produce fewer aftershocks than shallow continental and interplate events, and the largest

aftershock in a sequence in the oceanic lithosphere is more often roughly two magnitude units smaller than the mainshock.

Typically, the relative frequency of aftershocks decays rapidly at first, then slowly transitions into the normal background rate. Omori studied aftershocks in Japan in the late 1800's and early 1900's and developed an empirical formula to describe typical aftershock rates (*Omori's Law*). Let $n(t)$ represent the frequency of aftershocks at time t after the mainshock, then

$$n(t) = \frac{C}{(K+t)^P},\tag{7.27}$$

where $K, C,$ and P are constants that depend on the size of the earthquake. The P value is usually close to 1.0-1.4. The key features of the curve are a rapid decrease in the number of events followed by a very slow return to background activity levels. Fig. 7.9 is a plot of the time history of the aftershocks of the 2015 Nepal earthquake. The first large event in the sequence begins a classic Omori decay until a second cluster of events occurs two weeks later, which starts a second Omori decay pattern. Some recent studies have suggested that the decay rates for aftershocks depends on the rate of tectonic strain loading in the source volume. In this model, regions that accumulate strain more slowly host longer duration aftershock sequences. The implication is that relatively long aftershock sequences should be expected in regions of slow strain accumulation, such as the interiors of continents. At least some well known earthquake sequences are consistent with the idea.

Aftershock spatial distributions are often used to infer the mainshock rupture area. For most earthquakes, the rupture area (or aftershock area) scales with magnitude. Utsu and Seki (1954) developed an empirical relationship

$$\log A \sim 1.0\, M_S + 2.0, \tag{7.28}$$

where A is measured in m^2. Aftershock zones expand slightly during a month or more following the mainshock presumably involving changes in fluid pressures or aseismic deformation occurring in response to the mainshock. For this reason, when possible, rupture area is estimated from the extent of the aftershock zone after 1 to 2 days. Fig. 7.10 includes map and cross-sectional views of the aftershocks of the 1983 Borah Peak, Idaho, earthquake. The aftershocks define a zone approximately 70 km long and a dipping surface compatible with the southwest-dipping nodal plane of the focal mechanism. Note that the mainshock hypocenter is located at the southeast end and near the bottom of the rupture. Apparently the Borah Peak earthquake rupture mode was unilateral, the rupture propagated up and to the northwest. It is common for the hypocenter of a mainshock to be located near the bottom of the rupture; likely related to fault strength conditions in the crust.

Earthquake hypocenters outside subduction and collision zones are seldom deeper than 15-20 km. Along any fault, the deepest earthquakes define the base of the *seismogenic zone*.

Within the seismogenic zone the crust deforms primarily either by stable or unstable sliding on faults and more rarely by brittle failure when subjected to stresses greater than the strength of the material. Rock strength and frictional behavior depend on deformation rate, temperature, pressure, and composition. For most materials that could realistically make up a significant fraction of the crust and for realistic temperature profiles, rock strength increases to a depth of roughly 15 km and then decreases rapidly. Near the same depth frictional properties transition from those conducive to unstable sliding to those favoring stable sliding. The precise depth varies from place-to-place with details of composition and thermal structure. The maximum in the strength is often near the base of the unstable seismogenic zone. The geometry of the seismogenic zone has important consequences for earthquake size. The implication is that a *maximum* fault width is available for rupture, below which stable sliding or ductile deformation occur. Earthquakes that rupture this entire zone can be classified as *large*, and those that only rupture part of the seismogenic zone can be classified as *small*.

Thus far we have described fault slip, D, in terms of *average* slip on the part of the fault that ruptured. Within an earthquake rupture, slip varies considerably, and in fact the slip often appears to be concentrated on patches called *asperities* (by analogy to the microstructural protrusions in Amonton's early work on friction). Asperities are generally thought to represent zones of relatively high stress drop. *Average* stress drop appears independent of earthquake size, but parts of a rupture may have stress drops an order of magnitude higher than the average. The variation in frictional properties or slip history on the fault responsible for the nonuniform slip and/or stress drop remains an active area of research. Aftershocks on the fault hosting the rupture often concentrate around the edges of the mainshock asperities. Fig. 7.11 is a map of interplate aftershock density in the re-

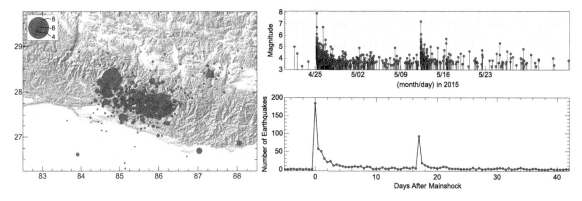

FIGURE 7.9 (left) Mainshock and aftershock locations for the 25 April, 2015 Gorkha, Nepal Earthquake (M_W 7.9). (right) Magnitude timeline and Omori-Law summary using the number of earthquakes in successive 12-hour time intervals. The aftershocks decay according to Omori's law until a large aftershock occurred just over two weeks later.

gion surrounding the 2011 M_W9.0 earthquake off the coast of Honshu, Japan. Regions of relatively intense aftershock activity surround the substantial area of mainshock slip centered on the star. Many additional aftershocks occurred within the upper and lower plates above and below the region of significant mainshock slip, but slip between the plates in that region was subdued.

7.6 Earthquake catalogs

Now that we have a basic understanding of how earthquakes are located, and how magnitude is used to estimate an earthquake's size, we can discuss the archives of this information, seismic catalogs. In a sense, an *earthquake catalog* is a table of information about earthquakes (when and where they occurred, how big they were, etc.). The most commonly used information in a catalog are the hypocentral locations and origin times of earthquakes, but seismic catalogs also include the arrival-time, amplitude and period measurements used to estimate the source parameters. The arrival time measurements are the basis of tomographic travel-time based investigations of Earth's interior. Modern catalogs

also include information on the earthquake size (seismic moment) and faulting geometry of the source if it is available, so are a rich source of information for tectonic investigations. The availability of faulting geometry information also allows seismologists to use the waveforms to constrain Earth's Interior by modeling seismograms directly.

The first lists of earthquakes (we know of) were constructed in the 1600's. Karl Ernst Adolf von Hoff of Germany (1771-1837) began publishing annual lists of worldwide earthquakes in 1826. The first global earthquake catalogs relied on historic accounts of shaking and earthquake damage. French seismologist, Alexis Perrey (1807-1882), compiled local, regional, and global earthquake lists. In the 1850's, Robert Mallet, assisted by his college-student son, John William (later a chemistry professor at the University of Alabama), compiled a comprehensive, global earthquake catalog for the British Association. Mallet (1810-1881), an Irish engineer and one of the founders of earthquake seismology, published broadly on earthquakes and used field observations of the 1857 Neapolitan Earthquake to investigate the origin and mechanics of earthquakes, seismic waves, and the damage they cause.

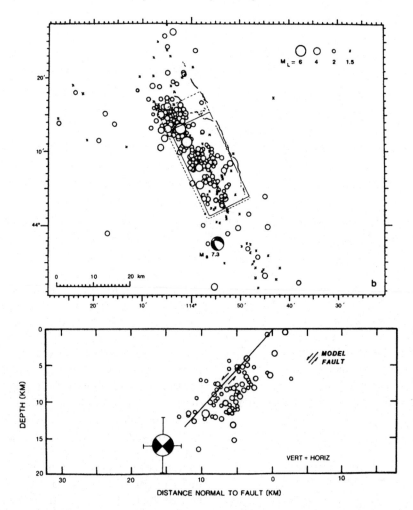

FIGURE 7.10 Aftershocks of the 1983 Borah Peak, Idaho, earthquake. The fault ruptured the surface for about 40 km. Note that the aftershocks define a plane that is consistent with the $N22°W$ plane of the focal mechanism (from Stein and Barrientos, 1985).

None of these earlier researchers had the benefit of data from seismometers, only historic records. Seismometer development began in the early 1800's and usable instruments that became available in the late 1800's allowed detection of more small earthquakes. In the late 1800's and early 1900's Fernand Jean Baptiste Marie Bernard comte de Montessus de Ballore (1851-1923) constructed perhaps the largest historical earthquake catalog and included some information from seismometer deployments. Volumes of his 21,000 event catalog once occupied 85 feet of shelf space in the Societe de Geographie in Paris (Davison, Charles. 1927. The Founders of Seismology. Cambridge [England]: The University Press).

Seismicity, in the modern sense, is a term used to indicate the spatial, temporal, and size distributions of earthquake activity in a given region. Gutenberg and Richter (1949, 1954) com-

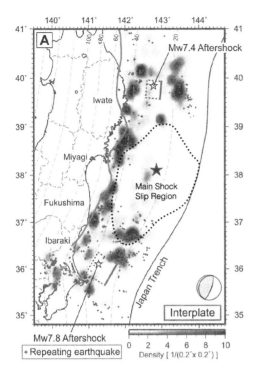

FIGURE 7.11 Map view of interplate aftershock density recorded throughout the one-year period immediately following the 2011 Tohoku-Oki mainshock. Typical reference focal mechanism are shown in the lower-right corners. The earthquake generated may aftershocks within the upper plate above the mainshock rupture, the focus here is on interplate events along the plate boundary. The black star indicates the epicenter of the 11 March 2011 M_W 9.0 Tohoku-Oki mainshock. Diamonds represent the epicenters of small repeating earthquakes. The gray curve identifies the down-dip limit of interplate seismicity. Large aftershock locations are identified. Most of the interplate aftershocks lie outside the region of large slip during the mainshock (modified from Kato and Igarashi, 2012).

TABLE 7.1 U.S. Geological Survey earthquake descriptors.

Magnitude Range	Descriptor
$M \geq 8.0$	Great
$7.0 \leq M < 8.0$	Major
$6.0 \leq M < 7.0$	Strong
$5.0 \leq M < 6.0$	Moderate
$4.0 \leq M < 5.0$	Light
$3.0 \leq M < 4.0$	Minor
$2.0 \leq M < 3.0$	Very Minor

Seismology (1958), tabulated and summarized much ground-breaking research of the 1930's 40's and 50's. Between 2007 and 2013, the U.S. Geological Survey produced a set of posters with a Seismicity-of-the-Earth theme. You can find and download them online.

As discussed in Chapter 1, earthquakes are commonly classified based on size of the event. Table 7.1 shows the U.S. Geological Survey earthquake descriptors based on magnitude. The descriptors can be used to quickly convey the potential impact an earthquake without specificy the exact magnitude.

7.6.1 Modern earthquake catalogs

Modern earthquake catalogs are for the most part available online. No catalog contains all the information on all the earthquakes that occur. We say a catalog is *complete* to a specified magnitude threshold if it has all the earthquakes of that size or larger. Even above the completeness threshold, the amount of information included in a catalog varies. Some catalogs are more suitable for some purposes than others. Catalogs serve users with a broad range of interest, from the public, to the press, and researchers. Many of the online interfaces provide a range of information formats that include different levels of information. Over long periods of time, catalogs are merged, revised, and sometimes redesigned. Many of the hypocenter and magnitude estimates for historical earthquakes come

pleted a comprehensive summary of earthquake geography including seismogram-based earthquake magnitude estimates. Their efforts established a number of regional seismicity classifications, and earthquake-depth categories (shallow, intermediate, and deep) that we continue to use. Their classic summary, Seismicity of the Earth, (1949, 1954) is available online in PDF format, and along with Richter's Book Elementary from analysis by Gutenberg and Richter in the 1950's (Abe, 1975; Pacheco and Sykes, 1992).

The USGS Common Catalog. The USGS maintains a number of catalogs including a real-time update of their Common Catalog, *COMCAT*. The term common indicates that the catalog integrates results obtained from the USGS global and regional networks. For example, an earthquake in east-central California may be located by three regional networks, northern and southern California, and Nevada as well as the global monitoring network. Some coordination is required to insure that duplicate events are not included and that the best estimate of the hypocenter parameters are archived. The information in *COMCAT* is dynamic and events are added and revised in near real time. During the weeks that follow an event, revisions may be performed as the data are reviewed (large events are more completely analyzed quicker).

COMCAT includes reasonably complete information on global activity with magnitudes greater than about 4.5. For smaller events, the results are probably complete for earthquakes in the conterminous US for earthquakes with magnitudes greater than 3.0. Regions with dense seismic networks such as California, Puerto Rico, etc include events complete to smaller magnitude ranges. The U.S. is not unique, and many countries and regions produce catalogs of smaller earthquakes that occur within their regions of focus. A thorough study including information on small earthquakes in a specific region always benefits from some communication with any local groups monitoring local activity. The UK's International Seismic Center (ISC) has historically aggregated many of the catalog results into the ISC Catalog. The task is substantial, particularly assessing the quality of solutions and arrival-time picks, but the result is a valuable resource.

The Global Centroid Moment Tensor Catalog. As mentioned in Chapter 1, researchers initially at Harvard and then Columbia University have developed the Global Centroid Moment Tensor Catalog (GCMT) - a valuable resource that includes not only consistent estimates of more than 50,000 post-1976 earthquake seismic moments, but also important information on the faulty geometry of the events. The GCMT is a major resource for seismologists and other geoscientists studying earthquakes and tectonics. The information is publicly available through a search interface or the entire catalog can be downloaded for more custom analyses. For a complete appreciation of the catalog information, details on the methods used to estimate the event parameters are documented in a series of scholarly articles (Dziewonski et al., 1981; Ekström et al., 2012).

The International Seismic Center, Global-Earthquake-Model Catalog. ISC-GEM is a historic earthquake catalog constructed to help the GEM group estimate seismic hazard and risk (Storchak et al., 2013). ISC-GEM includes many older events that have been relocated using modern algorithms and earth models to create a more consistent set of earthquake hypocenter and magnitude estimates. The variability in the quality and quantity of seismic travel time and amplitude measurements results in a variable, albeit improving quality of solutions with time. Since the seismological community continues to analyze information on historic and older earthquakes, the GEM catalog will continue to evolve for some time. Although global appears in the name, information from regional networks is integrated into the GEM catalog, so in some locales, the magnitude threshold is lower than in other global historic catalog.

The USGS Centennial Catalog. The USGS Centennial catalog is similar to the ISC-GEM, in that older events, with magnitudes larger than roughly 5.0 are reanalyzed to be more consistent with their modern equivalents. Again, the quality and quantity of the older data remains lower than their modern equivalents, so not all historic events can be located with the accuracy and precision of more recent earthquakes.

Catalogs of Large Historic Events. Onur and Muir-Wood (2014) constructed a catalog of the largest earthquakes ($M \geq 8.8$) believed to be complete back to 1700 and nearly complete back to the late 1500's. The catalog includes 18 events since 1700. From the information available, South America is the region most prone to the largest earthquakes. Paleoseismic methods have been used to construct paleocatalogs of earthquakes in many regions. The results are much less certain than instrumentally-based results, but provide an important look into the deeper history of fault behavior. No simple global catalog of all these events has been compiled - but histories for various subduction zone segments (Cascadia, Chile, etc.) have been estimated using geologic evidence related to tsunami and submarine landslide deposits. Trenching across and along continental faults also provides information on large fault systems such as the San Andreas in California, North Anatolian in Turkey, and many others.

Catalogs of Earthquake Impact. Several catalogs include information on the impact of earthquakes (and tsunami) on humans. Fatality catalogs spanning more than one century have been developed by Utsu (2002). The USGS' PAGER catalog was developed to use for rapid hazard estimation following large threatening modern events (Allen et al., 2009). The catalog was last updated in 2008. NOAA's National Centers for Environmental Information maintains catalogs of significant historic and modern earthquakes (deadly and damaging) and tsunami (NOAA, 2018). As with the hypocenter catalogs, precise casualty numbers are not available for all deadly earthquakes and the range of estimates for individual catastrophes can be dramatic. For example, the 2010 Haiti earthquake death toll estimates range from 90,000 to 300,000.

All seismic catalog information has uncertainty and as new information is uncovered and analyzed, some seismic catalogs are often updated. As more data are converted from paper records to digital, location analyses can be performed using consistent and modern earth models and magnitude formulas. This trend will likely continue for some time. Perhaps most important is the need to understand the history of an earthquake catalog and recognize changes in procedures used to locate and estimate the size of earthquakes, as well as acknowledge the history of seismic networks, which generally evolved to include more observations during the last century. Catalogs are an important source of information on background earthquake occurrence and earthquake sequences.

7.7 Summary

Richter developed magnitude with a modest goal of classifying earthquake into three strength levels (large, medium, and small). As measurements accumulated, he and Gutenberg realized that it worked much better than expected. In the years that followed, they extended the idea repeatedly and used magnitude to identify important patterns in earthquake size distribution and mapped earthquake geography with a more quantitative basis than ever before. They also empirically related magnitude to earthquake energy release, a fundamental advance in earthquake physics. But magnitude measures a complicated mix of static and dynamic effects associated with earthquake rupture, and the long history of measurement has resulted in a metric of earthquake size that is both extremely valuable for basic size classification, but that also includes subtle dependencies on the diversity of earthquake rupture characteristics. Using magnitude quantitatively requires care, but the success of the scale for communicating a first-order estimate of an earthquake size is without peer in seismology.

Earthquake sequences exhibit patterns, but not all follow the same script. Omori provided a quantitative model of the rate of the aftershock occurrence that indicates an initial rapid decay followed by an extended return to background levels. We have seen enough foreshock activity to know that while they are not as ubiquitous as aftershocks, they are often observed. Some optimism is warranted that a combination of geodetic and seismic observations may provide information on impending earthquakes in some (but not all) instances. But knowing when activity is foreshock in nature remains a challenge. The observations suggest that short-term earth-quake prediction remains an elusive goal. With more than a century of instrumental seismicity recorded in catalogs, including more than 50,000 consistently measured size and faulting geometry estimates, seismologists and other earth-quake scientists have a wealth of information to continue studying the complex and diverse earthquake processes.

8

Earthquake prediction, forecasting, & early warning

Chapter goals

- Introduce the concept of earthquake cycles.
- Review basic elements of earthquake prediction and forecasting.
- Discuss seismic hazard assessments.
- Describe earthquake interactions.

Support of scientific research is an important characteristic of our society and basic research is an important contribution to our culture. Seismology connects more directly with society on issues such as earthquake-hazard estimation and mitigation, nuclear-explosion monitoring, and shallow-earth imaging to identify natural resources and to monitor environment conditions. Our focus is earthquake seismology. Among the greatest societal challenges seismology faces in that domain are those related to earthquake prediction, seismic hazard estimation, and earthquake early warning. These are our focus in this chapter.

In common usage, *earthquake prediction* means defining the *precise* time, location, and size of an impending earthquake. Such specific predictions have been made, most often by non-earthquake scientists. But there is no reliable method for developing such earthquake predictions. Paraphrasing an old Danish saying, "prediction is difficult, especially about the future".

No scientists proffer specific earthquake times and precise locations. But research into ways to identify an impending event continues. Scientists study data from completed earthquake sequences to search for precursory clues (after the fact). Seismologists and geodesists monitor geophysical fields near large faults suspected to have the potential to produce damaging earthquakes – hoping to catch a precursory signal. Throughout the history of seismology, we have learned that Earth sometimes provides clues before a large event, sometimes doesn't, and sometimes provides the same clues without the large event. Earthquake prediction is certainly difficult (e.g. Hough, 2010).

But the difficulty or even impossibility of *precise* earthquake prediction doesn't mean we can't use what we know about earthquakes to prioritize our earthquake preparations. Instead of trying to predict specific events, seismologists and earthquake engineers have embraced probabilistic *forecasting* of the expected level of ground shaking. Useful forecasts provide the likelihood (and its uncertainty) of how strongly the ground may shake as a function of location and over a chosen time range, such as 30-50 years, the rough lifetime of a building. Seismic hazard estimation integrates information on fault structures, current and past earthquake patterns, regional and local geology, and seismic wave propagation. The potential level of seismic shaking expected is a measure of *earthquake hazard*. The potential loss resulting from that shaking, *earthquake risk*, is a combination of the likely strength of shaking and the exposure of humans and our infrastructure to the shaking (or indirect shaking hazards such as liquefaction, landslides, and tsunami). The forecasted likely ground shaking level can help direct efforts to reduce vulnerability of human-made structures and to increase resiliency of urban areas. Community preparations are a long-term undertaking that can take decades. On a shorter time scale, *earthquake early warning* systems use the automated analysis of near-real-time ground motion observations to detect and to identify large earthquakes as they initiate and grow. Such systems can provide tens of seconds, or in some cases minutes warning before the strong shaking arrives.

In this chapter, we outline the concepts underlying continuing efforts in scientific earthquake prediction (not precise times and places), strong-motion forecasting, and earthquake early warning. Accurate earthquake forecasts require an understanding of the relationship between earthquake occurrence and the tectonic processes that cause them (see Chapter 5). Perhaps most important in this regard are the active tectonic deformation rates frequently estimated using geological (long-term rates) and geodetic (current rates) observations. In addition, although it has a long history in seismology, human-triggered and induced seismicity (associated with damns and reservoirs), recent earthquakes caused by anthropogenic activity such as underground fluid injection, have drawn much attention and have been considered part of the seismic hazard in some regions.

8.1 The earthquake cycle

In our discussions of the earthquake process up to this point, we have considered earth-quakes as isolated, unique sequences. But nearly all shallow earthquakes occur on *preexisting faults*. Thus the vast majority of large earthquakes are repeat ruptures of an established fault. Initially unclear, the relationship between faults and earthquakes was a topic of substantial interest since the 1872 Owens Valley and 1891 Nobi (Mino-Owari), Japan earthquakes, which produced impressive dip-slip fault scarps. After the 1906 San Francisco earthquake, which produced a 400-500 km surface rupture with sometimes large lateral movements, Henry Fielding Reid used geodetic observations to develop the *elastic rebound model* of the earthquake process. Most of the time, a fault is "locked" and frictional resistance halts movement of the rocks immediately adjacent to the fault. But motion continues at a distance from the fault and strain accumulates in the fault-adjacent rocks. That strain exerts a stress on the fault, and when it reaches a critical value, on the order of 10^{-4}, the frictional resistance is overcome and sliding initiates. Once sliding begins, the frictional resistance, which is rate dependent, drops quickly and the strain is suddenly relaxed in an earthquake. When the rupture ceases, strain accumulation begins anew and the process repeats. The idea that faults fail repeatedly, with an irregular, but cyclic behavior, is central to earthquake forecasting and earthquake hazard estimation. The key aspect connecting elastic rebound and earthquake recurrence is the steady accumulation of strain ultimately caused by plate motions. If the tectonic motions are steady (as we believe they are on the relevant times scales), the strain accumulation should also be steady. If friction was also constant, in the ideal situation earthquake occurrence would be periodic and earthquakes would occur along faults at regular *recurrence intervals*.

Fig. 8.1A is a plot of the expected fault stress and slip patterns for such an ideal system. τ_1 represents the strength of the fault; once the stress reaches this value, the fault fails, and the stress drops to a minimum value τ_2, which de-

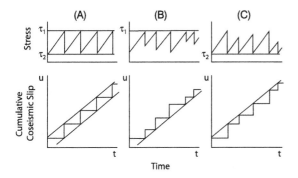

FIGURE 8.1 Various scenarios for buildup and release of stress on a fault. (A) Regular stick-slip faulting. (B) Time-predictable model. (C) Slip-predictable model (modified from Shimazaki and Nakata, 1980).

pends on the fault properties. In the ideal case, the displacement across the fault, or fault offset, is the same for each event (the earthquakes are the same size), and the recurrence interval is constant. The similarity of each event gives rise to the concept of a *characteristic earthquake*, in which earthquakes rupture the same fault segment, with roughly the same size at more or less regular intervals. Implicit in the characteristic earthquake model is the idea that faults are composites of *fault segments*, and individual segments behave in a reasonably predictable fashion. The ideal model is both *time and slip predictable*.

Unfortunately, although the time and slip predictable model is conceptually important, it is rarely observed. The reasons are many: fault mechanical behavior is not uniform, faults are not isolated, but interact, stress accumulation on a fault is not purely elastic, and even plate tectonic processes may not be steady in the short term. In Fig. 8.1A the two unchanging stresses, τ_1 and τ_2, control the behavior of the fault. Given the complex nature of fault friction and the occurrence of nearby earthquakes that load or unload adjacent faults, and the non-uniformity of strain release in each earthquake, it would be remarkable for τ_1 and τ_2 to be constant. Figs. 8.1B and 8.1C show two more general types of be-

havior in which either τ_1 or τ_2 is allowed to vary. When τ_1 is constant, the earthquake behavior is said to be *time predictable*. Since the stress drop may vary from event to event, the time to the next earthquake will vary. But at any stage in the sequence, we can predict the time to the next earthquake if the strain accumulation rate is known, and we can estimate the stress on the fault. In the second model, earthquakes occur at a range of τ_1, but the fault always relaxes to a constant τ_2. This is referred to as *slip predictable*. In the slip-predictable model, if the strain accumulation rate is known, we also know the size of the earthquake that might occur at a particular time.

If the time-predictable model is correct, then the amount of displacement in an event will specify the time interval to the next event. On the other hand, if the slip-predictable model is correct, the lapse time since the last event specifies the potential fault displacement at any given time. In both cases, an estimate of the long-term relative motion is required. Numerous faults have been investigated to determine whether either of these models is applicable. It is clear that neither model is perfect, a situation some seismologists jokingly refer to as the "unpredictable" model. More faults have a weak tendency to exhibit time-predictable behavior.

8.2 Paleoseismology

One of the biggest difficulties in determining the characteristics of fault behavior is that we rarely have more than one or two cycles in the historical record, and often historical data are less complete. For most regions, "significant" seismicity in the instrumental record (post-1960) is insufficient to characterize adequately the nature of earthquake occurrence over time. Archeologic observations can be used to identify and to characterize large events in historical times. To look farther back in time, we must use the information within the Earth. *Paleoseismology*

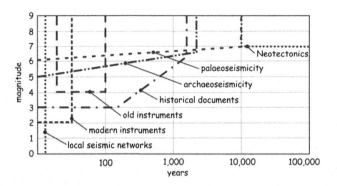

FIGURE 8.2 Diagram showing earthquake magnitude versus time of occurrence. Smaller magnitude earthquakes do not rupture the surface and will go undetected without sensitive seismic instruments. Large earthquakes have signatures in historic, archeologic, and geologic records. Large earthquake paleoseismic records can reach back thousands of years (from Caputo and Helly, 2008).

is the study of the geologic signature of earthquakes and tsunami that have occurred in the recent geologic past (the Holocene). A large earthquake may produce a fault scarp, trigger landslides, cause soil liquefaction, offset geomorphic features, or produce broad regions of uplift or subsidence. Such disturbances can be preserved in the geologic record and can be recognized hundreds, and sometimes thousands, of years later. Paleoseismologists date scarps or disruptions in faulted soil or rock using isotopic analysis, and correlate observations spatially and temporally to identify large-scale disruptions, and hence large earthquakes. The results can extend earthquake timelines back several earthquake cycles. Fig. 8.2 is a schematic magnitude versus timeline of what can be detected using modern local networks, modern seismic instruments, older generation seismic instruments, historic records, archeoseismic studies, and paleoseismic studies. Modern instrumentation has captured only a handful of large event recurrences, while the paleoseismic record can record several cycles. Fig. 8.3 is an example of the paleoseismic results for a section of the San Andreas fault in the Carrizo Plain (California, United States) at the Bidart Fan site. At that site, radiocarbon dates of 33 charcoal samples constrain the ages of the last six earthquakes that ruptured the southern San

Andreas fault in the Carrizo Plain (Akciz et al., 2010). For the time range between ca. 1360 A.D. and 1857, the estimated mean large earthquake recurrence rate is 88 ± 41 years. The last event was the 1857 Fort Tejon earthquake (estimated $M_W \sim 7.9$). But the variable intervals between large earthquakes indicate that the simple time- and slip-predictable models are simplifications. A *mean recurrence time* may be well defined, but significant stochastic fluctuations in the time between large earthquakes occur. Models of earthquake occurrence that include models of stress loading often exhibit *chaotic behavior*, which on the short time scale may be unpredictable. This has important consequences for those who desire to predict earthquake time, place, and size precisely.

Fault scarps are an excellent example of paleoseismic earthquake indicators. Consider a vertical dip-slip earthquake that ruptures the surface and produces an abrupt offset. The scarp may be steep and well-defined immediately following the earthquake, but as time passes, the scarp erodes and changes shape. Fig. 8.4 is an example based on the fault history of the Pajarito fault, which forms the western boundary of the Rio Grande rift near Los Alamos, New Mexico (McCalpin, 2005). The schematic diagram on the right is an integration of information from 14

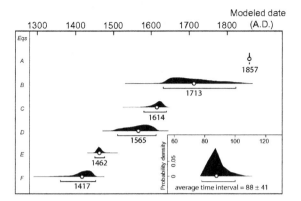

FIGURE 8.3 Probability density distributions quantifying the estimated age of large earthquakes (Eqs) along the San Andreas fault in the Carrizo Plain. The 1857 event was recorded in historical documents. The mean age estimate and its standard error (account for dating uncertainties) are indicated. The inset is a plot of the probability density function of average time interval between earthquakes that occurred in Carrizo Plain between ca. A.D. 1360 and 1857 (from Akciz et al., 2010).

trenches in the Pajarito fault zone. The data suggest a complex history for the 50 to 120 m high fault scarp on Bandelier Tuff (1.2 Ma), and yield a long-term average slip rate of $ca.\,0.1$ mm/yr. The last event occurred at least 20–40 ka, and the average estimated recurrence interval over the past $ca.\,300$ ka is about 20–40 kyr.

Although average earthquake recurrence intervals are difficult to estimate precisely and faults appear to exhibit substantial variability in inter-event intervals, we can make some basic generalizations. First, the long-term slip along a fault must be consistent with regional tectonic motions, generally plate motions. Thus, the total slip on all segments of a fault, over many cycles, must be equal to the tectonic slip. From this perspective, a reasonable initial estimate of an average recurrence interval can be estimated using measured and modeled plate-motion rates and an estimate of the event size (seismic moment). The data, however, have shown that this estimate may be crude and that paleo-investigations are essential to construct a reliable estimate of the uncertainties on the recurrence interval estimate.

8.3 Earthquake prediction

Earthquake prediction is a long-standing goal of earthquake seismology, listed as an important problem since the late 1800's. The goal remains unfulfilled. However, there has been some progress in understanding effects that *might* be expected to provide hints of an impending earthquake. We also can use earthquake and fault-rupture histories and plate-motion rates to define average expected large earthquake rates, but earthquake prediction is also a research field with a checkered reputation. Examples are too numerous in which the public has been made to fear an impending earthquake on the basis of a "prediction" that was either observationally or theoretically flawed (or both). For this reason, many earthquake scientists prefer to distinguish between predictions (associated with the more specific) and forecasts (associated with the more general).

We will continue to use the term prediction, but note the difference between scientific investigations based on an accepted physical model of the earthquake process, from the frequent and widely variable, non-scientific approaches too often reported in the media. Still, earthquake prediction is inherently a *social* exercise, and it is important not to couch predictions as purely a scientific endeavor. Predictions have social consequences. They may lead to reduction in property values, business losses, and general/or economic depression. The high stakes put extraordinary pressure on earthquake scientists to be correct in a subject that is intrinsically imprecise, but it also focuses attention on the societal importance of earthquake sciences such as seismology.

We must start any discussion of earthquake prediction with a clear definition. A **prediction**

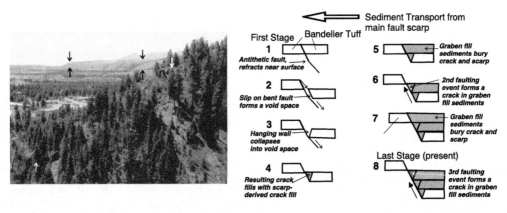

FIGURE 8.4 (left) Photograph of the main scarp of the Pajarito fault zone in the area of the paleoseismic trenches, looking south. In distance, fault scarp (between black arrows) trends southeast; in foreground, main scarp (between white arrows) is 100 m high. Buildings at the left center are part of Los Alamos National Laboratory. (right) Schematic cross-sections showing a crack fill-graben aggradation model for the Pajarito Fault. The complex history shows that repeated surface rupture exposes a free face composed of bedrock (white), which does not shed appreciable colluvium. Some rock fragments and soil from the free face and upthrown block fill crack at the base of the scarp (crack fill, stage 4). Subsequent fluvial deposition fills the graben and buries the scarp face (light gray unit, stage 5). The next faulting event (stage 6) drops the graben and creates a second tension crack that cuts the graben sediments. Then the graben aggrades again (medium gray unit, stage 7). Successively younger crack fills form up-dip against the fault surface. Older fills may rotate toward the downthrown block and be cut by subsidiary faults (modified from McCalpin, 2005).

is *successful* when it provides an accurate assessment of the *time*, *place*, and *size* of an earthquake. We categorize predictions using the precision or ranges allowed on these three parameters. The basic categories are (1) *long-term* (made years or decades in advance), (2) *intermediate-term* (made weeks in advance), and (3) *short-term* (made hours or days in advance) (see Box 8.1). Society responds to predictions from the different categories differently: long-term predictions can contribute to urban planning, helping to establish programs such as reinforcing buildings, to mitigate earthquake impacts. Intermediate-term predictions could help promote emergency preparedness and planning. Finally, short-term predictions can lead to responses from duck-cover-hold to evacuation and to a shutdown of pipelines and industries that may be critically damaged during an earthquake.

Long-term predictions are largely based on identification of fault characteristics such as segmentation, recurrence interval, and the time since the last earthquake. Identification of *seismic gaps* (a strain-accumulating region in a mature stage of its earthquake cycle surrounded by other regions that have slipped recently) is a form of long-term prediction. The utility of seismic gaps in earthquake prediction is controversial, as it has been shown that many earthquakes may occur before a seismic gap is filled. Complex interactions of earthquakes along the strike and dip of large faults makes identifying true seismic gaps challenging. Regardless of their predictive value, the concept of gaps is useful to discuss specific cases where areas of stress accumulation have resulted in an earthquake.

Intermediate- and short-term predictions, on the other hand, are based on *precursory phenomena*. By this, we mean observable changes caused by strain accumulation and nearby and perhaps slow strain release that may lead to an earthquake. Laboratory measurements suggest that volumetric properties of rock change as a function of strain. We can explain these changes

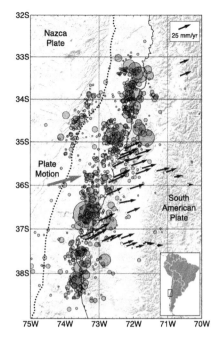

FIGURE 8.5 Map of pre-earthquake GPS vectors from Ruegg et al. (2009) and the 27 February 2010 M_W 8.8 Maule earthquake aftershocks. Motion of the Nazca Plate relative to the South American Plate in the ITRF2000 reference model is indicated by the large gray arrow (not on the same length scale as the GPS observations). The trench is identified with the dotted curve. Black arrows indicate the pre-event surface motion with a reference frame fixed on South America (i.e. motion relative to the rest of South America). The pre-event surface motion, which occurs roughly parallel to the plate motion and is largest along the coast, is dominated by deformation associated with the locked plate boundary and accumulating strain. The 2010 Maule aftershocks (within one week of the mainshock) indicate the rough spatial range of the earthquake rupture that released much of the accumulated strain.

using **dilatancy models**. Bridgman (Bridgman, 1949) first noted in the laboratory that rocks that are subject to uniaxial loads exhibit changing stress-strain behavior. Initial loading produces more volumetric compaction than would be expected from simple elasticity as microfractures close. When the cracks close, the stress-strain relation is linear, as expected in a solid. At stresses of approximately one half the fracture stress,

rocks typically dilate or expand. The most plausible explanation for this nonelastic volume increase is an increase in void space caused by the development of microfractures throughout the rock. In the laboratory, with continued increase in load, these microfractures coalesce into a cross-cutting localized fault and the rock fractures.

8.3.1 Long-term deformation and earthquake migration patterns

Geodetic observations such as those from the Global Positioning System (GPS) or Interferometric Synthetic Aperture Radar (InSAR) provide the best data to identify areas that accumulate strain and thus may be regions to host future earthquakes. For example, Ruegg et al. (2009) used GPS observations to observe the long-term accumulation in the region adjacent to the 2010 Maule, Chile (M_W 8.8) earthquake (Fig. 8.5). The area had hosted no large subduction zone earthquake since 1835, suggesting that this area was a seismic gap. Fig. 8.5 is a map with GPS observations before (Ruegg, 2002) and after (Vigny et al., 2011) the 2010 Maule earthquake. Ruegg et al. (2009) identified interseismic strain accumulation above the Nazca-South America subduction zone caused by a locked thrust zone extending to 60 km depth. Continuous GPS measurement during and/or following the 2010 event as used to estimate the co-seismic displacement that occurred during this large earthquake.

InSAR observations can be used to identify areas of strain accumulation and are used routinely to estimate the slip distribution caused by shallow earthquakes that deform the surface. InSAR *interferograms* are computed using satellite-based observations of the radar phase differences between two or more times. They provide an estimate of surface deformation during the time interval between satellite frames. An advantage of this approach is that we can get high-resolution spatial maps of deformation, as opposed to point measurements from GPS

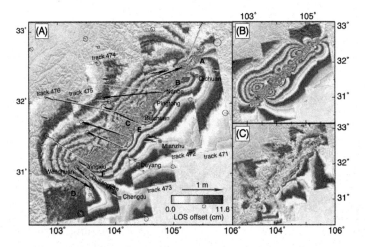

FIGURE 8.6 (A) InSAR interferograms and GPS observations of coseismic ground deformation produced during the 2008 M_W 7.9 Wenchuan earthquake. Vectors identify the horizontal surface offsets predicted by the preferred slip model of Feng et al. (2010), the epicenter of the Wenchuan earthquake, locations of nearby cities, and surface traces of the fault segments in the fault model. (B) Predicted interferogram. (C) Residuals (misfits) between the observed and predicted interferograms. Decorrelated regions of the interferograms are masked out (modified from Feng et al., 2010).

stations. An early application of this approach was to map the surface deformation caused by the 1993 Landers earthquake (Massonnet et al., 1994). The same methods are applied widely to map other processes, from subsidence to seasonal fluctuations of the water table. Fig. 8.6 is a map of observed and modeled InSAR and GPS observations from the 2008 M_W 7.9 Wenchuan earthquake ground deformation. Note the spatial resolution provided by this type of analysis, revealing the heterogenous nature of surface deformation from a major earthquake.

Simple models of strain accumulation often include the assumption that the rate of strain increase is steady. GPS and InSAR measure such steady accumulations and also measure more transient slow deformations that indicate that slow and steady is not the whole story. A well documented complication is the occurrence of nearby earthquakes that can advance or retard the steady march of a fault to its failure stress. Such triggering occurs at all scales (discussed below), but some notable sequences of large earthquakes have exhibited a clear "migration"

of the activity. Fig. 8.7 is a summary of observations related to a progression of large earthquakes along the large strike-slip North Anatolian Fault System in Turkey. Over the last century the earthquake activity has migrated roughly from east to west. Less well constrained by earlier progressions recorded historically and other examples suggest that at least for some sequences, such progressions are not unusual. The idea is that each event contributes to triggering the next adjacent one, by the static transfer of stress. The western extension of the progression includes the fault segments beneath the Sea of Marmara, just south of Istanbul. No precise information is available on the time we should expect a large event in this region, but the progression provides a clear warning to prepare.

8.3.2 Precursory phenomena

Short term predictions generally rely on precursory phenomena, which has included many different observations such as changes in electric and magnetic fields, gas emissions, groundwa-

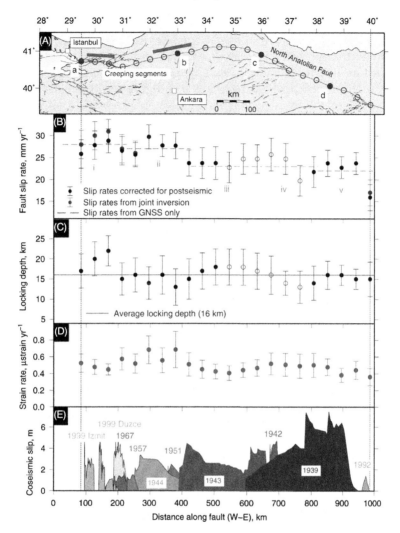

FIGURE 8.7 Along strike variation in fault parameters on (A) the North Anatolian Fault (NAF) determined from historic data and modeling of GPS and InSAR data. The variation in fault slip rate (B) and locking depth (C) along strike at the locations indicated by the black circles in (A). Also shown are the (D) calculated strain rate and (E) surface coseismic slip distributions of major earthquakes ($M_w > 6.5$) along the NAF since 1939 (modified from Hussain et al., 2018).

ter level changes, temperature changes, surface deformations, and seismicity. For example, some have attempted to use changes in the level of emission of the radioactive gas radon, a *geochemical precursor* as a precursor. Most geochemical precursors are associated with ground water and involve dissolved ions or gases. Ground-

water anomalies were among the earliest and most frequently reported earthquake precursors. Changes in taste and temperature in wells and springs prior to large earthquakes were documented in Japan more than 300 years ago. Most seismologists believe that microfracturing prior to major earthquakes in late stages of dilatancy causes increases in ion and gas concentrations in ground water. Laboratory studies of dilatant rocks show that porosity increases from 20% to as much as 40% prior to rupture. Dissolution or alteration of fresh rock surfaces could significantly increase the ion concentrations in the ground water. In the case of radon, which is produced by the radioactive decay of radium, the fresh fractures allow more of the gas to escape. Most precursory radon anomalies have appeared a few weeks to days before the impending earthquake. Radon is commonly used because it is easily measured. Other gas anomalies include hydrogen and helium. Fig. 8.8 shows the precursory changes of radon, water temperature, water level, and strain before the 1978 Izu-Oshima-Kinkai earthquake ($M_s = 7.0$). Although the precursor is clear and well defined, unfortunately, there are numerous counter-examples of earthquakes with *no* geochemical precursors, and we measure comparable fluctuations that are not associated with an earthquake.

The dilatancy model provides a physical framework to investigate expected precursory behavior. Fig. 8.9 shows changes of a rock under compressive loading. A volume increase ($\Delta V / V_0$) occurs dramatically at about 50% of the failure stress. This represents the development of cracks, which in turn causes variations in many of the physical properties of the medium. The behavior of rocks during dilatancy depends strongly on whether the cracks are dry or wet. In a wet model, the diffusional water plays an important role. Fig. 8.9 shows the expected physical parameter changes for wet and dry dilatancy models. Dilatancy experiments in which a rock is fractured are not exact analogs for the rupture of an existing fault. But the experiments provide a guide for the types of measurements that might be possible. Seismologists should be able to measure some of the predicted changes. For example, both models predict changes in seismic velocity. In the beginning stages of dilatancy, the development of cracks causes this because

they reduce the elastic modulus. From laboratory measurements, these cracks appear to affect α much more than β, which results in a decrease of α / β by $10 - 20\%$. During this stage, land uplift and ground tilt are also expected providing targets for geodesy. In late stages, the velocity ratio returns to normal due to either water saturation of the cracks or closing of the porosity. Seismologists have observed velocity changes within large fault zones, particularly the healing that occurs after a large earthquake (not suitable for prediction). The steady accumulation of tectonic strain is not the only signal that can be used to explore fault properties. Crack density in a fault's damage zone may vary with seasonal and tidal loading and such changes have been measured in attenuation (Malagnini et al., 2019). The seismological ability to extract these observations from seismograms is impressive, and more work is needed to map out the potential and limitations of these observations to detect precursory changes.

For seismologists, the easiest precursory indicator to measure is a seismicity pattern. Within a source region of a large earthquake, numerous small faults or heterogeneities on the main fault probably exist that can produce earthquakes in response to the loading cycle. A familiar pattern is foreshock activity (Chapter 7, which occur for many earthquakes. In some instances, we have observed that after the aftershock sequence of a major earthquake is complete, the seismicity level drops to a low or even very-low background level. This phenomenon is known as *seismic quiescence* and is fairly common. Quiescence is *sometimes* broken by a buildup of activity, with swarms of activity before the mainshock, which is known as a preshock sequence. At least two processes may contribute pre-shock activity. First an overall increase in the regional strain triggers earthquakes on subsidiary faults. An example is the relative deficiency of activity along the northern San Andreas Fault (the 1906 segment) that is hard to see on seismicity maps constructed with recent earthquakes. The

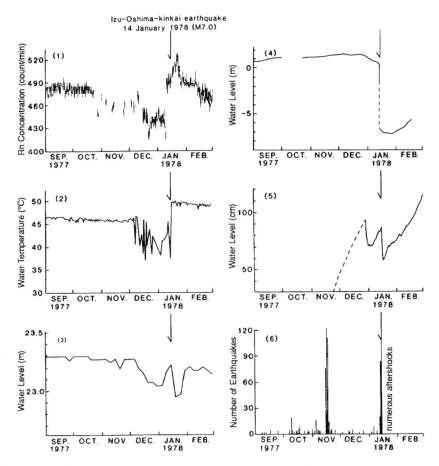

FIGURE 8.8 Precursory changes in radon concentrations, water temperature, and number of minor earthquakes prior to the Izu-Oshima earthquake of 1978 (from Wakita, 1981). As with all precursory phenomena, the fluctuations are neither unique to earthquake occurrence nor commonly observed before all earthquakes.

1906 San Francisco earthquake was preceded by decades of moderate and strong earthquakes on the surrounding faults of the Bay Area. Whether the regional pattern is robust is of course unknown. Second, the occurrence of significant slow slip or creep on adjacent segments of a large fault may load the locked regions to the point of failure. The 2011 M_W 9.0 Tohoku earthquake is believed to have been preceded by a slow deformation event that triggered small seismicity for months earlier and initiated an M_W 7.3 foreshock two days before the great

event that had a sequence of aftershocks that migrated to the eventual hypocentral region of the M_W 9.0 event (e.g. Kato et al., 2012). Other possible large subduction zone earthquakes exhibiting the same process are shown in Fig. 7.8. As is common, not all events exhibit this behavior, but if even some do, we may be able to develop an earthquake "watch" system similar to those used for hurricanes, tornadoes, and floods.

The 1975 Haicheng earthquake (M_s = 7.3) in northeast China was the first major earth-

Box 8.1 The Crustal Deformation Cycle

To understand the cyclic nature of earthquakes we must consider the entire process of crustal deformation. The *crustal deformation cycle* is often divided into four phases: (1) *interseismic*, (2) *preseismic*, (3) *coseismic*, and (4) *postseismic*. If Reid's model of elastic rebound were completely correct, only two phases of deformation would occur, coseismic and interseismic. The deformation cycle is more complex because the Earth is not perfectly elastic. A more complete representation of the deformation process is that of an elastic layer, which contains the fault, over a viscoelastic sublayer. The elastic layer is driven by a remote tectonic process such as plate motion. In the model, the fault has two parts: (1) the seismogenic zone, where strain is released only in the sudden slip of an earthquake, and (2) a deep region where the fault slips continuously. As the seismogenic zone cycles to failure, a strain is imparted to the rest of the system. Immediately after failure, strain is concentrated close to the fault. As time increases after the rupture, stress diffuses outward in the viscous material. Fig. B8.1.1 is a plot of the vertical land movement estimated using leveling in the region near the 1947 Nankai ($M = 8.1$) earthquake.

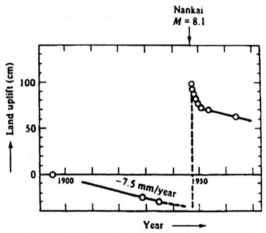

FIGURE B8.1.1 Vertical ground motion preceding and following the 1947 Nankai earthquake (from Okada and Nagata, 1953).

For three years after the earthquake, deformation occurred at a much higher rate than that expected from plate motion. This is a manifestation of the viscous relaxation of the stress.

quake believed to have been predicted. The event was preceded by many different kinds of precursors: foreshocks, ground-water changes, tilting, and strange animal behavior. Two official intermediate-term predictions were issued, but no official short-term prediction was made. An unofficial but nonetheless effective prediction of an imminent earthquake, based principally on the foreshocks (Fig. 8.10), was issued by the Provincial Seismological Bureau on the day of the earthquake, and buildings and communes were evacuated. The earthquake was destruc-

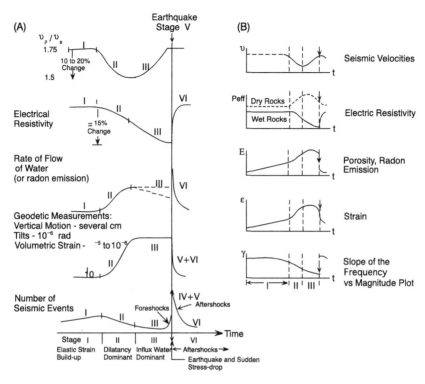

FIGURE 8.9 Conceptualized behavior of rocks under loading. (A) changes in physical quantities during different stages (identified by Roman numerals) of compressional loading and failure (earthquake). (B) Potentially measurable changes expected from dilatancy models for wet and dry conditions (from Kasahara, 1981).

tive, but relatively few died. The success caused great excitement in the earthquake prediction community. Unfortunately, a year later a magnitude 7.7 earthquake destroyed Tangshan, a city only 200 km southwest of Haicheng, and killed more than a quarter of a million people. In that case no precursors were detected, and no prediction was issued. Earthquake scientists clearly have their work cut out for them.

A systematic analysis of precursory phenomena by Cicerone et al. (2009) confirmed that precursory phenomena do occur in *some* cases. In those cases, the precursory anomalies tend to occur close to the eventual epicenter of the earthquake, perhaps caused by deformation that occurs prior to the main rupture. Yet, the uncertainty in predictions remains very high. An at-

tempt at short-term prediction of earthquakes in the U.S. occurred for the Parkfield, CA region of the San Andreas fault. Bakun and Lindh (1985) noted that this section of the San Andreas fault had earthquakes in 1881, 1901, 1934, and 1966 of approximately M 6.0. The 1966 event had observed precursory seismicity and slip, which led the U.S. Geological Survey (USGS) to launch the Parkfield, CA earthquake prediction experiment. Because of the expectation of another Parkfield earthquake in the near-future (around the year 1988), the USGS heavily instrumented the region, monitoring seismicity, tilt, strain, and other measurable quantities in order to detect any precursors that might occur. On 28 September 2004, a Parkfield (M 6.0) earthquake occurred in the same region as the pre-

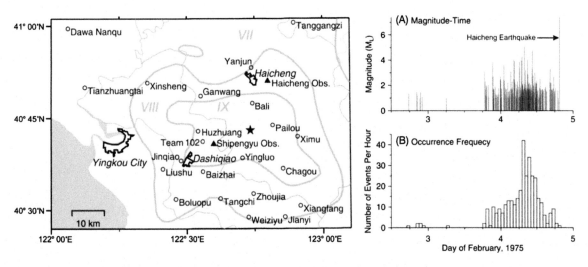

FIGURE 8.10 (left) Seismic intensity map for the 1975 Haiching, China earthquake. The earthquake epicenter is shown with a star, thick gray lines are isoseismals, and the thin lines are county boundaries. Dense urban areas are outlined in black. (right) Seismicity frequency before the earthquake showing the significant foreshock sequence that preceded the event (from Wang et al., 2006).

vious events. However, the 2004 earthquake did not have obvious precursors. Fig. 8.11 is a map of the locations of the 1934 and 1966 earthquakes, along with the rupture extent of the 2004 event. Although the magnitude and rupture extent of the 2004 earthquake were anticipated, the timing of the earthquake was years off, highlighting that reliable short-term earthquake prediction is currently not achievable (Bakun et al., 2005).

In summary, many phenomena appear precursory to earthquakes, so some level of earthquake prediction may be possible. Unfortunately, each earthquake shows extraordinary variability in its precursory behavior, including many events with a total absence of any known precursors. While work on earthquake prediction continues, we instead work toward mitigating earthquake hazards, focusing our efforts based on earthquake probabilities, and developing early warning systems that can warn the public on impending shaking created by an earthquake of significant size.

8.4 Earthquake forecasting and hazard estimation

The 1971 San Fernando, California earthquake caused more than $1.2 billion of damage and killed 64 people. It was painfully evident that the United States was not addressing the tremendous potential economic disaster associated with earthquakes. Congress created the National Earthquake Hazards Reduction Program (NEHRP) to address scientific, engineering, and social issues associated with earthquake occurrence to "to reduce the risks of life and property from future earthquakes in the United States through the establishment and maintenance of an effective earthquake hazards reduction program." One of the goals of NEHRP was to broaden and deepen our understanding of earthquakes and earthquake hazards, including earthquake prediction. Although much progress has been made toward understanding the earthquake process, as described above, accurately forecasting earthquakes remains elu-

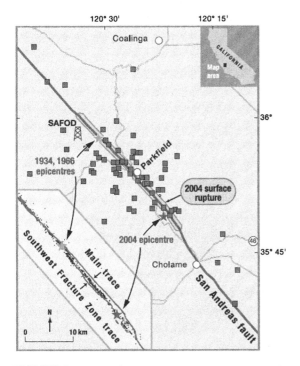

FIGURE 8.11 Map showing location of the 2004 Parkfield earthquake, the surface rupture, and the San Andreas fault. Seismographs, strainmeters, creepmeters, magnetometers, and continuous GPS stations shown as squares. Lower inset shows epicenters of 2004 aftershocks (black dots) plotted relative to fault. Upper inset, map location (modified from Bakun et al., 2005).

sive. What emerged was an approach to assign "earthquake probabilities" to different regions that is used to estimate the probable shaking levels caused by future earthquakes in each region of the country.

Earthquake probabilities are based on estimates of the recurrence interval and the time since the last major earthquake on each potential seismogenic fault. The conditional probability for earthquakes, defined as the likelihood that a given earthquake will occur within a specified time period, needs to be updated as time, T, increases. For example, a section of the San Andreas fault may have a probability of an earthquake occurring in the next decade of 0.3; if

the earthquake does not occur within 5 yr, the process can be repeated, and the probability will *increase* for the next 10-yr interval. In 1988, the Working Group on California Earthquake Probabilities produced a conditional probability map for the San Andreas fault for the ensuing three decades, that continues to be updated with improved fault movement estimates, shown in Fig. 8.12. A key assumption in this was the segmentation of the fault into independent rupture zones. The map reflects the seismic hazard as it relates to fault movement on specific faults or fault segments. For example, the area near Parkfield, CA currently has a high probability of a $M_W \approx 6.0$ event occurring. Fault rupture in this region has a recurrence interval of approximately 22 years, with the last event occurring in 2004 (Fig. 8.11). Several other segments have a $20 - 30\%$ probability of magnitude 7 to 8 earthquakes over the next 30 yr.

Hazard mitigation lies on the foundation of seismic hazard analysis, and in the United States, the USGS and NSF participate in the NEHRP program, which funds research that directly addresses reducing seismic hazard. As part of this program, the USGS produces hazard maps using Probabilistic Seismic Hazard Analysis (PSHA), based on the methodology developed by Algermissen and Perkins (1976), Algermissen et al. (1990), Frankel (1995) and Frankel et al. (1996). The USGS hazards maps are generally based on historical seismicity, known faults, geodesy, and b-value, with statistical framework to forecast predicted ground motion within a specified time window. These maps can inform building codes, insurance rate structures, risk assessments, and other public policy. The hazard maps are based on the annual rate, $\lambda(u > u_0)$, of exceeding ground motion, u_0 at a specific site, which is determined by summing over distance and magnitude (Frankel, 1995):

$$\lambda(u > u_0) = \sum_k \sum_l 10^{(\log N_k / T - b(M_l - M_{ref}))}$$
$$\times P(u > u_0 \mid D_k M_l), \qquad (8.1)$$

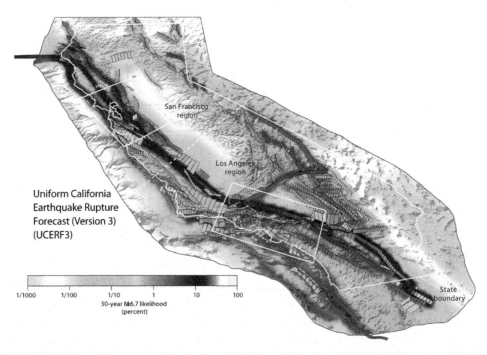

FIGURE 8.12 USGS 30-year forecast model showing the likelihood of an earthquake in each region of California that may experience a magnitude 6.7 or larger ($M \geq 6.7$) earthquake in the next 30 years (6.7 matches the magnitude of the 1994 Northridge earthquake). The shaded area represents greater California, and the white line across the middle defines northern versus southern California (modified from USGS factsheet, Field et al., 2014).

where k is the index for the distance bin, l is the index for the magnitude bin, T is the time in years of the earthquake catalog used to determine N_k. The first factor in the summation is the annual rate of earthquakes in the distance bin k and magnitude bin l. The b-value is taken to be uniform throughout most of the area. $P(u > u_0 | D_k, M_l)$ is the probability that u at the site will exceed u_0, for an earthquake at distance D_k, with magnitude M_l. This probability is dependent on the attenuation relation and the standard deviation (variability) of the ground motion for any specific distance and magnitude.

Fig. 8.13 illustrates the latest results from the USGS showing the 2% probability of exceedance of ground acceleration in %g (gravity) over the next 50 years throughout the contiguous U.S. The maps allow for the probabilistic approach for damaging shaking to be easily interpreted by a wide range of stakeholders, including policy makers, urban planners, emergency personal, and the general public. Since the inception of these maps, better seismic and fault characterization continue to improve these maps, and as new information emerges from the NEHRP program, new research findings can be easily incorporated. Regions such as the New Madrid fault zone in the central U.S. are recognized as having a high hazard (see Chapter 3), and well as much of the western U.S. It should be noted that the USGS includes all seismicity, which likely includes documented induced seismicity. Thus, Oklahoma now has a higher seismic hazard than previous versions of this map, as does western Texas (discussed below). Generally, areas away from tectonic

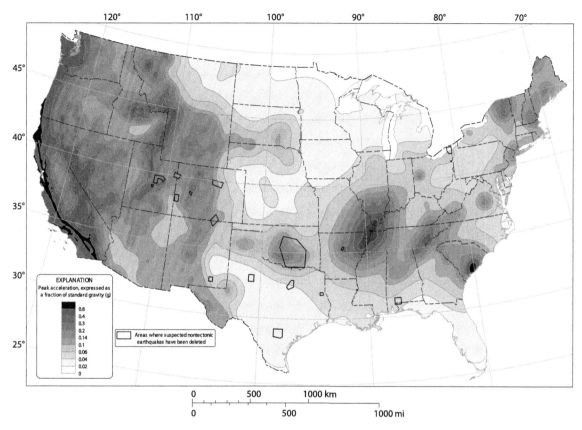

FIGURE 8.13 Map of the continental U.S. showing a 2% probability of exceedance of indicated levels of peak ground acceleration in the next 50 years. Note that the hazard is high in tectonic regions, but also in regions including the New Madrid fault zone in the central U.S. (see Chapter 3), and Oklahoma, where it has been recognized that earthquakes have been induced because of oil and gas exploration and exploitation (modified from USGS, 2018).

boundaries (intercontinental) have the lowest hazard.

8.5 Earthquake interactions and triggering

A foundation of earthquake prediction, forecasting, and probability is to assume that earthquakes occur independently. Although aftershocks, foreshocks, and mainshocks have long been recognized as connected in and around a fault that has ruptured, new studies have shown that the reach of an earthquake can be global. Earthquake triggering occurs when stress changes associated with an earthquake can induce or retard seismic activity at all distances from an earthquake. Static triggering occurs as a result of fault movement that creates additional stress that can promote triggering of another earthquake, and the stresses reduce to background levels usually 2 to 3 fault lengths from an earthquake rupture. Dynamic stresses created by the passage of seismic waves from a large earthquake can trigger a suite of phenomena at remote distances. Induced seismicity results from disturbing the ambient stress state

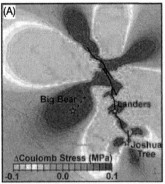 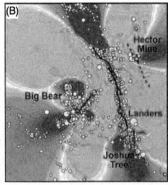

FIGURE 8.14 (A) The Landers and Joshua Tree earthquakes increased Coulomb stress (dark regions) where the Big Bear earthquake occurred three hours later. Coulomb stress calculations in (A) are for left-lateral strike-slip faults aligned with the Big Bear rupture surface. (B) Stress changes induced by the Joshua Tree, Landers, and Big Bear earthquakes increased Coulomb stress in regions where the vast majority of aftershocks occurred over the next seven years, culminating in the 1999 Hector Mine earthquake. Coulomb stress calculations in (B) are for right-lateral strike-slip faults aligned with the Hector Mine rupture surface (also reasonable for most aftershocks in the region) (modified from Freed, 2005).

of a fault through some external activity, thereby triggering an earthquake. Induced seismicity has been recognized as being a major contributor to seismicity in specific regions in the U.S., most notably in the state of Oklahoma.

8.5.1 Static triggering

When the shear stresses (driving stress) are large enough to overcome the normal (or clamping) stresses, an earthquake can occur, which is called the *Coulomb Failure Criteria* and will be discussed in **Chapter 18**. When a large earthquake occurs, it will deform the surrounding region with permanent deformation (called static deformation), which will increase or decrease the stress in the immediate surrounding region. This stress can either advance or retard the earthquake cycle, and is called *static triggering*. A *stress shadow* refers to a region that has had a reduction of stress from static stresses created by fault movement in a specific region, and thus can create regions of quiescence (fewer earthquakes than normal). In regions of increased stress, the earthquake cycle will advance, and the chances of an earthquake will increase be-

cause the stress that promotes earthquakes has increased. A widely accepted example is shown in Fig. 8.14, where the April 22, 1992 $M6.1$ Joshua Tree strike-slip earthquake preceded the June 28, 1992 $M7.3$ Landers strike-slip earthquake to the north. The Landers earthquake then was followed by the June 28, 1992 $M6.5$ Big Bear aftershock on a different fault, triggered by an increase of static stress caused by the motion and curvature of the fault. Seven years later, these earthquakes were then followed by the October 16, 1999 $M7.1$ Hector Mine earthquake, which occurred in a lobe of increased static stress from the Landers earthquake (Fig. 8.14). The July 5, 2019 $M7.1$ Ridgecrest, CA earthquake occurred over 170 km north-northwest of the Hector Mine earthquake (over two fault lengths, which had a fault length of ≈ 40 km). Thus these earthquakes are unlikely related through statics stress, which would be very small at these distances.

Modeling tools can be used to map the static stress caused by ground deformation following a large earthquake, with various computer tools that are openly available, for example *Coulomb* (Toda et al., 2005; Lin and Stein, 2004). The tool *Coulomb* calculates static displacements, stresses, and strains at a selected depth caused

---

Box 8.2 Operational Earthquake Forecasting

When a substantial earthquake occurs, there is a high likelihood that aftershocks will follow. The Omori law characterizes the typical temporal rate of aftershock occurrence, and the Gutenberg-Richter magnitude distribution b-value gives an expectation of the relative size distribution of events that may occur during the aftershock sequence. Each seismogenic region can also be characterized by an aftershock productivity law: $N(M) = k 10^{\alpha \cdot M}$, where N is the number of aftershocks, M is the mainshock magnitude, α is a parameter typically ~ 1, and k is a constant of proportionality that depends on the number of aftershocks per mainshock above the catalog completeness level (Reasenberg and Jones, 1989). These statistical models can be combined to give parametric representations of seismicity sequences, including epidemic-type sequences (ETAS) (Ogata, 1988), that capture the empirical history of activity in a region to give guidance on likely activity to follow a recent event. Formalization of this parameteric approach as Operational Earthquake Forecasting (OEF) distills the collective seismological observations into a probabilistic statement of likely number and size of events in the aftershock population as a function of time. This guides emergency responders and the public on likely aftershock sequence activity. The statistical models also allow for the low probability occurrence of aftershocks larger than an earlier event, which then become new mainshocks.

The USGS now reports an Aftershock Forecast for the region of large events for a time frame of one month after the mainshock. For an M 6.5 earthquake on March 31, 2020 in Idaho, the following information was provided: The USGS estimates the chance of more aftershocks as follows: Within the next one month until 2020-08-07 00:00:00 (UTC):

- The chance of an earthquake of magnitude 3 or higher is > 99%, and it is most likely that as few as 6 or as many as 22 such earthquakes may occur in the case that the sequence is re-invigorated by a larger aftershock.
- The chance of an earthquake of magnitude 5 or higher is 12%, and it is most likely that as few as 0 or as many as 2 such earthquakes may occur.
- The chance of an earthquake of magnitude 6 or higher is 1%, such an earthquake is possible but with a low probability.
- The chance of an earthquake of magnitude 7 or higher is 1 in 800, such an earthquake is possible but with a low probability.

---

by fault slip, magmatic intrusion, or dike expansion/contraction. This type of modeling has shown to be very effective for documenting static stress triggering. For example, this type of modeling has been used to link the Joshua Tree-Landers-Big Bear-Hector Mine earthquake sequence (Fig. 8.14) that occurred in southern California, plus events along the North Anatolian Fault in Turkey, another transform fault that has been documented to have static triggering from motions of large strike-slip earthquakes.

8.5.2 Dynamic triggering

Seismic waves from large earthquakes create dynamic stresses that can trigger local earthquakes in a variety of tectonic environments, tectonic (non-volcanic) tremor (movement in the brittle–plastic transition zone along major plate-boundary faults), and activity changes in hydrothermal and volcanic systems, plus changes in spring discharge, water well levels, soil liquefaction, and the eruption of mud volcanoes. This type of earthquake triggering is termed *dynamic triggering* or *remote triggering*. Surface waves with periods of 15–200 s trigger most events, yet body-waves, especially *S*-waves, have been shown to trigger phenomena. The first widely accepted case of dynamic triggering occurred immediately following the 1992 $M7.3$ Landers, CA earthquake, where seismicity increased throughout the state of California. The dynamic triggering was caused by strong surface wave arrivals. Since then, many studies have identified many examples of dynamic triggering globally, in a variety of tectonic provinces, and by different seismic waves. Another notable example is the 2002 $M7.9$ Denali Fault, AK earthquake, which triggered extensive seismicity in the continental U.S. In this case, the surface waves were amplified, on the order of 100 times larger in amplitude, due to both directivity and source process effects (the largest pulse of energy occurred at the end of the rupture) (Velasco et al., 2004), explaining the $M_S 8.5$ reported by USGS. Generally, triggering dynamic stresses can be < 1 kPa, but a range of values has been documented that contribute to triggering.

Fig. 8.15 shows an example of dynamic triggering caused by the dynamic stresses of the surface waves from the Dec. 26, 2004 Sumatra - Andaman Islands ($M9.2$) earthquake. High-pass filtering broadband seismograms reveals the local, triggered event hidden in the energy of the large surface wave arrivals, highlighting the utility of broadband instrumentation. The type of dynamic triggering in this example shows a small earthquake that is triggered at the same time as the arrival of the seismic waves, which can be referred to as coincident dynamic triggering. Researchers have identified triggering that occurs hours to days after the passage of the

seismic waves, referred to as delayed dynamic triggering. However, delayed dynamic triggering has proven to be more controversial and difficult to prove, and the mechanisms and observations remain highly debated. Despite these controversies, recent work in dynamic triggering highlights that large earthquakes have more of a global influence on smaller earthquakes ($M < 5.0$) than previously recognized, and perhaps can trigger larger earthquakes in the hours to days following an event.

8.5.3 Other triggering

Earthquakes have been proposed to be triggered from a variety of non-traditional sources, including anthropogenic activities such as reservoir-filling, mining, *hydraulic fracturing* (high-pressuring water injection intended to fracture rock) and fluid injection and/or withdrawal. *Reservoir induced seismicity* and seismicity created by the injection of fluids have been known since the 1970's. Fluids injected into the ground can reduce the normal stress that locks a fault, and any fault that is near failure can then rupture. Earth tides loading can increase the occurrence of shallow thrust earthquakes (Cochran, 2004), as can seasonal meteorological factors such as snow loading and ground water recharge (Christiansen et al., 2005).

Human-induced earthquakes or simply *induced earthquakes* resulting from oil and gas exploration and exploitation have recently become important topics of political and scientific debate. Many studies are currently being conducted to understand this phenomena, especially since the occurrence of > $M3$ earthquakes has increased in the central U.S. since 2010 (Fig. 3.29). The increase in seismicity correlates with increased industry activities for extraction of oil and gas resources. For example, the state of Oklahoma had a significant increase in seismicity related to waste water injection, making the state at one time more seismically active than California, reaching over 900 > $M3$

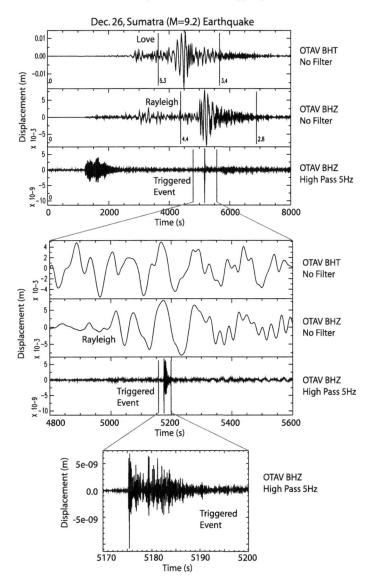

FIGURE 8.15 Displacement seismograms recorded at station OTAV ($\Delta = 173°$) illustrating unfiltered Love and Rayleigh waves, plus the high pass filtered (at 5 Hz) vertical component. (middle) Same as (top), but with a shorter time window showing the strong Rayleigh wave arrival and the triggered event. (bottom) Blowup of triggered event (from Velasco et al., 2008).

earthquakes in 2015. Intervention by various agencies to restrict injection has reduced the number of earthquakes in the state to 304 > $M3$ earthquakes in 2017.

The debate on *induced earthquakes* centers around the seismic hazards poised by industry practices, the importance of extraction of energy resources, and public safety. In particular, *hydraulic fracturing* stimulation, a non-conventional approach for oil extraction, has evolved into a very efficient approach for extracting fossil energy from the ground that would otherwise be unavailable. The by-product of "fracking" includes a large volume of fluids that need to be disposed, which have generally been injected deep into the ground. Both "fracking" and water disposal have been shown to trigger earthquakes, and there is significant exploration on how to reduce the amount of water that needs disposal, from recycling to re-use in injection. Furthermore, earthquakes created by *hydraulic fracturing* generally will cease following stimulation, while earthquakes attributed to injection can take months to years to reduce in frequency, as mentioned in the case of Oklahoma.

8.6 Earthquake early warning

Earthquake early warning (EEW) systems are designed to detect, analyze, and transmit information about an earthquake rapidly after an earthquake initiates. Such systems have existed in different forms since the late 1980's. The idea is even older. In 1868, J.D. Cooper envisioned a system whereby San Francisco could be alerted to impending shaking caused by earthquakes on the Hayward Fault using a telegraph-based system. Following the 1989 Loma Prieta (M7.1), which collapsed freeway structures near Oakland, CA, rescue workers used pager alerts sent via radio from USGS operated seismic instruments near the epicenter to alert them of impending aftershock shaking. Such shaking of already damaged and unstable structures threatened the lives of rescue workers and victims. Because the aftershock were roughly 90 km south, strong shaking from seismic waves did not reach the area until $\sim$ 30 s after the origin time, allowing some time for workers to vacate the unstable structures if needed.

The 1985, M_W 8.1, Michoacán earthquake that occurred along the Middle-American Trench killed more than 20,000 people roughly 300 km away in Mexico City. The large source-to-city distance allows up to a minute warning of the impending shaking. Mexico established an EEW for the region by 1991. Using radio receivers in schools, government offices, and TV and radio stations, broadcast alerts provide warning of imminent shaking. The initial system targeted earthquakes along the Guerrero Gap of the subduction zone specifically to warn residents of Mexico City. The current system, the Mexican Seismic Alert System (SASMEX, Fig. 8.16), covers portions of central and southern Mexico, and can provide up to a minute warning to Mexico City, Acapulco, Chilpancingo, Morelia, Puebla City, Oaxaca City and Toluca. In 2017, the system provided significant warning for Mexico City residents to expect shaking from the 2017 Tehuantepec (M8.2) earthquake. Less warning time was provided for the much closer Puebla-Morelos (M7.0), which caused significant damage in areas of high hazard zones within the city limits. Japan's 1995 Kobe earthquake, which killed 6000, initiated EEW development in Japan and lead to an operational system in 2007. Since their initial establishment, both of these EEW systems have been expanded and refined. Today, Mexico, Japan, South Korea, Taiwan have several years of experiences providing public alerts. India, Romania, Turkey, and the United States have deployed systems in focus regions, and systems are under development or newly established in of Chile, China, Costa Rica, El Salvador, Israel, Italy, Nicaragua, and Switzerland. Efforts to capture and use shaking information in mobile phones, which almost all contain reasonable-quality accelerometers are also underway.

Successful EEW faces a number of challenges related to the mix of scientific and social systems that must be integrated to establish and to operate a reliable system. Allen and Melgar (2019)

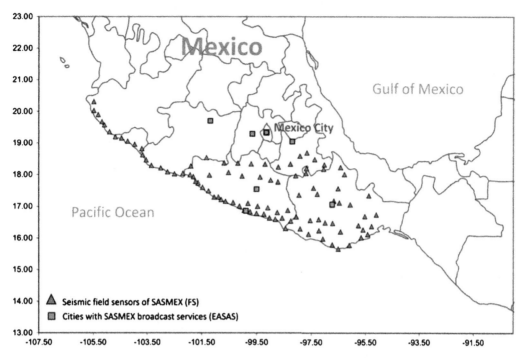

FIGURE 8.16 Distribution of Seismic Alert System of Mexico (SASMEX) strong-motion seismometers (triangles) and alternate emitters of seismic warnings designed to disseminate the alert in various Mexican cities (squares) (modified from Cuéllar et al., 2017).

list the key questions to be addressed to create a successful EEW system:

> How do you detect an earthquake? How do you determine the size (magnitude or otherwise) of the event and the distance to which ground shaking will be felt or damaging? How quickly can you do this, and how accurately? How do you choose the right trade-off between speed and accuracy?

> Who should receive alerts? How should the alerts be communicated (both message content and delivery technology) to different classes of users? How accurate do the alerts need to be? Since no system can be perfect, what is the tolerance for false and missed alerts? Who should pay for the system (government versus private sector/users)? Who is responsible for its successes and failures?

We focus on the first block of questions, which relate more directly to the underlying science – but the second block is no less important. Inter-

esting discussion of all the issues are included in Allen and Melgar (2019).

The foundation for all EEW systems is a real-time transmission of ground-motion observations and the development of algorithms that can use those data to produce a rapid and reliable estimate of the expected level of ground shaking as a function of location and time. This challenge integrates everything we know of earthquake processes and seismic-wave propagation with a requirement that decisions be made quickly. The establishment of real-time, dense ground-deformation monitoring networks in regions of substantial earthquake risk is essential. Two general approaches are adopted: first, characterize the source and then forecast the shaking based on the source model; second, detect the shaking and forecast future shaking based on current observations. The forecasts can be extrapolated over local-to-regional distances, or simply local using onsite observations.

Prediction of expected seismic ground motions is not a new problem, so once a source model is constructed, ground-motion predictions can be estimated. Most existing ground-motion models (GMM's) provide estimates of the average shaking levels as a function of distance from the rupture. But the actual strong shaking is quite variable and the models generally produce estimates that are reliable to perhaps a factor of two (or roughly one seismic intensity level). Since P waves are the first seismic waves to arrive, most analyses rely on this phase. Once an event is detected and signals from the event are associated (which is not trivial during vigorous aftershock sequences), the event size must be estimated. Many early algorithms used a point-source model and estimated the event size using the band-limited P-wave amplitude.

Full exploitation of the information in the early part of the P wave is perhaps best approached with a frequency-dependent analysis. Such approaches exploit characteristics of seismic signals such as the fact that very large, high-frequency observations indicate proximity to the source because at greater source-to-station distances, attenuation and scattering reduce the high-frequency signal components. At the other end of the spectrum, low-frequency signal components are an indication of a larger event. Adaptive, real-time, frequency-dependent analyses of observed ground motions can provide valuable early information as an earthquake initiates and grows in a region with a relatively dense seismic network. Of course an outstanding question, investigated for the last few decades and still debated, is whether the initial few seconds of P-wave motion provide reliable information on the ultimate size of an earthquake. Given that large earthquakes grow for tens to several hundred seconds, this could be considered a remarkable idea. But the research results on the issue are mixed. Some studies suggest that the event size is communicated in the initial ground motions and others suggest that with roughly ten seconds of data we can extract the answer. However, notable strong-motion studies indicate no difference in the onset of motion for large and small earthquakes located in the same region. Work will no doubt continue since the result is not only important for EEW but also addresses questions related to fundamental earthquake physics.

One important limitation of point-source approaches is magnitude saturation, which introduces an artificial cap on an EEW earthquake magnitude estimate (a $M\,9$ event might register as a $M\,8$ event). In addition, point-source approaches may underestimate shaking if the rupture is large and an incorrect distance (station-to-epicenter in place of station-to-rupture distance) is used to forecast the shaking. Except near the epicenter, the strength of shaking during large earthquakes is more dependent on the rupture properties immediately adjacent to a location than it is on the properties of the rupture near the hypocenter. Finite-fault models (FFM's) have been developed to minimize saturation and point-source prediction errors. In FFM's, the rupture finiteness is estimated as the earthquake grows. Accurately estimating the rupture size rapidly requires more than one seismometer. The more dense the seismic network, the more accurate earthquake growth can be tracked in real time. Constructing a reliable model of earthquake rupture in real time is a challenge. Several algorithms have been developed to approach the problem, and early results using the integration of seismic and geodetic observations are encouraging across a range of earthquake sizes. FFM's improve the ground motion predictions at distant from the event origin, but they require more data and take more time to compute.

Seismologists in Japan have developed approaches that avoid the need to construct a source model. In a region of dense seismic station coverage, the current, observed seismic

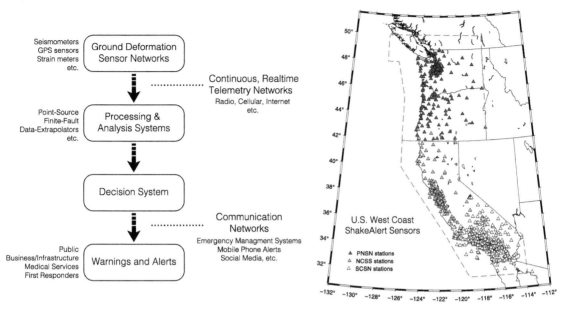

FIGURE 8.17 Summary diagram (left) of the main components of an earthquake early warning system (map from the US Geological Survey). Data flow is continuous as an earthquake develops and grows and the analysis and decisions are adapted to the most recent data. Map of USGS ANSS seismic stations used in the initial ShakeAlert System established along the west coast of the conterminous United States. Each triangle represents and observation point, the different shades correspond to different subnetworks.

shaking intensity is propagated 10 to 20 seconds into the future using numerically efficient techniques (based on radiative transfer theory). The propagation of local undamped motion (PLUM) approach includes rupture finiteness effects because they are part of the observations. As the rupture grows and shaking forecasts are updated, the region of expected large shaking increases. The result may be less warning time at large distances than a rapidly constructed and accurate model-based approach, but constructing an accurate source model is not trivial. The PLUM approach is operational in Japan and performed well during the 2016, M 7 Kumamoto earthquake. The empirical basis of the PLUM approach also works well during vigorous aftershock sequences that can confuse source-model based approaches. Waves arriving from more than one aftershock in different locations in-

crease the difficulty of constructing a source model in real time.

At this time, the state of the art in EEW is a rapidly developing competition of algorithms and approaches to produce the most timely and robust estimates of seismic shaking during an earthquake. Each large earthquake within one of the existing systems provides important information for system developers. Each approach has advantages and disadvantages that change with the nature of the earthquake and with the characteristics of the monitoring network. Even the metrics used to evaluate the performance of EEW systems are evolving. Given the differences, it is likely the most effective approach will include a combination of different algorithms and observations to perform well across a broad spectrum of earthquake types. The effort is sure to contribute to new observations and improvements in our understanding of earthquake nucleation and rupture.

Fig. 8.17 is a summary of the main components in an EEW. The arrows indicate a continuous flow of observations and data that must occur as a large earthquake grows. The essential communication systems transmitting the data and the alert information are a substantial effort. The decision system represents the algorithm used to decide what group of users should receive the alert at what stage in the process, balancing accuracy and timeliness. All aspect of the system are adaptive and continuously updating. The right side of Fig. 8.17 is a map of the seismic component of the western U.S. ShakeAlert System. Maintaining such large networks with realtime telemetry is a large investment, but the potential returns are greater. The challenge is substantial, but early deployments of EEW systems have significant public support and patience with regards to alarm accuracy. No one is tolerant of a system that produces false alarms. But when the cost is only a few moments of people's time, many are willing to accept an imperfect warning system that alerts them of possible imminent seismic shaking. If the strong shaking does not reach them is less important as long as the alert was based on an actual event.

8.7 Summary

This chapter is a review of elements for earthquake hazard assessment, forecasting, and the current state of scientific earthquake prediction. As stated previously, prediction has yet to be fully realized, but hazard mitigation based on systematic assessment of earthquake potential has advanced significantly these past decades. Furthermore, advances in data collection and analysis are providing a solid foundation for early warning systems that are being placed throughout the globe to warn of imminent shaking caused by large earthquakes.

9

Tsunami and tsunami warning

Chapter goals

- Define tsunami and briefly review recent and historic tsunami.
- Survey the process of tsunami excitation.
- Survey tsunami propagation physics.
- Survey tsunami measurement in deep water and coastal regions.
- Introduce systems used to provide timely tsunami warnings.

A *tsunami* is a series of sea waves that inundate coastal regions with heights ranging from a few centimeters to tens of meters. The term tsunami is derived from the Japanese words *tu*, for harbor, and *nami* for long-wave, and literally translated is long-wave-in-harbor (Imamura, 1937). Earlier in the history of seismology, the term used to refer to the same phenomena was *seismic sea-waves*. The signature characteristic of tsunami is the long wavelength of the waves. Each wave of a large tsunami can surge inland for 10-30 minutes then drain offshore for the same length of time. Tsunami have plagued coastal regions throughout history. Large tsunami demonstrate that the power of flowing water is seldom matched in its destructive capacity. The historical record includes numerous accounts of seismic and volcanic events that produced devastating tsunami. Many have lost their lives and livelihoods as a result of these phenomena. The longest historical records of tsunami are from Japan, which also suffers the greatest number of such events.

The last few decades have witnessed several devastating tsunami produced by large megathrust earthquakes. The 2004 Sumatra-Andaman Islands (M_W 9.2) earthquake created a large tsunami that impacted coastal communities throughout the Indian ocean, and caused more than 225,000 fatalities. In 2011, the Tohoku, Japan (M_W 9.0) earthquake created a large tsunami that caused nearly 20,000 fatalities, extensive damage, and resulted in the Fukushima Daiichi nuclear power plant meltdown. The 2010 Maule, Chile (M_W 8.8) earthquake produced a major tsunami along the coast of Chile, but caused many fewer deaths. The tsunami's disturbance of coastal waters was measurable for more than 12 hours, but not all impactful tsunamis are the result of large megathrust events. In 2018, a shallow, large (M_W 7.5) earthquake occurred along the Minahasa Peninsula in Indonesia, near a mountainous region in central Sulawesi. As a result of the event, the nearby provincial capital Palu was struck by a large, local tsunami that led to over 4000 deaths. Also in 2018, an eruption and collapse of Anak Krakatau (Anak meaning "child") volcano in Indonesia caused a deadly tsunami with waves up to five meters high that produced a death toll of over 400 and more than 30,000 injuries

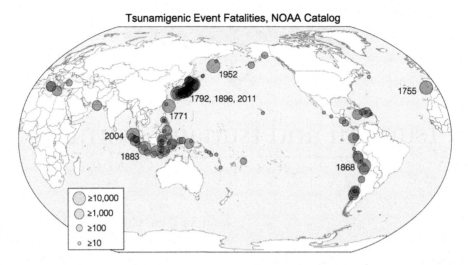

FIGURE 9.1 Distribution of tsunami from the Global Historical Tsunami Database of NOAA (doi:10.7289/V5PN93H7). All events with more than 10 fatalities, event validity scores (see the data base) greater than three, and sources attributed to earthquake and/or volcanic processes are included. The nine events with more than 10,000 fatalities are labeled using the year of the event.

(National Oceanic and Atmospheric Administration, NOAA). Anak Krakatau is the resurgent volcano that has formed since the devastating Krakatau eruption of 1883 in the Sunda Strait between the islands of Java and Sumatra. That eruption produced a large tsunami that wreaked havoc across the same region and killed more than 30,000.

Fig. 9.1 is a map of the distribution of historic tsunami listed in NOAA's Global Historical Tsunami Database (Table 9.1). Source locations for each event are identified by circles and the estimated tsunami death tolls are represented by four categories of circle size. Some quality control factors have been applied to limit the historic results to the most reliable events with earthquake or volcanic causes. As with earthquake catalogs, the data become less complete and less certain the farther back in time we look. The most reliable historic data are from Japan, which as you can see has experienced a large number of tsunami. The devastating impacts of tsunami have caused substantial human and economic losses throughout history. A

great challenge to understanding and preparing for tsunami is that it requires multiple disciplines, including seismology, basic hydrodynamics, oceanography, coastal engineering, and coordination with and among civil defense and emergency managements, as well as political officials from local and national governments.

Since water is relatively incompressible, tsunami attenuation is low and tsunami can propagate to great distances. Large tsunami are a hazard across entire ocean basins and historically, tsunami have arrived at many locations with no warning because the source was distant. Even today, warning systems are incomplete and imperfect, and although it may take hours for a large tsunami to reach distant shores, human losses are frequent in large tsunami, as is evident in the recent examples. The purpose of this chapter is to introduce basic tsunami-related concepts that overlap with seismology and to highlight current tsunami monitoring efforts, in which seismology plays an important role.

TABLE 9.1 Deadliest Tsunami in NOAA Global Historical Catalog.

Date	Location	Cause	Estimated Deaths
2004-12-26	Off W. Coast of Sumatra	Earthquake	227,899
1755-11-01	Lisbon	Earthquake	50,000
1883-08-27	Krakatau	Volcano	34,417
1896-06-15	Sanriku	Earthquake	27,122
1868-08-13	Southern Peru	Earthquake	25,000
2011-03-11	Honshu Island	Earthquake	18,431
1792-05-21	Shimabara Bay, Kyushu Island	Volcano	15,000
1771-04-24	Ryukyu Islands	Earthquake	13,486
1952-11-04	Kamchatka	Earthquake	10,000

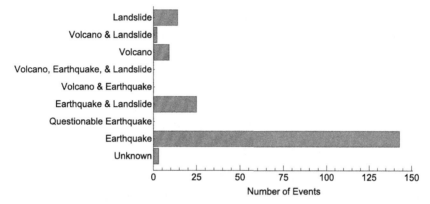

FIGURE 9.2 Bar chart of the number of different causes listed in the Global Historical Tsunami Database of NOAA (doi:10.7289/V5PN93H7). Labels correspond to the NOAA tsunami cause codes. Only events with event validity scores of 3 are included in the count, but even tsunami that produced no fatalities are included.

9.1 Tsunami excitation

Tsunami are produced by a sudden displacement of a large volume of water. The most common cause of the displacement is an uplift or subsidence of the ocean floor by large, shallow earthquakes (Fig. 9.2). The larger the earthquake (in terms of rupture area) the larger the potential tsunami. However, volcanic eruptions that produce relatively sudden water displacements also have resulted in large and deadly tsunami. Additionally, landslides crashing into water or submarine landslides have produced damaging tsunami. The November 18, 1929 M_W 7.2 Grand Banks earthquake produced a substantial submarine slump that triggered a deadly tsunami along the eastern Coast of Canada. The tsunami arrived roughly two and one-half hours after the earthquake and included several waves with heights between 2 and 8 m along the Burin Peninsula, Newfoundland (Natural Resources Canada).

Tsunamis caused by earthquakes are generally the result of vertical uplift and subsidence (e.g. Okal, 1988), although it is believed that at least part of the waves in regions of rugged bathymetry may result from sudden lateral offsets of sea-floor topography. The change in surface bathymetry induced by a reverse or normal faulting earthquakes can be computed if the rupture area and distribution of slip can be estimated or known. Strike-slip motions seldom directly excite large tsunamis, although they may trigger submarine landslides or cause tsunamis associated with lateral motion of rugged seafloor features. A particularly simple approach to estimating the sea-floor deformation is to ignore geologic complexity and use the solutions for surface deformation produced by a buried fault (see Box 9.1). The idea is illustrated conceptually in Fig. 9.3. In the car-

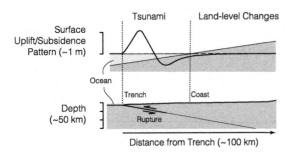

FIGURE 9.3 Conceptual illustration of tsunami excitation by a shallow-dipping subduction-style plate boundary thrust fault. The lower panel shows the physical model of plate-boundary. The upper panel shows the sea-floor and the vertical perturbation pattern caused by the earthquake. The event may produce uplift and subsidence of the sea floor and the coastal region (after Carvajal et al., 2017).

toon, slip on the fault is mostly near-horizontal, but includes vertical uplift and subsidence on the order of centimeters to meters. Larger displacements can be caused by steeper-angle splay faults in the overlying plate. Given a large enough area so disturbed, the sudden vertical water perturbation may excite a tsunami. Shallower earthquakes (closer to the trench) produce larger sea-floor deformation but the width of the rupture zone has strong influence on tsunami excitation. The specific pattern of uplift and subsidence depends on the fault dip and the distribution and direction of slip during the earthquake.

An example of the sea-floor deformation produced by a shallow subduction plate-boundary earthquake that occurred off the west coast of Sumatra in 2010 is shown in Fig. 9.4. Slip on the fault was estimated using seismic observations

and included a broad, slightly large, asymmetric regional shallow slip (U.S. Geological Survey). The slip distribution produced a roughly $A_u \sim$ 40 × 100 km^2 region of uplift near the trench and a smaller region of minor subsidence. Small

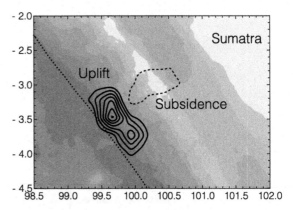

FIGURE 9.4 Estimated surface deformation produced by the M_W 7.8 25 October, 2010 Mentawai earthquake computed by the U.S. Geological Survey. The USGS routinely estimates finite-fault models (see Chapter 19) for large earthquakes. The Mentawai earthquake ruptured the shallow regions of the plate boundary and produced a significant and deadly tsunami. The surface deformation contour interval is 0.05 m, starting from ±0.1 m (regions of small uplift or subsidence do not produce tsunami). The maximum uplift was roughly 0.4 m, the maximum subsidence (on the Mentawai Islands) was roughly 0.1 m. Data from USGS website, (see Hayes, 2017, for details on methods).

details in the deformation pattern are averaged during tsunami excitation, but uplift of the region by an average amount of $\delta z \sim 0.2$ m displaces roughly 800 km^3, or 0.8 billion cubic meters of water. The potential energy in the uplifted water is roughly $E_u \sim M g \delta z = (\rho_w \cdot A_u \cdot z_w) g \delta z$, where M is the mass of the water, ρ_w is the density of water, and z_w is the water depth. Assuming an average water depth of $z_w = 3$ km, the energy transmitted to the water column in this instance is on the order of 2×10^{16} J. The event's seismic moment was 5.4×10^{20} Nm, which assuming typical earthquake properties would correspond to an energy radiated as seismic waves, $E_S \sim M_0 \cdot 5 \times 10^{-5} \sim 3 \times 10^{16}$ J, which

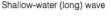

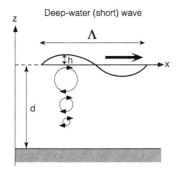

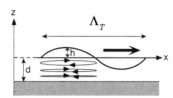

FIGURE 9.5 The terms shallow-water and deep-water refer to the relative sizes of the wave's wavelength and the water's depth. A deep-water wave has a wavelength that is small compared to the water depth (the *water* is relatively deep). A shallow water wave has a wavelength that is large compared to the water depth (the *water* is relatively shallow). (Left) Deep-water-wave, or short-wave, approximation. (Right) Shallow-water-wave, or long-wave, approximation. Energy of a long wave is distributed throughout the water column; energy of a short wave decays quickly with depth into the water. Tsunami are long-waves or shallow-water waves (modified from Satake and Tsunamis, 2015).

is on the same order as the potential energy change in the water. Calculations with more detailed slip distributions (e.g. Yue et al., 2014) that resolve larger slip near the trench, produce water-energy estimates on the same order as the smoother estimate in Fig. 9.4.

The tsunami is excited by the vertical displacement of the water away from its equilibrium position. Since water is relatively incompressible, the vertical change in seafloor position causes a relatively quick disturbance (uplift or subsidence) of the water column (and surface). The exact form of the water surface displacement depends on the earthquake deformation time history (slip and rupture), the properties of water, and the equation of fluid flow. The disequilibrium in water height is immediately assaulted by gravitational forces that operate to return the water surface to its equilibrium position (the ocean surface is a surface of equal gravitational potential). To first order and conceptually, we can assume that the water disturbance pattern is similar to the sea-floor deformation and the uplifted water becomes an initial condition for the fluid response that creates and propagates the tsunami. Perhaps most importantly, since the entire water column is raised (or lowered) the energy in a tsunami, a shallow-

water wave, is distributed throughout the entire water depth (Fig. 9.5). This is in contrast to the wind-generated, deep-water waves in which the deformation decreases quickly with depth (typically limited to depths of a few meters).

Tsunami are distinguished by particularly long periods (200-2000 s) and wavelengths of tens of kilometers. Tsunami wave amplitudes in the deep ocean range from centimeters to 5-10 m (close to the source) in height, but as the water shallows near the coast, the energy within the tsunami is focused into a smaller water column and the wave amplitude grows. Run-up (defined specifically later) of these long-period waves on shorelines can cause enormous destruction, overwhelming the standard storm-wave coastal defenses designed for much shorter-period waves. Truly large, damaging tsunamis are relatively rare, about one major event occurs per decade; but several large tsunami associated with earthquakes with $M \geq$ 8.5 have occurred in recent times, unleashing ocean-basin wide tsunami that killed hundreds of thousands of people in the Indian Ocean (2004 Sumatra) and tens of thousands in the Pacific Basin (2010 Chile, and 2011 Japan). We discuss efforts to reduce tsunami risk and impacts later in this chapter.

Box 9.1 Okada's Expressions for Halfspace Surface Deformation

The deformation of Earth's surface is a primary observation in earthquake science and static coseismic deformations contribute greatly to our understanding of earthquake ruptures and are required to compute earthquake-generated tsunami excitation. On land, these are the observations of InSAR and GPS, and they provide important near-field constraints on earthquake parameters and processes. For first-order work, surface observations are often modeled using simple, uniform elastic halfspace. For special cases, expressions for the surface deformation caused by faulting have been around since the 1950's. Several more general formulations were developed through the 1960's and 1970's. Yoshimitsu Okada (1985) developed consistent and clear expressions for surface deformation from point sources and rectangular faults. He separated the solutions into strike-slip, dip-slip, and tensile slip components (the components of a fault slip vector). The results can be combined to compute the deformation from an earthquake with arbitrary fault and slip orientations. The geometry is Cartesian with horizontal coordinates x, and y, and vertical coordinate z (positive upward). For a *point source* at the origin and depth $z = -d$, in a uniform isotropic halfspace with Lame parameters λ and μ, surface displacements produced by strike-slip offset on a surface striking in the x-direction and a dip of δ, are

$$u_x = -\frac{U_1}{2\pi}\left[\frac{3x^2q}{R^5} - I_1\sin\delta\right]\Delta\Sigma$$

$$u_y = -\frac{U_1}{2\pi}\left[\frac{3xyq}{R^5} - I_2\sin\delta\right]\Delta\Sigma \qquad (B9.1.1)$$

$$u_z = -\frac{U_1}{2\pi}\left[\frac{3xdq}{R^5} - I_4\sin\delta\right]\Delta\Sigma\;,$$

displacements produced by dip-slip offset are

$$u_x = -\frac{U_2}{2\pi}\left[\frac{3xpq}{R^5} - I_3\sin\delta\cos\delta\right]\Delta\Sigma$$

$$u_y = -\frac{U_2}{2\pi}\left[\frac{3ypq}{R^5} - I_1\sin\delta\cos\delta\right]\Delta\Sigma \qquad (B9.1.2)$$

$$u_z = -\frac{U_2}{2\pi}\left[\frac{3dpq}{R^5} - I_5\sin\delta\cos\delta\right]\Delta\Sigma\;,$$

displacements produced by tensile-slip offset are

$$u_x = \frac{U_3}{2\pi}\left[\frac{3xq^2}{R^5} - I_3\sin^2\delta\right]\Delta\Sigma$$

$$u_y = \frac{U_3}{2\pi}\left[\frac{3yq^2}{R^5} - I_1\sin^2\delta\right]\Delta\Sigma \qquad (B9.1.3)$$

$$u_z = \frac{U_3}{2\pi}\left[\frac{3dq^2}{R^5} - I_5\sin^2\delta\right]\Delta\Sigma\;,$$

where

$$p = y \cos \delta + d \sin \delta$$

$$q = y \sin \delta - d \cos \delta \tag{B9.1.4}$$

$$R^2 = x^2 + y^2 + d^2 = x^2 + p^2 + q^2 \, ,$$

and

$$I_1 = \frac{\mu}{\lambda + \mu} \, y \left[\frac{1}{R(R+d)^2} - x^2 \frac{3R+d}{R^3(R+d)^3} \right]$$

$$I_2 = \frac{\mu}{\lambda + \mu} \, x \left[\frac{1}{R(R+d)^2} - y^2 \frac{3R+d}{R^3(R+d)^3} \right]$$

$$I_3 = \frac{\mu}{\lambda + \mu} \left[\frac{x}{R^3} \right] - I_2 \tag{B9.1.5}$$

$$I_4 = \frac{\mu}{\lambda + \mu} \left[\qquad - xy \frac{2R+d}{R^3(R+d)^2} \right]$$

$$I_5 = \frac{\mu}{\lambda + \mu} \left[\frac{1}{R(R+d)} - x^2 \frac{2R+d}{R^3(R+d)^2} \right] \, ,$$

and $\Delta \Sigma$ represents the fault area and the slip vector, $D \hat{\mathbf{s}} = (U_1, U_2, U_3)$.

Okada's expressions are exceptionally convenient considering the number of terms and complexity of the calculation. They are widely used in earthquake science. The original work also includes explicit expression for a buried rectangular fault as well as the surface strains and tilts. A followup paper included expressions for the kinematic quantities at depth within the halfspace. Often, near-surface includes layering of geologic materials with a significant variation in properties – a useful reference for deformations from a point-source in layered media is Zhu and Rivera (2002).

9.2 Tsunami propagation

Roughly 70% of the Earth's surface is covered by water. P and Rayleigh waves transmit elastic energy through fluids. As noted earlier, fluids do not transmit shear stresses and thus they interact insignificantly with Love waves and horizontally (boundary parallel) polarized S waves. Vertically polarized S waves can transmit energy into and out of a fluid across a fluid-solid boundary where they couple with P and Rayleigh waves. We can model these interactions using elastic wave theory, which describes the interaction of inertial and elastic-restoring forces. However, fluids can also transmit energy via *gravity waves* that involve the interaction of inertial and gravity forces. Tsunami are gravity waves. To analyze tsunami mechanics, we adopt an *Eulerian* formulation, in which we consider the behavior of a material element according to its position at a particular time. Individual particles may flow into or out of the material element.

The Eulerian formulation is more convenient to describe gravity waves that may produce large particle motions in the fluid.

In tsunami generation, gravitational energy may constitute more than 95% of the energy (the rest is compressional energy in the slightly compressible water and compressional and shear energy in the underlying rock). Results from the field of fluid mechanics (e.g. Okal, 1982; Ward and Tsunamis, 1989) provide a wave equation for tsunami wave height, $h(\mathbf{x}, t)$,

$$\frac{\partial^2 h}{\partial t^2} = g \, \nabla \cdot (d \, \nabla h) \, , \qquad (9.1)$$

where $d(\mathbf{x})$ is the depth of the water and g is the acceleration due to gravity, which is roughly constant near Earth's surface. Eq. (9.1) behaves differently for end-member values of the ratio of the water depth to the wavelength (Fig. 9.5). At relatively short wavelengths ($\Lambda << d$), the wave velocity is $c(\Lambda) = (\Lambda_T \, g/2\pi)^{1/2}$, which corresponds to a dispersive wave with motions that decay exponentially away from the water surface (Fig. 9.5). In contrast, tsunami are dominated by components with long wavelengths, Λ_T, where, $\Lambda_T >> d$. In the long-wave regime, Eq. (9.1) can be reduced to the wave equation

$$\frac{\partial^2 h}{\partial t^2} = c_T^2 \nabla^2 h \, , \qquad (9.2)$$

where the velocity $c_T = \sqrt{g\,d}$ is non-dispersive and depends solely on water depth. In this case, the displacements in the vertical and radial directions vary linearly with depth. The tsunami velocities for periods of 200-2000 s are on the order of 700-900 km/hr in the open ocean, which is about the speed of a large commercial passenger jet.

If we measure the average speed a tsunami traveled (distance/time), we can use $c_T = \sqrt{g\,d}$ to estimate the average depth of the ocean along the path. If we combine observations corresponding to many paths, we can estimate the mean ocean depth. Fig. 9.6 is a plot of roughly

27,000 tsunami travel times in the Global Historical Tsunami Database of NOAA. A visual fit to

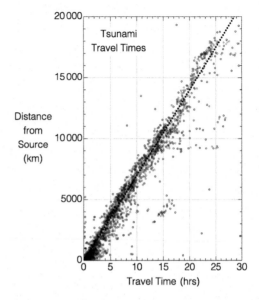

FIGURE 9.6 Observed tsunami travel times versus distance from source contained in the Global Historical Tsunami Database of NOAA (doi:10.7289/V5PN93H7). Most of the 26,948 are close to the source, but enough are distant to estimate a reasonable average slope. The dotted line shows a line corresponding to an average speed of 700 km/hr (a visual fit).

the observations (shown in the figure) suggests an average depth of about 3.9 km. A modern estimate of the mean ocean depth from more precise methods is about 3700 m. The tsunami result is reasonably accurate – the 200 m difference is easily within the uncertainty of the visual fit. Tsunami travel times were used long ago to estimate ocean depths, when the depths of the deep oceans were less well known and required substantial effort to measure.

We do not have to restrict our attention to wavelengths that are much larger or much smaller than the water depth. We can numerically model signals to include all wavelengths of interest. Such calculations directly include the effects of dispersion. A *dispersion curve* is a plot of the speed of a wave versus frequency or period (see Chapter 14). Tsunami can be considered an interference pattern that propagates through the ocean (much like surface waves within the Earth and oceans). Both tsunami and surface waves also have two period-dependent speeds that refer to the propagation of the packet as a whole (*group velocity*), and to the rate of change in phase within the interference packet (*phase velocity*). The two speeds are related and when dispersion is minor, the two speeds are roughly equal. Theoretical tsunami group and phase-velocity curves for a homogeneous self-gravitating Earth model covered by oceans that are 2, 4, and 6 km deep are shown in Fig. 9.7. The change in speed with ocean depth produces propagation speeds that range from roughly 150 m/s for a 2 km deep to 250 m/s for a 6 km deep ocean model. Short-period tsunami os-

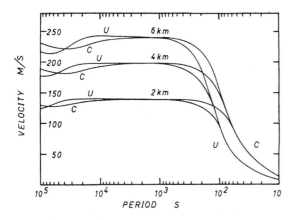

FIGURE 9.7 Tsunami dispersion curves for oceans 2, 4, and 6 km deep on a spherical planet (from Ward, 1989). When the curves are flat dispersion is minor (and the phase and group velocity are roughly equal). At the very longest periods and for periods less than about 200 s periods, the dispersion is substantial.

cillations travel slower (and arrive later), with speeds for periods between 100 and 10 s averaging roughly 50 m/s. The transition from a relatively flat dispersion curve (little dispersion) to a relatively rapid decrease in speed with decreasing period occurs at slightly longer periods for the deeper ocean models.

The lateral variation in ocean depth results in a laterally varying velocity field for tsunami waves – this produces tsunami-wave refractions. Focusing and defocusing occur, resulting in nonuniform tsunami amplitudes. Modern methods account for this by either computing tsunami waves with numerical methods for a laterally varying ocean model or by tracing rays through the velocity field to determine where strong focusing occurs. Fig. 9.8 illustrates the effects of ocean depth variations on the tsunami travel times (wavefronts) and the amplitude for the 2004 Sumatra–Andaman and the 2011 Tohoku earthquakes. Wavefront distortion and variations in the amplitude patterns (shown by shading) are the result of tsunami refraction. For example, islands may disrupt the wavefront as tsunami diffract around them. The hours needed for a tsunami to cross the ocean basins permits the development of ocean-wide tsunami warning systems. Following a series of large earthquakes and tsunami in the 1950's and 1960's the Pacific Ocean (and several regional) tsunami warning centers were established (discussed later in this chapter).

In addition to the warning applications, knowledge of the average tsunami speed from one location to another allows us to locate the source region that produced a tsunami using the arrival time of tsunamis on tide gauges that are azimuthally distributed around the source. The ideas are a two-dimensional version of earthquake location presented in Chapter 6. The decay of tsunami amplitude with distance is approximately represented by $1/\sqrt{r}$, corresponding to two-dimensional spreading of energy. The seafloor motion in the source region can be estimated by correcting the observed tsunami amplitudes for geometric spreading and possible local bathymetric effects near the observation location. In detail, tsunami excitation depends on the geometry of faulting, the depth of faulting, and the time history of faulting. Numerical calculation of the full propagation effects allows complete modeling of tsunami waveforms to es-

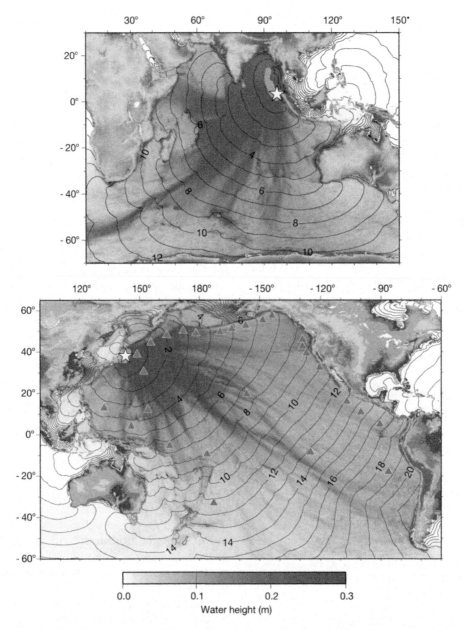

FIGURE 9.8 Tsunami propagation computed from the 2004 Sumatra–Andaman earthquake (top) and the 2011 Tohoku earthquake (bottom). The curves indicate tsunami travel times, which are plotted in one-hour increments from the tsunami source. Stars indicate the epicenters of the mainshock (data from USGS). The shaded background indicates the model-computed maximum water height (or energy) distribution. The triangles in the right panel show the locations of the ocean-bottom pressure gauge stations (deep-ocean assessment and reporting of tsunamis) (modified from Satake and Tsunamis, 2015).

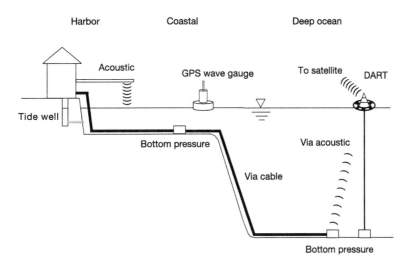

FIGURE 9.9 Various types of instruments used to measure tsunamis (from Satake and Tsunamis, 2015).

timate fault slip on submarine earthquakes (See Box 19.2).

9.3 Tsunami observation and monitoring

Most historical tsunami are known by the damage that they produced on land. Tsunami can be measured in the ocean using a suite of different instruments (Fig. 9.9). Measurements far from coastal regions are made using DART© (Deep-ocean Assessment and Reporting of Tsunamis) sensors that measure changes in pressure near the seafloor associated with changes in water height. Following the 2004 Sumatra earthquake, the number of DART sensors was increased and many of the sensors operated were made available online in near real-time. Closer to shore measurements can be made with pressure sensors, sea-height monitoring systems, and water-well based technology. Modern satellites can measure large tsunami from space, as was the case following the 2004 Indian Ocean tsunami.

Water-pressure sensors in the deep ocean can record the passage of tsunami larger than a few

millimeters, which is impressive given that the ocean depth is measured in kilometers. Such observations are used to track the wave as it traverses ocean basins. Most tsunami records are from tide gauges in harbors. The shallowing of the water in the harbor, along with geometric focusing, influences the peak amplitude of a tsunami. This effect is caused by the decrease in velocity as the depth shoals and the kinetic energy of the wave is transformed into gravitational energy by increasing the wave height.

Fig. 9.10 shows a tide gauge record of the 2004 tsunami on a 11 day tidal oscillation. The observations are from Colombo, Sri Lanka about 12° from the 2004 Sumatra-Andaman Island (M_W 9.2) earthquake. Colombo is on the west coast of Sri Lanka (the far side of the island from the earthquake) but still recorded a water height of over 2.5 m (Merrifield, 2005). Many currently operating tide stations are part of the Global Sea Level Observing System (GLOSS), an international sea level monitoring program. GLOSS station locations are shown in Fig. 9.11. GLOSS was established by the UNESCO Intergovernmental Oceanographic Commission (IOC) in 1985, and is designed to produce high-quality in situ

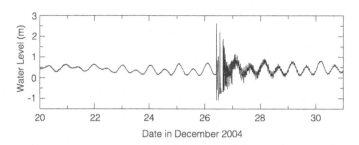

FIGURE 9.10 Example water level tide-gauge record from Colombo, Sri Lanka, from 20 to 31 December, 2004 showing the scale of the 2004 Sumatra-Andaman Island (M_W 9.2) tsunami with respect to the tidal range and also the presence of relatively high frequency oscillations subsequent to the tsunami (modified from Pattiaratchi and Wijeratne, 2009).

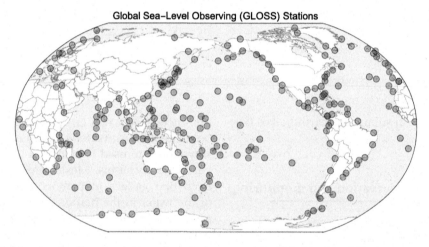

FIGURE 9.11 Map showing the current tide stations that are part of the Global Sea Level Observing System (GLOSS), an international sea level monitoring program (locations from https://www.gloss-sealevel.org/operational-status).

sea level observations. As with any natural phenomenon, observations of water-level variations caused by tsunami are essential for deepening our understanding of the processes that produce and propagate these waves.

Deep ocean observations are key to providing information on the status of a propagating tsunami. Although tsunami can be detected using satellites, to date, satellite methods cannot measure tsunami as accurately, reliably, and within time constraints required to forecast tsunami in real time. As part of the U.S. National Tsunami Hazard Mitigation Program (NTHMP), deep-ocean instruments (called

tsunameters) were developed at NOAA's Pacific Marine Environmental Laboratory (PMEL) for the early detection, measurement, and real-time reporting of tsunamis in the open ocean (Project DART). The original six DART sensors were placed strategically on the Pacific Ocean floor near regions of historic tsunami. By 2008, NOAA had expanded the DART network from 6 to 39 stations in the Pacific and Atlantic oceans. Following the development of the 3rd generation sensors, the network was expanded to 40 stations by 2010. Other countries have joined the effort – Fig. 9.12 is a map of recently operating DART stations shared by seven nations.

International DART Stations (from NOAA)

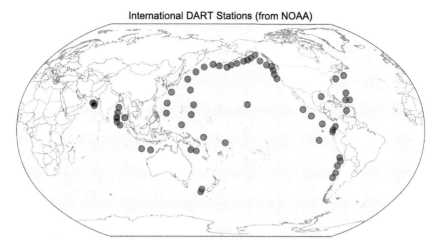

FIGURE 9.12 DART©(Deep-ocean Assessment and Reporting of Tsunamis) sensor network reporting on 25 March, 2020 (locations from https://nctr.pmel.noaa.gov/Dart/). Station shown include those operated by the United States, Australia, Chile, Colombia, Ecuador, India, and Thailand.

DART observations are rich in information. Fig. 9.13 shows a deep-water DART record of the 2011 M_W 9.0 Tohoku earthquake tsunami. The time range in each panel decreases from two weeks at the top to just over four hours at the bottom. The relatively regular, roughly 12 hour oscillations in the top panel are tide signals (which exhibit an interesting but expected change in period across the time range). The earthquake-generated signals pass during the time including the rapidly changing amplitude pattern starting abruptly on 11 March. The middle panel shows the Rayleigh Wave and tsunami signals. The dashed line is the earthquake origin time. The tsunami signals are clearly visible for more than a day as gravity waves propagate throughout the Pacific Basin. The bottom panel shows a more clear separation of the seismic signals (primarily $R1$) and the tsunami, which appears as a slightly dispersed pulse. The first dashed line is the earthquake origin time, the second corresponds to a velocity of 3.5 km/s (the Rayleigh wave speed) and the third corresponds to a velocity of 0.28 km/s (the tsunami speed). The sampling rate of the DART signal is insuf-ficient to resolve details in the seismic waves, but represents the tsunami signal well. These observations contain information on tsunami size and time important for real-time warning, and information on the tsunami excitation that can be used to help constrain the earthquake characteristics.

9.3.1 Onshore tsunami measurements

From a hazard mitigation point of view, the onshore measures of tsunami are most important. The *flow-depth* of a tsunami is the height of the water measured from the ground level, the *inundation height* is the height measured relative to the sea level at the time of the tsunami's arrival (Fig. 9.14). The *run-up* is the maximum vertical height above sea level reached by a tsunami. In other words, the run-up is equal to the maximum inundation height. The *inundation distance* is the maximum horizontal distance a tsunami reaches inland, measured from the coast. All measurements are adjusted to account for tide levels during the tsunami. Since the sea surface of the inundated tsunami is not flat in most cases, the run-up height generally

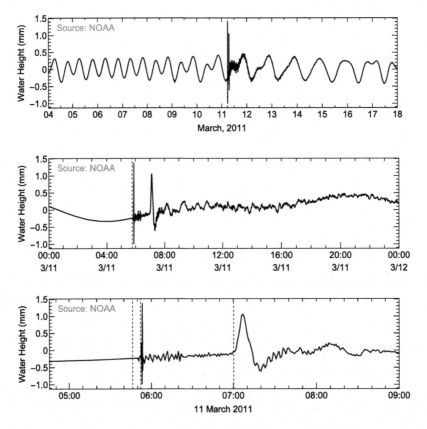

FIGURE 9.13 Deep sea DART deformation record recorded at Station 21413, roughly 1000 km southeast of Honshu, Japan. The mean value has been subtracted from the observations. Three complementary time ranges are used to progressively focus on the tsunami signal produced by the Great M_W 9.0 Tohoku Earthquake of 2011-03-11 05:46:24 (UTC). The dashed lines show the origin time, and travel times corresponding to Rayleigh- and tsunami-wave propagation speeds of 3.5 km/s and 280 m/s respectively. Data from NOAA's National Data Buoy Center (https://www.ndbc.noaa.gov/dart.shtml).

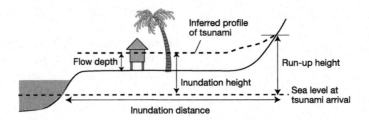

FIGURE 9.14 Definitions of tsunami heights and inundation distances on land (from Satake and Tsunamis, 2015).

does not equal the inundation height or water depth at the shore. The inundation height can be measured from water marks on buildings, the run-up distance is often measured using the maximum distance material is carried landward by the tsunami.

Box 9.2 Cascadia Subduction Zone and the Potential for a Tsunami

The Cascadia Subduction Zone (CSZ) stretches from northern Vancouver Island to Cape Mendocino in California, roughly 1000 km along the northwest coast of the conterminous United States and southwestern Canada (Fig. B9.2.1). Along the CSZ, the relatively young Explorer, Juan de Fuca, and Gorda plates subduct beneath the North America plate. In the south, the CSZ terminates in the Mendocino Triple Junction region, where the CSZ and the San Andreas and the Mendocino Transform faults meet. In the north, the CSZ terminates where it meets the Explorer-Pacific plate boundary and the Queen-Charlotte Fault. The CSZ is remarkable and unique in that it has not hosted any notable earthquakes outside the triple-junction regions for many decades (Fig. B9.2.1). The subducted plate has hosted several damaging strong and major events beneath Washington.

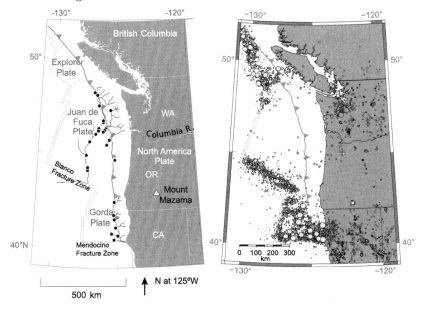

FIGURE B9.2.1 (left) Map showing tectonic plates, faults, and the location of turbidites deposited after the eruption of Mount Mazama (CraterLake), Oregon, 7800–7500 sidereal years ago (modified from Atwater and Griggs, 2012). (right) Map of earthquakes with $M \geq 3$ from 1900 to 2020 illustrating the remarkable lack of seismicity along the CSZ. Lighter shaded symbols represent shallow (≤ 33 km) events, darker shaded symbols represent deeper events.

Historical tsunami records, surface tsunami sedimentary deposits, tree-ring analyses, and Native-American oral histories all indicate a large (M_W 8.7-9.2), megathrust earthquake ruptured the CSZ on 26 January, 1700 (Satake et al., 1996; Satake, 2003; Atwater, 2005). Geological evidence, including sedimentary deposits observed in drill cores from deep-sea channels and fans and produced by shaking-triggered subaqueous slumps or slides (turbidites) provides a less precise, but longer large-earthquake history. These observations suggest an average giant earthquake-recurrence interval of about 500 years, but partial, but still substantial, rupture of the boundary may recur with a shorter, 300-year average.

Although the CSZ seismogenic zone is locked, the deeper regions of the plate boundary are active. Recent geodetic and seismic work has revealed repeated slow earthquakes occur beneath the locked section (e.g., Dragert, 2001). The deformation, called **Episodic Tremor and Slip (ETS)** (see Box 3.5), suggests that slow slip on the deep subduction interface could be stress loading the CSZ. The "quiet" CSZ subduction zone could generate a highly impactful, great earthquake in the near future.

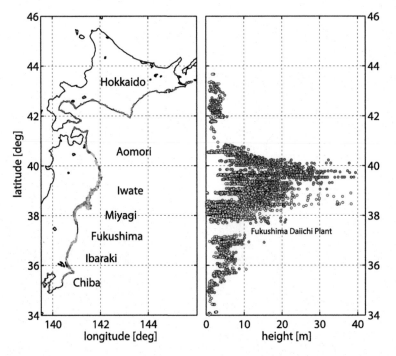

FIGURE 9.15 Maximum measured local tsunami heights for the Great M_W 9.0 2011 Tohoku Earthquake. Light symbols correspond to inundations, darker shading identifies run-up measurements (modified from Mori et al., 2011).

Fig. 9.15 is a plot of inundation heights and run-up heights measured along the coast of Honshu following the 2011 M_W 9.0 Tohoku earthquake. The observations are displayed as a function of latitude (the coast is oriented roughly north-south). The maximum run-up height was just under 40 m. Heights of more than ten meters span a range of 200-300 km of the coast. The gap in the data is the region of the Fukushima Daiichi nuclear power plant that suffered severe damage as a result of power failures caused by the tsunami. Radiation release prohibited measurement collection in the region.

9.4 Tsunami forecasting and warning

Tsunami have been a substantial hazard throughout history (Fig. 9.16). As populations have grown and coastal regions have attracted more people, the risk has increased. Recent

Box 9.3 Tsunami Earthquakes

An earthquake that generates a tsunami is called **tsunamigenic**. Most large shallow subduction zone earthquakes are in this class. They have large magnitudes and produce strong and sometimes prolonged shaking (Table B9.3.1). For many near the coast, the strong short-period shaking is the warning that a tsunami may be approaching. However, some moderate to large magnitude earthquakes, often located at the shallowest depths near the trench, generate large tsunami without strong short-period shaking. We call these events **tsunami earthquakes** and their existence creates a challenge for tsunami warning systems that rely on short-period body-wave and surface-wave magnitudes to detect events. The 1946 Aleutian earthquake (M_S 7.4) created an unusually large and deadly tsunami that struck the Hawaiian Islands. The event's M_S is substantially lower than the estimated M_W. A more recent example is the M_W 7.8 Mentawai earthquake of 25 October, 2010 (Lay et al., 2011), which had an M_S of 7.4 and was used to illustrate tsunami excitation in Fig. 9.4. Tsunami events must be analyzed using long-period signals to ensure that magnitude saturation does not influence the conclusions about potential tsunami. A number of methods have been developed for this purpose such as long-period P wave magnitudes, long-period surface wave magnitudes ($T \sim 20$ s), and W-Phase inversions.

TABLE B9.3.1   This table is a list of properties for unusual earthquakes that have occurred since 1896. These unique earthquakes share similar source characteristics: they are shallow, have long durations, and slow ruptures.

Region	Date	M_w	M_o (Nm)	Depth (km)	Duration (s)	NSD (s)
Japan	1896/06/15	8.0	1.2×10^{21}	0 – 10	100	9.9
Alaska	1946/04/01	8.2	2.3×10^{21}	0 – 10	100 – 150	10.0
Peru	1960/11/20	7.6	3.4×10^{20}	9	125	19.5
	1996/02/21	7.5	1.9×10^{20}	7 – 10	50	9.2
Kuriles	1963/10/20	7.8	6.0×10^{20}	9	85	10.5
	1975/06/10	7.5	2.0×10^{20}	5	80 – 100	16.1
Nicaragua	1992/09/02	7.7	4.2×10^{20}	0 – 10	125	17.5
Java	1994/06/02	7.8	5.2×10^{20}	15	80	-
	2006/07/17	7.8	6.7×10^{20}	20	185	-
Sumatra	2010/10/25	7.8	6.77×10^{20}	12	80	9.5

NSD is the normalized-source duration (divide the duration of the event by the cube-root of seismic moment). Ideally the scaling would result in an NSD of roughly one second (although scatter about that value is expected). The NSD of the events are an order of magnitude larger than those of typical earthquakes. That indicates a long duration, presumably associated with a slow rupture process, somehow related to the location of the event near the trench. One idea is that the low-rigidity of the material near the trench reduces the rupture speed and increases the earthquake duration, and decreases the amplitudes of shorter-period seismic waves. The presence of low rigidity material results in larger slip at shallow depth for a given seismic moment, and the large slip produces more seafloor deformation that enhances the tsunami excitation.

decades have witnessed tsunami catastrophes across the Indian Ocean Basin and along the coast of Japan, one of the most prepared regions of the world. The largest, ocean-basin

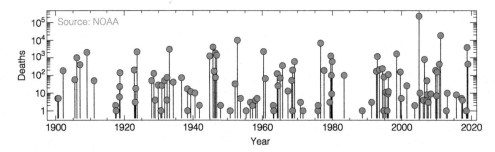

FIGURE 9.16 Timeline of deadly tsunami since 1990 listed in the Global Historical Tsunami Database of NOAA (doi:10.7289/V5PN93H7). All events with one or more fatalities and event validity scores greater than or equal to three, are included.

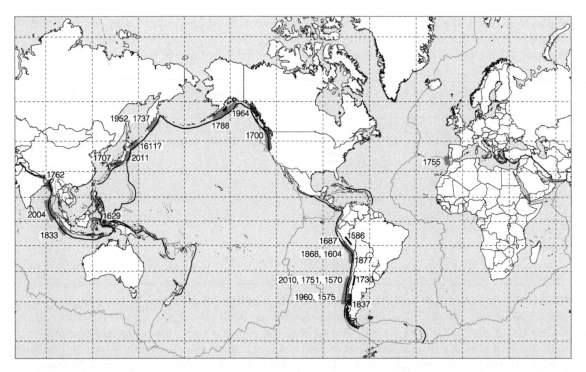

FIGURE 9.17 Location of giant earthquake ruptures from the Onur-Muir-Wood Catalog. Ruptures are idealized as roughly rectangular regions. Where ruptures substantially overlap, only the most recent event is shown and years of all ruptures listed (after Onur and Muir-Wood, 2014).

wide tsunami are generally caused by the largest seismic events. Onur and Muir-Wood (2014) constructed a list of 25 giant ($M_W \geq 8.8$) earthquakes spanning the last few centuries (Table 9.2, Fig. 9.17). The list is based on a number of observations (when known), including spatial extent of significant shaking damage, shaking duration, tsunami impacts, and information on coastal elevation changes. The list is believed to be complete back to 1700 AD and largely com-

TABLE 9.2 Giant ($M_W \geq 8.8$) Earthquakes (source: Onur and Muir-Wood (2014)).

Date	Estimated Magnitude	Location
2011 03 11	9.0	Tohoku, Japan
2010 02 27	8.8	Maule, Chile
2004 12 26	9.0	Sumatra-Andaman, Indonesia
1964 03 28	9.2	Alaska, US
1960 05 22	9.6	Valdivia, Chile
1952 11 04	9.0	Kamchatka, Russia
1877 05 10	9.1	Northern Chile
1868 08 13	9.2	Peru/Chile
1837 11 07	8.9	Southern Chile
1833 11 25	9.0	Southern Sumatra, Indonesia
1788 07 22	9.2	Alaska, US
1762 04 02	8.8	Bay of Bengal, Myanmar
1755 11 01	8.9	SW Iberia Atlantic Margin
1751 05 25	9.2	Concepción, Chile
1737 10 16	9.0	Kamchatka, Russia
1730 07 08	8.9	Valparaiso, Chile
1707 10 28	8.8	Nankai, Japan
1700 01 26	9.0	Cascadia, Canada/US
1687 10 20	8.8	Peru
1629 08 01	9.0	Banda Arc, Indonesia
1611 12 02 (?)	9.2	Hokkaido, Japan
1604 11 24	9.1	Peru/Chile
1586 07 10	9.1	Peru
1575 12 16	9.2	Valdivia, Chile
1570 02 08	8.8	Southern Chile

TABLE 9.3 Tsunami Impact and Warning Times (source: ITIC).

Tsunami Type	Typical Time to Impact	Target Response Time
Local	0-1 hr	2-5 min
Regional	1-3 hr	5-10 min
Distant	> 3 hr	10-20 min

plete back to 1550 AD (Table 9.2). Eighteen earthquakes in this size range have occurred since 1700, and the largest number in any particular region is along the west coast of South America.

First motivated by the 1946 Aleutian tsunami that struck the Hawaiian Islands and killed over 150 people, the United States has since developed, deployed, and refined tsunami warning systems. The large 1946 tsunami that hit the Big Island roughly five hours after the earthquake caught the region off guard. Although tsunami travel relatively fast (0.2 km/s), they are much slower than seismic waves ($5 - 10$ km/s), and for many years seismologists have analyzed seismic waves to detect large, possibly tsunamigenic earthquakes quickly and to issue a tsunami

warning when warranted. Many of the data used to detect large earthquakes are the same as those used in routine seismic analysis of large earthquakes (e.g., Fig. 4.1). For regional efforts, where the time required to issue a useful warning is much shorter, instruments have been deployed to detect large earthquakes quickly (e.g. the Cascadia region of western North America).

NOAA developed a Pacific Ocean-basin tsunami forecast system that produces estimates of tsunami watches and warning and tracks propagating tsunami. The National Tsunami Warning Center (NTWC), located in Alaska, issues tsunami warnings along the US and Canadian coasts (excluding Hawaii). At the NTWC and the Pacific Tsunami Warning Center (PTWC) in Hawaii, generally staffed 24 hours a day, 7 days a week, seismic data from the USGS are collected, processed, and analyzed after a large earthquake. Then the potential tsunami (size and inundation) is assessed and forecasts developed (including estimates of the tsunami arrival time). Data collected from tide gauges and tsunameters (DART buoys) is used to corroborate or update forecasts. For the procedure to work, data must be accessible in real time.

Many countries have since joined efforts to create comprehensive tsunami warning (and preparation efforts) across the globe. Many cooperate with the UN's International Tsunami Information Center (ITIC). Tsunami warning systems can be grouped into broad ocean-basin monitoring (such as the Pacific or Indian Ocean), regional, and local systems, each with different lead times to produce a useful warning (hours to minutes). The short response time for local warnings from tsunamis requires an educated local population to recognize the potential for a tsunami (following a nearby earthquake, that may, for example, have been felt). In fact, education is an essential component for any warning system to succeed. Similar to earthquakes (see Chapter 8), many details on who needs to be alerted, and how the information will flow from the warning centers to emergency managers, other government officials, and eventually the public, must be carefully planned and rehearsed.

9.5 Summary

This chapter is an introduction to the excitation, propagation, and measurement of tsunami, devastating waves that can propagate across entire oceans and that have killed hundreds of thousands of people in the last century. Forecast modeling allows for the size of a tsunami to be adequately estimated before the arrival of the wave far from the source. Local tsunamis remain difficult to quickly assess since they strike soon after the shaking stops. More time is available for regional and distance tsunami for which propagation across the ocean can take hours, leaving enough time for coastal communities to prepare for an imminent tsunami if accurate detection and warnings are issued, and response plans are in place and up to date.

Chapter goals

- Review basic structure of the Earth.
- Review how we image the Earth at different scales using seismic waves.
- Review how improved data collection and advanced techniques continue to change our understanding of Earth's processes.

Seismic waves provide a probe of the Earth's deep interior, and their predictable behavior makes it is possible to obtain high-resolution models of some of the Earth's internal properties, as will be discussed in later Chapters 12 and 19. A model is a simplified mathematical representation of the actual three-dimensional (3-D) material property variations within the planet. Seismology provides primary constraints on the variations of density, rigidity, and incompressibility and secondary constraints on the temperature field at all depths in the Earth. Interpretation of the actual chemistry, physical state, and dynamic behavior associated with the seismological structure requires experimental and modeling results from other disciplines such as mineral physics and geodynamics. Nonetheless, seismology has the primacy of providing our best resolution of the actual structure of the planet.

10.1 Global Earth structure

All different seismic wave types have been analyzed in determining Earth structure, ranging from free oscillations of the planet to high-frequency body waves reflected from shallow sedimentary layers. The waves reveal aspects of the Earth incorporated in Earth models, functional descriptions of how the material properties vary in the interior. A large number of body-wave travel times, free-oscillation eigenfrequencies, plus surface-wave and normal attenuation measurements were modeled in constructing the Preliminary Reference Earth Model (PREM) (Dziewonski and Anderson, 1981) (Fig. 1.21). The parameters of this model at a reference period of 1 s are given in Table 10.1, including density, P velocity, S velocity, shear attenuation coefficient (Q_μ), the adiabatic bulk modulus (K_s), rigidity (μ), pressure, and gravity. The PREM model includes anisotropic upper-mantle layers that are not listed here, and the velocities vary with reference period because of anelastic dispersion. Note that the variation of the attenuation coefficient with depth is very simple, with only a few-layer model being resolved. The core has very high values of Q, with almost no seismic wave attenuation, and attenuation of body waves, surface waves, and free oscillations requires a relatively low Q in the upper mantle. Although the PREM Q model is reasonable for globally averaging waves such as free oscilla-

TABLE 10.1 Parameters of the Preliminary Reference Earth Model (PREM) at a reference period of 1 s.

Radius (km)	Depth (km)	Density (g/cm^3)	V_p (km/s)	V_s (km/s)	Q_μ	K_s (kbar)	μ (kbar)	Pressure (kbar)	Gravity (cm/s^2)
0	6371.0	13.08	11.26	3.66	85	14253	1761	3638.5	0
200.0	6171.0	13.07	11.25	3.66	85	14231	1755	3628.9	73.1
400.0	5971.0	13.05	11.23	3.65	85	14164	1739	3600.3	146.0
600.0	5771.0	13.01	11.20	3.62	85	14053	1713	3552.7	218.6
800.0	5571.0	12.94	11.16	3.59	85	13898	1676	3486.6	290.6
1000.0	5371.0	12.87	11.10	3.55	85	13701	1630	3402.3	362.0
1200.0	5171.0	12.77	11.03	3.51	85	13462	1574	3300.4	432.5
1221.5	5149.5	12.76	11.02	3.50	85	13434	1567	3288.5	440.0
1221.5	5149.5	12.16	10.35	0	0	13047	0	3288.5	440.0
1400.0	4971.0	12.06	10.24	0	0	12679	0	3187.4	494.1
1600.0	4771.0	11.94	10.12	0	0	12242	0	3061.4	555.4
1800.0	4571.0	11.80	9.98	0	0	11775	0	2922.2	616.6
2000.0	4371.0	11.65	9.83	0	0	11273	0	2770.4	677.1
2200.0	4171.0	11.48	9.66	0	0	10735	0	2606.8	736.4
2400.0	3971.0	11.29	9.48	0	0	10158	0	2432.4	794.2
2600.0	3771.0	11.08	9.27	0	0	9542	0	2248.4	850.2
2800.0	3571.0	10.85	9.05	0	0	8889	0	2055.9	904.1
3000.0	3371.0	10.60	8.79	0	0	8202	0	1856.4	955.7
3200.0	3171.0	10.32	8.51	0	0	7484	0	1651.2	1004.6
3400.0	2971.0	10.02	8.19	0	0	6743	0	1441.9	1050.6
3480.0	2891.0	9.90	8.06	0	0	6441	0	1357.5	1068.2
3480.0	2891.0	5.56	13.71	7.26	312	6556	2938	1357.5	1068.2
3600.0	2771.0	5.50	13.68	7.26	312	6440	2907	1287.0	1052.0
3800.0	2571.0	5.40	13.47	7.18	312	6095	2794	1173.4	1030.9
4000.0	2371.0	5.30	13.24	7.09	312	5744	2675	1063.8	1015.8
4200.0	2171.0	5.20	13.01	7.01	312	5409	2559	957.6	1005.3
4400.0	1971.0	5.10	12.78	6.91	312	5085	2445	854.3	998.5
4600.0	1771.0	5.00	12.54	6.82	312	4766	2331	753.5	994.7
4800.0	1571.0	4.89	12.29	6.72	312	4448	2215	655.2	993.1
5000.0	1371.0	4.78	12.02	6.61	312	4128	2098	558.9	993.2
5200.0	1171.0	4.67	11.73	6.50	312	3803	1979	464.8	994.6
5400.0	971.0	4.56	11.41	6.37	312	3471	1856	372.8	996.9
5600.0	771.0	4.44	11.06	6.24	312	3133	1730	282.9	998.8
5650.0	721.0	4.41	10.91	6.09	312	3067	1639	260.7	1000.6
5701.0	670.0	4.38	10.75	5.94	312	2999	1548	238.3	1001.4
5701.0	670.0	3.99	10.26	5.57	143	2556	1239	238.3	1001.4
5771.0	600.0	3.97	10.15	5.51	143	2489	1210	210.4	1000.3
5871.0	500.0	3.84	9.64	5.22	143	2181	1051	171.3	998.8
5921.0	450.0	3.78	9.38	5.07	143	2037	977	152.2	997.9
5971.0	400.0	3.72	9.13	4.93	143	1899	906	133.5	996.8
5971.0	400.0	3.54	8.90	4.76	143	1735	806	133.5	996.8
6061.0	310.0	3.48	8.73	4.70	143	1630	773	102.0	993.6

continued on next page

TABLE 10.1 (*continued*)

Radius (km)	Depth (km)	Density (g/cm³)	V_p (km/s)	V_s (km/s)	Q_μ	K_s (kbar)	μ (kbar)	Pressure (kbar)	Gravity (cm/s²)
6106.0	265.0	3.46	8.64	4.67	143	1579	757	86.4	992.0
6151.0	220.0	3.43	8.55	4.64	143	1529	741	71.1	990.4
6151.0	220.0	3.35	7.98	4.41	80	1270	656	71.1	990.4
6186.0	185.0	3.36	8.01	4.43	80	1278	660	59.4	989.1
6221.0	150.0	3.36	8.03	4.44	80	1287	665	47.8	987.8
6256.0	115.0	3.37	8.05	4.45	80	1295	669	36.1	986.6
6291.0	80.0	3.37	8.07	4.46	80	1303	674	24.5	985.5
6291.0	80.0	3.37	8.07	4.46	600	1303	674	24.5	985.5
6311.0	60.0	3.37	8.08	4.47	600	1307	677	17.8	984.9
6331.0	40.0	3.37	8.10	4.48	600	1311	680	11.2	984.3
6346.6	24.4	3.38	8.11	4.49	600	1315	682	6.0	983.9
6346.6	24.4	2.90	6.80	3.90	600	753	441	6.0	983.9
6356.0	15.0	2.90	6.80	3.90	600	753	441	3.3	983.3
6356.0	15.0	2.60	5.80	3.20	600	520	266	3.3	983.3
6368.0	3.0	2.60	5.80	3.20	600	520	266	0.3	982.2
6368.0	3.0	1.02	1.45	0	0	21	0	0.2	982.2
6371.0	0	1.02	1.45	0	0	21	0	0.0	981.5

tions, it provides only a reference baseline for path-specific attenuation as sampled by body waves and surface waves. Later sections will thus emphasize three-dimensional attenuation variations.

In addition, different PREM models are provided for oceanic and continental lithosphere. As complete a model as PREM is, it still lacks a three-dimensional description of aspherical heterogeneity, and some details of the upper-mantle structure are undergoing revision. Ultimately, seismology will achieve a complete three-dimensional, anelastic (hence, frequency-dependent) anisotropic Earth model, but many aspects of such a complete model are still being resolved. Attaining such a detailed model will be critical for achieving a thorough understanding of the composition and dynamic processes inside the Earth. Since this is an ongoing process, the remainder of this chapter will traverse from the crust to the core, outlining major aspects of what we know about Earth structure and how it is determined by seismic-wave analysis.

10.2 Crustal structure

In terms of relative societal importance, seismological investigations of the structure of the shallow crust unquestionably have had the largest effort and the greatest impact. Much of that effort involves *reflection seismology*, the collection and processing of multichannel seismic data that record human-made explosive and vibrational sources. Although the principles involved in multichannel seismic processing have basic origins in the behavior of seismic waves and their reflection from boundaries, as described in previous chapters, the processing of the dense and now often two-dimensional recordings of the wavefield involves a multitude of specific procedures beyond the scope of this text. We instead focus our attention on whole-crustal-scale investigations of the shallow crust, which are typically performed with sparser seismic instrumentation than in shallow-crust reflection imaging.

On the larger scale of whole-crustal imaging, the main objective is to determine the basic layered structure of the crust; the P and S velocities as a function of depth, including the depth and contrasts across any internal boundaries; and the overall crustal thickness, or depth to the crust–mantle boundary. The effort to determine crustal thickness dates back to 1910, when Croatian researcher Andrija Mohorovičić first identified an abrupt increase in velocity beneath the shallow rocks under Europe. The boundary separating crustal rocks from mantle rocks is now called the **Moho** and is a ubiquitous boundary of highly variable character. Although we generally accept that the crust is chemically distinct from the upper mantle and that the Moho likely involves a chemical contrast, additional contributions to the seismically detectable boundary may arise from transitions in rheological properties, phase transitions in shallow mineral structures, and petrographic fabrics of the rocks. These complexities are combined with the complex tectonic history of the surface to provide a remarkably heterogeneous crustal layer. Fig. 10.1 shows a global contour map of crustal thickness variations. Generally speaking the crust is thickest within continental cratons, and thin near tectonic boundaries (see Chapter 3).

While we recognize the complexity of the crust, it is still useful to assess the basic seismological feature common to all crustal environments, which is that the shallow rocks have slower seismic velocities than the deeper rocks, usually approximating a low-velocity layer over a faster mantle. For example, continental and oceanic regions differ primarily in the thickness of the crust, which varies from 20 to 70 km beneath continents and from 5 to 15 km beneath oceans. Both regions have very low seismic velocity surface cover, with the water and mud layers on oceanic crust having a particularly low velocity. Both regions tend to have at least two crustal subdivisions, with the lower-velocity layer below the sediments typically having steep velocity gradients with depth down to a transition or midcrustal discontinu-

ity. At greater depths the average gradients and reflective properties of the deep crust vary substantially.

The travel-time curves for these generic layered crusts are thus composed of distinct primary-wave branches, with direct phases being called P_g in continental environments and P_2 in oceanic crust (layer 1 is the soft sediments; layer 2 is the shallow basaltic layer in the oceanic crust). For continental regions with a midcrustal discontinuity, often called the **Conrad discontinuity**, the head wave from this structure is called P^*, and P_n is the head wave traveling along the Moho. The analogous oceanic arrivals are P_3 and Pn, and both types of crust have PmP reflections from the Moho. As discussed in Chapter 5, S phases have corresponding labels (S_g, S^*, S_n, S_mS, etc.). The crossover distances of the various travel-time branches, the slopes of the branches, and their zero-distance intercepts reveal the layer thickness and velocities using the straightforward wave theory from Chapter 12. The variation in crustal thickness between oceanic and continental regions ensures a very different appearance of the seismograms as a function of distance for the two regions.

Fig. 10.2 shows a schematic of raypaths for different crustal phases and a sample record of a controlled source experiment in China. A continental-scaled seismic refraction/wide-angle reflection profile for crust requires large seismic sources (generally large explosives) that allows the seismic waves to penetrate the full crust and then be recorded by the receivers. This particular seismic profile shows that the crust–mantle boundary (Moho) with a typical seismic velocity structure found beneath a stable craton is a continuous feature, but intracrustal boundaries show significant heterogeneity, with laterally discontinuous layers.

A data profile of this type, covering several hundred kilometers from the source and allowing the head-wave branches to be identified, is called a **refraction profile**. Refraction seismology is quite straightforward and involves di-

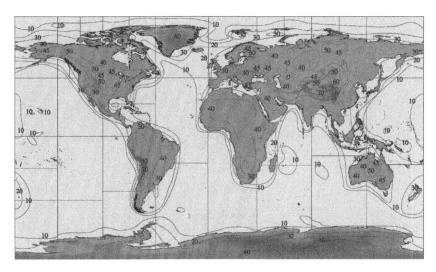

FIGURE 10.1 Global contour map of crustal thickness obtained from a great number of studies of regional-distance travel-time curves. Note the thinner crust in the Basin and Range extensional region under Utah and Nevada (modified from Mooney, 2015).

rectly identifying travel-time branches; measuring slopes, crossovers, and amplitude behavior along the branches; and relating these to one- or two-dimensional models of the primary layers in the structure. Note that at closer distances, less than 30 km, a weak Moho reflection, P_mP, occurs, as well as other strong arrivals. At close-in distances, direct body and surface waves dominate the seismograms, which display a long sequence of nearly vertically propagating waves composed of single or multiple reflections from crustal layers near the source region. This is the domain of *reflection profiles*. Reflection seismology strives to determine the *reflectivity* of the crust, meaning detailed layering and impedance contrasts below a localized region. The procedures for isolating energy associated with any particular reflector at depth involve unscrambling the many superimposed arrivals generated by the detailed rock layering. Usually, high-precision images of the shallow layering can be extracted from the data using the predictability of seismic-wave interactions with layering, just as for refraction work, but the signal environment is much more complex in re-

flection profiles. Modern whole-crustal imaging actually involves both classes of seismic recording, with refraction and reflection work in the same region giving the best overall model of the crustal structure. Still, the most extensive coverage of the subsurface has involved straightforward refraction modeling, so we emphasize results from those procedures.

One of the earliest, and most important, applications of refraction profiling was the systematic reconnaissance of uppermost mantle velocity variations, as directly measured by Pn velocity. Pn travels along the crust mantle boundary (Moho), and travels large horizontal distances. Assigning any single local velocity to it involves a lateral averaging along the path. Since these earlier studies, new large scale projects have collected new data, such as USArray, which placed a 400 station rolling array of seismometers across the US with 70 km spacing. Using tomographic inversions that explicitly model variations on each path, exploiting path overlap and intersection (see Chapter 20), this new data set has allowed for a more detailed investigations of Pn velocity variations than previously obtained

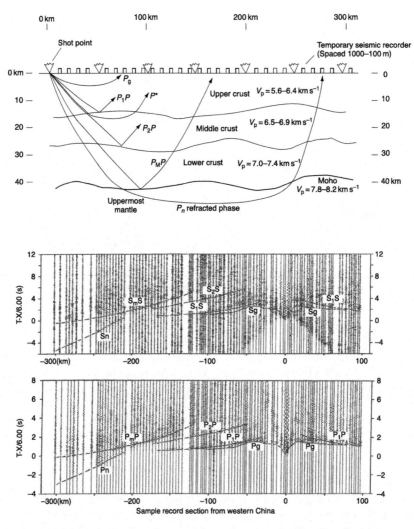

FIGURE 10.2 (Top) Ray paths for a continental seismic refraction/wide-angle reflection profile for crust with a typical seismic velocity structure found beneath a stable craton. The crust–mantle boundary (Moho) is a continuous feature, but intracrustal boundaries may be discontinuous laterally and several kilometers wide vertically. Commonly, shot points are spaced 50 km apart (or closer for higher resolution); temporary seismic recorders are spaced 1000–100m apart. (Bottom) Sample seismic record section from western China, showing S-wave record section with a reduction velocity of 3.46 km/s and P-wave record section with a reduction velocity of 6.0 km/s (modified from Mooney, 2015).

(Fig. 10.3). The data collection from USArray allows for unprecedented imaging at larger scales than previously available. Fig. 10.3 shows the Pn velocity variations obtained by seismic tomography for the western United States using US-Array data. The Basin and Range province Pn velocities are lower than beneath the Colorado Plateau (under northern Arizona and southeastern Utah) or under the stable continental platform underlying the Midwest, and there is sig-

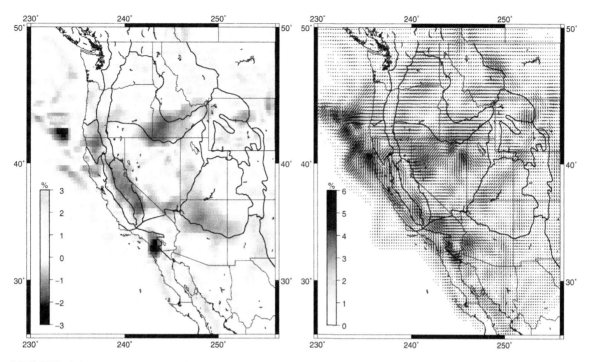

FIGURE 10.3 (left) Isotropic velocity perturbations resulting from a simultaneous inversion for isotropic and anisotropic parameters. Different shades indicate areas of lower and higher velocities, with the average velocity is 7.93 km/s. (right) The azimuthal anisotropy model. The black lines indicate the Pn fast axis with the length of the line proportional to the strength of the anisotropy. The anisotropy strength is also shaded by percent (modified from Buehler and Shearer, 2010).

nificant anisotropy for *Pn*. These variations are due to uppermost mantle differences in temperature and petrology, which are directly linked to crustal processes such as ongoing rifting in the Basin and Range, discussed in Chapter 3.

Refraction profiling has been extensively performed on several continents, and some gross continental characteristics are summarized for different continental provinces in Fig. 10.4. The velocities of the rocks at depth place important bounds on the petrology of the deep crust. Beginning in the 1970s, petroleum industry methods were utilized by university consortiums, such as COCORP and CALCRUST in the United States and similar groups in Europe, to examine the deep crust. Using high-frequency, near-vertical-incidence reflections, they have found that the crustal layers and boundaries defined by low-resolution refraction methods have highly variable, small-scale structure.

Seismic reflection data can reveal the nature of the Moho. For example, Fig. 10.5 shows a reflection profile from a large crustal experiment from an ancient a collision zone in the Superior Province of Canada. Dipping seismic reflections can be identified that extend 30 km into the mantle, representing a relict continental sutures associated with subduction. Although crustal structure, lithospheric thicknesses and convergence rates may have differed from those seen today, these seismic data provide direct evidence that plate tectonics was active in late Archaean times. In the Abitibi belt to the south, a comparatively non-reflective upper crust down to 3.0 s overlies a moderately reflective middle crust. The lower crust exhibits diffuse reflectivity that con-

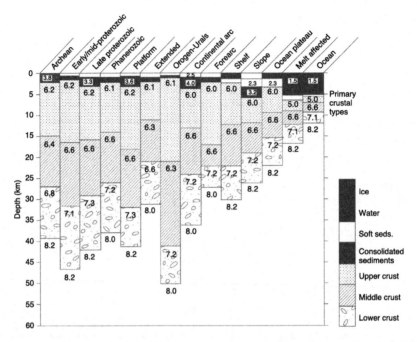

FIGURE 10.4 Fourteen primary continental and oceanic crustal types from previous studies. Typical P-wave velocities are indicated for the individual crustal layers and the uppermost mantle. Velocities refer to the top of each layer, and there commonly is a velocity gradient of 0.01–0.02 km/s/km within each layer. The crust thins from an average value of 40 km in continental interiors to about 7 km beneath oceans (modified from Mooney, 2015).

tinues to 13.5 s in places before dying out. In marked contrast, the lower crust of the Opatica belt contains well defined, high-amplitude subhorizontal seismic layering and a sharp reflection Moho. This variable character of crustal reflections even raises the question of what defines the crust. The Moho is more distinctive than intermittent midcrustal structures like the Conrad discontinuity, but does it strictly represent a chemical transition? Feldspar is the most abundant mineral in the continental crust, followed by quartz and hydrous minerals. The most common minerals in the mantle are ultramafic, such as olivine and pyroxene. A typical K-feldspar rock has a *P*-wave velocity of about 6 km/s, and dunite (olivine) may have a velocity of 8.5 km/s. The laminated Moho transition suggested by Fig. 10.5 is clearly not a single, sharp chemical boundary, and petrologists have

established a complex sequence of mineral reactions with increasing pressure and temperature. A combination of gross chemical stratification, mineralogical phase transformations and reactions, laminated sill injection and partial melting, and rheological variations occur near the crust–mantle boundary, and great caution must be exercised in interpreting limited-resolution seismological models for the transition in terms of associated processes and chemistry. The same caution holds for all models of deeper structure as well.

A key to making progress toward understanding the continental crust is the development of progressively higher-resolution, three-dimensional models. This can be achieved a number of ways. We can achieve this by merging the methodologies of refraction and reflection seismology with seismic tomography, or

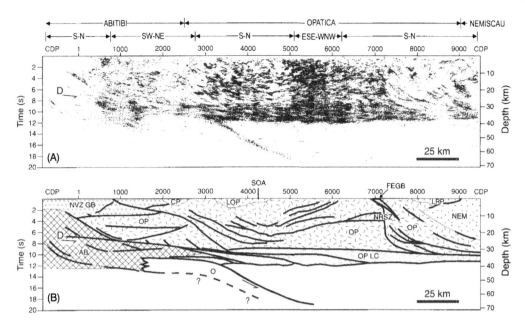

FIGURE 10.5 Seismic reflection line migrated using a stacking velocity of 6.5 km/s in the Superior Province of Canada. showing the complex nature of the Moho in a continental craton. Mantle reflections dip in a north to northwest direction beneath the Opatica belt, and intersect the Moho beneath the Abitibi Opatica boundary mapped at the surface. In the Abitibi belt to the south, a comparatively non-reflective upper crust down to 3.0 s, overlies a moderately reflective middle crust. The lower crust exhibits diffuse reflectivity that continues to 13.5 s in places before dying out. The lower crust of the Opatica belt contains well defined, high-amplitude subhorizontal seismic layering and a sharp reflection Moho. The upper and middle Opatica crust contains a broad zone of high-amplitude reflectivity with reflections of opposing dips between CDP 2500 and 6000 that extend down to 9 s (modified from Calvert et al., 1995).

develop new approaches based on the latest instrumentation and data collections approaches. For example, recent experiments have collected large amounts of data from the deployment of many sensors, such as USArray. This data collection allowed for unprecedented images of the upper mantle to be developed. The USArray has also allowed for new approaches to be developed and has been used extensively to image the Earth. Fig. 10.6 shows an example of using ambient noise (essentially surface waves created by diverse distributed processes) to image in 3-D the crust and upper mantle structure. The new images allow for the complex crust and mantle structure along with mantle flow to be mapped and interpreted.

Investigations of oceanic crust have followed a similar evolution, with primary reconnaissance refraction studies being used to map out gross crustal properties and more recent reflection and tomographic imaging experiments targeting localized regions of particular tectonic interest. As shown in Figs. 3.6 and 3.5, the topography and age of the ocean floors are well mapped and understood in the context of plate tectonics. Event with current instrumentation (with few ocean bottom seismometers (OBS)), the development of 3-D Earth models have allowed for probing the ocean crust and mantle structure. Fig. 10.7 shows an example of a comparison of different models between different isotropic V_S velocity models beneath the East Pacific Rise (EPR), with the local model using OBS obser-

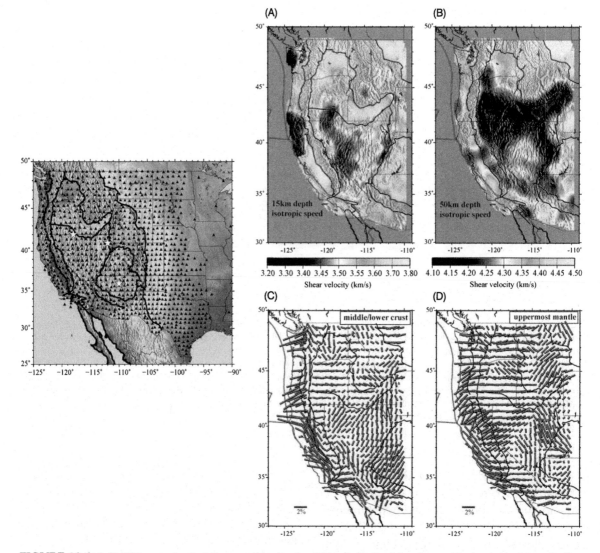

FIGURE 10.6 (left) USArray seismic stations used to develop 3-D velocity models using ambient noise tomography (top right, A and B) and anisotropic models (bottom right, C and D). Isotropic V_S velocity in the crust at depth (A) 15 km and (B) in the uppermost mantle at 50 km depth; Azimuthal anisotropy in the (C) middle to lower crust and (D) uppermost mantle (modifed from Ritzwoller et al., 2011).

vations (Harmon et al., 2011). All global models show the low-velocity zone (LVZ) beneath the EPR, but not all show the magnitude of the LVZ as the local model. Cross sections along the EPR show the strength and extent of the strong low-velocity anomaly.

Oceanic environments were also the first to provide convincing evidence for uppermost mantle anisotropy. Measurement of Pn velocities as a function of azimuth with respect to the strike of spreading-ridge segments has revealed higher velocities perpendicular to the

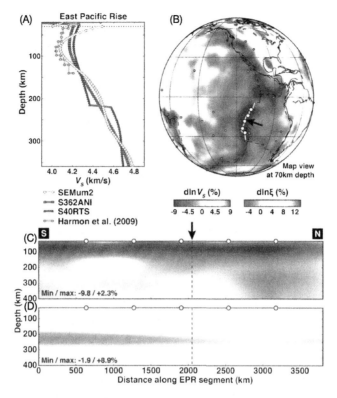

FIGURE 10.7 A comparison between different isotropic V_S velocity models beneath the East Pacific Rise (EPR). (A) a comparison between the mean 1D VS profile obtained from the high-resolution OBS based tomographic study of EPR structure with those sampled from global models SEMum2, S362ANI, and S40RTS in the same location. The pronounced low-velocity zone (LVZ) beneath the EPR inferred is captured by all models, but more defined by several models. (B) Relative variations in isotropic versus structure at 70 km depth in SEMum2, focused on the EPR, showing both the location of the comparison in (A) and the extent of the 3750 km line of section in (C). (C) and (D): a cross section following the portion of the EPR shown in (B), illustrating the strength and extent of the strong low-velocity anomaly imaged beneath the EPR in SEMum2 (C), nearly −10%. Black arrow and dashed line indicate approximate location of profiles in (A). SEMum2: second-generation global model (French et al., 2013); S362ANI: 3-D model from long-period waveforms and body wave travel times (Kustowski et al., 2008); S40RTS: 3-D model, spherical harmonics degree 40 model (Ritsema et al., 2011); local model (Harmon et al., 2011) (modified from French et al., 2013).

ridge. New models continue to be developed to highlight the nature of the oceanic and upper mantle. Fig. 10.8 shows 1-D profiles and three-dimensional imaging of anisotropy between V_{SV} and V_{SH} in oceanic crustal regions. Furthermore, the maximum difference between V_{SV} and V_{SH} occurs in the upper mantle, at around 125 km depth, where for PREM, this occurs just below the Moho. At 130 km depth, V_{SH} is 7% faster than V_{SV}, almost three times the value predicted by PREM. Furthermore, Fig. 10.8 shows that the anisotropic velocity variations are as large as the isotropic (thermal) variations, and that the isotropic S-wave variations correlate well with the age of the ocean floor. The largest deviation from this age correlation can be associated with the location of the Pacific Superswell, a broad area in the Pacific ocean floor marked by numer-

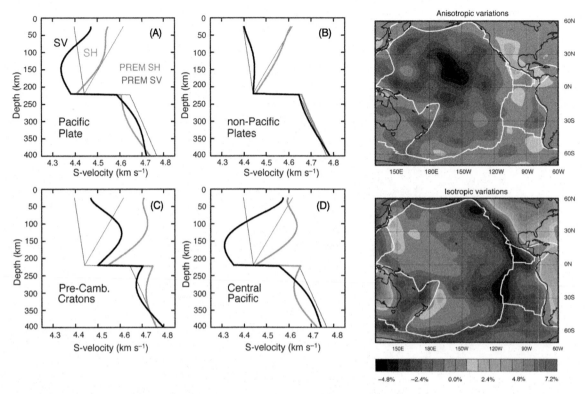

FIGURE 10.8 Velocity profiles of the shear-wave velocities V_{SV} and V_{SH} in the Pacific ocean. (Left) (A) Average velocities beneath the entire Pacific plate. The maximum difference between V_{SV} and V_{SH} occurs at around 125 km depth, in contrast to PREM, in which it occurs just below the Moho. (B) Average velocities calculated for all plates except the Pacific plate. (C), Average velocity profiles calculated for all Precambrian cratons within the Eurasia and North America plates. (D) Average velocity profiles calculated for a cap with 108 km radius centered on the anisotropy anomaly. At 130 km depth, V_{SH} is 7% faster than V_{SV}, almost three times the value predicted by PREM. In contrast with PREM, the radial anisotropy in the shallowest mantle is small. (Right) Maps showing the anisotropic and isotropic (thermal) variations velocity variations. The isotropic S-wave variations correlate better with the age of the ocean floor than either the dV_{SV} or dV_{SH}. The largest deviation from this age correlation is very clearly associated with the location of the Pacific Superswell (modified from Ekström and Dziewoński, 1998).

ous volcanic chains and very shallow seafloor compared to the depth predicted for its age (see Fig. 3.7.)

10.3 Upper-mantle structure

The material properties revealed by seismology play a dominant role in constraining both the composition and dynamics of the mantle.

The disciplines of mineral physics and geodynamics directly utilize seismological observations as boundary conditions or measurements that experimenters strive to explain. As we stated earlier, the entire Earth can be approximately viewed as a layered, one-dimensional, stratified, chemically differentiated planet composed of crust, mantle, and core. These major layers are separated by boundaries (the Moho and the core–mantle boundary, or CMB) across which seismically detectable material properties have strong contrasts. Additional stratification of the mantle is represented by global seismic-velocity boundaries at depths near 410, 520, and 660 km (Fig. 10.9) that define the *transition zone* in the lower part of the upper mantle. These boundaries give rise to reflections and conversions of seismic waves, which reveal the boundaries and allow us to model them in terms of depth and contrasts in velocity, density, and impedance. Once these values are reliably determined by seismology, high-pressure, high-temperature experiments on plausible mantle materials are used to assess the likely cause of the particular structure inside the Earth. Whereas major compositional contrasts probably underlie the Moho and CMB contrasts, the 410-, 520-, and 660-km discontinuities probably represent mineralogical phase transformations that involve no bulk change in composition but reflect a transition to denser lattice structures with increasing pressure.

The mantle also has localized boundaries at various depths associated with laterally varying thermal and chemical structure (Fig. 10.9). The upper 250 km of the mantle is particularly heterogeneous, with strong regional variations associated with surface tectonic provinces. The uppermost mantle just below the Moho is a region with high seismic velocities (*P* velocities of 8.0–8.5 km/s) that is often called the *lid* because it overlies a lower-velocity region. The base of the lid may be anywhere from 60 to 200 km in depth, and it is thought to represent the rheological transition from the high-viscosity lithosphere to the low-viscosity asthenosphere, or the Lithosphere-Asthenosphere Boundary (LAB) (Fig. 10.10). The LAB coincides with the top of a zone of decoupling between the lithosphere and asthenosphere, marked by an increased strain rate, a conductive *lid*, a change in the direction of anisotropy, and a transition layer that represents a near-surface region where temperature deviates from the adiabat. Beneath the lid is a region of reduced velocity, usually referred to as the low-velocity zone (LVZ); thus, LAB can difficult to seismically determine depending on the magnitude of the velocity contrast. Under very young ocean crust, the lid may actually be absent, but it is commonly observed under old ocean plates (Fig. 10.8). The thickness of the lid varies with tectonic environment, generally increasing since the time of the last thermotectonic event. Gutenberg first proposed the presence of the LVZ in 1959, and it is thought to correspond to the upper portion of the rheologically defined asthenosphere. The LVZ may be very shallow under ridges, and it deepens as the lid thickness increases. Beneath old continental regions the LVZ may begin at a depth of 200 km, with relatively low velocities usually ending by 330 km, where a seismic discontinuity is intermittently observed. A relatively strong seismic discontinuity is observed under some continental and island-arc regions near a depth of 220 km, which may be associated with LVZ structure or some transition in mantle fabric associated with concentrated LVZ flow structures.

Below 350 km, the transition zone has less pronounced lateral variations, although seismologists have reported intermittent seismological boundaries near depths of 710, 900, and 1050 km. A depth of 710 km or so is reasonable for the boundary between the upper and lower mantle, as known mineralogical phase transformations could persist to this depth. Few, if any, candidate phase transformations are expected in the lower mantle at greater depths; thus intermittent deeper boundaries may involve chemical heterogeneity. The only strong candidate for such a deep intermittent boundary is found about 250 km above the CMB, as discussed later. The only major boundary in the core is the inner core–outer core boundary at 5155 km depth, which is associated with freezing of the inner core as the geotherm crosses the solidus, giving rise to a solid inner core.

One of the most important ways of studying upper-mantle structure has been analysis of direct *P* and *S* waves at upper-mantle dis-

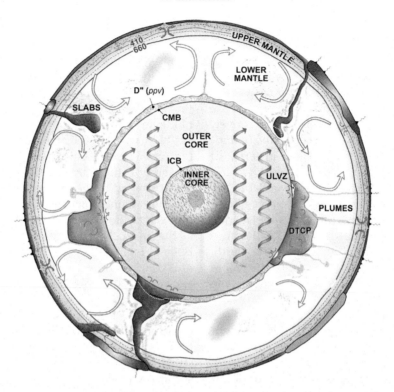

FIGURE 10.9 Schematic cross section through the Earth, showing depths of globally extensive and intermittent boundaries detected by seismic-wave analysis. Major mantle boundaries exist near depths of 410 and 660 km, bracketing the upper-mantle transition zone. A weak boundary near 520 km may exist globally as well. These are primarily phase boundaries and have minor topographic relief due to thermal variations. The core-mantle boundary is the major chemical boundary inside the Earth, separating the silicate mantle from the molten iron-alloy core. CMB: Core Mantle Boundary; ICB: Inner Core Boundary; ULVZ: Ultra-Low Velocity Zone; DTCP: Dense Thermo-Chemical Piles (from Ed Ganero, Arizona State University.)

tances. Beyond a distance of 10° to 13°, the first arrivals on seismograms in continental regions correspond to waves that have turned in the upper mantle rather than refracted along the Moho (Fig. 10.11). The precise depths to which the waves observed from 10° to 15° have penetrated depend strongly on the velocity structure in the lid; a strong lid-velocity increases refracting energy from relatively shallow depths, while lower velocity gradients or decreases cause the energy to dive deeply into the upper mantle. As early as the 1940s it was clear that at a distance of ∼ 15° the ray parameters associated with the first-arrival travel-time curves of P and S changed abruptly, but it was not until the 1960s that seismic data were adequate to reveal two triplications in the travel-time curves from 15° to 30°, leading to the discovery of the 410- and 660-km discontinuities (actually, reflections from the boundaries were being detected at around the same time). These triplications dominate the upper-mantle signals, and they arise from the wavefield grazing along the discontinuities with strong refractions, as shown by the raypaths in Fig. 10.11.

The collection of large volumes of high-quality, broadband seismic data over the last 30 years, plus advances in computational power and the advent of new approaches, have allowed for continuous advances in imaging the Earth at all

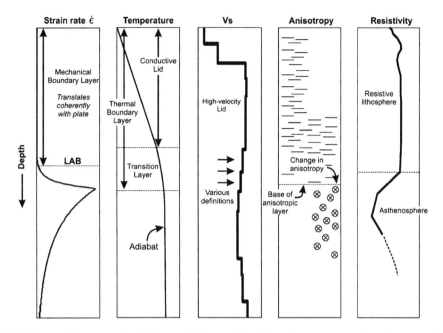

FIGURE 10.10 Diagram illustrating the different attributes of the lithosphere–asthenosphere boundary (LAB) coincides with the top of a zone of decoupling between the lithosphere and asthenosphere, marked by an increased strain rate. The thermal boundary layer (TBL), containing a conductive lid and a transition layer, represents a near-surface region where temperature deviates from the deeper adiabat. A zone of low seismic shear-wave velocity (V_s) is sometimes detected beneath a high-velocity lid; various definitions have been used to correlate this zone with the LAB (see text). The LAB may also correlate with a downward extinction of seismic anisotropy (e.g., Gaherty and Jordan, 1995) or a change in the direction of anisotropy (e.g., Debayle and Kennett, 2000). The electrical LAB is marked by a significant reduction in electrical resistivity (modified from Eaton et al., 2009).

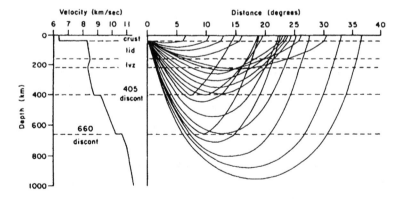

FIGURE 10.11 Complexity of seismic raypaths at upper-mantle distances produced by velocity variations with depth. Raypaths are shown for a velocity structure characteristic of a stable continental region, with a thick high-velocity lid above a weak low-velocity zone (LVZ). Oceanic or tectonically active continental structures will have different raypaths to each distance. Multiple arrivals at a given distance correspond to triplications caused by rapid upper-mantle velocity increases (from LeFevre and Helmberger, 1989).

scales and depths, especially the upper mantle with the deployment of the USArray. Fig. 10.12 shows the integration of results from a variety of geological and geophysical investigations that reveal new insight into upper mantle processes in the Great Basin region of the western United States. Specifically, mapping anisotropy (from *S*-wave splitting times) using the USArray identifies a rotational feature that surrounds a high velocity anomaly determined from travel time tomography (West et al., 2009). This feature, characterized as a vertical drip of cooler(denser) material with fast seismic velocities, flows downward as it plunges to the northeast. The mantle flow shifts rapidly from horizontal to vertical, which is counter to the motion of the North American plate, suggesting that regional mantle flows in the northeast direction. The lack of volcanism and the lower heat flow over the center of this drip supports this interpretation.

The velocity discontinuities near 410 and 660 km have contrasts of 4% to 8%, with comparable density increases. The 410- and 660-km discontinuities are observed rather extensively and appear to be global features, but they do vary in depth. Depths as shallow as 360 km and as deep as 430 km have been reported for the "410," with 20- to 30-km variations reported for the "660." The precision of depth estimates from wide-angle studies is not that great because of their limited resolution of lid and LVZ structure, which can baseline-shift the transition zone structure. Thus, depth estimates are usually based on near-vertical-incidence reflections from both above and below the boundary. Establishing the underlying cause of these changes in material properties continues to be a major effort. The primary features that must be established for any structure are (1) its global extent, (2) the velocity and density contrasts, (3) the depth (hence pressure) and variation in depth (topography) of the boundary, (4) the sharpness of the contrast, and (5) behavior of the contrast

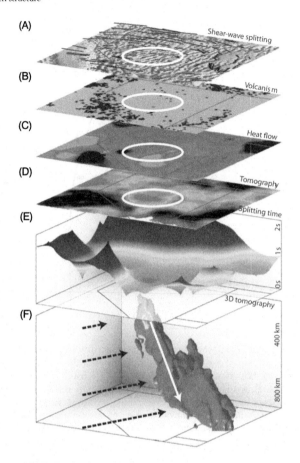

FIGURE 10.12 A layered summary of geological and geophysical constraints for the central Great Basin. (A) Shear-wave splitting with the topography background for reference. (B) Post-10-Myr volcanism. (C) Heat flow. (D) Seismic tomography horizontal slice at 200 km depth. (E) Shear-wave splitting times surface showing the strong drop in the central Great Basin. (F) Isosurface at +0.95% velocity perturbation for velocity model NWUS08-P2 showing the morphology of the drip, which merges with a larger structure at 500 km depth. The black arrows denote the inferred mantle flow direction; the white arrow denotes the flow direction of the Great Basin drip (modified from West et al., 2009).

in regions of known temperature structure, such as near downwelling slabs.

A classical procedure used for detecting and constraining the depth of mantle discontinuities is using underside reflections that arrive as precursors to mantle and core phases, such as *PKPPKP* (*P'P'* for short). The time difference between a main arrival and precursors is proportional to the depth of the discontinuity. The depth is determined using a model for seismic velocities above the reflector. The amplitude ratio and frequency content of the precursor provide a measure of the impedance contrast. We must account for the frequency-dependent propagation effects caused by focusing due to core velocity structure, but this technique, first developed in the 1960s, is reliable if the signals are rich in high-frequency energy. Unfortunately, the precursor approach is limited by the narrow distance ranges of favorable focusing by the core, but many analogous discontinuity interactions have been utilized to bound the depth, magnitude, and sharpness of upper-mantle discontinuities.

Many localized observations of the mantle-discontinuity phases have been examined; however, the availability of large digital data sets has created the opportunity to seek these phases on a global scale. Fig. 10.13 and Fig. 10.14 show results from S-wave receiver functions stacks, which map S-to-P conversions at interfaces, to image seismic discontinuities between the Moho, the 410 km seismic discontinuity, and the 660 km seismic discontinuity using USArray data. Each stack bin has over 1000 individual receiver functions, which increases the signal to noise ratio and highlights common features.

10.3.1 Discontinuities and anisotropy

We have seen a few of the ways in which seismology determines characteristics of the major transition zone mantle boundaries. This information is used by mineral physicists to constrain the mineralogy of the mantle materials. Since the 1950s it has been known that common upper-mantle minerals such as olivine $(Mg, Fe)_2 SiO_4$ and enstatite $(Mg, Fe)SiO_3$ undergo phase transformations with increasing pressure. In particu-

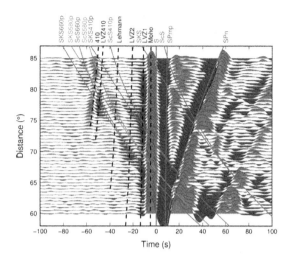

FIGURE 10.13 Binned S-receiver functions as a function of epicentral distance for the USArray. Each bin contains more than one thousand traces. Precursors of the *S* phase from *S*-to-*P* conversions in the upper mantle are marked with dashed black lines (410 conversion at the 410, LVZ410 conversion at a velocity reduction directly above the 410, Lehmann–Lehmann discontinuity, LVZ1, LVZ2 conversions at velocity reductions between Moho and Lehmann, Moho conversion at the Moho). Additional theoretical travel-time curves of ScS, SKS, S-to-P conversions of ScS and SKS at the 410 and 660 discontinuities and at a possible discontinuity at 580 km depth ($ScS_{410}p$, $ScS_{580}p$, $ScS_{660}p$, $SKS_{410}p$, $SKS_{580}p$, $SKS_{660}p$), crustal multiples ($SPmp$) and SPn are marked (modified from Kind et al., 2015).

lar, the low-pressure olivine crystal converts to the β-spinel structure at pressures and temperatures expected near 410 km depth, the β-spinel structure converts to γ-spinel structure near 500 km depth, and then the γ-spinel structure converts to Bridgmanite (Mg silicate perovskite) $(Mg, Fe)SiO_3$ and magnesiowüstite $(Mg,Fe)O$ at conditions near 660 km depth. These mineralogical phase transitions compact the crystal lattice and cause an increase in seismic-wave velocity. Because the phase transformations can occur over a fairly narrow range in depth (pressure), it is plausible that upper-mantle discontinuities may indicate internal phase boundaries.

The experimental and observational agreement in depth, sharpness, and topographic vari-

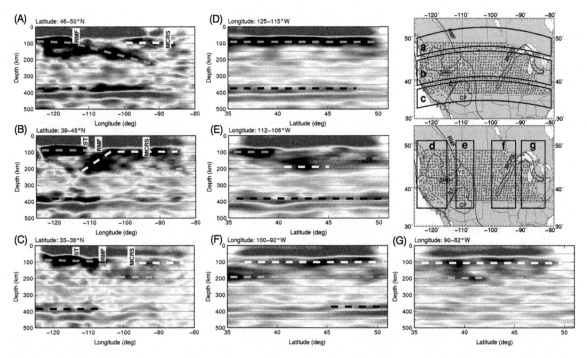

FIGURE 10.14 Depth-migrated west–east S-receiver function profiles summed over 2-degree latitude bins with 1-degree overlap (right maps). Region (A) (northernmost west–east profile) does not have any stations in Canada. However, due to the shallow incidence angle of the S-receiver functions, the mantle below southern Canada is also sampled by events from the northwest. The westernmost south–north profile (D) shows only the western LAB in the mantle lithosphere. In profile (E) we see at the southern and northern ends the western LAB, and in the central part the deep MLD. In profile (F) we see the cratonic MLD (yellow), the cratonic LAB (green) and in the north the deep western LAB (magenta). In the easternmost south–north profile (G) we see the MLD and weakly the cratonic LAB (modified from Kind et al., 2015).

ability of transition zone boundaries favors their interpretation as primarily the result of phase transitions. The olivine–β-spinel transition is widely accepted as the cause of the 410-km discontinuity, and this boundary should occur at shallower depth in cold regions, such as descending slabs, and at greater depth in warm upwellings. The 660-km discontinuity is most likely due to the dissociative phase transformations of olivine (γ-spinel) and enstatite to Bridgmanite and magnesiowüstite, with a similar transition in garnet perhaps occurring a bit deeper (~ 710 km). These endothermic transitions have pressure–temperature slopes opposite to those of the exothermic olivine–β-spinel

transition; thus cold downwellings should depress the boundary. This is consistent with the pattern of 660-km discontinuity topography seen in Fig. 10.15.

If these major transition zone discontinuities are predictable phase boundaries, does it follow that we know the upper-mantle composition? Unfortunately, the problem is complex, and many plausible mantle constituents are consistent with observed seismic models. The main seismic model parameters are compared with experimental results for a variety of plausible Earth materials in Fig. 10.16. Comparisons such as these are used to construct models of bulk-mantle composition compatible with the ob-

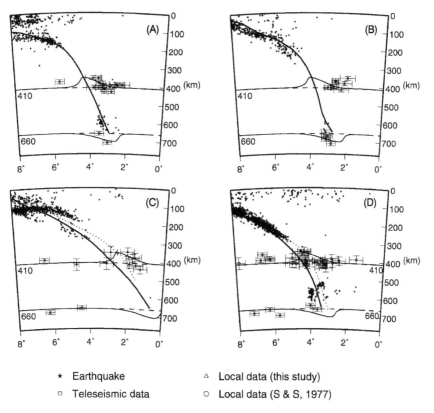

★ Earthquake △ Local data (this study)

□ Teleseismic data ○ Local data (S & S, 1977)

FIGURE 10.15 Cross-sections of the South American subduction zone. Squares show discontinuity observations made with the teleseismic records, circles the re-interpreted data, and triangles observations made from earthquake data. The nominal 410 and 660 km discontinuity depths are shown by the dashed lines. The 410 and 660 km discontinuity depths using Clapeyron slopes of 2.04 and -1.9 MPa $^{\circ}C^{-1}$ respectively are shown by the thin solid lines (modified from Collier and Helffrich, 2001).

served density and seismic velocities; however, this is a nonunique process. At this time it is still possible to conclude that either upper and lower mantles must be chemically stratified, even allowing for phase transformations in the transition zone, or that the upper and lower mantles have a uniform composition. We must refine both seismological models and mineral physics experiments before the extent of mantle chemical stratification can be resolved.

The upper mantle has laterally variable anisotropy, and we can use body waves to determine local anisotropic properties beneath three-component broadband stations. One tech-

nique that has been widely applied is analysis of teleseismic shear-wave splitting. Anisotropy causes shear waves to split into two pulses, one traveling faster than the other, with the differential time between the two pulses accumulating with the path length traversed in the anisotropic region (Fig. 10.17). The two pulses can most readily be observed for phases that have a known initial polarization before they enter the anisotropic region. One such phase is *SKS*, which has purely *SV* polarization following the conversion from *P* to *SV* at the core–mantle boundary. If the entire mantle along the upgoing path from the core to the surface

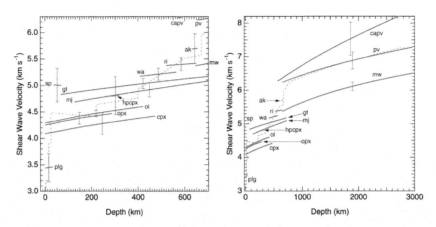

FIGURE 10.16 Calculated and observed (PREM: dashed line) shear wave velocities for a number of phases along a typical geotherm for (right) the whole mantle and (left) an expanded depth scale to show details of the upper mantle. Phases are shown over the approximate depth range that they are expected to occur in the mantle. Feldspar (plg), Spinel (sp), Olivine (ol), Wadsleyite (wa), Ringwoodite (ri), Orhopyroxene (opx), Clinopyroxene (cpx), HP-clinopyroxene (hpcpx), Ca-perovskite (cpv), Akimotoite (ak), Garnet (gt,mj), Stishovite (st), Perovskite (pv), Magnesiowüstite (mw) (modified from Stixrude and Lithgow-Bertelloni, 2005).

is isotropic, then the *SKS* phases will have no transverse component. If anisotropy is encountered, the *SKS* phases split into two polarized shear waves traveling at different velocities, and this generally results in a tangential component of motion when the ground motion is rotated to the great-circle coordinate frame. An example of *SKS* arrivals on horizontal broadband seismograms is shown in Fig. 10.18. By searching over all possible back azimuths, it is possible to find the polarization direction of the anisotropic splitting, which has fast and slow waves with the same shape (Fig. 10.18). Note that the ground motion has a complex nonlinear polarization.

By shifting the fast and slow waves to eliminate the anisotropic splitting, a linear polarization is retrieved. Rotation of the shifted traces into the great-circle coordinate system then eliminates the transverse component of the *SKS* phase. This procedure reveals the polarization direction of the fast and slow shear waves and the magnitude of the splitting. If paths from different azimuths give uniform anisotropic properties, the region where splitting occurs is likely to be in the shallow structure under the station.

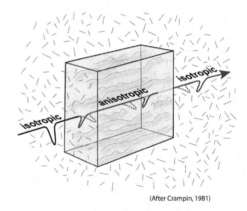

(After Crampin, 1981)

FIGURE 10.17 Example of an S-wave propagating in an isotropic medium, then intersecting an anisotropic region. As the S-wave travels through the anisotropic layer, the S-wave splits into fast and slow components, which travel at different speeds as dictated by the anisotropic medium (from Ed Ganero, Arizona State University modeled after Crampin, 1981).

10.4 Upper mantle heterogeneity

The variation in one-dimensional velocity models for different tectonic regions indicates

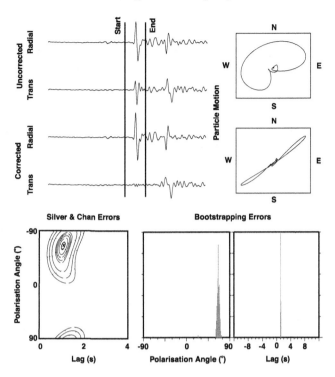

FIGURE 10.18 Example manual splitting measurement recorded at station WALA, from an event off New Britain. (Top left) Radial and transverse components of the incoming split SKS phase, the bottom two showing the waveforms after they have been corrected for splitting. (Top right) Corresponding particle motion plots oriented north–south/east–west. (Bottom left) Confidence intervals associated with the measurement calculated using a standard approach (Silver and Chan, 1991). (Bottom right) Histograms calculated using a bootstrapping method (Sandvol and Hearn, 1994) (modified from Evans et al., 2006).

upper-mantle heterogeneity, but tomographic techniques are required to determine the complete structure (see Chapter 20). Numerous seismic wave data sets have been used to investigate upper-mantle heterogeneity on a variety of scale lengths. Dense arrays of seismometers provide sufficient raypath coverage of the underlying lithosphere to develop tomographic images of the upper mantle to depths of several hundred kilometers. Because of the significant advances made over the last 25 years in the collection of high-quality broadband data, a large number of high-quality tomographic models of the upper mantle have been developed, with many including anisotropy. The Incorporated

Research Institute of Seismology have developed a tool for researchers to share their tomographic models, called the Earth Model Collaboration (EMC). The IRIS EMC openly provides access to various Earth models that have been offered to be shared by the researchers who developed them. The EMC also provides the framework for sharing: visualization tools, facilities to extract model data/metadata, and access to the contributed processing software and scripts.

An example of a large scale tomographic model, illustrated in Fig. 10.19 with a depth slice at 70 km depth, shows features that illuminate the inner workings of the Earth. Generally,

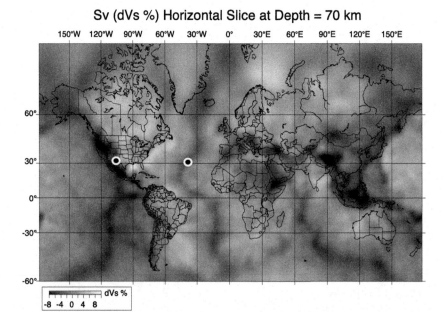

Sv (dVs %) Horizontal Slice at Depth = 70 km

FIGURE 10.19 Example of a depth slice at 70 km in the V_s global tomography model $3D2015_07Sv$ (Debayle and Kennett, 2000) derived from the IRIS EMC tool. The EMC allows a user to probe multiple models in various regions, and at various depths. Black dots represent the location of the depth profiles in Fig. 10.20.

slower velocities occur along oceanic ridges, even at 70 km depth, and cratonic regions (inner continents) have higher velocities. Probing even further, taking the depth profiles from various global models at the Mid Atlantic Ridge and more regional models at the southern Rio Grande Rift illustrate the compatibility of most models, all with similar results (Fig. 10.20). To compare the regional and global models, one would have to convert the globals models by adding the velocities from the reference models at each depth. Regardless, the global models, which can sample regions with poor local coverage, illustrate the depth of low velocity beneath the Mid Atlantic Ridge. The regional models of the southern Rio Grande Rift show remarkable similarities, with a limited low velocity zone under this rift, suggesting fundamental differences in the processes between the rifts.

In addition to being applied to regions of upwelling flow, seismic tomography has been extensively applied to determine upper-mantle structure beneath subduction zones, where old oceanic lithosphere sinks into the mantle (Fig. 3.20). Seismic observations have revealed that the sinking tabular structure of the plate provides a high-velocity, low-attenuation (high-Q) zone through which waves can propagate upward or downward through the mantle, and seismologists have expended extensive effort to determine properties of both the slab and the overlying mantle wedge. Up until the mid-1970s subducting slabs were always shown as cartoons, geometrically constrained by the intraplate earthquake locations in the Benioff zone, but as vast numbers of body-wave travel times began to accumulate, it became possible to develop tomographic images of the subducting high-velocity slab without *a priori* constraints. Current models, such as shown in Fig. 3.20 show details of orientation and depth of subducting slabs.

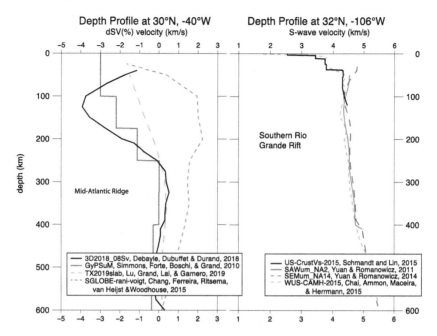

FIGURE 10.20 Depth slices from (left) $dVs\%$ global models at the Mid-Atlantic Ridge and (right) Vs regional models at the southern Rio Grande Rift derived from the IRIS EMC tool.

When surface-wave dispersion and body-wave travel-time observations are inverted for shear velocity, models like those in Fig. 10.19 are obtained, with shear velocity variations at shallow depths of 50 to 150 km having patterns that are clearly related to surface tectonics. However, variations at greater depths have weaker, tectonically unrelated patterns. These heterogeneities reflect both chemical and thermal variations in the mantle, with shear velocities varying moderately by ±5%. The associated thermal variations should be on the order of several hundred degrees, perhaps stronger in small-scale regions that are not resolved by the low-resolution models. Fluid-dynamics calculations show that such large thermal variations have density variations sufficient to drive solid-state convective motions of the upper mantle over long time scales. Thus, to the degree that the velocity variations map thermal heterogeneity, these images can be associated with dynamic flow in the interior. However, one must isolate the thermal

variations from possible petrological variations, which might, for example, contribute to the high-velocity deep roots of continents, before inferring dynamics from any seismic model.

One procedure for obtaining independent constraints on thermal variations in the mantle is to map the aspherical structure of seismic-wave attenuation. As described in Chapter 2, seismic attenuation structure in the Earth is only crudely approximated by any radially layered structure. With order-of-magnitude lateral variations in the quality factor, Q, at the upper-mantle depths, a simple stratified Q model like that for PREM (Table 10.1) is useful only to provide a baseline for body-wave attenuation or for global averages of surface-wave paths or free oscillations. Seismic attenuation is path specific in the Earth, and some general relationships with surface tectonics have been observed. Typically, paths through the mantle under stable cratons, which tend to be relatively high-velocity regions, have much lower attenuation, or lower t^*

values, than paths under tectonically active regions such as the western United States or midocean ridges. The lower mantle appears to have very little attenuation everywhere except possibly near the base of the mantle, so most of the regional variations in attenuation occur in the upper-mantle low-attenuation region, from 50 to 350 km depth. This has been demonstrated by comparing high-frequency attenuation between shallow and deep-focus earthquakes. Because most attenuation mechanisms at upper-mantle depths are expected to involve thermally activated processes, developing aspherical upper-mantle attenuation models can provide a mantle thermometer to complement velocity-variation and boundary-topography models.

In the 1980s much progress was made in determining high-frequency body-wave attenuation models and crude tectonic regionalizations. An important complication that was demonstrated is that attenuation does vary with frequency in the short-period body-wave band, leading to frequency-dependent t^* models. The high-frequency t^* determinations have relied mainly on comparison of teleseismic ground-motion spectra with theoretical source models or near-source recordings of the seismic radiation. At longer periods, many studies of body-wave attenuation have been conducted to determine average levels and lateral variations in t^* near a period of 10 s. Analysis of multiple ScS_n phases has been the most extensively applied procedure because of its intrinsic stability and source-effect cancellation properties. It is well documented that t^* varies regionally in the long-period body-wave band, but frequency dependence is not reliably resolved. Surface-wave attenuation coefficients can also be determined to develop regionalized and frequency-dependent models, although reliable separation of dispersive effects from attenuation effects is a complex procedure. Surface-wave analyses have confirmed the strong regional variations in upper-mantle attenuation across North America and Eurasia, and they constrain the vertical distribution of attenuation variations, which is difficult to attain with most body-wave studies.

Both the frequency dependence and regional variation of t^* near a period of 1 s contribute to variability of short-period body-wave magnitudes, m_b. This has actually had great political significance, since most nuclear test sites in the Soviet Union overlie shield-like mantle and the U.S. test site in Nevada (NTS) overlies a tectonically active region. As a result, explosions of the same yield have smaller teleseismic magnitudes for NTS explosions than for Soviet tests. This has complicated the verification of nuclear test yield limitation treaties, and this concern motivated many of the early studies of regional variations in seismic attenuation.

Much progress has been made for the development of an aspherical attenuation models. Specifically, the lateral variations in attenuation in the crust have been characterized well, especially under continents, and linking these models to tectonics has advanced. However, advances made with attenuation have not made as dramatic progress as with elastic velocity models. This is in part because it is more difficult to isolate effects of attenuation from propagation and source effects. Furthermore, characterizing lateral deep Earth attenuation remains difficult given the challenge in separating anelastic and scattering effects and the nonuniform sampling achieved with available data. In the case of surface-wave analyses, it is important to have a detailed velocity model first to account for focusing and dispersive effects. When this is done, the residual amplitude variations can be tomographically mapped into an aspherical model. Fig. 10.21 shows four models of Q for the mantle. The models show consistency with global tectonics. For example, global ridge and back-arc systems show low Q (high attenuation), whereas the Atlantic ridge is not as prominent in model. The Red Sea region shows low Q, much more than the Atlantic Ridge. This is likely the result of different parameterizations (not been

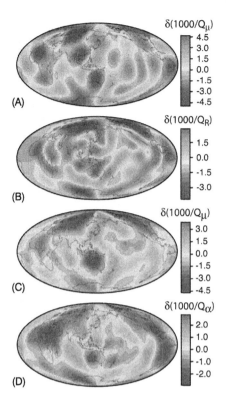

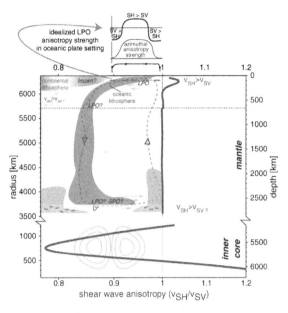

FIGURE 10.22 First-order anisotropic structure of the Earth with possible geodynamic interpretations (modified from Long and Becker, 2010).

FIGURE 10.21 Comparison of global Q models in the uppermost mantle. (A) QRLW8 model (Gung and Romanowicz, 2004), from inversion of three-component surface-wave and overtone waveform data, presented is Qm at a depth of 160 km. (B) Map at 150 km of the Rayleigh wave-based Q model QRFS12 of (Dalton et al., 2008). (C) Average Qm in the depth range 0–250 km from (Selby, 2002) and also from Rayleigh waves. (D) Average variations in Qa in the first 250 km of the mantle from amplitude ratios of P and PP (Shearer, 1991) (modified from Romanowicz and Mitchell, 2015).

inverted with depth), causing effects of deeper structure to merge into the results.

As with attenuation, mapping anisotropy of the upper mantle has been significantly advanced, especially with the USArray as part of EarthScope. Systematic analysis of SKS-splitting shows lateral variations that appear to relate to smaller scale mantle dynamics. Fig. 10.22 shows a geodynamic interpretation from the anisotropic structure of the Earth, showing that beneath continental regions, active horizontal flow and past deformational episodes ("frozen") will be present. Anisotropy beneath ocean basins, however, show primarily horizontal flow, leading to $V_{SH} > V_{SV}$, while deeper in the mantle vertical flow from upwellings and downwellings may lead to $V_{SV} > V_{SH}$. While teleseismic body waves largely provide point examples of upper-mantle anisotropic character, surface waves provide global path coverage required to constrain large-scale anisotropic properties. SKS-splitting results for the USArray (Fig. 10.23) show dramatic lateral variations suggesting that our ideas about laminar flow in the mantle are probably too simplistic. The results here are similar to those of Fig. 10.12, which suggested that there was rotational flow around a high velocity region beneath Nevada. These SKS splitting results are consistent with results from surface waves. A component of the surface-wave anisotropy appears to correlate with plate mo-

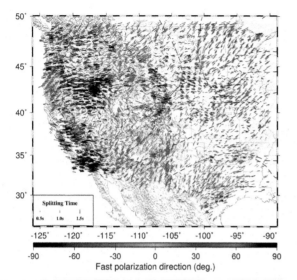

FIGURE 10.23 Shear-wave splitting measurements for western and central United States from the Missouri S&T western and central United States shear-wave teleseismic shear-wave splitting (SWS) database based hosted by IRIS (Services, 2012) that uses broadband seismic stations in North America (modified from Yang et al., 2016).

tion directions. This is illustrated in Fig. 10.6, which plots the orientation of horizontally fast directions of *azimuthal anisotropy* affecting surface waves determined from ambient noise. In general, fast directions tend to parallel plate spreading directions, yet variations illustrate the complexities of flow in the mantle.

10.5 Lower-mantle structure

The standard seismological procedure for studying the lower mantle has involved inversion of travel-time curves for smoothly varying structure, with array measurements of ray parameters as a function of distance playing a major role. This is due to the absence of significant boundaries throughout much of the vast region from 710 to 2600 km depth (Fig. 1.21). Both Herglotz–Wiechert and parameterized model inversions for lower-mantle structure suggest

smooth variations in properties compatible with self-compression of a homogeneous medium throughout the bulk of the lower mantle. Because all common upper-mantle materials have undergone phase transitions to the perovskite structure at depths from 650 to 720 km, it is generally believed that the lower mantle is primarily composed of $(Mg_{0.9}Fe_{0.1})SiO_3$ perovskite (Bridgemanite) plus $(MgFe)O$, with additional SiO_2 in the high-pressure stishovite form and uncertain amounts of calcium perovskite. No high-pressure phase transitions in the perovskite, stishovite, and MgO (rock salt) structures are expected over most of the pressure range spanned by the lower mantle, so the absence of structure is compatible with uniform composition. However, bulk composition is still uncertain, in particular the Fe and Si component relative to the upper mantle, so it remains unresolved whether a contrast in overall chemistry occurs that would favor stratified rather than whole-mantle convection. A critical parameter in this issue is the density structure of the lower mantle. Body-wave observations do not constrain the density structure in the lower mantle because it lacks detectable impedance contrasts. Instead, it is low-frequency free oscillations, which involve deep-mantle motions of large volumes, that we use to constrain the lower-mantle density structure. The free oscillations are not sensitive to the detailed density structure, but in combination with gross mass constraints from the Earth's moments of inertia, they can resolve the average density structure of the lower mantle rather well. The average density, along with density contrasts in the transition zone boundaries, provide the primary data modeled by mineral physicists in their effort to constrain the bulk composition of the lower mantle.

Although no major internal boundaries occur in the lower mantle above the D'' region, lateral heterogeneities occur throughout the region that can be modeled using seismic tomography. Several investigators have used body-wave

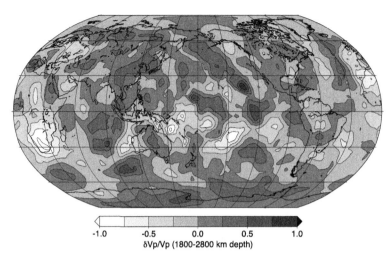

FIGURE 10.24 Relative P-wave velocity variations in the lower 1000 km of the mantle estimated using finite-frequency tomography. Perhaps most interesting are the relative low speeds beneath Africa (the African Superswell) and a series of low-speed features above the southwest Pacific Ocean "superplume'" that stretch (discontinuously) from north of eastern Australia to northeast of New Zealand. Model from Montelli et al. (2006).

travel times from thousands of events in three-dimensional inversions for lower-mantle structure. New approaches, such as finite frequency tomography, which allows for finite-frequency sensitivity kernels to account for wavefront propagation effects (wavefront healing) on low-frequency P waves travel times, plus the collection of large data sets allows for tomographic images that can detect subtle changes in the lower mantle. Fig. 10.24 illustrates one such inversion that uses the travel times of 66,210 P-waves, the differential long-period (20 s) travel times of 20,147 $PP - P$ and 2382 $pP - P$ waves, 1,427,114 short-period P, and 68,911 short-period pP travel times extracted from bulletins (Montelli, 2004). This figure shows a vertically averaged lower mantle (1800 to 2800 km depth) P velocity anomaly $\delta V_P / V_P$, illustrating that subtle features can be identified, such as the African *superplume* low-velocity anomaly beneath southern Africa. A *mantle plume* refers to a localized column of hot material that rises as a result of convection in the mantle, which can create hot spots away from plate margins (e.g.,

the Hawaiian Islands), while a *superplume* refers to a large plume or plume province that originates from the core-mantle boundary (Larson, 2013). If the velocity heterogeneity is caused by temperature variations, then temperature fluctuations of several hundred degrees occur in the deep mantle. This is sufficient to induce density heterogeneity that will in turn drive solid-state convection. Thus, slow-velocity regions are probably hot, relatively buoyant regions that are slowly rising, while fast-velocity regions are colder, sinking regions. Other features can be identified, such as a broad anomaly in the lower mantle beneath the Pacific Ocean, south of the equator, which may be a superplume that is responsible for the South Pacific *superswell*, a large region with shallow bathymetry, low surface wave velocities, and low effective elastic thickness relative to those predicted for its age. Other plumes, including Hawaii, are not well resolved but the models suggest that they reach the lowermost mantle. The observed/resolved plumes have diameters of several hundred kilometers, suggesting that a substantial fraction of

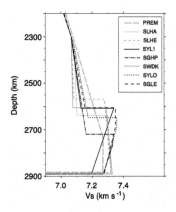

FIGURE 10.25 Radial S-wave velocity profiles for the lowermost mantle: PREM (Dziewonski and Anderson, 1981), SLHE, SLHA (Lay and Helmberger, 1983), SYL1 (Young and Lay, 1987a), SGHP (Garnero et al., 1988), SWDK (Weber and Davis, 1990), SYLO (Young and Lay, 1990), and SGLE (Gaherty and Lay, 1992). The discontinuity models represent localized structures, estimated for specific and distinct regions of the D'' layer. Shear velocity increases of 2.5–3.0% have been imaged in regions under Eurasia (SWDK and SGLE), Alaska (SYLO), Central America (SLHA), the Indian Ocean (SYL1), and the central Pacific (SGHP). The velocity increase is typically modeled as a sharp discontinuity, but the increase may be distributed over as much as 50 km. Decreased velocity gradients above the discontinuity may exist but are artifacts of the modeling in most cases. Several models include a reduction in velocity below the discontinuity. This may be real, or it may be an artifact of modeling a heterogeneous region with a one-dimensional model. The variations in depth of the discontinuity remain uncertain due to the lack of constraint on velocity above and below the discontinuity, but some variation appears to exist (from Lay (2015)).

the internal heat escaping from Earth originate from plumes (Montelli, 2004).

The D'' region at the base of the mantle has received much attention as well, since the core–mantle boundary is likely to be a major thermal boundary layer. If significant heat is coming out of the core, the mantle will be at least partially heated from below, and D'' may be a source of boundary-layer instabilities. It is often proposed that hot spots are caused by thermal plumes rising from a hot internal boundary layer, with D'' being a plausible candidate. Even

now, seismological investigations of D'' have not yet resulted in a complete characterization of the region, although it is clearly a more laterally heterogeneous region than the overlying bulk of the lower mantle.

The initial seismic models for the D'' region using travel-time studies, free-oscillation inversions, analysis of reflected phases, and studies of waves diffracted into the core-shadow zone showed remarkable diversity (Fig. 10.25). The low-resolution travel-time and free-oscillation inversions tend to give smooth models like the PREM structures, which simply showed a tapering of velocity gradients 175 km above the core–mantle boundary. Diffracted-wave studies tended to give models with stronger velocity gradient reductions, producing weak low-velocity zones just above the core (e.g., PEM-CLOI), while other studies detected abrupt velocity discontinuities near the top of D''. No single best radial model can be reliably interpreted in terms of boundary-layer structure of D''. Thus, there have been many efforts to determine the three-dimensional structure of the D'' region.

As for many tomography studies of the Earth in the last two decades, this region has been better characterized from specialized investigations of the deep mantle, again as advanced techniques have been developed, quantity and quality of data collection has been drastically improved, and much larger capacity computer access has evolved. Fig. 10.26 shows an example where a new approach was developed to image D'' (van der Hilst et al., 2007), where combined inverse scattering using a generalized Radon transform (GRT) of the wave field comprising core-reflected shear waves (ScS) along with statistical methods was used to produce images of D'' structure. Using $\sim 80,000$ ScS data points, the results show CMB reflections superimposed on tomography images of the mantle. Note that some of these reflections are not continuous, supporting the previous work from 1-D velocity models. Overall, the V_S velocity increase at 150

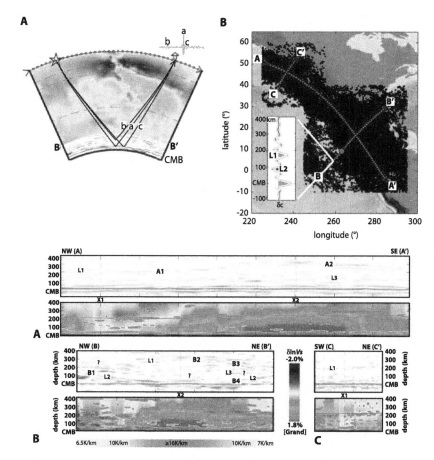

FIGURE 10.26 (Top) (A) In the mantle, tomography depicts smooth *P*-wave speed variations associated with deep subduction (lighter structure in center of section) beneath Central America; inverse scattering constrains deep mantle reflections in the lowermost 400 km. Superimposed on the tomography scattering image are schematic ray paths of *ScS* waves reflecting at and above the CMB: *a* depicts specular CMB reflections, which contribute to the main *ScS* arrival in the seismogram shown, *b* depicts scattering above the CMB, which produces precursors, and *c* depicts nonspecular reflections (at CMB or above it), which arrive mainly in the coda of ScS (we consider precritical reflections only). (B) Geographic map of the study region with bounce points of the ∼ 80, 000*ScS* data points (specular CMB reflections) used in our inverse scattering study. (Bottom) Reflectivity from inverse scattering, at more than 75% confidence (top) and S-speed(*dlnV$_S$*) from tomography (bottom). Scatter images are obtained by interpolation between GRT profiles calculated every 1° (∼ 60 km at CMB) along sections A-A', B-B', and C-C'. For the frequencies and incidence angles used, the radial resolution is ∼ 10 km. The gray scale for tomography is given between B-B' and C-C'. (A) L1, L2, and L3 label the scatter interfaces (thinly dashed). (B and C) The associated scatterers (visually enhanced) are superimposed on the tomography profiles. Interface L1 aligns increases of wave speed with increasing depth; L2 delineates a decrease; L3 is more ambiguous but generally coincides with a wave-speed increase. Whereas L1 and L2 are piecewise continuous, L3 has an intermittent, en echelon appearance. The solid (dashed) lines in the bottom panels depict the phase transition location. Points A1 and A2 and points B1 to B4 on L1 and L2 are used for temperature calculations. The gray scale below B-B' depicts the lateral variation in temperature gradient along the CMB. In the central portion of the section, *dT*/*dz* cannot be determined directly because the occurrence of the double crossing cannot be resolved (modified from van der Hilst et al., 2007).

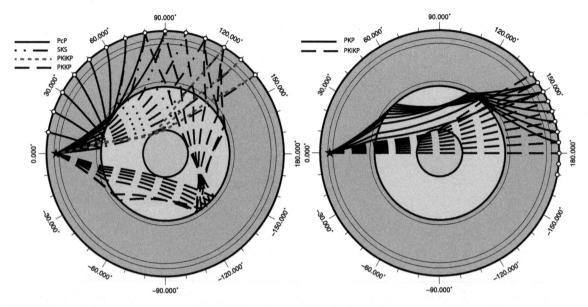

FIGURE 10.27 (Left) PREM theoretical raypaths between 0 to 140° for *PcP* reflecting off the outer core, *SKS* and *PKKP* waves traversing the outer core, *PKIKP* waves grazing the inner core. (Right) PREM theoretical raypaths between 140 to 180° for *PKP* and *PKIKP* waves (figure created with a modified script provided by Ed Ganero, Arizona State University, using the TauP package by Crotwell et al., 1999).

to 300 kilometers above the CMB aligns with a transition from Bridgmanite to postperovskite, while the internal D'' stratification may result from multiple phase-boundary crossings. These results can then be used for temperature calculations to estimate global average heat flux of 35 to 100 mW m^{-2}.

10.6 Structure of the core

The Earth's core was first discovered in 1906 when Oldham found a rapid decay of *P* waves beyond distances of 100°, and he postulated that a low-velocity region in the interior produced a shadow zone. Gutenberg accurately estimated a depth to the core of 2900 km in 1912, and by 1926 Jeffreys showed that the absence of S waves traversing the core required it to be fluid. The core extends over half the radius of the planet and contains 30% of its mass. The

boundary between the mantle and core is very sharp and is the largest compositional contrast in the interior, separating the molten core alloy from the silicate crystalline mantle. Seismologists have used reflections from the top side of the core–mantle boundary (*PcP*), underside reflections (*PKKP*), and transmitted and converted waves (*SKS, PKP,* and *PKIKP*) to determine topography on the boundary, which appears to be less than a few kilometers (Fig. 10.27). The contrast in density across the boundary is larger than that at the surface of the Earth; thus it is not surprising that little if any topography exists. The strong density contrast may be responsible for a concentration of chemical heterogeneities in the D'' region composed of materials that are denser than average mantle but less dense than the core. The material properties of the core are quite uniform (Fig. 1.21), with a smoothly increasing velocity structure down to a depth of 5150 km, where a sharp boundary separates

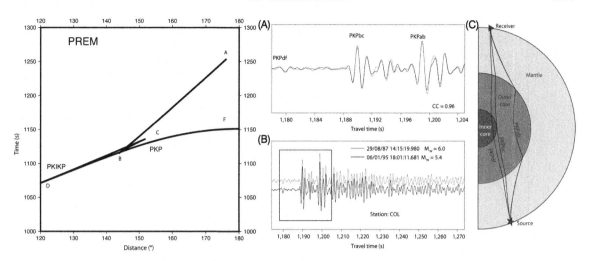

FIGURE 10.28 (Left) PREM theoretical travel times for core phases. (Right) Seismic recordings of two earthquakes (doublets) recorded at station COL (College station, Alaska). (A), 3 branches of *PKP* waveforms enlarged from the rectangle in (B). Cross-correlation coefficient CC between the two waveforms is shown in the lower right. (B), Traces are aligned on the *PKPbc* phase and filtered between 0.5 and 1.0 Hz. (C), A schematic representation of Earth's cross-section with ray paths of seismic waves referred to in this study; *PKPdf*, *PKPbc* and *PKPab*, shown. Note that only *PKPdf* waves traverse the Earth's inner core (modified from Tkalčić et al., 2013).

the outer core from the solid inner core. This 7–9% velocity increase boundary was discovered by the presence of refracted energy in the core shadow zone by Lehmann in 1936. It was not until the early 1970s that solidity of the inner core was demonstrated by the existence of finite rigidity affecting normal modes that sense the deep structure of the core.

The decrease in *P* velocity from values near 13.7 km/s at the base of the mantle to around 8 km/s at the top of the outer core profoundly affects seismic raypaths through the deep Earth. The principal *P* waves with paths through the core are shown in Fig. 10.27. As the takeoff angle from the source decreases from that of rays that just graze the core and diffract into the shadow zone, the *PKP* waves are deflected downward by the low velocities, being observed at distances of 188° to 143° (PKP_{AB}) and then again at 143° to 170° (PKP_{BC}) as the takeoff angle continues to decrease. Reflections from the inner-core boundary define the *PKiKP* (*cd*) branch (Fig. 10.28), after which the *P* wave penetrates the inner core

as *PKIKP*, which is observed from 110° to 180°. Note that most of this complexity of the core travel-time curve stems from the velocity decrease, the spherical geometry of the core, and the increase in velocity in the inner core.

These core phases are readily identified on teleseismic short-period recordings (Fig. 10.28), and the vast numbers of travel times reported by the ISC allow for the travel times, ray parameters, and positions of the cusps of the travel-time data to be used to invert for models of the core. Since the outer core does not transmit *S* waves, we believe it to be fluid, but the fact that the *S* velocity at the base of the mantle is slightly lower than the *P* velocity at the top of the core (Fig. 1.21) means that the core is not a low-velocity zone for the converted phase *SKS*. As a result, the *SKS* phase and attendant underside reflections off the core–mantle boundary, such as *SKKS*, are also used to determine the velocity structure of the outer core, particularly in the outermost 800 km where *PKP* phases do not have turning points. Some evidence from

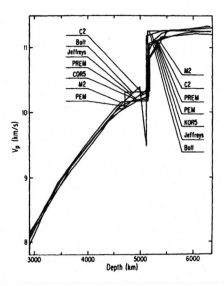

FIGURE 10.29 Various *P*-wave velocity models for the inner core–outer core boundary. Analysis of *PKiKP*, *PKIKP/PKP* waveforms and travel-time behavior underlies most of these models. Most recent models favor a relatively simple boundary, with a sharp velocity increase of 0.8–1.0 km/s, possibly overlain by a zone of reduced *P* velocity gradient at the base of the outer core (from Song and Helmberger, 1992).

SKKS phases suggests that the outermost 100 km of the core may have reduced seismic velocities relative to a uniform composition core, possibly representing a chemical boundary layer.

The inner core–outer core boundary has been extensively investigated for more than 50 years, in part due to the scattered arrivals preceding the B cusp, or caustic (Fig. 10.28). These arrivals were originally attributed to a complex transition zone at the top of the inner core, and a wide variety of models have been developed for this boundary. However, seismic arrays established that the *PKP* precursors are caused by scattering, probably in the *D″* region or at the core–mantle boundary. Fig. 10.29 shows an example of various *P*-wave velocity models for the inner core-outer core boundary.

Both the velocity and density contrasts and attenuation structure of the inner core have been

studied using *PKiKP* reflections and *PKIKP* refractions. An example of waveform comparisons used to determine inner-core properties is shown in Fig. 10.28. The different branches of the core phases *PKP* (*DF BC AB*) are readily visible on a single recording of two earthquakes close to each other (doublets). The *DF* branch arrival is broadened relative to the *BC* arrival, indicating a lower *Q* in the inner core. Waveform modeling can be used to match observations like these by considering a suite of Earth models and finding models that match the relative timing, amplitude, and frequency content of the core phases. Recent studies have used the travel times between the different branches of *PKP* to map long term changes in velocity of the inner core, suggesting inner core rotation. Specifically, the differential travel time between (PKP_{BC}–PKP_{DF}) is relatively insensitive to source location uncertainty and to 3-dimensional heterogeneities of the crust and mantle along the ray paths, allowing for the analysis of the $BC − DF$. By measuring $BC − DF$ and $AB − BC$ differential times by cross-correlation for a period of about 28, residuals between sources in the South Sandwich Islands recorded in seismic station COL in Alaska paths shows $BC − DF$ residuals increase from the years 1967–75, to 1980–85, and to 1992–95 (≈ 0.3 s) (Song and Richards, 1996). For this residual increase to occur, faster paths can be obtained by decreasing the angle between the inner core path and the axis of anisotropic symmetry. Since the ray path is essentially fixed, inner core rotation can be one mechanism to explain this observation (Fig. 10.30).

Mineral physics experiments demonstrate that the seismologically determined density of the core is lower than expected for pure iron, so the outer core is believed to have about 10% of a light alloying component such as Si, O, C, or S. The inner core may be almost pure iron, with the freezing process brought about by the geotherm dipping below the alloy melting temperature. The freezing process tends to concen-

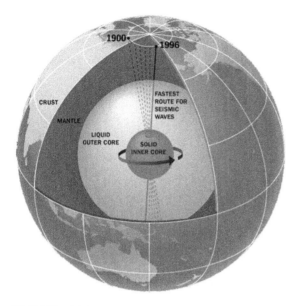

FIGURE 10.30 Mapping differential travel times of *PKP* core phases over 20 years has resulted in interpretation that the anisotropic symmetry axis of the inner core is moving owing to inner-core rotation. As the solid core and its symmetry axis rotates, a change in the angle between the symmetry axis and the inner-core leg of a ray for a fixed source and fixed station results in the systematic shift. The motion of the symmetry axis since AD 1900 moves at a rate of 1.1° per year eastward rotation about the north-south axis (modified from Song and Richards, 1996).

trate the lighter component in the fluid. Rise of this buoyant material provides compositionally driven convection in the core, which is believed to sustain core dynamics that produce the Earth's magnetic field. Proximity to the solidus is implied by the existence of the inner core; thus the outer core may actually contain suspended particles, up to 30% by volume. It is not known whether these impart any effective rigidity to the outer core, but anomalous, unexplained core modes may require a complex mechanism. Because the core rotates, the polar regions of the outer core may have a separate flow regime from the spherical annulus of material along the equator. Thus, cylindrical symmetry may play a role in core characteristics, perhaps with varying degrees of suspended particles in the polar regions.

The inner core is a very small region inside the Earth but appears to have surprising internal structure. Seismic waves traversing the inner core on paths parallel to the spin axis travel faster than waves in the equatorial plane, indicating the existence of inner-core anisotropy. This has been detected by travel times of *PKIKP* waves as well as by innercore–sensitive normal modes. The travel-time variations have ~ 1-s systematic differences with angle from the north-south axis. This axial symmetry may result from convective flow in the inner core that induces an alignment in weakly anisotropic crystals of solid iron. Thus, seismological measurements can reveal dynamic processes as deep as 6000 km into the Earth.

10.7 Summary

This chapter has reviewed the internal structure of the Earth, from the crust to the core. With the addition of high-volumes of high-quality data, increased computational capacity, and improved methodologies, great advances in imaging the Earth have occurred over the last 25 years. Large seismic experiments using thousands of seismometers are allowing for a new understanding of crustal structure. New discoveries of mantle and core processes have advanced due in a large part to improved imaging of the Earth. These advances are likely to continue, as data, computers, and methodologies continue to evolve.

PART II

Theoretical foundations

11

Elasticity and seismic waves

Chapter goals

- Introduce quantitative definitions of strain and stress.
- Introduce Hooke's Law and linear elasticity.
- Derive the equations of equilibrium and motion.
- Use deformation potentials to separate the equations of motion for a uniform material into *P*- and *S*-wave equations.
- Solve the homogeneous wave equation using separation of variables.
- Introduce plane waves and examine the particle motions of *P* and *S* waves.

In Chapter 2, we introduced the concepts of elasticity and continuum mechanics through examples that illustrated how we use elastic and nearly-elastic models to investigate earthquakes and seismic waves. In this chapter, we review the foundations of elasticity theory with a focus on seismic-wave propagation. We present a more rigorous examination of the concepts of strain and stress, the equations of equilibrium and motion, and the fundamental nature of solutions to the equations of motion, seismic waves. We derive, from first principles, the existence of *P* and *S* waves and examine their characteristics. In the chapters that follow, we apply these basic ideas to describe wave interactions and how seismologists investigate Earth's inte-

rior and seismic sources using observations of seismic deformations.

We adopt a continuum mechanics framework, which allows us to circumvent the need for detailed characterizations of microscopic phenomena. For seismology, this is critical, for we are, of course, ignorant of most of the detailed crystallographic and atomic-level structure inside the planet. Within a continuous material we can define mathematical functions for displacement, strain, and stress fields, which have well-behaved continuous spatial derivatives. Applying simple laws of physics to a continuum allows seismologists to understand nearly every arrival on a seismogram. Reliance on continuous mathematical quantities in elasticity analyses was not an immediately obvious approach. Continuum mechanics developed after decades of applied mathematics investigation that began with approximations to particle-based systems (Love, 1990).

Except in the immediate vicinity of the source, most of the seismic ground motions are small and ephemeral; the ground returns to its initial position after the transient motions subside. Vibrations of this type involve small elastic strains that correspond to small stresses. Thus, seismology, for the most part, is concerned with very small deformations (relative length changes of $\sim 10^{-6}$) over short periods of time (< 1 hr). This greatly simplifies the mathematical framework necessary, which is based on ***infinitesimal strain*** *theory*. In the immediate vicinity of seismic sources, or when we consider long-term, large-scale deformations of faults (as in structural geology), a more complete *finite-strain theory* must be employed. The relationship between forces and deformations in infinitesimal strain theory is largely empirically based and governed by a constitutive relation called Hooke's Law. Deformation is a function of material properties such as density, rigidity (resistance to shear), and incompressibility (resistance to volume change). The elastic material properties are known as moduli. When stress varies with time, strain varies correspondingly, and any imbalance between inertial forces and stresses produces seismic waves. Seismic waves transmit energy from a source (a sudden energy release) outward into the planet. These waves travel at velocities that depend on the elastic moduli and density and that are governed by the elastic equations of motion. We now proceed to show how these waves can be represented mathematically.

11.1 Deformation, deformation gradients, and strain

Seismology is so directly associated with measurement of motions that we begin by considering how motions within a material are quantified. We adopt a *Lagrangian* description, in which the motion of an infinitesimal volume of the material is tracked as a function of time and space. The Lagrange view is a natural system for seismology, because seismograms are essentially records of ground movements about an equilibrium position. Although we work with a continuum, seismic terminology sometimes refers to the motions of a small volume as particle-motions. Using the continuum model means that we use a continuous vector field, $\mathbf{u}(\mathbf{x}, t)$, to describe the motions of every location in the medium, and we are free to choose any convenient reference system for our purpose.

A rigorous foundation of the mathematics of continuum mechanics requires precise definitions of deformation and stress using limits. Only such an approach can allow us to treat the Earth as a continuum in which displacement, stress, and strain are defined at each point in the material. Thus we talk of displacement, stress, and strain *fields*. In reality, our observations are not made at infinitesimal points, but generally averaged over some fraction of a seismic wavelength. Thus for intuitive purposes, think of the points as a small volume of dimension small compared to the wavelength.

A material can undergo only a few fundamental types of motion: (1) whole-body translation, (2) whole-body rotation, (3) length, area, or volume changes (expansions and contractions in 1-, 2-, and 3-D) and (4) angular distortions or shear without a volume change (Fig. 11.1). Whole-body transformations such as translation or rotation do not change the shape of the material and hence are not resisted by the material's elastic properties, so they do not generate propagating elastic waves. So, although whole-body motions are important, they are not our focus. To produce static elastic deformation or to excite and propagate an elastic wave, we require a source that introduces a sudden imbalance in stress, and a material that resists the deformation with elastic restoring forces. Thus, we are interested in volume changes and shear distortions that produce spatially variable deformation gradients. To understand these processes quantitatively requires that we investigate spatial and temporal variations of the displacement field in deformed media.

We start with simple examples that show that the deformation of interest includes two type of distortion, length changes and angular distortions. Consider a body that is initially undeformed and unloaded with internal points O, P, and Q (Fig. 11.2A) connected by orthogonal lines. When forces are applied to deform the body, the points O, P, and Q move to point O', P', and Q' respectively. To describe the de-

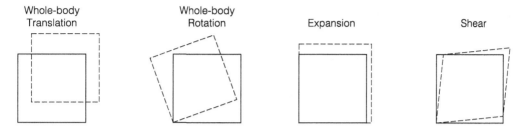

FIGURE 11.1 Imagine a region such as the solid square and the different ways that we can act on the shape. To quantify changes in the material suppose that we consider a position vector $\mathbf{x}$ that points at a particular point in the square. Suppose that after we act on the square, the material at position $\mathbf{x}$ moves to $\mathbf{x} + \boldsymbol{\xi}(\mathbf{x})$. For the first two deformations, a pure translation or rotation (or any combination of them), $\boldsymbol{\xi}(\mathbf{x})$ is constant – the same for all regions of the square. For the expansion and shear, $\boldsymbol{\xi}(\mathbf{x})$ is not constant, it depends on position. That means that the shape of the material has changed, and that shape change is resisted (by elastic restoring forces).

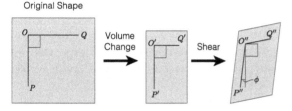

FIGURE 11.2 When a medium is deformed, we must describe both relative length changes and shearing rotations between portions of the medium. The cartoon shows the deformation of a square with two perpendicular reference lines undergoing deformation in two stages. First a volume change with principal strain directions in the directions of the lines illustrates normal strains (e.g. $\Delta\ell/\ell$). Next the material experiences a shear deformation that produces angular distortions of the original lines and the angle between.

formation of the material, we must characterize both the change in distance between points and any rotation of the reference lines relative to the surrounding material. For this purpose we introduce spatial gradients of the displacement field called **strains**. **Normal strains** are measures of elongation, defined as the change in length per unit length. For the vertical direction in the cartoon,

$$\varepsilon_{\text{normal}} = \frac{\overline{OP} - \overline{O'P'}}{\overline{OP}}, \qquad (11.1)$$

in the limit that the magnitude of the difference in length approaches zero. An analogous calculation can be performed for $\overline{OQ}$ and the value is clearly different in the case shown. Normal strains involve a fractional change in distance between points.

Shear deformations rotate line segments connecting nearby points with respect to the surrounding material. For example, the perpendicular line segments rotate such that the original perpendicular lines are no longer at right angles (Fig. 11.2). We define a **shear strain** as a measure of internal angular distortion, as

$$\varepsilon_{\text{shear}} = \frac{1}{2}\left(\frac{\pi}{2} - \angle P''O''Q''\right) = \phi. \qquad (11.2)$$

Keep in mind, these are small quantities formally defined for small distances between points, or in the limit that the line segment lengths approach zero.

Strain can be defined without need to reference a coordinate system, but in three dimensions, to describe all elongations and angular changes requires nine quantities, three normal strains, and six shear strains. In Cartesian coordinates, we represent these using index notation as ε_{11}, ε_{22}, ε_{33}, which represent relative length changes of line segments oriented in the coordinate directions. The six angular changes of each coordinate direction with respect to the other two directions are represented using $\varepsilon_{12}, \varepsilon_{13}, \varepsilon_{21}, \varepsilon_{23}, \varepsilon_{31}, \varepsilon_{32}$. For a continuum, these nine quantities are continuous throughout the medium and are functions of time. In the next section we define the strains in terms of displacement gradients.

11.1.1 Displacement gradients, strain, and rotation

We seek to establish general three-dimensional relationships between the nine Cartesian strain components and three Cartesian displacement components,

$$\mathbf{u}(\mathbf{x}, t) = u_1(\mathbf{x}, t)\,\hat{\mathbf{x}}_1 + u_2(\mathbf{x}, t)\,\hat{\mathbf{x}}_2 + u_3(\mathbf{x}, t)\,\hat{\mathbf{x}}_3. \tag{11.3}$$

The displacement, $\mathbf{u}(\mathbf{x}, t)$, is the motion of the material originally located at position $\mathbf{x}$, associated with the deformation. For example, $\mathbf{u}(\mathbf{x}, t)$, may represent the motion of the ground as a seismic wave passes, in which case we'll assume that $\mathbf{u}(\mathbf{x}, t)$, is zero before some specified time and that it eventually returns to zero as the vibrations cease. Or $\mathbf{u}(\mathbf{x}, t)$ may represent more gradual or permanent displacement of the ground before and after an earthquake.

Consider the deformation illustrated in Fig. 11.3. We identify and track two nearby (infinitesimally close) points in the material, P and Q. P is located at position $\mathbf{x}$ and Q is located at position $\mathbf{x} + \mathbf{dx}$ such that the initial, vector difference in positions is $\mathbf{dx}$. During the deformation, point P is moved to location P' and point Q is moved to location Q'. The displacement vectors describing the material's deformation are $\mathbf{u}$ for location P and $\mathbf{u} + \mathbf{du}$ for position Q.

A material is deformed when the nearby points move differently (otherwise the motion is a rotation or a translation). That is, when the displacement vector, $\mathbf{u}(\mathbf{x}, t)$ varies with position such that $\mathbf{u}(\mathbf{x}, t) \neq \mathbf{u}(\mathbf{x} + \mathbf{dx}, t)$. But since our interest in small (infinitesimal) deformations

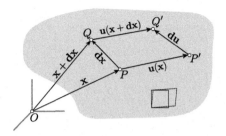

FIGURE 11.3 Deformation of a material can be quantified using the change in position between two nearby reference points. During the deformation the point P moves to P' and Q moves to Q'. The vector connecting the two points transforms from $\mathbf{dx}$ to $\mathbf{du}$. The change in position is quantified using the vector displacement, $\mathbf{u}(\mathbf{x}, t)$, produced by the deformation. When the displacement depends on position, the material undergoes a strain. The use of differentials indicates that all motions are small.

means that the difference is not too large, we can relate the two displacements with a Taylor Series expansion,

$$u_i(\mathbf{x} + \mathbf{dx}, t) = u_i(\mathbf{x}, t) + \frac{\partial u_i(\mathbf{x}, t)}{\partial x_j}dx_j$$
$$+ \frac{\partial^2 u_i(\mathbf{x}, t)}{\partial^2 x_j}dx_j^2 + \dots . \tag{11.4}$$

Note that we are using the standard summation convention – terms with repeated indices are summed over the three spatial dimensions. For small deformations, the first and second terms dominate the series, so

$$u_i(\mathbf{x} + \mathbf{dx}, t) \approx u_i(\mathbf{x}, t) + \frac{\partial u_i(\mathbf{x}, t)}{\partial x_j}dx_j, \tag{11.5}$$

or, using the vector definition of $\mathbf{du}$ in Fig. 11.3,

$$du_i(\mathbf{x} + \mathbf{dx}, t) \approx \frac{\partial u_i(\mathbf{x}, t)}{\partial x_j}dx_j. \tag{11.6}$$

The *deformation gradient tensor*, $\partial u_i / \partial x_j$, maps the difference in initial position to difference in final position (for nearby points). In other words, the deformation gradient tensor can quantify the local deformation of the material. Examine the tensor elements carefully,

$$\frac{\partial u_i(\mathbf{x}, t)}{\partial x_j} =$$
$$\begin{pmatrix} \dfrac{\partial u_1(\mathbf{x}, t)}{\partial x_1} & \dfrac{\partial u_1(\mathbf{x}, t)}{\partial x_2} & \dfrac{\partial u_1(\mathbf{x}, t)}{\partial x_3} \\ \dfrac{\partial u_2(\mathbf{x}, t)}{\partial x_1} & \dfrac{\partial u_2(\mathbf{x}, t)}{\partial x_2} & \dfrac{\partial u_2(\mathbf{x}, t)}{\partial x_3} \\ \dfrac{\partial u_3(\mathbf{x}, t)}{\partial x_1} & \dfrac{\partial u_3(\mathbf{x}, t)}{\partial x_2} & \dfrac{\partial u_3(\mathbf{x}, t)}{\partial x_3} \end{pmatrix} . \quad (11.7)$$

Each element is a ratio of lengths, so the elements have no physical units (think of them as change in length per unit length). Whenever we encounter a tensor in physics, it's a good idea to examine its irreducible parts to look for physical insight. For a second-rank tensor in Euclidean space, the irreducible parts are the symmetric and antisymmetric components of the tensor and the tensor's *trace* (sum of diagonal elements). We'll start by splitting the tensor into symmetric and antisymmetric parts

$$\frac{\partial u_i(\mathbf{x}, t)}{\partial x_j} = \frac{1}{2}\left(\frac{\partial u_i(\mathbf{x}, t)}{\partial x_j} + \frac{\partial u_j(\mathbf{x}, t)}{\partial x_i}\right)$$
$$+ \frac{1}{2}\left(\frac{\partial u_i(\mathbf{x}, t)}{\partial x_j} - \frac{\partial u_j(\mathbf{x}, t)}{\partial x_i}\right) \quad (11.8)$$
$$= \varepsilon_{ij} + \omega_{ij} .$$

The first term on the right-hand side, ε_{ij}, is symmetric, and the second term, ω_{ij}, is antisymmetric. The second term has zeros along the diagonal and includes three independent elements. This tensor (a vector) describes rotation. Our focus is the symmetric tensor, ε_{ij}, which is called the **strain tensor**. Since the strain tensor is symmetric, $\varepsilon_{ij} = \varepsilon_{ji}$, it contains only six independent elements (collectively, ε_{ij} and ω_{ij} combine to match the original nine elements of the displacement gradient tensor). The strain tensor elements are

$$\varepsilon_{ij} =$$
$$\begin{pmatrix} \dfrac{\partial u_1}{\partial x_1} & \dfrac{1}{2}\left(\dfrac{\partial u_1}{\partial x_2} + \dfrac{\partial u_2}{\partial x_1}\right) & \dfrac{1}{2}\left(\dfrac{\partial u_1}{\partial x_3} + \dfrac{\partial u_3}{\partial x_1}\right) \\ \dfrac{1}{2}\left(\dfrac{\partial u_2}{\partial x_1} + \dfrac{\partial u_1}{\partial x_2}\right) & \dfrac{\partial u_2}{\partial x_2} & \dfrac{1}{2}\left(\dfrac{\partial u_2}{\partial x_3} + \dfrac{\partial u_3}{\partial x_2}\right) \\ \dfrac{1}{2}\left(\dfrac{\partial u_3}{\partial x_1} + \dfrac{\partial u_1}{\partial x_3}\right) & \dfrac{1}{2}\left(\dfrac{\partial u_3}{\partial x_2} + \dfrac{\partial u_2}{\partial x_3}\right) & \dfrac{\partial u_3}{\partial x_3} \end{pmatrix} , \quad (11.9)$$

which is clearly symmetric. The diagonal elements of ε_{ij} represent elongations. The off-diagonal elements represent angular distortions of the material. We consider each in sequence.

Normal strains

Examine Fig. 11.4. The deformation shown includes an elongation (in the x_1 direction) and contraction (in the x_3 direction). Focus on the elongation. The deformation in the x_1 direction in this case depends only on the x_1 direction, so

$$du_1 = \frac{\partial u_1}{\partial x_1}dx_1 . \quad (11.10)$$

After deformation, the new length of the box is

$$dx_1 + du_1 = dx_1 + \frac{\partial u_1}{\partial x_1}dx_1 = \left(1 + \frac{\partial u_1}{\partial x_1}\right)dx_1$$
$$= (1 + \varepsilon_{11})dx_1 . \quad (11.11)$$

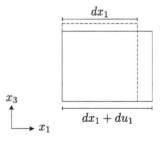

FIGURE 11.4 Example length changes used to clarify the geometric interpretation of the diagonal (normal) strains. The box be elongated in the x_1 direction and contracted in the x_3 direction. The original length along the x_1 direction, dx_1, is transformed into $dx_1 + du_1$.

This equation indicates that a geometric interpretation of ε_{11} is a change in length per unit length (in the x_1 direction). Consideration of similar examples demonstrates that the geometric interpretation extends to ε_{22} and ε_{33} for the simple case of elongation perpendicular to one axis, but the ideas hold for more complex geometry. We call the diagonal elements of the strain tensor *normal strains*. The normal strain terms involve volumetric changes. In our convention, compressional strains are negative and extensional strains are positive.

Next consider the sum of the three normal strains, $\theta = \varepsilon_{11} + \varepsilon_{22} + \varepsilon_{33}$,

$$\theta(\mathbf{x}, t) = \frac{\partial u_i(\mathbf{x}, t)}{\partial x_i} = \frac{\partial u_1(\mathbf{x}, t)}{\partial x_1} + \frac{\partial u_2(\mathbf{x}, t)}{\partial x_2}$$
$$+ \frac{\partial u_3(\mathbf{x}, t)}{\partial x_3}$$
$$= \nabla \cdot \mathbf{u}(\mathbf{x}, t) . \tag{11.12}$$

We call the divergence of the deformation-induced displacement the *cubic dilatation* and it corresponds to a fractional change in volume. For example, if we consider a simple cube with initial volume, $V_0 = dx_1 \, dx_2 \, dx_3$ and final volume, $V_1 = [(1 + \epsilon_{11})dx_1 \, (1 + \epsilon_{22})dx_2 \, (1 + \epsilon_{33})dx_3]$, then

$$\frac{\Delta V}{V_0} = \frac{V_1 - V_0}{V_0} \approx \varepsilon_{11} + \varepsilon_{22} + \varepsilon_{33} = \theta . \tag{11.13}$$

The idea generalizes to arbitrary shapes, but keep in mind that our expressions are appropriate for small deformations – for example, for deformations associated with large pressures, we must adopt a finite-strain theory to compute volume changes.

Shear strains

Examine Fig. 11.5. The deformation shown corresponds to a change in shape associated with only off-diagonal elements of the strain tensor. Focus on the deformation of the originally perpendicular lines of the lower-left corner of the box. If we follow the deformation along the

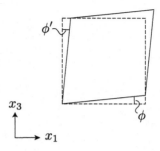

FIGURE 11.5 Example length changes and shear deformation used to define the geometric interpretation of diagonal and off-diagonal strains. On the left, the box is elongated in the x_1 direction and contracted in the x_3 direction. The original length along the x_1 direction, dx, is transformed into $dx + du$. On the right, the shear deformation produces a change in position such that $\partial u_3 / \partial x_1 > 0$ and $\partial u_1 / \partial x_3 > 0$.

horizontal axis, a change in position is associated with a constant value of $\partial u_3 / \partial x_1$. In terms of the angle ϕ, we have

$$\frac{\partial u_3}{\partial x_1} = \tan \phi . \tag{11.14}$$

Along the vertical axis, the deformation gradient has similar form,

$$\frac{\partial u_1}{\partial x_3} = \tan \phi' . \tag{11.15}$$

Since we are considering very small deformations (exaggerated in the figure), these quantities, including the angles ϕ and ϕ', are quite small. For small angles,

$$\tan \phi \approx \phi , \tag{11.16}$$

such that

$$\varepsilon_{13} = \frac{1}{2} \left(\frac{\partial u_1}{\partial x_3} + \frac{\partial u_3}{\partial x_1} \right) = \tfrac{1}{2} \left(\tan \phi + \tan \phi' \right)$$
$$= \frac{1}{2} \left(\phi + \phi' \right) , \tag{11.17}$$

and, of course, $\varepsilon_{31} = \varepsilon_{13}$. So we see that for small deformations, shear strains are geometrically associated with angular changes in the material.

Rigid-body rotation

The deformation gradient tensor (Eq. 11.7) also includes components and information on rotations (ω_{ij}), where ϵ_{ijk} is the permutation symbol (Box 11.1). This tensor represents a rigid-body rotation about a vector $\omega_k = \epsilon_{ijk}\omega_{ij}$. For reference we note that the rotations of the medium can be expressed using the curl of the displacement field,

$$\epsilon_{ijk}\,\omega_{ij} = \frac{1}{2}\left[\left(\frac{\partial u_3}{\partial x_2} - \frac{\partial u_2}{\partial x_3}\right)\hat{x}_1 + \left(\frac{\partial u_1}{\partial x_3} - \frac{\partial u_3}{\partial x_1}\right)\hat{x}_2 \right.$$
$$\left. + \left(\frac{\partial u_2}{\partial x_1} - \frac{\partial u_1}{\partial x_2}\right)\hat{x}_3\right]$$
$$= \frac{1}{2}\epsilon_{abc}\partial_a u_b$$
$$= \frac{1}{2}\nabla \times \mathbf{u}(\mathbf{x}, t) , \qquad (11.18)$$

which includes combinations of displacement gradients not in the strain tensor. Since the rotation is not resisted by the material's elasticity, our primary focus remains on the normal and shear strains.

11.2 Stress

When a continuum is acted upon by a force, either internal or external, that force influences every point in the body via a distribution of forces throughout the body. Two types of forces exist within a continuum, *body forces* and *contact forces* (what we call exertions in Chapter 2). Body forces, such as that caused by gravity, depend on the volume and density of material ($\mathbf{F}_g = \rho V\mathbf{g}$, where $m = \rho V$ is the mass, ρ is the density, and V is the volume). The net body force is proportional to the material volume. Contact forces operate across surfaces within and bounding a continuum. For example, the wind resistance a bicyclist experiences is a contact force because it depends on the area of the rider facing the wind. A net contact force is proportional to the area of the surface. Thus body

forces have physical dimensions of force per unit volume and contact forces have physical dimensions of force per unit area.

For a continuum that is acted on by external forces, internal contact forces operate within the medium. We can visualize the medium as having an internally distributed force system, as illustrated in Fig. 11.6. Imagine a plane that passes through the material and contains a point P within a small volume, ΔV. If we remove the material on one side of the plane, to keep the other side in equilibrium requires a distribution of forces. The surface forces must balance the forces within ΔV acting on the other side of the plane. The precise surface force required depends on the orientation of the plane. Call the net surface force $\Delta \mathbf{F}$, quantify the orientation using a normal vector, $\hat{n}$, and represent the area of the plane by ΔA. Then the *stress vector* or *traction vector* acting on surface defined by $\Delta A\hat{n}$, $\mathbf{T}(\hat{n})$, is defined as

$$\mathbf{T}(\hat{n}) = \lim_{\Delta A \to 0} \frac{\Delta \mathbf{F}}{\Delta A} = T_1\hat{x}_1 + T_2\hat{x}_2 + T_3\hat{x}_3 . \quad (11.19)$$

The limit is defined for the continuum model

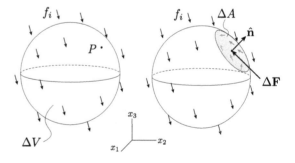

FIGURE 11.6 (Left) A small spherical volume embedded in a larger continuum acted upon by external forces, f_i. (Right) Imaginary plane with normal, $\hat{n}$, passing through an internal point P. A portion of the sphere has been removed and replaced by a distribution of forces acting on the surface (small gray vectors) that retain the remainder of the continuum in equilibrium (the sphere stays a sphere). The net force in the plane required to maintain the original shape, ΔF. The stress acting at point P is the ratio of the net force to the area, as the area becomes infinitesimally small.

Box 11.1 Useful vector relationships

Ground displacement is a vector, $\mathbf{u}(\mathbf{x}, t) = u_1(\mathbf{x}, t)\hat{\mathbf{x}}_1 + u_2(x, t)\hat{\mathbf{x}}_2 + u_3(x, t)\hat{\mathbf{x}}_3$, and the equations of motion (11.57) and (11.59) are vector equations. Thus, some review of basic vector operations is helpful.

(a) The *scalar product* (*dot product* or *inner product*) of two vectors

$$\mathbf{a} = a_1\hat{\mathbf{x}}_1 + a_2\hat{\mathbf{x}}_2 + a_3\hat{\mathbf{x}}_3$$
$$\mathbf{b} = b_1\hat{\mathbf{x}}_1 + b_2\hat{\mathbf{x}}_2 + b_3\hat{\mathbf{x}}_3$$

is given by

$$\mathbf{a} \cdot \mathbf{b} = a_1b_1 + a_2b_2 + a_3b_3 = a_ib_i = |\mathbf{a}|\,|\mathbf{b}|\cos\theta\;, \tag{B11.1.1}$$

where θ is the angle between the two vectors. The dot product is the length of each vector projected on the direction of the other vector, $\mathbf{a} \cdot \mathbf{b} = 0$ for perpendicular vectors ($\theta = \pi/2$), and clearly, $\mathbf{a} \cdot \mathbf{b} = \mathbf{b} \cdot \mathbf{a}$.

(b) The *vector product* (*cross product* or *curl*) of $\mathbf{a}$ and $\mathbf{b}$ is

$$\mathbf{a} \times \mathbf{b} = (a_2b_3 - a_3b_2)\,\hat{\mathbf{x}}_1 + (a_3b_1 - a_1b_3)\,\hat{\mathbf{x}}_2 + (a_1b_2 - a_2b_1)\,\hat{\mathbf{x}}_3$$

$$= \begin{vmatrix} \hat{\mathbf{x}}_1 & \hat{\mathbf{x}}_2 & \hat{\mathbf{x}}_3 \\ a_1 & a_2 & a_3 \\ b_1 & b_2 & b_3 \end{vmatrix}. \tag{B11.1.2}$$

In indicial notation we can define the permutation symbol

$$\epsilon_{ijk} = \begin{cases} 0 & \text{any two indices equal} \\ 1 & i, j, k \text{ in order} \\ -1 & i, j, k \text{ not in order} \end{cases} \tag{B11.1.3}$$

then,

$$\left(\mathbf{a} \times \mathbf{b}\right)_i = \epsilon_{ijk}a_jb_k\;. \tag{B11.1.4}$$

The cross product defines a new vector that is perpendicular to the two vectors. Properties of the dot and cross product include the following:

$$\mathbf{a} \times \mathbf{b} = -\mathbf{b} \times \mathbf{a}$$
$$\mathbf{a} \cdot \left(\mathbf{a} \times \mathbf{b}\right) = \mathbf{b} \cdot \left(\mathbf{a} \times \mathbf{b}\right) = 0$$
$$\mathbf{a} \times \left(\mathbf{b} + \mathbf{c}\right) = \mathbf{a} \times \mathbf{b} + \mathbf{a} \times \mathbf{c} \tag{B11.1.5}$$
$$\mathbf{a} \cdot \left(\mathbf{b} \times \mathbf{c}\right) = \mathbf{b} \cdot \left(\mathbf{c} \times \mathbf{a}\right) = \mathbf{c} \cdot \left(\mathbf{a} \times \mathbf{b}\right).$$

(c) The *gradient* of a scalar field uses the "del" operator

$$\nabla = \frac{\partial}{\partial x_1}\hat{\mathbf{x}}_1 + \frac{\partial}{\partial x_2}\hat{\mathbf{x}}_2 + \frac{\partial}{\partial x_3}\hat{\mathbf{x}}_3\;, \tag{B11.1.6}$$

applied to a scalar field, $\varphi(\mathbf{x})$,

$$\nabla \varphi(\mathbf{x}) = \frac{\partial \varphi}{\partial x_1} \hat{\mathbf{x}}_1 + \frac{\partial \varphi}{\partial x_2} \hat{\mathbf{x}}_2 + \frac{\partial \varphi}{\partial x_3} \hat{\mathbf{x}}_3$$

$$(\nabla \varphi)_i = \frac{\partial \varphi}{\partial x_i} = \varphi_{,i} \ .$$

(B11.1.7)

The gradient vector points in the direction of steepest slope, or rate of change, of the field $\varphi(\mathbf{x})$.

(d) The *divergence* of a vector field $\boldsymbol{\Psi}(\mathbf{x}) = [\psi_1(\mathbf{x}), \psi_2(\mathbf{x}), \psi_3(\mathbf{x})]$ is

$$\nabla \cdot \boldsymbol{\Psi}(\mathbf{x}) = \frac{\partial \psi_1}{\partial x_1} + \frac{\partial \psi_2}{\partial x_2} + \frac{\partial \psi_3}{\partial x_3} = \psi_{i,i} \ . \tag{B11.1.8}$$

This is a scalar field that measures the flux of the vector field through a unit volume. The integral over a volume V with surface area S is

$$\int_V \nabla \cdot \boldsymbol{\Psi}(\mathbf{x}) \, dV = \int_S \mathbf{n} \cdot \boldsymbol{\Psi}(\mathbf{x}) \, dS \ , \tag{B11.1.9}$$

where $\mathbf{n}$ is the outward-facing unit normal everywhere on S. This is *Gauss' theorem*. This states that the accumulation of the field $\boldsymbol{\Psi}(\mathbf{x})$ in the volume is equal to the flux through the surface.

(e) The *Laplacian* of a scalar field is the divergence of the gradient:

$$\nabla^2 \varphi(\mathbf{x}) = \nabla \cdot \nabla \varphi(\mathbf{x}) = \frac{\partial^2 \varphi}{\partial x_1^2} + \frac{\partial^2 \varphi}{\partial x_2^2} + \frac{\partial^2 \varphi}{\partial x_3^2} = \varphi_{,ii} \ , \tag{B11.1.10}$$

which is a scalar. The Laplacian of a vector field is a vector with components that are Laplacians of the original components (if Cartesian coordinates are used). Or, for any coordinate system,

$$\nabla^2 \boldsymbol{\Psi}(\mathbf{x}) = \nabla (\nabla \cdot \boldsymbol{\Psi}(\mathbf{x})) - \nabla \times \nabla \times \boldsymbol{\Psi}(\mathbf{x}) \ . \tag{B11.1.11}$$

(f) *Helmholtz's theorem* states that any vector field $\mathbf{u}(\mathbf{x})$ can be represented in terms of a vector potential $\boldsymbol{\Psi}(\mathbf{x})$ and a scalar potential $\varphi(\mathbf{x})$ by

$$\mathbf{u}(\mathbf{x}) = \nabla \varphi(\mathbf{x}) + \nabla \times \boldsymbol{\Psi}(\mathbf{x}) \tag{B11.1.12}$$

if $\nabla \times \varphi(\mathbf{x}) = 0$, $\varphi(\mathbf{x})$ is curl free, and $\nabla \cdot \boldsymbol{\Psi}(\mathbf{x}) = 0$, $\boldsymbol{\Psi}(\mathbf{x})$ is divergence free.

(g) Two useful vector identities are

$$\nabla \cdot (\nabla \times \boldsymbol{\Psi}(\mathbf{x})) = 0$$

$$\nabla \times (\nabla \varphi) = 0 \ .$$

and corresponds to a continuous distribution of internal forces. Intuitively, we can think of the surface as a small part of the material and the stress corresponds to an average of the microscopic interactions occurring within the surrounding small part of the overall volume. Macroscopically, stress corresponds to an exertion of one part of the material upon the other. The value of the traction changes with the orientation of each plane passing through the point P. However, we can represent the value for any plane using a tensor defined using the tractions operating on any three mutually orthogonal surfaces.

11.2.1 The stress tensor

To begin, choose a surface parallel to the $x_2 x_3$ plane ($\hat{\mathbf{n}} = \hat{\mathbf{x}}_1$) and define the stress components acting on this plane using

$$\sigma_{11} = \lim_{\Delta A_1 \to 0} \frac{\Delta F_1}{\Delta A_1}$$

$$\sigma_{12} = \lim_{\Delta A_1 \to 0} \frac{\Delta F_2}{\Delta A_1}$$

$$\sigma_{13} = \lim_{\Delta A_1 \to 0} \frac{\Delta F_3}{\Delta A_1}, \quad (11.20)$$

where

$$\Delta \mathbf{F} = \Delta F_1 \hat{\mathbf{x}}_1 + \Delta F_2 \hat{\mathbf{x}}_2 + \Delta F_3 \hat{\mathbf{x}}_3 . \quad (11.21)$$

The σ_{1j} are the components of the traction on the $\hat{\mathbf{n}} = \hat{\mathbf{x}}_1$ plane in the three coordinate directions. The first index of σ_{1j} in Eq. (11.20) corresponds to the direction of the normal to the plane acted on by the force, and the second index indicates the direction of the force (per unit area). Thus σ_{11} is a stress acting normal to the plane, and σ_{12} and σ_{13} are stresses acting in the plane. We can repeat that analysis for two mutually orthogonal planes containing the point P, one parallel to the $x_1 x_2$ and the other parallel $x_1 x_3$ planes, and define six additional stress components,

$$\sigma_{22}, \sigma_{21}, \sigma_{23} \quad \text{acting on the } \hat{\mathbf{n}} = \hat{\mathbf{x}}_2 \text{ plane}$$

$$\sigma_{33}, \sigma_{31}, \sigma_{32} \quad \text{acting on the } \hat{\mathbf{n}} = \hat{\mathbf{x}}_3 \text{ plane} .$$

All of these quantities are implicitly functions of position and time, i.e. σ_{ij} is shorthand for $\sigma_{ij}(\mathbf{x}, t)$ where the first index, i, indicates the plane upon which the stress acts, and the second index, j, indicates the coordinate direction of the stress. We adopt a sign convention such that positive stress components correspond to positively-directed forces acting on planes with normals in the $+x_i$ directions and negatively-directed forces acting on planes with normals in the $-x_i$ directions.

11.2.2 Cauchy's relation

Cauchy showed how to use these nine quantities to reconstruct the traction operating on any surface passing through the location where $\sigma_{ij}(\mathbf{x}, t)$ is known. We can demonstrate his result by considering a balance of forces on a tetrahedron with three faces parallel to the coordinate planes and a fourth face with an arbitrary orientation quantified using the area dA and unit-magnitude normal vector, $\hat{\mathbf{n}}$ (Fig. 11.7). For a body in equilibrium, the sum of the forces, and the sum of the moments, acting on the volume must vanish. Balancing the forces (stress × area)

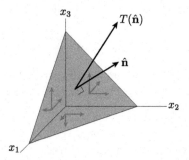

FIGURE 11.7 (left) Cauchy's formula uses a balance of forces on a tetrahedron bounded by the coordinate axes and a surface with normal vector $\hat{\mathbf{n}}$ to compute the traction on the surface from the stress tensor (components shown conceptually in gray). (right) Often for earthquake analysis we resolve the traction components into normal and shear components.

in the x_1, direction gives

$$T_1 \, dA = \sigma_{11} \, dA \, n_1 + \sigma_{21} \, dA \, n_2 + \sigma_{31} dA \, n_3 \,, \tag{11.22}$$

or

$$T_1 = \sigma_{11} \, n_1 + \sigma_{21} \, n_2 + \sigma_{31} \, n_3 \,. \tag{11.23}$$

Similarly, for the other directions, $\Sigma F_{x_2} = \Sigma F_{x_3} = 0$, leads to

$$T_2 = \sigma_{22} n_2 + \sigma_{12} n_1 + \sigma_{32} n_3 \,, \tag{11.24}$$

and

$$T_3 = \sigma_{33} n_3 + \sigma_{13} n_1 + \sigma_{23} n_2 \,, \tag{11.25}$$

or generally

$$T_i = \sigma_{ji} n_j \,. \tag{11.26}$$

The above formula, known as Cauchy's Formula, allows us to linearly combine our nine components of stress defined using the coordinate planes to calculate the stress on any arbitrarily oriented surface through the medium at location P. The nine components, σ_{ij} completely characterize the stress at the location P. We use these nine stresses to form a **stress tensor**, σ_{ij}

$$\sigma_{ij} = \begin{pmatrix} \sigma_{11} & \sigma_{12} & \sigma_{13} \\ \sigma_{21} & \sigma_{22} & \sigma_{23} \\ \sigma_{31} & \sigma_{32} & \sigma_{33} \end{pmatrix} . \tag{11.27}$$

The diagonal terms act perpendicular to the reference planes and are called **normal stresses**, and the off-diagonal terms act in the reference planes and are called **shear stresses**. Normal stresses with positive values (directed outward from positive or negative faces as defined above) are called **tensional stresses**, and negative values correspond to **compressional stresses**.

Representative absolute stresses within Earth

The *SI* units for stress are **Pascals** (*Pa*), where $10^6 \, Pa = 1 \, MPa$. Older units commonly used

TABLE 11.1 Pressure ranges within Earth (from PREM).

Region	Pressure Range	Pressure Range
The Crust	0.0001–0.6 GPa	1–6000 atm
The Upper Mantle	0.6–13 GPa	6k–125k atm
The Transition Zone	13–24 GPa	125k atm– 237k atm
Lower Mantle	24–136 GPa	237k–1.3M atm
Outer Core	136–329 GPa	1.3M–3.2M atm
Inner Core	329–364 GPa	3.2M–3.6M atm

in many classic seismology references are *bars*, $1 \, MPa = 10 \, bars$; in the *CGS* system a bar is $10^6 \, dyne/cm^2$. The state of stress at depth in the Earth is nearly always compressional, and therefore all three normal stresses in (11.27) are negative. The **maximum compressive stress** is the stress with the largest absolute value, and the **minimum compressive stress** is the stress with the smallest absolute value. To make the results more accessible intuitively, we can represent the pressure in terms of atmospheres, where on atmosphere is equal to the average atmospheric pressure on Earth. Near sea-level the atmosphere exerts a pressure of about 14.7 pounds per square inch (one psi, about 6.9 kPa), or about 100 kPa. Table 11.1 is a list of pressure values at selected depths within the Earth. The pressures are extremely high in the deep Earth – seismic waves are propagated by small stress perturbations around these very large values.

11.2.3 The conservation of linear momentum – the equations of equilibrium

In this section we combine the definition of stress and Newton's Second Law to develop the conditions of equilibrium for a material experiencing small stresses. Consider a small, cubic element in a continuum bounded by faces parallel to the coordinate planes (Fig. 11.8). Assume that the cube is in static equilibrium (the sum of

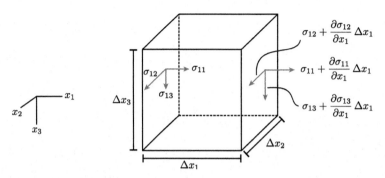

FIGURE 11.8 A cubic element in the continuum bounded by faces paralleling the coordinate planes. Balancing the stresses on each face acting in a given direction leads to the *equations of equilibrium*. Only the stresses acting on the $\pm x_1$ face are shown, but analogous stress terms act on the other four faces of the cube.

the forces acting on the cube vanish). First focus on the x_1 direction – since the cube dimensions are small, the stress on the right side of the cube can be represented in terms of the stress on the left side of the cube using a Taylor Series expansion.

Balancing the forces that act in the x_1 direction leads to

$$\sum F_{x_1} = \left(\sigma_{11} + \frac{\partial \sigma_{11}}{\partial x_1} \Delta x_1 - \sigma_{11} \right) \Delta x_2 \Delta x_3$$
$$+ \left(\sigma_{21} + \frac{\partial \sigma_{21}}{\partial x_2} \Delta x_2 - \sigma_{21} \right) \Delta x_1 \Delta x_3$$
$$+ \left(\sigma_{31} + \frac{\partial \sigma_{31}}{\partial x_3} \Delta x_3 - \sigma_{31} \right) \Delta x_1 \Delta x_2 = 0$$

or

$$\left(\frac{\partial \sigma_{11}}{\partial x_1} + \frac{\partial \sigma_{21}}{\partial x_2} + \frac{\partial \sigma_{31}}{\partial x_3} \right) = 0 . \qquad (11.28)$$

Similarly, letting $\Sigma F_{x_2} = \Sigma F_{x_3} = 0$ leads to

$$\left(\frac{\partial \sigma_{12}}{\partial x_1} + \frac{\partial \sigma_{22}}{\partial x_2} + \frac{\partial \sigma_{32}}{\partial x_3} \right) = 0 \qquad (11.29)$$

$$\left(\frac{\partial \sigma_{13}}{\partial x_1} + \frac{\partial \sigma_{23}}{\partial x_2} + \frac{\partial \sigma_{33}}{\partial x_3} \right) = 0 \qquad (11.30)$$

or compactly,

$$\frac{\partial \sigma_{ij}}{\partial x_i} = 0 \quad i, j = 1, 2, 3 . \qquad (11.31)$$

These are the **equilibrium equations** – equilibrium requires balanced spatial stress gradients.

11.2.4 Conservation of angular momentum stress tensor symmetry

A second condition of equilibrium requires that the conservation of momentum is satisfied, otherwise the cube will rotate. In other words, the sum of the *mechanical moments* operating on the cube must also sum to zero. Consider σ_{12} on either side of the elemental cube. The stresses are oppositely directed (no net force), but because of the cube's width, the two forces introduce rotational moment about the x_3 axis. Mechanical moment is the product of a force times the perpendicular distance of the force to the rotation axes. If we sum the moments about axes passing through the center of the cube in Fig. 11.8 and parallel to the coordinate axes, we obtain equations such as (for the x_3 axis)

$$\sum M_{x_3} = \left[\left(\sigma_{12} + \frac{\partial \sigma_{12}}{\partial x_1}\Delta x_1 + \sigma_{12}\right)\Delta x_2 x_3 \frac{\Delta x_1}{2}\right.$$
$$\left. - \left(\sigma_{21} + \frac{\partial \sigma_{21}}{\partial x_2}\Delta x_2 + \sigma_{21}\right)\frac{\Delta x_2}{2}\Delta x_3 \Delta x_1\right]$$
$$= 0,$$

or

$$2\sigma_{12} + \frac{\partial \sigma_{12}}{\partial x_1}\Delta x_1 - 2\sigma_{21} - \frac{\partial \sigma_{21}}{\partial x_2}\Delta x_2 = 0. \quad (11.32)$$

For the above condition to hold as $\Delta x_1, \Delta x_2 \to 0$, we must have $\sigma_{12} = \sigma_{21}$. Similarly, requiring that $\Sigma M_{x_2} = \Sigma M_{x_1} = 0$ requires, $\sigma_{13} = \sigma_{31}$ and $\sigma_{23} = \sigma_{32}$, or generally

$$\sigma_{ij} = \sigma_{ji} \quad (11.33)$$

The stress tensor is symmetric, and the symmetry reduces the number of independent components in the tensor to six.

11.2.5 Principal stresses

We can show that at each point in a continuum, three mutually perpendicular planes exist upon which no shear-stress components act. Directions normal to these planes are called the *principal axes* and the coordinate system defined by those axes is called the *principal-axes coordinate system*. The principal directions are determined by diagonalizing a tensor (stress or strain) as described in Box 11.2. The principal directions for the stress and strain tensors are known as principal stress and principal strain directions respectively. The trace of the stress tensor is invariant to choice of coordinate system and is related to the total stress state. *Hydrostatic stress* is defined as the average of the normal stresses:

$$P = \frac{\sigma_{11} + \sigma_{22} + \sigma_{33}}{3}. \quad (11.34)$$

The *deviatoric stress tensor* is that part of the stress tensor without the hydrostatic term

$$D_{ij} = \sigma_{ij} - P\delta_{ij}. \quad (11.35)$$

In addition to their physical significance of the principal stress and strain directions, many theoretical analyses are more easily analyzed in the principal axes coordinate system because the tensors are diagonal on those systems and diagonal matrices are convenient.

Tensors and tensor rotation

Tensors are important mathematical tools that we use to insure that our physical laws are independent of the particular coordinate system that we choose. Tensors have well-defined coordinate transformation properties. Scalars can be considered zeroth-order tensors (magnitude, no directional property), and vectors can be considered first-order tensors (in three dimensions, 3^1 elements, magnitude and directionality). Second-order tensors are used to quantify interactions between vectors and directional operators. Both the state of stress and state of strain can be represented using second-order tensors, which in three dimensions have $3^2 = 9$ elements, but in both cases, symmetry reduces the independent elements to six. Tensors obey specific transformation laws when a coordinate system is rotated, clearly a desirable property for our generally defined terms. If we let $a_{ij} = \partial x_i'/\partial x_j$ be the direction cosine between the rotated x_i' axes and the original x_j axes, then a vector, **u**, transforms according to

$$u_i' = a_{ij}u_j, \quad (11.36)$$

and the stress components transform according to

$$\sigma_{ij}' = a_{ip}a_{jq}\sigma_{pq}, \quad (11.37)$$

and the strain components transform according to

$$\varepsilon_{ij}' = a_{ip}a_{jq}\varepsilon_{pq}. \quad (11.38)$$

Box 11.2 Eigenvalues, eigenvectors, and tensor invariants

Stress and strain are symmetric tensors (i.e., $\sigma_{ij} = \sigma_{ji}$) and thus can be diagonalized, or rotated into a principal-axis coordinate system. Consider the stress tensor,

$$\sigma_{ij} = \begin{pmatrix} \sigma_{11} & \sigma_{12} & \sigma_{13} \\ \sigma_{21} & \sigma_{22} & \sigma_{23} \\ \sigma_{31} & \sigma_{32} & \sigma_{33} \end{pmatrix}. \tag{B11.2.1}$$

This tensor can be diagonalized by subtracting a constant value, λ (not Lamé's parameter), from the diagonal elements, setting the determinant of the resulting matrix equal to 0, and solving for λ,

$$\begin{vmatrix} \sigma_{11} - \lambda & \sigma_{12} & \sigma_{13} \\ \sigma_{21} & \sigma_{22} - \lambda & \sigma_{23} \\ \sigma_{31} & \sigma_{32} & \sigma_{33} - \lambda \end{vmatrix} = 0. \tag{B11.2.2}$$

The three roots of this equation, λ_k are called the *eigenvalues* and they are the values of σ_{ij} in a principal coordinate system where the tensor is diagonal. The symmetry of the matrix σ_{ij} ensures that the λ_k are real-valued quantities. Each eigenvalue has a corresponding *eigenvector*. Length-normalized eigenvectors provide the principal coordinate axis "directions". We compute the eigenvectors by solving

$$\begin{bmatrix} \sigma_{11} - \lambda_k & \sigma_{12} & \sigma_{13} \\ \sigma_{21} & \sigma_{22} - \lambda_k & \sigma_{23} \\ \sigma_{31} & \sigma_{32} & \sigma_{33} - \lambda_k \end{bmatrix} \begin{bmatrix} x_1 \\ x_2 \\ x_3 \end{bmatrix} = 0, \tag{B11.2.3}$$

for each λ_k. Eq. (B11.2.2) is equivalent to

$$\lambda^3 - \mathrm{tr}\left(\sigma_{ij}\right)\lambda^2 + \mathrm{minor}\left(\sigma_{ij}\right)\lambda - \det\left(\sigma_{ij}\right) = 0, \tag{B11.2.4}$$

where $\mathrm{tr}(\sigma_{ij}) = \sigma_{11} + \sigma_{22} + \sigma_{33}$ is the trace of the original tensor, $\mathrm{minor}(\sigma_{ij})$ is the sum of the minors of σ_{ij}, $(\sigma_{11}\sigma_{22} + \sigma_{22}\sigma_{33} + \sigma_{11}\sigma_{33} - \sigma_{21}^2 - \sigma_{32}^2 - \sigma_{31}^2)$, and $\det(\sigma_{ij})$ is the determinant of σ_{ij} $(\sigma_{11}\sigma_{22}\sigma_{33} + 2\sigma_{21}\sigma_{32}\sigma_{31} - \sigma_{11}\sigma_{32}^2 - \sigma_{22}\sigma_{31}^2 - \sigma_{33}\sigma_{21}^2)$. Because the eigenvalues of a matrix are unchanged by a coordinate transformation, the coefficients of the cubic polynomial in (B11.2.4) are *invariant* with respect to coordinate rotations. This means that the trace, minor, and determinant of the tensor are also independent of the coordinate system and, in general, have some special physical significance.

One way of defining tensors is to identify quantities that obey the above transformations. Physical fields that transform in this specific manner during coordinate rotations are represented using second-order tensors.

11.3 The equation of motion

Our primary interest in seismology is of course seismic waves, which are propagating deformations created by stress imbalances. Consider a force balance on a small cubic element in a continuum that is undergoing internal motions (not in equilibrium). Referring back to Fig. 11.8, the equilibrium Eq. (11.31) must now include inertial and any body-force contributions. For example, assume that the cube is acted on by a body force per unit volume $f = f_1\hat{\mathbf{x}}_1 + f_2\hat{\mathbf{x}}_2 + f_3\hat{\mathbf{x}}_3$ and represent the material density using ρ. Satisfying Newton's law now requires

$$\rho \frac{\partial^2 u_i}{\partial t^2} = f_i + \frac{\partial \sigma_{ij}}{\partial x_j} \ . \tag{11.39}$$

This set of three equations ($i = 1, 2, 3$) is called the *equations of motion* and they relate the continuum's accelerations to body forces and stress gradients. This is the most fundamental equation underlying the theory of seismology, it relates observable displacements to forces acting in the medium. We will see in Chapter 16 that many seismic sources can be represented using equivalent body forces that are introduced into Eq. (11.39) to fully describe seismic motions. For convenience, we will often denote temporal derivatives using Newton's dot notation, $\partial u/\partial t = \dot{u}$, $\partial^2 u/\partial t^2 = \ddot{u}$, so the equation of motion are compactly written

$$\rho \ddot{u}_i = f_i + \sigma_{ij,j} \ . \tag{11.40}$$

In the case in which sources or body forces such as gravity are not considered, the *homogeneous equation of motion* are

$$\rho \ddot{u}_i = \sigma_{ij,j} \ . \tag{11.41}$$

Homogeneous equations are important tools for exploring the fundamental solutions of a problem and homogeneous solutions are often used to construct solutions to the non-homogeneous equations that includes body-forces (sources).

11.3.1 Hooke's law and linear elasticity

To transform the equations of motion into a set of equations that can be solved for observ-able ground displacement, we require relationships between stress and displacement. Relationships called constitutive laws relate stress and strain (and hence displacement gradients). In any material, a complex relationship exists between stress and strain, depending on parameters such as pressure, temperature, stress rate, strain history, and stress magnitude. For example, nearly all Earth materials flow if small, steady stresses are applied for millions of years, or they fracture or fail plastically if high stresses are applied.

However, for the small-magnitude, short-duration stresses of interest in seismic-wave analysis, almost all Earth materials behave such that stress is linearly proportional to strain. The behavior has been often demonstrated empirically by applying controlled stresses to samples and observing the resulting stress-strain behavior. An example is shown in Fig. 11.9. Note that there is a *substantial*, nearly linear interval prior to failure of the rock and that for the small strains ($10^{-5} - 10^{-4}$) of interest here, the sample could well be represented by a linear elastic relationship, – removing the stress while in the elastic regime restores the material to its original state.

The most general form (i.e. the most general linear relationship between two second-order tensors) of a constitutive law for linear elasticity is *Hooke's law*

$$\sigma_{ij} = C_{ijkl} \, \varepsilon_{kl} \ . \tag{11.42}$$

The elastic constants of proportionality, C_{ijkl}, are known as *elastic moduli* (sometimes called elastic stiffnesses) and define the elastic material properties of the material. Since strain has no physical units, the physical units of the elastic moduli are the units of stress. We can think of the elements of C_{ijkl} as the intrinsic stresses with which the material resists deformation. C_{ijkl} is a fourth-order tensor with 81 terms relating the nine elements of the strain tensor to the nine elements of the stress tensor. Note the double repeated indices in Eq. (11.42) (and recall that we

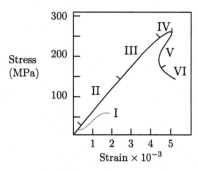

FIGURE 11.9 Stress-strain curve for a typical uniaxial compression test. Stage I involves closure of cracks; stage II is a linear elastic regime; stages III and IV involve dilatancy of the rock due to lateral expansion of the rock and micro-cracking; stage V involves loss of load-bearing capacity, strain localization, and development of a macroscopic shear failure; and stage VI has stress determined by residual friction on the shear zone (modified from Scholz, 2019).

are using the summation convention). The first term of Eq. (11.42) is

$$
\begin{aligned}
\sigma_{11} = {} & C_{1111}\varepsilon_{11} + C_{1112}\varepsilon_{12} + C_{1113}\varepsilon_{13} \\
& + C_{1121}\varepsilon_{21} + C_{1122}\varepsilon_{22} \\
& + C_{1123}\varepsilon_{23} + C_{1131}\varepsilon_{31} + C_{1132}\varepsilon_{32} + C_{1133}\varepsilon_{33} .
\end{aligned}
$$
(11.43)

There are nine such equations, but the symmetry of the stress and strain tensors ($\sigma_{ij} = \sigma_{ji}$ and $\varepsilon_{ij} = \varepsilon_{ji}$) reduces the number of independent equations to six and the number of independent coefficients to 36 ($\sigma_{ij} = \sigma_{ji} \rightarrow C_{jikl}$; $\varepsilon_{kl} = \varepsilon_{lk} \rightarrow C_{ijkl} = C_{klij}$). An additional symmetry relation ($C_{ijkl} = C_{klij}$) follows from consideration of a strain energy density function (see Malvern, 1969) and results in 21 elastic moduli for the most general elastic material, which has the most general *anisotropy*, meaning the stress-strain behavior depends on the orientation of the material.

Isotropic elastic materials

Fortunately, the elastic properties for many materials and material composites in the Earth are roughly independent of direction. An *iso-*

tropic elastic material has only two independent elastic moduli, called the *Lamé parameters*, μ and λ (sometimes called the first and second Lamé parameters, and μ is the shear modulus). These are related to C_{ijkl} by

$$
C_{ijkl} = \lambda \delta_{ij}\delta_{kl} + \mu \left(\delta_{ik}\delta_{jl} + \delta_{il}\delta_{jk} \right) ,
$$
(11.44)

where δ_{ij} is a Kronecker delta. For example, $C_{1111} = \lambda + 2\mu$, $C_{1122} = \lambda$, $C_{1212} = \mu$, etc. The symmetry in the first two and last two indices of the elastic stiffness tensor, C_{ijkl} allow us to represent its components using a two-dimensional matrix if we adopt the notation,

$$
11 \rightarrow 1, \ 22 \rightarrow 2, \ 33 \rightarrow 3, \ 23, 32 \rightarrow 4, \ 13, 31 \rightarrow 5,
$$
$$
12, 21 \rightarrow 6 .
$$
(11.45)

For example, we can write $C_{1111} \rightarrow C_{11} = \lambda + 2\mu$. For its convenience, the resulting 6×6 coefficient matrix is commonly used in discussions of anisotropy.

Inserting the isotropic relationship, Eq. (11.44), into Hooke's Law (11.42) results in

$$
\sigma_{ij} = \left[\lambda \delta_{ij}\delta_{kl} + \mu \left(\delta_{ik}\delta_{jl} + \delta_{il}\delta_{jk} \right) \right] \varepsilon_{kl} ,
$$
(11.46)

which reduces (e.g., $\delta_{kl}\epsilon_{kl} = \epsilon_{kk}$) to

$$
\begin{aligned}
\sigma_{ij} &= \lambda \varepsilon_{kk}\delta_{ij} + 2\mu\varepsilon_{ij} \\
&= \lambda \theta \delta_{ij} + 2\mu\varepsilon_{ij} .
\end{aligned}
$$
(11.47)

This form of Hooke's law for an isotropic linear elastic material was formulated by Navier in 1821 and Cauchy in 1823, 160 years after Hooke's original work. In the early 1800's, Cauchy and Poisson in particular applied the equations to a wide range of deformation problems. For completeness, we note that Hooke's law can also be expressed for strain components as well,

$$
\varepsilon_{ij} = \frac{-\lambda}{2\mu \left(3\lambda + 2\mu \right)} \sigma_{kk}\delta_{ij} + \frac{1}{2\mu}\sigma_{ij} ,
$$
(11.48)

which provides expressions useful for calculating the strains induced when the stress is

TABLE 11.2   Relationships between elastic moduli.

μ	κ	λ	E	v
$\frac{3(\kappa-\lambda)}{2}$	$\lambda + \frac{2\mu}{3}$	$\kappa - \frac{2\mu}{3}$	$\frac{9\kappa\mu}{3\kappa+\mu}$	$\frac{\lambda}{2(\lambda+\mu)}$
$\lambda\left(\frac{1-2v}{2v}\right)$	$\mu\left[\frac{2(1+v)}{3(1-2v)}\right]$	$\frac{2\mu v}{(1-2v)}$	$2\mu(1-v)$	$\frac{\lambda}{(3\kappa-\lambda)}$
$3\kappa\left(\frac{1-2v}{2+2v}\right)$	$\lambda\left(\frac{1+v}{3v}\right)$	$3\kappa\left(\frac{v}{1+v}\right)$	$\mu\left(\frac{3\lambda+2\mu}{\lambda+\mu}\right)$	$\frac{3\kappa-2\mu}{2(3\kappa+\mu)}$
$\frac{E}{2(1+v)}$	$\frac{E}{3(1-2v)}$	$\frac{Ev}{(1+v)(1-2v)}$	$3\kappa(1-2v)$	$\frac{3\kappa-E}{6\kappa}$

known. The coefficients in Eq. (11.48) have the units of inverse of pressure, such quantities are often called compliances.

Elastic moduli and parameters

Elastic moduli are often defined in terms of ratios of stress to strain for particular deformation geometry (that's how they are often measured). For any material, the moduli change with the pressure and temperature, both of which span large ranges within the planet. Laboratory measurements of the properties of rocks and minerals have provided a wealth of information on many geologic materials, particular at low pressures and temperatures. Designing experiments to measure the properties of materials in environments corresponding to the hot, deep earth remains a challenge. Advances in computational mineral physics (e.g. Stixrude and Lithgow-Bertelloni, 2005) are now provide useful estimates of material properties at the high pressures and temperatures within the lower mantle and core. The better we know the elastic moduli of earth materials, the more accurately we can interpret information from seismic imaging.

Commonly discussed elastic moduli involved in seismological analyses include the shear, bulk, and Young's moduli and of course the already mentioned second Lamé parameter, λ. The moduli are nonnegative and have units of stress. The *shear modulus*, or *rigidity* (μ) is a measure of a material's resistance to shear. For an isotropic material,

$$\sigma_{ij} = 2\mu\varepsilon_{ij} \Rightarrow \mu \simeq \frac{\sigma_{ij}}{2\varepsilon_{ij}} .$$

The larger μ, the less the material deforms under a particular stress. For a fluid, $\mu = 0$ because fluids have no resistance to shear. The shear modulus is the quantity in the formula for seismic moment (Chapters 1 and 2) and is the most important elastic modulus in terms of strain accumulation leading to earthquakes. Typical average values of μ in the shallow earth are 30 to 40 GPa. The *bulk modulus* or *incompressibility* (κ) is the material's resistance to a change in volume, and is defined as the ratio of an applied lthostatic pressure to the induced fractional change in volume,

$$\sigma_{ij} = -P\delta_{ij}, \quad \frac{\Delta V}{V} = \frac{-P}{\kappa} \Rightarrow -P = \kappa\,\varepsilon_{ij} \Rightarrow \frac{-P}{\varepsilon_{ij}} .$$

As a material becomes more rigid, κ increases. *Young's modulus* (E) is a measure of the ratio of uniaxial stress to strain in the same direction,

$$\sigma_{11} = E\left(\frac{\Delta L}{L}\right) = E\,\varepsilon_{11} ,$$

where L is the initial length and ΔL is the change in length of the material sample. Young's modulus often arises in one-dimensional analyses and was the first modulus discovered, when in the early 1800's, Young connected a physical definition of the quantity with a coefficient in the early forms of the equations of elasticity. Young was also one of the first scientists to recognize that a material's resistance to shear differed from its resistance to elongation or compression.

Two ratios that include elastic moduli are also important in seismology. **Poisson's Ratio** (ν) is the ratio of radial to axial strain when a uniaxial stress is applied ($\sigma_{11} \neq 0$, $\sigma_{22} = \sigma_{33} = 0$),

$$\varepsilon_{22} = \varepsilon_{33}, \quad \nu = \frac{-\varepsilon_{22}}{\varepsilon_{11}}.$$

Poisson's ratio is dimensionless and has a maximum value of 0.5. This is true for a fluid, when $\mu = 0$ (no shear resistance). Most Earth materials have a Poisson ratio between 0.22 and 0.35. Seismologists often use Poisson's ratio as a measure of the ratio of P- to S-wave speeds (V_P and V_S respectively) since the two quantities have a one-to-one, invertible relationship,

$$\frac{V_P}{V_S} = \sqrt{\frac{2(1 - \nu)}{1 - 2\nu}}, \quad \text{or} \quad \nu = \sqrt{\frac{(V_P/V_S)^2 - 2}{2(V_P/V_S)^2 - 2}}.$$
$$(11.49)$$

Another often-referenced ratio in seismology, the **seismic parameter** (Φ), is constructed using the ratio of bulk modulus to density,

$$\Phi = \frac{\kappa}{\rho} = V_P^2 - \frac{4}{3} V_S^2. \qquad (11.50)$$

Φ is related to the propagation speed of the dilatation associated with a seismic wave. The square root of the seismic parameter is often called the bulk velocity and this equals the seismic P-wave speed in fluids such as the outer core.

For most seismological applications, λ and μ or κ and μ are used for analytical calculations. Relationships between the common isotropic moduli are tabulated in Table 11.2. For many Earth materials, $\mu \approx \lambda$, and when they are exactly equal the material is called a **Poisson or Cauchy solid**, for which $\nu = 0.25$ and $\kappa = 5\mu/3$. Poisson's ratio is often referred to in studies of

Earth's materials since it has a one-to-one relationship with the ratio of P- to S-wave speeds. Values of κ and μ are often tabulated for Earth models (see Chapter 10). Table 11.3 is a list of near-surface (pressure and temperature) values of elastic moduli for selected Earth materials.

11.3.2 The equations of motion for linearly elastic materials

The substitution of Hooke's law into the equation of motion allows us to derive basic equations for displacement fields in linear elastic material. We will use the isotropic form of Hooke's Law (Eq. 11.47) in our development and the resulting equations are extremely useful. But before we proceed, it is important to note that many Earth materials are in fact not isotropic, and even average upper-mantle properties require anisotropic representations. This occurs mainly because olivine, a major mineral in the upper mantle, is intrinsically anisotropic, with elastic moduli varying by 10%, depending on orientation (relative to the olivine crystal structure). Some sedimentary rocks have fabrics that give rise to 25% anisotropy of elastic moduli. Although anisotropy can be fully analyzed, we develop our theory of seismic waves in the context of isotropic materials because it is algebraically simpler. We refer the interested readers to number of texts present a more general exposition, including Dahlen and Tromp (1998) and Chapman (2004).

To derive the homogeneous equations of motion for a linear, isotropic elastic material, we combine the homogeneous equation of motion, Eq. (11.41), Hooke's law, Eq. (11.47), and the strain-displacement relationship, Eq. (11.9). First, consider only the $i = 1$ term of Eq. (11.41),

$$\rho \frac{\partial^2 u_1}{\partial t^2} = \frac{\partial \sigma_{11}}{\partial x_1} + \frac{\partial \sigma_{12}}{\partial x_2} + \frac{\partial \sigma_{13}}{\partial x_3}. \qquad (11.51)$$

TABLE 11.3   Elastic Moduli for some common materials.

Material	κ (GPa)	μ (GPa)	λ (GPa)	ν	ρ (kg/m^3)
Water	2.1	0	2.1	0.50	1000
Sandstone	17	6	13	0.34	1900
Olivine	129	82	74	0.24	3200
Perovskite	266	153	164	0.26	4100

Combining the isotropic form of Hooke's Law and the strain-displacement relations, we find,

$$\sigma_{11} = \lambda\theta + 2\mu\varepsilon_{11} = \lambda\left(\frac{\partial u_1}{\partial x_1} + \frac{\partial u_2}{\partial x_2} + \frac{\partial u_3}{\partial x_3}\right)$$
$$+ 2\mu\frac{\partial u_1}{\partial x_1}$$

$$\sigma_{12} = 2\mu\varepsilon_{12} = \mu\left(\frac{\partial u_1}{\partial x_2} + \frac{\partial u_2}{\partial x_1}\right)$$

$$\sigma_{13} = 2\mu\varepsilon_{13} = \mu\left(\frac{\partial u_1}{\partial x_3} + \frac{\partial u_3}{\partial x_1}\right). \qquad (11.52)$$

Combining Eqs. (11.51) and (11.52) and assuming λ and μ are constant throughout the medium ($\partial\lambda/\partial x_i = \partial\mu/\partial x_i = 0$) results in

$$\rho\frac{\partial^2 u_1}{\partial t^2} = \lambda\frac{\partial\theta}{\partial x_1} + \mu\frac{\partial}{\partial x_1}\left(\frac{\partial u_1}{\partial x_1} + \frac{\partial u_2}{\partial x_2} + \frac{\partial u_3}{\partial x_3}\right)$$
$$+ \mu\left(\frac{\partial^2 u_1}{\partial x_1^2} + \frac{\partial^2 u_1}{\partial x_2^2} + \frac{\partial^2 u_1}{\partial x_3^2}\right). \qquad (11.53)$$

The first term in parentheses is the cubical dilatation, θ, and the second is the Laplacian, $\nabla^2 u_1$, so

$$\rho\frac{\partial^2 u_1}{\partial t^2} = (\lambda + \mu)\frac{\partial\theta}{\partial x_1} + \mu\nabla^2 u_1. \qquad (11.54)$$

A similar analysis for the u_2 and u_3 components produces

$$\rho\frac{\partial^2 u_2}{\partial t^2} = (\lambda + \mu)\frac{\partial\theta}{\partial x_2} + \mu\nabla^2 u_2, \qquad (11.55)$$

$$\rho\frac{\partial^2 u_3}{\partial t^2} = (\lambda + \mu)\frac{\partial\theta}{\partial x_3} + \mu\nabla^2 u_3. \qquad (11.56)$$

We can write all three equations in the equivalent vector form

$$\rho\,\ddot{\mathbf{u}} = (\lambda + \mu)\nabla(\nabla\cdot\mathbf{u}) + \mu\nabla^2\mathbf{u}, \qquad (11.57)$$

which is the three-dimensional vector homogeneous equation of motion for a uniform, isotropic, linear elastic medium. A common alternate form of this equation employs the vector identity (see Box 11.1)

$$\nabla^2\mathbf{u} = \nabla(\nabla\cdot\mathbf{u}) - (\nabla\times\nabla\times\mathbf{u}), \qquad (11.58)$$

allowing (11.57) to be written as

$$\rho\ddot{\mathbf{u}} = (\lambda + 2\mu)\nabla(\nabla\cdot\mathbf{u}) - (\mu\nabla\times\nabla\times\mathbf{u}). \qquad (11.59)$$

Eqs. (11.57) and (11.59) are complicated, three-dimensional, partial differential equations for displacements in a continuum (which we assume were initiated by an unspecified source). Although we can obtain solutions by numerical evaluation of these equations, we can gain insight into the solutions by using some standard analytical approaches.

11.4 Wave equations for P- and S-wave potentials

We can use Helmholtz's theorem (Box 11.1) to represent the displacement field as

$$\mathbf{u}(\mathbf{x}, t) = \nabla\varphi(\mathbf{x}, t) + \nabla\times\mathbf{\Psi}(\mathbf{x}, t), \qquad (11.60)$$

where $\varphi(\mathbf{x}, t)$ is a curl-free scalar potential field ($\nabla\times\varphi = 0$) and $\mathbf{\Psi}(\mathbf{x}, t)$ is a divergence-free vector potential field ($\nabla \cdot \boldsymbol{\Psi} = 0$). Physically, a curl-free field involves no shearing motion, and a divergence-free field involves no volume change. Substituting Eq. (11.60) into Eq. (11.59) and using the vector identity ($\nabla \times \nabla \times \boldsymbol{\Psi} = -\nabla^2 \boldsymbol{\Psi}$ since $\nabla \cdot \boldsymbol{\Psi} = 0$), we find

$$\nabla \left[(\lambda + 2\mu) \nabla^2 \varphi - \rho \ddot{\varphi} \right] + \nabla \times \left[\mu \nabla^2 \cdot \boldsymbol{\Psi} - \rho \ddot{\boldsymbol{\Psi}} \right]$$
$$= 0 . \tag{11.61}$$

We can clearly satisfy this equation if each term in brackets vanishes independently. Define

$$\alpha = \sqrt{\frac{\lambda + 2\mu}{\rho}} , \qquad \beta = \sqrt{\frac{\mu}{\rho}} \tag{11.62}$$

and note that (11.61) will be satisfied if

$$\nabla^2 \varphi - \frac{1}{\alpha^2} \ddot{\varphi} = 0 \quad \text{and} \quad \nabla^2 \boldsymbol{\Psi} - \frac{1}{\beta^2} \ddot{\boldsymbol{\Psi}} = 0 . \tag{11.63}$$

Eq. (11.63) lists a scalar wave equation for $\varphi(\mathbf{x}, t)$ and a vector wave equation for $\boldsymbol{\Psi}(\mathbf{x}, t)$. The quantity α is the velocity corresponding to the wave solutions, $\varphi(\mathbf{x}, t)$, and is called the P-wave velocity, and β is the S-wave velocity corresponding to solutions, $\boldsymbol{\Psi}(\mathbf{x}, t)$. Note that the equations of motion govern seismic displacements, and that the potentials used to decompose the displacements satisfy wave equations. For problems with certain symmetries, the displacement will satisfy a wave equation directly. But in general, the displacement satisfies the more general equations of motion. In fact in inhomogeneous media, the equations of motion involve gradients of the material properties and the expressions do not separate into two wave equations. Fortunately the equations of motion can be solved numerically in those instances, which is how we compute seismograms for more complicated media.

Thus we see that solving the equations of motion (11.59) generally involves solving wave equations such as (11.63), satisfied by wave potentials from which we can determine the displacement field using (11.60). In every case, the wholespace displacement field comprises two fundamental wave types, P and S waves, that propagate with distinct velocities determined by the properties of the medium. In both cases, the wave speed is a ratio of an elastic moduli to the density, which represents a ratio of elastic to inertial forces. The interactions of these two force types is what causes the wave to propagate. Later, we will show that P waves produce dilational and compressional motions and volumetric changes as the wave disturbance passes; S waves produce shearing motions without volume change. From (11.62) it is clear that $\alpha > \beta$ (for $\lambda \approx \mu$, $\alpha \approx \sqrt{3}\beta$). Since the P disturbance travels faster, it arrives before S disturbances – this is the original meaning of P for *primary* wave, and S for *secondary* wave. The existence of P and S wave solutions in a solid was first recognized by Poisson in 1829. An important additional result that will not be demonstrated here is that P and S waves are in fact the *only* transient solutions for the homogeneous elastic wholespace. Thus together they provide a *complete* solution to the equation of motion. Next, we extend our insight into these waves by considering one-dimensional and then three-dimensional cases.

11.4.1 The one-dimensional wave equation and solutions

We can demonstrate the essence of wave behavior using a simple one-dimensional case. Consider the longitudinal oscillations of a long, thin (infinitesimal), elastic bar extending in the $\pm x_1$, directions (Fig. 11.10). For this model, longitudinal oscillations involve displacements only in the x_1 direction, thus the displacement can be represented as $\mathbf{u}(\mathbf{x}, t) = (u_1(x_1, t), u_2 \approx 0, u_3 \approx 0)$. As in our general derivation, the equations of motion are derived by a balance between inertial terms and stress gradients. We assume that

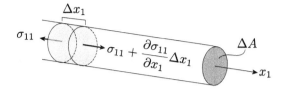

FIGURE 11.10 A long, thin elastic bar extending along the x_1 axis. An elastic disturbance produced by an unspecified source (say, a tap at the far end) propagates through the bar.

a source (unspecified, but known to exist) has created a stress imbalance in the bar. The sum of forces acting on a small volume of the bar, $\Delta V = \Delta A \, \Delta x_1$, is balanced by the acceleration of the volume

$$m \ddot{u}_1 = \sum F_{x_1}$$

$$\rho \, \Delta A \, \Delta x_1 \, \ddot{u}_1 = \left(\sigma_{11} + \frac{\partial \sigma_{11}}{\partial x_1} \Delta x_1 \right) \Delta A - \sigma_{11} \Delta A \,,$$

$$(11.64)$$

where ρ is the density and ΔA is the cross-sectional area of the bar. Thus

$$\rho \ddot{u}_1 = \frac{\partial \sigma_{11}}{\partial x_1} \,. \qquad (11.65)$$

We now assume that the bar is governed by an isotropic elastic constitutive law of the form $\sigma_{11} = E \, \varepsilon_{11}$, where E is Young's modulus. Then

$$\rho \frac{\partial^2 u_1}{\partial t^2} = E \frac{\partial^2 u_1}{\partial x_1^2} \,. \qquad (11.66)$$

If we define $c = (E/\rho)^{1/2}$, the result is the one-dimensional wave equation,

$$\frac{\partial^2 u_1}{\partial t^2} = c^2 \frac{\partial^2 u_1}{\partial x_1^2} \,. \qquad (11.67)$$

As before, the wave equation results from a simple combination of Newton's second law, an assumption of infinitesimal deformation, and Hooke's Law of linear elasticity. The equation

requires that the curvature of the deformation is proportional to the acceleration of the material and that the constant of proportionality is the wave speed squared. This derivation is, of course, approximate because in reality lateral strains in any realistic finite bar produce nonuniform stress across the cross section, but such variations are not resolvable with wavelengths much greater than the lateral dimension of the bar. But note that as a result of this approximation, in this case, the displacements directly satisfy the wave equation, in contrast with the case of our general elastic solutions, in which potentials satisfy the wave equations.

General solutions of the 1D wave equation

Producing the wave equation is only part of our analysis. The general solution of (11.67) is

$$u_1(x_1, t) = f(x_1 - ct) + g(x_1 + ct) \,, \qquad (11.68)$$

which is called **D'Alembert's solution**. The functions f and g are *arbitrary*, but the useful solutions are those that also satisfy the initial conditions associated with a particular source and any conditions on the boundary of the material. Disturbance f propagates in the $+x_1$ direction and disturbance g propagates in the $-x_1$ direction, both with velocity $c = (E/\rho)^{\frac{1}{2}}$. The arguments $(x_1 \pm ct)$ are called the *phase* of the solution. For a given value of phase, the translating functional shape is called a *wavefront*. For a bar with constant wave speed, c, the functions $f(x_1 - ct)$ and $g(x_1 + ct)$ maintain constant shapes as the waves propagate – a change in time is compensated by a change in position. The ideas are illustrated in Fig. 11.11.

The wave speed depends on the material properties, in this case the ratio of Young's modulus, E, to density, ρ, which corresponds to the ratio of the elastic to inertial forces that operate as the wave propagates. A stiffer material has a larger Young's modulus and thus hosts faster-traveling waves. Increasing the density

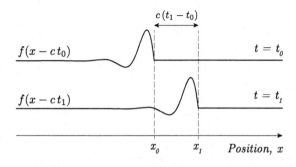

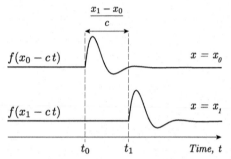

FIGURE 11.11 A propagating disturbance, $f(x - ct)$, shown above as a function of position for two times, t_0 and t_1, and below as a function of time for two positions, x_0 and x_1. Since the speed is constant, the disturbance maintains its shape as it moved from left to right, and exhibits the same time history at the two positions (ignoring the temporal shift). Dashed lines show the pulse's wavefront, which corresponds to the same phase value in each plot.

should reduce the speed, but for most materials, E tends to increase with increasing ρ, and denser materials usually exhibit a net increase in speed.

Harmonic solutions of the 1D wave equation

A general procedure that we can follow to solve partial differential equations such as (11.67) is to assume the solution has a form that separates the spatial and temporal dependences. This approach is called the method of *separation of variables*. For convenience, until needed, we ignore the subscript (1) on both position, x, and displacement, u. We assume a solution of

the form

$$u(x, t) = X(x) \cdot T(t) , \qquad (11.69)$$

(the product must have physical units equal to length) and insert this trial solution into (11.67), resulting in

$$c^2 \frac{1}{X(x)} \frac{d^2 X(x)}{dx^2} = \frac{1}{T(t)} \frac{d^2 T(t)}{dt^2} . \qquad (11.70)$$

The only way that the term on the left, a function only of x, can always (for all x, and all t) equal the term on the right, a function only of t, is if each of the two terms equal the same constant. We represent the constant, which has units of one over time squared, using $-\omega^2$. The separation approach has produced two coupled ordinary differential equations,

$$\frac{d^2 X(x)}{dx^2} + \frac{\omega^2}{c^2} X(x) = 0$$
$$\frac{d^2 T(t)}{dt^2} + \omega^2 T(t) = 0 . \qquad (11.71)$$

These equations can be solved by standard methods such as Fourier transforms, or in this simple case by recognizing that they have the form satisfied by simple *harmonic* functions. If we let

$$X(x) = A_1 e^{i(\omega/c)x} + A_2 e^{-i(\omega/c)x}$$
$$T(t) = B_1 e^{i\omega t} + B_2 e^{-i\omega t} , \qquad (11.72)$$

our solution will clearly satisfy Eq. (11.71). The solution for $u(x, t)$ given by (11.69) becomes

$$u(x, t) = C_1 e^{i\omega(t+x/c)} + C_2 e^{i\omega(t-x/c)}$$
$$+ C_3 e^{-i\omega(t+x/c)} + C_4 e^{-i\omega(t-x/c)} . \qquad (11.73)$$

This solution is a sum of harmonic terms that have the form of D'Alembert's solution, Eq. (11.68), and includes four arbitrary constants

TABLE 11.4 Relationships between wave variables.

Period	T	$T = 1/f = 2\pi/\omega$
Frequency	f	$f = \omega/2\pi = c/\Lambda$
Wavelength	Λ	$\Lambda = cT = 2\pi/k$
Wavenumber	k	$k = 2\pi/\Lambda = \omega/c$
Velocity	c	$c = \omega/k = f\Lambda$

(C_1, etc., with units of displacement) that are determined by the initial and boundary conditions in any particular problem.

Harmonic wave solutions such as those included in (11.73) are of fundamental importance in seismology and often are represented using sine and cosine functions,

$$u(x,t) = A e^{i\omega(t \pm x/c)}$$
$$= A \cos\left[\omega\left(t \pm x/c\right)\right]$$
$$+ iA \sin\left[\omega\left(t \pm x/c\right)\right] . \quad (11.74)$$

For a specified value of the **angular frequency**, ω, the harmonic terms represent a pattern that repeats with period, $T = 2\pi/\omega$. T is the time between of successive peaks of the harmonic deformation at a specific position (Fig. 11.12). When the wave is considered as a function of x, the **wavelength**, Λ, is the distance between peaks of the harmonic deformation. Since $\Lambda = cT$, the wavelength is also the distance the wave travels in a time equal to one period. The quantity $k = \omega/c = 2\pi/\Lambda$ is called the **wavenumber**. In general, seismic waves have frequencies between about 0.0003 and 100 Hz. For a typical seismic-wave velocity of 5 km/s, the corresponding wavelengths are between 15,000 and 0.05 km. Waves with such extremely different wavelengths sample very different characteristics of the Earth. Table 11.4 is a summary of the relationships between these fundamental quantities used to describe a harmonic signal.

The complex number representation of harmonic waves (11.74) does not imply the existence of "imaginary" waves or ground motions. Ground displacements are real-valued quanti-

ties, and whenever initial and boundary conditions are applied to general solutions such as (11.73), the complex quantities appear as parts of complex conjugates that eliminate the imaginary components from the solution. The complex-variable formalism is quite powerful, and so is commonly adopted in seismic wave-propagation analyses.

An approximate solution for an inhomogeneous 1D medium

We conclude our discussion of one-dimensional wave solutions by considering a case in which the thin bar in Fig. 11.10 has nonuniform (but smoothly varying) material properties. In other words, the density and Young's modulus are a function of position along the bar. The force balance equation becomes

$$\rho\left(x_1\right)\ddot{u}_1 = \frac{\partial}{\partial x_1}\left(E\left(x_1\right)\frac{\partial u_1}{\partial x_1}\right)$$
$$= E\left(x_1\right)\frac{\partial^2 u_1}{\partial x_1^2} + \frac{\partial E\left(x_1\right)}{\partial x_1}\frac{\partial u_1}{\partial x_1} . \quad (11.75)$$

If $\partial E(x_1)/\partial x_1$, the spatial gradient of Young's modulus, is sufficiently small, the rightmost term can be ignored (more precise criteria for this type of approximation are discussed in Chapter 12). In that case, we have

$$\frac{\partial^2 u_1}{\partial t^2} = c^2\left(x_1\right)\frac{\partial^2 u_1}{\partial x_1^2} , \quad (11.76)$$

which is similar to (11.67), except that

$$c\left(x_1\right) = \left[E\left(x_1\right)/\rho\left(x_1\right)\right]^{1/2}$$

varies with position. Applying the separation of variables produces two, coupled equations

$$\frac{d^2 T(t)}{dt^2} + \omega^2 T = 0$$

$$\frac{d^2 X(x_1)}{dx_1^2} + \frac{\omega^2}{c^2(x_1)} X(x_1) = 0 . \quad (11.77)$$

$x = x_0$ $T = \dfrac{2\pi}{\omega}$

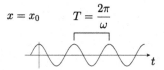

$t = t_0$ $\Lambda = cT$

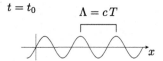

FIGURE 11.12 (left) Definition of period, T, and (right) wavelength, Λ, for a harmonic solution of a wave equation, $\cos[\omega(t \pm x/c)]$ where c is the wave speed and ω is the angular frequency.

Although the temporal dependence of (11.77) is satisfied by the same solution that we had in Eq. (11.72), we cannot choose $X(x_1) = Ce^{\pm i a \cdot x_1}$ where a is a constant (with physical units of one over length). Doing so results in

$$a^2 = \omega^2/c^2(x_1), \qquad (11.78)$$

which cannot be satisfied for all x_1 using constant values of a and ω. We instead choose $X(x_1) = Ce^{+i a(x_1)}$, where $a(x_1)$ is a function of position, which leads a nonlinear differential equation in $a(x_1)$,

$$i\frac{d^2a(x_1)}{dx_1^2} - \left(\frac{da(x_1)}{dx_1}\right)^2 + \frac{\omega^2}{c^2(x_1)} = 0. \qquad (11.79)$$

This nonlinear differential equation is very difficult to solve in general. To proceed, we assume that

$$\left|\frac{d^2a(x_1)}{dx_1^2}\right| \ll \frac{\omega^2}{c^2}, \qquad (11.80)$$

which allows us to drop the first term and solve

$$\frac{da}{dx_1} = \pm\frac{\omega}{c(x_1)}, \qquad (11.81)$$

such that

$$a(x_1) = \pm\omega \int_{-\infty}^{x_1} \frac{dx}{c(x)}, \qquad (11.82)$$

and the spatial component of the separation of variables solution is

$$X(x_1) = C \exp\left(\pm i\omega \int_{-\infty}^{x_1} \frac{dx}{c(x)}\right). \qquad (11.83)$$

Using Eq. (11.81), $d^2a/dx_1^2 = (\omega/c^2)(dc/dx_1)$ then condition (11.80) becomes $(\omega/c^2)(dc/dx_1) \ll \omega^2/c^2$, or $dc/dx_1 \ll \omega$. Physically, our approximation is valid when the spatial derivatives of speed are much smaller than the frequencies of interest. Or in other words, the gradient of speed, normalized by the speed, must be small compared to the wavenumber. The high-frequency approximate solution for the inhomogeneous bar is then

$$u(x_1, t) = A \exp\left[\pm i\omega\left(t \pm \int_{-\infty}^{x_1} \frac{dx}{c(x)}\right)\right]. \qquad (11.84)$$

Note that (11.84) still has a D'Alembert form, (11.68). If we define the wave phase, $\tau(x, t)$, as

$$\tau(x_1, t) = t \pm \int_{x_s}^{x_1} \frac{dx}{c(x)}$$

then it is clear that for a wavefront (a surface of constant phase) travel times are defined by

$$t(x_s, x_r) = \int_{x_s}^{x_1} \frac{dx}{c(x)},$$

where x_s is the source position and x_r is the receiver position. Travel time is equal to a path integral over the *slowness*, which is defined as the inverse of the speed. Solutions of this type lead to ray theory as described in Chapter 12.

11.4.2 Three-dimensional wave solutions

We return our attention to Eq. (11.59), the three-dimensional equations of motion, and Eq. (11.60), the decomposition of the displacement field into P-wave and S-wave potentials. In Cartesian coordinates, the displacements associated with the P wave are described by

$$u_p(\mathbf{x}, t) = \nabla \varphi(\mathbf{x}, t) = \frac{\partial \varphi}{\partial x_1} \hat{\mathbf{x}}_1 + \frac{\partial \varphi}{\partial x_2} \hat{\mathbf{x}}_2 + \frac{\partial \varphi}{\partial x_3} \hat{\mathbf{x}}_3 \,, \tag{11.85}$$

and $\varphi(\mathbf{x}, t)$ satisfies Eq. (11.63)

$$\frac{\partial^2 \varphi}{\partial t^2} = \alpha^2 \left(\frac{\partial^2 \varphi}{\partial x_1^2} + \frac{\partial^2 \varphi}{\partial x_2^2} + \frac{\partial^2 \varphi}{\partial x_3^2} \right) \tag{11.86}$$

where α is the P-wave speed (the square root of a ratio of elastic moduli to density). Based on our experience with one-dimensional solutions of the wave equation, we seek a solution in Cartesian coordinates using separation of variables,

$$\varphi(x_1, x_2, x_3, t) = X(x_1) \cdot Y(x_2) \cdot Z(x_3) \cdot T(t) \,. \tag{11.87}$$

Substitution of this form of solution into the equations of motion leads to four coupled equations (omitting explicit function arguments)

$$\begin{aligned}
\ddot{T} + \omega^2 T &= 0 \,, \\
\ddot{X} + k_1^2 X &= 0 \,, \\
\ddot{Y} + k_2^2 Y &= 0 \,, \\
\ddot{Z} + k_3^2 Z &= 0 \,,
\end{aligned} \tag{11.88}$$

where

$$k_\alpha^2 = k_1^2 + k_2^2 + k_3^2 = \omega^2/\alpha^2 \,, \tag{11.89}$$

is called the *wavenumber* of the P wave. The wavenumber has units of one over length.

Assuming harmonic solutions for each of Eq. (11.88) and multiplying terms as required by Eq. (11.87) results in a P-wave potential of the form

$$\begin{aligned}
\varphi(\mathbf{x}, t) \\
= A \exp[\pm i \, (\omega t \pm k_1 x_1 \pm k_2 x_2 \pm k_3 x_3)] \,,
\end{aligned} \tag{11.90}$$

which is the three-dimensional equivalent of Eq. (11.73). The physical units of the coefficient A, and hence the potential, are those of length squared.

The solution Eq. (11.90) has a D'Alembert style functional dependence (phase) on space and time and defines a set of *plane waves*, free to propagate in any direction in the medium. Although the values of wavenumber are restricted by Eq. (11.89), a general solution involves a sum of plane-wave solutions of the form Eq. (11.90) corresponding to a range of frequencies and a range of wavenumber components (k_1, k_2, k_3). In other words, the potential $\varphi(\mathbf{x}, t)$ corresponds to a system of waves that satisfy the equations of motion, or a *wavefield*. Seismologists use wavefield potentials to solve particular problems that include specific sources (initial conditions) and material boundaries (boundary conditions).

For a specific frequency, ω, and P-wave velocity, α, the requirement that $k_1^2 + k_2^2 + k_3^2$ is constant, defines a planar surface in Cartesian space with a normal vector called the *wavenumber vector*,

$$\mathbf{k}_\alpha = |k_\alpha| \hat{\mathbf{k}} = (\omega/\alpha) \hat{\mathbf{k}} \,, \tag{11.91}$$

where $\hat{\mathbf{k}}$ is a unit vector in the direction of $\mathbf{k}_\alpha$. The wavenumber vector quantifies the wave's propagation direction. In Chapter 12 we use wavenumber vectors to parameterize seismic rays. Using $\mathbf{k}_\alpha$, we can write a particular solution in the form of Eq. (11.90) as

$$\varphi(\mathbf{x}, t) = A \exp\left[i\left(\omega t - \mathbf{k}_\alpha \cdot \mathbf{x}\right)\right] , \qquad (11.92)$$

where $\mathbf{k}_\alpha \cdot \mathbf{x}$ is a dot product.

Corresponding solutions to the vector wave equation for the shear-wave potential, $\mathbf{\Psi}(\mathbf{x}, t)$ (Eq. 11.63),

$$\frac{\partial^2 \mathbf{\Psi}}{\partial t^2} = \beta^2 \left(\frac{\partial^2 \mathbf{\Psi}}{\partial x_1^2} + \frac{\partial^2 \mathbf{\Psi}}{\partial x_2^2} + \frac{\partial^2 \mathbf{\Psi}}{\partial x_3^2}\right) \qquad (11.93)$$

are similarly derived and represented by vector the potential

$$\mathbf{\Psi}(\mathbf{x}, t) = \mathbf{B} \exp\left[i\left(\omega t - \mathbf{k}_\beta \cdot \mathbf{x}\right)\right] , \qquad (11.94)$$

where $|\mathbf{k}_\beta| = \omega/\beta$ and

$$\mathbf{u}_s(\mathbf{x}, t) = \nabla \times \mathbf{\Psi}(\mathbf{x}, t) . \qquad (11.95)$$

Since plane waves are fundamental solutions of many seismological problems, we explore the wavenumber vectors that describe them in more detail.

Plane-wave phase and wavenumber vectors

Consider a plane P wave propagating with wavenumber vector $\mathbf{k}_\alpha$ contained in the $x_1 x_3$ plane (we can always orient our coordinate system so that this is the case). For this geometry, $k_2 = 0$ and $\partial\varphi/\partial x_2 = 0$. Now consider the part of the wave that has the constant phase, C. The plane-wave phase satisfies

$$\omega t - k_1 x_1 - k_3 x_3 = C . \qquad (11.96)$$

We also have the relationship from the variables separation, $k_1^2 + k_3^2 = \omega^2/\alpha^2$. For convenience, choose a reference such that $C = 0$ when $t = 0$. Then at that time, the locus of zero phase (a wavefront) is defined by $k_1 x_1 + k_3 x_3 = 0$, which is the equation of a line in the $x_1 x_3$ plane (in general equation form, $Ax + By + C = 0$). The wavenumber vector is perpendicular to the

plane wavefront, with components k_1 and k_3 along the x_1 and x_3 axes, respectively. The situation is illustrated in Fig. 11.13. If we define the angle between $\mathbf{k}_\alpha$ and the positive x_3 axis as i, we have

$$k_1 = \frac{\omega \sin i}{\alpha} = \omega p$$
$$k_3 = \frac{\omega \cos i}{\alpha} = \omega \eta_\alpha . \qquad (11.97)$$

The quantity $(p = \sin i/\alpha)$ is called the **horizontal slowness**, and $\eta = \cos i/\alpha$ is called the **vertical slowness** (slowness is the inverse of speed). Although we more naturally think in terms of seismic-wave speeds, seismic-wave slownesses are powerful tools for travel-time analysis. We will explore slownesses at great length in Chapter 12.

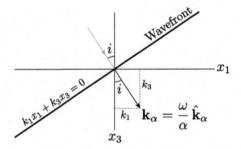

FIGURE 11.13 A wavefront is defined by a constant phase, which for a plane wave satisfies the equation of a plane (e.g. $Ax + By + Cz = D$). For simplicity, we choose to track the zero phase ($C = 0$) of a wavefront perpendicular to the $x_1 x_3$ plane. The choice results in $-k_1 x_1 - k_3 x_3 = \omega t$, which is the equation of a plane with a normal direction defined by $(k_1, 0, k_3)$ (in Cartesian coordinates, the normal to a plane defined by $Ax + By + Cz = D$ is (A, B, C)). If we further examine the time, $t = 0$, then the wavefront forms the locus of points along a line.

As time increases, to maintain the constant phase, the wavefront propagates in the direction specified by $\mathbf{k}_\alpha$. This is true regardless of the specific value of the phase, C. As time increases, the line of points with phase C must satisfy

$$\omega t - k_1 x_1 - k_3 x_3 = C , \qquad (11.98)$$

and so the lines sweep across the $x_1 x_3$ plane. Fig. 11.14 illustrates the situation and helps define a number of important geometric relationships for plane waves. From the geometry in the figure (focus on the triangles), the wave's apparent speed in the x_3 direction is

$$\frac{\Delta x_3}{\Delta t} = \frac{\ell/\cos i}{\Delta t} = \frac{\alpha}{\cos i} = \frac{1}{\eta_\alpha}, \qquad (11.99)$$

and the wave's apparent speed in the x_1 direction is

$$\frac{\Delta x_1}{\Delta t} = \frac{\ell/\sin i}{\Delta t} = \frac{\alpha}{\sin i} = \frac{1}{p}, \qquad (11.100)$$

which means that we can interpret the plane-wave slownesses as the inverse of the **apparent speeds** in the vertical and horizontal directions.

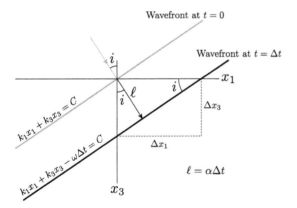

FIGURE 11.14 Variation of a wavefront's position with increasing time, Δt. To maintain the constant phase that defines the wavefront as time increases, the position of the wavefront advances in the direction of the wavenumber vector. The distance that the wavefront moves after time t is $\ell = \alpha \Delta t$.

Since the plane-wave's angle, i, can equal any value from $0°$ to $360°$, the general solution (11.92) corresponds to an infinite set of plane waves, with all possible orientations and spatial shifts filling the entire three-dimensional space. Each angle corresponds to a particular

wavenumber vector that points in the propagation direction. Plane waves are essential solutions to the wave equation when the distance from the source is large and the wavefront curvature is relatively small. More importantly, they are also basic tools that can be used to construct solutions for curved wavefronts such as spherical waves. Box 11.3 describes spherical wave solutions to the equations of motion in a whole space, which are relatively straightforward.

P- and S-wave displacements

If we want to determine the P-wave *displacements*, we must compute the gradient of φ,

$$\mathbf{u}_p(\mathbf{x}, t) = \nabla\varphi(\mathbf{x}, t) = \left(\hat{\mathbf{x}}_1 \frac{\partial}{\partial x_1} + \hat{\mathbf{x}}_2 \frac{\partial}{\partial x_2} + \hat{\mathbf{x}}_3 \frac{\partial}{\partial x_3} \right)$$
$$\times A \exp\left[\pm i \left(\omega t \pm \mathbf{k}_\alpha \cdot \mathbf{x} \right) \right]. \quad (11.101)$$

For the particular choice of φ given by

$$\varphi(\mathbf{x}, t) = A \exp\left[i \left(\omega t - k_1 x_1 - k_3 x_3 \right) \right], \quad (11.102)$$

we have as a solution

$$\mathbf{u}_p(\mathbf{x}, t) = (-ik_1 A) \times \exp\left[i \left(\omega t - k_1 x_1 - k_3 x_3 \right) \right] \hat{\mathbf{x}}_1$$
$$+ 0\, \hat{\mathbf{x}}_2$$
$$+ (-ik_3 A) \times \exp\left[i \left(\omega t - k_1 x_1 - k_3 x_3 \right) \right] \hat{\mathbf{x}}_3, \quad (11.103)$$

or

$$\mathbf{u}_p(\mathbf{x}, t)$$
$$= -iA\, (k_1, 0, k_3)\, \exp\left[i \left(\omega t - k_1 x_1 - k_3 x_3 \right) \right]. \quad (11.104)$$

The P-wave displacement field has the same spatial-temporal dependence as the P-wave potential field, but different multiplicative constants that allow it to satisfy the equation of motion rather than the wave equation. The P-wave displacements are contained in the $x_1 x_3$ plane and deform the material in a direction parallel to the wavenumber vector. Our analysis demonstrates (from first principles) that the *P-wave particle motion is perpendicular to the wavefront and*

Box 11.3 Spherical waves

Plane-wave solutions to the equations of motion receive most of our focus, but transient wave solutions with a concentrated source location are often more readily solved using spherical waves. The three-dimensional scalar wave equation

$$\nabla^2 \Phi(r, \theta, \phi, t) = \frac{1}{\alpha^2} \ddot{\Phi}(r, \theta, \phi, t) \qquad (B11.3.1)$$

can be solved by expressing the Laplacian operator in spherical coordinates:

$$\nabla^2 \Phi = \frac{1}{r^2} \frac{\partial}{\partial r} \left(r^2 \frac{\partial \Phi}{\partial r} \right) + \frac{1}{r^2 \sin\theta} \frac{\partial}{\partial \theta} \left(\sin\theta \frac{\partial \Phi}{\partial \theta} \right) + \frac{1}{r^2 \sin^2\theta} \frac{\partial^2 \Phi}{\partial \Phi^2}, \qquad (B11.3.2)$$

where the coordinates are defined in Fig. B11.3.1.

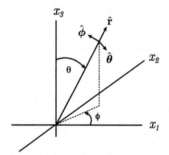

FIGURE B11.3.1 Standard spherical-geometry coordinate system.

For spherically symmetric solutions, $\Phi = \Phi(r, t)$, the homogeneous wave equation becomes

$$\frac{1}{r^2} \frac{\partial}{\partial r} \left(r^2 \frac{\partial \Phi}{\partial r} \right) = \frac{1}{\alpha^2} \frac{\partial^2 \Phi}{\partial t^2}. \qquad (B11.3.3)$$

This has solutions of the form

$$\Phi(r, t) = \frac{f(t \pm r/\alpha)}{r}, \qquad (B11.3.4)$$

where f is an arbitrary function, with the $(t - r/\alpha)$ phase indicating outward-propagating waves spreading spherically from the origin, and the $(t + r/\alpha)$ phase indicating inward-propagating spherical waves. The $1/r$ dependence differs from the Cartesian D'Alembert solution. The solution for the inhomogeneous wave equation with a source at $r = 0$, localized by the delta function defined by $\delta(r) = 0, r \neq 0$ and $\int_V \delta(r)dv = 1$,

$$\nabla^2 \Phi(r, t) = \frac{1}{\alpha^2} \ddot{\Phi} - 4\pi \delta(r) f(t) \qquad (B11.3.5)$$

is $\Phi(r, t) = -f(t - r/v)/r$. The displacements are $\mathbf{u}_p(\mathbf{r}, t) = \nabla \Phi = (\partial \Phi / \partial r)\hat{\mathbf{r}}$. We will use this solution in Chapter 16.

oriented in the direction that the wave is propagating. This characteristic of *P*-wave motion also holds for cylindrical and spherical waves. Because of the harmonic form of the motions, particles oscillate back and forth as the wave passes, as the wave alternately compresses and dilates the medium.

S-wave particle displacements are associated with vector plane-wave potential of the form Eq. (11.94). The displacements are found from Eq. (11.60)

$$
\begin{aligned}
\mathbf{u}_s(\mathbf{x}, t) &= \nabla \times \mathbf{\Psi}(\mathbf{x}, t) \\
&= \nabla \times (\psi_1(\mathbf{x}, t), \psi_2(\mathbf{x}, t), \psi_3(\mathbf{x}, t)) \\
&= \left(\frac{\partial \psi_3}{\partial x_2} - \frac{\partial \psi_2}{\partial x_3} \right) \hat{\mathbf{x}}_1 + \left(\frac{\partial \psi_1}{\partial x_3} - \frac{\partial \psi_3}{\partial x_1} \right) \hat{\mathbf{x}}_2 \\
&\quad + \left(\frac{\partial \psi_2}{\partial x_1} - \frac{\partial \psi_1}{\partial x_2} \right) \hat{\mathbf{x}}_3 .
\end{aligned}
$$
(11.105)

We simplify the algebra by again restricting our attention to plane waves with wavenumber vectors in the $x_1 x_3$ plane, so all $\partial \psi_i / \partial x_2$ quantities vanish, and

$$
\begin{aligned}
\mathbf{u}_s(\mathbf{x}, t) &= \left(-\frac{\partial \psi_2}{\partial x_3} \right) \hat{\mathbf{x}}_1 + \left(\frac{\partial \psi_1}{\partial x_3} - \frac{\partial \psi_3}{\partial x_1} \right) \hat{\mathbf{x}}_2 \\
&\quad + \left(\frac{\partial \psi_2}{\partial x_1} \right) \hat{\mathbf{x}}_3 .
\end{aligned}
$$
(11.106)

If we associate the $x_1 x_2$ plane with the Earth's surface and the x_3 axis with depth (a common convention), the $\mathbf{u}_{s_1}$ and $\mathbf{u}_{s_3}$ components comprise *S*-wave motions in the $x_1 x_3$ plane which we call the *SV* component because they include a component of vertical (x_3) motion. The x_2 component of motion involves purely horizontal (x_2) motions and is called the *SH* component of the shear-wave. Recall that for a comparable choice of coordinate system, the *P* waves had no x_2 component.

To examine the *SV* displacements we use a general plane-wave solution for $\psi_2(\mathbf{x}, t)$,

$$
\psi_2(x_1, x_3, t) = B' \exp[i(\pm \omega t \pm k_1 x_1 \pm k_3 x_3)] .
$$
(11.107)

Thus

$$
\begin{aligned}
\mathbf{u}_{SV}(\mathbf{x}, t) &= -\frac{\partial \psi_2}{\partial x_3} \hat{\mathbf{x}}_1 + \frac{\partial \psi_2}{\partial x_1} \hat{\mathbf{x}}_3 \\
&= \mp k_3 B' i \exp[i(\pm \omega t \pm k_1 x_1 \pm k_3 x_3)] \hat{\mathbf{x}}_1 \\
&\quad \pm k_1 B' i \exp[i(\pm \omega t \pm k_1 x_1 \pm k_3 x_3)] \hat{\mathbf{x}}_3,
\end{aligned}
$$
(11.108)

or,

$$
\begin{aligned}
\mathbf{u}_{SV}(\mathbf{x}, t) &= i B'(\mp k_3, 0, \pm k_1) \\
&\quad \exp[i(\pm \omega t \pm k_1 x_1 \pm k_3 x_3)].
\end{aligned}
$$
(11.109)

Careful examination of the above expression reveals that the *SV*-related displacement lies within the wavefront and in the $x_1 x_3$ plane, or,

$$
\mathbf{u}_{SV}(\mathbf{x}, t) \cdot \mathbf{k}_\beta \propto (\mp k_3, 0, \pm k_1) \cdot (k_1, 0, k_3) = 0 .
$$
(11.110)

For a particular case, $\psi_2 = B' \exp[i(\omega t - k_1 x_1 - k_3 x_3)]$, the wavefront is defined by the phase, $(\omega t - k_1 x_1 - k_3 x_3) = C$, which represents a line in the $x_1 x_3$ plane with a slope of $-k_1/k_3$. The ratio of the corresponding *SV* displacement components is the same,

$$
\frac{\mathbf{u}_{s_3}}{\mathbf{u}_{s_1}} = -\frac{k_1}{k_3} ,
$$
(11.111)

so the motion is parallel to the line of constant phase – i.e., the motion is contained in the plane-wavefront.

To examine the *SH* component of displacement, we let

$$
V(\mathbf{x}, t) = u_{s_2}(\mathbf{x}, t) = \frac{\partial \psi_1}{\partial x_3} - \frac{\partial \psi_3}{\partial x_1} ,
$$
(11.112)

where ψ_1 and ψ_3 are both solutions of the shear-wave equation

$$\frac{\partial^2 \psi_1}{\partial t^2} = \beta^2 \nabla^2 \psi_1$$

$$\frac{\partial^2 \psi_3}{\partial t^2} = \beta^2 \nabla^2 \psi_3 . \tag{11.113}$$

It turns out that in the case of SH waves, the displacement, $V(\mathbf{x}, t)$ also exactly satisfies the shear-wave equation

$$\partial^2 V(\mathbf{x}, t)/\partial t^2 = \beta^2 \nabla^2 V(\mathbf{x}, t) , \tag{11.114}$$

which can be demonstrated by direct substitution into Eq. (11.112). Thus, the SH wave displacements have the same form as the P potential, φ,

$$\mathbf{V}(x_1, x_3, t) = A' \exp\left[i\left(\pm \omega t \pm k_{\beta_1} x_1 \pm k_{\beta_3} x_3\right)\right]\hat{\mathbf{x}}_2 , \tag{11.115}$$

where $\left(k_{\beta_1}^2 + k_{\beta_3}^2\right) = (\omega^2/\beta^2)$, $k_{\beta_1}/\omega = \sin j/\beta = p$, and $k_{\beta_3}/\omega = \cos j/\beta = \eta_\beta$. Here j is the angle that the wavenumber vector makes with the positive x_3 axis. The wavefronts move in the $x_1 x_3$ plane as discussed before, but now with velocity β. The SH displacements are in the x_2 direction, and thus they lie in the plane of the wavefront, perpendicular to the direction of propagation. For the Earth reference system we described earlier, the SH motions are parallel to the surface.

The total displacement field is the sum of the P, SV, and SH waves,

$$\mathbf{u}(\mathbf{x}, t) = \mathbf{u}_P + \mathbf{u}_S$$

$$= \left(\frac{\partial \varphi}{\partial x_1} - \frac{\partial \psi_2}{\partial x_3}\right)\hat{\mathbf{x}}_1 + \left(\frac{\partial \psi_1}{\partial x_3} - \frac{\partial \psi_3}{\partial x_1}\right)\hat{\mathbf{x}}_2$$

$$+ \left(\frac{\partial \varphi_3}{\partial x_3} + \frac{\partial \psi_2}{\partial x_1}\right)\hat{\mathbf{x}}_3 . \tag{11.116}$$

For infinitesimal strains, the wave deformation is linear and the P and S waves do not interfere with one another. Expression (11.116)

emphasizes the separation of the SH components from the $P - SV$ components – which is a consequence of our assumption of isotropy. An anisotropic medium would couple all three components of motion. In an isotropic medium, as long as internal boundaries or free surfaces parallel the $x_1 x_2$ plane, the $P - SV$ / SH separation persists, as shown in later chapters. Our choice to discuss different components of the vector S motion in terms of SH and SV clearly has little significance for whole-space solutions; it sets the stage for subsequent discussions in coordinate systems that are defined with one coordinate orthogonal to Earth's surface. The total S-wave vector displacement lies in the plane of the wavefront, orthogonal to the direction of wave propagation. Splitting the wave into SH and SV components is simply for convenience.

Wave polarization on seismograms

The overall sense of particle motions associated with the P and S waves is shown in the block diagram in Fig. 1.2. The characteristic particle displacements associated with P and S waves result in predictable polarizations of the observed seismic displacements. Traditionally, most seismic stations record three components of ground motion: up-down (vertical), north-south, and east-west. The principal seismic waves arrive at the station propagating at some angle to the vertical in a horizontal direction defined to align with the great-circle connecting the source and receiver. We call that horizontal direction the **longitudinal** or **radial** direction. A P-wave arrival produces motions only in the direction of wave propagation, and vibrates the ground only in the vertical and radial directions, with relative strengths depending on the vertical angle of incidence (angle i in our earlier discussion). The vertically polarized S motion, SV, is also in the radial-vertical plane but parallel to the wavefront. On the other hand, the horizontally polarized S motion, SH, motion is entirely horizontal and perpendicular

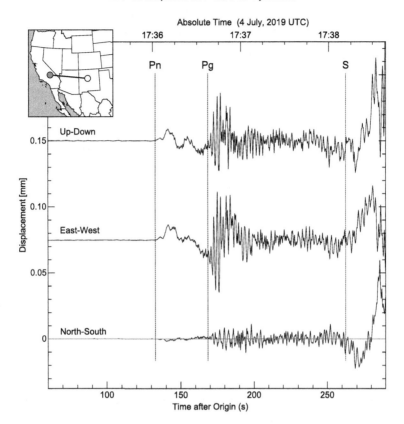

FIGURE 11.15 Three-component observations of eastern California earthquake (4 July, 2019, M_W 6.4) recorded at station ANMO, (Albuquerque, New Mexico). The source-to-station azimuth is $93°$. East is roughly the direction away from the source, and north-south is roughly orthogonal to the direction of propagation. P and SV motions are observed on the up-down and east-west components, SH motions are observed on the north-south component. Approximate arrival times of Pn and Pg and the S wave are identified. Prior to the S arrival, the transverse motions arise primarily from P-to-S and S-to-P scattering.

to the great-circle (vertical-radial) plane in what is called the ***transverse*** or ***tangential*** direction.

Although Earth is certainly not a uniform wholespace, the theoretical wave polarizations can be directly observed in seismograms when the direction from the source to the station aligned with one of the horizontal component directions. Fig. 11.15 is a plot of the three component seismograms recorded at seismic station ANMO (Albuquerque, New Mexico) that show the ground displacements produced by the 4 July, M_W 6.4 earthquake of the 2019 Ridgecrest,

California, earthquake sequence. The event was located roughly $1000\,km$ due west of the station, so the north-south seismogram records tangential motion, the east-west seismogram records the longitudinal. As predicted, the P and SV waves are recorded primarily on the vertical and longitudinal components, while the SH part of the S wave is recorded primarily on the transverse component. Since the Earth is not laterally uniform, the separation of motion is not perfect, but the observations certainly are notably consistent with the theoretical expectations.

Box 11.4 Seismic waves in anisotropic media

The P- and S-wave behavior in isotropic homogeneous media is remarkably simple, but greater complexity arises for anisotropic media. In an anisotropic, homogeneous medium, three independent body waves are generated that have orthogonal planes of particle motions. These are usually called quasi-compressional waves (qP) and quasi-shear waves (qSV and qSH), with names suggestive of the isotropic counterparts. In general, the propagation direction of these waves is not perpendicular to their wavefronts, so the particle motions differ from isotropic behavior. The velocities of these waves vary with the trajectory of the wave through the medium with respect to any axes of symmetry in the structure. For a wave propagating from an isotropic medium into an anisotropic medium, one of the primary effects is the separation of the isotropic S wave into two quasi-shear waves, which is called shear-wave splitting. .These properties arise from the general stress-strain relationship expressed by Hooke's law, for which the most general anisotropic medium has 21 independent elastic moduli. Increasing symmetry in the structure reduces the number of moduli. If the medium has symmetry about three orthogonal planes, the medium is *orthotropic*, and only nine independent constants exist. If it has axial symmetry, yielding a *hexagonal* medium, five independent constants exist. A common case relevant to some Earth structures occurs when the symmetry axis is vertical, which is called *transverse isotropy*. If the medium exhibits direction dependence of velocity in the horizontal surface, the behavior is called *azimuthal* anisotropy.

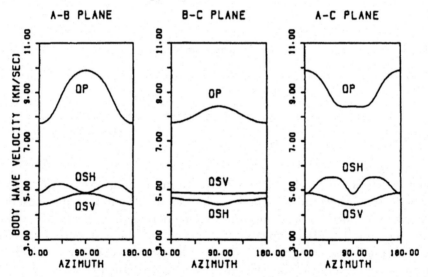

FIGURE B11.4.1 Variations of qP, qSH, and qSV wave velocities within planes of symmetry of single-crystal olivine. The labels A-B, B-C. and A-C denote symmetry planes that include the a and b axes, b and c axes, and a and c axes, respectively (from Kawasaki, 1989).

One of the major components of the Earth's mantle is olivine. A single olivine crystal has orthorhombic symmetry; thus the anisotropic seismic velocities have a complex behavior, as shown in Fig. B11.4.1. Since processes in the mantle may partially align crystal orientations on a macroscopic scale, net seismic wave anisotropy with reduced directional dependence is often observed, as is shear-wave splitting. Fig. B11.4.1 shows the variations of α and β in a single crystal of olivine.

Anisotropic behavior may also result from structural complexities rather than intrinsic crystallographic anisotropy. The presence of networks of flattened, possibly fluid- or magma-filled cracks causes directional wave-speed dependence, with the quasi-P and $-S$ waves being relatively slower in propagation directions perpendicular to the long axis of ellipsoidal cracks and relatively faster when propagating along the cracks' long axes, as shown in Fig. B11.4.2. Finely layered structures with alternating high- and low-velocity isotropic material can also give rise to effective anisotropic wave speeds. In later chapters, examples will be given of anisotropic body- and surface-wave observations in the Earth.

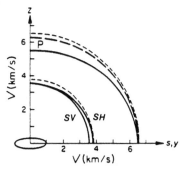

FIGURE B11.4.2 Velocities as a function of angle and fluid properties in granite containing aligned ellipsoidal cracks (orientation shown at origin) with porosity $= 0.01$ and aspect ratio $=0.05$. The short dashed lines are for the isotropic uncracked solid, the long dashes for liquid-filled cracks ($K_L = 100$ kbar), and the solid curves for gas-filled cracks ($K_L = 0.1$ kbar) (after Anderson et al., 1974).

11.5 Seismic-wave speeds in Earth materials

A more detailed review of the variations and ranges of seismic wave speeds in the Earth was presented in Chapter 10. We conclude this chapter with some brief comments and data regarding the seismic wave speeds in common earth materials. As noted earlier, P waves travel faster than S waves, and S waves do not propagate in fluids and gases. Most rock types exhibit a range of speeds depending on the pressure and temperature. At low pressures corresponding to the shallowest depths, laboratory samples show strong increases in wave speed with increasing pressure and cracks within the samples close. Usually, our interest lies in the properties of rocks at larger depths, where the rate of change in speed with respect to pressure is more modest. Within the atmosphere and ocean temperature affect the speed, as does ocean salinity. Within sedimentary rocks, the porosity and saturation levels are also important. Table 11.5 is a list of seismic-wave velocities for near-surface

TABLE 11.5 Seismic velocities (Clark, 1966, Christensen, 1982, 1996).

Material and source	P-wave (m/s)	σ	S-wave (m/s)	σ
Atmosphere	~333	–	–	–
Seawater	~1500	–	–	–
Sand	1800	–	500	–
Marine Sediments	1500–2500	–	–	–
Ice (sheets)	3850	–	1900	–
Clay	1100–2500	–	–	–
Limestone, Texas	6030	–	3030	–
Sandstone	1400–4300	–	–	–
Slate (avg)	6240	99	3351	96
Andesite (avg)	5712	227	3097	207
Granite-Granodiorite (avg)	6296	121	3692	104
Granite-Gneiss (avg)	6145	135	3553	143
Diorite (avg)	6566	144	3717	110
Basalt (avg)	5992	544	3246	293
Gabbro-norite-troctol (avg)	7200	125	3888	258
Serpentinite (avg)	5421	308	2610	173
Dunite (avg)	8352	83	4759	116

pressure and temperature conditions for a variety of materials.

For most rocks, when specific information is unavailable, a reasonable approximation is to assume that $\alpha \approx \sqrt{3}\beta \sim 1.73\beta$. But we sometimes assume relationships between P and S seismic wave speeds and density (related to what are often called Birch's Laws) and these are used to model changes within a particular rock type or for a range of similar rock types. For example, Brocher (2005) performed a regression of laboratory and field observations of the density and seismic wave speeds of common lithologies to develop

$$\rho(V_P) \sim 0.000106\, V_P^5 - 0.0043\, V_P^4 + 0.0671\, V_P^3$$
$$- 0.4721\, V_P^2 + 1.6612\, V_P, \qquad (11.117)$$

and

$$V_S(V_P) \sim 0.0064\, V_P^4 - 0.1238\, V_P^3 + 0.7949\, V_P^2$$
$$- 1.2344\, V_P + 0.7858. \qquad (11.118)$$

Such relationships are valuable when little other information is available, but we must keep in mind that the functions provide average values through uncertain observations that often span ranges of at least ± 0.25 km/s. When direct samples are unavailable, mineralogical averages are often constructed using the stiffnesses of compliances measured for the individual minerals. For the deepest earth, velocities based on computational estimates of the moduli and density at the appropriate depth and temperature provide model estimates.

Throughout our discussion of elasticity, and even in this section, we have assumed that the P- and S-wave velocities are independent of frequency or wavelength and depend only on the material properties of the continuum (this is the case for perfect elasticity). Anelastic effects lead to frequency dependent velocities, as introduced in Chapter 2. These effects are beyond our scope at this point, but the attenuation properties of the Earth are touched on in Chapter 10. In the next chapter, we adapt our wholespace solutions to investigate how seismic waves travel through an inhomogeneous structure like the Earth.

12

Body waves and ray theory – travel times

Chapter goals

- Derive and explore the Eikonal Equations.
- Develop formulas for travel times in vertically varying media.
- Develop formulas for travel times in spherically symmetric media.
- Develop formulas for travel times in a layered media.
- Compare global traveltime models for common seismic phases.

12.1 Wavefronts and rays

The only transient solutions to a stress imbalance suddenly introduced to a homogeneous elastic wholespace are P and S waves. P and S waves are known as *body waves* because they travel along paths throughout the continuum. The solutions for P and S waves, like those given in Eqs. (11.101) and (11.106), describe the locations of *wavefronts*, which are loci of points within the continuum that undergo the same motion at a given instant in time. A *ray* is a path normal to the wavefront that represents the route energy travels from the wavefront at that point. For a plane wave in a uniform material, the rays are parallel straight lines; for a spherical wave, the rays radiate outward from the

source like spokes on a wheel. Rays are a convenient means of tracking an expanding wavefront, and they provide an intuitive framework for extending elastic-wave solutions from homogeneous to inhomogeneous materials. If variations in seismic wave speed are not too extreme, rays corresponding to P or S waves behave very much as light traveling through a material with variations in the indices of refraction. Rays bend, focus, and defocus depending on the spatial distribution of velocity. Strictly speaking, ray-based solutions approximate solutions to the equations of motion and rays cannot be used to model all wave phenomena. These approximations are collectively known as *geometric ray theory* and are the basis for seismic body-wave interpretation.

In classical optics, the geometry of a wave surface is governed by Huygens' principle, which states that every point on a wavefront can be considered the source of a small secondary wave that travels outward with the velocity of the medium at that point. The wavefront at a later instant in time is found by drawing a tangent to the secondary wavelets, as shown in Fig. 12.1. Thus, given the location of a wavefront at a certain instant in time, we can predict future positions of the wavefront. Portions of the wavefront that are located in relatively high-velocity material produce secondary waves that travel farther in a specific time interval than those produced by points in relatively low-velocity

Foundations of Modern Global Seismology
https://doi.org/10.1016/B978-0-12-815679-7.00020-3

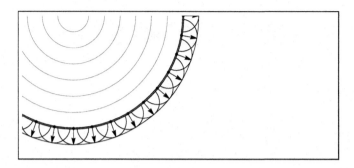

FIGURE 12.1 An expanding wavefront in a uniform material viewed with a Huygen's secondary source perspective. Huygens' principle models wavefronts by treating each point on the wavefront as a secondary source. The tangent surface of the expanding waves from the secondary sources provides the position of the wavefront at a later time.

material. This causes a temporal and spatial dependence in the shape of the wavefront. Since rays are normal to the wavefront, the ray orientation will also change with time and position. **Fermat's Principle** is a statement that a ray will follow a *minimum-time path* between a point, A, to a point, B, and is another way of saying that the wavefront travels from A to B in the shortest amount of time.

Since ray theory is an approximate solution to the equations of motion, we must know the conditions under which the ray-based solutions are valid. Recall the expression for a plane wave,

$$\varphi = Ae^{i\left(\pm\omega t \pm \mathbf{k}\cdot\mathbf{x}\right)}, \qquad (12.1)$$

where $\mathbf{k}$ is a vector that points in the direction of wave propagation and thus, by definition, is a ray. For homogeneous material, $\mathbf{k}$ does not change as the wave propagates (it is a straight line). Now if the seismic velocity varies smoothly in space (i.e., ρ, λ, and μ have small gradients), we must solve an equation analogous to Eq. (11.76):

$$\frac{\partial^2 \varphi}{\partial x_1^2} + \frac{\partial^2 \varphi}{\partial x_2^2} + \frac{\partial^2 \varphi}{\partial x_3^2} = \frac{1}{c^2(\mathbf{x})}\frac{\partial^2 \varphi}{\partial t^2}. \qquad (12.2)$$

The wave equation is an approximation of the equation of motion for heterogeneous media. As we did in the last chapter [see Eqs. (11.77)

and (11.79)], we will attempt to solve this partial differential equation by assuming a functional form

$$\varphi(\mathbf{x}, t) = A(\mathbf{x})\, e^{i\omega(T(\mathbf{x})-t)}, \qquad (12.3)$$

where $T(\mathbf{x})$ is a function of position with units of time. In two dimensions, contours of $T(\mathbf{x})$ are wavefronts, surfaces of identical travel time. Substitution of (12.3) into (12.2) yields

$$\nabla^2\left[A(\mathbf{x})e^{i\omega(T(x)-t)}\right] = \frac{1}{c^2(\mathbf{x})}\frac{\partial^2}{\partial t^2}\left[A(\mathbf{x})e^{i\omega(T(x)-t)}\right]. \qquad (12.4)$$

The required spatial derivatives are complex quantities. For example, $\partial^2\varphi/\partial x_1^2$ is

$$
\begin{aligned}
\frac{\partial^2\varphi}{\partial x_1^2} &= \frac{\partial}{\partial x_1}\left\{\frac{\partial A(\mathbf{x})}{\partial x_1}e^{i\omega(T(\mathbf{x})-t)}\right.\\
&\quad \left. + i\omega A(\mathbf{x})\frac{\partial T(\mathbf{x})}{\partial x_1}e^{i\omega(T(\mathbf{x})-t)}\right\}\\
&= \left[\frac{\partial^2 A(\mathbf{x})}{\partial x_1^2} - \omega^2 A(\mathbf{x})\left(\frac{\partial T(\mathbf{x})}{\partial x_1}\right)^2\right.\\
&\quad \left. + i\left(2\omega\frac{\partial A(\mathbf{x})}{\partial x_1}\frac{\partial T(\mathbf{x})}{\partial x_1} + \omega A(\mathbf{x})\frac{\partial^2 T(\mathbf{x})}{\partial x_1^2}\right)\right]\\
&\quad \times e^{i\omega(T(\mathbf{x})-t)}. \qquad (12.5)
\end{aligned}
$$

For $\partial^2 \phi / \partial x_2^2$ and $\partial^2 \phi / \partial x_3^2$, we obtain similar expressions. Equating the real and imaginary parts separately in Eq. (12.4) produces two equations,

$$\nabla^2 A(\mathbf{x}) - \omega^2 A(\mathbf{x})$$
$$\times \left[\left(\frac{\partial T(\mathbf{x})}{\partial x_1} \right)^2 + \left(\frac{\partial T(\mathbf{x})}{\partial x_2} \right)^2 + \left(\frac{\partial T(\mathbf{x})}{\partial x_3} \right)^2 \right]$$
$$= \frac{-\omega^2}{c^2(\mathbf{x})} A(\mathbf{x}) , \qquad (12.6)$$

and

$$2 \left(\frac{\partial T(\mathbf{x})}{\partial x_1} \frac{\partial A(\mathbf{x})}{\partial x_1} + \frac{\partial T(\mathbf{x})}{\partial x_2} \frac{\partial A(\mathbf{x})}{\partial x_2} + \frac{\partial T(\mathbf{x})}{\partial x_3} \frac{\partial A(\mathbf{x})}{\partial x_3} \right)$$
$$+ A(\mathbf{x}) \nabla^2 T(\mathbf{x}) = 0 . \qquad (12.7)$$

For non-zero frequency, we can rearrange the terms from Eq. (12.6) as

$$\left(\frac{\partial T(\mathbf{x})}{\partial x_1} \right)^2 + \left(\frac{\partial T(\mathbf{x})}{\partial x_2} \right)^2 + \left(\frac{\partial T(\mathbf{x})}{\partial x_3} \right)^2 - \frac{1}{c(\mathbf{x})^2}$$
$$= \frac{1}{\omega^2 A(\mathbf{x})} \nabla^2 A(\mathbf{x}) . \qquad (12.8)$$

The right-hand side of this equation is a ratio of the spatial Laplacian of the amplitude to the amplitude, scaled by the inverse of ω^2. For high frequencies (small wavelengths) this term is small – if we assume that it is approximately zero, then Eq. (12.8) reduces to

$$\left(\frac{\partial T(\mathbf{x})}{\partial x_1} \right)^2 + \left(\frac{\partial T(\mathbf{x})}{\partial x_2} \right)^2 + \left(\frac{\partial T(\mathbf{x})}{\partial x_3} \right)^2 = \frac{1}{c(\mathbf{x})^2} . \qquad (12.9)$$

This is called the *eikonal equation*, a nonlinear partial differential equation that relates wavefronts, rays, and travel times to the spatial distribution of seismic velocity. We can write the eikonal equations in the vector-calculus form,

$$\nabla T(\mathbf{x}) \cdot \nabla T(\mathbf{x}) = c^{-2}(\mathbf{x}) . \qquad (12.10)$$

The physical units of each term in the equation are the units of the inverse of velocity squared. We call the inverse of velocity, *slowness*.

Eq. (12.7), often called the *transport equation*, can be written as

$$2 \nabla T(\mathbf{x}) \cdot \nabla A(\mathbf{x}) + A(\mathbf{x}) \nabla^2 T(\mathbf{x}) = 0 . \qquad (12.11)$$

Now we consider some order-of-magnitude relationships developed from these equations and quantities. These are not algebraic results, they are useful only for comparing the relative scales of different quantities. For example, Eq. (12.10) suggests

$$\nabla T(\mathbf{x}) \sim c^{-1}(\mathbf{x}) , \qquad (12.12)$$

the travel-time spatial gradient scales with slowness. From the transport equation (12.11), the following ratios scale similarly

$$\frac{\nabla A(\mathbf{x})}{A(\mathbf{x})} \sim \frac{\nabla^2 T(\mathbf{x})}{\nabla T(\mathbf{x})} . \qquad (12.13)$$

In words, the curvature of a travel time surface (wavefront) normalized by the spatial rate of change of the travel time, scales as the wave amplitude gradient normalized by the amplitude. Combining these two order-of-magnitude relations, we have

$$\begin{aligned}
\frac{\nabla A(\mathbf{x})}{A(\mathbf{x})} &\sim \frac{\nabla^2 T(\mathbf{x})}{\nabla T(\mathbf{x})} \\
&\sim \frac{\nabla c^{-1}(\mathbf{x})}{c^{-1}(\mathbf{x})} \\
&\sim \frac{\nabla c(\mathbf{x})}{c(\mathbf{x})} .
\end{aligned} \qquad (12.14)$$

The last step is a simple application of the spatial derivative. We can use these relationships to provide some intuition on the conditions under which ray theory is appropriate.

Solutions to the eikonal equation are not exact solutions to the equations of motion because we ignored the term on the right-hand side in Eq. (12.8). But for many regions inside Earth, solutions of the eikonal equation are quantitatively useful approximations. They also provide insight into wave propagation that can be hard to glean from the partial differential equation. A solution of the eikonal equations will be a good approximate solution if the right-hand side of Eq. (12.8) is small compared to the gradients on the left-hand side,

$$\frac{1}{\omega^2} \frac{\nabla^2 A(\mathbf{x})}{A(\mathbf{x})} \ll \nabla T(\mathbf{x}) \cdot \nabla T(\mathbf{x}) . \qquad (12.15)$$

Since the spatial change in amplitude affects the reasonableness of the approximation, we use wavelength instead of frequency. At any location in the material, $\omega = 2\pi c(\mathbf{x})/\lambda(\mathbf{x})$, so we require that

$$\frac{\lambda^2(\mathbf{x})}{4\pi c^2(\mathbf{x})} \frac{\nabla^2 A(\mathbf{x})}{A(\mathbf{x})} \ll \nabla T(\mathbf{x}) \cdot \nabla T(\mathbf{x}) , \quad (12.16)$$

where we used Eq. (12.12). Considering only orders of magnitude, the condition can be written as

$$\lambda^2(\mathbf{x}) \frac{\nabla^2 A(\mathbf{x})}{A(\mathbf{x})} \ll 1 . \qquad (12.17)$$

Next, define the change in a quantity, say $F(\mathbf{x})$, over one wavelength distance as the product of the quantity's spatial derivative and the distance (one wavelength) and represent such wavelength-referenced changes in an order-of-magnitude sense using the notation $\delta[F(\mathbf{x})]$. The Laplacian of $A(\mathbf{x})$ can be thought of as the spatial rate-of-change in the rate-of-change of $A(\mathbf{x})$. If we multiply it by the wavelength,

$$\lambda(\mathbf{x}) \, \nabla^2 A(\mathbf{x}) = \lambda(\mathbf{x}) \, \nabla \cdot \nabla A(\mathbf{x}) \sim \delta[\nabla A(\mathbf{x})] \, (12.18)$$

Then Eq. (12.17) can be approximated as

$$\frac{\delta[\nabla A(\mathbf{x})]}{A(\mathbf{x})/\lambda(\mathbf{x})} \ll 1 . \qquad (12.19)$$

The result indicates that the *eikonal equation will approximate the wave equation well if the fractional change in the amplitude gradient over one seismic wavelength is small compared to the amplitude nor-*malized by the wavelength. Using Eq. (12.14) to derive another perspective,

$$\lambda(\mathbf{x}) \frac{\delta[\nabla c(\mathbf{x})]}{c(\mathbf{x})} = \frac{\delta[\nabla c(\mathbf{x})]}{c(\mathbf{x})/\lambda(\mathbf{x})} \ll 1 , \qquad (12.20)$$

which states that the *eikonal equation will approximate the wave equation well if the fractional change in velocity gradient over one seismic wavelength is small compared to the velocity normalized by the wavelength.* The conditions of validity require wavelengths smaller than a few hundred kilometers and slowly varying seismic wave speeds, criteria that apply to most body waves traversing the Earth's deep interior. Although the eikonal equations are not valid across a region with a sharp velocity gradient, we can, as we do with the wave equation, extend our solutions across boundaries by insuring the continuity of the appropriate stresses and displacements (the boundary conditions). We discuss this in detail later.

The eikonal equations (12.9) may appear complicated, however, they reduce to relatively simple equations for rays. The concept of rays is the basis of almost all body-wave interpretation. Rays allow us to track a displacement pulse from a source to a receiver, accounting for localized interactions along the specific path. Representing a portion of a seismic wavefield as a *ray* gives rise to the concept of *seismic phases* or *arrivals*. Seismic phases are transient disturbances at a receiver that correspond to P or S waves that have traveled a defined path between the source and receiver. Seismic phases have two primary characteristics: ***travel time*** and ***amplitude***. The eikonal and transport equations and their extensions can be used to quantify these characteristics. In this chapter we develop expressions for seismic-phase travel times, in the next, we develop expressions for amplitudes.

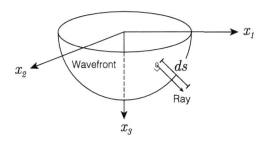

FIGURE 12.2 Three-dimensional wavefront with a ray or normal with length ds.

12.2 The Eikonal equations and seismic rays

Consider the wavefront, a surface of constant travel time, $T(\mathbf{x}) = t$, shown in Fig. 12.2. Rays are everywhere normal to the wavefront and can be characterized by their arc length, s, and travel time, t. Over a small time interval, we represent the local direction of the ray, $\hat{\mathbf{l}}$, using direction cosines,

$$\hat{\mathbf{l}} = (dx_1/ds,\, dx_2/ds,\, dx_3/ds)^T . \tag{12.21}$$

$\hat{\mathbf{l}}$ is parallel to the ray-path tangent, $d\mathbf{x}(s)/ds$. The direction cosines (components of $\hat{\mathbf{l}}$) satisfy

$$\left(\frac{dx_1}{ds}\right)^2 + \left(\frac{dx_2}{ds}\right)^2 + \left(\frac{dx_3}{ds}\right)^2 = 1 . \tag{12.22}$$

Since the ray is perpendicular to the wavefront, from calculus we know that it points in the direction with $\hat{\mathbf{l}} \propto \nabla T(\mathbf{x})$ and so dx_i/ds must be proportional to $\partial T(\mathbf{x})/\partial x_i$. Then,

$$\left(a\frac{\partial T(\mathbf{x})}{\partial x_1}\right)^2 + \left(a\frac{\partial T(\mathbf{x})}{\partial x_2}\right)^2 + \left(a\frac{\partial T(\mathbf{x})}{\partial x_3}\right)^2$$
$$= 1, \tag{12.23}$$

where a is a constant of proportionality. Eq. (12.23) is just the eikonal equation (12.9) if $a = c(\mathbf{x})$. Eqs. (12.22) and (12.23) can be com-

bined to form

$$\frac{1}{c(\mathbf{x})}\frac{dx_1}{ds} = \frac{\partial T(\mathbf{x})}{\partial x_1}$$
$$\frac{1}{c(\mathbf{x})}\frac{dx_2}{ds} = \frac{\partial T(\mathbf{x})}{\partial x_2}$$
$$\frac{1}{c(\mathbf{x})}\frac{dx_3}{ds} = \frac{\partial T(\mathbf{x})}{\partial x_3} . \tag{12.24}$$

Now let us consider how these equations change along the path of the ray. We can do this by taking the derivative of the equations with respect to ds. Examining the equation for x_1 equation only,

$$\frac{d}{ds}\left(\frac{1}{c(\mathbf{x})}\frac{dx_1}{ds}\right)$$
$$= \frac{d}{ds}\left(\frac{\partial T(\mathbf{x})}{\partial x_1}\right)$$
$$= \frac{\partial}{\partial x_1}\left(\frac{\partial T(\mathbf{x})}{\partial x_1}\frac{dx_1}{ds} + \frac{\partial T(\mathbf{x})}{\partial x_2}\frac{dx_2}{ds} + \frac{\partial T(\mathbf{x})}{\partial x_3}\frac{dx_3}{ds}\right)$$
$$= \frac{\partial}{\partial x_1}\left\{\frac{1}{c(\mathbf{x})}\left[\left(\frac{dx_1}{ds}\right)^2 + \left(\frac{dx_2}{ds}\right)^2 + \left(\frac{dx_3}{ds}\right)^2\right]\right\}$$
$$= \frac{\partial}{\partial x_1}\frac{1}{c(\mathbf{x})} . \tag{12.25}$$

The generalized form of this equation, called the *raypath equation*, is

$$\frac{d}{ds}\left(\frac{1}{c(\mathbf{x})}\frac{d\mathbf{x}(s)}{ds}\right) = \nabla\left(\frac{1}{c(\mathbf{x})}\right). \tag{12.26}$$

These expressions are second-order differential equations that constrain a particular path, $\mathbf{x}(s)$, through the medium (the ray path). Two initial conditions are required to solve (12.26): (1) an initial or reference position of the ray $\mathbf{x}(0)$, e.g. the source location; (2) the direction that the ray leaves that reference location.

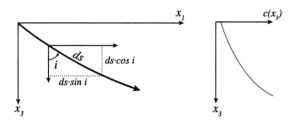

FIGURE 12.3 Raypath geometry for a medium in which the velocity is independent of the x_2 and x_1 directions.

12.3 Travel times in media with depth-dependent properties

We can develop some insight into the physics represented by (12.26) with a simple example. Consider a ray traveling through a material that has a change in velocity in only one direction, say depth. Then $c = c(x_3)$ and (12.26) reduces to

$$\frac{1}{c(x_3)}\frac{dx_1}{ds} = p_1$$

$$\frac{1}{c(x_3)}\frac{dx_2}{ds} = p_2 \qquad (12.27)$$

$$\frac{d}{ds}\left(\frac{1}{c(x_3)}\frac{dx_3}{ds}\right) = \frac{dc^{-1}(x_3)}{dx_3},$$

where p_1 and p_2 are constants and the constant ratio of p_1 to p_2 confines the raypath to a plane that is normal to the $x_1 x_2$ plane. The ray's azimuth does not change and the projection of the ray onto the $x_1 x_2$ plane is a straight line. Without loss of generality, but for simplicity, we choose a plane of propagation that coincides with the $x_1 x_3$ plane. Fig. 12.3 is an illustration of the geometry. Then $\hat{\mathbf{l}} = (dx_1/ds, 0, dx_3/ds)^T$ and Eq. (12.27) reduces to

$$\frac{1}{c(x_3)}\frac{dx_1}{ds} = p$$

$$\frac{d}{ds}\left(\frac{1}{c(x_3)}\frac{dx_3}{ds}\right) = \frac{dc^{-1}(x_3)}{dx_3}, \qquad (12.28)$$

where p is a constant. At any point along the ray, the ray's direction cosines are given by

$$\frac{dx_1}{ds} = \sin i$$

$$\frac{dx_3}{ds} = \cos i . \qquad (12.29)$$

The angle i, called the **angle of incidence**, quantifies the inclination of the ray relative to the vertical (x_3) direction. The "incidence" terminology arises from consideration of the ray orientation near Earth's surface, where most observations are made. Near the source, the same angle is often called the **take-off angle**.

12.3.1 The seismic ray parameter (horizontal slowness)

The first part of Eq. (12.28), written in terms of the angle, i, using Eq. (12.29), becomes

$$p = \frac{\sin i (x_3)}{c(x_3)} . \qquad (12.30)$$

We call the quantity, p, the **ray parameter** or **horizontal slowness**. The physical units of p are T/L (inverse of speed) and p can assume values between from 0 (for a vertical travel path) to $1/c$ (for a horizontal travel path). For a prescribed reference point and takeoff angle, a ray in a medium with a one-dimensional variation in speed will have a constant ray parameter along its entire path (hence the name). Eq. (12.30) is known as **Snell's Law**, which can also be derived from Fermat's principle, a statement that a raypath is a trajectory of stationary time (the time integrated along a ray path is an extremum, see Box 12.1). The ray parameter is a very important quantity in observational seismology.

12.3.2 Ray-path curvature

Careful consideration of Snell's Law (Eq. (12.30)) leads to the conclusion that a ray traveling into a medium with a positive velocity

Box 12.1 Geometric interpretation of Snell's law

Fermat's principle of least time can be used to derive *Snell's law* and the definition of seismic ray parameter. Consider a ray leaving point P in a medium of velocity α_1. What is the path the ray will take to arrive at point P' in a medium of velocity α_2? Fig. B12.1.1 is an illustration of the geometry.

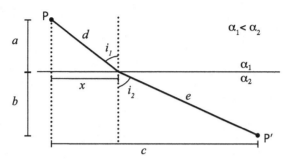

FIGURE B12.1.1 Raypath connecting two points on either side of a boundary.

The travel time on the path between P and P' is

$$T_{P-P'} = \frac{d}{\alpha_1} + \frac{e}{\alpha_2} = \frac{\sqrt{a^2 + x^2}}{\alpha_1} + \frac{\sqrt{b^2 + (c-x)^2}}{\alpha_2}. \tag{B12.1.1}$$

The minimum-time path must satisfy $dT/dx = 0$, which implies

$$\frac{dT}{dx} = 0 = \frac{x}{\alpha_1 \sqrt{a^2 + x^2}} - \frac{c-x}{\alpha_2 \sqrt{b^2 + (c-x)^2}}. \tag{B12.1.2}$$

Since $x/\sqrt{a^2 + x^2} = \sin i_1$, and $(c-x)/\sqrt{b^2 + (c-x)^2} = \sin i_2$,

$$\frac{\sin i_1}{\alpha_1} = \frac{\sin i_2}{\alpha_2}. \tag{B12.1.3}$$

This is the familiar expression from optics called Snell's law after Willebrod Snell (1591–1626). The generalization of Snell's law is $\sin i / v = p$, where p is called the *seismic parameter, ray parameter*, or *horizontal slowness*. The ray parameter is constant at any point along the entire travel path. A ray traversing material of vertically changing velocity, $v(x_3)$, must change its inclination angle, i, with respect to a horizontal reference plane. As a ray enters material of increasing velocity, the ray is deflected toward the horizontal. As a ray enters material of decreasing velocity, it is deflected toward the vertical. If the ray is traveling vertically, then $p = 0$, and the ray will experience no deflection.

gradient (increasing with depth) would eventually turn upward. We can better quantify this statement if we consider the second equation in (12.28),

$$\frac{d c^{-1}(x_3)}{dx_3} = \frac{d}{ds}\left(\frac{1}{c(\mathbf{x})}\cos i\right).$$

Rearranging and using the chain rule (for brevity we omit explicit notation indicating the x_3 dependence of velocity), we find,

$$\frac{d c^{-1}}{dx_3} = c^{-1}\frac{d\,\cos i}{ds} + \cos i\,\frac{d c^{-1}}{ds}$$

$$= c^{-1}(-\sin i)\frac{d i}{ds} + \cos i\,\frac{d c^{-1}}{dx_3}\frac{dx_3}{ds} \quad (12.31)$$

$$= c^{-1}(-\sin i)\frac{d i}{ds} + \cos^2 i\,\frac{d c^{-1}}{dx_3}.$$

Collecting and rearranging terms,

$$\frac{d c^{-1}}{dx_3}\left(1 - \cos^2 i\right) = c^{-1}(-\sin i)\frac{d i}{ds}$$

$$-c^{-2}\frac{d c}{dx_3}\left(1 - \cos^2 i\right) = c^{-1}(-\sin i)\frac{d i}{ds}$$

$$-c^{-2}\frac{d c}{dx_3}\sin^2 i = c^{-1}(-\sin i)\frac{d i}{ds} \quad (12.32)$$

$$-p^2\frac{d c}{dx_3} = -p\frac{d i}{ds},$$

or,

$$\frac{d i}{ds} = p\,\frac{d c(x_3)}{dx_3}. \quad (12.33)$$

Eq. (12.33) indicates that in a medium with speed dependent on depth, raypath curvature is directly proportional to the velocity gradient, $dc(x_3)/dx_3$ and the proportionality coefficient is the ray parameter. If velocity *increases* with depth, then the ray curves *upward* in proportion to the gradient and the take-off angle. If velocity *decreases* with depth, then the ray curves *downward* in proportion to the gradient and the take-off angle. Fig. 12.4 is an illustration of the concept of ray curvature using the

evolution of a wavefront in media with different velocity distributions. Fig. 12.5 is an example computed through a model with a linear, vertical velocity gradient equal to about 0.0375 km/s per *km*, which corresponds to a transition from roughly shallow crust speeds to the mantle speeds over a depth of about 40 *km*. The model does not include a crust-mantle transition, crustal properties continue with depth. We discuss the modifications introduced by the crust-mantle-boundary later. As required, the ray paths (curves with arrows) are orthogonal to the wavefronts and the wavefronts (gray curves) bend upward as a result of the positive velocity gradient.

Velocity gradients encountered in the continental crust span a relatively large range. The increase in pressure with depth causes a velocity increase, the increase in temperature with depth causes a velocity decrease. Composition and mineralogical changes can induce significant velocity variations. To estimate an order of magnitude crustal gradient based on rock properties, assume a crustal thickness of roughly 40 km, a total crustal pressure change, $\Delta P \sim$ 1 GPa, and a temperature change, $\Delta T \sim 500°C$. Assuming simple linear relationships between speed, pressure, and temperature,

$$\frac{dc}{dx_3} \sim \frac{\Delta c}{40\,\text{km}}$$

$$\sim \left(\frac{\partial c}{\partial P}\Delta P + \frac{\partial c}{\partial T}\Delta T\right)/40\,\text{km}. \quad (12.34)$$

The partial derivatives have been measured in the laboratory. For a granitic-like material, $\partial c/\partial P \sim 0.44\,\text{km s}^{-1}\,\text{GPa}^{-1}$ and $\partial c/\partial T \sim -0.00027\,\text{km s}^{-1}\,°\text{C}^{-1}$. These values suggest that the average gradient through a 40 km thick crust of pure granite should be on the order of 0.008 (km/s)/km. Since this value is much lower than typically observed (across the whole crust), the typical observed increase in speed is usually a result of composition and mineralogical changes. Early in the history of instrumental

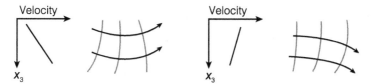

FIGURE 12.4 Conceptual illustration of wavefront and ray-path curvature for velocity gradients increasing and decreasing with depth. Gray lines are wavefronts, curves tipped with arrows are rays. After Officer (1974).

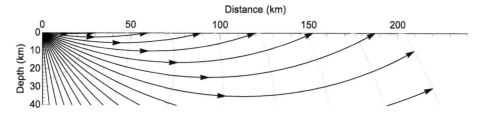

FIGURE 12.5 Calculated wavefronts and ray-paths for a steady increase in seismic velocity with depth. The source is at the surface and the take-off angles range from $5°$ to $80°$.

TABLE 12.1   Sample seismic-velocity gradients within earth materials and the PREM.

Material or Region	c_P [km/s]	dc_P/dz $[(km/s) \cdot km^{-1}]$	c_S [km/s]	dc_S/dz $[(km/s) \cdot km^{-1}]$
Unconsolidated Sand (0–5 m, wet-dry)	$400 - 1800$	$35 - 5$	$211 - 197$	$22 - 22$
Dry Sandstone	4.4	0.3	2.7	0.12
Granite, 0–4 km	5.95	0.1	3.52	0.05
Bulk Crust	6.2	0.04	3.6	0.02
Upper Mantle[*] (24–220)	8.0, 8.2	$-0.001, -0.0007$	4.4, 4.6	$0.0002, -0.001$
Upper Mantle (220–410)	8.8	0.002	4.7	0.0008
Transition Zone	9.8	0.004	5.3	0.002
Lower Mantle	12.3	0.001	6.6	0.0006
Outer Core	9.60	0.001	0.0	–
Inner Core	11.1	0.0002	3.4	0.0002

[*] *The PREM upper mantle is anisotropic, listed values correspond to vertically, horizontally propagating wave speeds respectively.*

seismology, crustal models consisted of a *granitic* layer above a basaltic layer. The seismic arrival notation, P_g, still used to refer to the crustal turning waves illustrated in Fig. 12.5, is a historical convention inherited from those times. Table 12.1 is a compilation of average seismic wave velocities and velocity gradients for earth materials (from laboratory measurements) and the

bulk crust (from simple models), regions of the mantle and core (from the PREM, Fig. 1.21). The sand, sandstone, granite gradients correspond to depths of a few kilometers; the gradient for granite near the surface is relatively large as cracks close with increasing pressure, but much more subdued deeper. Only averages are listed, most of the regions include a range of gradients. The crust and upper few hundred km of the mantle are the most heterogenous regions of the planet (Chapter 10). Regional analyses suggest the mantle just beneath the crust typically has a slight positive gradient.

12.3.3 Distance and travel-time formulas

The equations for rays in a vertically heterogeneous medium, Eq. (12.28), have several interesting aspects. For each specific angle i, a specific ray leaves the source and follows a specific raypath. Consider a medium such as Earth with an overall increase in speed with depth. The initial angle and the velocity structure determine the distance at which a ray will emerge at the surface. But that relationship is not unique. For a given pair of source and receiver positions, several possible rays may arrive at the same distance, which means that a *multiplicity* of arrivals can occur, all with different initial angles, paths, and travel times. We will discuss this more fully in later sections, as it is the basis for much seismic interpretation.

We can use (12.28) with initial conditions to predict where and when a ray will arrive. Consider Fig. 12.6. At any point along the path we have

$$\sin i = \frac{dx_1}{ds} = c(x_3)\,p$$

$$\cos i = \frac{dx_3}{ds} = \sqrt{1 - \sin^2 i} = \sqrt{1 - c^2(x_3)\,p^2}\,,$$

$$(12.35)$$

such that (using the geometry in Fig. 12.6)

$$dx_1 = ds\,\sin i = \frac{dx_3}{\cos i}\,c(x_3)\,p$$

$$= \frac{c(x_3)\,p}{\sqrt{1 - c^2(x_3)\,p^2}}\,dx_3.$$

$$(12.36)$$

For a surface source and receiver, Eq. (12.35) can be integrated over the depth range traversed by the ray to compute the distance, $X(p)$, at which

a ray with ray parameter p will emerge at the surface,

$$X(p) = 2\,p \int_0^{z_b} \frac{c(x_3)}{\sqrt{1 - c^2(x_3)\,p^2}}\,dx_3\,,\qquad (12.37)$$

where z_b is the maximum depth the ray descends. The ray parameter, p, can be moved outside the integral because it is constant along the path. The factor of 2 arises from the symmetry of the downgoing and upgoing portions of the raypath (Fig. 12.6). This is the *where* of ray equations; given the angle at which a ray leaves the source, we can calculate where (the distance) it will arrive. If we generalize this to a three-dimensional case, we also require specification of the azimuth of the raypath leaving the source to know where it will arrive.

An expression for the time it takes for the ray to arrive is obtained similarly. The ray parameter again defines the path, and the travel time is the integral of the slowness over the path,

$$T(p) = \int_{\text{path}} \frac{ds}{c(s)}\,.\qquad (12.38)$$

Using $ds = dx_3/\cos i$ to convert the path integral to an integral over depth, we have

$$T(p) = 2 \int_0^{z_b} \frac{dx_3}{c(x_3)\,\cos i}\qquad (12.39)$$

or

$$T(p) = 2 \int_0^{z_b} \frac{dx_3}{c^2(x_3)\sqrt{c^{-2}(x_3) - p^2}}\,,\qquad (12.40)$$

where $T(p)$ is the travel time along the raypath to the distance defined by Eq. (12.37). We can introduce some shorthand and rewrite Eqs. (12.37) and (12.40). Let $\gamma = 1/c(x_3)$ then

$$X(p) = 2\,p \int_0^{z_b} \frac{dx_3}{\sqrt{\gamma^2 - p^2}}\,,\qquad (12.41)$$

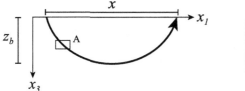

FIGURE 12.6 Geometry of the ray segment ds, along a path from a surface source to a surface receiver. The velocity of the medium varies only along the x_3 direction, so there will be symmetry of downgoing and upgoing legs of the raypath.

and

$$T(p) = 2 \int_0^{z_b} \frac{\gamma^2}{\sqrt{\gamma^2 - p^2}} dx_3. \qquad (12.42)$$

Noting the similarity in form of $X(p)$ and $T(p)$, we can relate the two:

$$T(p) = 2 \int_0^{z_b} \frac{\gamma^2}{\sqrt{\gamma^2 - p^2}} dx_3$$

$$T(p) = 2 \int_0^{z_b} \left(\frac{p^2}{\sqrt{\gamma^2 - p^2}} + \sqrt{\gamma^2 - p^2} \right) dx_3$$

$$T(p) = p X(p) + 2 \int_0^{z_b} \sqrt{\gamma^2 - p^2} dx_3. \qquad (12.43)$$

Eq. (12.43), the travel-time equation for a vertically heterogeneous medium, is a truly remarkable representation. The first term depends on x_1 and the second on x_3. The travel-time is *separable*. The *horizontal* travel time depends on the *horizontal* slowness, p; *vertical* travel time depends only on, $\eta(x_3) = \sqrt{c^{-2}(x_3) - p^2}$, which is known as the *vertical* slowness. Also note that $dT/dX = p$, the ray parameter is equal to the change in travel time with distance. We can observe the ray parameter by measuring the travel time as a function of distance and we use that quantity extensively when we work with seismograms.

12.3.4 Travel-time curves for continuous media

Fig. 12.7 is a plot of a travel-time curve for a continuous, increasing velocity distribution.

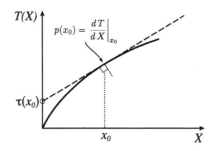

FIGURE 12.7 Travel-time curve in a continuous velocity structure. $\tau(X)$ is defined as the intercept of the tangent to the travel-time curve at any given X; the corresponding slope at that distance is p.

The horizontal slowness observed at a distance X is equal to the slope (dT/dX) of a line tangent to the travel-time curve. The intercept of the travel-time curve tangent lines, called τ, is a useful quantity in travel time analysis sometimes called the **delay time**. We can examine τ as a function of ray parameter. Starting with (12.43)

$$\tau(p) = T - p X = 2 \int_0^{z_b} \sqrt{\gamma^2 - p^2} dx_3, \qquad (12.44)$$

where $\tau(p)$ is the intercept ($X = 0$) of a line tangent to the travel-time curve at distance, $X(p)$, or horizontal slowness, p. As p increases, $X(p)$ increases and $\tau(p)$ decreases, hence

$$\frac{d\tau}{dp} = \frac{d}{dp} \left(2 \int_0^{z_b} \sqrt{\gamma^2 - p^2} dx_3 \right)$$

$$= 2 \int_0^{z_b} \frac{-p}{\sqrt{\gamma^2 - p^2}} \, dx_3$$

$$= -X(p). \tag{12.45}$$

Since the *tau function*, $\tau(p)$, is a single-valued function of p, its use can simplify the analysis of travel-time curves.

Fig. 12.8 includes plots of the raypaths, the travel-time curves, and p as a function of distance, and $\tau(p)$ for three major classes of continuous seismic models. In structure 1 (Fig. 12.8A) the seismic velocity increases smoothly with depth, and the travel-time curve is smooth and concave-downward. The ray parameter decreases monotonically with distance and $\tau(p)$ is smooth. In structure 2 (Fig. 12.8B), the velocity gradient changes with depth over a short depth interval. Seismic rays that turn above the gradient change are unaffected by it; hence the **travel-time branch** from A to B is identical to that for structure 1. Rays that enter the region of increased velocity gradient are turned, or deflected, toward the horizontal. If the gradient is strong enough, the rays will be turned such that they arrive at some distance C that is *smaller* than B. Rays that bottom well below the gradient zone will have a normal concave shape. If the velocity gradient change increases to the point where the transition becomes a velocity discontinuity, the travel-time curve will approach a discrete layered case (discussed in Section 12.5) branch AB will lengthen (point B will increase in distance X). For a layer over a halfspace model (Section 12.5), the AB branch is analogous to a direct arrival, the CD branch is analogous to the refracted arrival, and the BC branch is analogous to a reflected arrival.

Structure 2 produces a distinctive "bow tie" shape in the travel-time curve (Fig. 12.8B) that is called a **triplication**. The name comes from the fact that three distinct travel-time branches exist at certain distances. Seismograms at distances where the rays have passed through a structure such as that in Fig. 12.8B can be quite complicated. The three different arrivals will interfere,

and the character of interference will change very rapidly with distance. On the other hand, seismograms that are recorded across a triplication can be used to determine the character of the velocity change. The ray parameter is multivalued across triplication distances, corresponding to the different branches of the travel-time curve. However, $\tau(p)$ is single-valued, which is one of the advantages of using it to "unfold" a triplication curve.

Structure 3 (Fig. 12.8C) has a low-velocity zone beginning at a depth z_0. For rays that bottom above z_0, the travel-time curve is analogous to that of structure 1. As a ray penetrates below z_0, it is deflected toward the vertical, or bent downward, and a **shadow** is produced at distances where no arrivals occur (distance B to D). At a depth z_1, where the velocity is equal to that at depth z_0, the shadow is terminated. Below this depth, two arrivals result from an effect similar to the triplication. Theoretically this will result in a strong *cusp* at a distance D. The behavior of the slowness versus distance reflects the multiple values associated with two arrivals. Again, $\tau(p)$ is smooth and decreases monotonically, although it is discontinuous at the ray parameter corresponding to $1/c(z_0)$.

These three travel-time curves are important references when interpreting observed travel times. The three representations of the evolving wavefield [$T(X)$ $p(X)$, and $tau(p)$] are all equivalent. Depending on the circumstances, we can use any of the three representations to infer subsurface structure. We discuss the use of travel times to estimate Earth structure in Chapter 20.

12.4 Travel times in spherical Earth models

The travel-time equations derived in Section 12.5 are correct for problems in which the curvature of the Earth can be neglected. When curvature becomes important (at distances greater than about $12°$), we must mod-

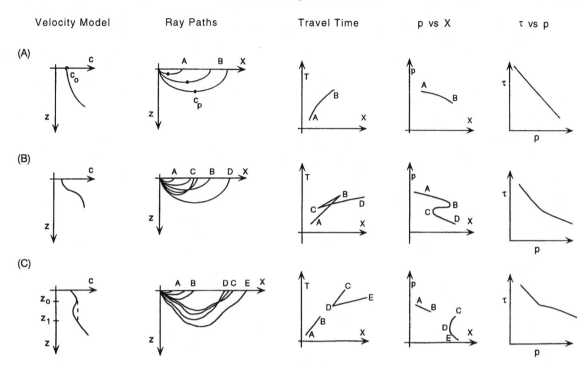

| Velocity Model | Ray Paths | Travel Time | p vs X | τ vs p |

FIGURE 12.8 Three different velocity structures and corresponding ray geometries, travel-time curves, ray parameter-versus-distance variations, and τ-versus-p variations (modified from Officer, 1974).

ify our approach. Fig. 12.9 shows a model of the Earth that is composed of concentric shells with uniform speeds (e.g. c_1 and c_2). For diagram clarity, thick shells are shown, but the approximation is better with much thinner shells. Within a constant speed shell, the rays travel along straight lines. Across each shell boundary is a discrete velocity change. On the local scale, the surface curvature is negligible and Snell's law is satisfied, so at position P,

$$\frac{\sin\theta_1}{c_1} = \frac{\sin\theta'_1}{c_2} . \quad (12.46)$$

Now consider the triangles highlighted in Fig. 12.9. Two right triangles ($\triangle OPS$ and $\triangle OQS$) share the side (OS), which has a length d. From trigonometry,

$$r_1 \sin\theta'_1 = r_2 \sin\theta_2 = d , \quad (12.47)$$

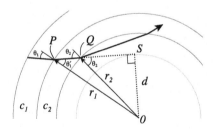

FIGURE 12.9 Ray geometry in a layered, spherical Earth.

or

$$\sin\theta'_1 = \frac{r_2 \sin\theta_2}{r_1} . \quad (12.48)$$

Combining these expressions requires that,

$$\frac{r_1 \sin\theta_1}{c_1} = \frac{r_2 \sin\theta_2}{c_2} . \quad (12.49)$$

Box 12.2 1st order ray tracing equations

The second-order form of the eikonal equations are useful for exploring solutions for simple models, including vertically inhomogeneous models that have wide application in seismology. The eikonal equations also apply to more general media – although the solutions must be approached numerically. Numerically integrating the eikonal equations is more easily accomplished using the first-order form of the equations. Expressions derived by Eliseevnin (1965), (cited in Julian, 1970) are particularly convenient since they parameterize the problem in terms of time and can be integrated through a 3D isotropic earth model.

Let a ray's direction angles be represented by α, β, and γ where $dx = ds\cos\alpha$, $dy = ds\cos\beta$, $dz = ds\cos\gamma$. The angle γ is the take-off/incidence angle. A ray is limited to the x_1x_3 plane when $\beta = \pi/2$ and $\alpha = 0$, or limited to the x_2x_3 plane when $\beta = 0$ and $\alpha = \pi/2$. Then the first-order form of the ray equations is

$$\frac{dx_1}{dt} = c(\mathbf{x})\cos\alpha$$

$$\frac{dx_2}{dt} = c(\mathbf{x})\cos\beta$$

$$\frac{dx_3}{dt} = c(\mathbf{x})\cos\gamma$$

$$\frac{d\alpha}{dt} = \frac{\partial c(\mathbf{x})}{\partial x_1}\sin\alpha + \frac{\partial c(\mathbf{x})}{\partial x_2}\cot\alpha\cos\beta + \frac{\partial c(\mathbf{x})}{\partial x_3}\cot\alpha\cos\gamma$$

$$\frac{d\beta}{dt} = \frac{\partial c(\mathbf{x})}{\partial x_1}\cos\alpha\cot\beta + \frac{\partial c(\mathbf{x})}{\partial x_2}\sin\beta + \frac{\partial c(\mathbf{x})}{\partial x_3}\cot\beta\cos\gamma$$

$$\frac{d\gamma}{dt} = \frac{\partial c(\mathbf{x})}{\partial x_1}\cos\alpha\cot\gamma + \frac{\partial c(\mathbf{x})}{\partial x_2}\cot\gamma\cos\beta + \frac{\partial c(\mathbf{x})}{\partial x_3}\sin\gamma$$

(B12.2.1)

You can start numerically integrating these equations at the source with a specified take-off angle using the derivatives to compute the position of the ray after a small time increment. More sophisticated approaches that adapt the step size for accuracy and efficiency of the integration can also be employed. For a one-dimensional velocity variation and a ray traveling in the x_1x_3 plane, the equations reduce to

$$\frac{dx_1}{dt} = c(x_3)\cos\alpha$$

$$\frac{dx_3}{dt} = c(x_3)\cos\gamma$$

$$\frac{d\alpha}{dt} = \frac{dc(x_3)}{\partial x_3}\cot\alpha\cos\gamma$$

(B12.2.2)

$$\frac{d\gamma}{dt} = \frac{dc(x_3)}{\partial x_3}\sin\gamma \,.$$

These equations can be integrated numerically to recover the ray path and travel time.

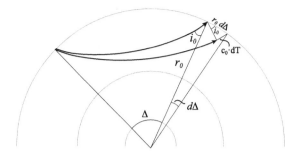

FIGURE 12.10 Raypaths for two adjacent rays in a spherical Earth.

Further, this relationship must hold along the entire raypath since r_1 and r_2 can be chosen anywhere on the path – this means that the ratios are *constant*. The relationship defines the spherical-earth ray parameter:

$$p = \frac{r \sin i}{c} . \tag{12.50}$$

Although the physical units of ray parameter in a spherical earth model differ from those we obtained for flat earth models, the meaning is consistent. The ray parameter, p, is the slope of the travel-time curve.

Consider Fig. 12.10, which traces the path of two adjacent rays. The parameters of the two rays are p, Δ, and T (ray parameter, angular distance, and travel time, respectively), and $p + dp$, $\Delta + d\Delta$, and $T + dT$. The distances $r_0 d\Delta$ and $c_0 dT$ are assumed small (the ray spacing is exaggerated for diagram clarity) such that c_0 represents the near-surface velocity.

From the problem geometry,

$$\sin i_0 = \frac{c_0 \, dT}{r_0 \, d\Delta} , \tag{12.51}$$

or

$$p = \frac{r_0 \sin i_0}{c_0} = \frac{dT}{d\Delta} . \tag{12.52}$$

The ray parameter, p, is precisely the slope of the travel-time curve, as it was for the flat-Earth case

– except that distance along the surface is measured in angular degrees. The ray parameter is still identified with the horizontal slowness or inverse *apparent velocity* along the surface, c_a. At the ray's turning point (maximum depth), r_t, we have $p = r_t \sin(\pi/2)/c(r_t) = r_t/c(r_t) = \xi(r_t)$. Unlike the flat-layered case, the spherical vertical slowness is a ratio of a velocity and a depth – in a spherical Earth, every ray will return to the surface, even if the velocity decreases with depth. The physical units of p in a spherical model are s/rad or s/deg (the "natural" units are s/rad, you must use these units when working with inversion formulas such as the Herglotz-Wiechert technique, see Chapter 20). Since one degree of arc distance along a great-circle on a sphere of radius 6371 km is 111.19 km, we convert a ray parameter in s/deg to a value in s/km, by dividing by 111.19 km/deg.

Fig. 12.11 is a plot of the P and S ray parameters (in s/km) for spherical earth model AK135 (Kennett, 2005). The complexity of the shallow earth and the influence of the transition zone are apparent. The one-degree distance sampling is insufficient to define the triplication from the crust-mantle boundary, but multiple arrivals are predicted for distances $\leq 10°$. The upper mantle triplications are apparent in the distance range from $15°$ to $30°$. Across this distance range, more than one arrival is observed at each distance – a result of the triplications associated with the rapid velocity increases in the mantle-transition zone. Beyond $\Delta \sim 30°$, the ray parameter variation is smooth and the effect of structure changes slowly. That is partly why P and S waves in this distance range are often used to investigate earthquake properties.

12.4.1 Travel-time expressions for spherical Earth models

We can derive a general equation for travel time in a sphere by considering the ray segment shown in Fig. 12.12. The length of a small seg-

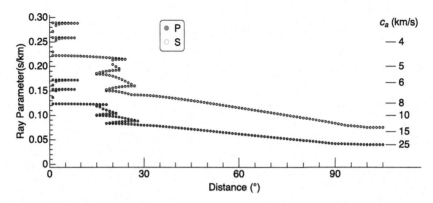

FIGURE 12.11 Ray parameters for P and S waves computed for spherical earth model AK135. Apparent speed $c_a = p^{-1}$ is shown along the right. Low-valued ray parameters correspond to high apparent speeds. A vertically incident wave has a ray parameter of zero. For distances beyond $\sim 100°$ the waves are diffracted and the ray parameter is constant.

ment of the ray (ds) satisfies

$$(ds)^2 = (dr)^2 + r^2(d\Delta)^2 . \tag{12.53}$$

Note that from the geometry, $\sin i = r\, d\Delta/ds$, so

$$p = \frac{r^2}{c}\frac{d\Delta}{ds} . \tag{12.54}$$

Eq. (12.54) can be used to eliminate ds from Eq. (12.53) to yield

$$(d\Delta)^2 = \frac{(dr)^2 p^2 c^2}{r^4 - r^2 p^2 c^2}$$

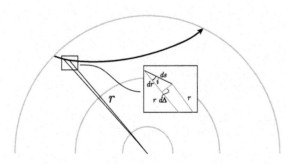

FIGURE 12.12 Geometry of ray segment ds in terms of radius r and angle $d\Delta$.

or

$$d\Delta = \frac{p}{r}\frac{dr}{\sqrt{\xi^2 - p^2}} , \tag{12.55}$$

where $\xi = r/c$. We can integrate (12.55) to obtain

$$\Delta = 2p \int_{r_1}^{r_0} \frac{dr}{r\sqrt{\xi^2 - p^2}}, \tag{12.56}$$

where r_0 is the radius of the Earth and r_t is the deepest sampled by the ray. This equation is analogous to Eq. (12.41) for a flat inhomogeneous model.

If instead, we choose to eliminate $d\Delta$ from (12.53) using (12.54) to obtain

$$(ds)^2 = (dr)^2 + \frac{p^2 c^2 (ds)^2}{r^2}$$

or

$$ds = \frac{dr}{\sqrt{1 - (p^2 c^2/r^2)}} = \frac{\xi\, dr}{\sqrt{\xi^2 - p^2}} . \tag{12.57}$$

The travel time along any path is the path length divided by the velocity,

$$T(p) = \int_{\text{path}} \frac{ds}{c} = 2 \int_{r_1}^{r_0} \frac{\xi^2}{r\sqrt{\xi^2 - p^2}}\, dr \ . \quad (12.58)$$

Eq. (12.58) is analogous to Eq. (12.42) for a flat, inhomogeneous model. Following the same logic that we used in the flat geometry, we can write (12.58) as separable travel-time equations:

$$T(p) = 2 \int_{r_1}^{r_0} \left(\frac{p^2}{r\sqrt{\xi^2 - p^2}} + \frac{\xi^2 - p^2}{r\sqrt{\xi^2 - p^2}} \right) dr$$

$$= p\,\Delta + 2 \int_{r_1}^{r_0} \frac{\sqrt{\xi^2 - p^2}}{r}\, dr. \quad (12.59)$$

For a given ray parameter, the first term on the right-hand side of Eq. (12.59) depends only on Δ, or *surface horizontal distance*, and the second term depends only on r, the vertical distance from Earth's center. This is analogous to (12.43), with the integral corresponding to the *tau* function, $tau(P)$, as in (12.44) for a spherical geometry.

The travel-time curves for a spherical geometry are very similar to those for a flat geometry, with the caveats that angular distance is used and ray parameter is scaled by the normalized radius. This implies that the qualitative behavior of the travel-time curves characterizing different velocity profiles in Fig. 12.8 can be used to infer the gross character of velocity structure in a spherical Earth. In the real Earth prominent triplications result from velocity increases at the crust-mantle boundary and near 410 and 660 km depth. The low-velocity core produces a major shadow zone (more on these in Chapter 10).

12.5 Travel times in layered Earth models

A common method to infer Earth's seismic velocity structure is to model seismic wave

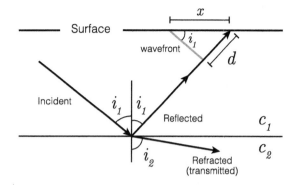

FIGURE 12.13 A wave incident on a boundary between contrasting materials. For clarity, we are ignoring potential reflected and transmitted converted waves (P-to-S or S-to-P), discussed in the next chapter.

travel times as a function of distance. The equations for travel time in a layered Earth model are a discretization of Eq. (12.43). We also can derive these equations from first principles. When a wave strikes a boundary marking a change in seismic velocity (see Fig. 12.13), the energy in the wave is partitioned between **reflected** and **transmitted** or **refracted** waves. To maintain the physical boundary conditions and be consistent with the eikonal equations (or equations of motion), these new, or *derivative*, rays inherit the ray parameter of the incident wave. Assume that the boundary is horizontal. The angle (i_1 or i_2) that the reflected and refracted rays make with a vertical plane is governed by Snell's law,

$$p = \frac{\sin i_1}{c_1} = \frac{\sin i_2}{c_2} \ . \quad (12.60)$$

Consider the wavefront associated with the reflected ray as shown in Fig. 12.13. The wavefront will advance a distance d in a time $\delta t = d/c_1$. The intersection of the wavefront with the surface will sweep along the surface at an **apparent velocity**, c_a, greater than the actual velocity of the layer,

$$c_a = \frac{x}{\delta t} = \frac{d}{\sin i_1} \frac{1}{\delta t} = \frac{c_1}{\sin i_1} = \frac{1}{p} \ . \quad (12.61)$$

The above expression is the reason *horizontal slowness* is synonymous with ray parameter. A horizontal ray corresponds to a wave with $p = 1/c_a$, or an apparent velocity equal to the physical velocity of the layer. A vertical ray corresponds to a wave with $p = 0$, or an infinite apparent velocity – the apparent speed, c_a has a range $c_1 \leq c_a \leq \infty$.

If the velocity in layer 2 is greater than the velocity in layer 1, angle $i_2 > i_1$. When the angle $i_2 = 90°$, we have a **critical refraction**,

$$\frac{\sin i_c}{c_1} = \frac{\sin 90°}{c_2} = \frac{1}{c_2} . \qquad (12.62)$$

The angle, i_c is called the **critical angle**. The critical refraction is associated with a wave that travels parallel to the interface between layers, immediately below the interface. This wave is usually referred to as a **head wave**, and has a unique property that it transmits energy back into layer 1 continually as it travels along the interface. This energy leaves the interface with the critical angle, i_c,

$$i_c = \sin^{-1}(c_1/c_2) . \qquad (12.63)$$

If $i_1 > i_c$, no energy transmits into layer 2, and all the energy is reflected back into layer 1 or travels along the boundary. If $c_2 < c_1$ there is no critical angle, and the refracted ray is deflected toward the vertical. Head waves in layered velocity structures are analogous to turning waves in continuous velocity structures. These waves are extremely important in determining the velocity structure of the Earth. The travel time of these seismic waves as a function of distance provides a direct measure of velocity at depth.

The layer-over-a-halfspace model

Consider the three rays in the layer-over-a-halfspace structure shown in Fig. 12.14. This model is canonical in observational seismology because it is a first-order approximation to problems that include the dry over saturated soil, unconsolidated sediment over rock, ice over rock,

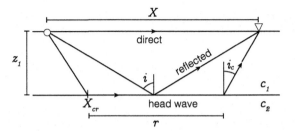

FIGURE 12.14 The three principal rays in a layer-over-a-halfspace velocity structure.

sedimentary rocks over basement (a basin), and the crust over the mantle. If $c_2 > c_1$, three *primary* waves travel between the source and the receiver: (1) the **direct arrival**, which travels in a straight line connecting source and receiver, (2) a **reflected arrival**, and (3) a **head wave**. Additional rays corresponding to multiple reflections and *P*-to-*S* conversions also propagate in the layer, but we focus on the easier-to-observe primary waves.

The expression for the direct-wave travel time as a function of distance is

$$T_D(X) = p X, \quad p = 1/\alpha_1 . \qquad (12.64)$$

The travel time for the reflected arrival is given by

$$T_R(X) = \frac{2 z_1}{\cos i} \frac{1}{c_1} , \qquad (12.65)$$

where z_1 is the layer thickness. Finally, the travel time of the head wave is given by

$$T_H(X) = \frac{r}{c_2} + \frac{2 z_1}{\cos i_c} \frac{1}{c_1} . \qquad (12.66)$$

These equations are for a surface source but only slight modifications are needed for sources within the layer. The second term in (12.66) is the same as (12.65) for the reflected arrival when $i = i_c$. The head wave first appears at $X = X_{cr}$, with a travel time equal to that of the critically reflected arrival. At closer distances only the direct and reflected waves exist. As X increases, only the r/c_2 term of (12.66) changes, thus, the wavefront travels along the surface with apparent velocity c_2. Because $r = X - 2z_1 \tan i_c$ and $\sin i_c = c_1/c_2$, (12.66) can be simplified,

$$T_H(X) = \frac{2z_1}{\cos i_c} \frac{1}{c_1} + \frac{1}{c_2}\left(X - \frac{2z_1c_1}{c_2 \cos i_c}\right)$$

$$= \frac{2z_1}{\cos i_c}\left(\frac{1}{c_1} - \frac{c_1}{c_2^2}\right) + \frac{X}{c_2}. \tag{12.67}$$

Recalling that $p = 1/c_2$ and $\cos i_c = \left(1 - \sin^2 i_c\right)^{1/2} = \left(1 - c_1^2 p^2\right)^{1/2}$, we can rewrite this as

$$T_H(X) = pX + 2z_1\eta_1, \tag{12.68}$$

where $\eta_1 = \sqrt{c_1^{-2} - p^2}$, is the vertical slowness. Equation (12.68) is the layer-over-a-halfspace equivalent to Eq. (12.43) and likewise is an extremely useful form of the travel-time equation because it separates the travel path into a *horizontal* term and a *vertical* term. No matter how complex a raypath in a layered structure becomes, it is possible to write the corresponding travel-time equation with a form similar to Eq. (12.68). The travel time is sum of the product of the horizontal slowness and horizontal distance, and the product of the vertical slowness and the vertical distance traveled by the wave.

Eqs. (12.64), (12.65), and (12.68) are mathematical formulas for *travel-time curves*. Fig. 12.15 is a plot of the travel-time curves for the principal rays for the structure in Fig. 12.14. At short distances only the direct and reflected arrivals exist. The direct arrival's curve is a straight line with a slope $dT/dX = p = 1/c_1$. The reflected-arrival travel time curve is a hyperbola. The $X = 0$ intercept has a travel time of $2z_1/c_1$ (the two-way travel time through the layer). At large distances the reflection travel time **branch** is asymptotic to the curve of the direct arrival. The travel-time branch associated with the head wave first appears at $X_{cr} = 2z_1 \tan i_c$. The head-wave arrival branch is a straight line with a

slope $dT/dX = p = 1/c_2$. The direct arrival is the first arrival until the **crossover distance**, X_X, after which the head wave is the first wave observed. At X_X the travel times of the direct arrival and head wave are equal,

$$T_{\text{direct}} = T_{\text{head}}$$

$$\frac{X_X}{c_1} = \frac{X_X}{c_2} + 2z_1\eta_1 \tag{12.69}$$

$$X_X = 2z_1 \frac{c_1 c_2}{c_2 - c_1}\eta_1$$

or

$$X_X = 2z_1\sqrt{\frac{c_2 + c_1}{c_2 - c_1}}. \tag{12.70}$$

Fig. 12.16 is a plot of a seismogram observed 314 km from an earthquake. Three prominent arrivals are noted, P_n, P_g, and S. Arrivals **Pn** and **Pg** correspond to the head wave (from the crust-mantle transition) and direct arrival (traveling within the crust), respectively, in Fig. 12.14. In 1909 a Croatian scientist named Mohorovičić first observed these two P-wave arrivals with different apparent velocities over a several-hundred-kilometer distance. P_g was observed to have a velocity of 5.6 km/s, P_n, a velocity of 7.9 km/s. The arrivals arise because, in a gross sense, the crust-mantle structure resembles a layer over a half-space. The head wave, P_n, is caused by the velocity increase at the crust-mantle boundary (known as the **Moho**). The reflection off the Moho is known as **PmP** but it is not readily identifiable in Fig. 12.16. The travel-time plot of Fig. 12.15 is not a complete view of the seismic wavefield because we have only considered three rays. The example seismograms include many more arrivals. Clearly, analogous shear-wave arrivals, **Sn**, **Sg**, and **SmS** also exist. In addition, multiple reflections, in which waves bounce between the surface and the Moho more than once with segments that are a mixture of P and S waves also traverse the layer. Many possible arrivals cause the complex oscillations in Fig. 12.16.

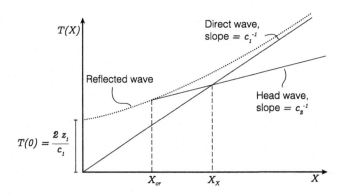

FIGURE 12.15 Travel-time curves for the primary waves in the layer-over-a-halfspace velocity structure.

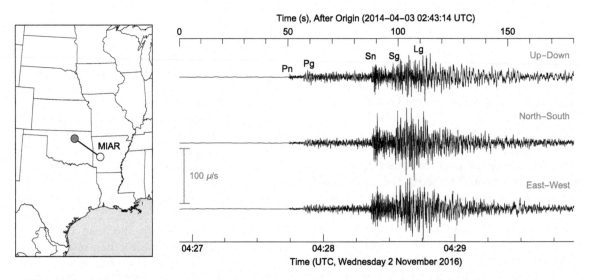

FIGURE 12.16 Ground velocity for an earthquake ($m_{b_{Lg}} = 4.4$) in north-central Oklahoma recorded at station MIAR, Mount Ida, Arkansas, USA. The head waves P_n and S_n and the crustal turning waves, P_g, S_g are labeled. The source-to-station distance is about 342 km The L_g phase is a crustal guided wave that propagates well through simple, stable regions of the continents. The wave is widely observed and used to estimate magnitudes in eastern North America. Distinguishing between S_g and L_g in this distance range is difficult as a result of interference.

The importance of travel-time curves is their interpretative power. If we consider a seismic station at a given distance from a seismic source, we expect a sequence of arrivals with predictable travel times. Suppose we have many seismic stations that record a seismic event. If we carefully measure the arrival times of the waves we can infer the structure from the pattern they form with distance. We can use our simple approximate model to estimate the layer velocity from the slope of the direct arrival and the halfspace velocity from the slope of the head-wave branch. The crossover distance or the zero offset ($X = 0$) reflection time constrains the layer thickness. All of this presumes of course that we know the time and location of the source.

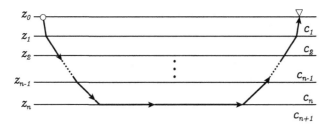

FIGURE 12.17 Head-wave raypath in a layered model.

Travel time curves are applied routinely to estimate those quantities as well (Chapter 6). Of course, Earth is more complex than a layer over a half-space. But the fact that we see a systematic variation in travel time with distance suggests that we can approximate the subsurface using one or more layers in which the velocity is constant.

The head-wave travel-time equations can be generalized to the case of n layers over a half-space. We can show the result geometrically, but instead, we use the results from the eikonal equations, Eq. (12.41) and Eq. (12.43). Consider the diagram in Fig. 12.17. Since the ray turns in the halfspace, the ray parameter is $p = 1/c_{n+1}$. The expression for the travel time for any distance to which the ray with that horizontal slowness arrives is

$$T(p) = p\,X + 2\int_0^{z_n} \sqrt{c^{-2}(x_3) - p^2}\,dx_3\,.$$

The integrand is constant within each of the layers in the model, so we can split the integral into a sum that includes one term for each layer. For example, the contribution of the top layer to the integral is

$$2\int_{z_0}^{z_1} \sqrt{c_1^{-2} - p^2}\,dx_3 = 2\sqrt{c_1^{-2} - p^2}\int_{z_0}^{z_1} dx_3$$
$$= 2\sqrt{c_1^{-2} - p^2}\,(z_1 - z_0)$$
$$= 2\eta_1\,(z_1 - z_0)\,. \tag{12.71}$$

If we represent the layer thicknesses using $H_i = z_i - z_{i-1}$, then the integral becomes a sum

$$\int_0^{z_n} \sqrt{c^{-2}(x_3) - p^2}\,dx_3 = \sum_{i=1}^n 2\,\eta_i\,H_i\,, \tag{12.72}$$

and the travel time for a wave that turns in the nth-layer becomes

$$T(p) = p\,X + \sum_{i=1}^n 2\,\eta_i\,H_i\,. \tag{12.73}$$

The formula is again of the form, horizontal slowness × horizontal distance plus the vertical slowness × the vertical distance. However, the vertical slowness assumes a different value in each layer.

Hidden layers and blind zones

Two special cases complicate the interpretation of travel-time curves. The first is the case of a low-velocity layer, which does not produce a head wave, and is sometimes called the hidden-layer problem. Consider the structure shown in Fig. 12.18, where the velocity of layer 2 is *less than* that of layer 1 and of the half-space. No head wave is excited along the interface between layers 1 and 2. Therefore, as first arrivals, we observe only a direct wave through layer 1, and a head wave from the interface between layers 2 and 3. We would also observe reflections from the top and bottom of layer 2, but they would be secondary arrivals. The corresponding travel-time equations for the direct wave and the head

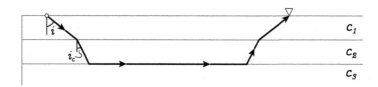

FIGURE 12.18 Raypath for a structure with a *low-velocity layer*, $c_2 < c_1 < c_3$. No head wave will exist on the interface between velocities c_1 and c_2. The critical angle satisfies $i_c = \sin^{-1} c_1/c_3$.

wave are

$$T_D(X) = p X, \quad \text{where} \quad p = 1/c_1$$

$$T_H(X) = p X + 2 H_1 \eta_1 + 2 H_2 \eta_2 \tag{12.74}$$
$$\text{where} \quad p = 1/c_3 .$$

Since the travel-time curve has only two first-arriving branches, unless we had clear information from the reflections, one would interpret the curve as a *single* layer of apparent thickness $\widehat{H}$ with velocity $\widehat{c_a}$, over a half-space of velocity c_3. The apparent thickness estimated from the crossover distance is $H_1 + H_2(\eta_2/\eta_1)$, which is in an *overestimate* of the actual depth to the half-space.

The second special case is called a *blind zone*, which arises when a layer is so thin that the head wave from it is never a first arrival. Consider the structure shown in Fig. 12.19. The travel times for the two rays shown are

$$T_1(X) = p X + 2 H_1 \eta_1 , \quad \text{where} \quad p = 1/c_2 ,$$

$$\tag{12.75}$$

and

$$T_2(X) = p X + 2 H_1 \eta_1 + 2 H_2 \eta_2 ,$$
$$\text{where} \quad p = 1/c_3 . \tag{12.76}$$

Note that η_1, for ray 1 *does not* equal η_1 for the second ray because p is different. For some combinations of c_2, c_3, and H_2, the travel-time curve will look like that shown in Fig. 12.19 (right), where the head wave with slope $1/c_2$ is never a

first arrival if c_2/c_1 is not much larger than one, or if H_2 is small.

12.6 Body-wave travel-time tables

The analysis of travel times has been an important part of seismology since the earliest days. Short-offset travel-time tables were used to study the speed of seismic waves at the shallow depths of Earth. Oldham (1900) produced the first global-scale travel-time curves and inferred the existence of the core from them. Beno Gutenberg and Inge Lehmann used travel times to better constrain first-order properties of the outer core and Lehmann used travel times to identify a reflection off the inner core. Jeffreys, Bullen, Macelwane, and Wadati, and their students studied seismic-wave travel times from shallow and deep earthquake sources. Much of this work, and early, globally-representative travel-time tables were completed using hand calculation. For many years, the standard reference for body-wave travel times were the tables of Jeffreys and Bullen (the JB Tables). In the 1980's and 1990's, as the number of observed travel times grew quite large, and computation power was sufficient, refinements of the tables were completed by Kennett and Engdahl (1991) and others. The **iasp91** model was developed in the early 1990's and later supplanted with **model AK135** in the late 1990's. Together with the PREM of Dziewonski and Anderson (1981), these are now the most-commonly used reference models for global travel-time calculations.

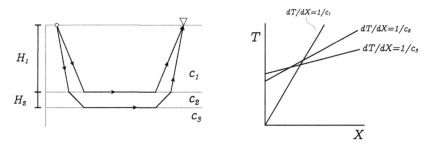

FIGURE 12.19 Travel path and corresponding travel-time curves for a blind zone scenerio. The observability of a first arrival with the slowness $1/c_2$ depends on the layer thickness and velocity contrasts involved.

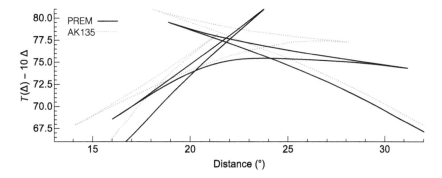

FIGURE 12.20 Travel-time curves for spherical earth models PREM and AK135 in the mantle-triplication distance range. The source is at the surface. Differences in the crust and in this range are the most dramatic. *P*-wave travel time differences of up to a second are shown.

Fig. 12.20 is a comparison of the PREM and AK135 earth models for *P* waves from a surface source, traveling to distances that are sensitive to the mantle triplication regions. The travel time curves are shown with a ***reducing velocity***, v_r, to remove the steep slope that exists at this distance range. Reducing velocities are common in seismogram displays – the time of arrival or travel time is modified by subtracting a zero-intercept line with a slope, x/v_r. The reducing velocity used in the figure, 0.1 s/deg corresponds to a speed of roughly 11 km/s. The plot indicates that the travel times for $AK135$ are later than those from the PREM. Although travel times were used to construct the PREM, it's primary purpose is not as a travel-time only model.

The travel time differences for direct *P* and *S* waves in the PREM, iasp91, and AK135 models is greatest in the distance range, $(15° \le \Delta \le 40°)$, where arrival times may differ by several seconds (greatest difference for *S*). The cause is differences in each model's upper mantle and mantle transition-zone structure. At distances greater than about 40°, predicted travel-time differences in *P* and *S* waves are small, on the order of 0.1 s. For core reflections, travel-time differences between all three models are all smaller still – the vertical integrals through the models are quite consistent. Outside the mantle triplication distances, the travel-time differences are significantly smaller than the natural variation in travel times arising from Earth's lateral heterogeneity. But systematic model differences may affect seismic location estimates. The USGS used the JB tables for earthquake location estimation until January of 2004, when they transitioned to the AK135 travel times. The change of models can result in difference in before and after origin-time patterns (Cleveland et al., 2018).

13

Body-waves and ray theory – amplitudes

Chapter goals

- Define propagation-related effects on body-wave amplitudes.
- Review geometric spreading and its relation to travel time curves.
- Review energy partitioning between waves interacting at elastic boundaries.
- Review scattering, attenuation and seismic attenuation models.

Now that we have fully developed the concept of travel time for a ray, we can return to *energy* associated with an arrival. The amplitude of a seismic wave depends on the strength of the wave as it leaves the source, the geometric spreading of energy that occurs as the wave travels away from the source, interactions along the path (reflections, transmissions, anelastic attenuation). We start our discussion with geometric spreading, which relates closely to the previous chapter on travel times and ray paths. We then discuss the interactions of waves with boundaries between elastic materials and conclude with some details on the physical models of attenuation (first introduced in Chapter 2) common in seismology.

13.1 Geometric spreading in vertically varying media

The decrease in amplitude caused by the geometric expansion of wavefronts is proportional to the distance traveled and the second derivative of the travel-time curve, d^2T/dX^2. Eq. (12.28) provides insight into the geometric spreading experienced by seismic body-waves. Consider a spherical wave a small distance from a surface seismic source in a region of uniform velocity. Let the energy of the disturbance be distributed uniformly on the spherical wavefront. As the wavefront expands, the total energy on the surface will remain constant, but the energy per unit surface area will decrease.

Define the total energy on the initially hemispherical wavefront (Fig. 13.1) as K, then the energy per unit area is $K/2\pi r^2$. Then consider a "bundle" of rays that leave the source between the angle i_0 and $i_0 + di_0$. The fraction of energy in a circular ring on the wavefront defined by these two takeoff angles is

$$E_0 = \frac{K}{2\pi r^2} (2\pi r \sin i_0) (di_0 r) , \qquad (13.1)$$

where $r \sin i_0$ is the radius of the strip, and $di_0 r$ is the width of the strip. Simplifying the expres-

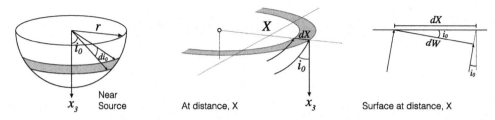

FIGURE 13.1 The area, which is inversely proportional to energy, for an expanding wavefront, near a surface source (left) and at the surface, a distance X away (center). Since the wavefront does not arrive normal to the surface, energy from a ray bundle with a width near the surface of $dW = dX \cos i_0$ is spread across the surface element of width dX (right).

sion, we find

$$E_0 = K \sin i_0 \, di_0 . \tag{13.2}$$

In a nonuniform medium, such a ray bundle expands or contracts as it propagates, depending on the characteristics of the wave velocity. Upon arrival at the surface, the energy spread out over the ring will be $2\pi X dX \cos i_0$ (Fig. 13.1). The factor, $\cos i_0$, adjusts for the non-normal incidence of the wavefront at the surface. The change in energy density, $E(X)$, is obtained by dividing (13.2) by the new area to obtain

$$E(X) = \frac{K \sin i_0 \, di_0}{2\pi X \cos i_0 \, dX} = \frac{K}{2\pi} \frac{\tan i_0}{X} \frac{di_0}{dX} . \tag{13.3}$$

This expression can be simplified, let c_0 be the speed near the source, then recall that

$$p = \frac{\sin i_0}{c_0} = \frac{dT}{dX}, \quad \text{or,} \quad i_0 = \sin^{-1}\left(c_0 \frac{dT}{dX}\right) . \tag{13.4}$$

Differentiating i_0 with respect to distance, X,

$$\frac{di_0}{dX} = \frac{c_0}{\sqrt{1 - c_0^2 (dT/dX)^2}} \frac{d^2T}{dX^2},$$

$$= \frac{c_0}{\cos i_0} \frac{d^2T}{dX^2} . \tag{13.5}$$

Thus Eq. (13.3) can be rewritten as

$$E(X) = \left(\frac{K}{2\pi}\right)\left(\frac{c_0 \tan i_0}{\cos i_0}\right)\frac{1}{X}\left(\frac{d^2T}{dX^2}\right) . \tag{13.6}$$

Wave amplitude is proportional to $\sqrt{E}$, so the amplitude of a seismic arrival is proportional to the *change* in ray parameter with distance. Velocity structures for which p changes rapidly produce large amplitude variations by geometric focusing or defocussing. Conversely, constant p implies small amplitudes. We've assumed that the source and receiver were at the same depth, so the speed was the same, and the take-off angle (at the source) was the same as the incidence at (at the receiver). If the source and receiver are at different depths, then the takeoff angle at the source will differ from the incident angle at the receiver and some minor adjustment of the formula is necessary. Let the subscript h identify quantities near the hypocenter and the subscript 0 represent quantities near the surface (receiver), then

$$E(X) = \left(\frac{K}{2\pi}\right)\left(\frac{c_h \tan i_h}{\cos i_0}\right)\frac{1}{X}\left(\frac{d^2T}{dX^2}\right) . \tag{13.7}$$

$E(X)$ is an energy density – specifically, it represents the energy per unit area. The units of K are those of energy, and the remaining terms combine to produce units of L^{-2}. These simple extensions of the ray equations for vertically varying media provide important insight. We next consider geometric spreading in spherical earth models, which are commonly used to account for the geometric effects in Earth.

13.2 Geometric spreading in spherical Earth models

To extend our results to spherical earth models requires a simple modification of (13.6). For the spherical problem, the bundle of rays leaving the source no longer illuminates a ring on a flat surface, they illuminate a spherical ring of width $d\Delta$ on the spherical surface. From the properties of a sphere, the area of this ring is given by

$$2\pi \, r_0^2 \sin \Delta \, |d\Delta| \, , \qquad (13.8)$$

where r_0 is the radius of the Earth and the $\sin \Delta$ factor arises from the form of the surface-area element in spherical coordinates. The geometry is summarized in Fig. 13.2. The ratio of the areas near the source and on Earth's surface, adjusted for the non-normal incidence of the wave, leads to

$$E(\Delta) = \frac{E_h \sin i_h \, di_h}{r_0^2 \cos i_0 \sin \Delta \, |d\Delta|} \, , \qquad (13.9)$$

where $E_h = K/2\pi$. Expressing $di_h/d\Delta$ in terms of ray parameter and travel time, changes Eq. (13.7) to

$$E(\Delta) = E_h \left(\frac{c_h}{r_0^2 \, r_h} \right) \left(\frac{\tan i_h}{\cos i_0} \right) \left(\frac{1}{\sin \Delta} \right) \left| \frac{d^2 T}{d\Delta^2} \right| \, , \qquad (13.10)$$

where we used

$$\frac{d^2 T}{d\Delta^2} = \frac{dp}{d\Delta} = \frac{d}{d\Delta} \left(\frac{r_h \sin i_h}{c_h} \right) = \left(\frac{r_h}{c_h} \right) \frac{d \sin i_h}{d\Delta}$$

$$= \left(\frac{r_h}{c_h} \right) \cos i_h \frac{di_h}{d\Delta} \, , \qquad (13.11)$$

which provides

$$\frac{di_h}{d\Delta} = \left(\frac{c_h}{r_h \cos i_h} \right) \frac{d^2 T}{d\Delta^2} \, . \qquad (13.12)$$

The quantity $E(\Delta)$ is an energy density (per unit area). The values of Δ used in these equa-

tions must be in radians for the physical units to make sense. For example, in the familiar arc length expression $ds = r \, d\Delta$, the angle is in radians and the units of length are associated with r. Thus, the physical units of $E(\Delta)$ are energy/area, which makes sense since the energy on the wavefront (a surface) is decreasing in proportion to the changes in area on the surface.

13.2.1 Seismic-wave energy and amplitude

Seismic waves propagate by inducing ground motions across the wavefront. As the material at the wavefront moves, kinetic energy is transferred to the non-moving particles across the wavefront as a compression, elongation, or shear that exerts a stress on the as-yet, unmoving material "down-stream" of the wave. The down-stream material moves in response to the up-stream strains and that elastic potential energy is later transferred to kinetic energy as a result of elastic restoring forces. The energy (wave) propagates.

The transfer of energy defines the wave – a wave *is* a disturbance that transmits energy. The strain energy in a small, deformed elastic volume is

$$E = \frac{1}{2} \int \sigma_{ij} \, \varepsilon_{ij} \, dV \, . \qquad (13.13)$$

Consider an SH plane wave propagating in the x_1 direction, with all motion in the x_2 direction,

$$u_2(\omega, x, t) = A(\omega) \, e^{i(\omega t - k x_1)} \, , \qquad (13.14)$$

where $A(\omega)$ is an amplitude factor, and the phase $i(\omega t - kx_1)$ represents the propagation of the quantity in the x_1 direction with a speed $c = k\omega$. The plane-wave's ground motion is generally a sum of similar terms over a range of frequency, ω. Because of the selected geometry, the only nonzero strains associated with $u_2(\omega, x, t)$ are

$$\varepsilon_{12} = \varepsilon_{21} = \frac{1}{2} \frac{\partial u_2}{\partial x_1} = -\frac{1}{2} i k A e^{i(\omega t - kx_1)} \, , \quad (13.15)$$

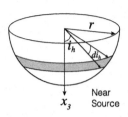 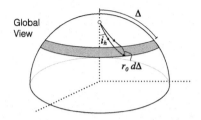 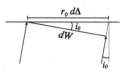

FIGURE 13.2 The area, which is inversely proportional to energy, for an expanding wavefront, near a surface source (left) and at the surface of the Earth, at a distance Δ (center). Since the wavefront does not arrive normal to the surface, energy from a ray bundle with a width near the surface of $dW = r_0 \sin \Delta \, d\Delta \, \cos i_0$ is spread across the surface element of width $r \, d\Delta$ (right).

and the elastic stress is (from Hooke's Law),

$$\sigma_{12} = \sigma_{21} = -i \, k \, \mu \, A(\omega) \, e^{i(\omega t - kx_1)}. \qquad (13.16)$$

The *average* strain energy integrated over a wavelength is

$$E = \frac{1}{\Lambda} \int_0^\Lambda \frac{1}{2} k^2 A^2(\omega) \, \mu \, dx_1 \,, \qquad (13.17)$$

where Λ is the wavelength. Recall that a wave travels a distance on one wavelength in a time equal to one oscillation period (T), $\Lambda = c\,T$. Further, for the shear wave speed, β, and density, ρ, $\mu = \rho\beta^2$, and using $k = (2\pi/\Lambda)$, we have

$$E = \frac{1}{2}\left(\frac{2\pi}{\beta T}\right)^2 A^2 \rho \, \beta^2 = 2\pi^2 \rho \frac{A^2}{T^2} \,. \qquad (13.18)$$

The simple example illustrates several important, general results. The *energy* carried in a seismic wave is proportional to the square of the displacement amplitude and inversely proportional to the square of the period. If the amplitude of two seismic signals is the same, the higher-frequency signal transports more energy.

Since seismic wave amplitude will change in proportion to the square root of the energy, using Eq. (13.10), the decrease in amplitude as a function of distance from geometrical spreading

is

$$G(\Delta, h) = \sqrt{\frac{E(\Delta)}{E_h}}$$

$$= \sqrt{\left(\frac{c_h}{r_0^2 \, r_h}\right)\left(\frac{\tan i_h}{\cos i_0}\right)\left(\frac{1}{\sin \Delta}\right)\left|\frac{d^2 T}{d\Delta^2}\right|} \,, \qquad (13.19)$$

where E_h represents the energy near the source and the spherical ray parameter, or angular slowness is $p = r \sin i / c$. We define a radial slowness as

$$q(r, p) = \sqrt{\frac{1}{c^2(r)} - \frac{p^2}{r^2}} \,. \qquad (13.20)$$

Note that the angular slowness has units *seconds-per-radians*, the radial slowness has units of *seconds-per-km*. Further, we have $q_0 = \cos i_0 / c_0$, $q_h = \cos i_h / c_h$, and $p = r_h \sin i_h / c_h$. Then,

$$G(\Delta, h) = \sqrt{\frac{c_h/c_0}{r_0^2 \, r_h^2} \frac{p}{q_h \, q_0}\left(\frac{1}{\sin \Delta}\right)\left|\frac{d^2 T}{d\Delta^2}\right|} \,. \qquad (13.21)$$

The slownesses (as did the angles) depend on the source-to-receiver angular distance, Δ. The units of $G(\Delta)$ are inverse length. In a homogeneous medium, the decrease in energy density with distance on an expanding spherical wavefront with radius, R, is proportional to $1/R^2$. The decay of amplitude is proportional to $1/R$. We can think of $G(\Delta)$ as an effective $1/R$ factor for the Earth, modified to account for the changes in ray path length and the effects of focusing and defocussing introduced by the variation of the wave speed with radius.

As described in the introduction, geometrical spreading is not the only contributor to the amplitude of a seismic wave. Seismic-wave amplitude is modified during propagation by several other phenomena. For example, $G(\Delta)$ does not take into account the loss of energy to non-elastic processes associated with seismic-wave attenuation. In the next section, we revisit anelastic attenuation, which was introduced in Chapter 2. We then conclude with a focus on wave reflection and transmission at a boundary between materials with differing elastic properties.

13.3 Body-wave attenuation

Thus far our discussion of body-wave amplitudes has been concerned with the *elastic* properties of the Earth. In an idealized, purely elastic Earth, geometric spreading and the reflection and transmission of energy at boundaries control the amplitude of a seismic pulse. As discussed in Chapter 2, Earth is not perfectly elastic, and propagating waves *attenuate* with time or distance traveled. Among other processes, movements along mineral dislocations or shear heating at grain boundaries taps the wave energy in a non-reversible manner. Energy is lost and amplitudes decrease. As noted earlier, we "model" these internal-friction effects with phenomenological descriptions because the microscopic processes are complex. Recall our discussion of seismic-wave attenuation from Chapter 2 where we quantified attenuation affects using the change in energy over a deformation cycle

with the quality factor defined by (Eq. (2.41)),

$$2\pi\, Q^{-1} = \frac{-\Delta E}{E} ,$$

where E represents the peak energy in the wave and ΔE the energy loss per deformation cycle. The change in amplitude with propagation time is (Eq. (2.38)),

$$A(t) = A_0\, e^{-\omega_0 t / 2Q} ,$$

or equivalently, the change in amplitude with propagation distance, x, is

$$A(x) = A_0\, \exp\left(-\frac{f\,\pi\,x}{c\,Q}\right) = A_0\, \exp\left(-\frac{x}{\lambda}\frac{\pi}{Q}\right) ,$$

where c is the wave speed, and f is frequency, the wavelength is λ, and we used $c = \lambda f$.

What we didn't discuss earlier is that the simple spring-slider model and the observed seismic attenuation exhibit different frequency dependencies. Fig. 13.3 is a summary of observed Q variations in Earth. For much of the low-frequency band, Q is relatively constant (but significant differences exist). At frequencies above $\sim 1\,\text{Hz}$, the energy a wave loses to intrinsic attenuation and to scattering becomes frequency dependent. *Scattering* is an elastic process, a redirection of energy, that differs from the absorption of energy by non-elastic processes. Seismic scattering is often treated using statistical approaches and the general patterns of the scattered waves (often called coda) are used to explore characteristics of Earth's heterogeneity. To first order, scattering can be modeled using an energy-loss phenomenology similar to that used to account for intrinsic energy loss and we can think of a scattering Q. Amplitude changes in the direct waves can be modeled using an effective quality factor,

$$\frac{1}{Q_{\text{effective}}} = \frac{1}{Q_{\text{intrinsic}}} + \frac{1}{Q_{\text{scattering}}} . \qquad (13.22)$$

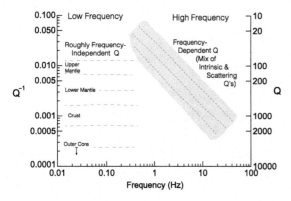

FIGURE 13.3 Conceptual view of Q-variations within Earth. Most of the high-frequency measurements corresponds to waves traveling through the lithosphere. Specific values are representative – Q values vary with position within the planet. Ranges include both Q_β and Q_α. Q in the outer core is very high, estimated to be greater than 4000; that just above and within the inner core is low. Recent core-phase P-wave analyses suggest Q_α values ranging from roughly 100–800.

Often, particularly at short periods, observationally separating the scattering and intrinsic amplitude attenuation is impossible. Ultimately, scattered waves, like direct waves, are absorbed by an intrinsic energy-absorption processes. Observations have led to models that at least partially explain the observed seismic attenuation. For shallow earthquakes, late-arriving scattered waves are often assumed to have well-averaged the properties of the bulk crust and the source and thus these waves carry valuable information on crustal and source properties. The coda-magnitude approach discussed in Chapter 7 is one such application of these ideas.

13.3.1 The standard-linear-solid attenuation model

The intrinsic quality factor, Q, for seismic waves is observed to be largely independent of frequency in the range from 0.001 to 1.0 Hz (Fig. 13.3). In the higher-frequency range, intrinsic Q generally increases with frequency. To explain the frequency dependence, we must modify our phenomenological model, the oscillating spring, as shown in Fig. 13.4. The springs represent elastic behavior, and the dashpot represents nonelastic, or *viscous*, losses. Many of the processes that contribute to seismic attenuation are consistent with a rheological model called a **standard linear solid** (SLS). Hooke's law, as written in Eq. (11.47), represents the model on the left, the center model represents our slight modification of the model to introduce the concepts of seismic attenuation in Chapter 2. The constitutive law of a standard linear solid (on the right) is

$$\sigma + \tau_\sigma \dot{\sigma} = M_r \left(\varepsilon + \tau_\varepsilon \dot{\varepsilon} \right) . \qquad (13.23)$$

M_r is called the **relaxed elastic modulus** (it dominates for low frequencies that deform the material over relatively long times), and τ_σ and τ_ε are called the stress and strain **relaxation times**, respectively. τ_σ represents the time it takes the stress to relax after an instantaneous step in strain, and τ_ε represents the time it takes the strain to relax after an instantaneous step in stress. To understand the physics of Eq. (13.23), consider the right panel in Fig. 13.4. If you deflect the mass a distance X, it is acted on by a restoring force F_0. But the force needed to hold the mass at X diminishes with time as the dashpot relaxes. The reduction in the system's restoring force is *not* recoverable. Hence the system behaves **anelastically**. If the relaxation times are zero (the material response is instantaneous), M_r represents the familiar elastic moduli.

The dynamics of (13.23) can be investigated by examining the ratio of stress to strain for a harmonic deformation ($\propto e^{i\omega t}$). Define the **complex elastic modulus**, M^* as

$$M^*(\omega) = \sigma(\omega)/\varepsilon(\omega) , \qquad (13.24)$$

which, for the SLS is

$$M^*(\omega) = M_r + \delta M \frac{\omega^2 \tau_\sigma^2}{1 + \omega^2 \tau_\sigma^2} + i \frac{\delta M \, \omega \tau_\sigma}{1 + \omega^2 \tau_\sigma^2} , \qquad (13.25)$$

Box 13.1 Seismic scattering

Some seismic arrivals can be explained in terms of reflections and mode conversions at boundaries within a simple layered model of the crust, but a one-dimensional structure cannot explain a significant amount of energy. These arrivals are produced by *scattering* caused by the wavefield's interaction with small-scale heterogeneities. Heterogeneities in material properties pervade the Earth and span many different length scales (see Chapter 10). Small-scale heterogeneity causes scattering that partitions the high-frequency wavefield into a sequence of arrivals that are often called coda waves. Fig. B13.1.1 shows seismograms produced by the impact of an Apollo Lander on the Moon's surface. These were recorded by a lunar seismometer installed during the Apollo 14 mission. The short-period three-component records ring on for more than 1 h, with waves being scattered from the highly heterogeneous region near the Moon's surface. The coda is spindle shaped, and analysis of the particle motions indicates that the energy is arriving from all directions. These differ from typical Earth recordings, for which the coda is weaker than the direct arrivals. This is because the seismic-wave attenuation on the Moon is much smaller, allowing strongly scattered waves to propagate for some time. The wave interactions with boundary irregularities and with volumetric gradients in rock properties all involve the conventional effects of refraction, conversion, reflection, and diffraction that we describe in this chapter, but the resulting overall wavefield is so complex that individual arrivals cannot be associated with a particular path through the medium given a limited number of surface recordings. Generally, seismologists attempt to characterize the statistical properties of the scattering medium in terms of the spectrum of spatial heterogeneities superimposed on any simple layered structure. Many techniques have been developed to relate the coda to the heterogeneity spectrum.

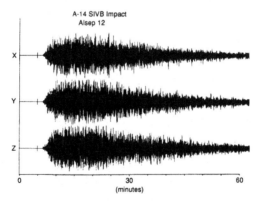

FIGURE B13.1.1 Three-component seismograms recording the impact of an Apollo lander on the Moon. Seismograms ring for more than 1 h (from Dainty et al., 1974).

Scattering can also decrease the amplitude of a seismic phase by shifting energy from the direct arrival back into the coda. This apparent attenuation is called scattering attenuation, and is often characterized by an exponential attenuation quality factor, Q_{sc}. Unlike Q defined for anelastic processes, Q_{sc} is not a measure of energy loss per cycle but, rather, a measure of energy redistribution. Q_{sc} depends very strongly on frequency and is very path dependent, since it depends on the particular heterogeneity spectrum encountered by a wavefield propagating through the Earth. Q_{SC} is usually modeled with stochastic operators, or randomization coefficients. Fig. B13.1.2 shows 3-D finite difference simulation of seismic wave propagation accounting for heterogeneous structure, which includes small-scale velocity inhomogeneity in subsurface structure and irregular surface topography (Takemura et al., 2015).

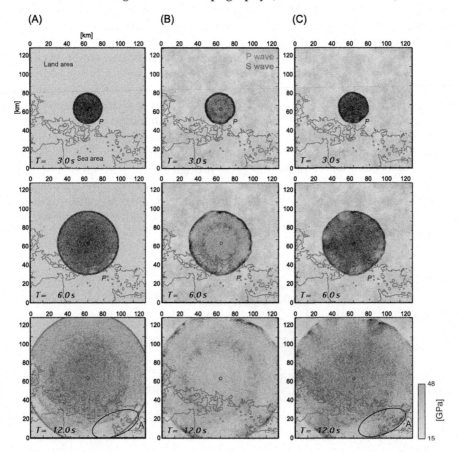

FIGURE B13.1.2 Snapshots of the seismic wavefield at the surface at $t = 3, 8$ and 12 s from the earthquake origin time showing P and S wave propagation derived from (A) the irregular topography model, (B) the velocity inhomogeneity model and (C) the composite model including both (modified from Takemura et al., 2015).

Simple Harmonic Motion   Damped Harmonic Motion   The Standard Linear Solid

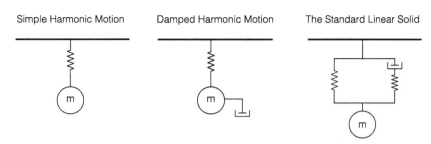

FIGURE 13.4 Phenomenological models of seismic attenuation (left) the simple elastic model (no energy loss); (center) simple damped harmonic motion (steady energy loss); (right) the standard linear solid (rate-varying energy loss).

where $M_u = M_r \tau_\epsilon / \tau_\sigma$ is the **unrelaxed elastic modulus**, the elastic response expected for high-frequency displacement applied over a short time – sort of like the initial deflection of the mass described above, and for brevity we defined $\delta M = M_u - M_r$. This complex elastic modulus has several significant differences from simple elastic moduli. Most important, the *behavior* of a standard linear solid depends on frequency (ω). Waves that travel through such a solid will be *dispersed*, different frequencies in a seismic wave will travel with different speeds.

We typically describe dispersion with an expression for (or observations of) the speed that different harmonic components of the signal travel. For small δM, the frequency-dependent phase velocity is

$$c_p(\omega) = \sqrt{\frac{M_r}{\rho}\left(1 + \frac{1}{2}\frac{\delta M}{M_r}\frac{\omega^2 \tau_\sigma^2}{(1+\omega^2\tau_\sigma^2)}\right)} . \quad (13.26)$$

Note that if $\delta M = 0$, then c_p is independent of frequency and is, of course, the velocity in the elastic case. For small δM we can also write an equation for Q:

$$\frac{1}{Q(\omega)} = \frac{\delta M}{M_r}\frac{\omega\tau_\sigma}{1+\omega^2\tau_\sigma^2} . \quad (13.27)$$

The foregoing expressions for phase velocity and Q can be understood by examining them as a function of $\omega\tau_\sigma$. Fig. 13.5 shows the behavior: attenuation is high when Q^{-1} is large; thus enhanced attenuation occurs over a limited range

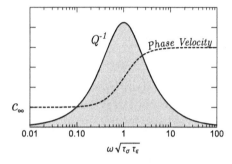

FIGURE 13.5 Response of a standard linear solid with a single relaxation mechanism. Q^{-1} as a function of frequency for a standard linear solid. The peak in Q^{-1} is known as a Debye peak. The dashed line shows phase velocity changes associated with the model. Phase velocity increases with frequency near the scaled frequency $\omega\sqrt{\tau_\epsilon\tau_\sigma}$.

of frequencies. The peak of attenuation is called a **Debye peak**. In general, each relaxation mechanism in the Earth has a distinct Debye peak. These relaxation processes include grain boundary sliding, the formation and movement of crystal lattice defects, and thermal currents. Relaxation times for processes of interest in earthquake seismology range from several thousand seconds to several thousandths of a second. The values τ_e and τ_σ differ by less than 1%, with τ_e the larger of the two quantities.

In the Earth we noted that measurements of seismic-wave attenuation indicate that Q is frequency *independent* over a large range in the seismic frequency band. How is this reconciled with the SLS Debye-Peak model? A great variety of attenuation processes operate within Earth and no single mechanism dominates. The sum or superposition of numerous Debye peaks for the various relaxation processes, each with a different frequency range, produces a broad, flattened *absorption band*. Fig. 13.6 is a plot of the superposition effect constructed using 12 different, coupled attenuation mechanisms. Note that Q^{-1} is basically constant for frequencies of 1.0 Hz (1.0 cycle/s) to 2.8×10^{-4} Hz (1.0 cycle/h).

Although Q is constant across the absorption band, the phase velocity is not and the response is dispersive with the functional form

$$c(\omega) = c_0 \left[1 + \frac{1}{\pi Q_m} \ln\left(\frac{\omega}{\omega_0}\right) \right], \qquad (13.28)$$

where ω_0 is a reference frequency (often chosen to be 1 Hz for body-wave analyses). For the period range shown, the change in phase velocity is on the order of 5% – higher frequencies travel faster. This result is important because it indicates that the velocity structure of the Earth, as estimated from long-period free oscillations and/or surface waves, will differ from that estimated by short-period body waves. In general, dispersion is minor for body waves of interest to earthquake seismology. But it can be important for very high frequencies, and it is *very* important for seismic surface waves, which we discuss in the next chapter.

13.3.2 Estimating Q in the seismic band

Measurements of Q vary laterally by an order of magnitude within the Earth, a much larger variation than that observed for seismic wave velocity ($\sim 10\%$). The mechanisms of intrinsic attenuation (grain-boundary and crystal-defect sliding) are very sensitive to pressure and temperature conditions. Tectonically active regions typically have relatively high heat flow and are more attenuating than "colder" tectonically stable regions. In this manner, Q variations often correlate with travel-time variations. Fast travel-time paths are typically high Q, slow paths typically low Q. This is a manifestation of the thermal activation of the attenuation mechanisms. Thus, mapping Q can reveal a snapshot of the thermal processes operating at depth. In Chapter 10 we discuss the lateral variation of Q and its consequences for tectonic processes.

Estimating Earth's attenuation properties is more difficult than traveltime analysis, but is an important goal of observational seismology. Q can be estimated from surface waves, free oscillations, and body waves. We focus on body-waves here, and discuss other signals in later chapters. A common approach to estimate body-wave Q is to compare the amplitude and frequency content of seismic waves that have traveled similar paths. Choosing observations with similar paths helps reduce unknown source effects. An example of such a comparison is shown in Fig. 13.7. The seismogram shown was produced by a 609 km deep source beneath southern Spain and recorded less than 4° to the northwest in Portugal. The transversely polarized horizontal component of motion (orthogonal to the great-circle path from source to station) is shown to isolate shear waves from P-motions. The focus is on multiple shear-wave core reflections (ScS, ScS_2, ScS_3) and their depth phases ($sScS_n$). The shear-wave reflections from the core and Earth's surface are strong, so the signals are clear for three bounces through the mantle and crust.

The ScS amplitudes are affected by geometrical spreading, reflection at the core and the surface, and attenuation as the waves traverse the mantle and crust. A crude approximation of the geometrical spreading difference between ScS and ScS_2 would suggest an expected decrease in amplitude of about a factor of two. The observed amplitudes show more than a factor of four decrease. Also apparent is a decrease in dominant frequency between the ScS and multiples ScS_2 and ScS_3. If we account for the reflection processes, we can estimate Q_β averaged over the entire mantle path by matching the

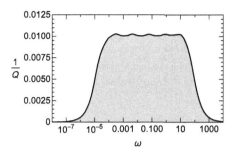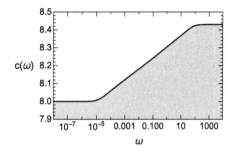

FIGURE 13.6 Superposition of the simple linear modelresponses for 12 attenuation mechanisms. The values of τ_e and τ_s are listed in Liu et al., 1976. The Q value is roughly constant across the seismic band lower than 10 Hz and then increases. The phase velocity is linearly proportional to the logarithm of frequency. An uppermost-mantle appropriate elastic value of 8.0 km/s is used to provide a concrete reference.

amplitude and frequency changes that occur in the multiply reflected signals. Values of mantle Q estimated from the procedure vary with tectonic setting and are typically in the range from 100–250 (a typical, average value $\sim$ 230) for the frequency ranges of 0.01 to 0.06 Hz. Estimates of P and S wave Q can be constructed by comparing other phases such as ScP and ScS. The key is to choose waves with similar source radiation patterns (described in Chapter 17) and source time functions (described in Chapter 18), but differences in propagation that allow isolation or comparison of the attenuation in different regions.

We will revisit attenuation when we discuss surface waves and Earth's Free oscillations in later chapters. In many ways, mapping the attenuation structure of Earth remains a frontier in seismology. Information on variations in seismic attenuation have the power to resolve ambiguities associated with variations in seismic wave speed. Additional work on the physical mechanisms of seismic attenuation is also important, particularly for the crust, where the role of fluids in absorption of energy is an important component of understanding and quantifying long-observed and clear differences in short-period seismic wave propagation between tectonically active and stable regions. We next turn our attention to the important interactions that occur at the boundaries between elastic materials. These

interactions play a central role in the amplitudes and shapes of seismic body waves and establish some background that is helpful when we discuss seismic surface waves in Chapter 14.

13.4 Seismic-wave reflection & transmission across geologic boundaries

When a body wave encounters a boundary or discontinuity at which the seismic velocity changes, the wave will reflect and refract. We now focus on the amplitudes of the waves, which are determined by the partitioning of energy at the boundary. We consider plane wave expressions, which are directly useful in many problems and which are integrated to represent spherical and cylindrical waves in problems where wavefront curvature is significant. Throughout our discussion, we will represent the P, S velocities with α and β respectively, and the density with ρ. We consider shear waves as consisting of two components, an SV that includes motion in the x_1x_3 plane, and an SH that includes motion perpendicular to the x_1x_3 plane. We focus on isotropic materials, the more general case of anisotropy is handled in an analogous manner (Chapman, 2004), but in anisotropic materials, the SH motions do not decouple from the P-SV.

When a P or vertically-polarized SV wave traveling in an $x_1 x_3$ direction impinges on a boundary ($x_1 x_2$ plane), four derivative waves result, as shown in Fig. 13.8 for a P wave: (1) P', the refracted or transmitted P wave (note that P head waves are a subset of P'), (2) SV', the refracted SV (it is possible to have P waves generate a SV head wave if $\beta_2 > \alpha_1$), (3) P, the

---

Box 13.2 The integrated path attenuation factor, t^*

In seismic body-wave studies we commonly account for the effects of attenuation by *convolving* an elastic pulse shape with an attenuation operator parameterized using the quantity t^*. In a region of uniform attenuation, t^* is the wave's travel time divided by the quality factor,

$$t^* = \frac{\text{travel time}}{\text{quality factor}} = \frac{t}{Q}. \tag{B13.2.1}$$

In the Earth, Q is a function of depth (and frequency), with the lowest Q values (highest attenuation) located in the upper mantle and inner core. Since $Q = Q(r)$, t^* is usually written as a path integral value

$$t^* = \int\limits_{\text{path}} \frac{dt}{Q} = \sum_{i=1}^{N} \frac{t_i}{Q_i}, \tag{B13.2.2}$$

where t_i and Q_i are the travel time and quality factor for the ith layer in a layered Earth model. Clearly, t^* equals the total travel time divided by the path-averaged value of Q. Observations suggest that t^* is approximately constant for body waves with periods longer than 1 s in the distance range $30° < \delta < 95°$. In this range, $t^*_\alpha \approx 1.0$ and $t^*_\beta \approx 4.0$. Thus, we can account for the effects of attenuation by replacing t/Q in Eq. (2.38),

$$A = A_0 e^{-\pi f t^*}. \tag{B13.2.3}$$

The value of t^* is roughly four times larger for S waves than for P waves – S waves attenuate much more rapidly with distance.

Fig. B13.2.1 is an illustration of the effects on a propagating seismic pulse traveling through materials corresponding to different values of t^*. The signals on the left correspond to an ideal pulse (a spike), those on the right correspond to a strong earthquake for which the rupture duration is modeled with a triangular pulse five-seconds wide. In each case the wave arrives at $t = 0$. In addition to amplitude attenuation, the signals include phase adjustments to insure the pulse is causal – all energy arrives after the temporal origin. The decrease in amplitude of high-frequency energy is vastly greater than that at low frequency, so the higher frequency pulse on the left shows the attenuation affects more dramatically. The difference is observationally important, but only relative. Attenuation affects the two signals identically in each case, the initial shape of the pulses produces the visually apparent relative differences.

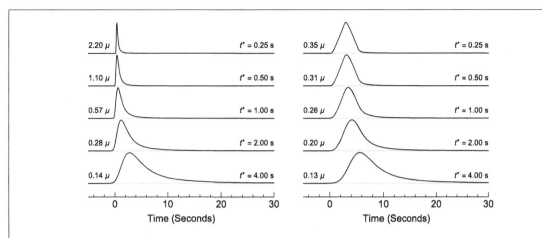

FIGURE B13.2.1 The effect of different t^* on body-waves traveling teleseismic distances. The left panel shows the attenuation filter with a perfectly impulsive source corresponding to an instantaneous earthquake rupture. The maximum amplitude is shown on the left of each seismogram and the amplitudes have been scaled to correspond to a P-wave from an earthquake with a magnitude of 6.3 observed at a distance of 60°. The right panel shows the same attenuation filters applied to a triangular pulse with a duration of 5 seconds, which better matches the duration of an earthquake of this magnitude.

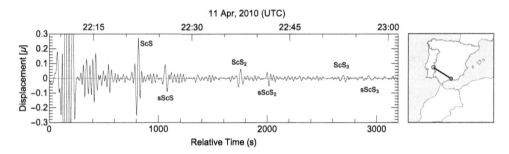

FIGURE 13.7 Transverse component displacements for the 11 April, 2014 earthquake deep beneath southern Spain observed at seismic station PESTR, Estremoz, Portugal, of the Portuguese Seismic Network. The record is focused on the signal behind the direct S arrival and shows the train of reflections from the core-mantle boundary.

reflected P wave, and (4) SV, the reflected SV wave.

Represent the ray parameter or horizontal slowness of the incident wave as p. Since all the waves must move along the boundary with the same apparent velocity, Snell's law requires

$$p = \frac{\sin i_1}{\alpha_1} = \frac{\sin j_1}{\beta_1} = \frac{\sin j_2}{\beta_2} = \frac{\sin i_2}{\alpha_2}. \quad (13.29)$$

When a horizontally polarized SH wave encounters a discontinuity surface parallel to the SH motion, only two waves are generated: (1) SH, reflected, and (2) SH', refracted (SH' can be a head wave). Then we have,

$$p = \frac{\sin j_1}{\beta_1} = \frac{\sin j_2}{\beta_2}. \quad (13.30)$$

The existence of multiple waves derived from a single incident wave implies that the energy of the incident wave must be partitioned. Although Snell's law and ray theory can predict the geometry of the wave interaction, we must return to a wavefield representation to determine the amplitude partitioning.

In Fig. 13.8, the interface separates two materials with different elastic properties. The equations of motion for homogeneous media are valid in each halfspace. The physics requires that *stresses* and *displacements* be "communicated" across the interface. A wave impinging on the boundary from halfspace 1 will induce a stress imbalance in halfspace 2, producing transmitted waves. We must consider several types of interface. At an interface between two solids, all the components of traction and all the components of displacement must be continuous across the interface. We call this a ***welded interface.*** At an interface between a solid and a perfect fluid, the fluid may slip along the interface, since it has no rigidity. Thus, the tangential displacements need not be continuous. The tangential tractions must vanish but the interface-normal traction and interface-normal displacements are continuous. At a ***free surface***, all the tractions must be zero, but no explicit restriction is placed on the displacements. Note that these conditions are on *tractions*, not stresses. For example, if the x_1x_2 plane is a free surface, then $\sigma_{31} = \sigma_{32} = \sigma_{33} = 0$, but the other components

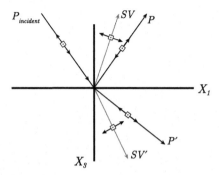

FIGURE 13.9 *P*- and *SV*-wave particle motion directions for the incident, reflected, and refracted waves. *SV* waves are required to match the boundary conditions of continuous stress and displacements. Project the *P*-motions back to the interface and observe that the P-motions alone cannot sum to zero.

of the stress tensor are not constrained. These displacement- and stress-continuity conditions are the basis for quantifying the partitioning of energy when a seismic wave interacts with a material boundary. In acoustics, only compressional waves need be considered, in electromagnetism, only transverse waves. In seismic-wave propagation, compressional, *P*, and vertically polarized shear, *SV*, waves couple at material boundaries and must be considered together. In Fig. 13.8, this is indicated by the generation of reflected and transmitted *SV* waves by an incident *P* wave.

When the *P* wave reaches the boundary, no *SH* wave is produced because the particle motion of the incident *P* wave is confined to the x_1x_3 plane, and thus neither reflection nor transmission will produce any motion in the x_2 plane. However, refraction of the *P* wave results in displacements that are not parallel to those on the incident side of the interface (see Fig. 13.9). Thus, the *P*-wave displacements *alone* do not combine to produce continuous displacements or tractions at the welded interface. Continuous displacements are provided by the addition of SV-type motion, which is also confined to the x_1x_3 plane. Hence, an incident *P* wave must introduce reflected and transmitted *SV* waves (and

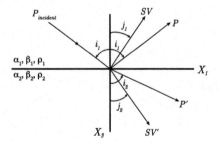

FIGURE 13.8 Ray for a *P* wave incident on a solid-solid boundary and the rays for waves generated at the interface.

vice-versa). In a fluid, where no S waves exist, the P waves reflect and transmit purely as P waves because only *normal* displacements and *normal* tractions need to remain continuous at the boundary.

13.4.1 P-waves at a fluid-fluid boundary

We can quantify the energy partitioning at the boundary using the plane-wave potentials introduced in section 11.4. The P-wave and SV-wave potentials are represented by

$$
\begin{aligned}
\varphi^-_{\text{(layer 1)}} &= \varphi_{\text{incident ray}} + \varphi_{\text{reflected ray}} \\
\varphi^+_{\text{(layer 2)}} &= \varphi_{\text{refracted}} \\
\psi^-_{\text{(layer 1)}} &= \psi_{\text{reflected}} \\
\psi^+_{\text{(layer 2)}} &= \psi_{\text{refracted}} ,
\end{aligned}
\tag{13.31}
$$

where φ and ψ are the P and SV potentials, respectively. Using $k_1 \omega^{-1} = p$ and $k_3 \omega^{-1} = \eta$, the plane-wave potentials can be expressed as

$$
\begin{aligned}
&\varphi_{\text{incident}}(\omega, p, \eta_{\alpha_1}, \mathbf{x}, t) \\
&= A_1 \exp\left[i\omega\left(p x_1 + \eta_{\alpha_1} x_3 - t\right)\right] .
\end{aligned}
\tag{13.32}
$$

We can write similar expressions for the other potentials in (13.31):

$$
\begin{aligned}
\varphi_{\text{reflected}} &= A_2 \exp\left[i\omega\left(px_1 - \eta_{\alpha_1} x_3 - t\right)\right] \\
\varphi_{\text{refracted}} &= A_3 \exp\left[i\omega\left(px_1 + \eta_{\alpha_2} x_3 - t\right)\right] \\
\psi_{\text{reflected}} &= B_2 \exp\left[i\omega\left(px_1 - \eta_{\beta_1} x_3 - t\right)\right] \\
\psi_{\text{refracted}} &= B_3 \exp\left[i\omega\left(px_1 + \eta_{\beta_2} x_3 - t\right)\right] .
\end{aligned}
\tag{13.33}
$$

The vertical slownesses subscripts correspond to layer velocities. Note that the sign of the x_3 term changes, depending on the direction (down or up) that the ray is traveling. The ratios of the post interaction amplitudes (A_2, A_3, B_2, B_3) over the incident amplitude (A_1) are called the **reflection** and **transmission** coefficients for potentials. These coefficients quantify the partitioning of

the incident wave's amplitude among the potentials. Recall that tractions and displacements can be calculated from the potentials by computing the spatial derivatives with respect to x_1 and x_3.

Deriving the reflected- and transmitted-wave amplitudes from the boundary conditions of a welded interface requires significant algebraic manipulation. The results are summarized in Table 13.1, *for displacements* (not potentials). We illustrate the process using the less laborious fluid-fluid boundary. A P wave incident on a fluid-fluid interface generates no S waves, so we need only consider reflected and refracted P waves. From (13.32) and (13.33) we can write expressions for the P-wave potentials in each layer,

$$
\begin{aligned}
\varphi_1 &= A_1 \exp\left[i\omega\left(p x_1 + \eta_1 x_3 - t\right)\right] \\
&\quad + A_2 \exp\left[i\omega\left(p x_1 - \eta_1 x_3 - t\right)\right]
\end{aligned}
\tag{13.34}
$$
$$
\varphi_2 = A_3 \exp\left[i\omega\left(p x_1 + \eta_2 x_3 - t\right)\right] .
$$

Within each layer, the P displacements are related to the potentials by Eq. (11.116):

$$
\mathbf{u}(x, t) = \frac{\partial \varphi}{\partial x_1} \hat{\mathbf{x}}_1 + 0 \cdot \hat{\mathbf{x}}_2 + \frac{\partial \varphi}{\partial x_3} \hat{\mathbf{x}}_3 .
\tag{13.35}
$$

The appropriate boundary conditions for the fluid-fluid boundary are continuity of normal stress and displacement (σ_{33} and u_3). The displacement condition is

$$
\left.\frac{\partial \varphi_1}{\partial x_3}\right|_{x_3=0} = \left.\frac{\partial \varphi_2}{\partial x_3}\right|_{x_3=0} .
\tag{13.36}
$$

Substituting (13.34) into this equation yields

$$
\begin{aligned}
&i\omega\eta_1 \left(A_1 - A_2\right) e^{i\omega(px_1 - t)} \\
&\quad = i\omega\eta_2 A_3 e^{i\omega(px_1 - t)}
\end{aligned}
\tag{13.37}
$$

or

$$
\eta_1 \left(A_1 - A_2\right) = \eta_2 A_3 .
\tag{13.38}
$$

The condition of stress continuity is given by

$$
\sigma^-_{33} = \lambda \nabla \mathbf{u} + 2\mu\varepsilon_{33} = \sigma^+_{33} ,
\tag{13.39}
$$

but $\mu = 0$ for the fluids. Thus,

$$\lambda_1 \nabla^2 \varphi_1 = \lambda_2 \nabla^2 \varphi_2. \tag{13.40}$$

We simplify (13.40) by using the fact that φ satisfies the wave equation:

$$\nabla^2 \varphi = \frac{1}{\alpha^2} \frac{\partial^2 \varphi}{\partial t^2} = \frac{-\omega^2}{\alpha^2} \varphi. \tag{13.41}$$

Therefore, for $x_3 = 0$,

$$\frac{\lambda_1}{\alpha_1^2}(A_1 + A_2) = \frac{\lambda_2}{\alpha_2^2} A_3. \tag{13.42}$$

Now, for a fluid, we have $\lambda_1 = \rho_1 \alpha_1^2$ and $\lambda_2 = \rho_2 \alpha_2^2$, so we can rewrite (13.38) and (13.42) as a system of equations

$$A_1 - A_2 = \frac{\eta_2}{\eta_1} A_3$$

$$A_1 + A_2 = \frac{\rho_2}{\rho_1} A_3, \tag{13.43}$$

that we can solve for ratios of the amplitudes,

$$T'_{PP} = \frac{A_3}{A_1} = \frac{2\rho_1 \eta_1}{\rho_1 \eta_2 + \rho_2 \eta_1}$$

$$R'_{PP} = \frac{A_2}{A_1} = \frac{\rho_2 \eta_1 - \rho_1 \eta_2}{\rho_1 \eta_2 + \rho_2 \eta_1}. \tag{13.44}$$

T'_{PP} and R'_{PP} are referred to as the potential transmission and reflection coefficients, respectively (we use the prime superscript to indicate a wave-potential coefficient). Note that T'_{PP} and R'_{PP} depend on the vertical slowness, $\eta = \cos i/\alpha$. Thus the partitioning of potential amplitudes depends on the angle at which the incident ray strikes the boundary. Consider the case of vertical incidence ($p = 0$, $\eta_1 = 1/\alpha_1$, $\eta_2 = 1/\alpha_2$):

$$R'_{PP_{i=0}} = \frac{\rho_2/\alpha_1 - \rho_1/\alpha_2}{\rho_1/\alpha_2 + \rho_2/\alpha_1} = \frac{\rho_2 \alpha_2 - \rho_1 \alpha_1}{\rho_1 \alpha_1 + \rho_2 \alpha_2} \tag{13.45}$$

$$T'_{PP_{i=0}} = \frac{2\rho_1/\alpha_1}{\rho_1/\alpha_2 + \rho_2/\alpha_1} = \frac{2\rho_1 \alpha_2}{\rho_1 \alpha_1 + \rho_2 \alpha_2}. \tag{13.46}$$

At this point, we have developed reflection and transmission coefficients for potentials. We can obtain displacement reflection and transmission coefficients by recalling $u_3 = \partial \phi / \partial x_3$, such that

$$R_{PP_{i=0}} = \frac{u_{\text{reflected}}}{u_{\text{incident}}} = \frac{-i\omega\eta_1}{i\omega\eta_1} \frac{A_2}{A_1} = \frac{\rho_1 \alpha_1 - \rho_2 \alpha_2}{\rho_1 \alpha_1 + \rho_2 \alpha_2}$$

$$= -R'_{PP_{i=0}} \tag{13.47}$$

$$T_{PP_{i=0}} = \frac{u_{\text{transmitted}}}{u_{\text{incident}}} = \frac{i\omega\eta_2}{i\omega\eta_1} \frac{A_3}{A_1}$$

$$= \frac{1/\alpha_2}{1/\alpha_1} \frac{2\rho_1 \alpha_2}{\rho_1 \alpha_1 + \rho_2 \alpha_2}$$

$$= \frac{\alpha_1}{\alpha_2} T'_{PP_{i=0}}. \tag{13.48}$$

The potential and displacement amplitude ratios are different quantities. We could also define ground velocity or acceleration ratios, or energy ratios. So it is important to keep track of what ratios are used in a particular application. One must also be careful to keep track of the vector displacement with respect to the direction the wave is propagating in defining the sign of the motion. The vector displacement reflection and transmission coefficients, R_{PP} and T_{PP}, have extensive use in geophysics despite their derivation for fluids and vertical incidence because these coefficients also hold for solid-solid interfaces at near-vertical incidence. The energy is partitioned quite simply: $T - R = 1$. The product $\rho\alpha$ is known as **acoustic impedance**, and depending on how the acoustic impedance changes across the boundary, the reflection coefficient can assume values from -1 to $+1$. The range of the transmission coefficient is 0 to 2. A free-surface boundary will have a vertical-incidence reflection coefficient of -1 (the displacement reverses direction with respect to the direction of propagation). The amplitude of transmitted displacement is zero.

Reflection variation with incidence angle / slowness

If we return to the general, potential form of T'_{PP} and R'_{PP} (non-vertical incidence, non-zero horizontal slowness), we can investigate the behavior of the system as the angle of incidence varies. If $\alpha_2 < \alpha_1$ (incidence from a high to low speed material) and $\rho_2\alpha_2 > \rho_1\alpha_1$, then $R'_{PP} > 0$ for normal incidence. But as i increases, R'_{PP} will decrease, reaching zero at an angle of incidence called the *intramission angle*, which occurs if i satisfies

$$\frac{\rho_2}{\rho_1} = \frac{\sqrt{(\alpha_1/\alpha_2)^2 - \sin^2 i}}{\sqrt{1 - \sin^2 i}} . \tag{13.49}$$

Beyond the intramission angle, the reflection coefficient decreases to a value of -1 at grazing incidence ($i = 90°$). If $\alpha_2 < \alpha_1$ and $\rho_2\alpha_2 < \rho_1\alpha_1$, the reflection coefficient is always negative and equals -1 for grazing incidence.

If $\alpha_2 > \alpha_1$ (incidence from a low to high speed material), a head wave is produced at the critical angle, $i_c = \sin^{-1}(\alpha_1/\alpha_2)$. At incident angles greater than the critical angle, no P waves will *propagate* in the lower medium because $p = \sin i/\alpha_1 = 1/c$ (where c is the apparent velocity) becomes greater than $1/\alpha_2$. Thus $\eta_2 = (\alpha_2^{-2} - p^2)^{1/2}$ become *imaginary*. We can write $\eta_2 = i\,\hat{\eta}_2 = \pm i\,(p^2 - \alpha_2^{-2})^{1/2}$, and we choose the positive sign so that the amplitude of the refracted potential (13.34) decreases exponentially away from the boundary. The choice insures that the wave energy is bounded because the head-wave displacement decays exponentially with depth into the half-space. The potential in the lower (higher-speed) medium takes the form

$$\varphi_2(\omega, p, \mathbf{x}, t) = T'_{PP} \exp\left[i\omega\,(p\,x_1 - t)\right]$$
$$\times \exp\left[-\omega\,\hat{\eta}_2\,x_3\right] . \tag{13.50}$$

The transmission coefficient, T'_{PP}, is now complex, and to keep the ray parameter constant, the angle i_2 also must become complex. Although

head wave travel times are relatively simple to compute with ray theory, the head-wave amplitudes are not.

We can rewrite the post-critical reflection coefficient in (13.44) as

$$R'_{PP} = \frac{\rho_2\,\eta_1 - \rho_1\,i\,\hat{\eta}_2}{\rho_2\,\eta_1 + \rho_1\,i\,\hat{\eta}_2} , \tag{13.51}$$

which is a complex number divided by its conjugate. Thus, the magnitude of R'_{PP} is 1, but the phase shift, θ, is nonzero. We have,

$$R'_{PP} = 1 \cdot e^{i\theta} , \tag{13.52}$$

where,

$$\theta = 2\tan^{-1}\left(\frac{\rho_1\hat{\eta}_2}{\rho_2\,\eta_1}\right) . \tag{13.53}$$

Since the modulus of the reflection coefficient is 1, the post-critical reflection is referred to as a *total reflection*, but post-critical reflections behave differently than precritical reflections. Fig. 13.10 shows a synthetic seismogram profile generated for increasing angles of incidence (increasing distance). Beyond 60 km, the reflected arrival has an angle of incidence that is greater than i_c. This is the distance at which a head wave first exists and begins to separate from the reflected arrival. At 450 km the reflected wave strikes the boundary at near-grazing incidence; the reflected waveform is very similar to that at 50 km, except the *polarity* is completely reversed. That the reflected wave shape changes as the source-receiver distance increases is clear from Fig. 13.10. Although the phase shift in Eq. (13.53) explains this shape change, it is instructive to return to the equation for the reflection potential. Noting that $A_2 = A_1 R'_{PP} = A_1 e^{i\theta}$, we can write the potential for the post-critical reflected arrival as

$$\varphi = A_1 \exp[i\theta]\exp\left[i\omega\,(px_1 - \eta_1 x_3 - t)\right] . \tag{13.54}$$

Now consider the behavior of θ:

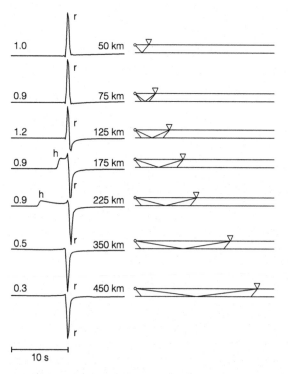

FIGURE 13.10 The change in reflected pulse shape (phase) as the incidence angle exceeds the critical angle. For the model shown, the head wave first appears at ~ 60 km. The absolute maximum amplitude is shown along the left, the source-to-station distance is shown on the right. The direct wave has been removed from the seismograms, but the head-wave (h) interferes with the reflection (r) at 175 and 225 km distances. The seismograms at 50 and 450 km show that the reflected wave polarity has reversed. Cartoons along the right show the reflected wave and head wave ray paths.

$$\theta = 0 \qquad \text{if } i = i_c$$
$$\theta < 0 \qquad \text{for } i > i_c$$
$$\theta = -\pi \quad i = \pi/2.$$

Now rewrite (13.54) as

$$\varphi = A_1 \exp\left[i\omega\left(px_1 - \eta_1 x_3 - t + \theta/\omega\right)\right]. \quad (13.55)$$

The quantity θ/ω is a **phase shift**. If we use a constant-phase argument to track the propagation of a particular wave, we have

$$p\,x_1 - \eta_1 x_3 - t + \theta/\omega = \text{constant}. \qquad (13.56)$$

We can interpret the term $-t + \theta/\omega = -(t - \theta/\omega) = -\hat{t}(\omega)$ is an *apparent* time that *depends on frequency*. The wavefront position is frequency dependent, lower frequencies (smaller ω) arrive *earlier* than higher frequencies (recall $\theta < 0$). As $\omega \to \infty, \hat{t} \to t$. This implies that the arrival of a post-critical reflection is "spread out" in time, each harmonic has a slightly different arrival time. This behavior is called **dispersion**, a phenomenon we will become very familiar with in the next chapter. A consequence of the dispersion is that the strongest reflection coefficient occurs exactly at i_c ($R'_{PP} = 1$, $\theta = 0$, and wavefronts do not disperse).

13.4.2 SH-waves at a solid-solid boundary

Reflection and transmission at a welded interface are much more complicated than at a fluid-fluid interface. However, for isotropic layers, the SH system remains fairly simple because interaction with the boundary does not produce any P or SV waves. We will briefly consider this case but rather than working with potentials as we did for the fluid-fluid boundary, we transition to displacement amplitude coefficients. The displacement coefficients represent the amplitudes of steady-state reflected and transmitted displacements relative to the amplitude of the incident displacements.

As with the fluid-fluid case, there are two boundary conditions: (1) continuity of tangential displacement ($v_2^+ = v_2^-$) and (2) continuity of shear stress ($\sigma_{23}^+ = \sigma_{23}^-$). Using these conditions yields the expressions for the SH-displacement reflection and transmission coefficients,

$$R_{SS} = \frac{\mu_1 \eta_{\beta_1} - \mu_2 \eta_{\beta_2}}{\mu_1 \eta_{\beta_1} + \mu_2 \eta_{\beta_2}}$$

$$T_{SS} = \frac{2\mu_1 \eta_{\beta_1}}{\mu_1 \eta_{\beta_1} + \mu_2 \eta_{\beta_2}}. \qquad (13.57)$$

Fig. 13.11 is a plot of the reflection and transmission coefficients from above and below an

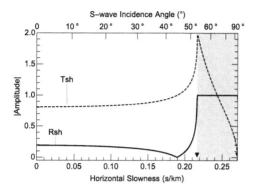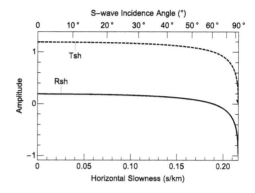

FIGURE 13.11 Reflection and refraction coefficients for an SH-polarized S wave incident from above (left) and below (right) a crust-mantle-boundary-like velocity change. The shear head-wave, S_n is initiated for a slowness of $1/\beta_2$ (indicated by the inverted triangle). For slownesses larger than that (shaded range), the coefficients are complex. Media properties: $\alpha_1 = 6.4\,\text{km/s}$, $\beta_1 = 3.7\,\text{km/s}$, $\rho_1 = 2800\,\text{kg/m}^3$ and $\alpha_2 = 8.0\,\text{km/s}$, $\beta_2 = 4.6\,\text{km/s}$, $\rho_1 = 3300\,\text{kg/m}^3$.

interface similar to a sharp crust-mantle boundary. When the incident wave originates in the crust, a head wave occurs for a horizontal slowness equal to $1/\beta_2 \sim 0.217\,\text{s/km}$. At that point the vertical slowness of the transmitted wave becomes imaginary and the coefficients become complex-valued quantities. In the figure, the slowness range of complex coefficients is shaded in gray. Complex coefficients correspond to phase shifts that distort the waveforms.

For the shear-wave model, the reflection coefficient, R_{SS}, has a zero near a slowness of roughly $0.19\,\text{s/km}$ (Fig. 13.11). For the corresponding angle of incidence, $\mu_1 \eta_{\beta_1} = \mu_2 \eta_{\beta_2}$ and no energy is reflected. Since the transmitted wave displacements are continuous with the incident, at that slowness, $T_{SS} = 1$, and transmission is perfect (for incidence from above or below).

The SH-polarized coefficients are nearly identical in form to fluid-fluid Eqs. (13.44). If we consider vertical incidence, then (13.57) reduces to

$$R_{SS} = \frac{\rho_1\beta_1 - \rho_2\beta_2}{\rho_1\beta_1 + \rho_2\beta_2}$$

$$T_{SS} = \frac{2\rho_1\beta_1}{\rho_1\beta_1 + \rho_2\beta_2} .$$

(13.58)

The quantity $\rho\beta$ is called the *shear impedance*. The SH critical-angle behavior for $\beta_2 > \beta_1$ is analogous to that described for the acoustic (fluid) case.

13.4.3 P- & S-waves at a solid-solid boundary

Analysis of the P-SV system includes every potential term in Eq. (13.31). Four derivative waves are created for an incident P or SV wave. The velocities may permit both P and SV head waves for incident P or S waves. The partitioning of a wave into *four* new waves at each boundary in the Earth results in seismograms that are rich in arrivals. We refer to the partitioning of P waves into SV waves or SV waves into P waves as **mode conversions**. Although mode conversions are often small in amplitude, they can provide important information about Earth structure. We discuss several examples in Chapter 10. For the welded interface, σ_{3i}, u_1, and u_3 must be continuous – this provides the boundary conditions. For an incident P wave, the displacement boundary conditions [using (13.33)], require (u_1, continuity)

$$p\,(A_1 + A_2) + \eta_{\beta_1} B_1 = pA_3 - \eta_{\beta_2} B_2 \quad (13.59)$$

TABLE 13.1   Solid-solid displacement reflection and transmission coefficients.

Coefficient	Formula
The P-SV System	
R_{PP}	$\left[(b\,\eta_{\alpha_1} - c\,\eta_{\alpha_2})\,F - (a + d\,\eta_{\alpha_1}\,\eta_{\beta_2})\,H\,p^2\right] \times D^{-1}$
R_{PS}	$-\left[2\,\eta_{\alpha_1}\,(a\,b + c\,d\,\eta_{\alpha_2}\,\eta_{\beta_2})\,p\,(\alpha_1/\beta_1)\right] \times D^{-1}$
T_{PP}	$2\,\rho_1\,\eta_{\alpha_1}\,F\,(\alpha_1/\alpha_2) \times D^{-1}$
T_{PS}	$2\,\rho_1\,\eta_{\alpha_1}\,H\,p\,(\alpha_1/\beta_2) \times D^{-1}$
R_{SS}	$-\left[(b\,\eta_{\beta_1} - c\,\eta_{\beta_2})E - (a + b\,\eta_{\alpha_2}\,\eta_{\beta_1})\,G\,p^2\right] \times D^{-1}$
R_{SP}	$-2\eta_{\beta_1}(ab + cd\eta_{\alpha_2}\eta_{\beta_2})\,p\,(\beta_1/\alpha_1) \times D^{-1}$
T_{SP}	$-2\,\rho_1\,\eta_{\beta_1}\,G\,p\,(\beta_1/\alpha_2) \times D^{-1}$
T_{SS}	$2\,\rho_2\,\eta_{\beta_1}\,E\,(\beta_1/\beta_2) \times D^{-1}$
The SH System	
R_{SS}	$\left(\mu_1\,\eta_{\beta_1} - \mu_2\,\eta_{\beta_2}\right) \times \left(\mu_1\,\eta_{\beta_1} + \mu_2\eta_{\beta_2}\right)^{-1}$
T_{SS}	$\left(2\mu_1\,\eta_{\beta_1}\right) \times \left(\mu_1\eta_{\beta_1} + \mu_2\eta_{\beta_2}\right)^{-1}$
where,	
$a = \rho_2(1 - 2\beta_2^2 p^2) - \rho_1(1 - 2\beta_1^2 p^2)$	$E = b\,\eta_{\alpha_1} + c\,\eta_{\alpha_2}$
$b = \rho_2(1 - 2\beta_2^2 p^2) + 2\rho_1\beta_1^2 p^2$	$F = b\,\eta_{\beta_1} + c\,\eta_{\beta_2}$
$c = \rho_1(1 - 2\beta_1^2 p^2) + 2\rho_2\beta_2^2 p^2$	$G = a - d\,\eta_{\alpha_1}\,\eta_{\beta_2}$
$d = 2(\rho_2\beta_2^2 - \rho_1\beta_1^2)$	$H = a - d\,\eta_{\alpha_2}\,\eta_{\beta_1}$
$D = EF + GHp^2$	

and (u_3 continuity)

$$\eta_{\alpha_1}(A_1 + A_2) + pB_1 = \eta_{\alpha_2}A_3 + pB_2 \,. \quad (13.60)$$

The continuity of stress (σ_{33} continuity) provides

$$\lambda_1 p^2\,(A_1 + A_2) + \lambda_1 p\eta_{\beta_1}\,B_1 + (\lambda_1 + 2\mu_1)$$
$$\times \left[\eta_{\alpha_1}^2\,(A_1 + A_2) - \eta_{\beta_1}pB_1\right]$$
$$= \lambda_2 p^2 A_3 - p\eta_{\beta_2}\lambda_2 B_2 + (\lambda_2 + 2\mu_2)$$
$$\times \left(\eta_{\alpha_2}^2 A_3 + \eta_{\beta_2}pB_2\right) \quad (13.61)$$

and (σ_{31} continuity)

$$\mu_1\left[2p\eta_{\alpha_1}(A_1 - A_2) + p^2 B_1 - \eta_{\beta_1}^2 B_1\right]$$
$$= \mu_2\left[2p\eta_{\alpha_2}A_3 + p^2 B_2 - \eta_{\beta_2}^2 B_2\right]. \quad (13.62)$$

Thus we have four equations with five unknowns. As before, it is sufficient to determine

the *ratios* with respect to A_1, thus obtaining R_{PP}, R_{PS}, T_{PP}, and T_{PS}. The algebra required to obtain these coefficients is extensive. A convenient way to compute and to verify coefficients is using a matrix formalism (Aki and Richards, 2009). Table 13.1 is a list of the standard reflection and transmission coefficients for solid-solid interfaces.

Figs. 13.12 and 13.13 show the reflection and transmission coefficients for P and SV-polarized waves incident from below and above a welded interface. In the first case, the wave is going from a *fast-* to a slow-velocity material, and there are no critical angles in the incident P-wave case. The energy partitioning is dominated by R_{PP} and T_{PP} from 0° to approximately 20°. Over this range, R_{PP} and T_{PP} are nearly identical to what would be obtained from the acoustic impedance mismatch (Eqs. (13.47) and (13.48)). For teleseismic P arrivals from a

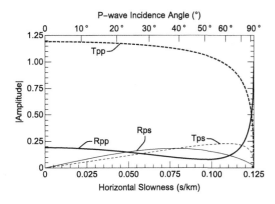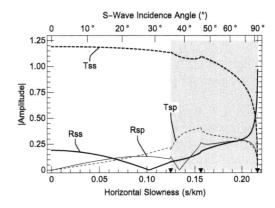

FIGURE 13.12 (left) Absolute reflection and transmission coefficients for a P wave incident on a boundary from a high-velocity region. The model approximates a sharp crust-mantle transition. For near-vertical incidence (angle $= 0°$), the reflected and transmitted P-wave amplitudes approximately equal those predicted by acoustic-impedance mismatches [(Eqs. (13.47)) and (13.48)]. There are no critical angles in this case. (right) Reflection and transmission coefficients for an SV-polarized shear wave incident on a boundary from a high-velocity region. For horizontal slownesses above 0.125 s/km (inverted triangle along axis) the reflected P wave is evanescent and all the coefficients are complex. For horizontal slownesses above 1/6.4 (inverted triangle along axis), the transmitted P-wave is also evanescent. Media properties: $\alpha_1 = 6.4$ km/s, $\beta_1 = 3.7$ km/s, $\rho_1 = 2800$ kg/m^3 and $\alpha_2 = 8.0$ km/s, $\beta_2 = 4.6$ km/s, $\rho_1 = 3300$ kg/m^3.

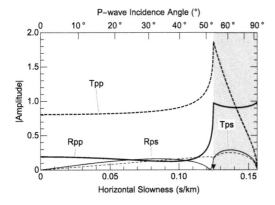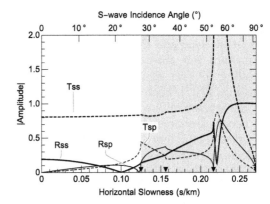

FIGURE 13.13 Reflection and refraction coefficients for a P wave incident on a boundary from a low-velocity region. i_c for the P wave occurs at 38.5°. Since the S velocity in the lower medium is lower than the upper P velocity, the refracted S wave never reaches a critical angle. Media properties: $\alpha_1 = 6.4$ km/s, $\beta_1 = 3.7$ km/s, $\rho_1 = 2800$ kg/m^3 and $\alpha_2 = 8.0$ km/s, $\beta_2 = 4.6$ km/s, $\rho_1 = 3300$ kg/m^3.

shallow source, $0.04 \lesssim p \lesssim 0.08$ s/km, and the behavior of the coefficients is simple. For an incident S wave, both the reflected and transmitted P waves can become evanescent. Their vertical slowness becomes imaginary for incident-wave horizontal slownesses greater than $1/\alpha_2$ and $1/\alpha_1$ respectively. All the coefficients are

complex and the evanescent wave amplitude(s) decrease exponentially away from the boundary. In the example, the effects begin for an incidence wave slowness of about 0.125 s/km, or an incidence angle of about 35°. This slowness is within the range of teleseismic S from shallow sources, $0.08 \lesssim p \lesssim 0.14$ s/km, and the effects on propagation are substantial. SV (radial component) waveforms, while rich in information because of these effects, are complex and difficult to unravel. Partly for this reason, P and SH-polarized waves are used for teleseismic earthquake modeling.

When a P wave is incident from the low-velocity medium, the critical slowness is $1/\alpha_2$ and the critical angle is $\sin^{-1}\alpha_1/\alpha_2$. The P coefficients are complex beyond this angle because the transmitted P vertical slowness is imaginary. The imaginary slowness produces an exponential decrease in the transmitted-wave amplitude below the interface (no wave is transmitted, it is trapped at the interface). As the angle of incidence approaches critical angle, the coefficients vary rapidly. In particular, T_{PP} gets very large before the wave become evanescent (rapidly decaying with depth) in the halfspace.

When an SV-wave is incident from the low-velocity medium (Fig. 13.13), the coefficients become complex when the incident wave slowness exceeds $1/\alpha_2$ and the transmitted P wave is evanescent. Two more critical values exist corresponding to $1/\alpha_1$ and $1/\beta_2$ (inverted triangles in Fig. 13.12). Each time the slowness exceeds the inverse of one of the speeds, another wave becomes evanescent as its vertical slowness becomes imaginary and the wave's amplitude decreases exponentially with distance from the boundary. Each change in propagation style causes a phase shift in all the waves, as is evident in the abrupt changes in the slope of the coefficient amplitude curves in Fig. 13.12. The behavior is richly complex and interesting.

Head-wave observations in Earth are often more commonly observations of waves turning in positive sub-interface velocity gradients (speed mostly increases with depth). P_g and S_g are turning waves within the crust. Their amplitudes depend on the nature of the gradient. The magnitude of the upper mantle velocity gradient beneath crust-mantle boundary can have a strong effect on P_n and S_n amplitudes. A steep positive gradient can strongly enhance these ar-

rivals, but a small decrease in speed with depth beneath even a sharp crust-mantle transition can weaken these signals substantially.

13.4.4 P- & S-waves at a solid-fluid boundary

The same approach used to examine the solid-solid boundary can be employed to examine the interactions and reflection and transmission coefficients from solid-fluid and fluid-solid boundaries, which approximates the ocean ocean-floor and the mantle-core boundaries within Earth. In the fluid, the continuity of stress and displacement applies only to the normal stresses and displacements. The problem is slightly simpler because there are no S waves in the fluid. The displacement reflection and transmission coefficients are summarized in Table 13.2. The incident medium is always medium 1.

For a P-wave vertically incident from the solid (medium 1), we have

$$R_{PP}^{\perp} = 1 - \frac{2\alpha_2\rho_1}{\alpha_2\rho_1 + \alpha_1\rho_2}$$

$$T_{PP}^{\perp} = \frac{2\alpha_1^2\rho_1}{\alpha_2^2\rho_1 + \alpha_1\alpha_2\rho_2} \qquad (13.63)$$

$$R_{PS}^{\perp} = 0 .$$

For vertically incident SV-wave from the solid (medium 1), we have

$$R_{SS}^{\perp} = -1$$

$$R_{SP}^{\perp} = 0 \qquad (13.64)$$

$$T_{SP}^{\perp} = 0 .$$

For a P-wave vertically incident from the fluid (medium 1), we have

$$R_{PP}^{\perp} = \frac{\alpha_2\left(\alpha_2\rho_2 - \alpha_1\rho_1\right)}{\alpha_1\left(\alpha_2\rho_1 + \alpha_1\rho_2\right)}$$

$$T_{PP}^{\perp} = \frac{\left(\alpha_1^2 + \alpha_2^2\right)\rho_1}{\alpha_2^2\rho_1 + \alpha_1\alpha_2\rho_2} \qquad (13.65)$$

TABLE 13.2 Fluid-solid displacement reflection and transmission coefficients.

Coefficient	Formula
Incident P-SV From Solid (Medium 1)	
R_{PP}	$\left[\eta_{\alpha_1} \left(4\beta_1^4 p^2 \rho_1 \eta_{\alpha_2} \eta_{\beta_1} + \rho_2 \right) - \rho_1 \eta_{\alpha_2} \left(1 - 2\beta_1^2 p^2 \right)^2 \right] \times D_F^{-1}$
R_{PS}	$4\alpha_1 \beta_1 p \rho_1 \eta_{\alpha_1} \eta_{\alpha_2} \left(2\beta_1^2 p^2 - 1 \right) \times D_F^{-1}$
T_{PP}	$2(\alpha_1/\alpha_2)\rho_1 \eta_{\alpha_1} \left(1 - 2\beta_1^2 p^2 \right) \times D_F^{-1}$
R_{SS}	$\rho_1 \eta_{\alpha_2} \left(1 - 2\beta_1^2 p^2 \right)^2 + \eta_{\alpha_1} \left(\rho_2 - 4\beta_1^4 p^2 \rho_1 \eta_{\alpha_2} \eta_{\beta_1} \right) \times D_F^{-1}$
R_{SP}	$4\alpha_1^{-1} \beta_1^3 p \rho_1 \eta_{\alpha_2} \eta_{\beta_1} \left(1 - 2\beta_1^2 p^2 \right) \times D_F^{-1}$
T_{SP}	$4\alpha_2^{-1} \beta_1^3 p \rho_1 \eta_{\alpha_1} \eta_{\beta_1} \times D_F^{-1}$

where,

$$D_F = \rho_2 \eta_{\alpha_1} + \rho_1 \eta_{\alpha_2} \left[4\beta_1^4 p^2 \eta_{\alpha_1} \eta_{\beta_1} + \left(1 - 2\beta_1^2 p^2 \right)^2 \right]$$

Incident SH From Solid (Medium 1)	
R_{SS}	1
Incident P From Fluid (Medium 1)	
R_{PP}	$\alpha_2 \eta_{\alpha_2} \rho_2 \left(4\beta_2^4 p^2 \eta_{\alpha_2} \eta_{\beta_2} + \left(1 - 2\beta_2^2 p^2 \right)^2 \right) - \alpha_1 \rho_1 \eta_{\alpha_2} \times D_{PF}^{-1}$
T_{PP}	$-(\alpha_1/\alpha_2)\rho_1 \left(\alpha_1 \eta_{\alpha_1} + \alpha_2 \eta_{\alpha_2} \right) \left(2\beta_2^2 p^2 - 1 \right) \times D_{PF}^{-1}$
T_{PS}	$-2\alpha_1 \beta_2 p \rho_1 \eta_{\alpha_2} \left(\alpha_1 \eta_{\alpha_1} + \alpha_2 \eta_{\alpha_2} \right) \times D_{PF}^{-1}$

where,

$$D_{PF} = \alpha_1 \rho_1 \eta_{\alpha_2} + \alpha_1 \rho_2 \eta_{\alpha_1} \left(4\beta_2^4 p^2 \eta_{\alpha_2} \eta_{\beta_2} + \left(1 - 2\beta_2^2 p^2 \right)^2 \right)$$

$T_{PS}^{\perp} = 0$.

For details on the derivation of the fluid-solid interactions, see Ben-Menahem and Singh (1981). In fact, we recommend that the matrix expressions in that reference be used to numerically evaluate coefficients, since the matrix expressions are easier to check. Seismology has a long history of typographic errors in reflection and transmission coefficient expressions.

Example solid-fluid reflection and transmission coefficients for waves incident on the boundary between Earth's surface and the atmosphere are shown in Fig. 13.14. At times, high-frequency waves from very shallow earthquakes transmit into the atmosphere and are audible as low-frequency sounding booms. The near-surface is modeled with $\alpha = 5.8$ km/s, $\beta = \alpha/\sqrt{3}$, and $\rho = 2500$ kg/m^3. The atmosphere is modeled using a fluid with $\alpha = 0.3$ km/s and $\rho = 1.2$ kg/m^3. Vertically incident P waves reflect with a coefficient just under one, and transmit a large displacement that carries little energy because of the elastic weakness of the atmosphere. As the angle increases, the conversion to SV motion increases as R_{PS} grows at the expense of R_{PP} and T_{PP}. S waves striking the fluid produce no transmission for vertical incidence but as the incidence angle increases, a transmitted P-wave is produced. For incident SV waves with horizontal slowness greater than $1/\alpha_1$, the reflected

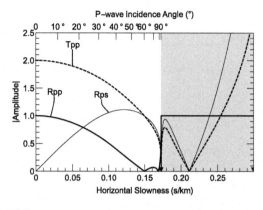
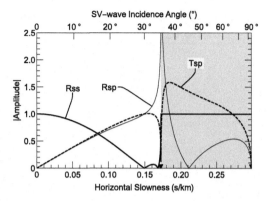

FIGURE 13.14 Reflection and transmission coefficients for a solid-fluid case. The solid has properties of a typical upper crust basement and the fluid has the properties of Earth's lowermost atmosphere.

P wave travels along the surface and the coefficients become complex valued – again resulting in a richly complex behavior corresponding to changes in pulse shapes. Seismologists usually do not model Earth's surface as a solid-liquid boundary, if we ignore the atmosphere, the surface is a free-boundary and the surface stresses vanish. This is a good approximation that simplifies the analysis, as we describe next.

13.4.5 P-S-wave reflection at a free surface

The last boundary that we must consider is the free surface. We often approximate Earth's surface as a free, meaning stress free, boundary. The atmosphere is more accurately modeled as a fluid with a low P-wave speed, $\alpha \sim 0.3\,km/s$. The free-surface boundary condition is examined in more detail in the next chapter, where we discuss seismic surface waves. At a free surface the tractions vanish (i.e. $\sigma_{1j} = 0$) and the displacements are unconstrained. The reflection coefficients for incident P and S waves are summarized in Table 13.3. Fig. 13.15 is a chart of the displacement plane-wave coefficients for P and S waves incident from a halfspace with a P-wave velocity of $\alpha = 5.8\,km/s$ and S-wave velocity of $\beta = \alpha/\sqrt{3}$. The magnitude of the ver-

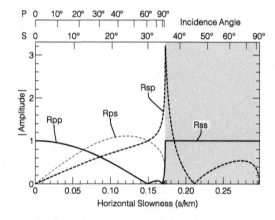

FIGURE 13.15 Free-surface displacement reflection coefficients for a halfspace with $\alpha = 5.8\,km/s$ and $\beta = \alpha/\sqrt{3}$. Expressions for the coefficients are listed in Table 13.3. P and S wave incidence angles corresponding to the slownesses are shown above. Note that $R_{PP} = R_{SS}$ and the coefficients are complex quantities for slownesses above $1/\alpha$ and that region is highlighted with gray. Compare with the coefficients in Fig. 13.14.

tical incidence R_{PP} and R_{SS} are one, and the mode-converted wave coefficients are zero. For an incident SV wave, the reflected P wave can become evanescent and the coefficients become complex. As the critical incidence is approached, much like the head-wave, the reflected P-wave displacements increase rapidly as the geometric focusing that occurs as the ray approaches

TABLE 13.3 Free-surface displacement reflection coefficients.

Coefficient	Formula
The P-SV System	
R_{PP}	$\left[-(\beta^{-2}-2p^2)^2+4p^2\,\eta_\alpha\,\eta_\beta\right]\times A^{-1}$
R_{PS}	$4(\alpha/\beta)\,p\,\eta_\alpha\,(\beta^{-2}-2p^2)\times A^{-1}$
R_{SP}	$4(\beta/\alpha)\,p\,\eta_\beta\,(\beta^{-2}-2p^2)\times A^{-1}$
R_{SS}	$\left[-(\beta^{-2}-2p^2)^2+4p^2\eta_\alpha\eta_\beta\right]\times A^{-1}$
where,	
	$A=[\beta^{-2}-2p^2]^2+4p^2\eta_\alpha\eta_\beta$
The SH System	
R_{SS}	1

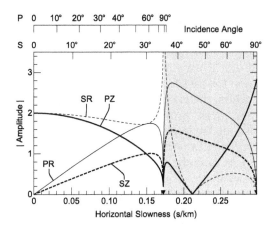

FIGURE 13.16 Free-surface displacement receiver functions for halfspace with $\alpha = 5.8\,\text{km/s}$ and $\beta = \alpha/\sqrt{3}$. P and S wave incidence angles corresponding to the slownesses are shown above. The quantity PZ corresponds to a P-wave on vertical (x_3); PR corresponds to a P-wave on radial (x_1); SZ corresponds to a SV-wave on vertical (x_3); SR corresponds to a SV-wave on radial (x_1). The horizontally polarize component of an S-wave, SH, has a receiver function is 2 for all angles. Compare with the coefficients in Fig. 13.15.

horizontal is substantial. The critically reflected P-wave, Sp, travels along the surface and can be quite large on the radial component of motion.

The free-surface receiver functions

The free surface is a special boundary in seismology because it is where most of our observations are collected. That our observations are collected *on a boundary* is an important point. Incident, reflected, and transmitted waves are coupled at boundaries. When a seismometer measures the arrival of a P or S wave, it registers the motions produced by all the waves that exist at the boundary – thus what we typically describe as the P-wave on a surface seismogram is the motion of the incident and reflected P waves and the reflected SV wave. We must account for all these waves if we are to quantitatively model the seismogram. We call the correction a free-surface receiver function (FSRF). To compute the FSRF, we must take the limit of the amplitudes of all the waves involved as the depth approaches the surface.

Assume that the surface is perpendicular to the x_3 direction. For an incident P-wave, the displacement FSRF for the vertical (x_3) and radial (x_1) directions of motion are

$$F_{P_{x_3}}(p)=\left[(2\alpha\eta_\alpha\beta^{-2})\left(\eta_\beta^2-p^2\right)\right]/R(p)$$

$$F_{P_{x_1}}(p)=\left[(4\alpha p\eta_\beta\eta_\alpha\beta^{-2})\left(\eta_\beta^2-p^2\right)\right]/R(p),$$
$$(13.66)$$

where

$$R(p)=\left(\beta^{-2}-p^2\right)^2+4p^2\eta_\alpha\eta_\beta.\qquad(13.67)$$

For an incident vertically polarized S-wave, the FSRF for the vertical and radial directions of motion are

$$F_{SV_{x_3}}(p)=\left[-4p\eta_\alpha\eta_\beta\beta^{-1}\right]/R(p)$$

$$F_{SV_{x_1}}(p)=\left[(2\eta_\beta\beta^{-1})\left(\eta_\beta^2-p^2\right)\right]/R(p).$$
$$(13.68)$$

For an incident horizontally polarized S-wave, the FSRF is equal to two. Example FSRF are shown in Fig. 13.16. The FSRFs provide the amplitudes of the surface displacement for an incident wave of unit amplitude. Since the incident S-wave can create an evanescent P wave (called the SP), the functions assume complex values for shear-waves with horizontal slowness equal to or greater than the inverse of the speed of a P wave in the shallow crust. This is common for teleseismic S waves in the mantle-triplication distance range. The denominator function, $R(p)$, is an important quantity in the analysis of Rayleigh waves and is often called the Rayleigh Denominator. We discuss the details in the next chapter.

The examples shown above only scratch the surface of the deep, beautiful, and complex behavior predicted when seismic waves interact with boundaries between elastic materials. Consider also that we have examined only the isotropic cases. Anisotropic materials involve a complete coupling of the P-S system – the expressions are more involved, but solvable (e.g. Chapman, 2004). Few, if any boundaries in Earth are perfectly sharp – but many are sharp on the scale of seismic wavelengths, and our expressions provide a first-order basis for interpretation of observations. More gradational transitions can be modeled by stacking steadily changing layers one upon another. The result can be viewed as a frequency-dependent reflection and transmission response that includes reverberations between the stacked layers (Kennett, 2009). Still, the expressions for reflection and transmission coefficients at an interface are central to many more sophisticated solutions of the equations of motion, for which the planes waves are a basis.

13.5 Body-wave energy flux factors

The steady-state reflection and transmission amplitude ratios insure that the boundary conditions at elastic boundaries are satisfied. To compute the amplitude of the waves transmitting through, or reflecting off boundaries requires that we also consider the energy flux at the boundary. No energy is trapped at the boundary – so the energy of the reflected and

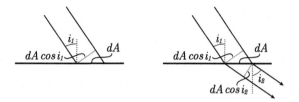

FIGURE 13.17 (left) The ratio of the incoming ray bundle to its projection onto the boundary. (right) Change in area of a transmitted wave ray tube. To conserve energy, we must account for this change in area and multiply the displacement amplitude ratios by the ratio of the area change to preserve energy.

transmitted waves must balance that of the incidence wave. If the displacement of a P-wave is $u(x_1, x_3, t) = A \cos \omega(px_1 + \eta_\alpha x_3 - t)$ then the energy flux across a unit area of the wavefront is $\rho \alpha A^2 \sin^2 \omega(px_1 + \eta_\alpha x_3 - t)$. If the wave strikes the interface with an angle i, we account for the wavefront projection onto the horizontal boundary by multiplying by ratio of the incident wave area to the boundary area, $\cos i$ (Fig. 13.17). For an S-wave incident with angle j, we must multiply by a factor of $\cos j$.

For the energy expression, we must include the density and wave-speed in the expressions. If we represent the speed of the incoming wave as v_{in} and that of the reflected or transmitted wave as v_{out}, and use a similar notation for the wave angles, then the complete energy flux factor is

$$\left(\frac{\rho_{in} \, v_{in}}{\rho_{out} \, v_{out}} \right)^{1/2} \left(\frac{\cos i_{out}}{\cos i_{in}} \right)^{1/2}$$

$$\text{or} \quad \left(\frac{\rho_{in} \, v_{in}}{\rho_{out} \, v_{out}} \right)^{1/2} \left(\frac{v_{out} \, \eta_{out}}{v_{in} \, \eta_{in}} \right)^{1/2} . \quad (13.69)$$

The square roots arise because the displacement is proportional the square-root of energy. For a non-mode converted reflection, the energy-flux factor is one, for transmitted and mode-converted waves, it is not.

For example, let $u_1(\mathbf{x}, t)$ represent a P-wave displacement pulse incident on a boundary, and let $u_2(\mathbf{x}, t)$ represent a transmitted pulse. The two are related by the product of the displacement transmission ratio discussed earlier and the energy flux factor,

$$u_2(\mathbf{x}, t) = u_1(\mathbf{x}, t) \times T_{PP}$$
$$\times \left(\frac{\rho_1 \alpha_1}{\rho_2 \alpha_2}\right)^{1/2} \left(\frac{\cos i_2}{\cos i_1}\right)^{1/2} . \quad (13.70)$$

For a symmetric turning ray (all downward and upward transmissions match), the upward leg of a ray will include the product of the inverses of all the flux factors accumulated on the downward leg so the product of all the factors will be one. Consider a P-wave reflection off the core mantle boundary (PcP) for a surface source and receiver. Let u_0, represent the displacement at the source. During propagation, the source amplitude is modified by the effects of geometric spreading, attenuation, and reflection and transmission using a sequence of multiplications to account for each effect or boundary interaction. For example, the vertical component of the core reflection PcP amplitude is

$$u_{PcP}(\mathbf{x}, t) \sim u_0(\mathbf{x}, t) \times G_{PcP}(\Delta) \times e^{-\pi f t^* PcP}$$
$$\times \prod_{i=1}^{n} (T_{PP})_i \times R_{PPcore} \times F_{P_{x_3}} ,$$
$$(13.71)$$

where the product of n transmission coefficients before and after reflection are included in the product operator. Because we have no mode conversions and a symmetric ray, the product of all the energy flux factors equals one. A more complete analysis of the wave amplitude would include the frequency dependence to include the phase changes associated with the attenuation operator. The point is that a ray-based amplitude analysis is intuitive. We follow the wave from the source to the seismometer and include the amplitude adjustments for each interaction along the path. For mode conversions, or for deep sources, we must also include the nonsymmetric part of the wave's energy flux factors. We discuss the factors that control $u_0(\mathbf{x}, t)$ (the amplitude leaving the source) in a later chapter. Next we investigate wave propagation from a completely different viewpoint – modes, as we discuss surface waves.

Box 13.3 Seismic diffraction

The analogy between seismic ray theory and optics extends to the concept of diffraction. *Diffraction* is defined as the transmission of energy by nongeometric ray paths. In optics, the classic example of diffraction is light "leaking" around the edge of an opaque screen. In seismology, diffraction occurs whenever the radius of curvature of a reflecting interface is less than a few wavelengths of the propagating wave. Fig. B13.3.1 shows a plane wave incident upon an opaque (acoustic impedance is infinite) boundary. Ray theory requires that waves arriving at seismometers at points F and G have identical amplitudes; *no* energy is transmitted to the right of point G. In fact, the edge of the boundary acts like a secondary source (Huygens' principle) and radiates energy forward in all directions. These diffractions can be understood from the standpoint of *Fresnel zones*, a concept that states that waves reflect from a large region rather than just a point. Thus, the Fresnel zone causes the ray traveling to F to "see" the edge of the reflector, although the geometric raypath clearly misses the boundary. The first Fresnel zone may be thought of as a cone with the edge of the reflector as its apex.

For a receiver that is a distance d beyond the reflector, the cone's radius is given by $r = d - \frac{1}{2}\lambda$, where A is the wavelength of the seismic wave. Fig. B13.3.2 shows the amplitude variation predicted for the experiment given in Fig. B13.3.1.

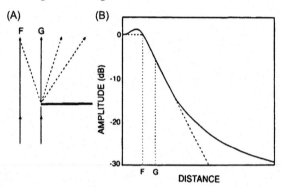

FIGURE B13.3.1 (A) Rays incident on a grating. Energy is *diffracted* around the edge. (B) Amplitude of energy as a function of distance into the diffraction zone (from Doornbos, 1989).

Diffraction is present at many scales within the Earth and has occasionally led to erroneous interpretations of structure. Fig. B13.3.2 shows an example from reflection seismology, which is a stacked section of a synthetic model containing five layers and four small lenses that simulate diffractors (Dell and Gajewski, 2011). The layer velocities are constant, with the fourth layer containing four small lenses with a lateral extension of 200 meters. The diffractors form a parabola known as a *diffraction frown*.

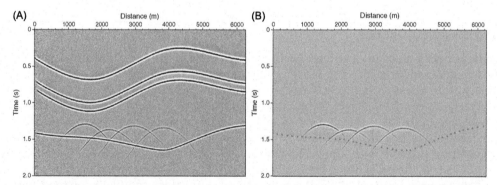

FIGURE B13.3.2 Synthetic example with four small lenses of 200 m lateral extension simulating diffractions. (A) Stacked section of the recorded wavefield and (B) diffraction-only data. Lateral extension of the seismic line is 6250 m. (modified from Dell and Gajewski, 2011).

Chapter goals

- Review the basic properties of surface waves.
- Introduce the mode-theory view of seismic-wave propagation.
- Review seismic approaches to dispersion analysis.
- Review surface-wave geometric spreading and attenuation.

Earth has two fundamental attributes that profoundly affect the seismic wavefield – a stress-free surface and a finite, spheroidal shape. P and S body waves form a complete solution to the equations of motion in a uniform, unbounded medium. In a bounded medium, body waves are only part of the solution. Near a shallow seismic source, where the wavefront curvature is significant, the interaction of P and S waves with Earth's surface and the shallow geologic structure can excite disturbances that have amplitudes that decay with depth and that travel along Earth's surface. We call these disturbances seismic *surface waves*. Analogous, smaller amplitude waves (Stoneley, solid-solid, or Scholte, liquid-solid) are produced near any elastic boundary, but they are seldom observed at the surface. Together with body waves, surface and interface waves form complete solutions to the equations of motion in a bounded medium. Our focus in this chapter is surface waves.

We divide seismic surface waves into two primary categories. The interaction of incident P and SV waves with Earth's (stress) free-surface produces disturbances that travel along the surface producing vertical and horizontal motions in the direction of propagation. To satisfy the equations of motion, the wave amplitudes must decrease with depth. We call these waves *Rayleigh waves* because they were theoretically introduced as a solution to the equations of motion for a uniform halfspace in 1889 by John William Strut, the 3rd Baron Rayleigh. Rayleigh waves were identified on seismograms by Oldham in 1900, in a paper that also included the earliest teleseismic body-wave travel time curves. Rayleigh-wave motions are in some ways similar to water waves, but they are elastic phenomena, whereas water waves are a gravity driven process. However, earthquake deformation can excite gravitationally driven oceanic waves called *tsunami*, which can propagate great distances across the ocean basins and often cause greater damage than the seismic waves (see Chapter 9). Rayleigh waves and tsunami are dispersive, different frequencies travel at different speeds. As the wave travels, the energy is dispersed over time (Fig. 14.1).

As seismic observations accumulated, a second type of strongly dispersed disturbance was observed, one that produced horizontal motion in a direction orthogonal to the direction

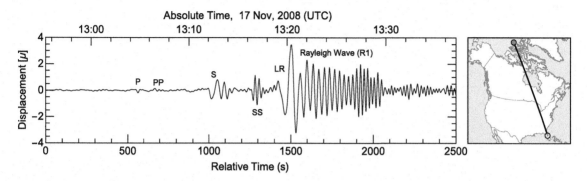

FIGURE 14.1 A characteristic vertical-component seismogram showing body-wave arrivals followed by a dispersed Rayleigh wave, the onset of which is labeled *L R* (long-period Rayleigh). The source was a shallow earthquake in the arctic, and station DWPF is located roughly 57° degrees distant at Disney World Park, Florida. The path lies within the North American continental lithosphere. The Rayleigh wave is dispersed and lower-frequency components that sampled deeper into the planet travel faster. Rayleigh-wave motions persist for roughly 12 minutes (from 13:18 to 13:30) and produce the largest ground motions on the seismogram. The speeds of Rayleigh-wave components range from about $3.0\,\mathrm{km/s} \leq c \leq 4.5\,\mathrm{km/s}$.

of propagation and no vertical motion. In 1911, A.E. Love demonstrated that these disturbances were produced by the total reflection of *SH* waves at the free surface and the general increase in shear-wave speed with depth. These wave amplitudes also decrease with depth, and we call them *Love waves*. For both Rayleigh and Love waves, the trapping of energy near Earth's surface reduces the geometrical spreading from a 3D to a 2D process, and surface waves are often the largest low-frequency waves observed at distance from a shallow earthquake (Fig. 14.1). This is one of the reasons Rayleigh waves were used for distant magnitude estimation (see discussion of M_S in Chapter 7).

The finiteness of the Earth and its internal layering, impose scale lengths and boundary conditions on the seismic wavefield. The planet's finiteness boundary conditions govern the solutions of the equations of motion in the medium. The essential role of the boundaries leads to an analysis of the systems *normal modes*, involving deformation at discrete frequencies for which the system can oscillate and match the boundary conditions. The mode view is analogous to the harmonic tones of an organ pipe or a vi-

brating guitar string. For internal sources, normal modes are called *free oscillations*. Whereas ray-based approaches construct approximate solutions to the equations of motion by tracking the propagation of energy through the planet; normal-mode approaches satisfy the equations of motion by combining solutions for a system's preferred frequencies of vibration. Each mode is a standing-wave solution to the equations of motion, propagating waves in this approach form as interference patterns in the time-varying standing-wave oscillations. Formally, the number of modes is infinite, practically, we consider only enough to provide a complete solution for a finite bandwidth. Normal modes are not as intuitive as ray theory approaches for body waves, but all body waves propagating in the Earth have counterparts in normal-mode oscillations. Modes provide an intuitive framework to analyze surface waves and are used extensively for this purpose. Both representations of seismic deformation have distinct advantages for studying Earth structure and seismic sources – seismologists adopt whichever view is more effective for a particular analysis.

14.1 Halfspace Rayleigh waves

For short distances, as we did for body waves, we can ignore Earth's curvature and examine surface waves in a cartesian system. For greater distances and longer periods we must include Earth's curved surface and finiteness and for the longest periods we must include the effects of gravity and Earth's rotation to understand our observations. We start with the simplest model and then advance to more complicated, but still symmetric models. For laterally heterogeneous media, we must rely on numerical approaches such as finite differences and elements.

We begin with Rayleigh waves, which can propagate along the top of a halfspace. We seek a solution to the equations of motion that propagates laterally, but deforms the halfspace over a finite depth range. Our familiar plane-wave potentials solution provides a starting point, but we seek solutions that decay with depth. Let the surface of the halfspace correspond to $x_3 = 0$ and the wave propagate in the x_1 direction. In other words, we adopt potentials of the form

$$
\begin{aligned}
\phi &= A \exp\left[i\omega\left(px_1 + \eta_\alpha x_3 - t\right)\right] \\
&= A \exp\left[-\omega\hat{\eta}_\alpha x_3\right]\exp\left[i\omega\left(px_1 - t\right)\right] \\
\psi &= B \exp\left[i\omega\left(px_1 + \eta_\beta x_3 - t\right)\right] \\
&= B \exp\left[-\omega\hat{\eta}_\beta x_3\right]\exp\left[i\omega\left(px_1 - t\right)\right],
\end{aligned}
\tag{14.1}
$$

where $(1/p) < \beta < \alpha$ and the horizontal apparent velocity $c = 1/p$. Such solutions confine the energy to propagate along the surface with exponential decay of the potentials away from the surface,

$$
\eta_\alpha = \sqrt{\frac{1}{\alpha^2} - p^2} = i\sqrt{p^2 - \frac{1}{\alpha^2}} = i\sqrt{\frac{1}{c^2} - \frac{1}{\alpha^2}} = i\,\hat{\eta}_\alpha
$$

$$
\eta_\beta = \sqrt{\frac{1}{\beta^2} - p^2} = i\sqrt{p^2 - \frac{1}{\beta^2}} = i\sqrt{\frac{1}{c^2} - \frac{1}{\beta^2}} = i\,\hat{\eta}_\beta,
\tag{14.2}
$$

where $1/p = c < \beta < \alpha$. If $\beta < c < \alpha$; we call c the *phase velocity*. Displacements are computed from the potentials using the Helmholtz Decomposition, Eq. (11.60), in this case,

$$
u_1(\mathbf{x}, t, \omega, p, \alpha, \beta) = \frac{\partial\phi}{\partial x_1} - \frac{\partial\psi}{\partial x_3}
$$

$$
u_3(\mathbf{x}, t, \omega, p, \alpha, \beta) = \frac{\partial\phi}{\partial x_3} + \frac{\partial\psi}{\partial x_1}.
\tag{14.3}
$$

Rayleigh (J. W. Strutt), in 1885, explored the system in Eq. (14.1) and found that the surface boundary condition can be satisfied, demonstrating the existence of a wave traveling along the surface with a velocity lower than the shear velocity and with amplitudes decaying exponentially away from the surface. We call these waves Rayleigh Waves and they were identified seismically about 15 years later. Following Rayleigh, we use the Helmholtz Decomposition (11.60) to compute displacements from the potentials and (14.1), Hooke's Law to compute the stresses and require the stress to vanish at $x_3 = 0$. The condition $\sigma_{33}\,|_{x_3=0} = 0$ gives

$$
A\left[(\lambda + 2\mu)\,\eta_\alpha^2 + \lambda p^2\right] + B\left(2\mu p\eta_\beta\right) = 0 , \tag{14.4}
$$

or, if we use $\rho\alpha^2 = (\lambda + 2\mu)$, $\rho\beta^2 = \mu$, to re-express in terms of β and slownesses,

$$
A\left(\beta^{-2} - 2p^2\right) + B\left(2p\eta_\beta\right) = 0, \tag{14.5}
$$

and $\sigma_{13}\,|_{x_3=0} = 0$ yields

$$
A\left(2p\eta_\alpha\right) + B\left(p^2 - \eta_\beta^2\right) = 0 . \tag{14.6}
$$

The coupled Eqs. (14.5) and (14.6) can be expressed in matrix form

$$
\begin{bmatrix}
\left(\beta^{-2} - 2p^2\right) & 2p\eta_\beta \\
2p\eta_\alpha & \left(p^2 - \eta_\beta^2\right)
\end{bmatrix}
\begin{bmatrix}
A \\
B
\end{bmatrix}
=
\begin{bmatrix}
0 \\
0
\end{bmatrix} .
\tag{14.7}
$$

The only solutions to these equations other than the trivial solution $A = B = 0$, are those for which the matrix determinant vanishes,

$$\left(\beta^{-2} - 2p^2\right)^2 + 4p^2 \eta_\alpha \eta_\beta = 0 , \qquad (14.8)$$

where we used $(p^2 - \eta_\beta^2) = -(\beta^{-2} - 2p^2)$. The term on the left is the denominator of the free-surface reflection coefficients, $R(p)$, (Chapter 13) and is called the *Rayleigh Denominator*.

14.1.1 Halfspace Rayleigh-wave speed

The vertical slownesses contain square roots, but we can rationalize Eq. (14.8) by multiplying each side of the equation by $\left(\beta^{-2} - 2p^2\right)^2 - 4p^2 \eta_\alpha \eta_\beta$, resulting

$$\left[\left(\beta^{-2} - 2p^2\right)^2 + 4p^2 \eta_\alpha \eta_\beta\right]$$
$$\left[\left(\beta^{-2} - 2p^2\right)^2 - 4p^2 \eta_\alpha \eta_\beta\right] = 0 \qquad (14.9)$$
$$\left(\beta^{-2} - 2p^2\right)^4 - 16p^4 \eta_\alpha^2 \eta_\beta^2 = 0 .$$

This equation is a constraint on the value of p for which a valid solution exists. Since the equation does not include frequency (remember, our displacements do), a halfspace Rayleigh wave is non-dispersive. If Eq. (14.8) is satisfied with real-valued η_α and η_β, then our displacements would not decay with depth and that would violate the implicit condition that energy is finite. Solutions to Eq. (14.9) that decay with depth are those with imaginary η_α and η_β, which limits our range of interest to the horizontal slowness range $p > \beta^{-1} > \alpha^{-1}$. If we define the speed, $c = 1/p$, then with a little algebra we can express Eq. (14.8) as

$$\frac{c^6}{\beta^6} - 8\frac{c^4}{\beta^4} + \left(24 - 16\frac{\beta^2}{\alpha^2}\right)\frac{c^2}{\beta^2} + 16\left(\frac{\beta^2}{\alpha^2} - 1\right)$$
$$= 0 . \qquad (14.10)$$

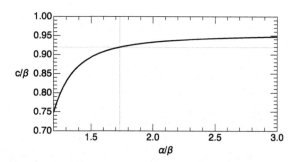

FIGURE 14.2 Half-space Rayleigh-wave phase velocity c as a function of the ratio of α/β. For *a* fluid, $\beta = 0$ and $c_R = 0$. For a Poisson solid, $\alpha = \sqrt{3}\beta$ and $c = 0.9194\beta$ (indicated by the gray lines). The Rayleigh speed changes by about $\pm 4\%$ for the range of P to S ratios typically found in rocks; the higher ratios ($\gtrsim 2$) generally correspond to soils and sediments.

Eq. (14.10) is a cubic polynomial in c^2/β^2 that depends only on the ratio of P- to S-wave speeds, α/β. If $c = 0$, then the polynomial value is $16\left(\beta^2/\alpha^2 - 1\right)$, which is negative since $\beta < \alpha$; if $c = \beta$, then the polynomial value is 1; the change in sign between the two values indicates that the polynomial has a root in the range $0 < c < \beta$.

Fig. 14.2 is a plot of solutions of (14.8) for different values of the ratio of P-to-S wave speeds. For typical values ($1.6 \leq \alpha/\beta \leq 1.9$), the Rayleigh-wave velocity is 0.9β to 0.95β. For soils and other materials with $2 \lesssim \alpha/\beta \lesssim 5$, the Rayleigh-wave speed approaches a value about 5% lower than that of the shear wave. If we assume that $\alpha/\beta = \sqrt{3}$, which is the condition for a Poisson solid (and close to the values for many rocks and regions of Earth), then

$$\frac{c^6}{\beta^6} - \frac{8c^4}{\beta^4} + \frac{56c^2}{3\beta^2} - \frac{32}{3} = 0 , \qquad (14.11)$$

which has a solution

$$\frac{c_R}{\beta} = \sqrt{2 - \frac{2}{\sqrt{3}}} \approx 0.9194 , \qquad (14.12)$$

where c_R is the root of the equation. This value is indicated by the gray lines in Fig. 14.2. The halfspace Rayleigh wave travels at a speed slightly less than that of a shear-wave. Equally important is the fact that c_R is not dependent on frequency. The Rayleigh wave for a halfspace does not disperse.

14.1.2 Halfspace Rayleigh-wave displacements

Now consider the nature of the halfspace Rayleigh-wave displacements. The Rayleigh-wave motion involves a mix of P and SV motions, combined with relative amplitudes A and B. We can rewrite (14.5) as

$$B = A \frac{\left(2p^2 - \beta^{-2}\right)}{2p\eta_\beta} = A \frac{\left(2 - c_R^2 \beta^{-2}\right)}{2c_R\eta_\beta} , \quad (14.13)$$

and then compute the Rayleigh-wave displacements using (14.3),

$$u_1(\omega, \mathbf{x}, t, p_R)$$
$$= i\omega p_R A e^{i\omega(p_R x_1 - t)}$$
$$\times \left[e^{-\omega\hat{\eta}_\alpha x_3} + \frac{1}{2}\left(\frac{c_R^2}{\beta^2} - 2\right) e^{-\omega\hat{\eta}_\beta x_3} \right]$$

$$u_3(\omega, \mathbf{x}, t, p_R)$$
$$= -\omega A e^{i\omega(p_R x_1 - t)}$$
$$\times \left[\hat{\eta}_\alpha e^{-\omega\hat{\eta}_\alpha x_3} + \frac{1}{2c_R^2\hat{\eta}_\beta}\left(\frac{c_R^2}{\beta^2} - 2\right) \times e^{-\omega\hat{\eta}_\beta x_3} \right].$$
$$(14.14)$$

Since the ground motion is a real-valued quantity, we use $\exp[i\omega(px_1 - t)] = \cos[\omega(px_1 - t)] + i\sin[\omega(px_1 - t)]$ and retain only real terms

$$u_1(\omega, \mathbf{x}, t, c_R)$$
$$= -A\omega p_R \sin\left[\omega(p_R x_1 - t)\right]$$
$$\times \left[e^{-\omega\hat{\eta}_\alpha x_3} + \frac{1}{2}\left(\frac{c_R^2}{\beta^2} - 2\right) e^{-\omega\hat{\eta}_\beta x_3} \right]$$
$$(14.15)$$

$$u_3(\omega, \mathbf{x}, t, c_R)$$
$$= -A\omega p_R \cos\left[\omega(p_R x_1 - t)\right]$$
$$\times \left[c_R\hat{\eta}_\alpha e^{-\omega\hat{\eta}_\alpha x_3} + \frac{1}{2c_R\hat{\eta}_\beta}\left(\frac{c_R^2}{\beta^2} - 2\right) e^{-\omega\hat{\eta}_\beta x_3} \right].$$

The $\sin(kx_1 - t)$ and $\cos(kx_1 - t)$ factors are associated with the Rayleigh wave's horizontal propagation. The depth-dependent factors are often called *eigenfunctions* (the problem can also be approached as an eigenvalue problem, the phase velocity is the eigenvalue). If we represent the eigenfunctions as $r_1(x_3, \omega)$ and $r_3(x_3, \omega)$, then

$$r_1(x_3, \omega)$$
$$= -A\omega p_R \times \left[e^{-\omega\hat{\eta}_\alpha x_3} + \left(\frac{c_R^2}{2\beta^2} - 1\right) e^{-\omega\hat{\eta}_\beta x_3} \right]$$

$$r_2(x_3, \omega)$$
$$= -A\omega p_R \times \left[c_R\hat{\eta}_\alpha e^{-\omega\hat{\eta}_\alpha x_3} + \frac{1}{c_R\hat{\eta}_\beta}\left(\frac{c_R^2}{2\beta^2} - 1\right) \right]$$
$$\times e^{-\omega\hat{\eta}_\beta x_3} .$$
$$(14.16)$$

In a problem with a source, the amplitude factor, A, is often dependent on the frequency and slowness. Here we ignore the numerical value of A, but its physical units, $[L^2]$, are essential to insure that the eigenfunction units are displacement.

The horizontal displacement eigenfunction, $r_1(x_3, \omega)$ has a zero crossing at a depth h defined by the equation

$$\exp\left[-\omega(\hat{\eta}_\alpha - \hat{\eta}_\beta)h\right] = 1 - \frac{c_R^2}{\beta^2} . \quad (14.17)$$

The vertical eigenfunction, $r_2(x_3, \omega)$ is always positive and both r_1 and r_2 approach zero at greater depths. For depths greater than h, the horizontal displacement sign change causes the motion to transition to prograde elliptical. The surface ellipticity, the ratio of the maximum amplitude in the horizontal and vertical directions, is

$$\frac{\max\left[r_1(0,\omega)\right]}{\max\left[r_3(0,\omega)\right]} = \frac{1 + \left(\frac{c_R^2}{2\beta^2} - 1\right)}{c_R\hat{\eta}_\alpha + \frac{1}{c_R\hat{\eta}_\beta}\left(\frac{c_R^2}{2\beta^2} - 1\right)}$$

$$= \frac{1 - \frac{c_R^2}{2\beta^2}}{\sqrt{1 - \frac{c_R^2}{\alpha^2}}}. \qquad (14.18)$$

The halfspace surface ellipticity depends on the ratio of P- to S-wave speeds directly and through c_r, but not on frequency.

Stresses introduced as the surface waves traverse a region are often associated with triggering small earthquakes and the strains at a particular source depth are an important component in the excitation of surface waves by earthquakes at different depths. For completeness and anticipating later discussions, we define stress eigenfunctions in terms of those for displacement,

$r_3(x_3, \omega)$

$$= \mu \left[\frac{dr_1(x_3, \omega)}{dx_3} - \omega p r_2(x_3, \omega)\right]$$

$$= \rho\beta^2 \left[\frac{dr_1(x_3, \omega)}{dx_3} - \omega p r_2(x_3, \omega)\right]$$

$r_4(x_3, \omega)$

$$= i\left[(\lambda + 2\mu)\frac{dr_2(x_3, \omega)}{dx_3} + \omega p\lambda r_1(x_3, \omega)\right]$$

$$= i\left[\rho\alpha^2\frac{dr_2(x_3, \omega)}{dx_3} + \omega p\rho(\alpha^2 - 2\beta^2)r_1(x_3, \omega)\right], \qquad (14.19)$$

where $r_3(z, \omega)$ and $r_4(z, \omega)$ correspond to the stresses σ_{13} and σ_{33} respectively. The relationships in Eq. (14.19) are general and not dependent on the halfspace model.

A Poisson solid

Thus far our expressions have been abstract. As always, a simple example can help clarify the key ideas. Our expressions simplify substantially for the reasonable assumption that the halfspace is a Poisson solid. Then, $c_R \approx 0.919\beta \approx 0.531\alpha$, and defining $k(\omega) = \omega p_R = \omega/c_R$ as the Rayleigh *wavenumber*, Eq. (14.15) becomes

$$u_1(\omega, \mathbf{x}, t, p_R)$$
$$= -Ak(\omega)\sin\left(k(\omega)x_1 - \omega t\right)$$
$$\times \left(e^{-0.85kx_3} - 0.58e^{-0.39kx_3}\right) \qquad (14.20)$$

$$u_3(\omega, \mathbf{x}, t, p_R)$$
$$= -Ak(\omega)\cos\left(k(\omega)x_1 - \omega t\right)$$
$$\times \left(0.85e^{-0.85kx_3} - 1.47e^{-0.39kx_3}\right).$$

At the surface of the halfspace, $x_3 = 0$ and

$$u_1(\omega, x_1, t, p_R) = -0.42Ak\sin\left(k(\omega)x_1 - \omega t\right)$$

$$u_3(\omega, x_1, t, p_R) = 0.62Ak\cos\left(k(\omega)x_1 - \omega t\right). \qquad (14.21)$$

The Rayleigh-wave displacements given by (14.20) depend harmonically on x_1 and exponentially on x_3 (depth). The displacements u_1 and u_3 are out of phase by 90° and therefore combine to produce ellipsoidal particles motion, as illustrated in Fig. 14.3. The surface maximum vertical motion is larger than the maximum horizontal motion by a factor of 1.5, consistent with Eq. (14.18). The ellipse eccentricity depends on the ratio of the P and S wave speeds. At the top of a deformation cycle (in the $-x_3$ direction) the surface horizontal motion is opposite the direction of propagation, and the elliptical motion is *retrograde* or counter-clockwise. On a seismogram the vertical and horizontal (radial) motion are shifted by 90°, the radial "leads" the vertical by one quarter cycle, as shown in the left panel. Along with arrival time and dispersion, the phase shift is one of the primary approaches to identifying Rayleigh waves on seismograms.

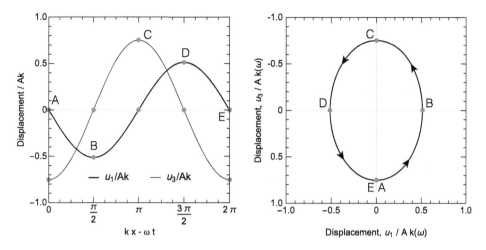

FIGURE 14.3 (left) Plot of Eq. (14.21), surface motion produced by a halfspace Rayleigh wave as a function of the phase argument $(k(\omega)x_1 - \omega t)$. (right) Trajectory of an individual particle as a function of time. The surface motion is retrograde elliptical.

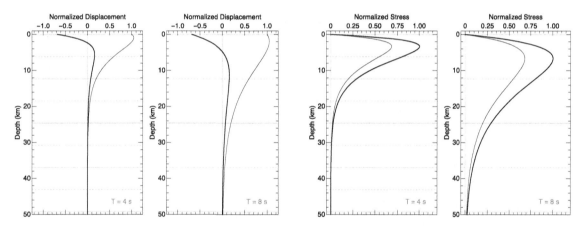

FIGURE 14.4 The left two panels show the displacement as a function of depth for Rayleigh waves with periods of four and eight seconds. The displacement eigenfunctions are normalized by the value of $r_2(\omega, 0)$. The thick line identifies the x_1 component, the thin line the x_3. The horizontal grid lines are spaced at one-half wavelength intervals. The right two panels show the stress eigenfunctions for Rayleigh waves with periods of four and eight seconds. The thick line identifies the σ_{13} component, the thin line the σ_{33}. The stress eigenfunctions are normalized by the value of $\sigma_{31}(\omega, \Lambda/4)$. The horizontal grid lines are spaced at one-half wavelength intervals.

Fig. 14.4 is a plot of the displacement and stress eigenfunctions for two sample periods. The halfspace used for the calculations had a P-wave velocity of 5.8 km/s ($\alpha/\beta = \sqrt{3}$) and a density of 2500 kg/m³. In this case, the shape of the eigenfunctions for different periods is identical, increasing the period only stretches the eigenfunctions across a larger depth range. The horizontal distance between surface particle motions at the same point in their elliptical trajectory defines the Rayleigh-wave wavelength, $\Lambda = 2\pi/k$. At a depth of about $\Lambda/5$ the horizontal motion goes to zero, and at greater depths, the elliptical motion has a **prograde** sense. By a depth of $\Lambda/2$, the horizontal particle motion is about 10% of the horizontal motion at the surface, and the vertical motion is about 30% of the surface vertical motion. Since the Rayleigh-wave amplitudes have exponential dependence in the form $e^{(-2\pi/\Lambda)x_3}$, long-wavelength Rayleigh waves have larger displacements at greater depth. In a homogeneous halfspace, the velocity of Rayleigh waves does not depend on frequency, but in a vertically inhomogeneous structure, a Rayleigh wave is **dispersive**. Because the velocity in the Earth increases with depth, the longer wavelength Rayleigh waves tend to sample faster material, giving rise to higher Rayleigh-wave velocities for large-wavelength, low-frequency wave components, which produces dispersion. The next most complicated structural model is a single layer over a halfspace, which for Rayleigh waves leads to the solution of a matrix determinant with six rows and columns. We usually do not approach the Rayleigh-wave analyses with the determinant based approach. Solution of Rayleigh-wave propagation in a layered or inhomogeneous elastic models are better investigated numerically. The methods are treated fully in advanced texts by Aki and Richards (2009), Kennett (2001), and Ben-Menahem and Singh (1981). The layer over a halfspace problem is much simpler for Love waves, which we examine in the next section. We will consider Rayleigh-wave motion in the Earth in the context of equivalent spheroidal free oscillations later in this chapter.

Surface-wave geometric spreading

Rayleigh waves only require a free surface to be a viable solution of the equations of motion, but only a half-space produces an undispersed Rayleigh pulse (see Box 14.1). A more characteristic Rayleigh waveform is shown in Fig. 14.1, where the Rayleigh wave is spread out over more than 10 min, and the lower-frequency energy arrives earlier in the waveform. We discuss such dispersion a little later. Note that the Rayleigh-wave motions are the largest of any arrivals on this seismogram, which results from the two-dimensional geometric spreading of the surface wave relative to the three-dimensional spreading that affects the body waves. Seismic sources near the surface tend to excite strong Rayleigh waves, whereas sources deep in the Earth excite only weak Rayleigh waves. In three dimensions, surface waves spread cylindrically outward from the source and since the energy is restricted in the vertical dimension, they exhibit a two-dimensional geometric decrease in amplitude with radius (distance) r from the source. Thus the geometric spreading for a surface wave, proportional to $1/\sqrt{r}$, is significantly lower than the three-dimensional ($1/r$) decay rate for body waves. As a result, Rayleigh Waves, tend to be much larger than body waves on long-period or broadband seismograms.

14.2 Love waves in a layer over a halfspace

The halfspace Rayleigh-wave analysis introduced some important surface-waves characteristics, deformation restricted near the surface, period-dependent depth sensitivity, elliptically polarized particle motion, 2D geometric spreading, etc. However, the most characteristic feature of seismic surface waves, strong frequency-dependent dispersion, is not part of the halfspace solution. In this section we explore dispersion from a using a Love-wave analysis, in the next, we quantify dispersion more generally.

Love waves are absent in a halfspace, to exist, they require an increase in speed with depth. The simplest such model includes a single, uniform layer resting on a uniform halfspace with a shear-wave velocity greater than that in the

Box 14.1 Lamb's problem

A complete theory for Rayleigh waves, even for a half-space, must include their excitation by a specific source. At this point we show a classic result, first obtained by H. Lamb (Lamb, 1904), which is the transient solution to an impulsive vertical point force applied to the surface of a half-space. Part (A) of Fig. B14.1.1. shows Lamb's (1904) calculations, which are believed to be the first theoretical seismograms. The motions begin with the P arrival. The small arrival prior to the large-amplitude pulse is the S wave, and the large pulse itself is a Rayleigh-wave pulse. The Rayleigh wave shows a clear phase shift between the radial (q_0) and vertical (w_0) components and is much larger than the body-wave arrivals. The experimental result shown in part (B) is a recording of a breaking pencil lead point-force source on a piece of brass, which has a vertical motion very similar to Lamb's prediction. Recordings of natural sources approximating Lamb's solution are shown in Box 17.1, but normally Rayleigh waves in the Earth are dispersed and resemble Fig. 14.1. Rayleigh-wave excitation varies substantially with source force system and depth.

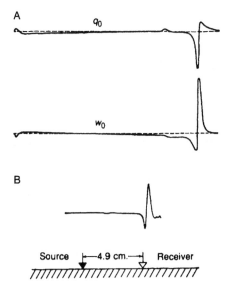

FIGURE B14.1.1 (A) Radial (q_0) and vertical (w_0) surface ground motions calculated by Lamb (1904) for an impulsive vertical force on the surface. (B) An experimentally recorded vertical ground motion for a vertical point source. The largest motion in each case corresponds to the undispersed Rayleigh pulse (from Ewing et al., 1957).

layer. This layer-over-a-halfspace model hosts both Rayleigh and Love waves, to focus only on Love, we consider horizontal motion perpendicular to the direction of wave propagation. In body-wave analyses, this corresponds to the SH motion. Indeed, Love waves in such a simple model can be envisioned as a superposition of post-critically reflected SH waves bouncing between the surface (a perfect reflector) and the top of the halfspace. The intuition of energy trapped by reflection is useful, but as the number of rays increases, simple intuition is overwhelmed. Thus we adopt a mode-based approach to the problem as we did for Rayleigh waves in a halfspace. The properties of an SH disturbance trapped in a near-surface waveguide were first explored by A.E.H. Love in 1911, and these waves are called *Love waves*. Love's work was motivated by observations of a "packet" of dispersed waves with no vertical motions and arriving earlier than Rayleigh waves.

In our mode-based approach, we investigate solutions to the equations of motion in the layer and in the halfspace that can also satisfy the boundary conditions at the surface and at the layer-halfspace boundary. The layer thickness, H, introduces a spatial characteristic to the problem that was absent in the halfspace Rayleigh-wave problem. As we will see, the inclusion of a layer results in a propagating, frequency-dependent dispersive interference pattern that we call a Love wave – an interesting result considering that the layer and halfspace shear-wave velocities are not frequency dependent.

Since our interest is in SH-type displacements, we use the result of Chapter 11; we do not need potentials because the SH displacements satisfy the wave equation. We represent the wavefield as a sum of plane-waves of the form

$$
\begin{aligned}
u_2(\omega, \mathbf{x}, t, p) = {} & A \exp\left[i\omega\left(px_1 + \eta_{\beta_1}x_3 - t\right)\right] \\
& + B \exp\left[i\omega\left(px_1 - \eta_{\beta_1}x_3 - t\right)\right], \\
& 0 \le x_3 \le H \\
= {} & C \exp\left[i\omega\left(px_1 + \eta_{\beta_2}x_3 - t\right)\right], \\
& x_3 \ge H
\end{aligned}
$$

$$(14.22)$$

for a particular horizontal slowness, p. From a body-wave perspective, the displacement in the layer is composed of upward- and downward-propagating plane waves, and the displacement in the half-space is composed of transmitted SH waves generated at the base of the layer. The complete solution is a sum of plane waves with

different values of p. If $\beta_1 > \beta_2$ (layer faster than halfspace), then energy will efficiently transmit into the halfspace and energy is not trapped near the surface. For $\beta_1 < \beta_2$, some energy is transmitted to the halfspace (for steep incidence angles), but for angles beyond critical, the energy is trapped in the layer and propagates laterally.

The boundary conditions are zero stress at the surface and continuous stress and displacement at the layer-halfspace boundary. Thus

$$
\sigma_{32}|_{x_3=0} = \mu_1 \frac{\partial u_2}{\partial x_3}\Big|_{x_3=0} = 0
$$

$$(14.23)$$

$$
\sigma_{32}|_{x_3=H^-} = \sigma_{32}|_{x_3=H^+}
$$

$$
u_2|_{x_3=H^-} = u_2|_{x_3=H^+},
$$

where the H^+ and H^- indicate locations just below and above the boundary respectively. Applying (14.23) to (14.22) yields three equations,

$$
A = B, \tag{14.24}
$$

$$
\begin{aligned}
A\mu_1\eta_{\beta_1}\left[\exp\left(i\omega\eta_{\beta_1}H\right) - \exp\left(-i\omega\eta_{\beta_1}H\right)\right] \\
= C\mu_2\eta_{\beta_2}\exp\left(i\omega\eta_{\beta_2}H\right),
\end{aligned} \tag{14.25}
$$

and

$$
\begin{aligned}
A\left[\exp\left(i\omega\eta_{\beta_1}H\right) + \exp\left(-i\omega\eta_{\beta_1}H\right)\right] \\
= C\exp\left(i\omega\eta_{\beta_2}H\right).
\end{aligned} \tag{14.26}
$$

For a specified frequency and earth model, these equations constrain the horizontal apparent speed or phase velocity, $c = 1/p = k_1/\omega$ for which solutions to the problem exist. We can rewrite the complex exponentials in terms of trigonometric functions, and take the ratio of Eqs. (14.25) and (14.26) to yield

$$
\tan\left(\omega\eta_{\beta_1}H\right) = \frac{\mu_2\eta_{\beta_2}}{i\mu_1\eta_{\beta_1}} = \frac{\mu_2\hat{\eta}_{\beta_2}}{\mu_1\eta_{\beta_1}}, \tag{14.27}
$$

where we assume the post-critical situation for which $c = 1/p < \beta_2$, $\eta_{\beta_2} = i\hat{\eta}_{\beta_2}$, and $\hat{\eta}_{\beta_2}$ is real-valued. Eq. (14.27) relates ω and c and must be satisfied to produce a stable horizontally propagating disturbance. Because the wave velocity explicitly depends on frequency, $c(\omega)$, Eq. (14.27) is called a *dispersion equation*. Rewriting (14.27) in terms of the material parameters μ_1, μ_2, β_1, and β_2 and the variables ω and c, we recover a transcendental equation

$$\tan\left(\omega H \sqrt{1/\beta_1^2 - 1/c^2}\right) = \frac{\mu_2 \sqrt{1/c^2 - 1/\beta_2^2}}{\mu_1 \sqrt{1/\beta_1^2 - 1/c^2}} \,. \tag{14.28}$$

Eq. (14.28) indicates that for real-valued solutions, $\beta_1 < c < \beta_2$. Solutions to the Love-wave dispersion Equation (14.28) are conventionally illustrated graphically. Fig. 14.5 is an illustration of four numerical solutions of the dispersion equation for four different periods (frequencies). Once a period or frequency is selected, the goal is to calculate the phase velocity. Phase velocities that satisfy the equations of motion and the boundary conditions are represented by the intersections of the left- and right-hand sides of the dispersion equation (Eq. (14.28)). Only the left-hand side (the tangent function) depends on frequency, so the right-hand side is the same in each panel of Fig. 14.5. At least one solution exists for each frequency. As the frequency increases, additional branches of the tangent function may enter the range of feasible solutions (from the right). Each additional branch adds another solution to the equations. For a given value of ω, a finite number of solutions exist, which we number from left to right using and integer, n, beginning with $n = 0$. The $n = 0$ solution is called the *fundamental mode* for that frequency, and larger values of n define the *higher modes* or *overtones* of the system. The fundamental mode corresponds to one-half a harmonic cycle distributed from the surface to the halfspace-layer boundary. The displacement in the layer can be derived from Eq. (14.22) and (14.24) to (14.26). The displacement has the form

$$u_2(\omega, \mathbf{x}, t) = l_1(\omega, x_3, t)\, e^{i(kx_1 - \omega t)}, \text{ where}$$

$$l_1(\omega, x_3, t)$$
$$= \begin{cases} 2A \cos\left(\omega x_3\, \eta_{\beta_1}\right), & 0 \leq x_3 \leq H \\ 2A \cos\left(\omega H\, \eta_{\beta_1}\right) e^{-\hat{\eta}_{\beta_2}(x_3 - H)}, & x_3 \geq H \end{cases} \tag{14.29}$$

where $\eta_{\beta_1} = \sqrt{\frac{1}{\beta_1^2} - \frac{1}{c^2(\omega)}}$, $\hat{\eta}_{\beta_2} = \sqrt{\frac{1}{c^2(\omega)} - \frac{1}{\beta_2^2}}$, and $l_1(\omega, x_3, t)$ is called the displacement eigenfunction. A stress eigenfunction, $l_2(\omega, x_3, t)$, can be computed using $l_2(\omega, x_3, t) = \mu\, dl_1/dx_3$. Over the entire depth range the sense of motion caused by the fundamental mode is uniformly in the $\pm x_2$ direction. Overtone solutions have n **nodes** (zero crossings) from $x_3 = 0$ to $x_3 = H$ that divide the layer into $n + 1$ vertical regions oscillating in the $\pm x_2$ direction and separated by nodal surfaces. (See Fig. 14.6.)

The nth overtone only exists as a horizontally propagating wave for frequencies greater than

$$\omega_{c_n} = \frac{n\pi}{H\sqrt{\left(1/\beta_1^2\right) - \left(1/\beta_2^2\right)}}, \tag{14.30}$$

where ω_{c_n} is the *cutoff frequency* for the nth mode. The phase velocity of the nth overtone at the cutoff frequency is β_2; for values equal to or greater than the cutoff, the displacement in the halfspace does not decrease with depth. For example, at the cutoff frequency, the halfspace eigenfunction is a constant value of $\pm 2A$ – a solution that is unacceptable because it violates the requirement that the displacement decrease with depth and the energy remain finite. The eigenfunction decay rate in the halfspace is proportional to $\hat{\eta}_{\beta_2}$, which is small when c is close to β_2. Thus for modes with phase velocities just below β_2, the displacement in the halfspace is large. Physically, the speed at which a wave travels depends on the elastic properties and density of the material that it deforms. Since these modes deform the halfspace significantly, the modes travel at a value near the

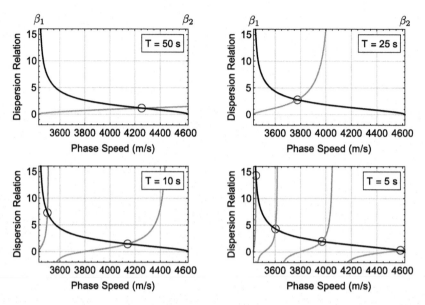

FIGURE 14.5 Graphical solutions of (14.28) for increasing frequencies (decreasing period). The gray curves represent the left-hand side, the black curves represent the right-hand side of the dispersion relation. The intersections, identified by circles, are solutions to the transcendental dispersion relation. For higher frequencies more branches of the tangent function lie in the range of real-valued solutions. The first higher mode appears by a period of 15 s, four modes exist for a period of 5 s. The model parameters used for the computation are $H = 40\,km$, $\beta_1 = 3.43\,km/s$, $\rho_1 = 2700\,kg/m^3$, $\beta_2 = 4.62\,km/s$, $\rho_2 = 3300\,kg/m^3$.

shear-wave speed of the halfspace. The effect is apparent in Fig. 14.6. The amount of halfspace deformation increases for modes with high phase velocities, as it must if the waves are to travel at speeds approaching that in the halfspace. For modes with displacements contained completely in the layer, the mode travels at a speed near the shear-wave velocity of the layer, as it must. As the frequency increases, the phase velocity of each mode decreases. At shorter periods, for example, the fundamental mode travels at a speed quite close to the layer velocity. The reason is that the displacement at short periods concentrates within the layer as frequency increases. At intermediate periods, the deformation is spread over the layer and halfspace, and the phase velocity is an average of β_1 and $\beta2$ (weighted by the eigenfunctions). At very long periods, the deformation

reaches deep into the halfspace, and the phase velocity approaches that of the halfspace. Love-wave propagation in a multilayered structure like the Earth can be analyzed in much the same fashion. Longer-wavelength, lower-frequency waves tend to have higher velocities, because velocity usually increases with depth; however, the actual velocity gradients in the mantle cause long-period Love waves to be less dispersive than Rayleigh waves of a corresponding period.

Love waves are always dispersive because they require at least a low-velocity layer over a half-space to exist. Because Love-wave particle motion is parallel to the surface, a complete separation of Love-and Rayleigh-wave surface motions occurs. Love waves travel faster and produce transverse motions ahead of the Rayleigh waves, which result in vertical and radial defor-

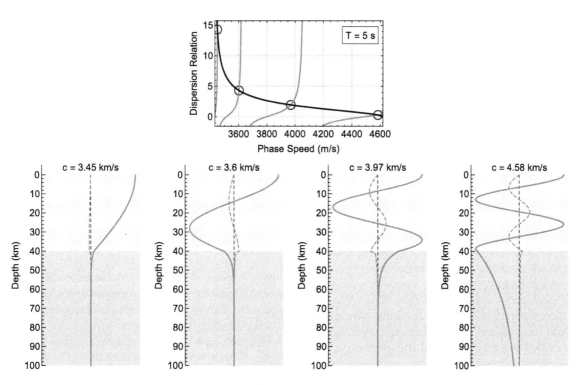

FIGURE 14.6 (top) Graphical solutions of (14.28) indicating four modes. The gray curves represent the left-hand side, the black curves the right-hand side of the dispersion relation. The intersections, identified by circles, are solutions to the transcendental dispersion relation. (bottom) Displacement (solid curve) and stress (dashed curve) eigenfunctions corresponding to the fundamental and first three higher modes identified above. Each function is self normalized. The light gray background identifies the layer, the darker background the halfspace. The model parameters used for the computation are $H = 40$ km, $\beta_1 = 3.43$ km/s, $\rho_1 = 2700$ kg/m^3, $\beta_2 = 4.62$ km/s, $\rho_2 = 3300$ kg/m^3.

mation. The block diagrams of Fig. 1.2 depict the sense of motion of body and surface waves. For short distances, dispersed Rayleigh and Love arrivals may overlap in time, but their particle motions allow us to separate them (the vertical component for Rayleigh and the transverse component for Love). The seismograms in Fig. 1.1 are an example of a (nearly) naturally rotated set of seismograms with clear body and surface-wave arrivals. The Love wave is "naturally" polarized on the transverse (E-W) component in this particular case.

The key result of our analysis is that Love wave phase velocity is frequency dependent and that phase velocity at each frequency is a weighted average of the depth-dependent shear-wave speeds and density. The same is true for the phase velocities of Rayleigh waves in media with density, and P- and S-wave speeds that vary with depth. The dependence of dispersion on the subsurface material properties is a fundamental result that allows seismologists to use observed patterns in dispersion to image the crust and upper mantle. Thus a deep appreciation of connection between surface-wave observations and phase velocities is essential for a seismologist. Our next step is to connect our phase-velocity analysis to seismogram-based observations of dispersion.

14.3 Dispersion

In general, surface waves disperse because the apparent velocity along the surface depends on frequency. Almost any seismic source excites waves across a continuous range of frequencies, and each harmonic has its own velocity, $c(\omega)$, which is called *phase velocity*. If a source somehow excited a monochromatic wave, the phase velocity for that frequency would characterize the disturbance completely. However, when waves spanning a range of frequencies are excited, the disturbances interfere to produce constructive and destructive ground-motion patterns. Constructive interference patterns behave as wave packets, which themselves propagate as disturbances along the surface with well-defined *group velocities*, $U(\omega)$. Phase velocity is directly controlled by the medium parameters (scale lengths of layering, intrinsic P and/or S velocities, rigidity, etc.) and the geometric "fit" of a particular harmonic component into the associated boundary conditions, as seen in the last section. Group velocity is indirectly controlled by the medium parameters through their influence on the phase velocity, but group velocity also depends on the variation of phase velocity with frequency, which controls the interference between different harmonics.

14.3.1 Discrete dispersion

To understand the relationship, consider two harmonic waves with the same amplitude but slightly different frequencies (ω', ω''), wavenumbers, and phase velocities $(k' = \omega'/c', k'' = \omega''/c'')$. If we add the two signals, the total displacement is

$$u(x, t) = \cos\left(\omega' t - k' x\right) + \cos\left(\omega'' t - k'' x\right).$$
$$(14.31)$$

Define ω as the average of ω'' and ω' such that $\omega' + \delta\omega/2 = \omega = \omega'' - \delta\omega/2$, and $k = \omega/c$ such that $k' + \delta k/2 = k = k'' - \delta k/2$, where the per-

turbations are small, $\delta\omega << \omega$, $\delta k << k$. Inserting these definitions into (14.31) and using the trigonometric relation, $2 \cos x \cos y = \cos(x + y) + \cos(x - y)$, we obtain

$$u(x, t) = 2 \cos\left(\omega t - k x\right) \cos\left[\frac{1}{2}(\delta\omega t - \delta k x)\right].$$
$$(14.32)$$

The combined signal is the product of two cosines, the second of which varies more slowly than the first (because the perturbations are small). Fig. 14.7 is an illustration of the idea. We call the phenomenon "beating" and view the signal as the modulation of the first, more rapidly varying signal, by the second, more slowly varying signal. The interference signal is periodic because we have summed two discrete harmonic signals – similar to the reconstruction of a signal with Fourier Series.

The envelope of the modulated signal propagates with a velocity different from the phase velocity of the average harmonic term c, which is defined as the *group velocity*,

$$U(\omega) = \frac{\delta\omega}{\delta k}.$$
$$(14.33)$$

In the limit as $\delta\omega$ and $\delta k \to 0$,

$$U(\omega) = \frac{d\omega}{dk} = \frac{d(kc)}{dk} = c + k\frac{dc}{dk} = c - \Lambda\frac{dc}{d\Lambda}.$$
$$(14.34)$$

The group velocity depends on both the phase velocity *and* the variation of phase velocity with wavenumber. If $dc/dk = 0$ (constant phase velocity), the phase and group velocities are equal. In the Earth, the phase velocity generally decreases with frequency, so $dc/dk < 0$ and $U < c$.

14.3.2 Continuous dispersion

An earthquake source excites surface waves with a continuum of frequencies rather than just two discrete frequencies. The total surface-wave

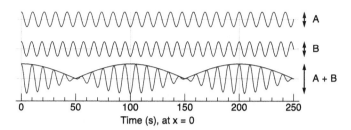

FIGURE 14.7 Two harmonics with equal amplitudes ($A = B$) and nearly equal frequencies and slightly different phase velocities. The lower trace shows the sum of the two signals and the modulating harmonic (absolute value) in Eq. (14.32).

displacement is a summation of all the propagating harmonic components. The sum of a continuum of harmonic terms with uniform amplitude over a finite frequency band $\delta\omega$ centered on average frequency ω_0 is

$$u(x,t) = \int_{\omega_0-\Delta\omega/2}^{\omega_0+\Delta\omega/2} \cos[\omega t - k(\omega)x]\,d\omega .$$

$$(14.35)$$

For small $\Delta\omega$, we expand $k(\omega)$ in a Taylor series:

$$k(\omega) = k(\omega_0) + \left(\frac{dk}{d\omega}\right)_{\omega_0}(\omega-\omega_0) + \cdots , \quad (14.36)$$

and evaluate the integral of the sum to first order in ω,

$$u(x,t) = \frac{1}{t - (dk/d\omega)_{\omega_0}\,x}$$
$$\left\{ \sin\left[\frac{+\Delta\omega}{2}\left[t - \left(\frac{dk}{d\omega}\right)_{\omega_0}x\right]\right.\right.$$
$$+\omega_0 t - k(\omega_0)x]$$
$$-\sin\left[\frac{-\Delta\omega}{2}\left[t - \left(\frac{dk}{d\omega}\right)_{\omega_0}x\right]\right.$$
$$\left.\left. + \omega_0 t - k(\omega_0)x\right]\right\} . \quad (14.37)$$

Then, using $2\sin\alpha\cos\beta = \sin(\alpha+\beta) - \sin(\beta-\alpha)$

$$u(x,t) = \frac{2}{t - (dk/d\omega)_{\omega_0}\,x}$$

$$\times \sin\left\{\frac{\Delta\omega}{2}\left[t - \left(\frac{dk}{d\omega}\right)_{\omega_0}x\right]\right\}$$

$$\times \cos(\omega_0 t - k(\omega_0)x) . \quad (14.38)$$

If we let $Y = (\Delta\omega/2)[t - (dk/d\omega)_{\omega_0}x]$, we have

$$u(x,t) = \Delta\omega\,\frac{\sin Y}{Y}\cos[\omega_0 t - k(\omega_0)x]. \quad (14.39)$$

The interference pattern is a cosine harmonic term with the reference parameters modulated by a *sinc* function, which is peaked at $Y = 0$ and has rapidly diminishing side lobes. The periodic modulations that we obtained by summing two cosine signals in Fig. 14.8 are replaced by a single, isolated wave packet when we consider a continuum of frequencies (Fig. 14.9):

Our simple analytic illustrations are limited to a narrow frequency range. Surface waves from large, shallow earthquakes, even at substantial distance, are dominated by a range of frequencies from about 0.01 to 0.1 Hz. For great earthquakes, the signals are clear for even lower frequencies (0.002 to 0.1 Hz). For moderate-magnitude regional earthquakes, intermediate-period signals (0.02 to 0.1 Hz) are typically well observed. For local earthquakes, surface waves with periods of a few seconds are commonly observed, but the excitation of such short-period observations is best for very shallow events. In hydrocarbon exploration, environmental, and engineering applications, even shorter period observations are commonly observed across

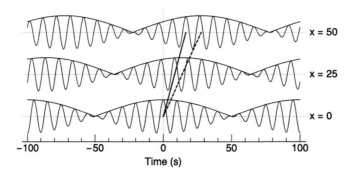

FIGURE 14.8 Sum of the two harmonics at three different distances. The solid line shows the position of the peak in the carrier wave, which travels at the phase velocity. The dashed line shows the position of the modulating signal, which travels at the group velocity.

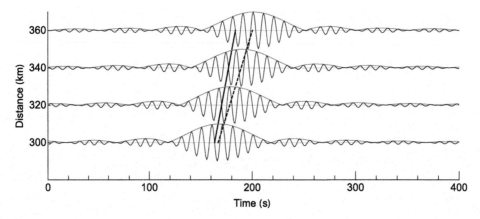

FIGURE 14.9 Example of continuous spectrum (uniform amplitude) interference pattern computed around a period of 10 seconds and using a phase velocity of $c = 3.0$ km/s and group velocity, $U = d\omega/dk = 2.0$ km/s. The dashed line tracks the zero crossing propagating at the phase velocity; the solid line tracks the peak of the envelope function (times measured at zero amplitude lines).

short distance ranges. The interference effects described above occur across every small frequency range in the signal band.

14.3.3 Calculating group velocity

We can calculate group velocity for models using a perturbation analysis based on the properties of the eigenfunctions of the equations of motion. The details of this approach are beyond our scope here, but can be found in more advanced texts such as Ben-Menahem

and Singh (1981) or Aki and Richards (2009). For simple models, when a dispersion equation can be obtained with an implicit form, $F(\omega, k) = 0$ (for example, from a vanishing determinant), the group-velocity can be estimated from $dF = (\partial F/\partial \omega)d\omega + (\partial F/\partial k)dk = 0$, from which,

$$U(\omega) = \frac{d\omega}{dk} = -\left(\frac{\partial F}{\partial k}\right)_\omega \Big/ \left(\frac{\partial F}{\partial \omega}\right)_k . \quad (14.40)$$

Consider the Rayleigh waves for a water layer over a halfspace. For this case, a potential-based analysis can be performed considering P-wave potentials in the water, and P and S potentials in the halfspace. Requiring vanishing stresses at the water surface (located at $x_3 = -h$) and continuous vertical displacement and vertical stress on the ocean floor (located at $x_3 = 0$), the following dispersion relation can be derived,

$$
F(\omega) = \tan\left(h\omega\sqrt{\frac{1}{\alpha_w{}^2} - \frac{1}{c^2}}\right)
$$
$$
- \left[\frac{\rho\beta^4\sqrt{c^2/\alpha_w{}^2 - 1}}{\rho_w c^4\sqrt{1 - c^2/\alpha^2}}\right]
$$
$$
\times \left[4\sqrt{1 - \frac{c^2}{\alpha^2}}\sqrt{1 - \frac{c^2}{\beta^2}} - \left(2 - \frac{c^2}{\beta^2}\right)^2\right]
$$
$$
= 0 .
$$

(14.41)

The phase velocity as a function of period can be computed numerically as it was for Love waves (a root-finding exercise). We can also rewrite Eq. (14.41) in terms of wavenumber using $c = \omega/k$, compute the derivatives, and then evaluate the group velocity for each phase velocity using Eq. (14.40). Fig. 14.10 is an example of the computation using Eqs. (14.41) and (14.40). The differentiations were computed symbolically and the group velocities evaluated numerically. Period sampling was more dense near a period of $10\,s$ in order to sample the area of relatively rapid increase in the phase velocity. In general, the group velocity is lower than the phase velocity, and as the rate of change in the phase velocity decreases, the two speeds approach one another. Large differences occur where the change in phase velocity with period is significant. We can understand the pattern using

$$
U(T) = c - \Lambda\frac{dc}{d\Lambda} = c - \Lambda\frac{dc}{dT}\frac{dT}{d\Lambda} = c - \frac{\Lambda}{c}\frac{dc}{dT}
$$
$$
= c - T\frac{dc}{dT} ,
$$

(14.42)

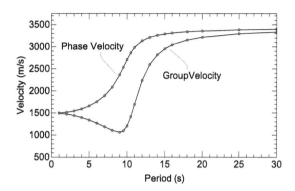

FIGURE 14.10 Phase and group velocities for a water layer with properties $\alpha_w = 1.5\,\text{km/s}$, and $\rho_w = 1000\,\text{kg/m}^3$ and an underlying halfspace with the properties $\alpha = 6.5\,\text{km/s}$, $\beta = \alpha/\sqrt{3}$, and $\rho = 2900\,\text{kg/m}^3$. The difference between the two curves depends on the phase velocity and the phase-velocity gradient.

where T represents the wave period. Clearly the larger the gradient in $c(T)$, the greater the difference between the two speeds.

For models with more than a few layers, the determinant-based approaches are unwieldy. Thus, eigenfunction perturbation analysis is the preferred approach for group-velocity computation for all but the simplest models. We can estimate the values using Eq. (14.42) directly with numerical differentiation, but numerical differentiation can be noisy. Next we explore the physical meaning of phase and group velocity, and understand their role in producing the distinct characteristics of surface waves on seismograms.

14.4 Dispersion on seismograms

Dispersion changes the overall appearance of a surface wave as it propagates. One can visualize the surface wave as having started from the source essentially as an undispersed pulse, with each frequency component having an amplitude $A(\omega)$ and initial phase, $\phi_0(\omega)$, determined by the excitation of the source and properties of the medium. As the wave travels outward from the source, dispersion modifies the interference pattern, spreading the energy out into a wavetrain, as shown in the Rayleigh waves of Fig. 14.1. The M_W 5.7 earthquake likely lasted only a few seconds but the Rayleigh-wave motions continue for roughly 10 minutes. Close to the source, the Rayleigh motion have a much shorter duration, but as the wave propagates to larger distances, the separation between the fastest and slowest components of the signal increases and the Rayleigh wave motion duration increases as the distances increases.

14.4.1 Measuring dispersion

Seismologists have long measured surface-wave dispersion to infer features of the subsurface geology. In the 1950's and 1960's as computer hardware and algorithms developed, solving the dispersion equations for plane-layered and spherical models allowed more quantitative modeling of the observations. A number of techniques were developed to measure group and phase velocity with more precision. Because earthquakes introduce frequency-dependent
Rayleigh and Love wave phase shifts related to source depth and faulting geometry, phase velocity is often measured using observations from multiple stations. Group velocity is less sensitive to the source properties, and it can be measured with single-station observations.

Group-velocity estimation

Group velocity is important because surface-wave energy propagates mainly in constructively interfering wave packets that propagate with group velocity. Given a single very well dispersed waveform from a source with known location and origin time, like that in Fig. 14.11, one can measure the arrival time of each period measured using peak-to-peak and trough-to-trough time measurements. The period is the difference in the times of successive peaks or

troughs, and the average of those two times, $\bar{t}$, can be used to compute the group velocity $U(T) = \Delta/\bar{t}$. Note that by using the epicentral distance, we assume that the surface-wave follows the great-circle arc from the source to the station. This may not always be the case, particularly for long paths, but it is a reasonable approximation.

Peak-to-peak methods are simple, useful for a quick estimate, but the results can be noisy and it may be hard to resolve the speeds of interfering periods. Filter-based methods are generally preferred for smoother and more robust group-velocity estimates. An observed surface wave can be thought of as a composite of interference packets. We can isolate packets by narrow-band filtering a seismogram. An example of the procedure, applied to the seismogram of Fig. 14.1 is shown in Fig. 14.12. The original unfiltered seismogram is shown at the top. Narrow-band filtered records with central periods shown on the right are shown below the original seismogram. The bottom axis shows time after origin, the top shows group velocity. Each narrowband-filtered trace includes one or more wave packets. The latest and largest is associated with the fundamental mode Rayleigh waves. The black circles mark the group delays and group velocities associated with components of the Rayleigh wave for each period. The path through tectonically stable eastern North America (Fig. 14.1) is relatively fast compared to the global average, and the group velocities for the period range from roughly 20-100s span from 3-to-4 km/s. Several higher modes are also apparent, they travel faster and arrive before the larger fundamental model signal. Finally, note the beating phenomenon in the coda of the shortest period Rayleigh waves. As we saw earlier, beating is a phenomenon when two similar-frequency harmonics are shifted slightly in phase. In this case, one possible explanation for the beating is the *multipathing* of short-period Rayleigh waves. Multipathing arises when parts of a signal follow slightly different paths, arrive at slightly dif-

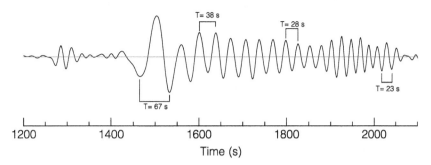

FIGURE 14.11 Observed Rayleigh waves shown Fig. 14.1 and filtered between 20 and 200 s period. The epicentral distance is 5, 963 km. Several measurements of the period using trough-to-trough and peak-to-peak times are shown.

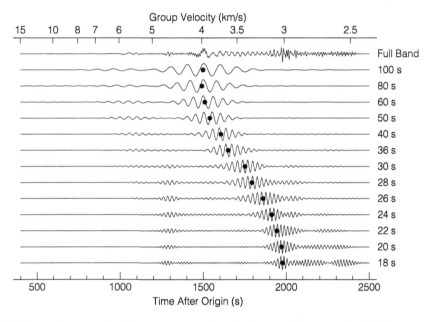

FIGURE 14.12 Observed propagating Rayleigh-wave packets for the signal shown in Fig. 14.1. The top trace is the original, unfiltered seismogram, the lower traces are narrow band-pass filtered versions of the signal with a central frequency corresponding to the period shown along the right. Each signal is self-normalized, so each filtered seismograms appear as large as the original. The period sampling is increased over the range where the velocity change is more rapid.

ferent times, and interfere, in this case producing a beating interference pattern.

If two stations are located along the same great-circle path, the group-velocities between the stations can be determined by measuring the difference in arrival times of filtered wave packets on the two seismograms. This is called a *two-station method*. A special application is the use of a single station to measure times between successive passes of surface waves traveling on the great circle (e.g., R_1, R_3). The measurement yields an average group velocity over the entire great-circle path containing the source and receiver. If information on the source depth and faulting geometry is available and reliable, we can compute a predicted (synthetic) seismogram and measure the phase difference between the observed and predicted surface waves to estimate the difference in phase velocity between the observation and the model. In theory, group velocity can be estimated from a phase-velocity dispersion curve by differentiation Eq. (14.34). Differentiation is a noisy calculation with actual data, so care is required for obtaining precise results with this approach.

Phase-velocity estimation

Several single- and two-station methods exist for measuring phase-velocity dispersion. We can obtain a crude measure using well-dispersed seismograms from two nearby stations, like those in Fig. 14.13. Each harmonic term at a given point in its cycle is associated with a peak or trough of a particular period of oscillation, and the differential time and propagation distance between corresponding cycles are used to estimate the phase velocity for each frequency. This simple procedure gives poor results unless the dispersion is so pronounced that the peaks are not envelopes of interfering frequencies. More typically, phase velocity is measured by taking the Fourier Transform of a seismogram to obtain a phase spectrum.

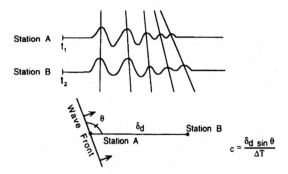

FIGURE 14.13 Examples of very well dispersed wave trains. Measurement of phase velocity between two nearby stations for which common cycles of a given phase can be reliably identified and differential travel time measured.

A surface wave can be represented using a continuous sum (integral) of sinusoids,

$$u(x, t)$$
$$= \frac{1}{\pi} \int_0^\infty \hat{u}(\omega, x) \times \cos\left(\omega t - \frac{\omega x}{c(\omega)} + \phi_0(\omega)\right) d\omega ,$$
$$(14.43)$$

where the harmonic's phase is $\phi(\omega) = \phi_0(\omega) - [\omega x / c(\omega)] + \omega t + 2n\pi$. The term $\phi_0(\omega)$ is the initial phase at the source, and the term $2n\pi$ accounts for the periodicity of the harmonic function (n is an integer). The amplitude spectrum, $\hat{u}(x, \omega)$, specifies the amplitude of each harmonic that contributes to the time-domain waveform. Consider a displacement seismogram that starts at time t_1 after the origin time, and that was observed at a distance x_1 from the source. The phase, $\psi_1(\omega)$, of the signal is

$$\psi_1(\omega) = \omega t_1 + \phi_0(\omega) - \frac{\omega x_1}{c(\omega)} + 2n\pi . \quad (14.44)$$

If the initial phase at the source, the origin time ($t = 0$), and the distance traveled (x_1) are known, then $c(\omega)$ can be determined to within the uncertainty arising from $2n\pi$. The appropriate value of n is usually identified by ensuring that the phase velocities for the longest-period signals agree with globally averaged values of $c(\omega)$; long-period phase velocities vary by only a few percent, which is sufficient to constrain the value of n.

The faulting geometry and source depth must be known to calculate $\phi_0(\omega)$. Additional corrections to the phase, particularly for long paths, are necessary to account for anelasticity and each sequential surface-wave passage of the source or the source's antipode (which adds $\pi/2$ to the phase). The most accurate procedure for estimating phase velocity is to take the difference in the phase spectra at two points on a great-circle path (as with group velocity, one can use a single station and look at successive surface-wave great-circle orbits). In this case the initial phase cancels out, leaving

$$\psi_1(\omega) - \psi_2(\omega)$$
$$= \omega(t_1 - t_2) - \frac{\omega}{c(\omega)}(x_1 - x_2) + 2\pi M,$$

$$(14.45)$$

or

$$c(\omega)$$
$$= \frac{x_1 - x_2}{(t_1 - t_2) + T\left[M - (1/2\pi)(\psi_1(\omega) - \psi_2(\omega))\right]},$$

$$(14.46)$$

where M, the difference in number of 2π cycles, is chosen to insure consistency with globally averaged values at long periods. Attenuation and polar passages corrections between the stations are needed for precise measurements.

14.4.2 Surface-wave dispersion and shallow Earth structure

There are two main applications of dispersion-curve measurements. The most critical is the determination of the subsurface velocity structure, and the second is correcting the observed phase back to the source so that the source radiation can be determined and used to constrain the source properties. Dispersion reflects the nature of the velocity gradients at depth. Stronger velocity gradients produce more pronounced dispersion.

Phase-velocity curves generally tend to be monotonic, whereas group-velocity curves often have a local minimum. The existence of a local minimum implies that significant energy will arrive at about the same time, producing an amplification and interference effect called an *Airy phase*. For continental paths an Airy phase with about a 20-s period often occurs, and long-period waves in the Earth have an Airy phase with approximately a 200-s period. Fig. 14.14 is a summary of the observed group velocities

for Rayleigh and Love waves in continental and oceanic regions. At periods longer than 80–100 s, regional near-surface differences have little effect since the waves are "seeing" deep into the upper mantle, where heterogeneity is less pronounced. The average oceanic crust is thinner than continental crust, resulting in a shift of the crustal Airy phase to periods of 10–15 s. Surface-wave dispersion sensitivity to crustal and upper-mantle velocity structure has led to extensive use of Rayleigh and Love waves to analyze three-dimensional Earth structure, which we describe in Chapter 10.

Rayleigh waves in a layered structure have overtones similar to those described for Love waves in the previous section (see Chapter 15). Both Love- and Rayleigh-wave overtones have their own dispersion curves. Generally the overtone group velocities are higher than velocities for the fundamental modes, causing overtones to arrive earlier. Fig. 14.15 shows examples of Rayleigh-wave overtones. The overtone wave packets are identified by X_n, where odd n correspond to initial minor-arc paths and even n to initial major-arc paths. The Rayleigh-wave overtone amplitudes tend to be stronger on the horizontal component than on the vertical component. Additional overtone observations were shown in Fig. 5.12 for a large strike-slip earthquake in Alaska. These Rayleigh-wave overtones are useful for probing deeper structure than that sampled by fundamental modes. Love-wave overtones are not well isolated from the fundamental modes in Fig. 5.12 but contribute to the long-period oscillations before the main Love-wave pulses.

14.5 Surface waves on a sphere

Earth's near-spherical shape has an important effect on surface-wave propagation – the waves spread over the spherical surface and hence converge at a point on the diametrically opposite side of the globe from the source, called

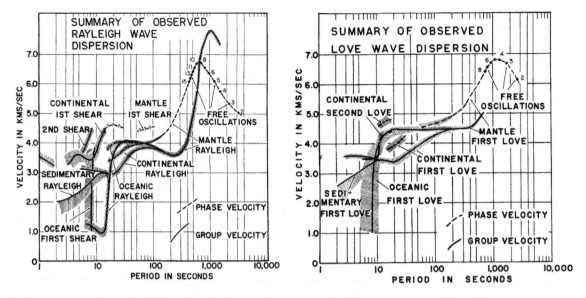

FIGURE 14.14 Observed phase and group velocity curves for Rayleigh (left) and Love (right) waves. These are observational patterns from the early 1960's (modified from Oliver, 1962).

the *antipode*. Waves converge from all directions at the antipode, and vertical-component Rayleigh waves constructively interfere and amplify, but the horizontal motions of Rayleigh and Love waves destructively interfere and defocus at the antipode. The waves "pass through" one another and diverge from the antipode, spreading over the surface again, eventually converging on the source and repeating the interference process. An example is presented in Fig. 14.16. At this distance, the Rayleigh and Love waves leaving the source in both directions along the great-circle arc containing the source and station (inset) arrive at the station simultaneously. The interference of the two wave packets greatly enhance the vertical component Rayleigh waves and suppress the horizontal component Rayleigh and Love waves. The interactions are complicated by Earth's lateral heterogeneity, which places some Love energy on the radial and some Rayleigh energy on the transverse components (although the instrument response should be checked to insure it is not an instrument effect). We can treat the motions

of the repeated passage of Rayleigh and Love waves on the Earth's surface as *traveling waves* or as patterns of **standing waves** or normal modes, which are discussed in the next chapter. Surface-waves obey Fermat's principle (Chapter 12) and follow the path of shortest travel-time across Earth's surface. For laterally uniform flat-layered structures, the surface-wave path is a straight line from source to receiver. Lateral variations in the medium require a curved trajectory (the surface waves refract) for the least-travel-time path.

On a sphere, the surface-wave path for a radially uniform structure is a **great-circle arc** (Box 14.2) connecting the source and receiver. Surface waves can travel in two directions along the great-circle arc containing the station. The shorter path is called the **minor arc** and the longer path is called the **major arc**. Since waves traveling along both arcs pass the station and continue to along the great circle, they eventually circuit the globe and pass by the station again, repeatedly. We denote long-period Rayleigh and Love waves by R and G (for Beno

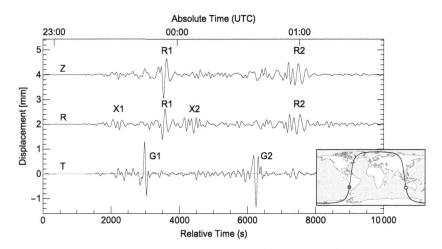

FIGURE 14.15 (A) Long-period seismograms of the 16 September, 2015, M_W 8.3, Illapel, Chile earthquake recorded at GSN station NWAO ($\Delta = 115°$). Channel labels are (Z), vertical component, (R), radial or longitudinal component, and (T), transverse component. All traces have been filtered to remove oscillations that have periods of less than 80 s. Long-period Rayleigh waves travel at a range of velocities with an average of roughly 3.5-3.9 km/s, Love waves travel at a central velocity of about 4.4 km/s. Higher-mode surface waves (X1,X2) are quite strong on the radial component, and travel with a speed of roughly $5.5 - 6.0$ km/s. Inset shows the great circle path containing the source (shaded symbol) and station (unshaded circle).

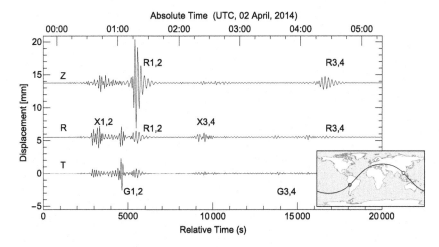

FIGURE 14.16 (A) Long-period seismograms of the 01 April, 2014, M_W 8.2, Iquique, Chile earthquake recorded at nearly antipodal seismic station QIZ, Qiongzhong, Hainan Province, China ($\Delta = 179°$). Channel labels are (Z), vertical component, (R), radial or longitudinal component, and (T), transverse component. All traces have been filtered to remove oscillations that have periods of less than 80 s. Inset shows the great circle path containing the source (shaded symbol) and station (unshaded circle).

Gutenberg, who studied Love waves), respectively. Minor-arc arrivals are indicated with odd-number subscripts that increase with the number of passages of the station (e.g., R_1, R_3, R_5), and major-arc arrivals are indicated by even-number subscripts (R_2, R_4, R_6, etc.). Fig. 14.15

shows an example of minor-arc and major-arc surface-wave arrivals recorded on a long-period channel of GSN station NWAO, located in Narrogin, Australia.

The horizontal ground motions are rotated to correspond to motion transverse to the great circle or along the great circle (radial positive away from the source). Note that the G_1 and G_2 arrivals at these periods (> 100 s) are more impulsive than the more dispersed Rayleigh waves. The Rayleigh-wave energy in R_1 and R_2 is stronger on the vertical component than on the longitudinal component by about a factor of 1.25, slightly lower than expected for Rayleigh waves in a Poisson half-space. The arrivals labeled X_1 and X_2 are a superposition of Rayleigh-wave overtones that have traveled on the minor- and major-arc paths with group velocities of about 5–7 km/s. Note the opposite direction of the G1 and G2 motions. The polarity of the transverse is defined using the minor-arc direction. R_2 is more dispersed than R_1 and has a lower amplitude because it has traveled farther. In general, one expects to see the amplitudes $|R_1| > |R_2| > \cdots > |R_n|$, but both propagation effects and source effects (Chapter 16) can produce more complicated amplitude behavior.

Fig. 14.17 is a plot of dispersion curves calculated for the PREM, which is an average of both continental and oceanic regions. As such, it's 23 km thick crust is inappropriate for short-period calculations almost everywhere on the planet. For periods longer than roughly ~ 60 s, the values are often reasonable on long paths, where sufficient averaging of Earth's shallow and heterogeneous geology has occurred. The Rayleigh-wave mantle Airy phase is clear at periods just over 200 s, as is the rapid increase for the long-period waves $T > 300$ s. The R1 Airy phase is often has a distinct pulse like character at the start of the Rayleigh wavetrain (e.g. Fig. 14.1, 14.11, etc.). Since the Rayleigh group velocity does not change much from about 60 s to more than 200 s, Rayleigh

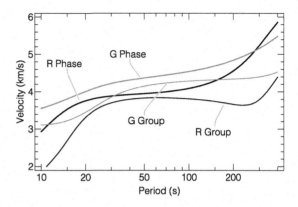

FIGURE 14.17 Group and phase velocity dispersion curves computed for the PREM (Fig. 1.21). R indicates Rayleigh waves and G indicates Love waves.

wave energy in this band arrives roughly at the same time, resulting in a relatively large pulse-like signal. The peak-and-trough method cannot separate the components with different periods in the pulse (Fig. 14.11), and that is one major advantage of the multiple filter approach to measuring group velocities.

14.6 Surface-wave amplitude and attenuation

Like body waves, surface-wave amplitudes carry substantial information on seismic sources and there is a long history of extracting and interpreting that information. We discuss more details in a later chapter, but some general comments are appropriate at this time. The surface-wave amplitude excited by a source is proportional to the eigenfunctions and their depth-dependent derivatives. Since the eigenfunctions are larger at shallow depths surface waves are better excited by shallow sources – and this is of course period dependent.

Box 14.2 Great-circle paths, azimuth, and back azimuth

Parameters of great-circle paths can be determined using spherical trigonometry. Consider the spherical triangle shown below. E is the epicenter, S is the seismic station, and N is the north pole. A, B, and C are the three internal angles of the spherical triangle.

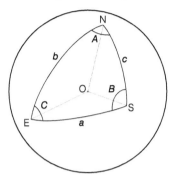

FIGURE B14.2.1 Spherical geometry for great-circle paths.

In general, $A + B + C \neq 180°$. Arc lengths a, b, and c are the sides of the triangle in degrees measured between radii from an origin in the center of the sphere. If A, b, and c are known, then

$$a = \cos^{-1}(\cos b \cos c + \sin b \sin c \cos A) , \tag{B14.2.1}$$

and

$$C = \cos^{-1}\left(\frac{\cos c - \cos a \cos b}{\sin a \sin b}\right) . \tag{B14.2.2}$$

The angular distance a is the **epicentral distance**, Δ. For most applications, A is the difference in longitude between E and S, and b and c are the source and station colatitudes, respectively (co-latitude is $90° -$ latitude). When measured clockwise from north, angle C is called the **azimuth** and gives the direction in which a ray must leave the source to arrive at the station. Source radiation patterns are usually expressed in terms of azimuth from the source. If the station was located to the left of the epicenter in Fig. B14.2.1, the azimuth would be $360° - C$ (remember, always measure *clockwise*). The **back azimuth** is the angle measured from north to the direction from which energy arrives at the station, is given either by

$$B = \cos^{-1}\left(\frac{\cos b - \cos a \cos c}{\sin a \sin c}\right) \tag{B14.2.3}$$

or by $360° - B$ (as in the case shown). Note that B is not simply related to C and they must be calculated separately. Back azimuth is used to determine the longitudinal and transverse directions for each source-station pair. The longitudinal component lies along the great circle connecting the source and receiver, and the transverse component is perpendicular to the great circle.

14.6.1 Geometric spreading

Once excited, surface-wave propagate, and their amplitudes decrease as the wavefront expands. In a cartesian system, or for short distances and wavelengths, the geometric spreading amplitude decrease, $G(\Delta)$, occurs at a rate

$$G_{flat}(\Delta) \propto \frac{1}{\sqrt{\Delta}} , \qquad (14.47)$$

where Δ is the source-to-station distance in km. On a sphere, the decay occurs as the wavefront expands as a small circle, which has results in an amplitude decay rate of

$$G_{sphere}(\Delta) \propto \frac{1}{\sqrt{a \sin(\Delta/a)}} , \qquad (14.48)$$

where Δ represents the minimum of the distances to the source and the antipode (in km) and a is Earth's radius. As the wave approaches the anti-pode, geometric spreading is reversed. Geometric spreading is not a factor at the antipode, all the energy is focused at the same location. These expressions, are of course ideal – in the real Earth, the fall of may not be exactly $1/\sqrt{r}$ or $1/\sqrt{\sin\Delta}$.

14.6.2 Attenuation

Anelastic losses also cause surface-wave motions to attenuate with time. For body waves we characterize anelastic properties of the Earth in terms of radial and lateral variations of the P-wave attenuation quality factor Q_α and the S-wave attenuation quality factor Q_β. Since, in general, both P- and S-wave motions contribute to surface waves and standing waves, there are separate Rayleigh (Q_R), Love (Q_L), spheroidal (Q_s), and toroidal (Q_t) quality factors, all depending on frequency as well as varying from path to path. The existence of anelasticity produces velocity dispersion, given by

$$c(\omega) = c_0 \left[1 + \frac{1}{\pi Q_m} \ln \frac{\omega}{\omega_0} \right], \qquad (3.132)$$

where subscripts indicate a reference frequency, ω_0, and reference phase velocity, c_0, and Q_m is the wave quality factor. Since surface-wave Q values are relatively low, on the order of 100 for short-period waves and a few hundred for long-period waves, the effects of physical dispersion become important. Thus, Q is studied for long-period waves both to understand attenuation processes in the Earth and to allow models of Earth structure consistent with both body waves and surface waves or normal modes to be derived.

Measurement of surface-wave attenuation is conceptually straightforward but difficult in practice. First, define a decay factor that accounts for amplitude, $A(\Delta)$, decays as the wave propagates a distance Δ,

$$A(\Delta) = A(0) e^{-\gamma \Delta} . \qquad (14.49)$$

Assuming that source effects, geometric spreading, etc. are accounted for, we can estimate a decay coefficient γ at a given period T using observations from two distances,

$$\gamma(T) = \ln\left(\frac{A_i}{A_{i+1}}\right) \bigg/ (\Delta_{i+1} - \Delta_i) . \qquad (14.50)$$

Some of the first measurements were made for sequential great-circle passages of R_i and R_{i+2} or G_i and G_{i+2} waves. For that geometry, the source effects and geometric spreading are the same for R_i and R_{i+2}. Also, $(\Delta_{i+1} - \Delta_i)$ is the circumference of the Earth. Call the spectral amplitudes at a specific period, A_i (for R_i) and A_{i+2} (for R_{i+2}). The corresponding inverse quality factor is then

$$Q(T) = \frac{\pi}{T U(T) \gamma(T)} , \qquad (14.51)$$

where $U(T)$ is the group velocity on the great-circle path. This approach has been used extensively to measure surface-wave attenuation values for periods less than $500 s$. We can estimate Q using the antipode $R1/R2$ and $R3/R4$ arrivals on the vertical component Rayleigh waves

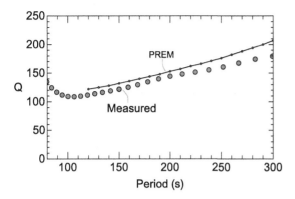

FIGURE 14.18 Attenuation quality factor, Q for long-period Rayleigh waves estimated using Eqs. (14.50) and (14.51) and the vertical-component seismogram in Fig. 14.16. The values for the PREM (from Chael and Anderson, 1982) are shown with small symbols and joined with a curve.

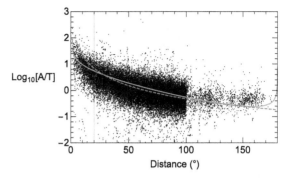

FIGURE 14.19 Logarithm of the 20 s period Rayleigh wave amplitude over period versus distance for 429 globally distributed earthquakes (34,750 individual observations) that occurred between 2010 and 2018. The mean logarithmic amplitude for each event is subtracted from the observations of that event. The curves show the geometric correction from the M_S formula (dashed curve) and the theoretical formula for $\gamma = 2 \times 10^{-4}$ km^{-1} (solid, light-gray curve). The magnitude formula is not used for distances closer than 20° (vertical line). The substantial change in the number of observations at $\Delta \sim 100°$ is a characteristic of the catalog and a result of measurement practices.

shown in Fig. 14.16. The $R1/R2$ and $R3/R4$ signals were isolated from the seismogram and the spectral ratio computed for a set of periods between about 80 s to 300 s (Fig. 14.18). Since the Rayleigh-wave amplitudes are sensitive to Earth's lateral heterogeneity, the estimated Q values using the spectral ratio approach are only approximate. Corrections for heterogeneity can be applied, but the measurements provide a reliable, rough estimate of the upper mantle Rayleigh-wave Q. The longer-periods (longer wavelengths) are probably less susceptible to velocity heterogeneity and so the increase is more interesting. An outstanding question in global seismology is a noted discrepancy between surface-wave and normal mode Q estimates in the period band from 15-300 s. Surface-wave-based Q estimates are systematically lower, and this is the case in Fig. 14.18. However, as with many seismological analyses, modern approaches synthesize many more observations than this simple, single-path example. But the concepts underlying the estimates remain the same.

Consider the empirical formulas developed for the surface-wave magnitude scale,

$$M_S(20) = \log A/T + 1.66 \log \Delta + 3.3 ,$$
$$20° \leq \Delta \leq 160° .$$

The M_S distance correction formula suggests distance amplitude decay in a form

$$A(\Delta) \sim A(0) \, \Delta^{-5/3} , \quad \Delta \text{ in degrees.} \quad (14.52)$$

This empirical form includes both geometric spreading, scattering, and attenuation (at $T \sim$ 20 s). The geometric spreading and attenuation factors developed in this chapter can be represented as

$$A(\Delta) \sim A(0) \, \frac{e^{-\gamma \Delta}}{\sqrt{a \sin (\Delta /a)}} , \quad \Delta \text{ in km.} \quad (14.53)$$

The two expressions correspond to curves with different curvatures, so it's not possible for them to agree exactly, but the overall agreement is

Box 14.3 Surface-wave amplitude anomalies

Surface-wave amplitudes in a flat-layered structure decrease with increasing propagation distance because of geometric spreading, anelastic attenuation, and (generally) dispersion. On a spherical surface, surface-wave amplitudes decrease progressively with propagation distance because of anelasticity and dispersion, but geometric spreading has a more complex form. It can be shown (e.g., Aki and Richards, 2009) that away from the source or its antipode, geometric spreading is given approximately by $(\sin \Delta)^{-1/2}$, where Δ is the angular distance between source and receiver. This spreading gives the lowest amplitudes near $\Delta = 90°$, i.e., when the surface wavefront is spread over the entire circumference of the planet. Curiously, R_1, R_2, and R_3, for example, all have the same geometric spreading at a given station ($\Delta = \Delta_0$). Generally, however, we expect $|R_1| > |R_2| > |R_3|$, etc. due to the dominating effects of attenuation and dispersion, as seen in Fig. B14.3.1. The seismograms in Fig. B14.3.2 show several stations with the normal behavior (HAL, RAR), but other stations (PFO, CMO, KIP) for which strong amplitude anomalies (e.g., $|R_5| >> |R_4|$ at PFO, KIP; $|R_4| >> |R_3|$ at CMO) are observed. Since this earthquake did not have any source complexity that could account for these anomalies, propagation effects are probably responsible. Surface waves propagating on the surface of a laterally heterogeneous sphere (like the Earth) are deflected from the great-circle path, and focusing and defocusing can result. Part (B) of the figure shows raypaths on the surface of the Earth for 200-s-period surface waves traveling through a model having a laterally varying phase velocity. The rays bundle up, enhancing the amplitude. Part (C) shows predicted Rayleigh-wave amplitude anomalies at different distances from sources in Japan and North America plotted as functions of azimuth from the sources. Amplitude ratios are predicted to vary by a factor of 3, comparable with observations. Deflection of Love-wave energy (G_2) from the great-circle path can also be observed in seismograms. While the long-period deflections are usually minor, large-amplitude anomalies can result, and one must be cautious in assuming the surface-wave energy has propagated on the great circle.

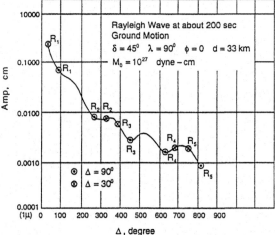

FIGURE B14.3.1 200-s-period Rayleigh-wave amplitude on the vertical component as a function of distance. Observations at two different distances, 30° and 90°, are marked for great-circle orbits. The source is 33 km deep and has a moment of 1×10^{20} Nm and a fault mechanism of strike = 0°, dip = 45°, rake = 90° (courtesy of H. Kanamori).

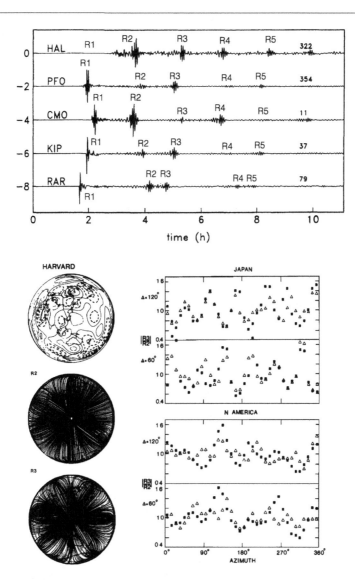

FIGURE B14.3.2 (A) Great-circle Rayleigh-wave arrivals at IDA stations for the September 1977 Tabas, Iran earthquake. (B) Projection of phase-velocity heterogeneity for 200-s-period Rayleigh waves on the hemisphere centered on Japan, along with surface-wave raypaths for R_2 arrivals at each point and R_3 arrivals at each point on the same hemisphere. (C) Calculated amplitude anomalies at different distances from two source regions for two models of surface-wave phase-velocity heterogeneity (boxes and triangles). **A** (from Masters and Ritzwoller, 1988); **B** (from Schwartz and Lay, 1985).

not bad with a $\gamma \sim 2\text{-}4 \times 10^{-4} \, \text{km}^{-1}$, which assuming a group velocity of $3.3 \, \text{km/s}$ for the 20 s period Rayleigh wave, corresponds to a $Q \sim 120\text{--}240$.

We compare the magnitude distance correction and spherical geometric spreading function with Rayleigh-wave amplitude observations in Fig. 14.19. The observations were extracted from the ISC Catalog and include shallow (depth $\leq 70 \, \text{km}$) earthquakes with magnitudes in the range from 5.8 to 7.5 that occurred between 2010 and 2018. We selected events that included 64 or more observations and subtracted the mean $\log_{10}(A/T)$ from each measurement. The resulting 429 earthquakes include almost 35,000 measurements (an average of about 80 measurements per event). The data show a trend, but also include significant scatter at each distance. The curves have been shifted vertically to align with the observations – the specific calibrations are part of the magnitude scale definitions. The curves agree reasonably well for most of the distance range, particularly given the first-order goals of magnitude measurements, which are not overly precise.

For shallow earthquakes surface waves are the largest intermediate- and long-period signals observed at distance. They produce some of the most interesting patterns on seismograms and are rich in information on Earth's interior and the seismic source that generated them. Generalization of our surface-wave discussions leads to the normal-mode analysis of the planet – a substantial extension of our analyses for the layer over a halfspace. The shape and major material boundaries within (and bounding) the planet constrain Earth's modes of vibration and mode observations provide important information on the seismic and density structure of Earth. We discuss these oscillations in some detail in the next chapter.

Chapter goals

- Examine mode-theory for spherical bodies.
- Discuss traveling- and standing-wave views of vibration.
- Review the characteristics of Earth's free oscillations.
- Discuss the importance and history of long-period seismology.

We have explored how vertical scale lengths in a layered medium constrain the character of motions that waves propagating along the surface can induce. In an analogous manner, the Earth's finite spherical shape introduces radial and circumferential constraints on solutions to the equations of motion for the planet. For example, only surface waves that constructively interfere after propagating around the Earth's surface will persist as long-term motions. The circumference provides a scale length into which an integral number of wavelengths can fit to produce persistent standing motions. Because only *discrete* wavelengths and frequencies fit the Earth's spherical boundary conditions, the corresponding standing waves are called the *free oscillations* or *normal modes* of the system.

15.1 A vibrating string

We can develop insight into normal modes by considering the case of a one-dimensional string held fixed at both ends (Fig. 15.1). We assume that a source excites small-amplitude motions of the string that propagate as waves away from the source in the $\pm x_1$ directions. The motions of the string, $u(x) = u_3(x)$ in the $\pm x_3$ direction. These motions must obey the one-dimensional wave equation,

$$\frac{\partial^2 u}{\partial x_1^2} = \frac{1}{c^2}\frac{\partial^2 u}{\partial t^2}. \tag{15.1}$$

In Chapter 11 we derived general solutions of this equation in the form

$$u(x,t) = C_1 e^{i\omega(t+x/c)} + C_2 e^{i\omega(t-x/c)}$$
$$+ C_3 e^{-i\omega(t+x/c)} + C_4 e^{-i\omega(t-x/c)}. \tag{15.2}$$

The boundary conditions for the string's fixed ends are $u(0,t) = u(L,t) = 0$. The first gives $C_1 = -C_2$ and $C_3 = -C_4$. The condition at $x = L$ then gives

$$\left(C_1 e^{i\omega t} + C_3 e^{-i\omega t}\right) 2i \sin(\omega L/c) = 0. \tag{15.3}$$

The nontrivial solutions are given by zeros of the sine function,

$$\omega L/c = (n+1)\pi, \quad n = 0, 1, 2, \ldots, \infty. \tag{15.4}$$

Thus, only motions with discrete frequencies, $\omega_n = (n + 1)\pi\ c/L$, called *eigenfrequencies*, exist and satisfy the boundary conditions. The eigenfrequencies have corresponding displacement patterns, $e^{i\omega_n t} \sin(\omega_n x/c)$, called *eigenfunctions* or *normal modes* of the system. The $n = 0$ mode is the fundamental mode and has no internal nodes (places where motion is zero) within the system; $n > 0$ corresponds to higher modes or overtones that each have n internal nodes. Fig. 15.1 includes a plot of the first three eigenfunctions that are allowed by the boundary conditions. The oscillatory motion of each eigenfunction such that the nodes are fixed, and the deformations are called *standing-wave patterns*, when viewed individually. The modes sum to a complete solution of the equations of motion, so any general propagating disturbance on the string can be represented by a weighted sum of the eigenfunctions (possibly an infinite number),

$$u(x,t) = \sum_{n=0}^{\infty} \left(A_n e^{i\omega_n t} + B_n e^{-i\omega_n t} \right) \times \sin\left(\frac{\omega_n x}{c}\right).$$

$$(15.5)$$

Thus, the standing-wave representation in terms of normal modes can equivalently represent traveling waves in the system. The Fourier power spectrum of (15.5) is discrete with spikes at the eigenfrequencies ω_n, with relative amplitudes given by the weighting functions, which are generally determined by the source-related boundary conditions.

Remarkably, this simple string system is a superb analog for Earth. The planet is more complex, but the basic ideas are the same. Earth can be set into global motions that make it ring like a bell at specific frequencies. At high frequencies, the resonances are so numerous the spectrum of shaking appears continuous. But at the longest periods, individual modes can be resolved directly from a long-duration seismogram. If we take a continuous-displacement recording that extends many hours or days after a large earthquake, like that in Fig. 15.2, we

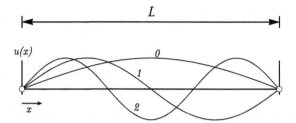

FIGURE 15.1 Geometry of a string under tension with fixed end points separated by distance L. Motions of the string excited by any source f comprise a weighted sum of the eigenfunctions, which are solutions that satisfy the boundary conditions with discrete eigenfrequencies. The first three eigenfunctions are shown – the fundamental mode ($n = 0$) has no zero crossings, the first higher mode ($n = 1$) has one, and the second higher mode ($n = 2$) has two.

can view the time-domain signal as a sequential passage of surface waves traveling along great circles. When the Fourier amplitude spectrum is computed for this signal (Fig. 15.2), we observe discrete peaks at different frequencies with variable relative amplitudes.

These correspond to eigenfrequencies of the Earth system, involving standing waves that fit into the layered spherical geometry of the planet. Interference of these coexisting standing-wave vibrations corresponds to disturbances that move along the surface as a function of time, which from a traveling-wave perspective are Love and Rayleigh waves. In fact, we can also represent all body-wave motions by summing a sufficient number of normal modes, for the infinite set of modes must represent all motions in the medium.

15.2 A vibrating sphere

The modes of a spherical body involve both radial and surface patterns that to match the physical boundary conditions, must fit the geometry of the system. The viable oscillations are of two basic types: (1) *spheroidal oscillations*, analogous to the *P*-, *SV*-, and Rayleigh-

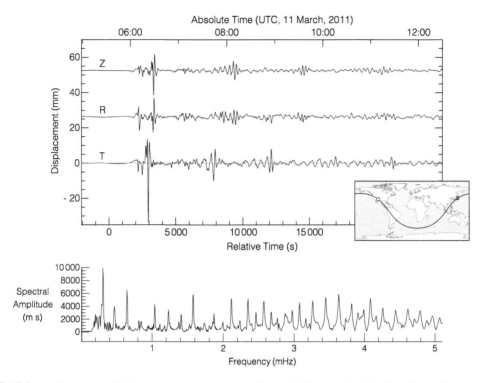

FIGURE 15.2 (top) Long-period displacement seismograms observed following the 2011 Tohoku earthquake (M_W 9.0). The origin time on the relative scale is roughly 800 s. The multi-orbit surface waves arrive as discrete packets. The inset shows the great-circle including the source and seismic station, ANMO, in Albuquerque, NM, USA. Following such a large event, Earth oscillates at long-periods for weeks. (bottom) Spectral analysis of long-duration time series illustrates the remarkable and discrete nature of the vibration frequencies. The spectral peaks center on Earth's preferred frequencies of long-period vibration. The spectrum was computed using 100 000 s of the vertical displacement high-pass filtered to exclude signals with periods longer than 4 000 s.

wave motions, which have a component of motion parallel to the radius (*radial* motion in *spherical* geometry) from the Earth's center; and (2) **toroidal** or **torsional** oscillations, involving shear motions parallel to the sphere's surface, analogous to SH- or Love-wave motions. Gravity does not influence toroidal motions, but long-period ($T > 500$ s) spheroidal motions involve significant work against gravity, and are more sensitive to the Earth's gross density structure than other seismic waves.

Fig. 15.3 is a summary of some of the characteristics of normal-mode motions for a spherical, elastic, nonrotating medium. The easiest modes

to visualize are the toroidal modes, which involve twisting motions of portions of the sphere. A nomenclature from spherical harmonics (Box 15.1) is used to identify patterns of motions. The toroidal modes are labeled $_nT_l$, where n is the number of zero crossing in the eigenfunction along the radius of the Earth, and where l is the number of nodal lines on the surface (the **angular order** number or **degree** of the spherical harmonic term). The radial order is analogous to the depth-dependent eigenfunctions in our discussion of surface waves. The angular order, l is a generalization of the horizontal wavenumber k, which was continuous in our surface-wave dis-

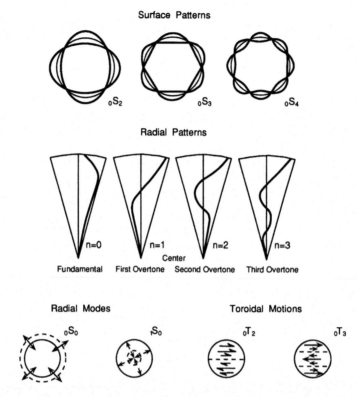

FIGURE 15.3 (Top) Surface and radial patterns of motions of spheroidal modes. (Bottom) Purely radial modes involve no nodal patterns on the surface, but overtones have nodal surfaces at depth. Toroidal modes involve purely horizontal twisting of the Earth. Toroidal overtones ($_1T_2$) have nodal surfaces at constant radii across which the sense of twisting reverses (after Bolt, 1982).

cussion, but not takes on a limited number of values because the medium is spherical and has a finite horizontal extent.

Mode deformation patterns are generally referenced to the source location, although rotational and aspherical effects are related to Earth's rotation axis. For our discussion, imagine the source is located at the pole. Both n and l can take on integer values up to infinity, but from a pragmatic perspective, for the Earth it is important to identify only the first few hundred values. Mode $_0T_0$ is undefined, and mode $_0T_1$ cannot exist because it would correspond to oscillation in the rate of rotation of the whole Earth, which for internal sources violates con-

servation of angular momentum. The lowest order toroidal mode is $_0T_2$, which corresponds to alternating twisting of the upper and lower hemispheres of the body. Mode $_1T_2$ corresponds to similar twisting of an interior sphere overlain by twisting in the reverse direction of the outer hemispherical shells.

Spheroidal modes are labeled $_nS_l$, where, in general, n and l have similar significance to their toroidal counterparts, although the poles are no longer positions of zero motion. Modes corresponding to $l = 0$ have no surface nodes and that subset is called *radial modes*, since all their motion is in the radial direction. Mode $_0S_0$ involves expansion and contraction of the sphere as a whole. Mode $_1S_0$ has one internal surface of zero motion separating alternating layers moving inward or outward. For $l > 0$, nodal lines occur on the surface along small circles parallel to the equatorial plane or along longitudinal great circles through the poles, which subdivide the surface into portions with alternating motion. Mode $_0S_1$ is undefined, as this would correspond to a horizontal shift of the center of gravity, which can happen only if the sphere is acted on by an external force. Mode $_0S_2$ is the longest-period normal mode of a sphere and is sometimes called the "football" mode. It involves alternating motion from a prolate to an oblate spheroid, as shown in Fig. 15.3. Mode $_0S_2$ has only two equatorial bands of zero motion, while $_0S_3$ and $_0S_4$ have three and four nodal lines, respectively.

15.3 Earth's free oscillations

The normal modes in Earth behave basically in this manner, but are complicated by the variation of material properties with depth and by departures from spherical symmetry caused by rotation, aspherical shape, and lateral material-property heterogeneity. For example, the Earth has a fluid outer core (see Chapter 10) in which the shear velocity is very small or zero. Torsional modes depend only on shear-velocity and are confined to solid shells of the mantle and crust. The inner core appears to be solid and may host inner-core toroidal motions, but if that is the case, some inner-core sources must excite the motions and they cannot be observed at the surface because they are decoupled from the mantle by the fluid outer core. Spheroidal modes are sensitive to both P and S velocity and den-

---

Box 15.1 Spherical harmonics

Analysis of the Earth's normal modes is most naturally performed in a spherical coordinate system. Here we consider basic mathematical solutions of the wave equation $\nabla^2 S = (1/c^2)(\partial^2 S/\partial t^2)$ in the spherical polar coordinate system (r, θ, ϕ) defined in Box 11.3. The wave equation for a homogeneous, nonrotating spherical fluid becomes (see Box 11.3)

$$\frac{1}{r^2}\frac{\partial}{\partial r}\left(r^2\frac{\partial S}{\partial r}\right) + \frac{1}{r^2\sin\theta}\frac{\partial}{\partial\theta}\left(\sin\theta\frac{\partial S}{\partial\theta}\right) + \frac{1}{r^2\sin^2\theta}\frac{\partial^2 S}{\partial\phi^2} = \frac{1}{c^2}\frac{\partial^2 S}{\partial t^2}.$$
(B15.1.1)

To solve this we use separation of variables, let $S(r, \theta, \phi, t) = R(r)\Theta(\theta)\Phi(\phi)T(t)$. We find the standard solution for the time-dependent term is a harmonic function, $T(t) = e^{-i\omega t}$, leaving

$$\frac{d^2\Phi}{d\phi^2} + m^2\Phi = 0,$$
(B15.1.2)

$$\frac{d}{dr}\left(r^2\frac{dR}{dr}\right) + \left[\frac{\omega^2 r^2}{c^2} - l(l+1)\right]R = 0,$$
(B15.1.3)

and

$$\frac{d}{d\theta}\left(\sin\theta\frac{d\Theta}{d\theta}\right) - \left[\frac{m^2}{\sin^2\theta} - l(l+1)\right](\sin\theta)\,\Theta = 0,$$
(B15.1.4)

where m^2 and $l(l+1)$ are constants of separation. Eq. (B15.1.2) is the harmonic equation, which has solutions $e^{im\phi} = \cos m\phi + i \sin m\phi$, where m must be an integer for the solutions to satisfy the spherical geometry. Eq. (B15.1.3) for $R(r)$ involves the frequency, ω, but not m; thus in the homogeneous, nonrotating system, ω will be independent of m but will depend on the constant l. This is a well-studied differential equation that has solutions called spherical Bessel functions, which have the form

$$j_1(x) = x^l \left(-\frac{1}{x}\frac{d}{dx} \right)^l \frac{\sin x}{x}, \tag{B15.1.5}$$

where $x = \omega\, r/c$. For $l = 0$ and $r R(r) \alpha \sin(\omega\, r/c)$, spherical Bessel functions have the form of decaying sinusoids, as shown in Fig. B15.1.1.

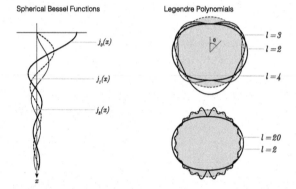

FIGURE B15.1.1 Functional behavior of spherical Bessel functions (left, x increasing downward) and Legendre polynomials shown in a polar projection (right). For clarity, the polynomials are shown by plotting $1 + 0.2 P_l (\cos\theta)$.

Eq. (B15.1.4) for $\Theta(\theta)$ is also in the form of a classic equation called the associated Legendre equation, which is usually given in terms of $x = \cos\theta$ with cases $m = 0$

$$\Theta(\theta) = P_l(\cos\theta) = P_l(x), \tag{B15.1.6}$$

where $P_l(x)$ are Legendre polynomials, which are computed using

$$P_l(x) = \left(\frac{1}{2^l l!}\frac{d^l}{dx^l} \right) \left(x^2 - 1 \right)^l, \tag{B15.1.7}$$

which gives $P_0(x) = 1$, $P_1(x) = x$, $P_2(x) = \frac{1}{2}(3x^2 - 1)$. Examples of Legendre polynomial functional dependence for $l = 2$ to 20 are shown in Fig. B15.1.1. For $m \neq 0$ the solutions are given by the associated Legendre functions $P_l^m(x)$, where

$$P_l^m(x) = \left(1 - x^2 \right)^{m/2} \left(\frac{d^m P_l(x)}{dx^m} \right) = \left(\frac{\left(1 - x^2\right)^{m/2}}{2^l l!} \right) \left(\frac{d^{l+m}}{dx^{l+m}} \left(x^2 - 1 \right)^l \right)$$

$$(-l \leqslant m \leqslant l). \tag{B15.1.8}$$

Many mathematical texts describe the properties of these functions in detail. For $m = 0$, Φ is a constant and S will have axial symmetry.

The product $\Theta(\theta)\Phi(\phi) = P_l^m(\cos\theta)e^{im\phi}$ is called a surface spherical harmonic of degree l and order m. The most common form used in seismology is the fully normalized spherical harmonic,

$$Y_l^m(\theta,\phi) = (-1)^m \left[\left(\frac{2l+1}{4\pi} \right) \frac{(l-m)!}{(l+m)!} \right]^{1/2} P_l^m(\cos\theta)\,e^{im\phi}, \qquad (B15.1.9)$$

where l and m are integers. The function $e^{im\phi}$ has zeros along $2m$ meridians of longitude (m great circles), while $P_l^m(\cos\theta)$ has zeros along $l - m$ parallels of latitude. Examples of the surface patterns produced by spherical harmonics are shown in Fig. B15.1.2. The *angular order number* l gives the total number of nodal lines on the surface with zero displacement. The parameter m gives the number of great circles through the pole with zero displacement. Thus, there are always $l - m$ nodal lines along latitude. Rotation of the coordinate system cannot change the order number l but can change m.

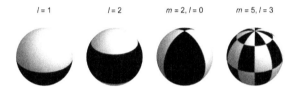

FIGURE B15.1.2 Examples of real-valued components of the surface spherical harmonics, $m = 0$ yields *zonal* harmonics of degree l. For $l = m$, the nodal surfaces are longitudinal lines giving *sectoral* harmonics. For $0 < |m| < l$, the combined latitudinal and longitudinal nodal patterns are called *tesseral* harmonics. The patterns show the polarity of the motion produced by the corresponding term in the modal sum.

Spherical boundary conditions constrain the eigenfrequencies in terms of the sphere geometry and constants n and l, $_n\omega_l$, where $n = 0$ corresponds to fundamental modes and $n > 0$ corresponds to overtones. Since $_n\omega_l$, does not depend on m for the case considered, all modes with angular order l (for a given n) will have the same frequency but different displacement patterns (eigenfunctions). Thus, a normal-mode power-spectrum peak for the system is actually a *multiplet*, composed of $2l + 1$ overlapping singlets with different displacement patterns. Overlap of all values of m is called normal-mode *degeneracy*.

Finally, for the elastic Earth we use vector surface harmonics:

$$R_l^m(\theta,\phi) = Y_l^m \hat{r} \qquad (B15.1.10)$$

$$S_l^m(\theta,\phi) = \frac{1}{\sin\theta}\frac{\partial Y_l^m}{\partial\phi}\hat{\phi} + \frac{\partial Y_l^m}{\partial\theta}\hat{\theta} \qquad (B15.1.11)$$

$$T_l^m(\theta,\phi) = \frac{1}{\sin\theta}\frac{\partial Y_l^m}{\partial\phi}\hat{\theta} + \frac{\partial Y_l^m}{\partial\theta}\hat{\phi} \qquad (B15.1.12)$$

to represent the total ground motion. The first two terms, R_l^m and S_l^m, are needed to describe spheroidal motion, while T_l^m describes toroidal motion.

sity structure, and the partitioning of compressional and shear energy with depth is complex. The displacement on and within Earth is expressed as a sum of time-dependent oscillations (modes) that can be written in terms of radially-varying eigenfunctions and spherical harmonic functions as

$$_n\mathbf{u}_l^m(\mathbf{r}, t)$$

$$= Re\left\{\left[{}_nU_l(r)\,Y_l^m(\theta, \phi)\,\hat{\mathbf{r}}+\right.\right.$$

$$\left.{}_nV_l(r)\,\nabla_1 Y_l^m(\theta, \phi) - {}_nW_l(r)\,\hat{\mathbf{r}} \times Y_l^m(\theta, \phi)\right]$$

$$\left.\exp\left(i\,{}_n\omega_l^m\,t\right)\right\},$$

$$(15.6)$$

where θ and ϕ are the co-latitude and longitude, ∇_1 is the horizontal gradient operator (only the θ, ϕ terms), ${}_n\omega_l^m$ are the eigenfrequencies, U, V, and W are the scalar radial eigenfunctions and the integer order numbers and the multiplet numbers satisfy $n \geq 0$, $l \geq 0$, and $-l \leq m \leq l$. The radial eigenfunctions quantify how each mode deforms Earth with depth – the larger n, the more oscillations in the function, and they are identical for each singlet in the multiplet defined by m. $nU_l(r)$ (radial) and $nV_l(r)$ (longitudinal) correspond to spheroidal motions, $nW_l(r)$ corresponds to toroidal motions. To compute the motion for each mode, we must solve an eigenvalue problem incorporating the earth-model properties and the boundary conditions to estimate the eigenfrequency and eigenfunctions. To compute a seismogram, we also compute an excitation factor for each mode, which depends on the source location, faulting geometry, and rupture properties, including the seismic moment, and we sum all the modes over the frequency range of interest.

Example eigenfunctions (spheroidal modes) are shown in Fig. 15.4. In general, *fundamental* modes ($n = 0$) have energy concentrated in the mantle (Fig. 15.4), and the shear energy is distributed deeper into the mantle than the compressional energy. Mode $_0S_2$ is sensitive to the entire mantle and gravity variations over the depth extent of motions caused by the mode. Including effects of self-gravitation changes the predicted period of this mode by almost 10 min. As l increases, the energy in both shear and compression is concentrated toward the surface. For $l > 20$, the fundamental spheroidal modes interfere to produce traveling Rayleigh-wave fundamental modes. The overtones ($n > 0$) of spheroidal motion generally involve energy sampling deeper in the Earth, including in the inner and outer core. In the Earth, radial-motion eigenfunctions of the spheroidal modes ($n, l > 0$) do not necessarily have n zero crossings along the radius, although this is true for the toroidal system.

15.3.1 Observing Earth's natural frequencies of vibration

Measuring Earth's eigenfrequencies is a substantial and important seismological activity. Earth's normal modes are identified primarily by computing ground-motion Fourier spectra, as seen in Fig. 15.2, and by associating the corresponding eigenfrequencies with those calculated for a model of the planet. The process is iterative, changes in the model can lead to re-identification of a particular mode peak. Mode identification began in 1882 when Horace Lamb first calculated the normal modes of a homogeneous, elastic, solid sphere, and discovered that $_0S_2$ must have the longest period. The search for this mode of the Earth required development of very sensitive ground-motion instruments. Hugo Benioff, a pioneer in development of ultra-long-period instrumentation, and colleagues used signals from the great 1952 earthquake in Kamchatka to observe a mode with a period of ~ 57 min, close to the ~ 1-h period expected for $_0S_2$. That observation was refined following the 1960 Chile earthquake ($M_w = 9.5$, the largest earthquake instrumentally observed). By

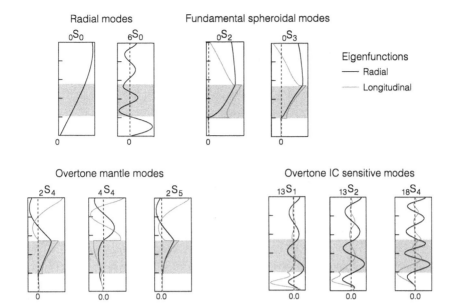

FIGURE 15.4 Low-order spheroidal normal modes for selected regions of Earth. The top of each panel is Earth's surface, the bottom is Earth's center. The gray region shows the range of the fluid outer core. IC indicates inner core, and modes such as these provide the best evidence on bulk inner core shear-wave properties (modified from Laske and Widmer-Schnidrig, 2015).

1960, reasonably good Earth models derived from travel times were available, and many modes could be confidently identified. In all, about 40 normal modes were observed using signals from the 1960 event, and the period of $_0S_2$ was estimated to be roughly 53.83 min. Subsequently, several thousand modes have been identified and their degenerate eigenfrequencies determined. Table 15.1 is a list of degenerate frequencies (see Box 15.1) of various observed modes of the Earth.

The process of isolating and identifying mode frequencies and refining Earth models using the information flourished in the 1970s. Fig. 15.5 is a plot of predicted and observed spheroidal mode eigenfrequencies ordered by angular order number l. Measurement of these observations has been a multi-decadal effort pioneered by Gilbert and Dziewonski (1975) and continued through to the present. The modes sort into distinct branches for different values of n, but

TABLE 15.1 Sample observed normal-mode periods.

Spheroidal modes	T (s)	Toroidal modes	T (s)
$_0S_0$	1227.52	$_0T_2$	2636.38
$_0S_2$	3233.25	$_0T_{10}$	618.97
$_0S_{15}$	426.15	$_0T_{20}$	360.03
$_0S_{30}$	262.09	$_0T_{30}$	257.76
$_0S_{45}$	193.91	$_0T_{40}$	200.95
$_0S_{60}$	153.24	$_0T_{50}$	164.70
$_0S_{150}$	66.90	$_0T_{60}$	139.46
$_1S_2$	1470.85	$_1T_2$	756.57
$_0S_{10}$	465.46	$_1T_{10}$	381.65
$_2S_{10}$	415.92	$_2S_{40}$	123.56

some branches come close together and have very similar eigenfrequencies. Precisely measuring the overlapping mode frequencies is a data intensive (years of recordings), iterative process. Groups of overtones along trajectories in the ω-l plot correspond to particular body-wave

15. Free oscillations

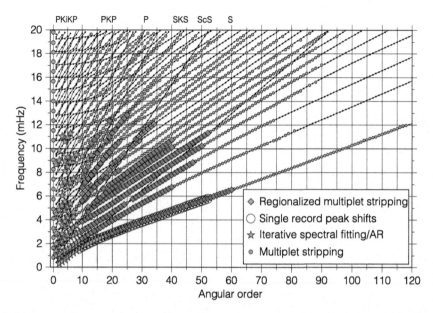

FIGURE 15.5 Spheroidal mode eigenfrequencies as a function of angular order number, l. Note that modes align on fundamental ($n = 0$) and overtone ($n > 0$) branches. The small solid circles identify modes not yet identified, but predicted using the PREM. Body-wave equivalent modes, which cross branches, are indicated for a few main body-wave phases (modified from Laske and Widmer-Schnidrig, 2015).

equivalent energy, a few of which are identified in the figure. This association is based on the modes that have appropriate phase velocities and particle motions. Given an Earth model that adequately predicts the observed eigenfrequencies, one can, of course, predict the eigenfrequencies of all modes. However, as we move to higher frequencies, three-dimensional heterogeneity begins to complicate the use of spherically symmetric models to disentangle overlapping branches. Depending on the method, modes up to $20\,mHz$ and $12\,mHz$ have been estimated, although some remain to be identified and measured individually (Fig. 15.5).

Mode splitting

This elegant approach of considering the entire Earth system in a boundary-value problem is complicated by asymmetry of the system. The most important factor is Earth's rotation, which produces an asymmetric Coriolis force.

The asymmetry disrupts eigenfrequency degeneracy and separates each mode's multiplet into $2l + 1$ distinct m **singlet** peaks (Box 15.1). We call this phenomenon **splitting**. Split eigenfrequencies are close and the relative split modes displacement patterns interfere with one another. The singlets are identified by the superscript m, so the multiplet $_0S_2$ is composed of singlets $_0S_2^{-2}$, $_0S_2^{-1}$, $_0S_2^{0}$, $_0S_2^{1}$, and $_0S_2^{2}$, each with a singlet eigenfrequency, $_n\omega_l^m$, and eigenfunction. Splitting of modes $_0S_2$ and $_0S_3$ was first observed for the 1960 Chile earthquake. Rotation splits the singlet eigenfrequencies according to the amount of angular momentum they possess about the Earth's rotation axis. A modern example is shown in Fig. 15.6, a spectrum of strain recorded at the Black Forest Observatory in Germany and enhanced by clever signal processing and corrections for known, interfering signals. The recovery is superb.

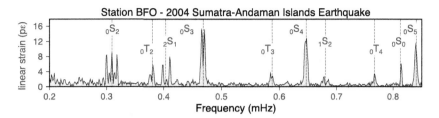

FIGURE 15.6 Normal-mode spectra computed using 220 hour-long time-series from the invar-wire strainmeters at the Black-Forest Observatory (BFO, Germany). Time-series started at 2:30:00.0 hr UTC on 26 December 2004 and the data were calibrated, instrument corrected, detided, corrected for air pressure effects, and tapered before the Fourier transform (modified from Zürn et al., 2015).

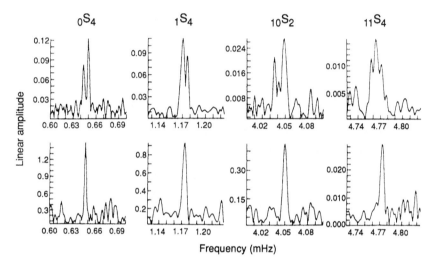

FIGURE 15.7 High-resolution spectra of four low-harmonic-degree multiplets recorded at nonpolar latitudes (top row) and polar latitudes (bottom row). The polar spectra are not obviously split, indicating effective cylindrical symmetry as produced by rotation. The low-latitude spectra are split. Rotation explains the splitting of the $_0S_4$ and $_1S_4$ modes well, but under-predicts the splitting of $_{10}S_2$ and $_{11}S_4$ (from Masters and Ritzwoller, 1988).

Earth's rotational splitting of modes manifests itself differently for each source-receiver combination. For example, rotational splitting decreases for stations near the rotation poles (high latitudes) (Fig. 15.7). The same mode has discrete multiplets at nonpolar stations (actually the $2l + 1$ multiplets are smeared together to give broadened, multiple peaks that do not resolve the individual eigenfrequencies for each l, m eigenvalue) but a single degenerate multiplet spike at high latitudes where the Coriolis force does not perturb the symmetry of the mode patterns. The strong splitting of the modes $_{10}S_2$ and $_{11}S_4$ is greater than expected due to rotation, and these modes are sensitive to the core; this is now attributed to anisotropy of the inner core aligned along the spin axis (see Chapter 10). In the time domain, split singlets produce a strong beating phenomenon (Section 14.4).

Rotation is not the only cause of multiplet splitting – aspherical structures associated with Earth's ellipticity and large-scale heterogene-

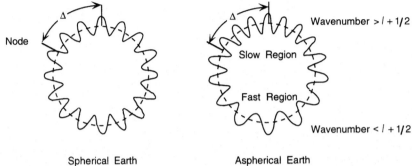

FIGURE 15.8 A cartoon illustrating the distortion of the standing-wave multiplet caused by lateral heterogeneity in velocity structure. Although the number of wavelengths around any great circle remains constant, the local wavenumber, k, varies with local frequency perturbation $\delta\omega_{local}$. The spatial shift of the phase at distance Δ perturbs the observed multiplet amplitude (modified from Park, 1988).

ity also can disrupt mode degeneracy. If we think of the standing-wave energy distributed over a great circle, lateral variations in velocity structure will distort the standing-wave pattern, locally perturbing the eigenfrequencies of the multiplet, as shown in Fig. 15.8. A mode will average the great-circle velocity structure, and different average great-circle velocities result in different multiplet frequencies for different paths. The local shift of phase at a particular distance affects the amplitude of the multiplet, and leads to variations of the mode spectral peaks that correspond to focusing and defocussing in the traveling wave-equivalent surface waves (see Box 14.3), although one must also account for lateral (off-path) averaging of the modes.

Mode coupling

So far we have considered how a mode can be affected or modified by rotation, geometry, and heterogeneity. In some circumstances, different distinct modes can interact or couple. Mode coupling is impossible in a simple, symmetric sphere, but Earth is more complex. The eigenfrequencies of some distinct sets of modes are roughly the same (Fig. 15.5). Oscillations with eigenfrequencies can result in interactions between singlets of a multiplet, adjacent modes on the same branch, modes on different branches, and even interactions between toroidal and spheroidal modes, induced by Coriolis asymmetry. We refer to such interactions as *mode coupling*. Rotation, aspherical structure, and possible anisotropy of the medium must all be included in the complex calculations of coupling effects, but these calculations are necessary to estimate accurately the eigenfrequencies and attenuation of the individual modes. Fig. 15.9 is a plot of observed and synthetic seismograms that demonstrate coupling between spheroidal and toroidal modes (most favorably observed in great-circle paths traveling near the poles, with tangential motion that is very strong and spheroidal motion that is very weak). The complex signals before the arrival of R_4 are a mix of toroidal and spheroidal energy on the vertical-component seismogram. Mode coupling remains an active avenue of research that provides important insight and information into Earth seismic heterogeneity.

15.4 Attenuation of free oscillations

Anelastic losses cause free-oscillation motions to attenuate with time. For body waves we characterized anelastic properties of the

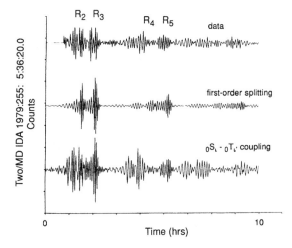

FIGURE 15.9 Data and synthetics for the September 12, 1979 New Guinea earthquake on a vertical-component recording at IDA station TWO. Coriolis coupling is high on this path, which goes within 5° of the rotation axis and leads to mixed spheroidal and toroidal motions on the seismogram. This is not included in the synthetics for first-order splitting, which account only for interactions within each multiplet, but is better accounted for when coupling between nearby fundamental-mode toroidal and spheroidal modes is calculated (from Park, 1988).

Earth in terms of radial and lateral variations of the P-wave attenuation quality factor Q_α and the S-wave attenuation quality factor Q_β. For surface waves we characterized attenuation in terms of period-dependent (modes) factors for Rayleigh (Q_R), Love (Q_L). Similar to surface waves, each spheroidal and toroidal mode has an associated quality factor, (Q_S), and toroidal (Q_T). Free-oscillation attenuation measurements can be made by a variety of procedures. For an isolated split multiplet, with mean eigenfrequency $\omega_0 = {_n\omega_l}$, the contribution to the displacements at the surface will have the form

$$_n u_l(\mathbf{r}, t) = \sum_{m=0}^{2l+1} a_m(\mathbf{r}) \exp\left[i\left(\omega_0 + \delta\omega_m\right)t\right]$$
$$\times H(t) \exp\left[-\frac{\omega_0 t}{2Q_m}\right], \qquad (15.7)$$

where t is time, $\mathbf{r}$ is a position vector, $\delta\omega_m$ is the difference between the singlet eigenfrequency, and ω_0 is the mean multiplet eigenfrequency, $a_m(\mathbf{r})$ is the amplitude of the singlet at the receiver. The function $H(t)$ is the Heaviside step function that is zero when its argument is less than zero and one when its argument is zero or greater. The amplitude, $a_m(\mathbf{r})$, is a function of frequency, the source and receiver location, earth model, and earthquake faulting parameters. The attenuation quality factor Q_m may or may not vary for each singlet.

If a multiplet is not split, then Q can be readily measured by narrowband filtering to isolate the mode and by using the temporal decay of the natural logarithm of the seismogram envelope. The displacement of an unsplit mode can be represented as the real part of Eq. (15.7),

$$_n u_l(\mathbf{r}, t) = Re\{_m a_l(\mathbf{r})\} \cos {_n\omega_l} t$$
$$\times H(t) \exp\left[-\frac{\omega_0 t}{2Q}\right]. \qquad (15.8)$$

The exponential function is an amplitude bound, or envelope, of the displacement. The logarithm of the envelope in slope intercept form is

$$\ln[_n u_l(\mathbf{r}, t)] = \ln[Re\{_m a_l(\mathbf{r})\}] - \frac{\omega_0}{2Q} t, \quad 0 \geq t.$$
$$(15.9)$$

Q is inversely proportional to the slope of the line. Smoothly decaying motions yield stable attenuation estimates. Fig. 15.10 is an example of the idea. The upper seismogram shows a multimodal (full) signal filtered only to remove long- and short-period noise. These motions are a sum of many modes and the long-period spectrum of the signal is shown in Fig. 15.2. The lower seismogram shows a narrow band filtered version of the same seismogram, with a filter designed to isolate the $_0 S_9$ mode. The $_0 S_9$ signal has been scaled by a factor of 30 compared to the original signal. After 10 hours of interference and focusing, the mode's oscillations begin a steady,

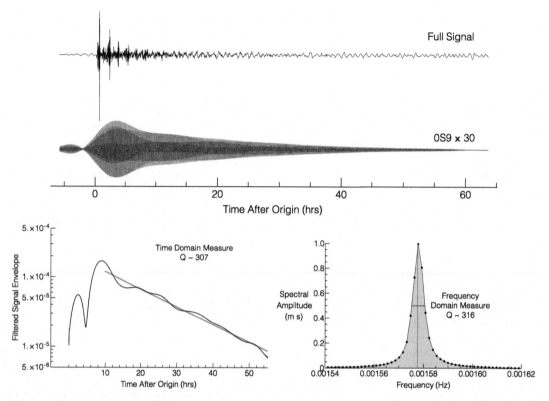

FIGURE 15.10 Q as estimated by narrowband filtering to isolate a single mode ($_0S_9$). The top seismogram shows a filtered vertical-component displacement seismogram recorded at station ANMO in the days following the 2011 Tohoku earthquake. The filter has removed periods longer than $4000\,s$ and shorter than $80\,s$ period. The bottom seismogram shows a narrow-band filtered signal centered on the period of the mode. The chart on the lower left shows a linear fit to the logarithm of the unsmoothed envelope of the seismogram mode as a function of time. The slope of the decay of amplitude is proportional to Q, and the estimated value is shown. The power spectrum (computed using just over 166 hrs of data) on the lower right shows the spectral peak associated with $_0S_9$. The width of the peak is proportional to Q, and the estimate from the spectrum is shown.

exponential decrease in amplitude. The panel on the lower left shows the logarithm of the $_0S_9$ signal envelope along with a linear fit in the time range of steady amplitude decay. The linear fit is reasonably good and the estimated $Q \sim 307$. This single-station estimate would be averaged with those of other stations to produce a robust estimate of the mode Q. The value listed in Dziewonski and Anderson (1981) is $319 \pm 5\%$, but several more recent estimates are slightly higher.

In the frequency domain, Q is estimated by the spread of the corresponding spectral peak at ω_0. Near the peak, the spectrum of a signal with the form of Eq. (15.9) has a shape that can be approximated by

$$|_nu_l(\mathbf{r}, \omega)|^2 \approx \frac{1}{4}\left[(\omega - \omega_0)^2 + \left(\frac{\omega_0}{2Q}\right)^2\right]^{-1}.$$

(15.10)

When $\omega = \omega_0 \pm \omega_0/2Q$, then which means that the full width of the peak at half of its maximum value, $\Delta\omega$, is related to Q by

$$Q^{-1} = \frac{\Delta\omega}{\omega_0}. \qquad (15.11)$$

We can measure Q by measuring the width of the power spectrum at half its peak value. The chart on the lower right of Fig. 15.10 is an example of a spectral peak Q estimation. The horizontal line is at half the power spectral value and the width of the peak suggests a Q value compatible with the value estimated from the time-domain approach (as it should, the data are the same, only the approach differs). The slight difference in value is within measurement uncertainty and both values are within the 5% uncertainty of the accepted value from a larger suite of measurements.

Clearly, if modes are split, both the frequency-domain and time-domain signals are complex, and simple Q measurements cannot be made. The analysis used to estimate Q then depends on the relative amount of pulse broadening due to attenuation versus multiplet splitting. If one can accurately predict the individual singlet eigenfrequencies, one can estimate Q by modeling the time-domain signal or the split spectral peaks and a sum of modes. As high-quality digital data have increased in abundance, seismologists have even measured separate singlet attenuation values for a few strongly split modes. More sophisticated time-series and spectral analysis approaches applied to thousands of waveforms recording hundreds of earthquakes have been combined to improve estimates of eigenfrequencies and Q values, and to isolate and measure the properties of additional modes including multiplets and singlets. The establishment and maintenance of long-period global seismographic networks continue to provide observations that enable important advances in the analysis of Earth's long-period oscillations. The work remains important because it provides crucial information on the nature of Earth's interior.

15.5 Building models of Earth's interior using normal modes

The 1960 Chile earthquake commenced the analysis of free-oscillation attenuation, and it was quickly recognized that Q is higher for longer-period fundamental modes. This indicates that Q increases with depth. The desire for quantitive approaches to extract the information contained in modes was a major motivation for the development of formal approaches to quantify model resolution and uniqueness that continue to play a central role in geophysical inverse modeling. We wish to relate the measurable quantities associated with a set of normal modes ($_n\omega_l^m$ and $_nQ_l^m$) to quantities that we would like to know – Earth's elastic parameters and density (or seismic wave speeds and density), and the attenuation properties represented by Q_κ and Q_μ or Q_α and Q_β. The quality factor for pure shear, Q_μ, and the quality factor in pure compression, Q_k, are related to the P and S-wave quality factors, Q_α and Q_β, by

$$Q_\beta = Q_\mu \qquad (15.12)$$
$$Q_\alpha^{-1} = L Q_\mu^{-1} + (1-L) Q_\kappa^{-1} \qquad (15.13)$$
$$Q_\kappa = [(1-L) Q_\alpha Q_\beta] / (Q_\beta - L Q_\alpha), \qquad (15.14)$$

where $L = \frac{4}{3}(\beta/\alpha)^2$.

Using a perturbation approach and Rayleigh's Principle, we can derive expressions that relate perturbations in model parameters to perturbations in the observed quantities. For example, for a radially symmetric earth model with known boundary positions (e.g. fixed core-mantle boundary), we can show that the perturbation of an eigenfrequency, $_n\delta\omega_l$ (ignoring splitting for notation convenience) can be expressed as

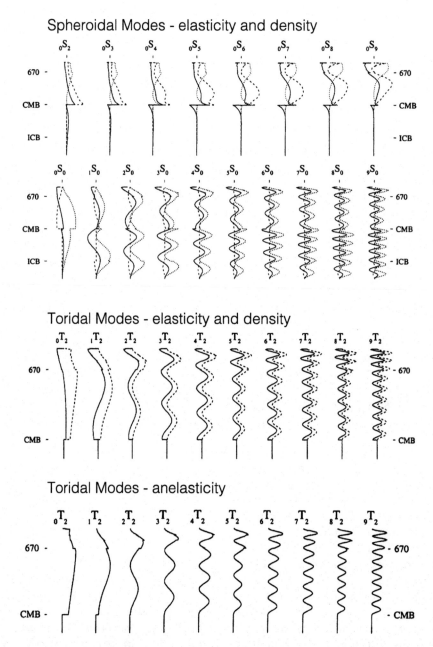

FIGURE 15.11 The top two panels are plots of eigenfrequency sensitivity kernels, $K_\alpha(r)$ (dotted line), $K_\beta(r)$ (dashed line), and $K'_\rho(r)$ (solid line) for select spheroidal modes (identified at the top of each plot). The vertical axis is depth and the location of the 670 discontinuity and the outer and inner core boundaries are indicated. The third panel from the top is a suite of sensitivity kernel, $K_\beta(r)$ (dashed line), and $K'_\rho(r)$ (solid line) plots for select toroidal modes (identified at the top of each plot). The bottom panel is a suite of the mode quality factor, Q, sensitivity kernel plots for the same toroidal modes shown in the panel above (from Dahlen and Tromp, 1998).

$$
_n\delta\omega_l = \int\limits_0^a [\delta\alpha(r)\, K_\alpha(r,n,l)
$$

$$
+\delta\beta(r)\, K_\beta(r,n,l) + \delta\rho(r)\, K'_\rho(r,n,l)\big]\, dr \;,
$$

$$(15.15)$$

where $K_{\alpha,\beta,\rho}(r,n,l)$, each a function of radius, are called *sensitivity kernels*, the perturbations to the P, S-wave speeds and density are functions of radius only, and a represents Earth's radius. Sensitivity kernels *for each mode* can be computed from the eigenfunctions and their derivatives (Dahlen and Tromp, 1998). This is an important expression since it says that we can use an existing model to relate changes in an observable to changes in model parameters. Let $\delta\omega$ represent the difference between observed and predicted $_n\omega_l^m$. Through Eq. (15.15), we can use that difference and an earth model estimate to compute model perturbations that would improve the fit to a set of observations. More generally, we can combine all our observations from many modes (and other data such as travel times) to "invert" the observational misfits to construct an improved model that reduces the amount of misfit. For attenuation, a similar perturbative approach produces

$$
(_n Q_l)^{-1} = \omega^{-1}\int\limits_0^a \Big[\delta\alpha(r)Q_\alpha^{-1}(r)K_\alpha(r,n,l)
$$

$$
+\delta\beta(r)Q_\beta^{-1}(r)\, K_\beta(r,n,l)\Big]\, dr \;,
$$

$$(15.16)$$

where $Q_\alpha^{-1}(r)$ and $Q_\beta^{-1}(r)$ represent the variations in P and S-wave quality factor inverses with depth.

Example sensitivity kernels computed using the transversely isotropic PREM are shown in Fig. 15.11. The sensitivity kernels indicate the depth range of eigenfrequency sensitivity for a small set of long-period spheroidal and toroidal modes. Each kernel is self normalized, so only the relative depth sensitivity is communicated.

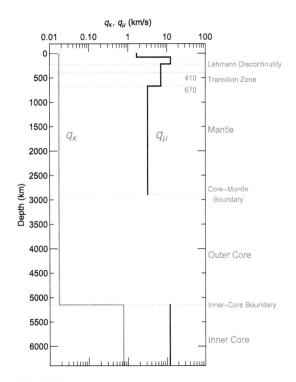

FIGURE 15.12 The PREM Q variations with depth ($q = 1\,000/Q$) represented in terms of bulk (κ) and shear modulus (μ). Q_μ is undefined in the outer core, since the fluid does not support shear motions.

The modes show interesting patterns in sensitivity and sample different regions of Earth's interior distinctly. The modes show interesting patterns in sensitivity. All modes sense density variations, spheroidal modes sense both P- and S-wave speeds, and toroidal modes sense S-wave speeds. The toroidal modes sample only the crust and mantle because the outer core is fluid. The bottom panel is a suite of plots of the attenuation sensitivity for toroidal modes. They resemble the elastic-property kernels because they are a weighted sum of the density and wave-speed kernels. The positive and negative variations in the kernels indicate *trade-offs* in model parameters that can occur between different parts of the model. We can increase the speed in one depth ranges and decrease the speed in one or more others without changing the eigenfrequency. Such trade-offs can be minimized if we combine observations across a broad frequency range using observations from many modes. But the smooth shape of the kernels ultimately limits the resolution of material property variations with depth.

If we model the planet using a set of uniform spherical shells (surrounding a uniform spherical core), with N layers, then Eq. (15.15) allows us to relate a change in eigenfrequency to a change in the model parameters in each layer. The complete integral becomes a sum of integrals across each layer and we can isolate the contributions of a layer. Although aging, the PREM is a model that is compatible with combined surface-wave, normal-mode, and body-wave attenuation measurements. The seismic wave speeds and densities are shown in Fig. 1.21, the corresponding attenuation model is shown in Fig. 15.12. Table 10.1 is a list of the model parameters (speeds, density, moduli, quality factors). Q_α is very high in the mantle and outer core, and seismic attenuation is dominated by shear. The low Q_β in the upper mantle is usually associated with the asthenosphere.

The low-Q upper mantle, affects surface waves significantly and the effect must be included in precise predictions of their amplitudes and phase. Q_β increases through the transition zone and then assumes higher values in the lower mantle. In the inner core, Q_β is relatively low, but Q_α is high, although not as high as in the mantle and crust.

Normal mode sensitivity to Earth's average structure and deviations from spherical symmetry provide important information on the planet's interior. Modes are used extensively to study Earth's interior. The problem is not easy, however, since the amplitude of modes is also sensitive to focusing and defocusing caused by lateral gradients in structure. Our estimates of Earth's attenuation variations are generally restricted to broader spatial scales than we estimate for seismic wave speeds (Box 14.3). We discuss models Earth's interior in more detail in Chapter 10. Finally, although we have focused on structure, the sensitivity of modes to earthquake depth, faulting geometry and long-period rupture characteristics also makes them invaluable for the analysis of moderate- and large-magnitude earthquakes.

16

Seismic point-source models

Chapter goals

- Introduce seismic wave polarities.
- Introduce simple seismic point-source models.
- Discuss physical approximations to explosion and earthquake sources.
- Introduce equivalent body forces for explosions and force couples.
- Introduce the seismic moment tensor.

Seismic waves excited by a source carry information on the nature of the source and accumulate information on the interior of the Earth along the paths they travel. As the wave propagates, the source and structure information is blurred together. An observed displacement seismogram is a convolution of source and propagation effects,

$$u(\mathbf{x}, t) = E(\mathbf{x}, t) * G(\mathbf{x}, t) , \qquad (16.1)$$

where $\mathbf{x}$ is a position vector, t is time, $u(\mathbf{x}, t)$ is displacement (a seismogram), $E(\mathbf{x}, t)$ represents source effects, and $G(\mathbf{x}, t)$ represents propagation effects (which we have discussed in other chapters). If your focus is on the source, you choose observations that minimize the unknown components in $G(\mathbf{x}, t)$, if your focus is on imaging Earth's interior, you choose observations that minimize the uncertainty in $E(\mathbf{x}, t)$. In some instances, with sufficient data, both quantities can be investigated simultaneously. But in general (and throughout much of the history of seismology), the problems are approached separately. Indeed, some research teams specialize in one or the other type of study and progress occurs as each group advances our understanding of seismic sources or earth structure.

We can split the source excitation function into two factors, one that depends on the fault geometry and near-source material properties, the other that depends on the rupture process. We let

$$E(\mathbf{x}, t) = A_F(\mathbf{x}, t, [\omega]) \cdot S(\mathbf{x}, t) , \qquad (16.2)$$

where $A_F(\mathbf{x}, t, [\omega])$ represents amplitude produced by fault orientation and $S(\mathbf{x}, t)$, the source time function (or moment-rate) function, represents the rupture process. The brackets around the radial frequency, ω, indicate that A_F may be frequency dependent – as is the case for surface waves, but not for body waves. Although we introduce simple forms for $S(\mathbf{x}, t)$ in the next chapter, our focus in this and the next chapter is on faulting-geometry and near-source material properties effects. We discuss more complex source time functions in Chapter 18.

We start our discussion with ideal explosive sources. We introduce a number of seismological concepts using the explosion model, which is historically and seismologically important. Then we turn our attention to simple earthquake models, focusing our analyses on spa-

439

tially localized sources, a reasonable approximation when the source is small compared with the wavelength of the seismic waves used to investigate the process (a common situation for small- to moderate-size earthquakes). We will see that elastic-wave theory provides simplified force systems that can be used to represent seismic sources and reveal important characteristics of seismic sources. We do not present a first-principles theory of seismic source phenomena, but review classic and key results used to model seismic sources of many types.

16.1 An ideal explosion

We first consider an ideal representation of underground explosions. Explosions are a common source of seismic waves and span a range of sizes. Small explosions are used for mining, quarrying, road excavating, and other construction applications, as well as in natural resource exploration and active-source seismic lithospheric study. Large explosions include underground nuclear tests, which produce waves strong enough to be observed on the opposite side of the Earth. Natural explosive or implosive sources are rare, but some may occur as a result of metastable mineralogical phase transitions or magmatic processes. The most general and ideal character that we associate with explosions is spherical symmetry – this is an approximation, but a useful place to start.

Consider an ideal explosion in a uniform isotropic wholespace (Fig. 16.1). Such an explosion can be modeled as a sudden increase in pressure inside of a finite spherically symmetric cavity. The explosion itself may produce a cavity by melting, vaporizing, and deforming the surrounding rock (underground nuclear explosions can produce 0.5-km-radius cavities in preexisting rock and shatter the surrounding rock). More generally, whatever the process that occurs immediately after the explosive energy release, at some distance from the source

called the **elastic radius**, r_e, nonlinear processes subside and small elastic deformations propagate the energy outward. We assume that at the elastic radius, the complex processes inside the nonlinear-region can be explained with a time-dependent pressure force, $F(t)$. The time history of $F(t)$ can range from an impulse (the pressure spikes and quickly drops back to its initial state) to a step (permanent strains in the inelastically distorted medium produce a permanent pressure change). Nuclear explosions often involve a combination of these, with an overshoot of pressure that decays to a static permanent step as gas pressure in the newly created cavity dissipates. The elastic radius for an underground nuclear explosion may be 1 km or larger.

Beyond the elastic radius, the equations of motion reduce to a one-dimensional inhomogeneous wave equation,

$$\frac{\partial^2 \phi(r,t)}{\partial r^2} - \frac{1}{\alpha^2}\frac{\partial^2 \phi(r,t)}{dt^2} = -4\pi F(t)\delta(r_e)\,, \quad (16.3)$$

where the P-wave displacement potential is $\phi(r,t)$, the P-wave velocity is α, and the effective force function applied at the elastic radius is $F(t)$. The solution for the displacement potential has the form

$$\phi(r,t) = \frac{-F(t - r/\alpha)}{r}\,, \quad (16.4)$$

where $F(t)$ is called the reduced displacement potential, and r is the distance from the elastic radius. Often the elastic radius is ignored and the source is treated as a spatial **point source**. A point source is localized and represented with a delta function such as $\delta(\mathbf{r})$. The point-source assumption is common and reasonable when the seismic wavelength is at least several times larger than the source dimension. In that case, the resolvable features of the source are a spatial average of the underlying unresolved processes. For an ideal explosion, seismic waves propagate outward from the point source with equal amplitude in all directions. A common

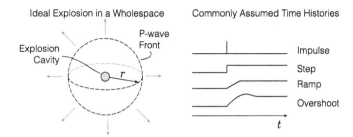

FIGURE 16.1 An ideal underground explosion source is conceptualized as a small volume on which a spherically symmetric pressure pulse is suddenly applied. The radius of the spherical source is called the elastic radius, r_e. At the instant of application a P wave is initiated in the surrounding elastic medium and propagates outward. The time dependence of the P-wave potential is the same as the time history of the pressure pulse, examples of which are shown in the inset. Displacements are computed from the potential by spatial differentiation (Eq. (16.5)).

characteristic of particular solutions of inhomogeneous wave equations for point sources is that the wave potential has a D'Alembert form with a potential time-history equal to the force-time history. The time history represents an effective integration of nonlinear processes inside the source radius. In a uniform wholespace, since the P-wave energy is spread over a spherical wavefront, the wave potential decays as $1/r$.

The spherically symmetric displacement field, $u(r, t)$, can be calculated by spatial differentiation, using $u(r, t) = \nabla \phi(r, t)$,

$$\mathbf{u}(r, t) = \frac{\partial \phi(r, t)}{\partial r} \hat{\mathbf{r}} = \frac{F[\tau(r)]}{r^2} \hat{\mathbf{r}} + \frac{1}{\alpha r} \frac{\partial F[\tau(r)]}{\partial \tau} \hat{\mathbf{r}},$$

(16.5)

where $\tau(r) = t - r/\alpha$ is called the **retarded time**. The surrounding medium is at rest until time $\tau(r)$ after the explosion, which corresponds to the time that it takes for the P wave to travel from the source to the observation point. The first term in (16.5) involves displacements that are directly proportional to the reduced displacement potential; these decay quickly with distance from the source (at a rate proportional to $1/r^2$). A term with such a rapid amplitude decay is called a **near-field term**. If any permanent step in the effective pressure is produced by a source, a static (permanent) deformation of the surrounding elastic medium will result. The

second term decays more slowly and dominates the displacement at large distances. We call such a term a **far-field term**. The time-history of the far-field displacement is proportional to a time derivative of the reduced displacement potential. Thus, a step in effective pressure at the source produces an impulsive far-field ground motion. This is a characteristic that we will find for far-field motions from other sources, and our ability to infer the time history of the forces acting at the source hinges on the sensitivity of far-field displacements to the temporal derivative of the source time history.

The solution in (16.5) indicates that the particle motions produced by an explosion in a whole space are outward along radial directions (both static deformations, if any, and transient P-wave motions). Such motions are observed in the Earth, which is clearly not a whole space, but the direct P waves from a large explosion generally exhibit **compressional** initial arrivals at all stations. Compressional P-wave motion is defined as P-wave particle motion away from the source, after allowing for direction changes along the seismic raypath (Fig. 16.2). Since elastic-wave propagation generally does not modify the shape of the wave initiated at the source, the near-source compressions can be observable at large distances. Although not part of the solution for an explosion, P-waves with the

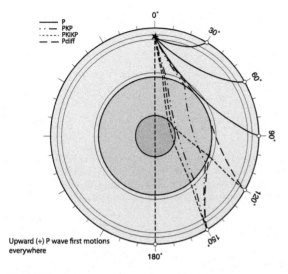

FIGURE 16.2 Initial *P*-wave motions from an ideal underground explosion are compressional, defined as being in the direction of wave propagation, or "away" from the source, allowing for the raypath perturbations caused by Earth structure. Thus, explosions produce upward ground motions for the first *P* or *PKP* arrivals everywhere on the surface of the Earth.

opposite motion, towards the source, are called *dilatational*. For reasons presented later, a motion leaving the source with small (ideally zero) amplitude is referred to as *nodal*.

An ideal explosion involves no shear deformation at the source, and hence no *S* wave is excited or directly generated by the explosion. However, even if the Earth had perfect radial symmetry, we expect vertically polarized shear (*SV*) waves and hence Rayleigh waves to be produced by the wave interactions with elastic boundaries. For example, we would expect to observe *P*-to-*SV* conversions at Earth's surface (*pS*, *PS*, *PPS*, etc.) or the core-mantle boundary (*PcS*, *PKS*, *PKIKS*, etc.). Theoretically, we would not expect to see either horizontally polarized shear (*SH*) or Love waves from an explosion in a radially symmetric Earth. But we commonly observe *SH* and Love waves from both large and small explosions. The transversely-polarized energy is produced by the departure

from spherical symmetry of the source, triggering of an earthquake or deviatoric strain release near the explosion, and/or conversion of *P* and *SV* energy to *SH* energy by Earth anisotropy and heterogeneity (Box 16.1).

16.2 Faulting sources

Most seismic sources lack the simple spherical symmetry of an ideal underground explosion. For example, an earthquake involves a sudden displacement offset across a fault (a dislocation) that is certainly not spherically symmetric. But the process produces a diagnostic seismic-wave amplitude pattern that is imprinted on the wavefronts emanating outward from the source. A *radiation pattern* is a geometric description of the wave amplitude and sense of initial motion. The symmetry of earthquake faulting provides predictable relationships between the radiation pattern of detectable wave motions and the faulting geometry, which allow us to investigate faulting processes remotely.

The Elastic Rebound Model was introduced in Chapter 1. In the model, sliding occurs when the elastic strain accumulation in the vicinity of the rupture overcomes the static frictional stress that had been resisting motion. Slip initiates in a small region or point called the *hypocenter*. As the instability grows, a *rupture front* expands outward over the fault, separating slipping regions (or perhaps regions that have slipped and then come to rest) from regions that have not yet slipped (Fig. 16.3). The area of rupture is a function of space and time, $A_R(\mathbf{x}, t)$, as is the vector slip, $\mathbf{D}(\mathbf{x}, t)$, which describes the offsets across the fault within the ruptured area. As the rocks slide, elastic strain energy stored adjacent to the fault (the source volume) is transformed – across-fault contacts are broken, frictional sliding under high stress heats the fault-bounding rocks, and the suddenness of motion excites seismic waves. Eventually the rupture stops, sliding arrests, and the earthquake ends. In some re-

Box 16.1 Explosion SH waves?

One of the societally important applications of seismology is monitoring for the occurrence of underground nuclear explosions. Seismic waves provide estimates of the explosion location and size, mainly by empirical calibration of *P*- and Rayleigh-wave amplitudes with explosions of known *yield*, or energy release, in equivalent kilotons of TNT. Distinguishing between an explosion and a natural source is the first analysis task. Usually explosions are identified by examining a variety of waveform characteristics that may distinguish earthquakes from explosions. Since an ideal explosion will not generate significant transversely polarized waves, it would seem reasonable to rely on whether or not significant *SH*- or Love-wave energy is observed. However, Fig. B16.1.1 compares *SH* and Love waves from an earthquake with the same component recorded for an explosion.

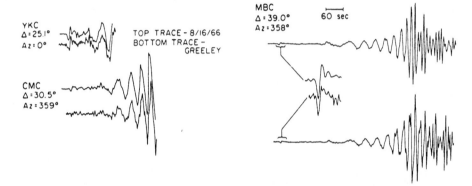

FIGURE B16.1.1 Comparison between *SH* and Love waves for the nuclear explosion GREELEY (December 20, 1966) and an earthquake in eastern Nevada (August 16, 1966). The seismograms are lined up on the *S* arrival, and the amplitude scale is the same for both the earthquake and explosion (from Wallace et al., 1983).

The *SH* components from both events are virtually identical, and this holds for the full azimuthal range surrounding the events. The large explosion GREELEY, presumably triggered nearby fault motions in the surrounding crust, a process called *tectonic release*, which accounts for the transverse-wave energy. Although not all explosions produce such clear transversely polarized radiation, it is clear that other waveform attributes are needed to discriminate explosions. The most successful discriminants prove to be location (e.g. source depths greater than humans can drill) and some quantity that can be conceptually related back to the ratio of compressional-to-shear-wave energy. For example, the ratio of short-period (1-s) *P*-wave energy (m_b) to 20-s-period Rayleigh-wave energy (M_s) is higher for explosions than for earthquakes larger than magnitude 3.5. Short-period (1-to-10 Hz) regional spectral amplitude ratios of *P*-to-*S* or L_g waveforms are used to discriminate events observed regionally (see for example Douglas 2013 and Maceira et al. 2017).

gions, the same process occurs at a much slower rate and the loss of suddenness results in only small seismic motions. We call that category of fault slip *slow earthquakes*.

As with an ideal explosion, for the long-wavelength waves excited by the earthquake, the rupture area and source volume participating in strain-energy release are relatively small, and the source can be approximated as a spatially localized *point source*. Shorter-wavelength waves are sensitive to the finite rupture extent, rupture-front expansion, and to the variation of slip across of the rupture. Their analysis requires consideration of *finite source* models – we discuss these effects in Chapter 18. For now, we concentrate on the geometric aspects of faulting (fault orientation and slip direction) and near-source material properties, which have a substantial effect on the amplitude patterns of seismic waves.

16.2.1 Shear-faulting nomenclature

The standard nomenclature that has evolved for describing fault orientation and slip direction was introduced in Box 3.1. During rupture, slip direction is constrained to lie on the fault surface, which precludes opening or closing cracks, but we discuss such sources later. To describe the orientation of the fault in geographic coordinates, we use two angles, the fault *strike*, ϕ_f, which is the azimuth (measured from north) of the projection of the top of the fault onto Earth's surface; and the fault *dip*, δ, which is the angle measured downward from Earth's surface to the fault in the vertical plane perpendicular to the strike. The strike direction is chosen such that the dip is $\leq 90°$ (i.e., if you orient the thumb on your outstretched right hand along the strike and rotate your hand downward from the horizontal to the fault plane, your hand will pivot through an angle less than $90°$). Strike direction is arbitrarily either orientation for a vertically dipping fault ($\delta = 90°$). The relative motion of the rocks on either side of the

fault is defined by the *slip vector*, which can have any orientation on the fault surface. The direction of the slip vector is specified by the angle of *slip*, also called the *rake*, λ. The rake is measured on the fault surface from the strike direction counter-clockwise to the slip vector. The rake quantifies the motion of the *hanging wall*, rocks above the fault, relative to the *footwall*, rocks below the fault. The magnitude of the slip vector is D, the total displacement of the rocks across the fault. In general, ϕ_f, δ, λ, and D vary over the rupture surface, so *average* values are used in simple models.

Three basic categories of slip are commonly used to characterize motions on faults (Box 3.2). When the relative slip of the rocks on either side of a fault is horizontal, the motion is called *strike-slip* ($\lambda \sim 0°$ or $180°$); if in addition, the dip is $\sim 90°$, the geometry is called *vertical strike-slip*. For $\lambda \sim 0°$, viewed from either side of the fault, the other side moves to the left and this type of fault movement is called *left-lateral*. If $\lambda \sim 180°$, for the similar reasons, the fault movement is called *right-lateral*. When the relative slip of the rocks on either side of a fault is vertical, the motion is called *dip-slip* because the slip vector parallels the dip direction. For $\lambda \sim 90°$, the hanging wall moves upward, resulting in *reverse faulting*. Reverse faulting with a shallow dip is often referred to as *thrust faulting*. For $\lambda \sim 270°$ the hanging wall moves downward, resulting in *normal faulting*. Dip-slip motion along a vertical fault is often called *vertical dip-slip* faulting. In general, λ will have a value different than these special cases, and the motion is then called *oblique slip*, and the faulting character is described by joining appropriate modifiers (e.g., *right-lateral oblique normal faulting*, for $180° < \lambda < 270°$).

16.3 Earthquake P-wave "first motions"

To motivate discussions that follow, let's apply some intuition and consider the pattern of

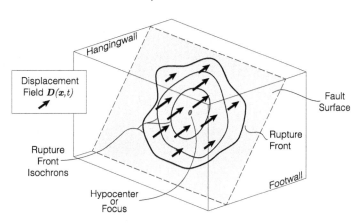

FIGURE 16.3 A schematic diagram of a fault rupture spreading from the hypocenter, or earthquake focus, over the fault surface. All regions that are sliding continually radiate outgoing P and S-wave energy. The displacement field, $\mathbf{D}(\mathbf{x}, t)$, varies over the fault surface. Note that the direction of rupture propagation does not generally parallel the slip direction (after Bolt, 1982).

P-wave displacement that we expect for an arbitrarily oriented earthquake (shear dislocation). Specifically, focus on the initial motion of the P-wave because over time, the motion may oscillate in the direction of wave travel (as a result of source characteristics, limited bandwidth, scattering, etc). Our interest is whether the initial motion is away from the source (in the direction of propagation) or towards the source (opposite the direction of propagation). Since the direction of the initial motion is not changed by the propagation of the wave through the Earth, a motion away from the source will result in an initial P-wave motion that is upward at Earth's surface. A motion towards the source will result in an initial P-wave motion that is downward at Earth's surface. Thus P-wave initial, or *first motions*, are often described in terms of the P-wave polarity (upward is positive and downward is negative). Example P-waveforms shown in Fig. 16.4 indicate motion towards the source at stations located to the northwest and southeast, and motion away from the source at stations to the northeast. When sufficient observations are available, a clear pattern emerges related to the earthquakes faulting geometry (strike, dip, and rake). P-wave polarity patterns were doc-

umented in the early 1900's by R. Labozetta in Italy and T. Shida in Japan. To understand the pattern and to use similar patterns to learn about earthquakes, we must consider the deformation produced during faulting.

Consider the deformation produced by motion along a horizontal fault (Fig. 16.5, fault viewed from side). Given the respective compression (pushing) and dilation (pulling) of rocks in the quadrants surrounding the rupture, we expect alternating quadrants in which the initial motions of the P arrival will be *compressional* (motion away from the source) or *dilatational* (motion toward the source). However, because the rupture is finite, away from the fault, changes in static deformations and the initial P-wave amplitude are not abrupt – a smooth three-dimensional radiation pattern is produced.

To begin to quantify the pattern of earthquake-radiated P-wave amplitudes, we introduce a local Cartesian coordinate system at the source, with the x_1 axis parallel to the slip direction and x_3 perpendicular to the fault (the fault is in the x_1-x_2 plane). We also introduce spherical coordinates r, θ, ϕ as illustrated in Fig. 16.6. P-wave displacements correspond to the dis-

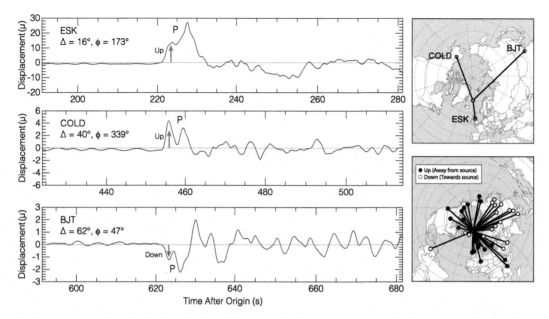

FIGURE 16.4 P-waveforms observed following the 29 Jan, 2011 earthquake along the northern Mid-Atlantic Ridge System. The initial direction of the *P*-wave motions indicate whether the initial motion near the earthquake was towards (BJT) or away (ESK, COLD) from the source region. Source-to-station distance (Δ) and azimuth (φ) are listed for each station (unshaded circles on map). Upper inset map is an azimuthal-equidistant projection centered on the source (shaded circle) showing the three sample station locations; grid lines identify the geographic grid. Lower inset map includes observed polarities for a larger set of data that exhibit a systematic pattern related to the earthquakes faulting geometry.

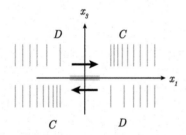

FIGURE 16.5 Conceptual view of the sense of initial *P*-wave motion with respect to the fault plane and auxiliary planes (*for P-waves traveling close to the fault*). The initial motion of the ground is a dilatation (D) in the upper-left and lower-right, and a compression (C) in the upper-right and lower-left. The sense of initial motion changes sign across two surfaces, an extension of the fault surface and across a surface orthogonal to the fault (the auxiliary plane).

placement in the radial direction, $u_r(\theta, \phi) \propto \sin 2\theta \cos \phi$ (the result is derived in the next

chapter). When $\phi = 0$ (i.e., in the $x_1 x_3$ plane) $u_r(\theta, 0) \propto \sin 2\theta$, which is a simple four-lobed azimuthal pattern reflecting the alternating quadrants in Fig. 16.5.

Away from the fault, *P* amplitudes vary smoothly, polarity reversals occur in directions where amplitudes have decreased to zero. No discontinuity in ground motion occurs outside the rupture region. The *P* wavefront includes a smooth transition from outward initial motions to inward initial motions. The largest *P*-wave motions are radiated from the middle of the four quadrants, at 45° angles to the rupture ($x_1 x_2$) (Fig. 16.6).

As suggested earlier (Fig. 16.4), the geometric patterns in *P*-wave amplitude/polarity are characteristic of seismic-wave radiation from faults. Seismologists routinely use this kind of information to infer earthquake faulting geome-

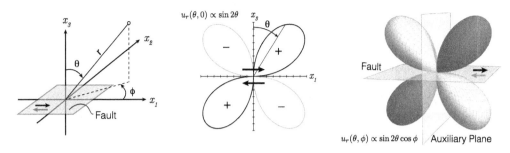

FIGURE 16.6 (left) The local source coordinate system (x_1, x_2, x_3) and spherical coordinate system (r, θ, ϕ). x_3 is perpendicular to the fault surface and slip is along the x_1 direction. (center) The radiation-pattern variation over the P wavefront in the $x_1 x_3$ plane. The directions of sign change are nodes – to change the P-wave polarity, the radiation pattern must pass through zero. (right) The full "radiation" pattern of P-wave amplitude shown in three dimensions. Positive radial motion (away from the source) is shown shaded.

try. Most of the complexity in the analysis arises from mapping the observations in Earth's geographic reference frame back to the source coordinate system, where radiation patterns have simple low-order symmetry. If observations of P-wave first motions from a sufficient number of directions are available, we can estimate the orientation of the fault and the direction of slip. To do so, we back-propagate the waves to a sphere just large enough to surround the source (called the *focal sphere*). The four-lobed pattern of P-wave polarity divides the sphere into four quadrants, two with motions away from the source, and two with motions toward the source. With observations sufficiently well distributed in azimuth (ϕ) and take-off angle ($\pi - \theta$), the observed polarity pattern constrains the orientation of the radiation lobes and thus the faulting geometry. Back-propagating body-waves from a station to the focal sphere requires assumptions about the ray paths – and we rely on ray theory (Chapter 12) to complete the analysis. When the observations are made close to the source, the case for small earthquakes, uncertainty in the location can map into uncertainty in azimuth and take-off angle that can complicate recognition of the polarity pattern. We discuss the details later.

A notable complication with a body-wave polarity analysis arises from the symmetry of the radiation pattern. If we exchange the slip and fault-normal directions and then reverse the direction of slip (i.e. exchange the slip and fault-normal vectors) we recover the exact same polarity pattern (Fig. 16.6). In other words, an *auxiliary plane*, orthogonal to the fault (Fig. 16.6) could have hosted an earthquake with an opposing sense of slip to produce the same polarity pattern. For example, right-lateral motion on the fault is indistinguishable from left-lateral motion on the auxiliary plane. Resolving the ambiguity requires additional information such as near-source static deformation, observed surface rupture, or in some instances, a definitive alignment of aftershocks. The ambiguity remains unresolved for most moderate-magnitude and smaller earthquakes, but the tectonic forces producing either style of faulting are the same, so the ambiguous results are valuable.

Body-wave polarity patterns are simple and useful, but a direct fit to the seismograms is a preferred approach to extract faulting information, particularly if the data are sparse. To model seismograms directly we must solve the equations of motion including the effects of seismic source geometry and rupture. Towards this end, we now consider how simple models of faulting can be represented in a form convenient for incorporation in the equations of motion. The physical models of explosions and faulting are replaced by mathematically simpler and seismically equivalent body forces. The result is not only a quantitative explanation for the observed amplitude patterns, but also underlies our ability to compute **synthetic seismograms** that can be directly compared with observations. In the next chapter, we examine specific solutions to the equations of motion that include the body-force equivalent of common seismic sources.

16.4 Equivalent body forces for seismic sources

Equivalent body forces (EBFs) are central to understanding much of the terminology and simple mathematical models used to quantify earthquakes and to model seismograms. EBFs are not physical sources, they are mathematical tools that simplify computation, but they also encapsulate some important characteristics of seismic sources. Their use has led to a deeper understanding of faulting and tectonics. Recall that the equations of motion for a uniform, elastic material, including body-forces, $\mathbf{f}$, are

$$\rho \ddot{\mathbf{u}} = (\lambda + 2\mu) \nabla (\nabla \cdot \mathbf{u}) - \mu \nabla \times \nabla \times \mathbf{u} + \mathbf{f}. \quad (2.52)$$

In earlier chapters, we set $\mathbf{f} = 0$ (no body forces) and set $\mathbf{u} = \nabla \phi + \nabla \times \boldsymbol{\Psi}$ to separate P- and S-wave disturbances. To include a physical model of a seismic source in the equations requires that we evaluate the deformation caused by a sudden pressure pulse over the surface of a sphere (for an explosion) or a sudden discontinuity in displacement across a fault rupture. Integrating the deformation over an explosion boundary or an earthquake rupture surface is instructive, but EBFs provide a simpler alternative to represent seismic sources. To derive EBFs, we look for simple force systems, $\mathbf{f}$ that can balance the deformations produced by a physical model of a seismic source; the opposite of these balancing forces can be used to replace the physical source, so they are called **equivalent body forces**. We start with the obvious, a point force, then

show how we can represent the motions produced by slip on a fault using two force couples. We generalize that result using a set of force couples combined to form a seismic moment tensor.

16.4.1 Seismic point-force sources

The equivalence of our first EBF is self evident, since the first source we consider is a point force. Point forces have applications in a number of important problems that include volcanic eruptions, landslides, nuclear-explosion spall, and the extraction of propagation patterns from inter-sensor comparisons of ambient ground motions. Point-force solutions are also the fundamental building block for constructing EBF representations for explosions, earthquakes, and more exotic seismic point-sources. We present explicit solutions of the equations of motion for a directed point force later. Here, we introduce definitions and ideas to help make subsequent analyses more concrete.

Consider first the solution of a static problem of a force, $\mathbf{F}$, applied at a point in a homogeneous elastic medium (Fig. 16.7). When the force is imposed, the material deforms and reaches equilibrium (accelerations stop). The displacement field is the change in position of each point in the medium in response to the force. The magnitude

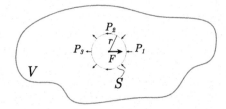

FIGURE 16.7 A planar cut through a three-dimensional volume, V, in which a point force, $\mathbf{F}$, is acting. We consider the nature of the vector displacement field, $\mathbf{u(r)}$, on a spherical surface of radius r centered at the point of application of the force. Small vectors show the stresses acting on S to balance the point force F when the system is in equilibrium.

of the displacement, $\mathbf{u(r)}$, on a spherical surface S with radius r centered on the point source can be approximated as follows. For a system in equilibrium, the body force, **F**, must be balanced by the stresses acting on S. Intuitively, the balancing stress will be compressional at point P_1, purely shear at point P_2, a dilation at P_3, but more generally a mix of shear, compression, or dilatation. If we represent the stress acting on S by σ, a first-order force balance requires that $|\mathbf{F}| \approx 4\pi r^2 \sigma$. For linear elasticity we know that $\sigma = C\varepsilon = C\nabla\mathbf{u}$, where C represents a general elastic modulus and ε represents the strain. If we let $u(\mathbf{r})$ represent the magnitude of the displacement at r, then $du(\mathbf{r})/dr$ provides the magnitude of σ. Using F to represent $|\mathbf{F}|$, we have

$$\sigma = C\frac{du(\mathbf{r})}{dr} = \frac{F}{4\pi r^2} , \qquad (16.6)$$

so that

$$du(\mathbf{r}) \approx -\frac{F\,dr}{4\pi r^2 C} . \qquad (16.7)$$

Integrating the above provides an estimate of the radial variation in the magnitude of the deformation produced by the point force,

$$u(\mathbf{r}) = \frac{1}{C}\frac{F}{4\pi r} . \qquad (16.8)$$

The static displacement field from a point force is expected to diminish with distance from the source, and the displacements are directly proportional to the force divided by an elastic modulus. This analysis lacks directional information on the displacement field, which involves a combination of radial and shear components. We remedy this analysis shortcoming later, after we show how we can build models of seismic sources using combinations of point forces.

16.4.2 An ideal explosion

As discussed earlier, the physical source model of an ideal explosion is a rapid pressurization of a sphere. To balance the outward deformation on the sphere requires an equal and opposite force distributed over the sphere. Under a point source approximation (small source region compared with the seismic wavelength), as the sphere becomes small, the surface forces can be replaced with three orthogonal inward pointing dipoles (pairs of forces acting in opposite directions). Thus, an explosion can be represented using three outward-pointing, equal-strength mutually perpendicular force *dipoles* (Fig. 16.8). When used in the equations of motion, the three force dipoles produce a seismic wavefield identical to that from the physical, ideal explosive point-source model. The strength of a dipole, or force couple, is quantified using its moment, $M = f\,dx$, where f is the strength of the force and dx is the distance separating the forces. Source time-dependence can be accommodated by requiring the forces to balance the deformation for each instant of time as the source evolves. Then, for an ideal explosion source with time history of $F(t)$, we have $M(t) = -4\pi\rho\alpha^2 F(t)$, where $F(t)$ is defined in terms of *force per unit mass*. Both a sudden spherical expansion and a set of three orthogonal, symmetric dipoles produce a spherically symmetric pattern of strain, so the equivalence of the body force dipoles is not unreasonable outside the explosion's effective elastic radius.

16.4.3 An ideal earthquake

At first thought, it seems unlikely that any simple body-force system could adequately represent an earthquake rupture. To do so requires that we identify a set of forces that can produce the same deformation (strain change) as that produced by slip across a bounded rupture of an elastic material. As described earlier, an earthquake rupture initiates at the hypocenter and spreads over the fault producing slip behind an expanding rupture front (Fig. 16.3). As the rupture expands, all of the enclosed sliding regions continually radiate P and S waves. Fault offset across different parts of the rupture may have smooth or irregular time histo-

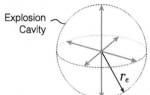
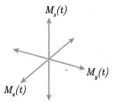

FIGURE 16.8 The physical model of an idealized earthquake, a sudden pressure increase on the surface of a spherical cavity is mathematical equivalent to the application of three equal-strength dipoles (force per unit mass) at the center of the sphere. The P-wave radiation from both source models is identical. The equivalent body forces provide a more convenient way to include the source in the equations of motion, so are commonly used in forward and inverse seismological computation.

ries. The total amount of slip varies spatially within the rupture, and it must decrease to zero at the rupture edges. Each location on the rupture may also have a slightly different slip direction. During sliding, many phenomena, such as local heating and perhaps melting of rock, hydrologic pressure variations, and rock fracture occur – these are an important part of the earthquake energy budget and influence the time history of energy release. Despite this rich, interesting complexity, we *can* successfully approximate this process well enough to explain many details in seismograms.

We begin by "standing back" from the fault and considering the *average properties* of the rupture. For seismic waves with periods longer-than or comparable-to the duration of the rupture and for wavelengths that are large relative to the fault dimensions, most of the details of the fault rupture are unresolvable, only the average properties substantially influence the character of the seismogram. This allows us to work with a simpler representation. We focus on first-order earthquake characteristics such as the total rupture area, A, the average rupture offset, $\overline{\Delta u}$, and the average velocity and direction of rupture propagation, v_r. In its simplest form this **dislocation model** of earthquakes can be reduced to a point source (i.e., no spatial extent) with a parameterized dislocation time dependence that approximates the slip history and

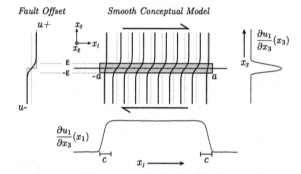

FIGURE 16.9 A conceptual model demonstrating the equivalence of body-force equivalents with a simple model of faulting as slip across an interface.

rupture expansion. The model complexity has to increase as the ratio of seismic-wavelength to fault-length decreases. More complex models can be built by summing such point-sources (superposition).

A clear and rigorous derivation of the earthquake EBFs starting with first principles is available in Aki and Richards (2009). We follow Kasahara (1981) and Maruyama (1973) and present a conceptual model outlining the relationship between slip across a fault and an EBF system appropriate for a linear, isotropic elastic material surrounding a small fault (small compared with the seismic wavelength). The model is summarized in Fig. 16.9. The fault is assumed to be orthogonal to the x_3 direction and the slip is parallel to the x_1 direction. Slip only occurs in the gray region, $-a \leq x \leq a$ (the rupture). The fault offset is $\Delta u = u^+ - u^-$ as shown in the cartoon on the far left. We seek to replace the rupture deformation with a set of forces acting in an unbroken elastic material, so rather than allow a discontinuity in slip, we replace the offset with a smooth change in position across an infinitesimal distance $-\epsilon \leq x \leq \epsilon$. As $\epsilon \to 0$, the conceptual model offset will more closely resemble that of the physical dislocation model. We construct the rupture so that the amount of offset smoothly vanishes at the rupture edges over a distance c. All motion in the system is in the x_1 direction.

Now using the smooth-curve approximations, we examine the equations of equilibrium for an isotropic linear elastic material in the vicinity of the fault (we are interested in forces that balance the fault's deformation field – their opposites produce a deformation field equivalent to the rupture). We start with Hooke's Law,

$$\sigma_{ij} = \lambda \varepsilon_{kk} + 2\mu \varepsilon_{ij} , \qquad (16.9)$$

where the strains can be written in terms of spatial derivatives of the displacement,

$$\varepsilon_{ij} = \frac{1}{2} \left(\frac{\partial u_i}{\partial x_j} + \frac{\partial u_j}{\partial x_i} \right) . \qquad (16.10)$$

In our simple model, which has displacement only in the x_1 direction, the only substantial spatial displacement gradient is $\partial u_1 / \partial x_3$. Thus we must be concerned with only two strains ε_{13} and ε_{31}, or two stresses defined by Hooke's Law,

$$\begin{aligned} \sigma_{13} &= 2\mu \varepsilon_{13} = \mu \, \partial u_1 / \partial x_3 \\ \sigma_{31} &= 2\mu \varepsilon_{31} = \mu \, \partial u_1 / \partial x_3 . \end{aligned} \qquad (16.11)$$

The rate of change of u_1 with respect to x_3 (as a function of x_1 and as a function of x_3) is shown conceptually below and to the right of the fault in Fig. 16.9. Keep in mind that the spatial quantities ϵ, a, and c are all small compared to the seismic wavelengths for which our approximation will hold.

The equations of motion,

$$\rho f_i = -\sigma_{ij,j} + \rho \ddot{u}_i , \qquad (16.12)$$

where ρ is density, and f_i represents a force per unit mass, require that stress gradients are balanced by accelerations and body-forces (ρf_i). Whereas the strains and stresses were proportional to the displacement gradient, the forces are proportional to the stress gradients, or displacement *curvature*. To be equivalent, the body-forces must balance the stress gradients exactly, no accelerations are allowed. Thus, the equations of motion lead to two non-zero relationships for the equivalent body forces,

$$\begin{aligned} \rho f_1 &= -\sigma_{13,3} = -\mu \, \frac{\partial \left(\partial u_1 / \partial x_3 \right)}{\partial x_3} \\ \rho f_2 &= 0 \qquad\qquad\qquad\qquad (16.13) \\ \rho f_3 &= -\sigma_{31,1} = -\mu \, \frac{\partial \left(\partial u_1 / \partial x_3 \right)}{\partial x_1} . \end{aligned}$$

To match the deformation produced by our smooth fault offsets will require forces in both the direction normal to and parallel to the slip direction.

The EBFs (f_1 and f_3) are proportional to the displacement curvature. The differentiation is illustrated conceptually in Fig. 16.10, where the spatial gradients of $\partial u_1 / \partial x_3$ in the x_1 and x_3 directions are shown. In the limit that ϵ, a, and c become small, the distance between regions of opposite curvature decreases and the amplitude of the curvature increases. The result is the development of two force couples or a *double couple*. Thus for ruptures small compared to the seismic wavelength, fault slip produces deformation equivalent to that produced by a double couple, one couple in the direction of slip (above and below the rupture) and the other couple perpendicular to the direction of slip. In the EBF double-couple system, the couple parallel to the fault includes forces with directions that match the sense of offset across the fault. The force couple perpendicular to the rupture looks like a 90°

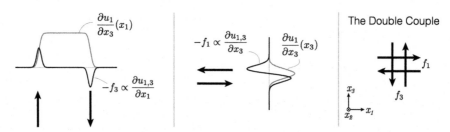

FIGURE 16.10 Equivalent body forces for a small earthquake on a fault normal to the x_3 direction and slip in the x_1 direction. The left panel shows the origin of the force couple with forces in the x_3 direction (EBF f_3); the center shows the origin of the force couple with forces in the x_1 direction (EBF f_1). Both couples create a moment about the x_2 axis, but the moments sum to zero and angular momentum is conserved.

degree rotation, then reversal of slip directions, of the fault-parallel couple. This is the origin of the fault-plane auxiliary-plane ambiguity discussed earlier in our discussion of *P*-wave first motions.

In addition to the equations of equilibrium (which are the conservation of linear momentum), the body-forces must also satisfy the conservation of angular momentum. Each of our single couples introduces a moment about the x_2 axis, and to avoid rotation, the moments of the two couples must balance. The moment acting on an element of area dA exerted by body forces offset in the x_3 direction (forces in the x_1 direction) can be computed from

$$-dA \int_{-\epsilon}^{\epsilon} x_3 \times \rho f_1 \, dx_3 = -\mu \, \Delta u \, dA \ . \qquad (16.14)$$

The couple offset in the x_1 direction (forces in the x_3 direction) introduces an opposing moment about x_2, equal to $+\mu \, \Delta u \, dA$. Although the net moment is zero (the fault does not rotate) the EBF model is the conceptual origin of quantifying earthquake strength using the seismic moment,

$$M_0 = \mu \, \overline{\Delta u} \, A \ . \qquad (16.15)$$

Of course, the physical intuitiveness of seismic moment, a product of how large the rupture was, how large the offset was, how stiff the rocks

are, makes it a physically appealing measure of earthquake size.

The above discussion is meant only to show that representing seismic radiation from a fault using force couples is reasonable. The level of approximation implied depends on the sensitivity of the seismic waves to the details of the rupture complexity, which is frequency and wavelength dependent. Later in this chapter we will introduce a more general system of body forces called a *seismic moment tensor*, which includes the double couple as a special case. The model is routinely combined with a *time-varying* force dependence to simulate seismograms and to constrain earthquake size and faulting geometry (fault strike and dip, and rake). But EBFs and even dislocation models at some point must fail to express the physics of faulting – i.e. what goes on as the rocks slide. Perhaps the main limitation of the models is that they do not explicitly include physics for initiating or terminating rupture, and hence they leave many fundamental issues unresolved. The limitation is important to keep in mind, but point-source and kinematic finite-source analyses constructed from point sources have revealed many characteristics of earthquake processes.

16.4.3.1 Equivalent body force system non-uniqueness

From the 1920s until the mid-1960s seismologists debated whether a single- or double-

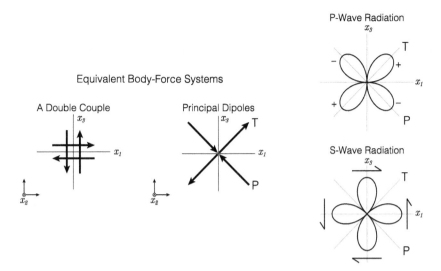

FIGURE 16.11 The double-couple force system in the x_1x_3 plane for a shear dislocation in the x_1x_2 plane. An equivalent set of dipole forces without shear (aligned along the principal axes), is shown in the center. On the right are the patterns of P- and S-wave radiation distributed over the respective wavefronts in the x_1x_3 plane. The + indicates motion away from the source, the − indicates motion towards the source.

couple model was appropriate for earthquakes. Although the single-couple model makes less sense physically, the main reason for not ruling it out was that the P-wave radiation from both models is indistinguishable. However, the S-wave and surface-wave radiation is not, and when sufficient data became available, it was unambiguously demonstrated that the double-couple model was more appropriate. In addition, elastodynamic solutions for actual stress and displacement discontinuities in the medium (an alternate way of modeling shear faults, which we do not consider in this text) confirm the equivalence of double-couple body forces and shear dislocations. Having established the appropriateness of the double-couple, the result leads to an interesting conclusion – *EBF systems are not unique.*

For example, our double-couple force system can be transformed into an equivalent pair of orthogonal dipoles at 45° angles to the fault (the x_1-x_2 plane), in a plane perpendicular to the fault (the x_1-x_3 plane). These axes are often called the

principal axes associated with the earthquake. The dipole directed toward the source is called the *compressional* or **P**-axis and lies in the quadrants of dilatational P-wave first motions *toward* the source. The dipole directed outward from the source is called the *tensional* or **T**-axis, and lies in the quadrants of compressional P-wave first motions *away* the source. P- and S-wave radiation patterns are shown in Fig. 16.11. The sense of shearing motion on the wavefront reflects the orientation of the force on the near side of the source double couple. The P and S radiation patterns are rotated by 45° from one another, but the symmetry still precludes distinguishing the fault plane and the auxiliary plane using only first motions. But since an S-wave radiation for a single couple produces a two-lobed radiation pattern, the pattern provides the basis for distinguishing it from the double-couple mechanism. Note that the principal directions, **P**, and **T** are directions of no shear, hence no shear wave radiates in those directions.

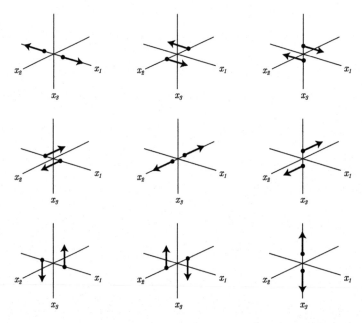

FIGURE 16.12 The nine couples represented in the seismic moment tensor. Component M_{ij} corresponds to a couple with forces oriented in the $\pm i$th direction offset in the $\pm j$th direction. The moment tensor corresponding to a source that conserves momentum must be symmetric.

The non-uniqueness of source representations is quite apparent when we generalize our force systems using a seismic moment tensor. Moment tensors are flexible mathematical tools for representing a broader class of seismic sources and offer distinct advantages for seismogram modeling. However, the flexibility and advantages sometimes carry added complexity during interpretation.

16.5 Seismic moment tensors

We can combine the three dipoles needed to represent explosions and the six single-couples needed to represent earthquakes on three orthogonal surfaces, to create a set of nine force couples (Fig. 16.12). Each dipole or couple's strength is quantified using a moment (force × distance), M_{ij}, where $\pm x_i$ are the directions

of the couple's forces and x_j is the direction of the offset between the forces (the moment-arm). This collection of forces can be used to represent a wider class of sources that includes ideal explosions and double-couple earthquake models. For example, as we did above, we can combine single couples to form double couples. To represent an earthquake with slip in the $\pm x_1$ direction across a fault in the $x_1 x_2$ plane, we sum M_{12} and M_{21} (with equal strengths). We can also sum equal strength couples M_{13} and M_{31} or M_{23} and M_{32} to create double couples acting in remaining orthogonal coordinate planes. An ideal explosion source can be represented by the sum of three equal-strength dipoles, M_{11}, M_{22}, and M_{33}. The full set of couples can be summed to produce a wide range of effective source deformations.

We can also arrange the force-couple moments to create a symmetric second-order *mo-*

ment tensor,

$$\mathbf{M} = \begin{bmatrix} M_{11} & M_{12} & M_{13} \\ M_{21} & M_{22} & M_{23} \\ M_{31} & M_{32} & M_{33} \end{bmatrix} . \quad (16.16)$$

The moment tensor satisfies the established rules for coordinate transformations and can be analyzed using standard tensor mathematics. The moment-tensor components are the moments (possibly time-dependent moment histories) corresponding to the force couples. In tensor form, the double-couple solution given by (16.13) has a corresponding moment tensor

$$\mathbf{M} = M_0 \begin{bmatrix} 0 & 0 & 1 \\ 0 & 0 & 0 \\ 1 & 0 & 0 \end{bmatrix} , \quad (16.17)$$

where a scalar factor (the seismic moment), M_0, has been factored out the tensor. The ideal explosion is also called an isotropic source because of its symmetry – its moment tensor is

$$\mathbf{M} = M_0 \begin{bmatrix} 1 & 0 & 0 \\ 0 & 1 & 0 \\ 0 & 0 & 1 \end{bmatrix} . \quad (16.18)$$

The sum of the diagonal elements of a tensor is called its *trace*. The moment-tensor trace is non-zero for a volumetric source such as an explosion, implosion, mine collapse, etc. The moment-tensor trace for a double-couple is zero because a double-couple produces no volume change. Observations suggest that most faulting sources have little isotropic component, and moment tensors for faulting events are often estimated using a constraint that the moment-tensor trace vanishes (i.e. $M_{11} + M_{22} = M_{33}$).

16.5.1 Moment tensors and shear faulting

We can use the moment tensor to conveniently represent earthquakes of all orientations.

An arbitrarily oriented shear-faulting event is defined by a fault normal vector $\hat{\mathbf{n}}$, and a slip vector, $D\hat{\mathbf{s}}$, where D is the magnitude of the slip. In terms of the fault strike, ϕ_f, dip, δ, and rake, λ, the vector components (north, east, and down) of these vectors are

$$\hat{\mathbf{n}} = \left(-\sin\delta \, \sin\phi_f, \; \sin\delta \, \cos\phi_f, \; -\cos\delta \right)^T, \quad (16.19)$$

and

$$\hat{\mathbf{s}} = \begin{pmatrix} \cos\delta \, \sin\lambda \, \sin\phi_f + \cos\lambda \, \cos\phi_f \\ \cos\lambda \, \sin\phi_f - \cos\delta \, \sin\lambda \, \cos\phi_f \\ -\sin\delta \, \sin\lambda \end{pmatrix} .$$

$$(16.20)$$

The slip vector, $D\hat{\mathbf{s}}$, lies on the fault surface and is orthogonal to the fault normal vector. By convention, the slip vector indicates the direction of hanging-wall movement relative to the footwall.

The corresponding moment-tensor components, M_{ij}, can be defined in terms of the fault-slip and fault-normal unit vectors since, as described above, these are part of the definition of the couples in a double-couple. We have,

$$M_{ij} = \mu \, A \, D \left(s_i n_j + s_j n_i \right) , \quad (16.21)$$

where the source strength is the seismic moment, $M_0 = \mu \, A \, D$. Note the symmetry of the slip vector $\hat{\mathbf{s}}$ and fault normal $\hat{\mathbf{n}}$ in the expressions for the moment-tensor elements (16.21). The two quantities can be exchanged without changing the moment tensor. The symmetry corresponds to the ambiguity of the fault and auxiliary planes of a point double couple (discussed in Section 16.3).

Finally, we can define the moment tensor in term of the faulting geometry angles and the average slip and rupture area using the expressions for $\hat{\mathbf{n}}$ and $\hat{\mathbf{s}}$ in geographic coordinates. Substituting Eqs. (16.19) and (16.20) in Eq. (16.21), results in

$$M_{NN} = M_{11}$$
$$= -M_0 \left(\sin\delta \cos\lambda \sin 2\phi_f + \sin 2\delta \sin\lambda \sin^2\phi_f \right)$$
$$M_{EE} = M_{22}$$
$$= M_0 \left(\sin\delta \cos\lambda \sin 2\phi_f - \sin 2\delta \sin\lambda \cos^2\phi_f \right)$$
$$M_{ZZ} = M_{33}$$
$$= M_0 \left(\sin 2\delta \sin\lambda \right) = -(M_{11} + M_{22})$$
$$M_{NE} = M_{12}$$
$$= M_0 \left(\sin\delta \cos\lambda \cos 2\phi_f + \tfrac{1}{2}\sin 2\delta \sin\lambda \sin 2\phi_f \right)$$
$$M_{NZ} = M_{13}$$
$$= -M_0 \left(\cos\delta \cos\lambda \cos\phi_f + \cos 2\delta \sin\lambda \sin\phi_f \right)$$
$$M_{EZ} = M_{23}$$
$$= -M_0 \left(\cos\delta \cos\lambda \sin\phi_f - \cos 2\delta \sin\lambda \cos\phi_f \right) ,$$
$$(16.22)$$

where N, E, Z represent the directions north, east, and down respectively. Note that the trace of a double-couple moment tensor, $Tr(\mathbf{M}) = M_{11} + M_{22} + M_{33}$, vanishes because the source-region volume change during shear faulting is zero. These expressions can be used to convert strike, dip, and rake into M_{ij} components.

Shear-faulting moment tensors in principle-axis coordinates

Since a moment tensor is symmetric, it can be rotated into a principal-axis coordinate system using its real-valued eigenvalues (M_i) and their associated orthonormal eigenvectors ($\hat{e}_i$). For example, the moment tensor for the source described by (16.17) when rotated into the principal-axis coordinate system takes a diagonal form,

$$\mathbf{M}' = \begin{bmatrix} M_0 & 0 & 0 \\ 0 & -M_0 & 0 \\ 0 & 0 & 0 \end{bmatrix}, \qquad (16.23)$$

where the dipoles orient along directions defined by the principal axes, $\mathbf{P}$ (maximum compressional deformation), $\mathbf{T}$ (minimum compressional deformation), and $\mathbf{B}$ (intermediate or null

axis). The result is a statement of the fact that a double-couple force system is equivalent to two orthogonal force dipoles, as illustrated in Fig. 16.11. Since shear faulting includes no net volume change, the trace of $\mathbf{M}'$ is zero. In addition, one of the eigenvalues for a double-couple will be zero since the fault normal, the slip vector, the $\mathbf{P}$ and $\mathbf{T}$ axes all lie in a single plane. Given our conventions, the eigenvector corresponding to the positive eigenvalue is aligned along the tension axis, $\mathbf{T}$; the eigenvector for the zero eigenvalue is aligned along the intermediate stress axis, $\mathbf{B}$; and the eigenvector for the negative eigenvalue is aligned along the compressional axis, $\mathbf{P}$.

Computing fault-normal and slip vectors from a moment tensor

The relationships between the moment-tensor principal axes and the double-couple stress axes allow us to map a moment tensor corresponding to a double couple into the strike, dip, and rake of the fault and auxiliary planes. Let $\hat{e}_t$ and $\hat{e}_p$ represent the moment-tensor eigenvectors associated with the $\mathbf{T}$ and $\mathbf{P}$ axes respectively. The key relationships are

$$\hat{u} = \frac{1}{\sqrt{2}} \left(\hat{e}_t - \hat{e}_p \right)$$
$$(16.24)$$
$$\hat{s} = \frac{1}{\sqrt{2}} \left(\hat{e}_t + \hat{e}_p \right) .$$

Once the numerical components of $\hat{u}$ and $\hat{s}$ are computed, Eqs. (16.19) and (16.20) can be inverted for strike, dip, and rake of both the fault and auxiliary planes, provided that some care is given to the multi-valued nature of trigonometric functions.

Computing seismic moment from a moment tensor

Two different formulas are used to compute an event's size, or seismic moment, from moment tensor representations. One formula is a scaled Frobenius norm (Silver and Jordan, 1992),

$$M_0 = \frac{1}{\sqrt{2}} \sqrt{\sum_i \sum_j M_{ij}^2} = \frac{1}{\sqrt{2}} \sqrt{\sum_k M_k^2} . \quad (16.25)$$

The other formula originates from the construction of a best-double couple fit to the moment tensor and is equal to the average of the two largest (in an absolute sense) eigenvalues. Assuming that the eigenvalues, M_i, are ordered such that the two absolute largest are M_1 and M_2,

$$M_0 = \frac{1}{2} (|M_1| + |M_2|) . \quad (16.26)$$

For a double couple, the results of both formulas are equal since the intermediate eigenvalue (smallest in an absolute sense, M_3) is zero. For an ideal explosion, the two formulas provide the same result because all three M_i are equal. For most events, the two formulas provide very similar values.

16.5.2 Non-double-couple seismic sources

Purely shear-faulting is not the only style of faulting. A volcanic injection-related event could induce a component of motion perpendicular to the fault (and a local volume change). These and other deformation processes are more complicated than simple double-couples, but they can be represented with moment tensors. The moment tensor for a fault with a component of motion perpendicular to the fault surface has the form

$$\begin{aligned} M_{ij} &= \mu \, A \, D \left(s_i n_j + s_j n_i \right) \\ &+ \lambda \, A \, D \, \delta_{ij} \, (\hat{\mathbf{s}} \cdot \hat{\mathbf{n}}), \end{aligned} \quad (16.27)$$

where λ represents Lame's parameter and δ_{ij} is a Kronecker delta. The component of slip in the direction normal to the fault adds a constant value to the moment-tensor diagonal (and changes the trace). Since in all cases the moment tensor is symmetric, it is always diagonalizable and a linear combination of three orthogonal dipoles can represent any moment-tensor force system. The eigenvalues of the moment tensor, M_1, M_2, M_3, and associated orthonormal eigenvectors $\hat{\mathbf{e}}_i$ provide an alternate and often more convenient representation of the tensor (particularly for theoretical or source-type discussions). The eigenvalues of the moment tensor provide information on the nature of the source – we have already seen that for an ideal explosion, all the eigenvalues are equal, none are zero; for a double-couple, two eigenvalues are equal in magnitude and opposite in sign, the third is zero. When you want to identify the type of source described by a moment tensor, you can examine its eigenvalues.

Moment-tensor decompositions

Generally, the first step in interpreting a moment tensor is to decompose it into physically meaningful deformation components. We'll assume that we've already rotated the tensor into its principal-axis coordinate system, which makes decomposition much easier. The rotation can always be reversed using the eigenvectors on the components of the decompositions described below. Let $\mathbf{E}$ be a matrix with columns constructed using the three orthonormal eigenvectors, $\hat{\mathbf{e}}_i$, of $\mathbf{M}$. Then from standard tensor algebra, we can diagonalize $\mathbf{M}$ using (the diagonal elements are the eigenvalues),

$$\begin{bmatrix} M_1 & 0 & 0 \\ 0 & M_2 & 0 \\ 0 & 0 & M_3 \end{bmatrix} = \mathbf{E} \mathbf{M} \mathbf{E}^T , \quad (16.28)$$

and we can reverse the transformation (return to the original coordinate system) using,

$$\mathbf{M} = \mathbf{E}^T \begin{bmatrix} M_1 & 0 & 0 \\ 0 & M_2 & 0 \\ 0 & 0 & M_3 \end{bmatrix} \mathbf{E} . \quad (16.29)$$

Usually we start with a moment tensor in a known geographic coordinate system, trans-

---

Box 16.2 Moment tensor geographic conventions

Not all seismologists used the same coordinate system to define the moment tensor. Two common reference frames are the local, north-east-down system, (N, E, Z), convention clearly documented in Aki and Richards (2009), and a spherical coordinate system, (θ, ϕ, r), commonly used in global, normal-mode analyses – most notably the system used by the Global Centroid Moment Tensor (GCMT) Project. In the normal-mode system, the co-latitudinal direction, θ is positive towards the south, the longitudinal direction, ϕ, is positive towards the east, and the radial direction, r, is positive outward. The opposite directions of the north and co-latitudinal, and the radial and downward-vertical only slightly complicate transformation between the two reference frames. Reverse the sign of any element that includes *one* of N or Z or θ or R to convert to the other system

$$
\begin{aligned}
M_{NN} &= M_{\theta\theta} \\
M_{NE} &= -M_{\theta\phi} \\
M_{NZ} &= M_{\theta r} \\
M_{EE} &= M_{\phi\phi} \\
M_{EZ} &= -M_{\phi r} \\
M_{ZZ} &= M_{rr}.
\end{aligned}
\qquad \text{(B16.2.1)}
$$

Other coordinate systems are often employed, so it is important to always check the conventions when working with moment tensors.

---

form it to and decompose it in the principal axis system, and then transform the quantities back to geographic coordinates using Eq. (16.29).

Seismologists have developed a suite of moment-tensor decompositions. Which decomposition you choose depends on the suspected nature of the source – an analysis in a volcanic region may not proceed exactly as an analysis in a dominantly shear-faulting environment; a suspected underground explosion may be analyzed differently than a tectonic event. Sometimes investigating more than one decomposition is necessary to ensure that all reasonable source types are considered.

Generally, the first step is to separate a diagonalized moment tensor into isotropic and deviatoric tensors components, $\mathbf{M} = \mathbf{M}^{(I)} + \mathbf{M}^{(D)}$, or, in the principal-axis coordinate system,

$$
\begin{bmatrix} M_1 & 0 & 0 \\ 0 & M_2 & 0 \\ 0 & 0 & M_3 \end{bmatrix} = \begin{bmatrix} M^{(I)} & 0 & 0 \\ 0 & M^{(I)} & 0 \\ 0 & 0 & M^{(I)} \end{bmatrix}
$$
$$
+ \begin{bmatrix} M_1^{(D)} & 0 & 0 \\ 0 & M_2^{(D)} & 0 \\ 0 & 0 & M_3^{(D)} \end{bmatrix},
$$

(16.30)

where $M^{(I)} = (M_1 + M_2 + M_3)/3$ and the $M_i^{(D)} = M_i - M^{(I)}$ are the *deviatoric* eigenvalues of $\mathbf{M}$. Many routine earthquake modeling efforts (e.g. GCMT, USGS, and regional efforts) include a constraint that the isotropic component of the solution equals zero, because doing so improves the stability of the seismogram modeling procedures used to estimate the moment tensors.

That requirement is less common in analyses of events mining regions, explosion studies, etc.

The simplest decomposition of a *deviatoric moment tensor* is into three vector dipoles,

$$
\begin{bmatrix} M_1^{(D)} & 0 & 0 \\ 0 & M_2^{(D)} & 0 \\ 0 & 0 & M_3^{(D)} \end{bmatrix} = \begin{bmatrix} M_1^{(D)} & 0 & 0 \\ 0 & 0 & 0 \\ 0 & 0 & 0 \end{bmatrix}
$$
$$
+ \begin{bmatrix} 0 & 0 & 0 \\ 0 & M_2^{(D)} & 0 \\ 0 & 0 & 0 \end{bmatrix}
$$
$$
+ \begin{bmatrix} 0 & 0 & 0 \\ 0 & 0 & 0 \\ 0 & 0 & M_3^{(D)} \end{bmatrix}. \tag{16.31}
$$

The dipoles act in the directions defined by the moment-tensor eigenvectors, $\hat{\mathbf{e}}_i$. If the moment tensor represents a double couple, the eigenvalue associated with the null, B, axis is zero. As noted earlier, this is seldom exactly true when moment tensors are estimated from observations, but for many earthquakes, the intermediate eigenvalue is small compared to the others.

Alternatively, a deviatoric moment tensor can be split into three double couples

$$
\begin{bmatrix} M_1^{(D)} & 0 & 0 \\ 0 & M_2^{(D)} & 0 \\ 0 & 0 & M_3^{(D)} \end{bmatrix}
$$
$$
= \frac{1}{3} \begin{bmatrix} M_1^{(D)} - M_2^{(D)} & 0 & 0 \\ 0 & M_2^{(D)} - M_1^{(D)} & 0 \\ 0 & 0 & 0 \end{bmatrix}
$$
$$
+ \frac{1}{3} \begin{bmatrix} 0 & 0 & 0 \\ 0 & M_2^{(D)} - M_3^{(D)} & 0 \\ 0 & 0 & M_3^{(D)} - M_2^{(D)} \end{bmatrix}
$$
$$
+ \frac{1}{3} \begin{bmatrix} M_1^{(D)} - M_3^{(D)} & 0 & 0 \\ 0 & 0 & 0 \\ 0 & 0 & M_3^{(D)} - M_1^{(D)} \end{bmatrix}. \tag{16.32}
$$

But we aren't restricted to decompositions using only dipoles and double couples. We can also decompose a deviatoric moment tensor into a sum of three *compensated linear vector dipoles* (CLVDs). A CLVD has one dipole with a strength equal to twice the other two. For example,

$$
\begin{bmatrix} M_1^{(D)} & 0 & 0 \\ 0 & M_2^{(D)} & 0 \\ 0 & 0 & M_3^{(D)} \end{bmatrix}
$$
$$
= \frac{1}{3} \begin{bmatrix} 2M_1^{(D)} & 0 & 0 \\ 0 & -M_1^{(D)} & 0 \\ 0 & 0 & -M_1^{(D)} \end{bmatrix}
$$
$$
+ \frac{1}{3} \begin{bmatrix} -M_2^{(D)} & 0 & 0 \\ 0 & 2M_2^{(D)} & 0 \\ 0 & 0 & -M_2^{(D)} \end{bmatrix}
$$
$$
+ \frac{1}{3} \begin{bmatrix} -M_3^{(D)} & 0 & 0 \\ 0 & -M_3^{(D)} & 0 \\ 0 & 0 & 2M_3^{(D)} \end{bmatrix}. \tag{16.33}
$$

A common earthquake-based decomposition is one with a *major double couple* and a *minor double couple*. Since the trace of a deviatoric moment tensor is zero (that defines deviatoric), if we arrange the deviatoric eigenvalues such that

$$
\left| M_1^{(D)} \right| \geqslant \left| M_2^{(D)} \right| \geqslant \left| M_3^{(D)} \right|, \tag{16.34}
$$

then

$$
\begin{bmatrix} M_1^{(D)} & 0 & 0 \\ 0 & M_2^{(D)} & 0 \\ 0 & 0 & M_3^{(D)} \end{bmatrix}
$$
$$
= \begin{bmatrix} M_1^{(D)} & 0 & 0 \\ 0 & -M_1^{(D)} & 0 \\ 0 & 0 & 0 \end{bmatrix} \tag{16.35}
$$
$$
+ \begin{bmatrix} 0 & 0 & 0 \\ 0 & -M_3^{(D)} & 0 \\ 0 & 0 & M_3^{(D)} \end{bmatrix},
$$

where the first term on the right, the major double couple, includes the largest eigenvalue, and the second term, the minor double couple, includes the smallest eigenvalue. The major double couple has the same principal axes as the original moment tensor. In the minor double couple, we swapped the **P** and **T** axes, which might correspond to an unusual tectonic situation. We can work out decompositions that preserve one or the other axis in both double couples, if we have reason to expect compression or tension dominates local tectonics.

Another common deviatoric-tensor decomposition is the sum of a double couple, and a CLVD. For eigenvalues sorted by absolute value as in Eq. (16.34), we can write

$$
\begin{bmatrix} M_1^{(D)} & 0 & 0 \\ 0 & M_2^{(D)} & 0 \\ 0 & 0 & M_3^{(D)} \end{bmatrix}
$$

$$
= \left(1 - 2F'\right) M_3^{(D)} \begin{bmatrix} 0 & 0 & 0 \\ 0 & -1 & 0 \\ 0 & 0 & 1 \end{bmatrix} \quad (16.36)
$$

$$
+ F' M_3^{(D)} \begin{bmatrix} -1 & 0 & 0 \\ 0 & -1 & 0 \\ 0 & 0 & 2 \end{bmatrix},
$$

where

$$
F' = -M_1^{(D)} / M_3^{(D)} . \quad (16.37)
$$

A measure of the size of the CLVD component relative to the size of the double couple can be defined using the ratio of the smallest eigenvalue (in an absolute sense) to the largest eigenvalue (in an absolute sense),

$$
\epsilon = \left| \frac{M_3^{(D)}}{M_1^{(D)}} \right| . \quad (16.38)
$$

For a pure double couple, $\epsilon = 0$; for a pure CLVD, $\epsilon = \pm 0.5$. A measure often quoted is $200 \times \epsilon$, which is the percent non-double component contained in a moment-tensor source representation. An alternate definition of ϵ that retains the sense of the stress state is

$$
\epsilon_{\pm} = -\frac{m_2}{\max(|m_1|, |m_3|)} , \quad (16.39)
$$

where for clarity we have relabeled the deviatoric moment-tensor eigenvalues $m_1 \geq m_2 \geq m_3$ (no absolute values). Assuming that the non-double component is caused by the tectonic environment, if $\epsilon_{\pm} < 0$, the stress state is compressive, if $\epsilon_{\pm} > 0$, the stress state is tensional.

Box 16.3 Non-double-couple sources

Moment tensor inversions are routinely performed using a variety of seismic phases for each earthquake. Most natural events appear to be primarily a double-couple shearing source, with any minor double-couple or CLVD component [measured by ϵ as defined in the text for Equation (16.36)] being attributed to noise in the inversion arising from inaccurate propagation corrections. However, some events have $\epsilon\pm$ values larger than expected to be caused by noise. An example is the January 1, 1984 deep earthquake under Japan. Fig. B16.3.1 shows the non-double-couple moment tensor solutions found using waves with different frequencies. Note that these projections, which show the P-wave nodal surfaces separating compressional (dark regions) and tensional P-wave motions, do not have the simple orthogonal planes expected for a double couple. The $\epsilon\pm$ values range from -0.2 to -0.33, indicating that the **B** axis has a positive value, contributing to the tensional character of the source (essentially making the P-wave compressional motion zone larger). The question is then, What is the significance of this non-double-couple component?

Kuge and Kawakatsu (1990) analyzed the broadband *P*-wave motions and found evidence for complex rupture. This can be well modeled assuming two double couples shifted in space and time (procedures for this modeling are described in Chapter 19). The sum of the two double couples equals the total moment tensor determined at long periods. Thus, in this case the deep event involved faulting on two separate fault planes, with only high-frequency body-wave signals having sufficient resolution to determine the detailed nature of the source.

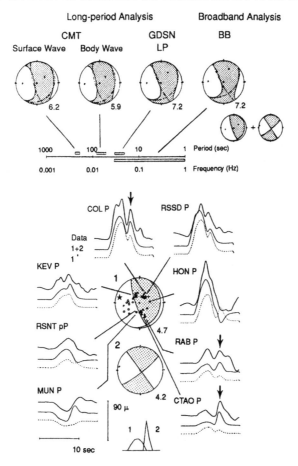

FIGURE B16.3.1 (Top) Projection of *P*-wave radiation nodes of moment tensor solutions for the source of a deep earthquake on January 1, 1984. (Bottom) Observed *P* waves (top traces) and predictions for two subevent models with different mechanisms (middle) or the same mechanism (modified from Kuge and Kawakatsu, 1990).

Moment tensors are flexible tools for representing seismic point sources – understanding them is essential if you want to develop a deep understanding of earthquake seismology. We have approached the subject from a perspective of seismic source representation. An important additional characteristic of seismic moment tensors is that their components, for a fixed depth and source time function, are linearly related to seismogram amplitudes (this is not the case for parameters such as strike, dip, and rake). The linear relationship allows relatively simple modeling approaches to be used to estimate earthquake moment tensors, provided that we can handle the nonlinearity associated with source depths and time histories. For many events, the time history can be conveniently approximated (especially when the seismic wave periods are much longer than the source duration) using boxcars and triangles, and source depth can be estimated using systematic searches or iterative approaches. The

process is applied routinely both globally and regionally. The Global Centroid Moment Tensor Catalog (GCMT) includes seismic moment tensors for more than 40,000 earthquakes that have occurred since the onset of routine digital seismic recording in the mid-1970's.

Next, we develop the relationships between seismic-source models and seismic-wave amplitudes. The key to constructing solutions of the equations of motion for a complex set of body forces such as a double couple or moment tensor is to solve first for the displacement field due to a single point force. Then we can use the linearity of elastic solutions to superimpose point force deformations to represent deformations caused by force couples and their combinations. We introduce the approach using static (time-independent) equilibrium solutions, and then present the elastodynamic solutions that apply to propagating waves.

17

Seismic point-source radiation patterns

Chapter goals

- Derive expressions for static deformations produced by point forces and point-force couples.
- Derive expressions for seismic body-wave radiation from explosions, point forces and point-force couples.
- Discuss seismic-wave radiation patterns for a whole space and surface-wave radiations patterns for simple earth models.
- Review the theoretical basis for seismic modeling of *P* wave first motions to estimate faulting geometry.

This chapter is an extension of the previous chapter in which we described simple mechanical models of seismic point sources that we can use to represent explosions, earthquakes, and more exotic sources. The physical model of an earthquake as slip along a fault surface was replaced with equivalent body forces that produce the same seismic deformation patterns provided that the source dimension was small compared with the seismic wavelength. Our most general description was in terms of combinations of force couples and dipoles (the moment tensor). Pure shear faulting was represented using a sum of two single couples called the double couple. Point forces can be used to represent other pro-

cesses such as volcanic eruptions and landslides, etc. Now we focus on the explicit expressions for the *seismic-wave radiation patterns* produced by common seismic sources. We start our discussion with time-independent static problems, which are easier to solve but provide a useful view of the deformation patterns produced by point forces, couples, and double couples. Then we outline and present the key results for the dynamic, time-dependent seismic radiation from point sources.

17.1 Elastostatics

Earlier (Section 16.4.1) we discussed the magnitude of the deformation produced by a point force. In this section we develop explicit relationships for the static displacement $\mathbf{u}(\mathbf{r})$ at point P in an isotropic, infinite, homogeneous elastic medium with density, ρ and elastic parameters λ and μ, caused by a point force at the origin. At large distances from the source, $\mathbf{u}(\mathbf{r}) = 0$. We define a point force $\mathbf{F}$ using a limit,

$$\mathbf{F} = \lim_{\delta V \to 0} \rho \, \mathbf{f} \, \delta V \,, \qquad (17.1)$$

where $\mathbf{f}$ is the force per unit mass, $\rho \, \mathbf{f}$ is the body force per unit volume, and δV is the infinitesimal volume element acted on by this idealized force. Before discussing details, we also introduce the three-dimensional spatial delta function $\delta(r)$,

$$\delta(r) = \begin{cases} 0, & r \neq 0 \\ \int_V \delta(r)\, dV = 1, & r = 0. \end{cases} \qquad (17.2)$$

Using Gauss' Theorem (covered in Chapter 11) we can represent a delta function using spatial derivatives of the radial coördinate r^{-1}

$$\delta(r) = -\frac{1}{4\pi} \nabla^2 \left(\frac{1}{r}\right). \qquad (17.3)$$

17.1.1 Static displacement field due to a single force

We now introduce a point force, $\mathbf{F}$, into the elastic equations of motion for equilibrium conditions ($\ddot{\mathbf{u}} = 0$),

$$\mathbf{F} + (\lambda + 2\mu)\nabla(\nabla \cdot \mathbf{u}) - \mu\nabla \times \nabla \times \mathbf{u} = 0. \quad (17.4)$$

Represent the force's magnitude with F and locate the force at the origin ($r = 0$), then

$$\mathbf{F} = \rho\mathbf{f} = F\mathbf{a}\,\delta(r) = -F\,\nabla^2\left(\frac{\hat{\mathbf{a}}}{4\pi r}\right)$$

$$= -F\left[\nabla\left(\nabla \cdot \frac{\hat{\mathbf{a}}}{4\pi r}\right) - \nabla \times \nabla \times \left(\frac{\hat{\mathbf{a}}}{4\pi r}\right)\right], \qquad (17.5)$$

where $\hat{\mathbf{a}}$ is a unit vector defining the direction of the force and we have used the vector identity $\nabla^2\mathbf{v} = \nabla(\nabla \cdot \mathbf{v}) - \nabla \times \nabla \times \mathbf{v}$. The equations of equilibrium become

$$-F\nabla^2\left(\frac{\hat{\mathbf{a}}}{4\pi r}\right)$$

$$= -F\left[\nabla\left(\nabla \cdot \frac{\hat{\mathbf{a}}}{4\pi r}\right) - \nabla \times \nabla \times \left(\frac{\hat{\mathbf{a}}}{4\pi r}\right)\right]$$

$$= -(\lambda + 2\mu)\nabla(\nabla \cdot \mathbf{u}) + \mu\nabla \times \nabla \times \mathbf{u}. \quad (17.6)$$

With this source form, any displacement field can be represented by a sum of solenoidal (divergence free) and irrotational (curl free) fields,

so we seek a solution of the form

$$\mathbf{u}(\mathbf{r}) = \nabla(\nabla \cdot \mathbf{A}_P) - \nabla \times (\nabla \times \mathbf{A}_S), \qquad (17.7)$$

where

$$\begin{cases} \nabla \times \mathbf{A}_P = 0 & \therefore \quad \nabla^2\mathbf{A}_P = \nabla(\nabla \cdot \mathbf{A}_P) \\[2mm] \nabla \cdot \mathbf{A}_S = 0 & \therefore \quad \nabla^2\mathbf{A}_S = -\nabla \times \nabla \times \mathbf{A}_S. \end{cases} \tag{17.8}$$

We call $\nabla \cdot \mathbf{A}_P$ and $\nabla \times \mathbf{A}_S$ Helmholtz potentials and $\mathbf{A}_P$ is irrotational and $\mathbf{A}_S$ is solenoidal. Substitution of this form of solution into the equations of equilibrium leads to

$$\nabla\left\{\nabla \cdot \left[\frac{-F\hat{\mathbf{a}}}{4\pi r} + (\lambda + 2\mu)\nabla^2\mathbf{A}_P\right]\right\}$$

$$+ \nabla \times \nabla \times \left(\frac{F\hat{\mathbf{a}}}{4\pi r} - \mu\nabla^2\mathbf{A}_S\right) = 0, \qquad (17.9)$$

which is satisfied if

$$(\lambda + 2\mu)\nabla^2\mathbf{A}_P = \frac{F\hat{\mathbf{a}}}{4\pi r}$$

$$\mu\nabla^2\mathbf{A}_S = \frac{F\hat{\mathbf{a}}}{4\pi r}. \qquad (17.10)$$

If we choose the directions of the potential fields such that $\mathbf{A}_P = A_P(r)\,\hat{\mathbf{a}}$, and $\mathbf{A}_S(r) = A_S\,\hat{\mathbf{a}}$. Then we obtain two Poisson's Equations

$$\nabla^2 A_P(r) = \frac{F}{4\pi(\lambda + 2\mu)r}$$

$$\nabla^2 A_S(r) = \frac{F}{4\pi\mu r}. \qquad (17.11)$$

In spherical coordinates, $\nabla^2 r = 2/r$, so we can integrate these expressions to obtain

$$A_P(r) = \frac{Fr}{8\pi(\lambda + 2\mu)}$$

$$A_S(r) = \frac{Fr}{4\pi\mu}. \qquad (17.12)$$

The potential functions, $A_P(r)$ and $A_S(r)$ solve the inhomogeneous equilibrium equations, (17.10) and the corresponding displacements can be computed using (17.7). Adopting a cartesian system for the spatial gradient operations, substituting Eq. (17.12) into (17.7), and expressing the vector operations using indicial notation yields the following expression for the ith component of displacement produced by a unit magnitude force ($F = 1$) in the jth direction, $u_i^j(\mathbf{x})$,

$$u_i^j(\mathbf{x}) = \frac{1}{8\pi(\lambda + 2\mu)}\frac{\partial}{\partial x_i}\frac{\partial r}{\partial x_j} - \frac{1}{8\pi\mu}\frac{\partial}{\partial x_i}\frac{\partial r}{\partial x_j}$$

$$+ \delta_{ij}\frac{1}{8\pi\mu}\nabla^2 r$$

$$= \frac{1}{8\pi\mu}\left(\delta_{ij}\nabla^2 r - \frac{\lambda + \mu}{\lambda + 2\mu}\frac{\partial^2 r}{\partial x_i \partial x_j}\right),$$

or

$$u_i^j(\mathbf{x}) = \frac{1}{8\pi\mu}\left(\delta_{ij}r_{,kk} - \Gamma r_{,ij}\right), \qquad (17.13)$$

where,

$$\Gamma = \frac{\lambda + \mu}{\lambda + 2\mu} = \frac{\lambda + 2\mu - \mu}{\lambda + 2\mu} = 1 - \beta^2/\alpha^2, \quad (17.14)$$

α is the P-wave speed, β is the S-wave speed, and $r = \sqrt{x_1^2 + x_2^2 + x_3^2}$ (Fig. 17.1).

For $\lambda = \mu$ (a Poisson solid), $\Gamma \approx 2/3$. The Cartesian coordinate system is referenced to the source (i.e., the force acts along one of the axes). The quantities, $u_i^j(\mathbf{x})$, can be used to form a symmetric ($u_i^j = u_j^i$) tensor called the **Somigliana Tensor**. For a set of forces in each coordinate direction, each with strength F, the six independent components in the tensor are

$$u_1^1(\mathbf{x}) = \frac{F}{8\pi\mu}\left[\frac{2}{r} - \Gamma\left(\frac{1}{r} - \frac{x_1^2}{r^3}\right)\right]$$

$$u_1^2(\mathbf{x}) = \frac{F}{8\pi\mu}\left(\Gamma\frac{x_1 x_2}{r^3}\right)$$

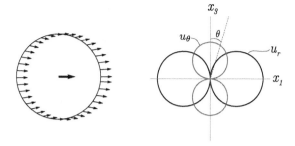

FIGURE 17.1 Vector displacements produced at equal distance from a point force (thick arrow) directed in the x_1 direction. Variation in radial and azimuthal component of $\mathbf{u}(r)$ of the deformation with azimuth, θ.

$$u_1^3(\mathbf{x}) = \frac{F}{8\pi\mu}\left(\Gamma\frac{x_1 x_3}{r^3}\right)$$

$$u_2^2(\mathbf{x}) = \frac{F}{8\pi\mu}\left[\frac{2}{r} - \Gamma\left(\frac{1}{r} - \frac{x_2^2}{r^3}\right)\right]$$

$$u_2^3(\mathbf{x}) = \frac{F}{8\pi\mu}\left(\Gamma\frac{x_2 x_3}{r^3}\right)$$

$$u_3^3(\mathbf{x}) = \frac{F}{8\pi\mu}\left[\frac{2}{r} - \Gamma\left(\frac{1}{r} - \frac{x_3^2}{r^3}\right)\right]. \qquad (17.15)$$

Or, in one compact, general form,

$$u_i^j(\mathbf{x}) = \frac{F}{8\pi\mu}\left[\frac{2 - \Gamma}{r}\delta_{ij} + \Gamma\frac{x_i x_j}{r^3}\right], \qquad (17.16)$$

where δ_{ij} is the Kronecker delta.

For a force applied in the x_1 direction, the displacements in spherical polar coordinates (r, θ, ϕ) (Fig. 17.1), can be computed using the Jacobian coordinate transformation,

$$\begin{bmatrix} u_r(\mathbf{r}) \\ u_\theta(\mathbf{r}) \\ u_\phi(\mathbf{r}) \end{bmatrix} = \begin{bmatrix} \sin\theta\cos\phi & \sin\theta\sin\phi & \cos\theta \\ \cos\theta\cos\phi & \cos\theta\sin\phi & -\sin\theta \\ -\sin\phi & \cos\phi & 0 \end{bmatrix}$$

$$\times \begin{bmatrix} u_1^1(\mathbf{x}) \\ u_2^1(\mathbf{x}) \\ u_3^1(\mathbf{x}) \end{bmatrix}, \qquad (17.17)$$

where $\mathbf{r}$ and $\mathbf{x}$ represent the same position in the two different coordinate systems. For example, in the $x_1 x_3$ plane, $\phi = 0$, we have

$$u_r(r, \theta) = \sin\theta\, u_1^1(x_1, x_3) + \cos\theta\, u_3^1(x_1, x_3)$$

$$= \frac{F}{4\pi\mu r}\sin\theta$$

$$u_\theta(r, \theta) = \cos\theta\, u_1^1(x_1, x_3) - \sin\theta\, u_3^1(x_1, x_3)$$

$$= \frac{F}{4\pi\mu r}\left(1 - \frac{\Gamma}{2}\right)\cos\theta . \qquad (17.18)$$

These rigorous solutions, with their simple trigonometric patterns of static deformation as illustrated in Fig. 17.1, resemble the intuitive solution described in Section 16.4.1. The magnitude of the displacements falls off as $1/r$, which indicates a stress decrease proportional to $1/r^2$. The static radial motions have a simple two-lobed sinusoidal distribution around the source, with maxima toward and away in the direction of the force. The shearing deformations in the $x_1 x_3$ plane also have a two-lobed pattern of shearing parallel to the direction of the force with maximum and minima perpendicular to the force direction. Axial symmetry requires that these patterns, rotated around the x_1 axis, produce the full, three-dimensional patterns.

17.1.2 Static displacement field due to a force couple

If we apply a force at position (ξ_1, ξ_2, ξ_3) instead of at the origin, the displacement at $P(x_1, x_2, x_3)$ will still be given by the Somigliana tensor, with all distances adjusted by the offset of the source location, e.g., $r^2 = (x_1 - \xi_1)^2 + (x_2 - \xi_2)^2 + (x_3 - \xi_3)^2$. If we apply a force $\mathbf{F}$ in the x_1-direction at $\xi_2 + \frac{1}{2}d\xi_2$ and an equal-strength force in the $-x_1$-direction at $\xi_2 - \frac{1}{2}d\xi_2$, we have constructed a single force couple as shown in Fig. 17.2. The displacement at $P(x_1, x_2, x_3)$ is the sum of the displacements from the two individual point forces,

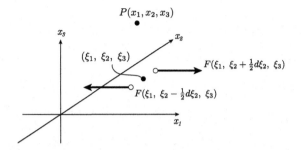

FIGURE 17.2 A force couple acting parallel to the $x_1 x_2$ plane in the x_1 direction at position (ξ_1, ξ_2, ξ_3).

$$u_i^1\left(\xi_1, \xi_2 + \frac{1}{2}d\xi_2, \xi_3 : x_1, x_2, x_3\right)$$

$$- u_i^1\left(\xi_1, \xi_2 - \frac{1}{2}d\xi_2, \xi_3 : x_1, x_2, x_3\right)$$

$$= \frac{\partial u_i^1}{\partial\xi_2}d\xi_2 + O(d\xi_2)^2 ,$$

where we used the definition of a spatial derivative. Since the source is not at the origin, the displacement arguments include the positions of both the source and the observing location. The single-couple response is the difference between the displacement fields due to the two single forces, accounting for the infinitesimal spatial offset $d\xi_2$. Using

$$r^2 = (x_1 - \xi_1)^2 + (x_2 - \xi_2)^2 + (x_3 - \xi_3)^2 ,$$

we have

$$\frac{\partial r}{\partial\xi_i} = -\frac{\partial r}{\partial x_i} ,$$

which leads to

$$\frac{\partial u_i^j}{\partial\xi_k} = -\frac{\partial u_i^j}{\partial x_k} . \qquad (17.19)$$

Then the displacement for the force couple in Fig. 17.2 is equal to

$$-\frac{\partial u_i^1}{\partial x_2}d\xi_2 + O(d\xi_2)^2 . \qquad (17.20)$$

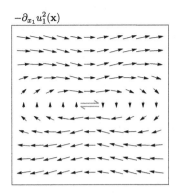 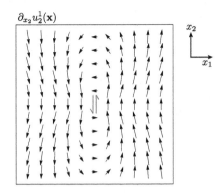

FIGURE 17.3 Static displacements in the x_1 and x_2 plane produced by single couples acting the x_1 (left) and x_2 (right) directions. The material is uniform and isotropic with $\lambda = \mu$. The vector lengths have been scaled by the square of the distance from the origin to allow distant deformations to be visible.

If we let $d\xi_2 \to 0$ and $F \to \infty$ such that $Fd\xi_2 \to M$, a finite moment, we can compute expressions for the static displacement field for the single couple with moment M. The displacement (u_i) due to a single couple *at the origin* with forces acting in the $\pm x_1$ direction offset in the x_2 direction can be obtained by replacing F by M in the Somigliana tensor, taking the spatial derivative of all the terms with respect to x_2, and changing the sign. The result is

$$u_1(x_1, x_2, x_3)$$
$$= -\frac{M}{8\pi\mu}\left[\frac{-2x_2}{r^3} - \Gamma\left(\frac{-x_2}{r^3} + 3\frac{x_1^2 x_2}{r^5}\right)\right]$$
$$u_2(x_1, x_2, x_3) = -\frac{M}{8\pi\mu}\Gamma\left(\frac{x_1}{r^3} - 3\frac{x_1 x_2^2}{r^5}\right)$$
$$u_3(x_1, x_2, x_3) = -\frac{M}{8\pi\mu}\Gamma\left(-3\frac{x_1 x_2 x_3}{r^5}\right). \quad (17.21)$$

If a single couple is oriented along the x_2 axis with offset arm along the x_1 direction, we obtain

$$u_1(x_1, x_2, x_3) = -\frac{M}{8\pi\mu}\Gamma\left(\frac{x_2}{r^3} - 3\frac{x_1^2 x_2}{r^5}\right)$$

$$u_2(x_1, x_2, x_3)$$
$$= -\frac{M}{8\pi\mu}\left[\frac{-2x_1}{r^3} - \Gamma\left(\frac{-x_1}{r^3} + 3\frac{x_2^2 x_1}{r^5}\right)\right]$$
$$u_3(x_1, x_2, x_3) = -\frac{M}{8\pi\mu}\Gamma\left(-3\frac{x_1 x_2 x_3}{r^5}\right). \quad (17.22)$$

In general, for a couple oriented with forces in the jth direction and offset in the kth direction, the displacements are $u_i = -\partial_{x_k}(u_i^j)$, where ∂_{x_k} indicates partial differentiation with respect to x_k. The moment is defined to be positive for clockwise rotation around the axis perpendicular to the force direction and offset arm, and negative for counterclockwise rotation.

Fig. 17.3 includes plots of the equilibrium static displacements produced in the $x_1 x_2$ plane for single couples located at the origin. The left panel corresponds to displacements produced by a single couple acting horizontally (x_1) and offset vertically (x_2). The right panel is a plot of the displacements produced by a single couple acting vertically (x_2) and offset horizontally (x_1). The computations were performed assuming a Poisson solid ($\lambda = \mu$). The single-couple displacements decrease as r^{-2} away from the location of the forces. Displacements shown in the figure are scaled by a factor of r^2 so that the more distance deformations are visible.

$$-\partial_{x_2} u_2^1(\mathbf{x}) + \partial_{x_1} u_1^2(\mathbf{x})$$

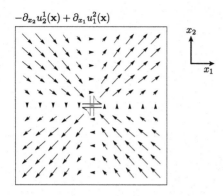

FIGURE 17.4 Static displacements in the x_3 for the double couple included two single couples along the x_1 and x_2 axes. The source is located at the center of the grid and the double-couple force orientations are shown in gray. The vector magnitudes are scaled by r^2 to insure that distant deformations are visible in the display.

17.1.3 Static displacement field due to a double couple

The principle of superposition that holds for point forces also holds for couples. To compute the displacement field due to a double couple, we simply sum the displacements for the individual couples. If we combine the single couples in Fig. 17.3, we simply add the displacement patterns from the two panels in the figure.

$$u_i(\mathbf{x}) = -\partial_{x_2} u_2^1(\mathbf{x}) + \partial_{x_1} u_1^2(\mathbf{x}) . \qquad (17.23)$$

The numerical results using the single-couples in Fig. 17.3 is shown in Fig. 17.4. The displacements decrease rapidly away from the source (which is at the center of the plot), so for display, the vector magnitudes are again scaled by r^2 to make distant deformations visible. The pattern indicates four deformation quadrants, two with motion towards the source and two with motion away from the source. The maximum displacement is at 45° angles from the orientation of each single couple and the minima are along the directions of each couple, where the deformation direction changes – these are nodes. The **T**-axis is oriented with an azimuth (relative to the x_2 di-

rection) of 45° and the **P**-axis is orthogonal to **T**. The intermediate axis passes vertically through the origin (grid-center).

Analytic expressions for the Cartesian displacements of a double-couple have the form,

$$
\begin{aligned}
u_1(\mathbf{x}) &= \frac{M}{8\pi\mu}\left(\frac{2x_2}{r^3}\right) - \frac{2M\Gamma}{8\pi\mu}\left(\frac{x_2}{r^3} - \frac{3x_2 x_1^2}{r^5}\right) \\
&= \frac{M}{4\pi\mu r^2}\frac{x_2}{r}\left[-1 - \Gamma\left(1 - \frac{3x_1^2}{r^2}\right)\right]
\end{aligned}
$$

$$
\begin{aligned}
u_2(\mathbf{x}) &= \frac{M}{8\pi\mu}\left(\frac{2x_1}{r^3}\right) - \frac{2M\Gamma}{8\pi\mu}\left(\frac{x_1}{r^3} - \frac{3x_2^2 x_1}{r^5}\right) \\
&= \frac{M}{4\pi\mu r^2}\frac{x_1}{r}\left[1 - \Gamma\left(1 - \frac{3x_2^2}{r^2}\right)\right]
\end{aligned}
$$

$$u_3(\mathbf{x}) = \frac{M}{8\pi\mu}\left(6\Gamma\frac{x_1 x_2 x_3}{r^5}\right) = \frac{M}{4\pi\mu r^2}\left(3\Gamma\frac{x_1 x_2 x_3}{r^3}\right). \qquad (17.24)$$

In spherical polar coordinates (r, θ, ϕ), the displacements can be written in terms of $u_r(\mathbf{r})$, $u_\theta(\mathbf{r})$, and $u_\phi(\mathbf{r})$:

$$u_r(\mathbf{r}) = \frac{M}{4\pi\mu r^2}\left(1 + \frac{\Gamma}{2}\right)\sin^2\theta \sin 2\phi$$

$$u_\theta(\mathbf{r}) = \frac{M}{4\pi\mu r^2}\left(\frac{1}{2} - \frac{\Gamma}{2}\right)\sin 2\theta \sin 2\phi \qquad (17.25)$$

$$u_\phi(\mathbf{r}) = \frac{M}{4\pi\mu r^2}\left(1 - \Gamma\right)\sin\theta \cos 2\phi .$$

The displacement field for the double couple decreases proportional to the inverse of distance squared, much more rapidly than for the point force, which decreased in proportion to the inverse of distance. On the $x_1 x_2$ plane, $\theta = \pi/2$, $u_\theta = 0$, and

$$u_r \approx (1 + \Gamma/2)\sin 2\phi$$
$$u_\phi \approx (1 - \Gamma)\cos 2\phi. \qquad (17.26)$$

Fig. 17.5 is a view of the displacements for the same double-couple illustrated in Fig. 17.4 with

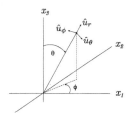 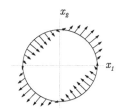 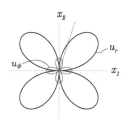

FIGURE 17.5 Displacements produced by a double-couple at the origin in the $x_1 x_2$ plane. (left) Spherical polar coordinate system (center) The total displacement pattern in the $x_1 x_2$ plane on a circle around the source, involving a combination of u_r and u_ϕ components. (right) Azimuthal pattern of radial (u_r) and tangential (u_ϕ) displacements in the $x_1 x_2$ plane.

a focus on azimuthal variations of both the vector and polar-coordinate displacement magnitude. The coordinate system and unit vectors (which change orientation with position) are shown on the left. Vector displacements in the $x_1 x_2$ plane sampled a fixed distance from the source are shown in the center panel. The right most panel shows the magnitude of the components in the radial and tangential directions, both of which exhibit a four-lobed (quadrant) variation in amplitude. The factor-of-four difference in component amplitude results from the difference in coefficients in Eq. (17.26).

Keep in mind that our results, while illustrative of the deformation fields associated with point forces, single- and double-couples, are for an isotropic uniform elastic wholespace. Using superposition of point-source solutions, we can model more spatially distributed sources. Since static deformations decay rapidly with distance from the source, most observations of geodetic deformation are observed close to the source. Indeed, they provide critical information on the nature of the strain release during earthquakes. However, to model the measurements, most of which are made at the surface, we must use a solution that includes a stress-free surface. Although the derivation is beyond our scope, a halfspace solution can be constructed using the wholespace solutions. Expressions for the surface deformation in a uniform, isotropic halfspace are described in Box 9.1. Derivations and details are available in the references, and more

general solutions and discussions can be found in Segall (2010). We proceed now to problems that allow for time dependence of the source, which gives rise to both transient waves and permanent static deformations.

17.2 Elastodynamics

We now consider the application of a time-varying force system to a homogeneous elastic whole space. Mathematically this is a much more difficult problem than the statics case. The temporal dependence of the force and displacement introduces an extra set of integrations with respect to both time and space. But the logical framework used for elastodynamics follows the same path we used for elastostatic problems. We construct the full solution for a double-couple force system by summing solutions of the single-force response. But because some of the mathematics is beyond the scope of this text, we omit some of the mathematical details. A more complete analysis can be found in advanced texts such as Aki and Richards (2009) or Kasahara (1981). The elastodynamic equations for the displacements in a uniform isotropic elastic material are

$$\rho \ddot{\mathbf{u}} = \rho \mathbf{f} + (\lambda + 2\mu) \nabla (\nabla \cdot \mathbf{u}) - \mu \nabla \times \nabla \times \mathbf{u} .$$

Our goal is to introduce expressions for the displacement field, $\mathbf{u}(\mathbf{r})$, produced by a point source, single-couple, double-couple, and moment tensor.

17.2.1 Elastodynamic point-force displacements

We assume a body force per unit volume with the time-dependent from $\rho\,\mathbf{f}(t) = F(t)\,\delta(\mathbf{r})\,\hat{\mathbf{a}}$, where $F(t)$ is the time history of the applied force and $\mathbf{r}$ is a position vector (the source is at the origin). Commonly assumed force-time histories include delta functions, $\delta(t)$, step functions, $H(t)$, and ramp functions, $R(t)$, similar to those shown in Fig. 16.1 for explosions. As before, we use a vector identity to write

$$F(t)\,\delta(\mathbf{r})\,\hat{\mathbf{a}} = -F(t)\,\nabla^2\left(\frac{\hat{\mathbf{a}}}{4\pi r}\right)$$

$$= -F(t)\left\{\nabla\left[\nabla\cdot\left(\frac{\hat{\mathbf{a}}}{4\pi r}\right)\right] \quad (17.27)\right.$$

$$\left. -\nabla\times\nabla\times\left(\frac{\hat{\mathbf{a}}}{4\pi r}\right)\right\},$$

where r represents the distance from the origin. We seek a solution of the form [see (17.7)]

$$\mathbf{u}(\mathbf{r}, t) = \nabla(\nabla\cdot\mathbf{A}_P) - \nabla\times\nabla\times\nabla\mathbf{A}_S$$

$$\text{where} \quad \begin{cases} \nabla\times\mathbf{A}_P & = 0 \\ \nabla\cdot\mathbf{A}_S & = 0 \end{cases} \quad (17.28)$$

Using this decomposition, the elastodynamic equation separates into two vector equations,

$$(\lambda + 2\mu)\,\nabla^2\mathbf{A}_P = \frac{F(t)}{4\pi r}\hat{\mathbf{a}} + \rho\frac{\partial^2\mathbf{A}_P}{\partial t^2}$$

$$\mu\nabla^2\mathbf{A}_S = \frac{F(t)}{4\pi r}\hat{\mathbf{a}} + \rho\frac{\partial^2\mathbf{A}_S}{\partial t^2}. \quad (17.29)$$

Choosing $\mathbf{A}_P = A_P\,\hat{\mathbf{a}}$ and $\mathbf{A}_S = A_s\,\hat{\mathbf{a}}$, these reduce to scalar equations,

$$\nabla^2 A_P = \frac{F(t)}{4\pi(\lambda + 2\mu)r} + \frac{1}{\alpha^2}\frac{\partial^2 A_P}{\partial t^2}$$

$$\nabla^2 A_S = \frac{F(t)}{4\pi\,\mu r} + \frac{1}{\beta^2}\frac{\partial^2 A_S}{\partial t^2} \quad (17.30)$$

where the P-velocity, α, and S-velocity β are

$$\alpha = \sqrt{\frac{\lambda + 2\mu}{\rho}}, \qquad \beta = \sqrt{\frac{\mu}{\rho}}.$$

Our equations for A_P and A_S are inhomogeneous wave equations of the form

$$\nabla^2\phi(\mathbf{x}, t) - \frac{1}{c^2}\frac{\partial^2\phi}{\partial t^2}(\mathbf{x}, t) = g(\mathbf{x}, t). \quad (17.31)$$

An important form of $g(\mathbf{x}, t)$ is a point source in both space and time,

$$g(\mathbf{x}, t) = -\delta(\mathbf{x})\,\delta(t), \quad (17.32)$$

which corresponds to a solution of the form

$$\phi(\mathbf{x}, t) = \frac{1}{4\pi}\frac{\delta(t - r/c)}{r}. \quad (17.33)$$

This result indicates that the solution to the inhomogeneous wave equation with a impulsive, symmetric point-source is an outward-propagating wave that ultimately decays in amplitude as $1/r$ (our equations are for displacement potentials). The amplitude decrease keeps the total energy on the spreading wavefront constant. A similar solution was invoked in the initial discussion of explosion sources at the start of this chapter. As we will see, the form of displacement time history more than a few wavelengths from an impulsive force is a delta-function.

Using (17.33) and the properties of delta functions, we can develop additional solutions. For a point force acting at location $\boldsymbol{\xi} = (\xi_1, \xi_2, \xi_3)$ and time τ, we have

$$\nabla^2\phi - \frac{1}{c^2}\frac{\partial^2\phi}{\partial t^2} = -\delta(\mathbf{x} - \boldsymbol{\xi})\,\delta(t - \tau), \quad (17.34)$$

which has a solution

$$\phi(\mathbf{r}, t) = \frac{1}{4\pi}\frac{\delta(t - \tau - |\mathbf{x} - \boldsymbol{\xi}|/c)}{|\mathbf{x} - \boldsymbol{\xi}|}. \quad (17.35)$$

For a more general time history of force, $f(t)$,

$$\nabla^2 \phi - \frac{1}{c^2} \frac{\partial^2 \phi}{\partial t^2} = -\delta(\mathbf{x} - \boldsymbol{\xi}) f(t) , \qquad (17.36)$$

which has a solution

$$\phi = \frac{1}{4\pi} \frac{f(t - |\mathbf{x} - \boldsymbol{\xi}|/c)}{|\mathbf{x} - \boldsymbol{\xi}|} . \qquad (17.37)$$

If a source is extended throughout a volume, V_{ξ}, and in time

$$\nabla^2 \phi - \frac{1}{c^2} \frac{\partial^2 \phi}{\partial t^2} = -\Phi(\mathbf{x}, t) , \qquad (17.38)$$

then the solution is

$$\phi(\mathbf{x}, t) = \frac{1}{4\pi} \iiint\limits_{V_{\xi}} \frac{\Phi(\boldsymbol{\xi}, t - |\mathbf{x} - \boldsymbol{\xi}|/c)}{|\mathbf{x} - \boldsymbol{\xi}|} dV_{\xi} .$$

$$(17.39)$$

The potential at $(\mathbf{x}, t)$ is sensitive to source activity in the element dV_{ξ} (at $\boldsymbol{\xi}$) only at the *retarded time*, $t - |\mathbf{x} - \boldsymbol{\xi}|/c$. Thus, we can write solutions to Eq. (17.30) for the P- and S-wave potentials,

$$A_P(\mathbf{x}, t) = \frac{1}{4\pi} \iiint\limits_{V_{\xi}} \frac{-F(t - |\mathbf{x} - \boldsymbol{\xi}|/\alpha)}{4\pi (\lambda + 2\mu) r |\mathbf{x} - \boldsymbol{\xi}|} dV_{\xi}$$

$$A_S(\mathbf{x}, t) = \frac{1}{4\pi} \iiint\limits_{V_{\xi}} \frac{-F(t - |\mathbf{x} - \boldsymbol{\xi}|/\beta)}{4\pi \mu r |\mathbf{x} - \boldsymbol{\xi}|} dV_{\xi} ,$$

$$(17.40)$$

where $\xi = 0$ for a point force at the origin.

The next steps are rather messy – we must integrate over the volume around the source position $\boldsymbol{\xi}$. Let the distance $|\mathbf{x} - \boldsymbol{\xi}| = \alpha \tau$, where τ is the transit time. Then it can be shown,

$$A_P(\mathbf{x}, t) = \frac{1}{4\rho\pi r} \left[\int_0^{\infty} F(t - r/\alpha - \tau) \tau \, d\tau \right.$$

$$\left. - \int_0^{\infty} F(t - \tau) \tau \, d\tau \right]$$

$$(17.41)$$

$$A_S(\mathbf{x}, t) = \frac{1}{4\rho\pi r} \left[\int_0^{\infty} F(t - r/\beta - \tau) \tau \, d\tau \right.$$

$$\left. - \int_0^{\infty} F(t - \tau) \tau \, d\tau \right] .$$

The displacement field is obtained from the potentials using

$$\mathbf{u}(\mathbf{x}) = \nabla (\nabla \cdot \mathbf{A}_P) - \nabla \times \nabla \times \mathbf{A}_S . \qquad (17.42)$$

For a body force with time history $F(t)$ applied in the x_1 direction at the origin, the total displacement field is

$$u_i(\mathbf{x}, t) = \frac{1}{4\pi\rho} \left(\frac{\partial^2}{\partial x_i \partial x_1} \frac{1}{r} \right) \int_{r/\alpha}^{r/\beta} \tau F(t - \tau) d\tau$$

$$+ \frac{1}{4\pi\rho\alpha^2 r} \left(\frac{\partial r}{\partial x_i} \frac{\partial r}{\partial x_1} \right) F\left(t - \frac{r}{\alpha}\right)$$

$$+ \frac{1}{4\pi\rho\beta^2 r} \left(\delta_{i1} \frac{-\partial r}{\partial x_i} \frac{\partial r}{\partial x_1} \right) F\left(t - \frac{r}{\beta}\right) .$$

$$(17.43)$$

For solutions corresponding to point forces in the x_2 or x_3 directions, replace each ∂x_1 with ∂x_2 or ∂x_3, respectively. Thus, for a point force with time history $F(t)$ in the x_j direction, located at the origin, we have the classic *Stokes* solution,

$$u_i(\mathbf{x}, t) = \frac{1}{4\pi\rho} \left(3\gamma_i\gamma_j - \delta_{ij} \right) \frac{1}{r^3} \int_{r/\alpha}^{r/\beta} \tau F(t - \tau) d\tau$$

$$+ \frac{1}{4\pi\rho\alpha^2} \gamma_i\gamma_j \frac{1}{r} F\left(t - \frac{r}{\alpha}\right)$$

$$+ \frac{1}{4\pi\rho\beta^2} \left(\gamma_i\gamma_j - \delta_{ij} \right) \frac{1}{r} F\left(t - \frac{r}{\beta}\right) ,$$

$$(17.44)$$

where γ_i is the direction cosine; $\gamma_i = x_i/r = \partial r/\partial x_i$ and δ_{ij} is a Kronecker delta. The first term decreases as $1/r^2$ for short-duration sources, the later two terms decrease as $1/r$. The first term is called the **near-field** term because it decays so quickly it is usually only important near the source. Near-field displacements include contributions from both the P and S wavefields, and these cannot be easily separated. For analogous reasons, the latter two terms are **far-field** terms. The first far-field term is a P wave,

$$u_i^P(\mathbf{x}, t) = \frac{1}{4\pi\rho\alpha^2} \gamma_i \gamma_j \frac{1}{r} F\left(t - \frac{r}{\alpha}\right), \quad (17.45)$$

with the following properties: (1) amplitude attenuates as $1/r$, (2) the wave propagates with speed $\alpha = [(\lambda + 2\mu)/\rho]^{1/2}$, (3) the displacement waveform is a scaled version of the applied force, and (4) the displacement is parallel to the direction from the source to the observation point. The second far-field term is an S wave,

$$u_i^S(\mathbf{x}, t) = \frac{1}{4\pi\rho\beta^2} (\delta_{ij} - \gamma_i\gamma_j) \frac{1}{r} F\left(t - \frac{r}{\beta}\right) \quad (17.46)$$

with the following properties: (1) amplitude attenuates as $1/r$, (2) the wave propagates with velocity $\beta = (\mu/\rho)^{1/2}$, (3) the displacement waveform is a scaled version of the applied force, and (4) the direction of displacement is perpendicular to the direction from the source to the observation point.

17.2.2 Elastodynamic single-couple displacements

As we did for the static fields, the displacement field for single couples can be obtained by differentiating the point-force results. Differentiating the Stokes solution with respect to each coordinate direction produces the displacements for a couple paralleling each direction. The complete solution including near-field terms is described in Aki and Richards (2009). For brevity, we focus on the far-field P wave, other results are arrived at using similar analyses. P waves in the far field from a force couple (Fig. 17.6) with time history $h(t)$ are given by summing the

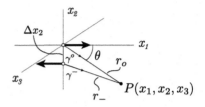

FIGURE 17.6 A couple for which far-field P-wave displacements are computed. Eventually our focus is on a geometry for which $\Delta x_2 \to 0$, but slight differences in the position and travel time from two point-force displacement fields to the observation point P require the limit be taken at the end of the analysis.

motions due to individual point forces. We label the forces F_o and F_-, and define directions and distances from the (0) and $(-)$ forces as γ^o, r_o, γ^-, and r_- respectively. The distances to an observing point are $r_o = \sqrt{x_1^2 + x_2^2 + x_3^2}$ and $r_- = \sqrt{x_1^2 + (x_2 - \Delta x_2)^2 + x_3^2}$. The vector from F_o to the observation location is $r_o \gamma_i^o$. The displacements corresponding to each of the point forces are

$$u_i^o(\mathbf{x}, t) = \frac{1}{4\pi\rho\alpha^2} \gamma_i^o \gamma_1^o \frac{h(t - r_o/\alpha)}{r_o}$$

$$u_i^-(\mathbf{x}, t) = -\frac{1}{4\pi\rho\alpha^2} \gamma_i^- \gamma_1^- \frac{h(t - (r_-/\alpha))}{r_-},$$
$$\quad (17.47)$$

where the negative sign in front of the second term arises from the force direction. The total displacement due to the force couple is $u_i^o + u_i^-$,

$$u_i^c(\mathbf{x}, t) = \frac{1}{4\pi\rho\alpha^2} \left[\gamma_i^o\gamma_1^o \frac{h(t - r_o/\alpha)}{r_o} - \gamma_i^-\gamma_1^- \frac{h(t - r_-/\alpha)}{r_-} \right].$$
$$\quad (17.48)$$

Now consider the direction cosines, γ^o and γ_i^-. The first is

$$\gamma_i^o = \frac{x_i}{r_o} = \frac{\partial r_o}{\partial x_i}. \quad (17.49)$$

We would like to express γ_i^- in terms of γ^o.

$$\gamma_i^- = \frac{\partial r_-}{\partial x_i} = \frac{\partial \sqrt{x_1^2 + (x_2 - \Delta x_2)^2 + x_3^2}}{\partial x_i}$$

$$= \begin{cases} x_i/r_- , & i \neq j \\ \\ (x_2 - \Delta x_2)/r_- & i = 2 \end{cases} \quad (17.50)$$

$$= \frac{x_i}{r_-} - \frac{\Delta x_2}{r_-} \delta_{2i} ,$$

where δ_{2i} is the Kronecker delta. When the difference in distances is small, $\Delta r = |r_- - r_o| \ll r_o$, we can approximate

$$\frac{x_i}{r_-} \approx \frac{x_i}{r_o} \sim \gamma_i^o , \quad (17.51)$$

although Δx_2 is not necessarily $\ll x_2$, so we must retain the Δx_2 term when $i = 2$. Dropping the subscript on the direction cosines ($\gamma^o \to \gamma$) to produce a simpler notation leads to

$$u_i^c(\mathbf{r}) = \frac{\gamma_1}{4\pi\rho\alpha^2} \left[\gamma_i \frac{h(t - r_o/\alpha)}{r_o} - \left(\gamma_i - \frac{\Delta x_2}{r_-}\delta_{2i} \right) \frac{h(t - r_-/\alpha)}{r_-} \right]. \quad (17.52)$$

We can approximate $1/r_-$ as $1/r_o$ for $\Delta r \ll r_o$, but what about $h(t - r_o/a)$ compared with $h(t - r_-/\alpha)$? For notational convenience, replace r_o with r and examine the function

$$h(t - r_-/\alpha) = h(t - r/\alpha - \Delta r/\alpha) . \quad (17.53)$$

The time shift $\Delta r/\alpha$ corresponding to the difference in arrival time from each point force is not small enough to ignore and has a fundamentally important observable consequence. Define a difference from the arrival time as $t_0 = t - r/\alpha$ and use a Taylor series to expand $h(t)$ about t_0,

$$h(t) = h(t_0) + \frac{\partial h(t)}{\partial t}\bigg|_{t_0} (t - t_0) + \cdots . \quad (17.54)$$

Then use this expansion to evaluate $h(t - r/\alpha - \Delta r/\alpha)$ such that to first order,

$$\begin{aligned} h(t &- r/\alpha - \Delta r/\alpha) \\ &\approx h(t_0) + \dot{h}(t_0) \times (t - r/\alpha - \Delta r/\alpha - t_0) \\ &= h(t_0) + \frac{\Delta r}{\alpha} \dot{h}(t_0) \\ &= h(t - r/\alpha) + \frac{\Delta r}{\alpha} \dot{h}(t - r/\alpha) \end{aligned}$$
$$(17.55)$$

Then, to first order

$$\begin{aligned} u_i^c(\mathbf{x}, t) &= \frac{\gamma}{4\pi\rho\alpha^2} \Bigg\{ \gamma_i \left[\frac{h(t - r/\alpha)}{r} \right] \\ &\quad - \gamma_i \left[\frac{h(t - r/\alpha)}{r} - \frac{\Delta r}{\alpha} \frac{\dot{h}(t - r/\alpha)}{r} \right] \\ &\quad + \frac{\Delta x_2}{r}\delta_{2j} \left[\frac{h(t - r/\alpha)}{r} - \frac{\Delta r}{\alpha} \frac{\dot{h}(t - r/\alpha)}{r} \right] \Bigg\} \\ u_i^c(\mathbf{x}, t) &= \frac{\gamma_1}{4\pi\rho\alpha^2} \Bigg\{ \gamma_i \frac{\Delta r \, \dot{h}(t - (r/\alpha))}{\alpha r} \\ &\quad + \frac{\Delta x_2 \delta_{2j}}{r^2} \left[h(t - r/\alpha) - \frac{\Delta r}{\alpha} \dot{h}(t - r/\alpha) \right] \Bigg\} . \end{aligned}$$
$$(17.56)$$

The second group of terms decays as $1/r^2$ which implies that these terms are part of the near-field response that we can dismiss relative to the far-field terms (for our analysis), which decay as $1/r$. Thus the far-field displacements are

$$u_i^c(\mathbf{x}, t) = \frac{\gamma_1}{4\pi\rho\alpha^3} \left[\frac{\gamma_i \, \Delta r \, \dot{h}(t - r/\alpha)}{r} \right] . \quad (17.57)$$

The spatial offset of the forces leads to far-field displacement sensitivity to near-source particle velocities rather than to near-source particle displacements. This was also found for the explosion described at the start of the chapter.

We now need to consider $\Delta r = r_- - r$

$$\Delta r \approx \frac{\partial r}{\partial x_2} \Delta x_2 = \gamma_2 \Delta x_2 , \quad (17.58)$$

where we used the definition of the direction cosines $\gamma_i = x_i/r = \partial r/\partial x_i$. Thus,

$$u_i^c(\mathbf{x}, t) = \frac{\gamma_1 \gamma_2 \gamma_i}{4\pi\rho\alpha^3}\left[\frac{\Delta x_2}{r}\dot{h}\left(t - \frac{r}{\alpha}\right)\right]. \qquad (17.59)$$

Finally, we consider the limit as $\Delta x_2 \to 0$ and $h \to \infty$ such that $\Delta x_2 h \to M$, which is the moment of the couple. We then have

$$M\left(t - \frac{r}{\alpha}\right) = \lim_{\substack{\Delta x_2 \to 0 \\ h \to \infty}} \Delta x_2 h(t - r/\alpha)$$

$$\dot{M}\left(t - \frac{r}{\alpha}\right) = \lim_{\substack{\Delta x_2 \to 0 \\ h \to \infty}} \Delta x_2 \dot{h}(t - r/\alpha).$$

$$(17.60)$$

The far-field P-wave displacements for a single couple with forces in the x_1 direction and offset in the x_2 direction are

$$u_i^c(\mathbf{x}, t) = \frac{\gamma_1 \gamma_2 \gamma_i}{4\pi\rho\alpha^3}\frac{\dot{M}(t - r/\alpha)}{r}. \qquad (17.61)$$

For a uniform material, the motion is inversely proportional to the P velocity cubed and inversely proportional to the density. Perhaps the most important physical insight is that the shape of the observed far-field ground motion is proportional to the near-source velocities during the deformation. Since the double-couple sources are sum of single-couple sources, the same will be true for an earthquake model. Thus, the motion at great distance (more than a few wavelengths) from the fault is proportional to an earthquake's slip velocity history.

More generally, the far field motion in the n direction for a couple in the pq plane (force in the p direction, offset in the q direction) is

$$u_n^c(\mathbf{x}, t) = \frac{\gamma_n \gamma_p \gamma_q}{4\pi\rho\alpha^3}\frac{\dot{M}_{pq}(t - r/\alpha)}{r}. \qquad (17.62)$$

From a similar analysis, it can be shown that the general form of the far-field S-wave displacements for a couple in the pq plane is given by

$$u_n^c(\mathbf{x}, t) = \frac{-\left(\gamma_n \gamma_p - \delta_{np}\right)\gamma_q}{4\pi\rho\beta^3}\frac{\dot{M}_{pq}(t - r/\beta)}{r}.$$

$$(17.63)$$

These are key results, albeit they remain in an abstract Cartesian reference frame.

17.2.3 Moment-tensor radiation patterns

The seismic moment tensor was defined in terms of nine force couples (Section 16.5) and we use the results of the previous section to compute the far-field response for each. If we assume each moment tensor component shares the same time history, $\dot{M}(t)$, then we can write, for example,

$$u_n^c(\mathbf{x}, t) = \frac{R^P}{4\pi\rho\alpha^3}\frac{\dot{M}(t - r/\alpha)}{r}, \qquad (17.64)$$

where $R^P = \gamma_n \gamma_p \gamma_q M_{pq}$, and M_{pq} represents the moment-tensor elements. For an S-wave,

$$u_n^c(\mathbf{x}, t) = \frac{-\left(\gamma_n \gamma_p - \delta_{np}\right)\gamma_q M_{pq}}{4\pi\rho\beta^3}\frac{\dot{M}(t - r/\beta)}{r}.$$

$$(17.65)$$

We can define analogous quantities R^{SV} and R^{SH}, if we want to split the S motion into vertically and horizontally polarized components. To do so, requires that we use a spherical coordinate system. We can use the geographic reference system to render the expressions more practical. If we assume our coordinate system (x_1, x_2, x_3) represents (north, east, and down), as we did in our moment-tensor definition in terms of strike, dip, and rake, we can define the spherical coordinate polarization vectors as

$$\boldsymbol{\gamma} = \hat{\mathbf{r}} = (\sin\theta\cos\phi, \sin\theta\sin\phi, \cos\theta)$$
$$\hat{\boldsymbol{\theta}} = (\cos\theta\cos\phi, \cos\theta\sin\phi, -\sin\theta) \quad (17.66)$$
$$\hat{\boldsymbol{\phi}} = (-\sin\phi, \cos\phi, 0) \ .$$

$\hat{\mathbf{r}}$ is the direction of P-wave polarization, $\hat{\boldsymbol{\theta}}$ is the direction of SV polarization, and $\hat{\boldsymbol{\phi}}$ is the direction of SH polarization. The angle θ is measured from downward (x_3 is positive down) and thus is also known as the take-off angle since seismic waves observed at great distance take-off from the source downward. For a wave leaving the source region in a direction $\boldsymbol{\gamma}$, the amplitudes of P, SV, and SH can be calculation using

$$R^P = \hat{\boldsymbol{\gamma}} \cdot \mathbf{M} \cdot \hat{\boldsymbol{\gamma}}$$
$$R^{SV} = \hat{\boldsymbol{\theta}} \cdot \mathbf{M} \cdot \hat{\boldsymbol{\gamma}} \quad (17.67)$$
$$R^{SH} = \hat{\boldsymbol{\phi}} \cdot \mathbf{M} \cdot \hat{\boldsymbol{\gamma}}$$

Each formula is the projection of the displacement from the source, $\mathbf{M} \cdot \hat{\boldsymbol{\gamma}}$, onto the appropriate polarization vector. These formulas are a convenient way to compute radiation patterns, but an examination of the explicit formulas for double-couple sources provides insight into the seismic wavefield.

17.2.4 Elastodynamic double-couple displacements

A double-couple response is constructed by superposition of two single couple responses. If we sum two couples with forces oriented along the x_1 and x_2 directions and offset along the x_3 direction, we create the double couple shown in Fig. 17.4. The far-field P waves response is

$$u_i^{DC}(\mathbf{x}, t) = 2 \left[\frac{\gamma_1\gamma_2\gamma_i}{4\pi\rho\alpha^3} \frac{\dot{M}(t - r/\alpha)}{r} \right]. \quad (17.68)$$

If the displacements are transformed into the spherical coordinate system (Fig. 16.1), the far-

field P and S displacements are

$$\mathbf{u}_P(\mathbf{r}, t) = \frac{1}{4\pi\rho\alpha^3} (\sin 2\theta \cos\phi \, \hat{\mathbf{r}}) \frac{\dot{M}_0(t - r/\alpha)}{r}$$

$$\mathbf{u}_S(\mathbf{r}, t) = \frac{1}{4\pi\rho\beta^3} \left(\cos 2\theta \cos\phi \, \hat{\boldsymbol{\theta}} - \cos\theta \sin\phi \, \hat{\boldsymbol{\phi}} \right)$$
$$\times \frac{\dot{M}_0(t - r/\beta)}{r},$$

$$(17.69)$$

where $M_0(t) = \mu A_R(t)D(t)$ is the time-dependent *moment function* ($A_R(t)$ is the rupture area as a function of time and $D(t)$ is the fault slip history). The far-field displacements are proportional to the time derivative of the moment function, $\dot{M}_0(t)$, which is called the *moment-rate function*. The moment-rate is the part of the signal that includes information on rupture processes – wave amplitude factors are governed by the faulting geometry and near-source properties. Both P and S have the four-lobed radiation patterns shown in Fig. 17.7. Radial (P-wave) motions are maximum in the directions of the P and T axes, transverse motion (S-wave) is minimal in those directions. The azimuthal patterns are identical to those of the static case (Fig. 17.4).

17.3 Double-couple radiation patterns in geographic coordinates

Radiation patterns expressed in terms of geographical references are more convenient. The transformation requires mapping the source description into the spherical coordinate systems shown in Fig. 17.8 (note that we choose the x_3 direction as downward). We follow Aki and Richards (2009) and define a ray coordinate system with directions $\hat{\mathbf{l}}, \hat{\mathbf{p}}, \hat{\boldsymbol{\phi}}$ along the P, SV, and SH polarization directions at the source.

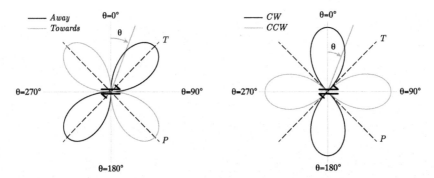

FIGURE 17.7 Far-field radiation patterns in the x_1x_3 plane ($\phi = 0$) for radial components of displacement (left) and transverse components of displacement (right), caused by a double couple in the x_1x_3 plane. The full vector displacements are given by Eq. (17.69). The slip vectors for motion on a fault normal to the x_3 plane are shown. Radial motion is positive away from the source, the transverse motion is positive in the direction of increasing θ (clockwise, CW); counter-clockwise (CCW) motion is negative. The corresponding pressure **P** and tension (**T**) are also shown.

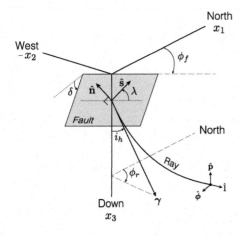

FIGURE 17.8 Definition of a geographic coordinate system, with x_3 positive downward. The fault strike, ϕ_f, is measured clockwise from north; the dip, δ, is measured from the surface to the fault in the direction of dip. The observing station is at azimuth ϕ_s, and the raypath to that station has a takeoff angle, i_h, relative to the x_3 axis.

17.3.1 Body-waves

Using the coordinates systems defined in Fig. 17.8, the equations for the far-field P and S waves from a point double-couple described using standard fault orientation angles strike, ϕ_f,

dip, δ, and rake, λ, are

$$u_P(\mathbf{r}, t) = \frac{\mathcal{F}^P}{4\pi\rho\alpha^3} \frac{1}{r} \dot{M}\left(t - \frac{r}{\alpha}\right)$$

$$u_{SV}(\mathbf{r}, t) = \frac{\mathcal{F}^{SV}}{4\pi\rho\beta^3} \frac{1}{r} \dot{M}\left(t - \frac{r}{\beta}\right) \qquad (17.70)$$

$$u_{SH}(\mathbf{r}, t) = \frac{\mathcal{F}^{SH}}{4\pi\rho\beta^3} \frac{1}{r} \dot{M}\left(t - \frac{r}{\beta}\right),$$

where the radiation patterns, $\mathcal{F}^P$, $\mathcal{F}^{SV}$, and $\mathcal{F}^{SH}$ can be computed using the moment-tensor formulation (Eq. (17.67)) and the moment-tensor definition for a double couple in terms of the fault normal and slip vectors (Eq. 16.21). Representing the receiver-azimuth using ϕ_r, and the take-off angle using i_h (Fig. 17.8), we have

$$\mathcal{F}^P = \left[\cos\lambda \sin\delta \sin 2(\phi_r - \phi_f)\right.$$
$$\left. - \sin\lambda \sin 2\delta \sin^2(\phi_r - \phi_f)\right] \sin^2 i_h$$
$$\left[\sin\lambda \cos 2\delta \sin(\phi_r - \phi_f)\right.$$
$$\left. - \cos\lambda \cos\delta \cos(\phi_r - \phi_f)\right] \sin 2i_h$$
$$+ \sin\lambda \sin 2\delta \cos^2 i_h \qquad (17.71)$$

$$\mathcal{F}^{SV} = \left[\sin\lambda \cos 2\delta \sin(\phi_r - \phi_f)\right.$$
$$\left. - \cos\lambda \cos\delta \cos(\phi_r - \phi_f)\right] \cos 2i_h$$

$$+ \tfrac{1}{2} \cos\lambda \sin\delta \sin 2(\phi_r - \phi_f) \sin 2i_h$$

$$- \tfrac{1}{2} \sin\lambda \sin 2\delta \sin 2i_h \left[1 + \sin^2(\phi_r - \phi_f) \right] \tag{17.72}$$

$$\mathcal{F}^{SH} = \left[\cos\lambda \cos\delta \sin(\phi_r - \phi_f)\right.$$
$$\left. + \sin\lambda \cos 2\delta \cos(\phi_r - \phi_f)\right] \cos i_h$$
$$+ \left[\cos\lambda \sin\delta \cos 2(\phi_r - \phi_f)\right.$$
$$\left. - \tfrac{1}{2} \sin\lambda \sin 2\delta \sin 2(\phi_s - \phi_f)\right] \sin i_h . \tag{17.73}$$

The azimuthal quantities, ϕ_f and ϕ_r are both measured clockwise from north. These are the equations from Aki and Richards (2009) and can be used to compute body-wave radiation patterns. An alternate formulation for radiation patterns is also common (Ben-Menahem and Singh, 1981; Stein and Wysession, 2003, etc.), in which the receiver azimuth relative to the fault strike is measured in a counterclockwise direction. In terms of the standard azimuths (used above), the alternate definition of relative azimuth is $2\pi - (\phi_r - \phi_f)$ which corresponds to a change the sign of the relative azimuth. The difference in the relative-azimuth definitions results in a sign difference for terms in $\mathcal{F}^P$, $\mathcal{F}^{SV}$, $\mathcal{F}^{SH}$ that include the factors $\sin(\phi_r - \phi_f)$ or $\sin 2(\phi_r - \phi_f)$. We note these differences because later we use the alternative relative azimuth definition to describe surface-wave radiation patterns.

The factor $1/r$ in Eq. (17.70) accounts for geometric spreading in a whole space. To use these equations to analyze observations, we must modify this factor to account for realistic geometric spreading in the Earth. Earlier, in Section 13.2 we developed an expression for the geometric spreading in a spherically symmetric earth model,

$$G(\Delta, h) = \sqrt{\frac{c_h/c_0}{r_0^2 r_h^2} \frac{p}{q_h q_0} \left(\frac{1}{\sin\Delta}\right) \left|\frac{d^2 T}{d\Delta^2}\right|} ,$$

and so a P-wave amplitude, for example, can be represented by

$$u_P(t) = \frac{\mathcal{F}^P \, G(\Delta, h)}{4\pi\rho\alpha^3} \, \dot{M}\left(t - \frac{r}{\alpha}\right) . \tag{17.74}$$

Analogous expressions exist for SH and SV waves. The factor, $\dot{M}(t)$, is the time derivative of the moment function – so we need the time history of $M(t)$. Earlier, we represented the static moment, $M_0 = \mu A \bar{D}$, where $\bar{D}$ is the average displacement over the rupture area A after the rupture arrests. Here we define $M(t) = \mu A_R(t) D(t)$ such that $\dot{M}(t) = \mu \, \partial[A(t) D(t)]/\partial t$. If the slip during an earthquake is so fast it can be approximated as instantaneous, then $M(t) \propto H(t)$ and $\dot{M}(t) \propto \delta(t)$. We consider the nature of the moment rate function $\dot{M}(t - r/\alpha)$ in more detail in the next chapter.

17.3.2 Surface-waves

Our focus on wholespace solutions has restricted our attention to P and S waves. Consideration of more complex models leads to the idea of surface-wave radiation patterns. Their derivation is more complex (see for example, Aki and Richards, 2009), but the ideas are analogous and can be discussed in a similar manner. Most importantly, the existence of surface-wave radiation patterns means that surface-wave observations also carry information about the fault orientation and slip direction and that they can be used to estimate faulting geometry. Both time- (seismogram fitting) and frequency-domain (amplitude and phase spectral fitting) approaches have been developed to extract information on seismic sources from surface-waves. We discuss some of the methods in Chapter 19.

Consider surface-wave displacements expressed using a Fourier Transform,

$$u(t) = \frac{1}{2\pi} \int_{-\infty}^{+\infty} U(\omega) \, e^{i\omega t} \, d\omega, \tag{17.75}$$

where $U(\omega)$ is the Fourier spectrum of $u(t)$. We can represent one frequency component the surface-wave spectrum, $U(\omega)$, using

$$U(\omega) = E(\omega) \cdot G(\omega) , \qquad (17.76)$$

where $E(\omega)$ represents the source excitation effects and $G(\omega)$ represents surface-wave propagation effects. For example, for a Love-wave observed at a distance Δ from the earthquake, we have

$$U_L(\omega) = \left(p_L P_L^1 + i \, q_L Q_L^1 \right)$$
$$\times \left[\frac{1}{\sqrt{\sin \Delta}} e^{-i\pi/4} e^{i\omega a(\Delta/c)} e^{-\omega \Delta (a/2 Q v_g)} \right]$$
$$\times \dot{M}(\omega) , \qquad (17.77)$$

and for a Rayleigh wave at the same distance,

$$U_R(\omega) = \left(s_R S_R^1 + p_R P_R^1 + i \, q_R Q_R^1 \right)$$
$$\times \left[\frac{1}{\sqrt{\sin \Delta}} e^{-i\pi/4} e^{i\omega a(\Delta/c)} e^{-\omega \Delta (a/2 Q v_g)} \right]$$
$$\times \dot{M}(\omega) , \qquad (17.78)$$

where a is Earth's radius, $c(\omega)$ and $v_g(\omega)$ are phase and group velocity, and $Q(\omega)$ is the corresponding attenuation quality factor. $\dot{M}(\omega)$ is the spectrum of the moment-rate function. The propagation phase is $\exp[i\omega\Delta/c]$, and $1/\sqrt{\sin \Delta}$ is the geometric spreading. The *horizontal radiation patterns* are defined as

$$p_L = \sin \lambda \cos \delta \sin \delta \sin 2\phi' + \cos \lambda \sin \delta \cos 2\phi'$$
$$q_L = -\cos \lambda \cos \delta \sin \phi' + \sin \lambda \cos 2\delta \cos \phi'$$
$$s_R = \sin \lambda \sin \delta \cos \delta$$
$$q_R = \sin \lambda \cos 2\delta \sin \phi' + \cos \lambda \cos \delta \cos \phi'$$
$$p_R = \cos \lambda \sin \delta \sin 2\phi' - \sin \lambda \sin \delta \cos \delta \cos 2\phi',$$
$$(17.79)$$

where $\phi' = \phi_f - \phi_s$ (the opposite definition than what we used earlier for body-waves). The frequency and depth-dependent excitation functions, $P_L^1(z,\omega)$, $Q_L^1(z,\omega)$, $S_R^1(z,\omega)$, $P_R^1(z,\omega)$, and

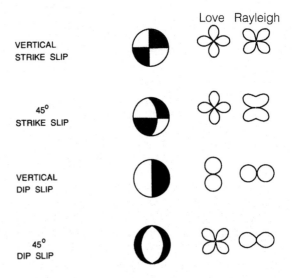

FIGURE 17.9 *P*-wave focal-mechanism projections and surface-wave radiation patterns for four basic fault orientations. The relative amplitude of the Love-wave (left) and Rayleigh-wave [right] radiation patterns is not drawn to scale (adapted from Kanamori, 1970).

$Q_R^1(z,\omega)$, are combinations of the surface-wave eigenfunctions, their spatial derivatives, and near-source material properties.

Source information in near-source surface waveforms is represented by the product of horizontal radiation patterns, vertical excitation functions, and the moment-rate function. We define surface-wave amplitude radiation patterns for a given frequency as

$$E_L(\phi, \omega) = \dot{M}(\omega) \sqrt{\left(p_L P_L^1 \right)^2 + \left(q_L Q_L^1 \right)^2}$$
$$E_R(\phi, \omega)$$
$$= \dot{M}(\omega) \sqrt{\left(s_R S_R^1 \right)^2 + \left(p_R P_R^1 \right)^2 + \left(q_R Q_R^1 \right)^2} .$$
$$(17.80)$$

Fig. 17.9 includes the surface-wave radiation pattern azimuthal variation plots for several canonical fault orientations represented using *P*-Wave focal mechanisms. Surface-wave phase spectra carry important information on faulting geometry as well, particularly changes in the nodal directions.

Before concluding this section, we note that the horizontal radiation patterns (Eq. (17.79)) can also be used to express the body-wave radiation patterns. Specifically, we can rewrite the expressions for the P, SV, and SH radiations patterns as

$$\mathcal{F}^P = s_R(3\cos^2 i_h - 1) - q_R \sin 2i_h$$
$$\quad - p_R \sin^2 i_h$$
$$\mathcal{F}^{SV} = s_R(3/2)\sin 2i_h + q_R \cos 2i_h \qquad (17.81)$$
$$\quad - p_R(1/2)\sin 2i_h$$
$$\mathcal{F}^{SH} = -q_L \cos i_h - p_L \sin i_h \ .$$

Thus we see that the body waves can also be represented as the product of a horizontal and vertical radiation (i_h-dependent) patterns.

17.4 Estimating faulting geometry

We conclude with a brief discussion of how faulting geometry (fault orientation and slip direction) is estimated using the radiation patterns contained in seismic observations. We restrict our attention to a method based on P-wave first-motion polarity. After exploring earthquake rupture processes in the next chapter we discuss (in Chapter 19) how both the fault orientation and slip patterns are imaged using more complete seismogram analysis.

17.4.1 *P*-wave first motion modeling

In section 16.3 we motivated our entire discussion of seismic wave excitation using observed patterns in initial *P*-wave motions (Fig. 16.4). Our radiation pattern expressions provide a quantitative explanation for the observations – what is left to be developed is the construction of a relatively simple approach to using radiation patterns to estimate earthquake faulting geometry. Early seismologists

recognized that propagation leaves the sense of motion (towards or away) from the source unchanged. Over the years they developed graphical analysis methods more convenient than using maps such as that in Fig. 16.4. First, we project each observation back to the focal sphere using the receiver azimuth (ϕ_r) and wave takeoff angle (i_h) (Fig. 17.10) and then we identify two planes (the fault and auxiliary planes) passing through the center of the sphere and separating observations of different polarity into four quadrants.

We display and analyze the focal-sphere patterns using cartographic projections. Although some use a *stereographic projection*, which preserves angles at the expense of distorting areas, we will use the *Lambert azimuthal equal-area projection*, which accurately represents the focal-sphere area at the expense of some angular distortion. Each *P*-wave polarity is displayed in the equatorial circle that represents either the upper- or lower-hemisphere of the focal sphere. We focus on teleseismic observations, which leave the source downward, so we'll project data onto the lower hemisphere. The upper hemisphere is often used in local-distance analyses, where the waves leave the source upward. In all applications it is important to specify which hemisphere was used. Both upward- and downward-departing waves can be projected onto either hemisphere. A lower hemisphere projection is illustrated in Box 3.2 and Fig. 17.10. The center of the equatorial circle is the center of the focal sphere and the location of the source. The projected coordinates, x_s and y_s, of a polarity observation are

$$r_s = \sqrt{2} \sin(i_h/2)$$
$$x_s = r_s \sin \phi_r \qquad (17.82)$$
$$y_s = r_s \cos \phi_r \ .$$

For a stereographic projection, use $r_s = \tan(i_h/2)$. Rays that take off upward and intersect the upper hemisphere are projected onto the lower hemisphere by adding π to the station azimuth (exploiting the low-order symmetry of radiation patterns). The fault and auxiliary planes intersect the focal sphere, and the intersections project to the equatorial plane as curves that separate regions of compressional and dilatational P-wave motions (Fig. 17.11). The curves are great circles on the equatorial plane, and their orthogonality produces characteristic appearance of the different faulting types (See Box 3.2).

For each observation, we calculate the receiver azimuth (ϕ_r) and great-circle distance (Δ) using the source and receiver locations and then compute the takeoff angle (i_h) using the slope of the travel time curve at the receiver's great-circle distance using

$$\frac{dT}{d\Delta}(\Delta) = p(\Delta) = \sin i_h / v_h \,, \qquad (17.83)$$

where p is the ray parameter (in s/km, already corrected for spherical geometry), and v_h is the wave speed at the hypocenter (in km/s). Table 17.1 is a list of P-wave takeoff angles appropriate for a shallow source. Note that the more distant the station, the lower the takeoff angle. Thus, observations from more distant stations plot closer to the center of the focal mechanism than observations from closer stations.

Fig. 17.11 is a plot of the expected P-wave amplitudes and S-wave polarizations for the dip-slip faulting geometry. An equal-area projection is used for P, a stereographic projection for S

Box 17.1 Surface-wave radiation from surface point-force sources

Seismic-wave radiation from some natural phenomena have been found to be well explained by equivalent point-force sources. The simplest examples include vertical volcanic eruptions, which can be modeled as a point force representing the eruption's reaction force. Fig. B17.1.1 shows the theoretical horizontal and vertical ground motions for a vertical point force action on Earth's surface. The solution for this system was first provided by Lamb (1904) and represents the ground-motion calculation of a transient wave. The figure on the right shows observed and synthetic ground motion (mainly Rayleigh wave) for a station 67 km from an eruption of Mt. St. Helens, on June 13, 1980 (a minor eruption after the main blast). Kanamori and Given (1983) estimated an equivalent force strength of $5.5 \times 10^8 \, N$ by matching the observed amplitude.

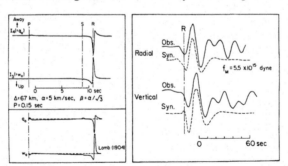

FIGURE B17.1.1 (Left) Theoretical radial and vertical ground motions for a point force on the surface of a half-space. (Right) Comparison of observed and synthetic ground motions for the June 13, 1980 eruption of Mt. St. Helens. The synthetics are computed for a vertical downward point force at the source, with strength $f_m = 5.5 \times 10^8 \, N$ (from Kanamori and Given, 1983).

Another process that has be modeled by a point force quite successfully is a nearly horizontal landslide. In this case the BFE represents the reaction force on the surface due to laterally moving the slide mass off the hillside. Kanamori and Given (1983) modeled the large landslide that occurred during the May 18, 1980 Mt. St. Helens eruption with a horizontal force, which they found best matched the event's long-period Love and Rayleigh waves. Another example is the 1975 Kalapana, Hawaii event, which involved either slip on a very shallow dipping fault or a large slump (Fig. B17.1.2). The observed radiation pattern of Love waves from the event is two-lobed, consistent with the single-force model as shown. From the strength of the point force one can estimate the peak acceleration of the slide block as 0.1–$1.0 \mathrm{m/s}^2$.

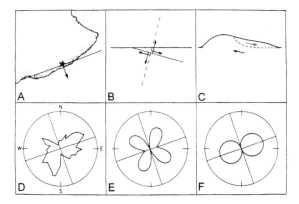

FIGURE B17.1.2 Observations and interpretations of the source mechanism for the 1975 Kalapana, Hawaii event. Observations are shown in (A) and (D), where subhorizontal surface ground motions were observed southeast of the hypocenter and the teleseismic Love-wave radiation has the observed pattern in (D). Interpretation of the source as a shallow double couple (B) predicts a four-lobed Love wave radiation, if the dip is larger than about 10° (as dip goes to 0°, the pattern becomes two-lobed, Lay et al. (2018)) (E); interpretation as a reaction point force (C) due to land sliding produces a Love-wave radiation pattern (F) in better agreement with the data (from Eissler and Kanamori, 1987).

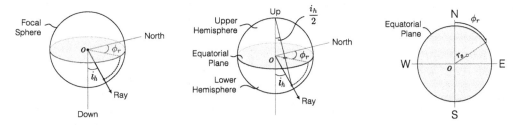

FIGURE 17.10 (left) The focal sphere near the source, which can be thought of as the initial outgoing P (or S) wavefront. Each ray to a seismic station at a particular location cane be parameterized in terms of the receiver azimuth and takeoff angle at the source. (center) three-dimensional geometry used to project the sphere to an azimuthal projection. (right) For analysis, the focal sphere is projected to preserve area elements. Azimuth is unchanged, the takeoff angle is mapped to a distance from the circle's origin, r_s.

TABLE 17.1 *P*-wave takeoff angles for source at 15 km depth (model AK135).

Δ (°)	10	15	20	25	30	40	50	60	70	80	90
p (s/km)	0.123	0.101	0.085	0.088	0.079	0.075	0.068	0.062	0.055	0.049	0.042
i_h (°)	46	36	30	31	28	26	23	21	19	16	14

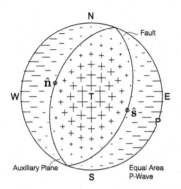
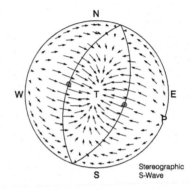

FIGURE 17.11 Focal mechanisms for a pure reverse fault with a dip of 45° and strike of 20°. *P*-wave polarities and relative amplitudes are shown on the left, *S*-wave polarizations and amplitudes are shown on the right. Plus signs (+) indicate motion away from the source. The fault and auxiliary planes are shown as curves, the projections of the **P** and **T** are shown using labels, the **B** axis is located at the intersection of the planes. The two dots represent the normal and slip vectors, which are co-linear on the projection, but in three dimensions, they are orthogonal.

(we are showing angles). Note the smooth variation in both *P* and *S* radiation patterns. The fault and auxiliary planes are nodal directions for *P* waves, the **P** and **T** axes directions are maxima. *S*-waves have no nodal planes but the stress axes are nodal points. *S*-wave polarizations converge on the direction of the **T** axis and diverge from the direction of the **P** axis. Note that the *S* polarizations are orthogonal to the fault and auxiliary planes. The **B** axis lies on the intersection of the two planes, and this is also the "pole" (perpendicular) to the plane containing the **P** and **T** axes. The intersection of the plane orthogonal to the **B**-axis with the fault plane is the slip-vector direction – n̂, ŝ, **P**, and **T** all lie in that plane. Fig. 17.12 is a plot of the radiation patterns associated with an oblique reverse faulting event. The patterns are rotated but the same relationships hold. An investment of time studying these radiation-pattern projections in concert with the original equations describing the patterns can help develop intuition into geometric

patterns associated with earthquake-generated seismic waves.

For decades, focal mechanisms were introduced and estimated using graphical approaches. The analysis uses a projection grid (a Wulff or a Schmidt stereonet) and an overlay that includes the projection's bounding circle and coordinate directions, on which we plot the observations. The overlay's center must be pinned to the center of the grid, but free to rotate so that we can identify two planes that separate observations such as *P*-wave polarities into four quadrants. One of the planes is the fault, the other is the auxiliary plane. The graphical procedure replaces tedious calculations associated with the data and the projections. Nowadays, these numerical calculations are quite modest, so we generally estimate faulting geometry using simple computer-based tools to match observed *P*-wave amplitudes and/or *S*-wave polarizations. In fact, computational approaches allow us to match any combination of the obser-

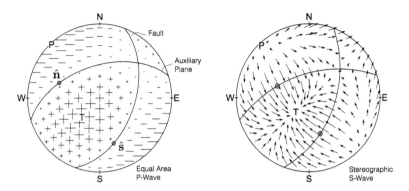

FIGURE 17.12 Focal mechanisms for an oblique reverse fault with a dip of 50°, a rake of 50°, and strike of 20°. *P*-wave polarities and relative amplitudes are shown on the left, *S*-wave polarizations and amplitudes are shown on the right. Plus signs (+) indicate motion away from the source. The fault and auxiliary planes are shown as curves, the projections of the **P** and **T** are shown using labels, the **B** axis is located at the intersection of the planes. The two dots represent the normal and slip vectors.

vations including polarity, polarization, *P*-to-*S* amplitude ratios, etc. Of course automating the procedure requires that we define sensitive and stable misfit norms, which is not always easy. Observations from more than one event can be combined to construct composite focal mechanisms (when the data are sparse for any one event) or to estimate faulting geometries of a suite of events that may differ, but should be consistent with a single tectonic stress orientation.

An example focal mechanism is shown in Fig. 17.13, which was constructed using the data for a strike-slip event along the northern Mid-Atlantic Ridge (sample data are shown in Fig. 16.4). The solution (the planes) were estimated interactively by varying the strike, dip, and rake of the "fault". Teleseismic data cluster near the center of the mechanism, reflecting the narrow range of takeoff angles spanned by teleseismic *P* (and *SH*). Only regional and local-distance data will locate far from the projection origin. The example is an unusually well-constrained *P*-wave first-motion result (angles are probably resolved to ±5° or better). Only one observation polarity is inconsistent, and since it is near a node, the signal amplitude is small. The location of the event between two continents

hosting many seismic observatories results in superb azimuth coverage. This is not always the case. Small events generally produce sparse and ambiguous observations. Often one plane may be well constrained, but the other can lie within a large range of possible orthogonal orientations. Combined modeling of *P* and *S* polarities can reduce the ambiguity. Even in this case, discriminating which plane corresponds the actual fault plane, requires information on surface rupture, aftershocks preferentially distributed along one plane, or analysis of seismic signals to resolve any source finiteness effects. Since this particular event occurred along a well-understood mid-ocean ridge, based on the orientation of the nearby transform faults, the east-southeast striking structure likely corresponds to the fault. In that case, the motion represents left-lateral strike-slip motion (North American moving west-northwest relative to Eurasia).

The ISC phase-pick catalog includes polarity measurements from thousands of earthquakes. These can be used to both practice accessing data in the catalog and to explore *P*-wave first motion polarities. However, in any large catalog, you must recognize that the data are not perfect. Polarities for many *P* waves are often ambiguous when the arrival is emergent. Still, first-

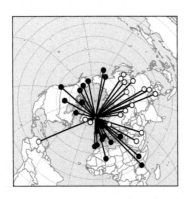
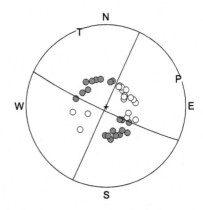

FIGURE 17.13 (left) Geographic distribution of P-wave polarity observations for the earthquake along the 2011 northern Mid-Atlantic Ridge (also show in Fig. 16.4) (Right) Focal mechanism constructed using the P-wave polarities shown on the left ($\phi_f = 115°$, $\delta = 85°$, $\lambda = 0°$). The fault and auxiliary planes are shown as well as projections of the **P** and **T** axes. The **B** axis is located at the intersection of the two planes. Data for this northern hemisphere earthquake located between North America and Europe constrain the faulting geometry well.

motion method is based on some fundamental relationships that contribute to a deeper understanding of more modern waveform-modeling approaches. At a minimum, understanding these concepts provides a valuable tool to check results produced by other methods or even to validate seismic instrumentation (checking orientations).

Classic stereonet focal-mechanism analysis emphasizes identification of the fault and auxiliary planes and their associated slip vectors. Computational approaches are more general and can be used for example to estimate moment-tensor elements for sources that may not be consistent with shear-faulting described by a simple double-couple (i.e. two planes). Moment-tensor approaches are more appropriate for explosive and mining-related seismic sources. The basis of moment-tensor estimation approaches using wave amplitudes and polarities is Eq. (17.67). Note that formulating the problem to estimate

the moment tensor increases the number of free parameters from three (strike, dip, rake) to six (the number of independent moment-tensor elements). Thus, a moment-tensor analysis may require more data to produce a robust result than does an analysis incorporating a double-couple constraint.

First-motion modeling is one of the oldest approaches used to investigate the earthquake sources. For small local earthquakes and other seismic events, it remains an attractive approach. For most moderate-to-large earthquakes we include much more of the information in the seismogram to constrain seismic source properties. To model seismic waveforms or their spectra, we must account for the temporal characteristics of an earthquake's rupture. We return to source modeling approaches in Chapter 19, after we discuss rupture processes and seismic moment-rate functions and their spectra in the next chapter.

Earthquake rupture and source time functions

Chapter goals

- Define the source-time function and its Fourier Transform, the source spectrum.
- Introduce fundamental concepts involved in earthquake rupture processes.
- Introduce rupture finiteness effects on seismograms using the Haskell source rupture model.
- Relate simple earthquake rupture models to seismogram properties.
- Introduce deconvolution approaches used to estimate seismic source time functions.

At the start of the last chapter, we noted that an observed displacement seismogram, $\mathbf{u}(\mathbf{x}, t)$, is a convolution of source and propagation effects (Eq. (16.1)),

$$u(\mathbf{x}, t) = E(\mathbf{x}, t) * G(\mathbf{x}, t) \,,$$

where $\mathbf{x}$ is a position vector, t is time, $E(\mathbf{x}, t)$ represents source excitation effects, and $G(\mathbf{x}, t)$ represents propagation effects (which we have discussed earlier). We then split the source excitation function into two factors, one that depends on the fault geometry and near-source material properties, the other that depends on

source-time history such as an earthquake rupture process. Specifically, we let (Eq. (16.2))

$$E(\mathbf{x}, t) = A_F(\mathbf{x}, t, [\omega]) \cdot S(\mathbf{x}, t) \,,$$

where $A_F(\mathbf{x}, t, [\omega])$ represents amplitude patterns that result from fault orientation, which we examined in detail last chapter. In this chapter, our focus is on the source time function (or moment-rate) function, $S(\mathbf{x}, t)$, which includes all effects of the earthquake rupture process (or source time history in general).

The spatiotemporal history of fault slip affects the temporal variation and frequency content of seismic waveform, which means that we can use seismograms to study earthquake rupture processes. Earthquake slip involves three main stages: (1) initiation of the sliding (or crack formation), (2) growth of the rupture zone, or rupture-front expansion, (3) and rupture termination. The complementary laboratory and theoretical study of rock failure and the fracture processes is called *fault mechanics*. A simple review of the basics of fault mechanics provides some context for a discussion of the dynamics of earthquakes.

18.1 Rock fracture and fault rupture

The earliest roots of fault mechanics can be traced to the work of Amonton in 1699 and Coulomb in 1773. Coulomb introduced a simple theory for rock failure that states that the shear strength of a rock is equal to the initial strength of the rock, plus a constant times the normal stress, σ_n, on the plane of failure:

$$|\tau|_{\text{failure}} = c + \mu_i\,\sigma_{\text{n}}, \qquad (18.1)$$

where c is the strength of the rock, sometimes called *cohesion*, and μ_i is called the *coefficient of internal friction* (not the shear modulus). Numerous variations on (18.1) have been proposed, but this simple equation predicts failure surprisingly well and is usually referred to as the *Coulomb failure criterion*. A similar relationship, known as *Amonton's second law* (Amonton's first law states that frictional forces are independent of the size of the fault surface), is used to describe frictional sliding on an existing crack:

$$\tau = \mu_s\,\sigma\,. \qquad (18.2)$$

This equation has the same form as (18.1) without the cohesion term. In (18.2) μ_s is called the *coefficient of friction*, and it does not have the same value, μ_i, required for failure of unfaulted rock. The coefficient of friction, μ_s, generally has a larger value before sliding takes place, called the *static friction coefficient*, and a smaller value once sliding commences, called the dynamic or *kinetic friction coefficient*. Amonton proposed that μ_s was related to roughness or protrusions on the fault surface. These protrusions, known as *asperities*, are welded contacts between the two sides of a fault. These welds must be overcome to allow sliding, and once these welds are broken, sliding proceeds at a reduced friction level.

Much experimental work has been done on rock friction, and in general it has been shown that Amonton's law is applicable over a wide range of normal stress values. Byerlee (1978) compiled data from a large number of rock-friction experiments and found that maximum friction was nearly independent of rock type. For normal stresses greater than 200 MPa (Fig. 2.5), the relationship is

$$|\tau| = 50 + 0.6\,\sigma_n \ \text{MPa} \qquad (18.3)$$

and for normal stresses less than 200 MPa, the relationship is

$$|\tau| = 0.85\sigma_{\text{n}}. \qquad (18.4)$$

If we assume that normal stress is approximately equal to the overburden pressure, then Eq. (18.3) is valid for all depths greater than about 6 km. This is known as *Byerlee's law*. That Byerlee's law is relatively independent of rock type is rather remarkable, and it suggests that the details of fault roughness are relatively unimportant in the frictional behavior associated with earthquake rupture.

The dynamics of frictional sliding are more complex than Eq. (18.3) indicates. In the laboratory the relationship between sliding displacement and applied shear stress is not smooth. In general, no slip occurs on the fault surface until the critical value of τ is reached, and then sudden slip occurs followed by a drop in stress. This causes a time interval of "no slip" during which the stress builds up again to the critical value, and then the sudden-slip episode is repeated. This type of frictional behavior is known as *stick-slip*, or *unstable sliding*. For extremely smooth fault surfaces the slip may be continuous, or nonepisodic. This is referred to as *stable sliding*. Brace and Byerlee (1966) proposed stick-slip behavior as a possible explanation for shallow earthquakes, building on ideas developed by Bridgman in the 1930s and 1940s. Earthquakes are generally thought to be recurring slip episodes on preexisting faults followed by periods of no slip and increasing strain. This was a profound change in the understanding of faults in earthquake processes. The emphasis is not on the *strength* of the rock but rather on a friction-controlled stress/stability cycle. The difference between the shear stress just before the slip episode and just after the slipping has ceased is known as the **stress drop**. The stress drop observed in an earthquake may represent only a fraction of the total stress supported by the rock.

In detail, a number of factors influence rock friction. These include temperature, slip rate, and slip history. Many materials become weaker with repeated slip and may eventually transition to a stable sliding mode. This behavior is known as **slip weakening**. Also, most Earth materials exhibit an inverse dependence of friction on slip velocity. This type of behavior is known as **velocity weakening**. Finally, for most materials, stick-slip behavior is observed only at temperatures below about 300°C. Almost all seis-

mogenic faults involve a shallow, near-surface zone where stable sliding normally occurs. Occasional ruptures that nucleate in the deeper unstable sliding regime may extend throughout much of the crust (and breach the surface). At yet greater depth, higher-temperatures induce a transition to stable sliding and grading into a zone of continuous ductile deformation. The slip-rate dependence of friction can cause normally stable sliding regimes to rupture. The key point is that it is frictional behavior rather than internal strength that controls earthquake rupture processes.

Earthquake rupture dynamics

The Coulomb failure criteria and the Anderson Theory of Faulting provide a *static* framework for understanding the earthquakes, but the *dynamics* of the sudden slip are much more dif-

---

Box 18.1 Anderson's theory of faulting

Application of Eq. (18.1) to geologic materials and tectonic stresses in the Earth to predict newly created fault orientations is known as **Anderson's theory of faulting**. Represent the principal stresses using σ_1 (maximum), σ_2 (intermediate), and σ_3 (minimum). The display Eq. (18.1) in a Mohr diagram, is shown in the figure below. The upper line identifies the *failure envelope*. Failure in an un-faulted rock occurs when the difference between the maximum and minimum compressive stresses intersects the envelope. The orientation of the failure surface (fault) is perpendicular to the failure envelope (the angle 2θ in the figure).

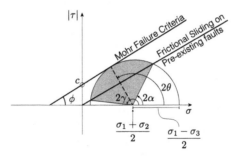

FIGURE B18.1.1 Mohr-Coulomb failure envelope and its relation to the difference between σ_1 and σ_3. The cohesion is represented by c. The stress state defines the semi-circle, the lines define failure criteria. Two failure criteria are shown – one for rock fracture and the other for sliding on pre-existing faults.

The slope of the failure envelope is related to the internal friction by $\mu_i = \tan\phi$ where ϕ is called the angle of internal friction. Ideally, a fault will form at an angle $(90° - \theta)$ from the axis of maximum compressive tectonic stress,

$$\theta = \pm\left(45° + \phi/2\right) . \tag{B18.1.1}$$

Only when $\mu_i = 0$ will the principal stresses align with **P** and **T** axes of earthquake focal mechanisms (which intersect the fault surface at 45° angles). For most rocks, $\phi \sim 30°$, and failure will occur on either fault of a conjugate pair oriented at $\pm30°$ to σ_1.

Near the surface of the Earth one of the principal stresses is almost always vertical, thus we can predict the orientations of faults for a particular stress environment. In a compressive region, σ_1 and σ_2 are horizontal and σ_3 is vertical, resulting in $\sim 30°$ dipping reverse faults that strike parallel to σ_2. In an extensional region, σ_1 is vertical and σ_3 is horizontal, resulting in $\sim 60°$-dipping normal faults. In regions where σ_3 and σ_1 are both horizontal, vertical strike-slip faults form, oriented at $\sim 30°$ from the σ_1 direction.

Anderson's Theory describes the basic characteristics of observed faults reasonably well. But the theory only relates the orientation of the faults to the stress field at the time of fault formation. The frictional strength of faults is less than the strength of unbroken rock, so once they form, faults are a weak surface that will slip even when the stress field is not optimally aligned. The line through the origin in Fig. B18.1.1. represents the condition for frictional sliding of preexisting faults. If a region contains preexisting faults with normal vectors that lie between the angles defined by 2α and 2γ, those faults with slip, probably preventing creation of a new fault at angles θ. The complications associated with friction make inferring principal stress orientations from isolated seismic focal mechanisms difficult. However, a regional stress field orientation can be constrained using an analysis of many focal mechanisms in a region.

ficult to understand. An earthquake rupture can be described as a two-step process: (1) formation of a crack and (2) propagation, or growth, of the crack. A crack tip serves as a stress concentrator – if the stress at the crack tip exceeds a critical value, then the crack grows unstably (sudden slip). Fig. 18.1 is a plot of the stress at a point, P, along part of a fault ruptured in an earthquake. Before the rupture reaches the neighborhood of P, the stress on the fault is τ_0, which is less than the frictional resistance stress, τ_s. As the rupture approaches, the stress at P rises as stress concentrates ahead of the propagating crack tip. When the stress reaches τ_s (at time t_0), slip at point P initiates. As the rocks at location P slide, the fault stress diminishes to a dynamic fric-

tional value τ_f. When sliding ceases, the stress level adjusts up or down slightly (depending on whether velocity weakening or strengthening occurred) to τ_1 the final stress. The **static stress drop** at the location on the fault is $\Delta\sigma = \tau_0 - \tau_1$. The dynamic stress that drives the offset across the fault is $\tau_s - \tau_f$.

The fault stress reduction connected to the observable seismic-wave radiation acts through the strain energy released as the rocks on either side of the fault move. In Fig. 18.1 the fault at location P slipped from its initial to its final value during a finite period of time, $t_1 - t_0$. Fig. 18.2 is a hypothetical slip history for location P, along with its corresponding time derivative. The moment rate function, $\dot{M}(t)$, is directly proportional

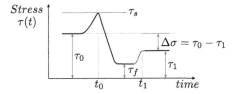

FIGURE 18.1 Schematic of the stress history at a point on a rupture surface. As the rupture front approaches the point (with a finite rupture speed), stress increases to a value of τ_s, after which failure occurs at the point. The point slips to a displacement D, and stress is reduced to some value τ_f. The difference between the initial stress and the final stress, $\Delta\sigma$, is defined as the stress drop (after Yamashita, 1976).

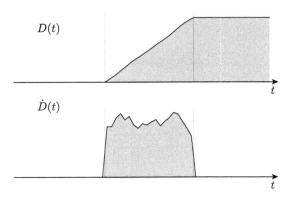

FIGURE 18.2 The relation between the displacement history of a particle on the fault, $D(t)$, and the far-field source time function. The top panel shows the displacement history for a small region of the rupture. In this case slip occurs with small (hard to even see on the plot) fluctuations about a nearly constant average rate. The shape of a far-field seismic wave is proportional to the time derivative of the on-fault displacement history, $\dot{D}(t)$. For nearly constant slip rates, the far-field source time function has a boxcar-like shape.

to the slip velocity on the fault. The slip function may have substantial irregularity, but it can often be approximated by a simple ramp function such that the derivative is boxcar shaped. Thus the history of the fault slip will map directly into the moment rate, which in turn will give the shape of the *P*- or *S*-wave energy radiated by the earthquake. The time derivative of slip at a point, called the **particle velocity**, depends on the frictional constitutive law, the determination of which is an active area of research. The seismic moment of an entire fault is the sum of the slip of all the particles on the fault. The moment rate is a convolution of the slip history at each point and the history of propagation of the rupture front along the fault.

Most dynamic crack models predict that after a rupture front passes a point, that point will continue to slide until information is received that the rupture front has stopped. In these models, information on rupture cessation comes in the form of a **healing front** initiated from the rupture's boundary, which sweeps inward across the faulted surface, halting slip. This idea requires, for example, that for a circular, radially outward growing crack, the center of the fault is the first and last place to be sliding. Fig. 18.3 illustrates the continued slip and changing slip velocity on an expanding crack surface for such a model. Each point on the fault experiences a stress history like that in Fig. 18.1

as the crack tip passes by and produces far-field radiation like that in Fig. 18.2, but with the particle time history having a prolonged tail corresponding to "slow slip" following the passage of the slip front, that terminates only when the healing front passes. The average slip duration of a point for such crack models is approximately $t_r \approx 2\sqrt{s}/3v_r$, where s is the slip area and v_r is the rupture velocity. This clearly is on the same order as the time required for the entire fault to rupture, $t_r \approx \sqrt{s}/v_r$.

Alternatively, it is possible that rupture surfaces heal locally before the broader crack expansion terminates. Heaton (1990) proposed the slip-pulse model, in which the dynamic friction is inversely proportional to the slip velocity. In this model the friction is greatly reduced as the crack tip passes, but when the slip velocity slows, the friction rises (Fig. 18.3). In such a case, the slip duration at any particular point on the fault, t_r, may be only a small fraction of the total earthquake rupture duration. Because seismic-wave radiation is sensitive

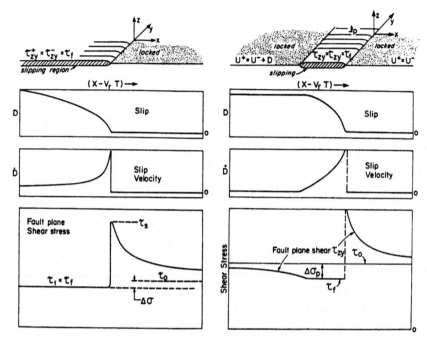

FIGURE 18.3 Two idealized models of rupture. On the left is the crack model of Kostrov (1966) for a crack propagating at rupture velocity v_r in the x direction. Fracture initiates when the stress at the crack tip exceeds the static strength of the fault, τ_s, and the stress on the slipping fault is a constant frictional level, τ_f. The ambient stress before rupture is τ_0, and the stress drop, $\Delta\sigma$, is $\tau_0 - \tau_f$. Every point on the rupture surface continues to slide until the rupture front stops and sends a "healing pulse" back across the fault. On the right is the self-healing slip pulse model of Heaton (1990) in which slip-velocity-dependent friction heals the rupture surface behind the slip pulse. The sliding friction is very low and returns to a high level after the rupture. The static stress drop, $\tau_0 - \tau_1$, may be very low (~ 10 bars), but the average stress drop within the rupture pulse may be an order of magnitude larger (from Heaton, 1990).

to the derivative of the particle dislocation history (as well as to growth of the rupture area), $\dot{M}(t) = \mu(\partial/\partial t)[A(t) \cdot D(t)]$, far-field seismic signals have limited resolution of any small long-duration tails on the fault displacement history at any particular location. Many seismic observations favor short particle slip durations and we can certainly explain most seismic observations without long-duration slip at any particular location on the rupture. But that may be partly because our data are most sensitive to the early part of the slip history – long tails could go unnoticed, buried within the noise. Most kinematic models of earthquakes include localized slip histories that are short compared with the

overall duration of a large-earthquake rupture. The "shape" is often referred to as the **source time function**.

18.1.1 Simple moment-rate function shapes

For wavelengths larger than the rupture area (a point source), we can approximate the moment rate function with parameterized shapes that integrate and average the spatial variations within the small rupture. Introduction of some simple parameterized functions commonly used to represent seismic moment-rate functions can make our discussion more concrete. For exam-

Box 18.2 Crack failure modes

The brittle failure of rocks has long been studied both experimentally and theoretically. The field of modern *fracture mechanics* grew out of a major discrepancy between the two approaches. In general, the theoretical strength of rock, the stress required to break the atomic bonds in a crystal lattice, is several orders of magnitude greater than the stress required to break rock in the laboratory. In the first quarter of this century, A.A. Griffith suggested an explanation based on the fact that all materials contain defects or microscopic cracks. *Griffith's theory* is based on a balance of energy in the system. Creating a new crack requires *work* to break molecular bonds, which increases the potential energy of the system. This increase is compensated with a reduction in the strain energy. If the rate of strain energy supplied to a crack tip equals the energy required to extend the crack, stable crack growth occurs.

The growth of a crack depends on the tip of the crack serving as a stress concentrator. The ability of a crack to concentrate stress depends on the type of displacement at the crack tip. There are three fundamental types of cracks. The simplest, referred to as mode I, or tensile mode, is a crack that opens normal to the direction of crack propagation. The other two types of cracks include shear displacements. Mode II has displacements in the plane of the crack and along the direction of crack propagation (referred to as the *sliding* mode). Mode III has displacements in the plane of the crack but normal to the direction of crack growth (referred to as the *tearing mode*). In general, a crack can include combination of the three fundamental types, but in general, earthquake rupture is modeled with mode II cracks. (See Fig. B18.2.1.)

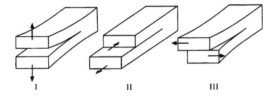

FIGURE B18.2.1 Cartoon illustration of the style of deformation associated with each crack mode (from Lawn, 1993).

The crack-tip stress depends on the mode type and is proportional to $K_n r^{-1/2}$, where r is the distance from the crack tip and K_n is called the **stress intensity factor** (n is the mode type). Griffith's theory requires an energy balance for cracks to grow, which can be used to define the crack extension force, G. For mode II cracks,

$$G = K_n^2 \left(1 - v^2\right)/E .$$

(B18.2.1)

When G exceeds some critical value G_c, unstable fracturing occurs. The dynamics of fracture growth remains an area of active research. A number of modifications to Griffith's theory are now used and are applied in seismology to predict rupture velocity and to model the way faults grow.

ple, if all the fault displacement occurred instantaneously in time, then a reasonable model of the moment function would be

$$M(t) = H(t), \tag{18.5}$$

where $H(t)$ is a Heaviside Step function (ignoring, for convenience, an overall scale factor of M_0). Then

$$\dot{M}(t) = \delta(t) , \tag{18.6}$$

as shown in Fig. 18.4. More realistically, it takes a finite length of time for any given region of the rupture to achieve its total offset, even for a spatial point source, for which $A_R(t)$ is a step function. In this case, a ramp function can be used to represent the moment function (Fig. 18.4),

$$M(t) = R(t) \tag{18.7}$$

The temporal derivative of $R(t)$ is a boxcar function, $B(t)$

$$\dot{M}(t) = B(t) . \tag{18.8}$$

The area A under the boxcar function is equal to M_0:

$$A = \int_{-\infty}^{\infty} \dot{M}(t) \, dt = M' \tau = M_0. \tag{18.9}$$

Thus, the details of the *particle-dislocation time history* affect the far-field body-wave signals. In the next chapter we will consider a few simple rupture time histories, including both finite particle dislocation and areal expansion histories. The boxcar function has some rather sharp edges that remain unrealistic. Brune (1970) introduces a smoothed version based on a dynamic model of faulting that is more smooth, but still relatively simple. The model is parameterized with a seismic moment, and an effective source duration, τ_r,

$$M_B(t) = M_0 \left[1 - (1 + t/\tau_r) \exp(-t/\tau_r) \right] H(t) . \tag{18.10}$$

The corresponding moment rate is

$$M_B(t) = M_0 \frac{\exp(-t/\tau_r)}{\tau_r^2} H(t) . \tag{18.11}$$

These two functions are shown in Fig. 18.4. Brune's physical model was introduced in Section 2.4.5, and the model is revisited a little later in this chapter. When we cannot simply ignore the effects of the finite size of rupture, but the rupture isn't so large (compared with the wavelength) that we have to explicitly introduce the temporal and spatial variation in slip and slip rate, we still can employ some relatively simple earthquake rupture models to account for the source effects in far-field seismograms.

18.2 The one-dimensional Haskell source model

Four kinematic faulting source parameters can be used to represent earthquake rupture kinematics (these parameters of course depend on the underlying faulting dynamics) in far-field seismic waveforms (observations at distances beyond a few fault lengths). These are (1) the final (average) displacement across the rupture ($\bar{D}$), (2) the dimensions of the rupture (L, length, and w, width), (3) the rupture velocity (v_r), and (4) the particle velocity (the average rate at which an individual location on the rupture travels from its initial to its final position). From Eq. (17.71) let us consider a far-field P-wave displacements,

$$u_r(r, t) = \frac{1}{4\pi\rho\alpha^3} \frac{R^P}{r} \dot{M} \left(t - \frac{r}{\alpha} \right) , \tag{18.12}$$

where R^P represents the P-wave radiation pattern.

In the simplest case, the fault can be considered a single point source. Then the seismic moment rate is just the temporal derivative of the displacement history of that point. If the displacement occurred instantaneously as a step,

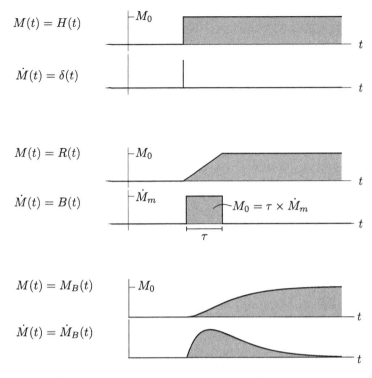

$$M(t) = H(t)$$
$$\dot{M}(t) = \delta(t)$$

$$M(t) = R(t)$$
$$\dot{M}(t) = B(t)$$

$$M(t) = M_B(t)$$
$$\dot{M}(t) = \dot{M}_B(t)$$

FIGURE 18.4 Far-field P- and S-wave displacements are proportional to $\dot{M}(t)$, the time derivative of the moment function $M(t) = \mu A_R(t)D(t)$. That is the temporal derivative of the moment function dictates the shape of the body-wave. An instantaneous step moment function (top) results in an impulse moment-rate function, and a more gradual, uniform increase in moment over a characteristic source time τ (middle), results in a box-car moment-rate function. The still relatively simple, but smoother model of Brune (1970) is illustrated by the lower two functions. In all cases, the area under the curve of the moment-rate function is equal to the seismic moment.

the moment rate would be a delta function. More realistically, it takes a finite time for the fault to achieve its total offset. The simplest case would be represented by a ramp history, as in Fig. 18.2. The source time function corresponding to a ramp-shaped displacement history is a boxcar of length τ_r, where τ_r is called the **rise time**. The rise time is the time it takes a single particle on the fault to achieve its final displacement. Thus if a fault could be described by a single point with a ramp particle time history, the far-field $P-$ and S-wave displacements would be shaped like boxcars. Their amplitudes would vary with azimuth depending on the radiation pattern, but the pulse shape would be identical everywhere. A simple ramp displacement history describes faults as viewed in the far field surprisingly well.

Clearly, all faults involve displacement at more than a single point. Consider a simple kinematic fault model to include the role of rupture expansion in the response. Assume that all points along the finite rupture follow a similar displacement history, but the slip at different locations occurs at different times, triggered as the rupture front expands. The signals from all these point sources must be summed with the appropriate time lags to construct the complete time function. We can see how to do this by considering a simple "ribbon" fault (see Fig. 18.5), where

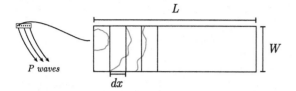

FIGURE 18.5 Geometry of a one-dimensional fault of width W and length L. The individual segments of the fault are of length dx, and the moment of a segment is $m\,dx$. The fault ruptures with velocity v_r. Gray contours show a more realistic rupture kinematics, starting from a hypocenter and expanding across the fault surface.

the fault ruptures initially at one end and the rupture propagates with a finite velocity to the other end. The fault is long and narrow and can be treated as a series of small segments that we individually approximate using point sources. We use the principle of linear superposition to determine the far-field displacement. The summation of the subevent point-source displacements is

$$u(r,t) = \sum_{i=1}^{N} u_i \left(r_i, t - \frac{r_i}{\alpha} - \Delta t_i \right), \qquad (18.13)$$

where Δt_i is the lag between subevents. We proceed with it being understood that time for the displacement field always involves the delay time r/α, which is the P-wave propagation time. Using (18.12), to represent the far-field P wave response from each subevent and $\dot{M}_i(t) = \mu A_i \dot{D}_i(t) = \mu W \, dx \, \dot{D}_i(t)$, we can rewrite (18.13) as

$$u_r(r,t) = \frac{R_i^P \mu}{4\pi\rho\alpha^3} W \sum_{i=1}^{N} \frac{\dot{D}_i}{r_i} (t - \Delta t_i)\,dx. \quad (18.14)$$

Now consider the geometry appropriate for the far-field. For a station a large distance and perpendicular to the fault, then r_i is approximately constant, as is R_i^P. If the rupture front expands with a constant rupture velocity, v_r, and the displacement history is the same everywhere on the

fault, then Δt_i, is the distance along the fault divided by the rupture velocity,

$$u_r(r,t) = \frac{R^P \mu}{4\pi\rho\alpha^3} \frac{W}{r} \sum_{i=1}^{N} \dot{D}\left(t - \frac{x}{v_r}\right) dx. \quad (18.15)$$

We can rewrite (18.15) in a more useful form by using the shift property of the delta function

$$\dot{D}\left(t - \frac{x}{v_r}\right) = \dot{D}(t) * \delta\left(t - \frac{x}{v_r}\right), \qquad (18.16)$$

where $*$ denotes convolution. Now assume that the particle velocity is everywhere the same on the fault. Note that this assumption is not fully consistent with dynamic crack models, but it is reasonable for the slip pulse model or for high-frequency radiation from a crack model (ignoring the long tails). This simple, approximate model provides insight into the effects of rupture finiteness that will hold qualitatively for all rupture models.

If we substitute (18.16) into (18.15) and take the limit of the sum as $dx \to 0$, we obtain an integral expression

$$u_r(r,t) = \frac{R^P \mu}{4\pi\rho\alpha^3} \frac{W}{r} \int_0^x \dot{D}(t) * \delta\left(t - \frac{x}{v_r}\right) dx, \qquad (18.17)$$

where x represents position along the rupture. Since $\dot{D}(t)$ is independent of x, it can be taken outside the integral

$$u_r(r,t) = \frac{R^P \mu}{4\pi\rho\alpha^3} \frac{W}{r} \dot{D}(t) * \int_0^x \delta\left(t - \frac{x}{v_r}\right) dx . \qquad (18.18)$$

Evaluation of the integral requires the integration of the delta function. Let $z = t - (x/v_r)$, then $x = tu_r - zv_r$, and $dx = (dx/dz)dz = -v_r dz$. Thus

$$\int_0^x \delta\left(t - \frac{x}{v_r}\right) dx = - \int_t^{t-(x/v_r)} v_r\, \delta(z)\, dz. \quad (18.19)$$

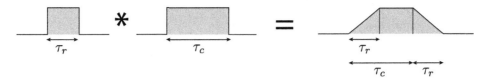

FIGURE 18.6 The convolution of two boxcars, one of length τ_r and the other of length τ_c ($\tau_c > \tau_r$). The result is a trapezoid with a rise time of τ_r, a top duration of $\tau_c - \tau_r$, and a fall of duration τ_r.

The integral of $\delta(z)$ is the Heaviside step function, $H(t)$, where

$$H(t) = \begin{cases} 0 & t \leq 0, \\ 1 & t \geq 0. \end{cases} \tag{18.20}$$

Thus

$$u_r(r, t) = \frac{R^P \mu W}{4\pi\rho\alpha^3 r} \dot{D}(t) * v_r \left. H(z) \right|_{t-x/v_r}^{t}$$

$$= \frac{R^P \mu W}{4\pi\rho\alpha^3 r} \dot{D}(t) * \left[H(t) - H(t - x/v_r) \right]$$

$$= \frac{R^P \mu W}{4\pi\rho\alpha^3 r} v_r \dot{D}(t) * B(t; \tau_c), \tag{18.21}$$

where $B(t; \tau_c)$ is a boxcar that starts at time t, and has a duration equal to the total rupture time ($\tau_c = x/v_r$). Thus the far-field displacement pulse shape is the convolution of two boxcars, one representing the displacement history at one location the rupture (which we assumed was everywhere the same), and the other representing the effects of a rupture propagation along the finite-length fault. The convolution of two boxcars is a trapezoid (Fig. 18.6). This trapezoid has two fundamental dimensions: (1) a duration equal to the sum of the two boxcars and (2) the *rise* and *fall* times of the trapezoid, which are equal to the duration of the shortest boxcar, which for large earthquakes is the one corresponding to slip across the fault.

This simple line source, or *Haskell fault model* (Haskell, 1964), predicts that the far-field P- and S-wave displacements should have a roughly

trapezoidal shape. Consider the integral of a far-field P wave displacement pulse,

$$\int_{-\infty}^{\infty} u_r(r, t)\, dt = \int_{-\infty}^{\infty} \frac{R^P \mu}{4\pi\rho\alpha^3} \frac{v_r W}{r} \dot{D}(t) * B(t; \tau_c)\, dt, \tag{18.22}$$

or, rearranging terms,

$$\frac{4\pi r\rho\alpha^3}{R^P} \int_{-\infty}^{\infty} u_r(r, t)\, dt$$

$$= \int_{-\infty}^{\infty} \dot{D}(t) * \mu W v_r B(t; \tau_c)\, dt. \tag{18.23}$$

The right-hand side of (18.23) is the integral of a convolution or two functions (the area under the curve), which is equal to the product of the individual areas (under the curves) of the two functions. The area of $\dot{D}(t)$ is D, the total slip. The area of $\mu w v_r \times B(t; \tau_c)$ is μWL, or μA. Thus, the right-hand side is equal to the seismic moment, $M_0 = \mu DA$. The left-hand side is the area under the displacement pulse corrected for propagation, the radiation pattern, and the near-source material properties. This equality provides a procedure for determining the seismic moment from far-field displacements. Fig. 18.7 is a plot of an observed SH displacement waveform from an earthquake near Parkfield, California. Note that its shape is roughly trapezoidal.

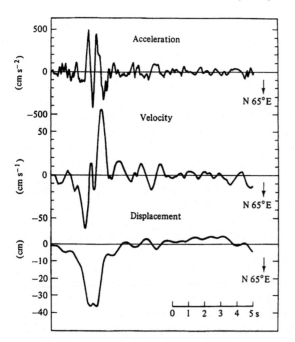

FIGURE 18.7 A recording of the ground motion near the epicenter of the 1966 Parkfield, California earthquake. The accelerometer was located within a few km of the rupture (or less) and very close to the fault. The top trace shows the observed accelerations, the velocity and the displacement seismograms were obtained by integrating the acceleration. The station is located along a nodal direction P waves and a maximum for SH. The displacement pulse is the SH wave. Note the trapezoidal shape on the displacement pulse (from Aki, 1968).

18.2.1 Rupture directivity

In our initial consideration of the Haskell source model, the boxcar associated with the propagation of the rupture had a duration, $\tau_c = L/v_r$, for a station at an azimuth perpendicular to the strike of the ribbon fault. Only stations perpendicular to the rupture observe the true rupture duration. Other stations observe an apparent rupture duration that depends on the azimuth of the observer relative to the rupture direction. In general, the rupture velocity is less than the S-wave (and usually the surface-wave) velocity near the source, so the body waves from ruptured fault segments arrive at the station in

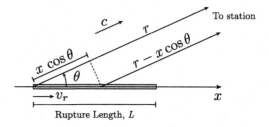

FIGURE 18.8 Geometry of a rupturing fault and the path to a remote recording station. Gray region identifies the rupture with the x origin at the left end. Rupture propagates in the x direction with speed v_r and the wave leaves the source region with an apparent speed c. The difference in distance to the station for a location along the rupture (relative to the origin) is $x \cos \theta$ (after Kasahara, 1981).

the order that the segments ruptured. But when the path to the station is not perpendicular to the fault, the body waves generated from different segments of the fault travel different distances to the recording station and thus have relative travel times that depend on azimuth. Fig. 18.8 shows a rupture of length L, for which the rupture propagates from left to right, starting at position $x = 0$. Let the distance to the recording station, r, be such that $r \gg L$, and call the arrival time of a ray from the rupture initiation point, $t = r/c$, where c is the velocity corresponding to the wave type. Then the arrival time of waves radiated from a rupture segment at location x on the fault is given by

$$t(x, \theta) = \frac{x}{v_r} + \frac{(r - x \cos \theta)}{c} . \qquad (18.24)$$

The first term represents the time it takes the rupture to propagate a distance x, the second accounts for the difference in path length for positions $x = 0$ and x. The time difference for energy leaving the origin and the end of the rupture (L) defines an apparent rupture duration for the direction θ (the azimuth relative to the

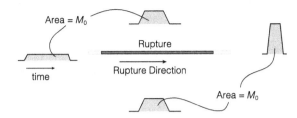

FIGURE 18.9 Azimuthal variability of the source time function for a unilaterally rupturing fault. The duration changes, but the area of the source time function is constant (independent of azimuth) and equal to the seismic moment.

rupture) is

$$\tau_c(\theta) = \left[\frac{L}{v_r} + \left(\frac{r - L\cos\theta}{c} \right) \right] - \left(\frac{r}{c} \right)$$

$$= L \left[\frac{1}{v_r} - \frac{\cos\theta}{c} \right] .$$

$$(18.25)$$

The component of the source time function associated with the fault finiteness is still a boxcar, but its duration, or **apparent rupture duration**, depends on the azimuth measured relative to the rupture direction. Thus, the Haskell model predicts a trapezoidal source time function for all stations, but the duration varies with direction. The azimuthal dependence resulting from rupture propagation is called **directivity**. If a seismic station is located in the direction of rupture propagation ($\theta = 0$), the trapezoid is very narrow and has a high amplitude. If a seismic station is located such that the fault ruptures away from it, the source time function will be spread out and have a small amplitude (see Fig. 18.9). For all azimuths, the time-function area equals the seismic moment. Thus the ratio of the rupture (v_r) and wave phase velocity (c) can strongly affect the amplitude of a particular phase. As the rupture velocity approaches the phase velocity, the directivity effects become more pronounced. Amplification of the signal in the rupture direction is frequency dependent. Long-period observations are less sensitive to directivity than shorter periods. At the shortest

periods, directivity effects may be disrupted by variations in slip, slip rate, and rupture propagation.

The simple Haskell line source representation that we have considered involved **unilateral rupture**, or rupture in only one direction. For some earthquakes, unilateral rupture is a sufficient model of the faulting process, but many earthquakes nucleate in the center of a fault segment and expand in both directions. This is known as **bilateral rupture**. The source time function for bilateral rupture varies less with azimuth, and it is often difficult to distinguish bilateral rupture from a point source. Some ruptures appear to expand radially, as **circular ruptures**. Generally, it is much easier to observe the horizontal component of source finiteness than the vertical component because the temporal delays are more pronounced for horizontal offsets. However, in some rare cases, the effects of **vertical directivity** are observed. This is possible when the length of the fault is short compared with its width and if the rupture is mainly up or down dip. In this case, the source time function for direct arrivals (P and S) will differ from those for the depth phases (pP, sS, sP, pS). Directivity can also be detected for complex ruptures with multiple subevents that are spatially and temporally offset. Directivity effects for nonuniform faulting allow spatially variable-displacement models to be estimated for some earthquakes.

Examination of Eq. (18.25) leads to following observationally useful formula. The maximum duration for a unilateral rupture is

$$\tau_{max} \sim L \left(\frac{1}{v_r} + \frac{1}{c} \right) , \qquad (18.26)$$

where the approximation is to account for the facts that the model is quite simple compared with actual earthquakes, and we are considering the rise time to be a second-order modification. The corresponding minimum duration is

$$\tau_{min} \sim L \left(\frac{1}{v_r} - \frac{1}{c} \right) . \qquad (18.27)$$

Adding and differencing these two relations yields (for unilateral ruptures)

$$\tau_{max} - \tau_{min} \sim \frac{2L}{c} ,$$

$$\tau_{max} + \tau_{min} \sim \frac{2L}{v_r} . \quad (18.28)$$

The left-hand sides in the above expressions are observable and the phase velocity of the wave, c is usually reasonably well known. Provided a sufficient number of observations from a sufficient range in azimuth are available, the first relationship can provide a quick estimate of a large unilateral earthquake's rupture length. Experience suggests that the estimated rupture speed, which require first an estimate of the length, are less precise, but still a useful check on results based on other observations.

18.3 Seismic source spectra

The equivalent time- and frequency-domain representations of a seismogram can provide valuable insight into source characteristics. The source time function (18.23) for the faulting model developed in the last section can be written

$$M(t) = M_0 \, B(t; \tau_r) * B(t; \tau_c) , \quad (18.29)$$

where the boxcar of width τ_r represents the effects of the slip dislocation history, and the boxcar of width τ_c represents the effects of fault finiteness. The boxcar heights are normalized to $1/\tau_r$ and $1/\tau_c$, respectively. The Fourier transform of a boxcar is given by

$$F[B(t; \tau_r)] = \hat{B}(\omega) = \frac{\sin(\omega\tau_r/2)}{\omega\tau_r/2} . \quad (18.30)$$

Recall that convolution of two functions in the time domain has a frequency-domain representation equal to the multiplication of the Fourier transforms of the two functions. Thus we can express the spectral density of the source time function as

$$\widehat{M}(\omega) = M_0 \left| \frac{\sin(\omega\tau_r/2)}{\omega\tau_r/2} \right| \left| \frac{\sin(\omega\tau_c/2)}{\omega\tau_c/2} \right| . \quad (18.31)$$

Eq. (18.31) indicates that the displacement amplitude decreases with increasing frequency. Fig. 18.10 (left) shows that we can approximate the boxcar amplitude spectrum (for normalized area, $M_0 = 1$) as

$$\left| \frac{\sin(\omega\tau_r/2)}{\omega\tau_r/2} \right| \approx \begin{cases} 1 & \omega < \frac{2}{\tau_r} \\ \frac{1}{\omega\tau_r/2} & \omega > \frac{2}{\tau_r} . \end{cases} \quad (18.32)$$

The spectrum of a boxcar has a plateau at frequencies less than $2/\tau_r$ and for higher frequencies, decays in proportion to $1/\omega$. The crossover frequency between the plateau and ω^{-1} behavior, defined by the intersection of the asymptotes to the low- and high-frequency spectra, is called a *corner frequency*. For a convolution of two boxcars, as in Eq. (18.31), and assuming $\tau_r < \tau_c$, the peak-amplitude spectra will have three distinct trends:

$$\widehat{M}(\omega) \approx \begin{cases} M_0 & \omega < \frac{2}{\tau_c} \\ \frac{M_0}{\omega\tau_c/2} & \frac{2}{\tau_c} < \omega < \frac{2}{\tau_r} \\ \frac{M_0}{\omega^2(\tau_r\tau_c/4)} & \omega > \frac{2}{\tau_r} \end{cases} \quad (18.33)$$

Physically, the model suggests that the spectral amplitude of a seismic pulse should be flat at periods longer than the rupture time of the fault. For periods between the rise time and rupture time, the spectra will decay as $1/\omega$, and at high frequencies the spectra will decay in proportion to $1/\omega^2$ (see Fig. 18.10, right). This is called an *omega-squared source model*. The decay of the far-field displacement spectra with increasing frequency is consequence of the interference patterns of waves leaving different parts of the rupture, caused by the rupture's temporal and spatial finiteness – components of the signal with periods less than the rupture duration interfere destructively.

The Haskell model can be modified to include the effects of bilateral rupture, finite-fault width, oblique rupture propagation (Udías et al., 2014). For example, accounting for bilateral propagation in a rectangle of finite length, L and width, W results in a function form similar to our one-dimensional results

$$\widehat{M}(\omega) \propto \dot{D}(\omega) \left| \frac{\sin(\omega\tau_L/2)}{\omega\tau_L/2} \right| \left| \frac{\sin(\omega\tau_W/2)}{\omega\tau_W/2} \right| ,$$

(18.34)

where we've associated τ_L with the rupture propagation along strike direction and τ_W includes the effects of propagation along the dip direction. We moved the particle rise time effects into a slip rate spectrum, $\dot{D}(\omega)$ (it could be modeled as a Brune-like source or another boxcar spectrum). Although the Haskell model and its variants are simple kinematic approximations of earthquake rupture, they do introduce important concepts including rise time, rupture propagation and directivity, that both affect the duration of seismic signals – any physical finite

source model will give rise to qualitatively comparable behavior of the far-field radiation.

18.3.1 Simple earthquake spectra models

For smaller earthquakes, the directivity effects are less important and the spectral model can be related to average rupture propagation (duration) and earthquake slip processes. In practice we usually can identify only one corner, which is defined by the intersection of the asymptote to the plateau and the asymptote of the $1/\omega^2$ decay (see Fig. 18.10 (right)). A simpler spectral shape, introduced by Aki (1968) and adopted by Brune (1970), is often assumed to model seismic source spectra,

$$\widehat{M}(\omega) = M_0 \frac{f_c^n}{f_c^n + f^n} ,$$

(18.35)

where M_0 is the seismic moment, f_c is the corner frequency of the spectrum, and $n = 2$ is usually assumed (an omega-squared model). The model is appealing because the spectrum is a function of two parameters, a long-period seismic moment, and a single corner frequency. The shape can be fit to seismic observations quite easily

---

Box 18.3 Earthquake rupture propagation

Earthquake fault rupture is a complex process that includes significant variations in the rate of rupture-front expansion. These fluctuations are important because they influence the strong shaking close to large earthquakes. The spatially variable process is hard to constrain in detail so we characterize the process in terms of statistical fluctuations about an average expansion process. Exploring the broad spatial and temporal features of earthquake rupture has been a research focus attention for many decades. Information from local, regional and teleseismic observations has been used to define the range of speeds at which an earthquake rupture propagates. Estimates of earthquake rupture speed are derived from a variety of observations, strong motion observations such as the timing of S-wave arrival at near-fault stations, and directivity patterns observable in seismic waveforms. Even average rupture speed can be difficult to resolve in detail unless the rupture is large. Fig. B18.3.1 is an example of using seismic array observations of slowness as a function of time to map spatial and temporal variations in seismic energy emitted by a propagating rupture.

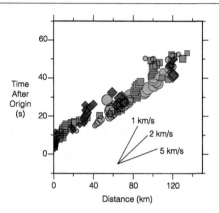

FIGURE B18.3.1 Rupture front expansion estimates for the 2015 M_W 7.9 Gorkha, Nepal earthquake. *P*-wave back-projection estimates from three seismic networks are shown with different symbols. A linear-fit produces an optimal speed estimate of about 2.7 km/s (modified from Meng et al., 2016).

Most average rupture-speed ($\bar{v}_r$) estimates are *roughly* in the range $0.7\beta \le \bar{v}_r \le 0.9\beta$, but are not limited to that range. We refer to rupture fronts with $\bar{v}_r < \beta$, as **sub-shear** and those that travel faster as **super-shear**. For more specificity, we use **subsonic** ($\bar{v}_r < \beta$), **intersonic** ($\beta \le \bar{v}_r < \alpha$), and **supersonic** ($\bar{v}_r \ge \alpha$). The heterogeneity of rupture-front expansion at small spatial scales likely decreases the directivity effects at high frequencies, but at longer periods and large ruptures, directivity effects can be substantial. The effects are maximal for speeds near the wave speed but directivity-associated amplification decreases for waves traveling at speeds much faster or slower than the rupture. Since the intensity of shaking adjacent to a large earthquake increases with rupture speed, the phenomena is important. A lack of aftershocks has been correlated with rupture segments associated with the super-shear rupture. Resolving the maximum speed of a rupture front is a challenge and some examples (Imperial Valley) have been debated for decades. Recent examples include the Denali (Alaska, USA), Izmit (Turkey), Kunlun (Tibet, China), Haida Gwaii (Canada), and Palu (Indonesia).

(e.g., Allmann and Shearer, 2009; Denolle and Shearer, 2016)

Table 18.1 is a list of a few model-based estimates of corner frequencies that can be used for rough estimates or to compare with observations and identify unusual observations. Note that models suggest that on average *P*-waves are richer in high frequencies than *S*-waves, a result of the greater similarity of the rupture and shear-wave speeds (compared with the *P* wave speed). Commonly used estimates developed for circular ruptures with an assumed rupture speed of $0.9\,\beta$ are

$$f_c^P = 0.32\,\beta/r$$
$$f_c^S = 0.21\,\beta/r\,, \tag{18.36}$$

where r is the rupture radius and the superscript identifies the body-wave type.

These corner-frequency relations are useful as typical values or as starting points for modeling, but both observed *and* rupture-model corner frequencies are complicated, so simple aver-

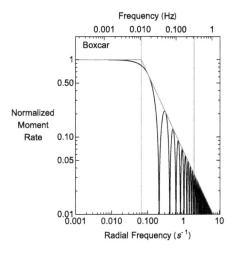 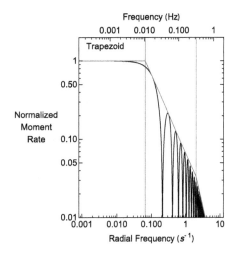

FIGURE 18.10 (left) Amplitude spectrum of a boxcar computed for $\tau_r = 1\,s$. The spectrum has two distinct regions, one where the spectral density is flat, and a second where the spectral density decays as ω^{-1}, the transition occurs at the corner frequency, $\omega_{c_r} = 2/\tau_r$. (right) Seismic spectrum of a trapezoid computed as the convolution of two boxcars with durations $\tau_r = 1\,s$ and $\tau_c = 30\,s$. The spectrum has two transitions, one at $\omega_{c_r} = 2/\tau_c$ and another at $\omega_{c_r} = 2/\tau_r$ where the falloff is ω^{-2}. The vertical lines identify the corner frequencies.

ages must be used with care. More robust is the fact that since the corner frequency is inversely proportional to an earthquake's duration, larger earthquakes, which take longer to rupture, have lower corner frequencies.

Corner frequency is useful to parameterize seismic-wave spectra, but more valuable is the estimate of the rupture dimension that it provides. Together with the seismic moment, that rupture dimension can be related to stress drop. For circular faults, the combination of relationships between stress-drop, moment and rupture radius, with a relationship to corner frequency leads to a notional estimate of

$$f_c = 0.42\,\beta \left(\frac{\Delta\sigma}{M_0} \right)^{1/3}, \qquad (18.37)$$

where all quantities are in *SI* units. Note that the corner frequency is larger for higher-stress drop events and lower for earthquakes with larger moment. Thus higher stress-drop events are relatively enriched in higher frequencies and larger

TABLE 18.1 Model-based *P*- and *S*-wave corner frequency estimates.

Reference	*P* Wave	*S* Wave
Savage (1972) (rectangle)	$\omega_1 = 1.2\,\alpha/L$	$\omega_1 = 3.6\,\beta/L$
Savage (1972) (rectangle)	$\omega_2 = 2.4\,\alpha/W$	$\omega_2 = 4.1\,\beta/L$
Madariaga (1976) (circle)	$\omega_c = 1.16\,\alpha/a$	$\omega_c = 1.32\,\beta/a$

events are relatively enriched in lower frequencies. These are relative differences.

Fig. 18.11 is a plot of the omega-squared model earthquake source spectra for a range of earthquake sizes. Despite the variability in specific corner frequency values, the spectral shape represented by Eq. (18.35) explains some first-order observations of classic seismology. We already mentioned the relative enrichment of signals from larger events in long-periods. In addition, seismograms corresponding to small earthquakes are dominated by short-period signals because of the high corner frequencies and because their long-period signals are generally below the long-period noise levels. The spectral view of seismic wave radiation provides a quan-

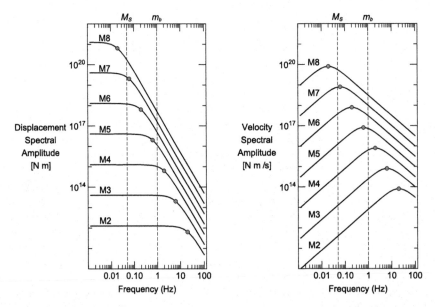

FIGURE 18.11 (left) Source amplitude spectra for ground displacement produced by an omega-squared source spectral model computed using a range of seismic moments and a constant stress drop of 3 MPa (Eq. (18.37)). Corresponding M_W values are shown on the left of each curve, circles identify f_c, computed using a shear-wave speed of 3.6 km/s. The dashed lines indicate the typical frequencies at which magnitude measurements for M_S and m_b are made. (right) Source amplitude spectra for ground velocity for the same models. Spectra with an omega-curbed decrease with frequency show less difference in spectral amplitude at high frequencies.

titative explanation for magnitude saturation described qualitatively in Section 7.2, Fig. 7.4.

Both M_S and m_b were designed to be as compatible as possible with M_L and sometimes all three magnitudes have the same value for an earthquake. Unfortunately, this is rarely the case. Each of these three magnitudes depends on a frequency-dependent amplitude measurement at roughly 1.2, 1.0, and 0.05 Hz for M_L, m_b, and M_S, respectively. Only for small earthquakes (very short fault lengths) with corner frequencies well above 1 Hz will the amplitude be the same for all three frequencies. For earthquakes above a certain size, for example, the frequency at which we measure m_b will be located on the ω^{-2} decay slope, and the event size will be under-estimated by m_b – this is magnitude saturation. For the spectral models shown, m_b saturation begins between $m_b \sim 5$ and $m_b \sim 6$ and saturation is significant for $m_b \gtrsim 6.5$. M_S satura-

tion begins between values of M_S above 7 and below 8; saturation is significant for $M_S \gtrsim 8.0$. The spectra in Fig. 18.11 are computed for specific parameters and don't apply exactly to all earthquakes, so there are frequent examples of reported m_b larger than 6.0 and M_S larger than 8.0. The same is true for other magnitudes such as M_L. But the concept of magnitude saturation is robust.

Magnitude saturation complicates the interpretation of a single magnitude, but can also provide insight into source characteristics. For example the m_b/M_S ratio is a way of identifying earthquakes deficient in high-frequency signals which may be an indication of a shallow rupture in subduction regions – where shallow ruptures often trigger tsunami. Still, having a magnitude that does not saturate is desirable for communicating earthquake size as a single, easy to appreciate quantity.

TABLE 18.2 Typical corner frequencies for simple earthquake models.

M_W	M_0(N m)	f_c (1 MPa)	f_c (3 MPa)	f_c (10 MPa)
1	3.5×10^{10}	43 Hz	63 Hz	94 Hz
2	1.1×10^{12}	14 Hz	20 Hz	30 Hz
3	3.5×10^{13}	4 Hz	6 Hz	9 Hz
4	1.1×10^{15}	1 Hz	2 Hz	3 Hz
5	3.5×10^{16}	0.4 Hz	0.6 Hz	0.9 Hz
6	1.1×10^{18}	0.14 Hz	0.2 Hz	0.3 Hz
7	3.5×10^{19}	0.04 Hz	0.06 Hz	0.09 Hz

Numerous studies of earthquake spectra have been combined to produce estimates of the *typical* rupture size and infer slip offsets for earthquakes of different size. The numbers are not precise, but are useful quantities for order of magnitude calculations about earthquakes. Table 18.2 is a summary of the seismic moment and corner frequencies for events of different sizes and stress drops. The corner frequencies were computed using Eq. (18.37) with a shear-wave speed of $3400 \, \text{m/s}^2$. We often use the inverse of the corner frequency to estimate a characteristic duration of a particular size earthquake. This is a rough average of rise-time (slip) processes and rupture propagation, and it's not a bad approximation for quick calculations. Table 18.3 is a list of the values of $\frac{1.5}{f_c}$, which provides reasonable estimate of the typical earthquake durations (assuming a standard stress drop of 3 MPa). The values are certainly well within a factor of two of what is usually observed for moderate and large earthquakes.

TABLE 18.3 Typical durations for simple earthquake models.

M_W :	1	2	3	4	5	6	7
τ (s) :	0.02	0.08	0.24	0.76	2.4	7.5	25

18.3.2 Earthquake self similarity

Earthquakes are often viewed as *scale invariant* processes, which means that small and large earthquakes exhibit the same patterns and the physical properties of earthquakes scale in specific ways. Fig. 18.12 is a summary cartoon based on simple models, used by Walter et al. (2006) to summarize the implications of scale invariance. Implicit in the self similarity model is that average stress drop and rupture speed are constant. Some earthquakes sequences appear to follow self-similar scaling patterns, but other data sets show contradictory evidence. Walter et al. (2006) provide a nice review – and note that the large scatter in the observations (primarily from estimates of seismic radiated energy, E_S) allow a number of interpretations – but some careful work on aftershock sequences suggests that earthquakes are not always self-similar. Walter et al. (2006) note that the nature of earthquake scaling is reflected in variations of scaled energy, E_S/M_0. If the scaled energy is independent of earthquake size, earthquakes are self similar and small and large earthquakes differ primarily only in scale. If the scaled energy increases with earthquake size, then large earthquakes release more strain energy per unit fault slip and differ fundamentally from small earthquakes. The ideas continue to be investigated and as a result of the variability of the data will require large, multiple data sets to resolve issues of self-similarity conclusively.

18.4 Earthquake-slip heterogeneity

Thus far in our discussion we have assumed that the rupture process is fairly smooth, which is fine for average properties, but in detail, slip is not smoothly distributed, and source time functions can deviate significantly from the simple shapes that we have considered. Physically, complexity in the moment-rate function corresponds to variations in slip, slip rate, or rupture front propagation. Separating the effects of all these processes is a challenge and we often discuss models in simple terms of variations in slip with position. Regions of very high slip, known

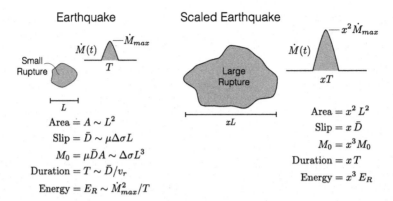

FIGURE 18.12 Summary of self-similar relationships based on simple earthquakes models. The cartoon on the left shows a small rupture and corresponding source time function. The list below includes relationships based on simple earthquake models and observed patterns. The cartoon on the right shows a larger rupture (dimensions increased by a factor x) and the corresponding list of how the quantities change if the earthquake is simple and self similar (modeled after figure in Walter et al., 2006).

as ***asperities***, are extremely important in earthquake hazard analysis because the failure of the asperities radiates most of the high-frequency seismic energy. The concentration of slip on asperities implies they are regions of high moment release, which, in turn, implies a fundamental difference in the fault behavior at the asperity compared with that of the surrounding fault. Mapping slip to stress drop suggests that asperities are regions of apparent high strength (very large stress drop). The reason for the high relative strength could be heterogeneity in the frictional strength of the fault contact or variations in geometric orientation of the fault plane. Rupture slip roughness and the concept of asperities seems to apply to earthquakes at all scales.

A geometric explanation for asperities reflects the fact that faults are not perfectly planar. On all scales, faults are rough and contain jogs or steps. The orientation of the fault plane as a whole is driven by the regional stress pattern. Segments of the fault that are subparallel to this orientation can have significantly higher normal stresses than surrounding regions, making them "sticking" points that resist steady, regular slip. Fig. 18.13 shows a geometric irregularity that could serve as an asperity. The size and apparent strength of the asperity depend on d_s and θ_s (see Fig. 18.13). At high frequencies, failure of

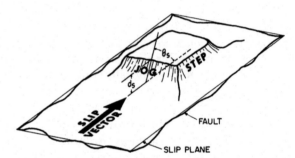

FIGURE 18.13 Geometric irregularity that could serve as an asperity. The protuberance increases the relatively normal stress on the top of the irregularity resulting in an increase in effective stress (from Scholz, 2019).

discrete asperities may be manifested as distinct seismic arrivals. This implies that the details of source time functions may correspond to seismic radiation on particular segments of the fault.

An important question is, What causes the rupture to stop? Along with the concept of asperities, the concept of *barriers* has been introduced for regions on the fault that have exceptional strength and impede or terminate rupture. Alternatively, barriers may be regions of low strength in which the rupture "dies out." This type of barrier is known as a *relaxation barrier*. The concepts of strength and relaxation barrier are generally consistent with the asperity model if adjacent segments of the fault are considered. A strength barrier that terminates rupture from an earthquake on one segment of the fault may serve as an asperity for a future earthquake. Similarly, the high-slip region of a fault during an earthquake may act as a relaxation barrier for subsequent earthquakes on adjacent segments of the fault. Aseismic creep may also produce relaxation barriers surrounding asperities that limit the rupture dimensions when the asperity fails. Unfortunately, there are also inconsistencies between the barrier and asperity models of fault behavior. Consider a region of moderate slip located between the two asperities. Is this reduced slip caused by a region of previous failure, or is this a region of the fault that is primed for a future earthquake? It may be possible to resolve this question by studying the detailed spatial distribution of aftershocks. If the regions adjacent to the asperities have a concentration of aftershocks but the asperities themselves are aftershock-free, this would be inconsistent with strength barriers. There is some indication that aftershock distributions outline asperities, but there are still problems with spatial resolution that preclude strong conclusions. Aftershocks are clearly a process of relaxing stress concentrations introduced by the rupture of the mainshock, but there remains an active debate as to their significance in terms of asperities and barriers. The only thing that is certain is that, averaged over long periods of time, the entire fault must slip equal amounts.

18.5 Source-spectrum estimation

We can use the simple source time function shapes and source spectrum parameterizations to quantify certain characteristics of seismic sources by directly fitting the ideal shapes to observations. For example, consider a direct wave, which is often modeled using a relatively simple propagation factor that includes geometric spreading and a free-surface receiver function

$$\mathbf{u}(\mathbf{x}, t) = A_F(\mathbf{x}, t, [\omega]) \cdot S(\mathbf{x}, t) * G(\mathbf{x}, t).$$

Recall that $A_F(\mathbf{x}, t, [\omega])$ represents amplitude patterns that result from fault orientation, $G(\mathbf{x}, t)$ represents propagation effects, and $S(\mathbf{x}, t)$ is the source time function (or moment-rate) function. To isolate the source time function from the seismogram requires that we isolate the source contribution from the convolution (Eq. (16.2)). We must also account for the propagation of the wave and the radiation patterns (at least in an average sense).

18.5.1 Source-spectrum estimation

A commonly used approach to estimate small event corner frequencies and stress drops is to fit the spectrum of a seismic waveform with the two-parameter omega-squared spectral model. The approach requires corrections for radiation patterns and wave propagation, but for short distances, or relatively simple paths, corrections can be developed. Consider the spectrum of an observed seismic wave (a single component), $u(\mathbf{x}, t)$,

$$\widehat{U}(\mathbf{x}, \omega) = A_F(\mathbf{x}, \omega) \cdot \widehat{S}(\mathbf{x}, \omega) \cdot \widehat{G}(\mathbf{x}, \omega).$$

Time-domain convolution corresponds to frequency-domain multiplication. We'll take the logarithm to transform the multiplication into summation of the logarithms of the various quantities that make up A_F, S, and G. For local and regional propagation, we can express a body-wave amplitude using the form

$$\log \widehat{U}(\omega, \mathbf{x}) \approx \log A_F(\mathbf{x}) + \log M_0$$
$$- \log \left(1 + \frac{\omega^2}{\omega_c^2}\right) + \log \widehat{G}'(r, \omega)$$
$$+ \widehat{P}(\omega, \mathbf{x}) \qquad (18.38)$$

Box 18.4 The spectrum of seismic slip

Although the source duration of most earthquakes scales directly with seismic moment (see Fig. 18.15), there are some exceptions. In particular, *slow earthquakes* have unusually long source durations for the seismic moments associated with them. Slow earthquakes typically have an m_b that is small relative to M_s. Fig. B18.4.1 shows the effect of duration on short- and long-period body waves. The slow rise time presumably results from a very low stress drop, which controls the particle velocity. Variability in the source function occurs on all scales, from rapid events to slow creep events. Fig. B18.4.2 compares the seismic recordings of several aftershocks of the 1960 Chile earthquake. The upper two recordings are normal earthquakes, with typical fundamental mode excitation. The May 25 event has some greater complexity in the surface wave train, while the June 6 event is incredibly complex, with over an hour-long interval of surface wave excitation.

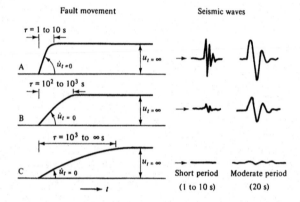

FIGURE B18.4.1 The effect of different rise times on teleseismic signals (from Kanamori, 1974).

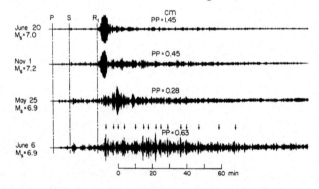

FIGURE B18.4.2 Recordings of four aftershocks of the 1960 Chile earthquake. The upper two traces are conventional in appearance, with well-concentrated R_1 wavepackets. The lower two events have much more complex surface waves intervals, indicative of long, complex source radiation, extending over more than an hour for the June 6 event (from Kanamori and Stewart, 1979).

Kanamori (1986) noted that some subduction zone earthquakes produce extraordinarily large tsunamis but have moderate surface-wave amplitudes. In these cases M_s is small for the actual moment, and very slow *rupture* propagation reduces short-period wave excitation. The physical mechanism responsible for such a slow rupture process is unknown, but in the extreme, it could produce a "silent" earthquake devoid of short-period body and surface waves. Two investigators, G. Beroza and T. Jordan, suggested that several silent earthquakes occur each year that can be identified only because they produce free oscillations of the Earth. However, several sources, including large atmospheric storms and volcanic processes, can excite low-frequency oscillations, so the source of the free oscillations observed by Beroza and Jordan is somewhat uncertain, but likely to be associated with unusual earthquake dynamics.

where $\widehat{U}(\omega, \mathbf{x})$ is the observed spectral amplitude, ω is the angular frequency $(2\pi f)$, $\mathbf{x}$ is the station position vector, r is the source-to-station distance, M_0 is the seismic moment, ω_c is the angular corner frequency, $A_F(\mathbf{x})$ is an excitation factor that depends on the earth model (properties near the source and receiver) and faulting geometry, and $\widehat{G}'(r, \omega)$ is a function of distance used to account for geometric spreading and frequency-dependent attenuation, and the site-amplification filter, $\widehat{P}(\omega)$ accounts for near-station focusing and defocussing, but has no physical units. For simplicity, we assumed that the moment-rate, or source-time-function spectrum can be represented with an omega-squared model,

$$\widehat{S}(\omega) = M_0 \left(\frac{\omega_c^2 + \omega^2}{\omega_c^2} \right)^{-1} = M_0 \left(1 + \frac{\omega^2}{\omega_c^2} \right)^{-1} . \tag{18.39}$$

The moment-rate function, $\widehat{S}(\mathbf{x}, t)$, has physical units of $[Nm \ s^{-1}]$, so it's Fourier Transform, $\widehat{S}(\mathbf{x}, \omega)$, has units of Nm. We ignore directivity-related differences and use a simple corner frequency-based proxy that is effective in the standard practice for body waves. This is not a bad assumption for small events. The corner frequency, ω_c depends on the seismic moment and

stress-drop,

$$\omega_c = K^{1/3} \left(\frac{\Delta \sigma}{M_0} \right)^{1/3} . \tag{18.40}$$

In Eq. (18.37), we assumed that $K = 2\pi \, 0.42 \, \beta$, and several alternate values are listed in Table 18.1. Although often assumed to be equal, P and S wave corner frequencies may not be identical (Table 18.1). Taylor and Hartse (1998) discuss some details and parameterize the situation as

$$\omega_{cP} = \zeta \, \omega_{cS} , \tag{18.41}$$

where $1 \le \zeta \le v_p / v_s$.

For a model with near source and near-receiver speeds of c_s and c_r, respectively,

$$A_F(\mathbf{x}) = \frac{R_{\theta \phi}}{4\pi \sqrt{\rho_s \, \rho_r \, c_s^5 \, c_r}} , \tag{18.42}$$

where the subscripts s and r correspond to the source and receiver respectively. The speed, c, represents the P or S wave speed, and ρ represents the density. The inclusion of c_r and ρ_r here accunts for some of the propagation effects. When the faulting geometry is unknown, we replace the specific radiation pattern factor with the square root of the RMS value of the radiation pattern. Using the expressions from Aki and Richards (2009), and representing the RMS using angular brackets, we have

$$A_F(\mathbf{x}) = \frac{\langle R_{\theta\phi} \rangle_{P,S}}{4\pi \sqrt{\rho_s \rho_r c_s^5 c_r}} \, , \qquad (18.43)$$

where

$$\langle R_{\theta\phi} \rangle_P = \sqrt{4/15} \, , \qquad \langle R_{\theta\phi} \rangle_S = \sqrt{2/5} \, . \qquad (18.44)$$

If $c_r \sim c_s$ and $\rho_r \sim \rho_s$, then

$$A_F(\mathbf{x}) = \frac{\langle R_{\theta\phi} \rangle_{P,S}}{4\pi \rho_s c_s^3} \, . \qquad (18.45)$$

We have combined the geometric and attenuation amplitude factors in our expression. The units of $G'(r)$ for body waves are the units of r^{-1}. If an explicit form for the two factors is required, we can assume something similar to

$$\widehat{G}'(\omega, r) \sim G(r) \, \exp\left(-\frac{\omega r}{2c\, Q(\omega)}\right) \, , \qquad (18.46)$$

where $Q(\omega)$ is the frequency-dependent attenuation quality factor, and c represents the average speed from the source the receiver.

If we assume that we can adequately approximate the source and station locations, the earth properties, and the faulting geometry factors in Eq. (18.38), we can construct a fit to an observed spectrum by adjusting the two parameters M_0 and ω_c until an optimal match is obtained. This can be performed using one spectrum or a suite of spectra (from the same earthquake). Obviously, the more data, the more reliable the result, although even with many observations, as detailed above, the assumptions on source geometry, source spectral shape, and propagation are substantial. The results are often capable of extracting relative differences across events within the same region and provide some information with which to compare results from one area to another.

When data are plentiful – such as when we have observations of a suite of earthquakes observed at a substantial number of seismic stations, we can forego many of the assumptions and construct a regression with which we can estimate properties of each source, a common geometrical-spreading/attenuation as a function, and individual site effects. The source spectra, site effects, and attenuation are functions of frequency, the geometrical spreading is not. Most often a regression (linear inverse problem) is performed for each frequency of interest using the observed spectra. Even with many data, assumptions of a simple and single geometric spreading and attenuation function and site effect independent of azimuth and incidence angle remain significant. Observations are often complex and much of the variability in the data remains unexplained by a regression-based model. But on average, such models have provided important and valuable information on earthquakes and other seismic sources. Again the most reliable information is obtained for the relative differences of the events included, but comparisons from region to region show enough consistency to compare events over a broad region, and even globally.

18.6 Source-time function estimation

Another approach to estimate the source spectrum is to **deconvolve** a predicted seismogram computed for an impulsive source time history (a Green's function) from the observed seismograms. Deconvolution is the undoing of convolution. We start with our familiar convolution-based description of seismic deformation in the frequency domain,

$$\widehat{U}(\mathbf{x}, \omega) = A_F(\mathbf{x}, \omega) \cdot \widehat{S}(\mathbf{x}, \omega) \cdot \widehat{G}(\mathbf{x}, \omega).$$

Now represent the Fourier transform of a synthetic seismogram computed for an impulsive source, $S(\mathbf{x}, t) = \delta(t)$, as $\widehat{Y}(\omega, \mathbf{x})$. Because the

Fourier spectrum of the delta function is one for all frequencies,

$$\widehat{Y}(\omega, \mathbf{x}) = \widehat{A}_{F_0}(\mathbf{x}, \omega) \cdot \widehat{G}_0(\mathbf{x}, \omega) , \qquad (18.47)$$

where the subscript on $\widehat{A}_{F_0}$ and $\widehat{G}_0$ indicate simply that our model of source excitation and propagation are approximate. Then, to the extent that approximation is adequate, we can estimate the source spectrum using

$$\widehat{S}(\mathbf{x}, \omega) \approx \frac{\widehat{U}(\mathbf{x}, \omega)}{\widehat{Y}(\omega, \mathbf{x})} = \frac{\widehat{A}_F(\mathbf{x}, \omega) \cdot \widehat{S}(\mathbf{x}, \omega) \cdot \widehat{G}(\mathbf{x}, \omega)}{\widehat{A}_{F_0}(\mathbf{x}, \omega) \cdot \widehat{G}_0(\mathbf{x}, \omega)} . \qquad (18.48)$$

The better our knowledge of the excitation (faulting geometry, near-source properties) and the propagation effects on the seismogram, the better our estimate of $\widehat{S}(\omega, \mathbf{x})$. Eq. (18.48) is an effective way to think about deconvolution – but it is seldom an effective way to perform deconvolution. Within the limitations imposed by the signal sample rate (Chapter 5), the Fourier transform relationships are numerically stable and invertible. But computational division is difficult when the denominator is a small number that may contain significant noise. Consider Eq. (18.48) in a form such that the denominator is a real-valued quantity,

$$\widehat{S}(\mathbf{x}, \omega) \approx \frac{\widehat{U}(\mathbf{x}, \omega) \, \widehat{Y}^{\dagger}(\omega, \mathbf{x})}{\widehat{Y}(\omega, \mathbf{x}) \, \widehat{Y}^{\dagger}(\omega, \mathbf{x})} . \qquad (18.49)$$

In many cases with actual data, the spectral ratio in Eq. (18.49) is numerically unstable because the value in the denominator becomes quite small, and may be dominated by noise. Still, the information in the source time functions is so important to understanding earthquake physics, that we accept the limitations of deconvolution and employ a broad range of methods to solve these often poorly-posed inverse problems. For example, when the situation is not too severe, a common approach in earthquake seismology is to use a simple stabilization approach to avoid division by small numbers. We modify the denom-

inator by replacing values smaller than a threshold (often expressed as a fraction of the maximal value in the denominator spectrum).

$$\widehat{S}(\mathbf{x}, \omega) \approx \frac{\widehat{U}(\mathbf{x}, \omega) \, \widehat{Y}^{\dagger}(\omega, \mathbf{x})}{\Upsilon(\omega)} , \qquad (18.50)$$

where the denominator for a particular frequency, ω_i is defined as

$$\Upsilon(\omega_i) = \max \left\{ \widehat{Y}(\omega_i, \mathbf{x}) \, \widehat{Y}^{\dagger}(\omega_i, \mathbf{x}), \right.$$
$$\left. w \cdot \max_{all\ \omega} \left[\widehat{Y}(\omega, \mathbf{x}) \, \widehat{Y}^{\dagger}(\omega, \mathbf{x}) \right] \right\} , \qquad (18.51)$$

where w is an adjustable parameter ($0 \leq w \leq 1$) often called the water-level. This method is called the water-level approach. More sophisticated and optimal modifications of the denominator exist if you know something about the spectral content of the noise, or in our case, synthetic seismogram errors. Other, more sophisticated methods have been developed in both the frequency domain and in the time domain. A comprehensive review is beyond our scope, but the seismological literature is replete with discussions of deconvolution.

18.6.1 Body waves

In general, there are no near-field recordings for most earthquakes of interest, and we must infer any faulting heterogeneity from details of the far-field time function alone. This is possible when the source orientation is known independently and we simply want the source time function. The major problem with this procedure is that it maps uncertainty in the Earth transfer function and source orientation into the time function. This is a problem for analysis of large earthquakes unless the Earth transfer function correctly includes the effects of fault finiteness. One way to allow for finiteness is to produce a suite of Earth transfer functions for a given geometry and write the time-domain displacement response as

$$u(x,t) = \sum_{j=1}^{M} B_j \left[b(t - \tau_j) * g_j \right], \quad (18.52)$$

where g_j is the Earth transfer function from the jth element of the fault that "turns on" at some time τ_j, which is prescribed by the rupture velocity; $b(t - \tau_j)$ is the parameterization of the time function as described in Eq. (19.19); and B_j is the variable of interest in the inversion, namely the strength of element $b(t - \tau_j)$ in the source time function. A separate source time function is found for each element of the fault by solving for $B_j(t)$. Fig. 18.14 shows an example of the forward problem for a fault that ruptures from 15 to 36 km depth. It is obvious that unless the variability in timing of the depth phases is accounted for, an inversion for the time function will be biased. Chapter 19 reviews and shows examples of inversions for source time functions based on Eq. (18.52).

Meier et al. (2017) investigated source time function (STF) data sets for subduction zone earthquakes with moment magnitude $M_w \geq 7$ and identified some similarities in the STFs of large ruptures. Fig. 18.15 shows median STFs in seven magnitude bins and for 12 STFs with $M_w \geq 8$. Specifically, the median STF scales with different event size, grows linearly, and is shaped like a triangle. Because of this similarity, the STFs can be approximated with the functional form: $y_{fit} = \mu t e^{1/2\lambda t^2}$ (Meier et al., 2017), where μ and λ are determined through inversion. Since STFs are proportional to the product of slip and rupture area, the scalability does not require the ratio of slip over length be constant, a classical concept of earthquake similarity. Furthermore, the centroid times (the time at which the 1/2 of energy has been released), shows a linear progression with size of the earthquake. Despite these similarities, many

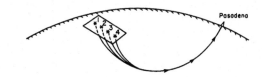

Velocity Structure			
α(km/sec)	β(km/sec)	ρ(gm/cm³)	h(km)
5.00	2.88	2.40	8.0
5.80	3.34	2.50	9.0
7.00	4.04	2.75	13.0
8.00	4.62	2.90	—

FIGURE 18.14 Earth transfer functions for a four-point source representation of a thrusting earthquake. The sum of the $g(t)$ convolved with time functions appropriate for each point source will give the synthetic seismogram (from Hartzell and Heaton, 1985).

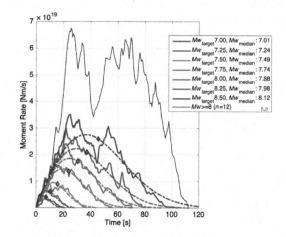

FIGURE 18.15 Median STFs in seven magnitude bins and for 12 STFs with $M_w \geq 8$, with the approximate range of when the median normalized STFs reach peak amplitudes is indicated in gray (35 to 55% of full-rupture duration). The median STFs on the physical scale are accurately approximated with the functional form (dashed lines). Diamonds in show centroid times for fitted functions (modified from Meier et al., 2017).

subduction zone earthquakes show varied complexities in their STFs, reflecting differences in rupture processes.

18.6.2 Surface waves

The earth models that we use to compute predicted seismograms are much more accurate for long periods than for short periods. Resolution of the details in the source time function increases with bandwidth, and often the details we would like to know are beyond our ability to predict with simple seismic wave-speed models. But for large events, where interesting variations in the source time function extends to longer periods (where the earth models are adequate), we can estimate source time functions directly from the data using synthetic corrections. Fig. 18.16 is an example of a set of time functions estimated from observed Rayleigh waves excited by the 01 April, 2007, M_W 8.1, Solomon Islands earthquake. The observations show a straight-forward example of the directivity pattern expected from Eq. (18.25), and conceptually illustrated in Fig. 18.9. The time functions are displayed as a function of a *directivity parameter*, defined as $\Gamma(\theta, p) = p \cos \theta$, where θ is the azimuth of the station measured relative to the rupture direction and p is the horizontal slowness measured along the fault surface.

The directivity parameter transforms the sinusoidal variation in time-function duration (Eq. (18.25)) to a linear pattern using a reference direction, in this case 305°, the direction of the plate-boundary strike (rupture along the shallow dipping subduction interface is approximately along strike). $\Gamma < 0$ corresponds to rupture away from the station, $\Gamma > 0$ corresponds to rupture towards the station. STF estimates in the direction of rupture are noisier, but have shorter durations and (larger amplitudes, not shown). Dashed reference lines show the onset ($t = 0$) and the non-directivity affected duration (corresponding to the $\Gamma = 0$ observations). Fitting a line along the apparent duration of the STFs,

projected to limiting values of Γ ($\pm 0.25 \, \text{s/km}$), suggests minimum (50 s), maximum (180 s), and average (115 s) STF durations. In this case, the removal of propagation effects using EGF deconvolution also isolates an early aftershock. The slope of the aftershock signals indicates that it is not collocated with the position used to compute the point-source synthetic seismograms

Earth's propagation often includes wave interference effects that produce spectral variability that may or may not be sufficiently modeled in the predicted spectra.

Empirical Green's functions

Although Earth models have become quite sophisticated, there are many instances where our ability to compute accurate theoretical Green's functions is inadequate to allow source information to be retrieved from particular signals. This is common for broadband recordings of secondary body waves with complex paths in the Earth (*PP, SSS,* etc.), as well as for short-period surface waves ($T = 5$–$80 \, \text{s}$). A strategy for exploiting these signals is to let the Earth provide the propagation effects for these signals, which are usually complex. Consider the seismograms from a small earthquake located near a larger event of interest. If the source depth and focal mechanism of the two events are identical, the Earth response to each station will be the same. If the small event has a relatively short (impulse-like) source time function, its recordings approximate the Earth's Green's functions, including attenuation, propagation, and radiation pattern effects, with a corresponding seismic moment (and instrument). We can use these signals to model the signals for a larger event, any differences in the signals is attributed to the greater complexity of the source time function for the larger earthquake.

We deconvolve the "empirical" Green's functions (EGFs) from the corresponding records for the larger event. The result is an approximation of the source time function for the larger event, normalized by the seismic moment of

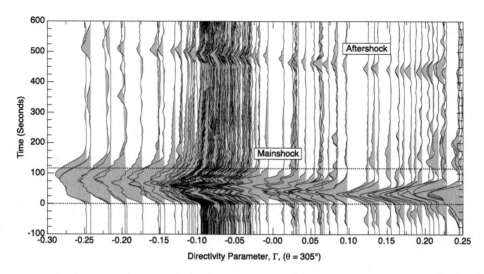

FIGURE 18.16 Relative source time function estimates deconvolved from R1 Rayleigh waves generated by the 01 April, 2007 Solomon Islands earthquake (modified from Furlong et al., 2009).

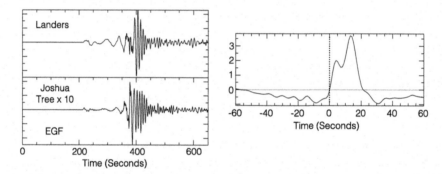

FIGURE 18.17 (left) Observed transverse-component seismograms for the 1992 Landers (M_W 7.3) and Joshua Tree (M_W 6.1) earthquakes that occurred in southern California observed at seismic station CCM, Cathedral Cave, MO. (right) Source time function estimated from the seismograms on the left using a water-level deconvolution (data from Velasco et al., 1994).

the smaller event (Fig. 18.17). Isolated phases with single ray parameters are usually used in this type of analysis because the source time functions depend on the wave phase velocity. The azimuthal and ray parameter (takeoff angle) variations in the relative source time functions can be used to constrain the finiteness of the larger event. Formally, interpretation of deconvolution as a source-time function is valid for frequencies below the corner frequency of the

smaller event, otherwise, the results are better known as a relative source time function. Rupture processes of both small and great events have been studied using empirical Green Functions. Fig. 18.17 is an example EGF computation. The signals are dominated by dispersed Love waves and the larger, Landers earthquake is relatively enriched in longer-period signals. Finer details in the source time functions are often better resolved using an EGF. An obvious limitation is the availability of a nearby event with similar faulting geometry and depth, but aftershocks often provide the needed data. However, small-magnitude EGFs often have low signal-to-noise ratios at long periods. The Landers earthquake time function in Fig. 18.17 has broad long-period side-lobes caused by a lack of sufficient long-period signal in the Joshua Tree EGF. The time function appears to sink into a broad trough.

Source time functions are valuable observations of the radiation of seismic energy during the rupture process. Independent of faulting geometry, they relate directly to the physics op-

erating during earthquakes. The methods we reviewed in this chapter provide either constrained or first-order properties of these important quantities. To extract more information from the signals we often construct parameterized models of the earthquake process that can be directly and quantitatively compared with observed seismograms to provide constraints on the temporal and spatial distribution of earthquake slip and slip-rate. We discuss these models and approaches in the next chapter on waveform modeling.

19

Imaging seismic-sources

Chapter goals

- Introduce methods used to estimate seismic point-source parameters.
- Introduce moment-tensor estimation using body waveforms, surface-wave spectra, and full waveform modeling.
- Introduce approaches to estimating source finiteness using sub-fault modeling and iterative pulse stripping.
- Introduce finite-fault modeling methods used to estimate the rupture characteristics and slip distributions of earthquakes.

The preceding chapters have dealt with phenomena that influence the patterns that we observe on seismograms: source-excitation effects, propagation effects, and characteristics of the seismometer/seismic-recording system. These effects can be combined to *predict* what we expect to see on a seismogram. Such a mathematical construction is known as a ***synthetic seismogram***. The formalism of comparing synthetic and observed seismograms is known as ***waveform modeling***. Waveform modeling is one of the most powerful tools available to seismologists for refining models of Earth's interior and seismic sources. In general, waveform modeling is an iterative process in which differences between the observed and synthetic seismograms are minimized by adjusting the Earth model or source representation to improve the match

between the observations and predictions. We refer to the process of estimating the characteristics of earthquakes, explosions, and other seismic sources as seismic source imaging – using "imaging"' as a generic term to represent the processes by which we extract information from seismograms to investigate the source processes. Seismic waves carry information on their source and accumulate information on the interior of the Earth along the paths they travel. As waves propagate, the source and structure information is blurred together. An observed seismogram is a convolution of source and propagation effects,

$$u(\mathbf{x}, t) = E(\mathbf{x}, t) * G(\mathbf{x}, t) , \qquad (19.1)$$

where $\mathbf{x}$ is a position vector, t is time, $u(\mathbf{x}, t)$ is displacement (a seismogram), $E(\mathbf{x}, t)$ represents source effects, and $G(\mathbf{x}, t)$ represents propagation effects (for an ideal, impulsive source). If your focus is on the source, you choose observations that minimize the unknown components in $G(\mathbf{x}, t)$, if your focus is on imaging Earth's interior, you choose observations that minimize the uncertainty in $E(\mathbf{x}, t)$.

For example, $G(\mathbf{x}, t)$ accounts for the multiplicity of arrivals due to reflections and refractions at material boundaries within the Earth along with a filter that accounts for the seismic-wave attenuation. If only a few rays are important, they may be modeled separately and combined. If the number of rays is large, we may use a more complete representation of the

515

equations of motion in a stratified structure computed by directly solving the differential equations. We may sum long-period free oscillations that well describe the vibrations that follow an earthquake, or we integrate Laplace- or Fourier-transformed differential equations in cylindrical coordinates for shorter-period observations.

Source effects, $E(\mathbf{x}, t)$, include both radiation pattern and rupture effects. Radiation pattern effects are included in synthetic seismograms in two ways. If a ray-based framework is adopted, the expressions for the wholespace are assumed to govern at the source and the radiation patterns are calculated as described earlier (Chapter 17) and the amplitudes are propagated to the observing distance (Chapters 12 and 13). For more complex stratified media we often represent the source radiation as a discontinuity in stress (or strain) at the source location. Rupture effects are generally included by convolution of an impulsive response with a source time function (Chapter 18). To model more complex faulting scenarios we combine the results of multiple point sources to represent the rich detail involved in the earthquake rupture process.

The ability to model seismograms opens the door to interpretation of the details about earthquakes that influence seismogram characteristics. The idea is an extension of using first-motion patterns to constrain the orientation of faulting geometry, extending the work to use more information in the seismogram. Most of what is known about seismic source processes has been learned by applying such a procedure. In this chapter we will explore waveform modeling and provide some examples of seismogram modeling to estimate simple point and finite-source parameters. We start with a simple discussion of body-wave modeling using rays, then discuss surface-wave spectral modeling. We conclude with an introductory discussion of modeling source finiteness with body waves.

19.1 Body waveform modeling – a point source

We can readily construct the quantities on the right-hand side of Eq. (19.1) for a point source. If the source is small (compared to our wavelengths of interest), directivity effects are minimal, and we can approximate the moment-rate function for all directions with a single time history. For example, the source-time function could be approximated with a trapezoid (see Chapter 18). The most complex filter in Eq. (19.1) is $G(\mathbf{x}, t)$, which is sometimes called an *earth transfer function*. $G(\mathbf{x}, t)$ accounts for all propagation effects such as reflections, triplications, diffractions, scattering, attenuation, mode conversions, as well as geometric spreading. The usual procedure is to separate $G(\mathbf{x}, t)$ into a filter that accounts for elastic phenomena, $G_E(\mathbf{x}, t)$, and a filter that accounts for attenuation, $G_A(\mathbf{x}, t)$. For body waves, $G_E(\mathbf{x}, t)$ is a time series with a sequence of impulses temporally distributed to account for the variability in arrival times. The amplitudes depend on the direction the wave left the source and any interactions that wave had with Earth's structure (reflections, transmission, mode conversions, etc.).

At teleseismic distances ($0° \leq \Delta \leq 90°$), the most important P-wave arrivals are P, pP, and sP, and $G_E(\mathbf{x}, t)$ is a "spike train" with three pulses. The amplitude of each spike depends on the wave's angle of incidence, i_h at the surface and the seismic radiation pattern. At these distance ranges for shallow sources, the ray parameter of each of these waves is essential the same, so the take-off angles differ only in terms of up-going (pP and sP) and down-going (P) waves. In Chapters 13, mathematical expressions were developed to calculate the amplitudes of impulse P waves. For example, the expression for the far-field P waveform is

$$u_r(r, i_h, \phi, t) = \frac{R^P}{4\pi\rho\alpha^3} \cdot \frac{1}{r} \cdot \dot{M}\left(t - \frac{r}{\alpha}\right), \quad (19.2)$$

where R^P is the radiation pattern in terms of fault geometry and takeoff angle. Analogous expressions can be written for the horizontally (SH) and vertically (SV) polarized components of shear waves.

19.1.1 Fundamental fault responses

For convenience, we exploit the relationships among the radiation patterns for double-couples corresponding to different faulting geometry. Any double couple can be represented using combinations of three different elementary moment tensors (Aki and Richards, 2009), and in the simplest combination, the elementary moment tensors correspond to shear dislocations or *fundamental faults*. The ability to express the displacement for an arbitrary double-couple as the sum of fundamental faults or moment-tensor elements (described later) is the basis for many seismogram modeling algorithms and tools.

We rewrite (19.2) as a weighted sum of three fundamental faults,

$$u_r\,(r, i_h, \phi, t)$$

$$= \frac{1}{4\pi\rho\alpha^3} \cdot \frac{1}{r} \cdot \sum_{i=1}^{3} A_i\,(\phi, \lambda, \delta)\, C_i \cdot \dot{M}\left(t - \frac{r}{\alpha}\right),$$

$$(19.3)$$

where the A_i are called the *horizontal radiation patterns*, and the C_i are called the *vertical radiation patterns*, one for each fundamental fault. An analogous expression can be written for the SV component, with vertical radiation patterns, SV_i. The SH displacement requires only two fundamental faults, using vertical radiation patterns, SH_i, we have

$$u_\phi\,(r, i_h, \phi, t)$$

$$= \frac{1}{4\pi\rho\beta^3} \cdot \frac{1}{r} \cdot \sum_{i=4}^{5} A_i\,(\phi, \lambda, \delta)\, SH_i \cdot \dot{M}\left(t - \frac{r}{\beta}\right).$$

$$(19.4)$$

The five horizontal radiation patterns are functions of the fault geometry and station azimuth.

$$A_1 = \sin 2\phi \cos \lambda \sin \delta + \tfrac{1}{2} \cos 2\phi \sin \lambda \sin 2\delta$$

$$A_2 = \cos \phi \cos \lambda \cos \delta - \sin \phi \sin \lambda \cos 2\delta$$

$$A_3 = \tfrac{1}{2} \sin \lambda \sin 2\delta$$

$$A_4 = \cos \phi \cos \lambda \sin \delta - \sin 2\phi \sin \lambda \sin \delta \cos \delta$$

$$A_5 = -\sin \phi \cos \lambda \cos \delta - \cos \phi \sin \lambda \cos 2\delta\,,$$

$$(19.5)$$

where ϕ is the station azimuth, ϕ_s, relative to the fault strike, ϕ_f, i.e. $\phi = \phi_s - \phi_f$, and the first three are used for P and SV motion, and the last two are used for SH motion. The three fundamental faults are (1) a vertical strike-slip fault, (2) a vertical dip-slip fault, and (3) a 45° dipping thrust fault ($\lambda = 90°$) evaluated at an azimuth of 45°. By plugging in the appropriate strike, dip, and rake, you can see that A_2 and A_3 vanish for the first fundamental fault, A_1 and A_3 vanish for the second fundamental fault, and so on, which means that the three fundamental responses are proportional to the C_i. In terms of the horizontal radiations patterns defined for surface-wave analysis (Eq. (17.79)), we have $A_1 = -p_R$, $A_2 = q_R$, $A_3 = s_R$, and $A_4 = p_L$, $A_5 = -q_L$.

The vertical radiation patterns are functions of the take-off angle, i_h, or the horizontal slowness, p, and vertical slownesses $\eta_{\alpha,\beta}$,

$$p = \frac{\sin i_h}{\alpha}, \qquad \eta_\alpha = \frac{\cos i_h}{\alpha} = \sqrt{\alpha^{-2} - p^2}\,,$$

$$\eta_\beta = \frac{\cos i_h}{\beta} = \sqrt{\beta^{-2} - p^2}\,,$$

$$(19.6)$$

the vertical pattern terms for P waves are

$$C_1 = -p^2\,, \qquad C_2 = 2\varepsilon\, p\, \eta_\alpha\,, \qquad C_3 = p^2 - 2\eta_\alpha^2\,,$$

$$(19.7)$$

for SV, we have

$$SV_1 = -\varepsilon\, p\, \eta_\beta\,, \qquad SV_2 = \eta_\beta^2 - p^2\,,$$

$$SV_3 = 3\varepsilon\, p\, \eta_\beta$$

$$(19.8)$$

and for SH we have

$$SH_1 = \beta^{-2}, \qquad SH_2 = -\frac{\varepsilon}{\beta^2}\frac{\eta_\beta}{p}, \qquad (19.9)$$

where

$$\varepsilon = \begin{cases} +1 & \text{if ray is upgoing} \\ -1 & \text{if ray is downgoing}. \end{cases} \qquad (19.10)$$

Eqs. (19.3) and (19.4) provide a simple methodology for the calculation of the displacements for arbitrary faulting geometries. We calculate the response for each of the three fundamental faults (i.e. the C_i, SV_i, SH_i), weight them by the source-geometry dependent A_i, and sum. The use of fundamental fault responses extends to more general responses (than just a far-field body wave) and allow us to combine a few fundamental responses to predict the response for sources with any faulting geometry. But these equations, as written, are only accurate in a uniform wholespace. If the wave interacts with a more realistic structure, the interaction will introduce reflected and possibly mode-converted waves. To model a seismogram, we must account for these interactions.

We'll adopt an intuitive ray-based view where each arrival is tracked individually. The ground displacement is the sum of k ray-responses, which for a ray leaving the source as a P waves take the form

$$u_k(r, i_h, \phi, t)$$
$$= \frac{F_{P_c}}{4\pi\rho\alpha^3} G_k(\Delta, h)$$
$$\times \sum_{i=1}^{3} A_i \, C_i \, \widetilde{RT}_k \, \delta(t - T_k) * Q_\alpha(t) * \dot{M}(t), \qquad (19.11)$$

for a ray leaving the source as n SV wave,

$$u_k(r, i_h, \phi, t)$$
$$= \frac{F_{SV_c}}{4\pi\rho\beta^3} G_k(\Delta, h)$$

$$\times \sum_{i=1}^{3} A_i \, SV_i \, \widetilde{RT}_k \, \delta(t - T_k) * Q_\beta(t) * \dot{M}(t), \qquad (19.12)$$

and for and ray leaving the source as n SH wave,

$$u_k(r, i_h, \phi, t)$$
$$= \frac{F_{SH_c}}{4\pi\rho\beta^3} G_k(\Delta, h)$$
$$\times \sum_{i=4}^{5} A_i \, SH_i \, \widetilde{RT}_k \, \delta(t - T_k) * Q_\beta(t) * \dot{M}(t). \qquad (19.13)$$

F_{M_c} is the free-surface receiver function (M identifies the mode type, P or S wave, c is the component of motion, radial or vertical) of the kth ray, which was described in Chapter 13. T_k is the arrival time of the kth ray (usually only the time relative to the direct arrival is needed) and $\widetilde{RT}_k$ is the product of all the transmission and reflection coefficients that the kth ray experiences as it travels from the source to receiver. The geometric spreading factor, $G_k(\Delta, h)$ replaces the $1/r$ in a uniform material. Attenuation is included using a linear filter $Q_{\alpha,\beta}(t)$, usually parameterized in terms of the t^*, a path integral through the attenuation model. See Box 13.2 for examples of the filter response. $\dot{M}(t)$ is the average moment-rate function, assumed here to be the same for P and S.

Eqs. (19.11)–(19.13) have been extensively employed modeling sources at teleseismic and regional distances. Purely ray-based approaches have become less common because many rays may be required to model the complete earth response with suitable accuracy. Although rays are easy to compute, it becomes a challenge to just list enough rays to produce all the details in a complete solution of the equations of motion. Today, most approaches directly and completely solve the equations of motion in a stratified material (at least near the source) to compute the displacements, see for example, Hudson (1980) or Herrmann (2013) Computer Programs in Seismology. But even in those applications, at least part of the propgation is treated using ray theory. In any case, a ray-based view is essential for understanding the underlying physics and is often sufficient to extract first-order information from the signals.

19.1.2 Teleseismic body-wave modeling

Although Eqs. (19.11)–(19.13) look complicated, they are straightforward to compute for teleseismic distances because for most rays arriving close in time to the P or S wave, the horizontal slowness changes relatively slowly and the structure in the lower mantle produces relatively simple propagation effects. Fig. 19.1 includes a plot of $G_E(\mathbf{x}, t)$ for a dip-slip fault in a half-space. The amplitudes of the depth phases are affected by both the source radiation pattern and the surface reflection. In the example, P and pP both leave the source with a compressional motion. Upon reflection at the free surface, pP is inverted (reflection coefficient is negative). The combined effects of the SV radiation pattern and free-surface reflection also invert the polarity of the sP arrival relative to P. For observations at teleseismic distances $G_k(\Delta, h)$ is approximated as described in Chapter 13, and t^* is nearly constant over much of the body-wave frequency band (for P waves, $t^* \sim 1$ and for S waves, $t^* \sim 4$). The larger t^*, the more the high frequencies are attenuated. Thus the *amplitude* of the short-period signals is more sensitive to t^* than is the amplitude of long-period signals. The *relative* arrival times (ΔT_k) of the phases depend on the earthquake depth, h, and the distance between the source and receiver (through its control on the horizontal slowness). The travel time delay for each ray, relative to the direct-wave, is the sum of the vertical slowness × the vertical distance for any segment in the ray that is not also in the direct-wave path. Halfspace surface-

reflection delay times are

$$\Delta T_k \approx d\,\eta_{\mathrm{u}} + d\,\eta_{\mathrm{d}}, \qquad (19.14)$$

where η_{u} and η_{d} are the vertical slownesses of the upward and downward travel rays.

The relative amplitudes of the spikes in $G_E(\mathbf{x}, t)$ vary greatly depending on faulting geometry, so seismic waveforms are strongly diagnostic for different fault orientations. Waveform modeling is much more powerful for constraining fault orientation than first-motion focal mechanisms because it provides more complete coverage of the focal sphere and uses the wave relative-amplitude information. A realistic $G_E(\mathbf{x}, t)$ generally requires more than just the three halfspace wave arrivals. For a stratified earth structure, multiple reflections and conversions occur both near the source and beneath the receiver. In general, these multiples are much less important than the primary three rays at teleseismic distances unless the earthquake occurred beneath the ocean floor. In this case *water reverberations*, rays bouncing between the surface and ocean floor, can produce significant coda on the signal.

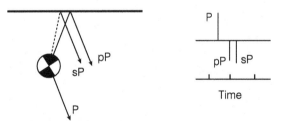

FIGURE 19.1 Primary raypaths corresponding to direct P and surface reflections pP and sP that arrive at a teleseismic station. For a shallow source these arrivals arrive close together in time, and together they comprise the P "arrival". The relative amplitudes of the arrivals are influenced by the source radiation pattern and the free-surface reflection coefficients. Small-amplitude differences due to extra attenuation or geometric spreading for the upgoing phases also can be accounted for if the source is deep, but aren't that important for crustal sources.

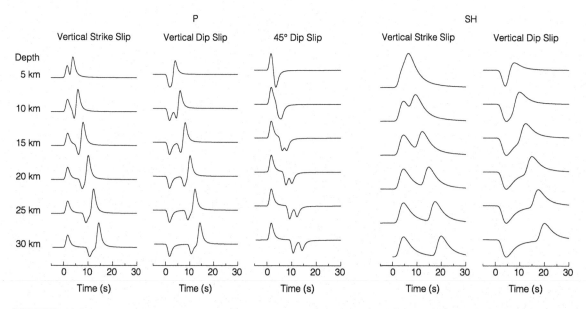

FIGURE 19.2 *P* and *SH* point source, halfspace seismograms for crustal sources computed with a ray parameter corresponding to a distance of roughly 60°. Only the direct and primary near-source reflected waves are included. The signals have been convolved with the two-second duration triangle and attenuation filters with $t_P^* = 1$ and $t_S^* = 4$. The response for each fundamental fault type is shown for six crustal depths.

Fig. 19.2 is a plot of *P* and *SH* synthetic seismograms for the relevant fundamental faults and a suite of source depths. Each signal has been convolved with a relatively short-duration source time function (a triangle with a width of two seconds). The source depth, time function, fault orientation, and seismic moment are known as the *seismic source parameters*. The goal of *waveform modeling* is to recover the source parameters by "fitting" observed waveforms with synthetics. *P* and *SH* waveforms for different fault orientations differ enough to be diagnostic of the source type, although some care is required because trade-offs exist. For low-frequency observations (e.g. shear waves) the strongest trade-off is between source depth and source time-function duration. A deeper source with a shorter-duration source function may be similar to a shallower source with a longer source function. Broadband *P* waves alleviate the depth-duration trade off, but they

can be quite complex from vertical strike-slip earthquakes for which they are often nodal and variable. For very shallow sources (near ocean ridges, explosions), interference of the direct and surface-reflected waves can reduce the information carried in the waveform. The key to minimizing uncertainty, including trade offs, is combining observations that provide sufficient azimuthal coverage and data multiplicity.

19.1.3 Moment-tensor inversion

The separation of source and propagation effects is the fundamental notion underlying waveform modeling (for seismic source-parameter estimation). For a double couple, (19.11) provides the separation. Now consider a full moment tensor source for which all moment tensor elements share an identical source time history, $S(t)$. Represent the earth response corresponding to each moment tensor element by $G_{in}(t)$, where i is an index representing each moment tensor element and n is the component of motion. Then the displacement can be modeled using

$$u_n(\mathbf{x}, t) = S(t) * \sum_{i=1}^{6} m_i \cdot G_{in}(\mathbf{x}, t) \qquad (19.15)$$

where the m_i represent the six independent components of the moment tensor, $m_1 = M_{11}$, $m_2 = M_{22}$, $m_3 = M_{12}$, $m_4 = M_{13}$, $m_5 = M_{23}$, and $m_6 = M_{33}$, and u_n is the vertical, radial, or tangential displacement. If the source is a double-couple, then we can reduce the sum to five terms using $M_{33} = -M_{11} - M_{22}$. The displacement is the product of the seismic moment tensor elements and their corresponding $G_{in}(t)$, which we call *Green's functions*. Relationships between the moment-tensor Green's functions and the fundamental-fault responses can be found in Langston (1981). For source modeling, the moment tensor elements are values we wish to estimate. The Green's functions are impulse displacement responses for a seismic source with orientation indicated by the corresponding moment tensor element. Each moment tensor element excites *three* components of displacement. The Green's functions may be simple teleseismic body-wave responses, or they may represent local, regional, or teleseismic body and surface-wave responses. Because any arbitrary faulting geometry can be represented by a specific linear combination of moment tensor elements (see Section 16.5.1), the corresponding waveform can be constructed as a *linear* combination of moment-tensor Green's functions. This is an extremely powerful representation of the seismic waveform because it requires the calculation of at most six Green's functions to produce a synthetic waveform for an arbitrary moment tensor at a given distance.

Eq. (19.15) is the basis for **moment-tensor inversion** procedures to recover the seismic source parameters. The purely double-couple representation is included (19.11) as a special case. In the simplest case, let us assume that the source time function and source depth are known. Then $S(t)$ can be directly convolved with the Green's functions, yielding a system of linear equations:

$$u_n(\mathbf{x}, t) = \sum_{i=1}^{5} m_i \cdot G_{in}(t) , \qquad (19.16)$$

which we can write in matrix form

$$\boldsymbol{u} = \boldsymbol{G}\,\boldsymbol{m} , \qquad (19.17)$$

where the data are arranged to form a vector, $\boldsymbol{u}$ and the Greens functions are stored as columns of the matrix $\boldsymbol{G}$, and the moment-tensor elements are stored in the vector $\boldsymbol{m}$. For a seismogram with k time samples, the equations would look like

$$\begin{bmatrix} u_1 \\ u_2 \\ u_3 \\ u_4 \\ \vdots \\ u_k \end{bmatrix} = \begin{bmatrix} G_{11} & G_{12} & \cdots & G_{15} \\ G_{21} & G_{22} & \cdots & G_{25} \\ G_{31} & G_{32} & \cdots & G_{35} \\ G_{41} & G_{42} & \cdots & G_{45} \\ \vdots & \vdots & & \vdots \\ G_{k1} & G_{k2} & \cdots & G_{k5} \end{bmatrix} \begin{bmatrix} m_1 \\ m_2 \\ \vdots \\ m_5 \end{bmatrix} . \qquad (19.18)$$

When more than one seismogram is used, individual $\boldsymbol{u}$ vectors and $\boldsymbol{G}$ matrices can be appended to create a single matrix expression for the five moment-tensor elements in $\boldsymbol{m}$. The most stable procedure is to simultaneously fit many seismograms from stations with distinctive Green's functions. In practice, the system must include observations from a distribution of azimuths to resolve $\boldsymbol{m}$. Most approaches rely on least-squares optimal fits to the seismograms, although more robust metrics are also commonly employed.

Of course, we usually do not know the source time function or source depth *a priori*, so when needed, we can recast the linear problem as an *iterative* inversion. In this case we parameterize the source time function and invert both the moment-tensor elements and source-time function parameters. Two common parameterizations of the time function are a series of box-cars, or overlapping triangles (Fig. 19.3).

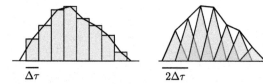

$\overline{\Delta\tau}$ $\overline{2\Delta\tau}$

FIGURE 19.3 Two alternative parameterizations of an arbitrarily shaped source function. The box-car width and triangle half-width (τ), the number of basis elements, and the start times of each are specified so that the unknowns are only the basis shape's amplitude. Curves show the interpretive view (smooth) of the corresponding moment-rate functions.

Consider the case where we represent $S(t)$ as a series of M boxcars,

$$S(t) = \sum_{j=1}^{M} B_j\, b(t - \tau_j)\,, \qquad (19.19)$$

where $b(t - \tau_j)$ is a boxcar of width $\Delta\tau$ that begins at time τ_j, and ends at $\tau_j + \Delta\tau$. B_j is the boxcar height and $M\Delta\tau$ is the total length of the time function. Eq. (19.19) can be used to rewrite (19.15) as

$$u_n(t) = \sum_{j=1}^{M}\sum_{i=1}^{5} B_j\, m_i \left[b(t - \tau_j) * G_{in}(t) \right].$$

$$(19.20)$$

This equation has two sets of unknowns, the boxcars heights (the source time function), B_j, and the moment-tensor elements, m_i. Eq. (19.20) is a nonlinear function of the unknowns (products of the model parameters) and its solution is more challenging, but, an iterative, linearized least-squares approach works well. We assume an initial model, compute the initial predictions, and then solve for a set of model-parameter corrections that will reduce the misfit, $\mathbf{\Delta d}$, which is

equal to the observed – prediction at each time sample in the seismogram. We then solve

$$\mathbf{\Delta d} = A\,\mathbf{\Delta P}, \qquad (19.21)$$

where A is a matrix of partial derivatives ($A_{ij} = \partial u_i/\partial P_j$) computed for each sample in the observed waveform ($u_n(t)$) with respect to a given model parameter P_j. The problem unknowns are the ΔP_j which are the corrections to apply to the model parameters, P_j, in order to reduce the difference between the observed and predicted seismograms. The linearization is valid for small ΔP_j, but with sufficient data, convergence of the iterative procedure is reasonably quick.

Eq. (19.21) can be solved with the techniques described in Section 6.4.1 and Aster et al. (2013). The approach generally leads to stable results but issues can arise if the data are noisy, azimuthal coverage is modest, or trade-offs are not resolved by the signal bandwidth. Moment tensors estimated from waveform inversion hardly ever correspond to pure double couples. The estimate is usually diagonalized and decomposed into a major and minor double couple or into a major double couple and a CLVD (Section 16.5.2). The minor double couple is usually small and is ignored because it is usually assumed that the minor double couple is the result of noise or of mapping incomplete or inaccurate Green's functions into the source parameters.

The effectiveness of waveform modeling for determining seismic source parameters depends on our ability to calculate the Green's functions accurately. At teleseismic distances this is usually not a problem, since the rays P, pP, and sP and SH and sSH have simple structural interactions and turn in the lower mantle where the seismic velocity structure is smooth. Although strong reverberations can occur in a sedimentary basin, for the most part teleseismic Green's functions for isolated body-wave arrivals are simple. The same is not true for observations at upper-mantle-triplication and regional distances.

At upper-mantle distances, triplications from the 400- and 670-km discontinuities make the body-wave Green's functions complex. Further, the mantle above the 400-km discontinuity has tremendous regional variability (Chapter 10). In general, beyond 14°, the first-arriving P wave has turned in the upper mantle, and the 400-km triplication occurs between 14° and 20°. The triplication from the 670-km discontinuity usually occurs between 16° and 23°. The complexity and regional variability of upper-mantle-distance seismograms diminish their utility in seismic source parameter studies. Only when an earthquake occurs where the upper-mantle structure is very well known are the records of use for source analysis.

At regional distances the crust acts as a waveguide, and hundreds of reflections between the surface and crust-mantle boundary are important for the waveform character. But local and regional-distance analysis is extremely important in the study of small or moderate-sized earthquakes ($m_b \leq 5.5$), which are less often well recorded at teleseismic distances. In many instances, lower frequency arrivals at local to near-regional distance are modeled using Green's functions computed using stratified earth models. Such inversions include both the intermediate-period (~ 5 to $\sim 80\,\text{s}$) body and surface waves. The bandwidth is adjusted depending on the source size, but incorporation of periods less than $10\,s$ generally requires relatively close observations (less than a few hundred kilometers).

19.1.4 Time-dependent moment-tensor inversion

Seismic source may exhibit changes in faulting geometry as the rupture progresses, or interactions between explosive sources and the local tectonic strain field and fault systems may lead to changes in the source character with time (e.g. Box 19.1). We can investigate these processes by allowing the moment-tensor to change with time. That is, we can model the observations using a *time-dependent* moment tensor – each of the six independent moment tensor elements is allowed to have an individual time history. We rewrite Eq. (19.15) as

$$u_n(\mathbf{x}, t) = \sum_{i=1}^{5} m_i(t) * G_{in}(\mathbf{x}, t), \qquad (19.22)$$

where each moment tensor element is an independent time series. We can solve these equations for the moment-rate as a function of time (or using parameterized time functions). A time-domain approach allows us to constrain the roughness of the resulting moment-rate functions, limit their duration, etc. We can also approach the problem from a frequency-domain perspective. Transforming Eq. (19.22) to the frequency domain, we have

$$\widehat{u}_n(\mathbf{x}, \omega) = \sum_{i=1}^{5} \widehat{m}_i(\omega)\, \widehat{G}_{in}(\mathbf{x}, \omega), \qquad (19.23)$$

where $\widehat{m}_i(\omega)$ is the only unknown. We can solve for $\widehat{m}_i$ at a set of discrete frequencies and use the inverse Fourier transform to obtain a time-dependent moment tensor. To transform Eq. (19.23) to a matrix form suitable for a standard inverse analysis, we have to recognize that each spectral value is a complex valued quantity with a real and imaginary part. Consider a single frequency, f. Let $\widehat{u}_1^R$ and $\widehat{u}_1^I$ correspond to the observed real and imaginary components of the spectrum observed at the first station, and $\widehat{u}_n^R$ and $\widehat{u}_n^I$ correspond to the spectrum observed at the nth station. The Green's function spectral values are organized into a matrix composed of 10 columns corresponding to the real- and imaginary-valued components of five moment tensor spectra for each station. The character of the matrix equations is dictated by the complex multiplication,

$$(\widehat{m}^R + i\,\widehat{m}^I)(\widehat{G}^R + i\,\widehat{G}^I)$$
$$= (\widehat{m}^R\,\widehat{G}^R - \widehat{m}^I\,\widehat{G}^I) + i\,(\widehat{m}^R\,\widehat{G}^I + \widehat{m}^I\,\widehat{G}^R). \qquad (19.24)$$

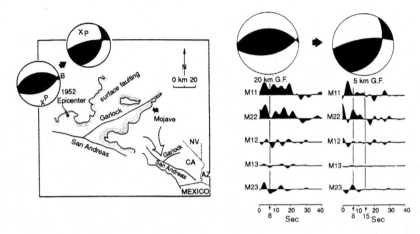

FIGURE 19.4 Results of a time-dependent moment-tensor inversion of the 1952 Kern County earthquake. Source time functions for each moment tensor element are shown for two depths. The preferred solution involves a pure thrust at 20 km depth in the first 8 s and a shallower oblique component in the next 7 s (from Kim, 1989).

For one frequency and n observed seismic spectra, the matrix form of the equations are

$$
\begin{bmatrix} \widehat{u}_1^R \\ \widehat{u}_1^I \\ \vdots \\ \widehat{u}_n^R \\ \widehat{u}_n^I \end{bmatrix} = \begin{bmatrix} \widehat{G}_{11}^R & -\widehat{G}_{11}^I & \widehat{G}_{15}^R & -\widehat{G}_{15}^I \\ & & \cdots & \\ \widehat{G}_{11}^I & \widehat{G}_{11}^R & \widehat{G}_{15}^I & \widehat{G}_{15}^R \\ \vdots & & \ddots & \vdots \\ \widehat{G}_{n1}^R & -\widehat{G}_{n1}^I & \widehat{G}_{n5}^R & -\widehat{G}_{n5}^I \\ & & \cdots & \\ \widehat{G}_{n1}^I & \widehat{G}_{n1}^R & \widehat{G}_{n5}^I & \widehat{G}_{n5}^R \end{bmatrix}
$$

$$
\times \begin{bmatrix} \widehat{m}_1^R \\ \widehat{m}_1^I \\ \vdots \\ \widehat{m}_5^R \\ \widehat{m}_5^I \end{bmatrix}. \tag{19.25}
$$

These equations are a linear relationship between the observed spectral values, the assumed (known) earth model and Green's functions, and the desired moment-tensor elements. Analogous equations for a suite of other frequencies are appended and a linear system can be solved using standard approaches of geophysical inverse theory. Inversion of Eq. (19.25) is unstable at high frequencies due to inaccuracies of the

Green's functions, so only the lower frequencies are used typically.

Fig. 19.4 shows the results of a time-dependent moment-tensor inversion for the 1952 Kern County earthquake. The results show a temporal evolution of rupture from primarily northwest-southeast thrusting to east-west oblique strike-slip motion. The geologic interpretation of the Kern County earthquake is that it started at the southwest corner of the fault at a depth of approximately 20 km. The fault ruptured to the northeast, where the fault plane became much shallower and the slip became partitioned into shortening (thrusting) and strike-slip components. For the entire rupture, the **P** axes remained nearly constant, but the **T** axis rotated from being nearly vertical to a much more horizontal position.

19.2 Surface-wave modeling for the seismic source

In addition to time-domain modeling, we often model seismogram spectra, particularly the surface waves, in the frequency domain. Spec-

Box 19.1 Tectonic release from underground nuclear explosions

Theoretically, the seismic waves generated by an underground nuclear explosion should be very different from those generated by an earthquake. An explosive source creates an isotropic stress imbalance without the shear motion that characterizes double-couple sources. Therefore, the seismograms from an explosion should not have *SH* or Love waves, but many explosions do have *SH-type* energy. This energy is thought to be generated by a "tectonic" component, namely the release of preexisting strain by the detonation of an explosion. There are three possible mechanisms for generation of the nonisotropic seismic radiation, known as *tectonic release*: (1) triggering of slip on prestressed faults, (2) release of the tectonic strain energy stored in a volume surrounding the explosion, and (3) forced motion on joints and fractures. For all three of these mechanisms for tectonic release, the long-period teleseismic radiation pattern can be represented by an equivalent double-couple source. Depending on the orientation and size of the tectonic release, the seismic waveforms from underground explosions can be significantly modified from those we expect for an isotropic source (an explosion). Waveform modeling can be used to constrain the size and orientation of the tectonic release. For large explosions, it appears that tectonic release is associated with a volume of material surrounding the detonation point, and the volume is related to the size of the explosion. If an explosion is detonated within the "volume" of a previous explosion, the tectonic release is dramatically reduced.

Fig. B19.1.1 shows two large underground nuclear explosions at the Nevada Test Site (NTS). BOXCAR (April 26, 1968, $m_b = 6.2$) was detonated 7 yr before COLBY (March 14, 1975, $m_b = 6.2$); the epicenters are separated by less than 3 km. Although the *P* waveforms recorded at LUB are similar, there are some distinct differences. Below the BOXCAR waveform is a synthetic seismogram constructed by "adding" the waveform of a strike-slip earthquake to the waveform of COLBY. The near-perfect match between the observed and synthetic waveform for BOXCAR supports the double-couple interpretation for tectonic release.

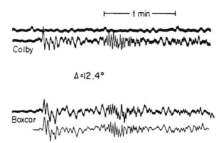

FIGURE B19.1.1 A comparison of the *P* and P_L waveforms for BOXCAR and COLBY at LUB. Also shown is a synthetic waveform constructed by summing the COLBY waveform and a synthetic calculated for a strike-slip double couple (moment is 1.0×10^{17} N m) (from Wallace et al., 1983).

tral modeling allows a more even weight of the information at different frequencies or signals recorded at different distances, which may have a substantial difference in duration as a result of dispersion (Chapter 14). Large earthquakes rupture large regions of one or more faults that may include changes in strike, dip, and slip direction. To assess the average faulting geometry of a large, shallow event, we often rely on intermediate- and long-period Love and Rayleigh-wave spectral amplitudes. Observed azimuthal patterns in surface-wave amplitude and phase can be used in either a formal geophysical inversion or with forward modeling approaches. The restriction to shallow events is a result of the surface-wave excitation depth dependency (Chapter 14). Both spectral amplitude and phase carry information on the source. In general, amplitude measurements are easier to correct for propagation than the phase measurements, so in some cases, only amplitude observations are employed.

For global analyses of large events, we generally focus on low-frequency observations for which we have models to calculate reasonably good propagation corrections. Large events also excite surface-wave signals that rise above the long-period noise levels (small events do not). Once we account for distance and attenuation (and propagation phase shifts if we are also using the source phase information), we can examine surface-wave spectral measurements for radiation patterns and model those patterns using the equations described in section 17.3.2. The information in Love and Rayleigh wavetrains complement one another and double the information available in each direction. Patterns seen at each period are similar, but not identical, and this also includes information on such parameters as the average fault dip, etc. Azimuthal phase variation provides relatively precise information on azimuthal orientations of polarity reversal (a phase shift of π). These same ideas can be applied to smaller, regionally observed earthquakes if we use shorter-period observations, which are often quite well observed from shallow, moderate-magnitude earthquakes. To do so requires earth models adequate to compute the phase and amplitude corrections needed to correct the surface-wave spectral measurements back to the source region. The results are quite effective in revealing tectonic activity across broad regions such as eastern and western North America, etc.

Geometric information from surface waves can provide very good constraints on the average faulting geometry but surface-wave *resolving* power for source depth and source time function is limited. In addition, surface-wave amplitude and phase is strongly dependent on the earth structure along the travel path. This means that we must correct for the effects of velocity and attenuation heterogeneity precisely for an inversion scheme to be robust. This is equivalent to knowing the earth transfer function in body-wave inversion procedures, but the body-waveforms are less sensitive to absolute travel time than are surface waves. For this reason, global surface-wave inversions are best performed at very long periods ($> 100\,\text{s}$) for which the heterogeneity is relatively well mapped. These periods are so long compared to most source durations that we can usually consider the far-field time function simply as a boxcar function with duration τ. In this case we can write the source *spectrum* of an earthquake source as

$$\widehat{V}(\omega, h, \varphi) = \widehat{A}(\omega, h, \varphi) + i\,\widehat{B}(\omega, h, \varphi)\,, \quad (19.26)$$

where ω is angular frequency, h is source depth, ϕ is the station azimuth and $\widehat{A}$ and $\widehat{B}$ are real-valued quantities. The observed spectrum, $\widehat{V}(\omega)$, is calculated directly from a surface-wave seismogram corrected for instrument and propagation effects. For Rayleigh waves $\widehat{A}$ and $\widehat{B}$ are

$$\widehat{A}(\omega, h, \phi) = \big[-P_R(\omega, h)\,M_{12}\sin 2\phi$$
$$+ \tfrac{1}{2} P_R(\omega, h)\,(M_{22} - M_{11})\cos 2\phi$$
$$- \tfrac{1}{2} S_R(\omega, h)\,(M_{22} + M_{11})\big]\,M(\omega)\,, \quad (19.27a)$$

$$\widehat{B}(\omega, h, \phi) = [Q_R(\omega, h)\,M_{23}\sin\phi$$
$$+ Q_R(\omega, h)\,M_{13}\cos\phi]\,M(\omega)\,, \quad (19.27b)$$

and for Love waves

$$\widehat{A}(\omega, h, \phi) = \left[-\tfrac{1}{2} P_{\mathrm{L}}(\omega, h)(M_{22} - M_{11}) \sin 2\phi \right.$$
$$\left. - P_L(\omega, h) M_{12} \cos 2\phi \right] M(\omega) ,$$
$$(19.28\mathrm{a})$$

$$\widehat{B}(\omega, h, \phi) = [-Q_L(\omega, h) M_{13} \sin \phi$$
$$+ Q_{\mathrm{L}}(\omega, h) M_{23} \cos \phi] M(\omega) .$$
$$(19.28\mathrm{b})$$

The M_{ij} are assumed to be frequency-independent and we'll assume that they all share a known source spectrum, $M(\omega)$. For long periods a simple moment-rate shape is often assumed, such as a box-car or triangle of duration τ. The P_R, S_R, Q_R, P_L, and Q_L terms are called *surface-wave excitation functions* (analogous to the body-wave Green's functions) and depend on the elastic properties of the source region and the source depth. The excitation functions depend on the material properties, frequency, and depth-dependent eigenfunctions (and their spatial derivatives) near the source. Fig. 19.5 is a plot of P_L as a function of depth and period for different types of near-source structures.

We can express Eq. (19.26) as a matrix equation. For Rayleigh waves

$$V = \Phi D , \qquad (19.29)$$

where

$$V = [A, B]^T , \qquad (19.30)$$

$$\Phi = \begin{bmatrix} -\sin 2\phi & \tfrac{1}{2} \cos 2\phi & -\tfrac{1}{2} & 0 & 0 \\ 0 & 0 & 0 & \sin \phi & \cos \phi \end{bmatrix} ,$$
$$(19.31)$$

and

$$D = [P_R M_{12}, \; P_R (M_{22} - M_{11}), \; S_R (M_{11} + M_{22}),$$
$$Q_R M_{23}, \; Q_R M_{13}]^T . \qquad (19.32)$$

The matrix Φ depends only on known source-station geometry. D contains the unknown moment-tensor elements. Eq. (19.29) can be extended to include spectral observations from N

stations. Then Φ is a $2N \times 5$ real-valued matrix, and V is a real-valued vector (dimension $2N$) containing the real and imaginary components of the observed spectra. The system of equations can be solved for $D(\omega)$ at several frequencies. Typically the optimal choice of source duration τ is determined by grid-searching (testing values systematically over a range of values) and choosing the value that minimizes the misfit to the spectral observations. Once $D(\omega)$ is determined, we decompose it into two vectors, one containing the excitation functions and the other containing the elements of the moment tensor:

$$\widehat{\Lambda}(\omega) = \left[\widehat{D}^T(\omega_1), \widehat{D}^T(\omega_2), \dots, \widehat{D}^T(\omega_n) \right]^T$$
$$\widehat{\Lambda}(\omega) = \widehat{E} M ,$$
$$(19.33)$$

where

$$\widehat{E} = [E_1, E_2, \dots, E_n]^T$$
$$\widehat{E}_i = \mathrm{diag}\,[P_R(\omega_i), \; P_R(\omega_i), \; S_R(\omega_i), \; Q_R(\omega_i),$$
$$Q_R(\omega_i)]$$
$$M = [M_{12}, M_{22} - M_{11}, M_{22} + M_{11}, M_{23}, M_{13}]^T .$$
$$(19.34)$$

Eq. (19.33) is a standard overdetermined problem that can be solved by least squares approaches. No fit to real data is perfect and the spectral misfit can be used as a measure of the uncertainty and used to calculate the uncertainties on the model parameters. The excitation functions in $\widehat{E}_i$ are, of course, dependent on depth, so the inversion must be repeated for several depths. A comparison of the errors for the different depths should result in a *minimum error*, which yields the source depth and thus the preferred moment tensor. To the extent that the source can be approximated as a point source with a boxcar source function, the source spectrum will not affect the long-period surface-wave spectra. But for large events, the effective source duration will have an azimuthal pattern, as can be seen by considering the equation for

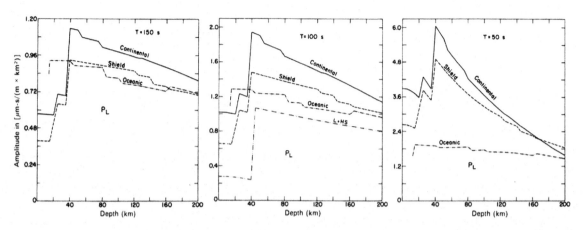

FIGURE 19.5 Dependence of the fundamental Love-mode displacement spectrum on source depth for a vertical strike-slip source. Excitation functions are shown for three different upper-mantle models, representative of shield, continental, and oceanic regions. Variations in the excitation coefficients as a function of period provide information about the source depth. The curve marked L+HS corresponds to a layer over a halfspace solution (from Ben-Menahem and Singh, 1981).

source finiteness (18.25). Directivity effects are more apparent in surface waves than in body waves because their phase velocity is much closer to the typical rupture speed of earthquakes. Source finiteness not only introduces an azimuthal phase dependence, it also reduces the amplitude of the shorter-period components of the signal. In this case, the spectrum for a large event should be corrected for source finiteness. Despite these complexities, surface waves often can better constrain total seismic moment and total rupture duration of larger earthquakes than shorter-period body-waves.

19.3 Global centroid moment-tensor solutions

In 1981 the seismology research group at Harvard headed by Adam Dziewonski began routinely estimating the seismic source parameters of all earthquakes with $M_s > 5.5$ using the centroid moment tensor (CMT) method (Dziewonski et al., 1981). Originally, the analysis simultaneously fit (1) the long period ($T > 40$ s) body wave train from the P-wave arrival until the

onset of the fundamental modes and (2) mantle waves ($T > 135$ s). Since 2004, improvements in Earth models have allowed the inclusion of intermediate period surface waves, which are often the largest arrival at teleseismic distance for shallow earthquakes (Ekström et al., 2012). Since that time, more earthquakes per year have been modeled as the initial modeling threshold, $M_W \sim 5.5$, was reduced by the inclusion of shorter-period surface waves to $M_W \sim 5.0$. Centroid-moment-tensor refers to the fact that the estimates provided are spatial and temporal averages of the earthquake's distribution of seismic moment.

The normal-mode based inversion procedure (with fundamental mode corrections for aspherical earth structure) provides estimates of the best point-source hypocentral parameters (centroid latitude, longitude, and depth, and the centroid time) and the six independent moment-tensor elements (but the inversion includes a constraint that the trace of the moment tensor is small, and for most events in the catalog, it is zero). The triangular time function (originally a box-car was used) has a half-duration set using an empirical relation,

$$\tau_h = 1.05 \times 10^{-8} \, M_0^{1/3} \,, \qquad (19.35)$$

and centered on the centroid time. The centroid spatial coordinates are ideally centered on the centroid of the spatial distribution of seismic moment characterizing the source. The seismogram modeling approach is based on an equation very similar to Eq. (19.15),

$$u_n(\mathbf{x}, t) = \sum_{i=1}^{6} \psi_{in}(\mathbf{x}, \mathbf{x}_s, t) \cdot m_i, \qquad (19.36)$$

where ψ_{in} represents an *excitation kernel* – the complete seismogram Green's function for each of the moment tensor elements. The receiver is at $\mathbf{x}$, and the source is at $\mathbf{x}_s$ (which is unknown). The problem is nonlinear, so a standard local, gradient-driven search for a solution is used to provide optimal source-parameter estimates. The procedure starts with an initial location (e.g. USGS hypocenter) and moment-tensor estimates, m_i, and an iterative procedure that adjusts both the spatial and temporal parameters and the moment-tensor elements

$$u_n(\mathbf{x}, t) - u_n^0(\mathbf{x}, t) = a_n \, \delta r_s + b_n \, \delta \theta_s + c_n \, \delta \phi_s$$

$$+ d_n \, \delta t_s + \sum_{i=1}^{6} \psi_{in}^0(\mathbf{x}, \mathbf{x}_s, t) \cdot \delta m_i,$$

$$(19.37)$$

where u_n^0 and ψ_{in}^0 are based on the initial location and moment-tensor estimate. δr_s, $\delta \theta_s$, and $\delta \phi_s$ are the changes in spatial coordinates of the hypocenter, and a_n, b_n, and c_n are the partial derivatives of the seismogram fit with respect to perturbations in the hypocentral coordinates. The change in fit corresponding to a time shift is d_n and δt_s is the change in the origin time. The kernels are obtained by summing normal modes. The inversion process can be efficiently performed for a large number of seismograms (currently roughly 178 stations are examined as

part of the routine processing). A sample fit of the GCMT solution for the 28 January, 2020 M_W 7.7 earthquake north of Jamaica and west of Cuba (observed in northern Greenland) is shown in Fig. 19.6. The earthquake ruptured a large roughly east-west striking strike-slip fault. The radiation to the north includes strong Love ($G1$) and weak Rayleigh ($R1$). The signals have been filtered to be dominated by a narrow band with periods between 100–200 s. Still the information on the faulting geometry is quite clear as indicated by the large Love waves. More subtle patterns can be extracted when data from different directions are incorporated. The constraints provided by the waveform shape allow the routine (and important) quantification of the average faulting parameters of all moderate- and large earthquakes. Fig. 1.16 shows more than fifty-thousand GCMT solutions. The catalog constructed beginning in early 1980's is one of the most widely used and consulted resources in earthquake seismology.

19.4 Iterative sub-event identification

Faulting heterogeneity can be described using the concept of **subevents**. In other words, for some large earthquakes the source process can be thought of as a series of moderate-sized earthquakes. When the roughness of a rupture is so strong that the source time functions becomes sufficiently complicated to suggest failure of a collection of strong nearly individual subevents, the event is known as a *complex earthquake*. All earthquakes are complex in detail, we must reference fault complexity to the passband of observation.

In our previous discussions of estimating source parameters, we assumed that the rupture front progressed in a smooth and predictable manner. However, we have no *a priori* reasons to expect that a rupture is smooth. We can develop a generalized waveform inversion in which the *temporal* and *spatial* distribution of moment can

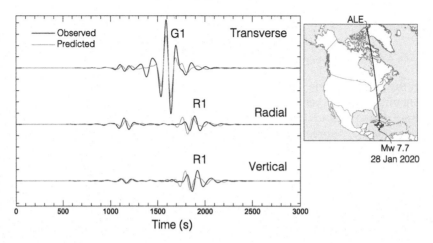

FIGURE 19.6 Left observed and predicted seismograms recorded at seismic station ALE (Alert, Canada, doi.org/10.7914/SN/II) for the 28 January, 2020, M_W 7.7 earthquake along the northern boundary of the Caribbean Plate (between Jamaica and Cuba). The east-west striking strike-slip rupture radiated strong Love waves to the north (in the direction of the station). The predictions were computed using PREM (no phase corrections) and the quick GCMT solution for the event.

be recovered. In the simplest case, a fault can be parameterized as a series of sub-faults with known spatial coordinates but with unknown moment and/or rupture time. Express the least-squares difference between an observed and predicted waveform is

$$\epsilon_2(m_1, t_1) = \int_0^\infty [u(t) - m_1 \cdot w(t - \tau_1)]^2 \, dt ,$$

$$(19.38)$$

where $u(t)$ is the observed seismogram, $w(t)$ is a predicted seismogram calculated for a point source [$w(t)$ is computed with a specified time function, $S(t)$ for a "unit" earthquake of moment m_0]. The sub-event scale factor is m_1 and the sub-event rupture initiation time is τ_1. The least-square misfit, ϵ_2 can be minimized in terms of m_1 and τ_1 to provide and estimate of the rupture time and size of a subevent.

More importantly, we can generalize Eq. (19.38) for the problem including multiple sub-events and observations by successively "stripping away" the contribution of each subevent to the original signal. In this procedure a wavelet is correlated with the observed signal, an optimal time lag and amplitude for the subevent is calculated from the correlation. The contribution of that subevent is subtracted from the seismogram to construct a residual waveform. This residual is correlated with another wavelet, stripped, and so on until the original observed seismogram is adequately reproduced. This problem is usually severely underdetermined, so a "search procedure" is used to find the minima in $\epsilon_2(m_i, t_i)$. The generalized form of (19.38) is given by

$$\epsilon_k = \sum_{j=1}^M \int [x_{jk}(t) - m_k \, w_{jk}(t - \tau_k, f_k)]^2 dt ,$$

$$(19.39)$$

where M is the number of seismic waveforms, x_{jk} is the residual (observed minus current predicted seismogram) at the jth station after $k - 1$ iterations, m_k is the moment chosen for the kth iteration, w_{jk} is the predicted waveform for the jth station from the kth subevent, τ_k is the time of the kth subevent, and f_k are the source parameters for the kth subevent (epicenter, focal mechanism, etc.). This process is repeated for a prescribed number of iterations, and the results from each iteration are combined to produce a model of the overall rupture process.

The pulse-stripping procedure is a useful exercise to identify the time and location of substantial seismic moment. The procedure includes minimal constraints on the rupture process and extracts the largest features first, producing the most robust results early in the analysis. An alternative, and quite common approach is to include a simple rupture model (fault orientation, segmentation, rupture propagation, maximum slip duration, etc.) using a planar model of an earthquake. The inclusion of geometric information provides *a priori* information that can help stabilize the inversion of seismic waveforms (as well as InSAR, GPS, and tsunami information).

19.5 Earthquake finite-fault models

As the preceding discussions have indicated, seismologists use numerous methodologies to extract the details of faulting from seismic waveforms. Since earthquakes involve faulting with a finite spatial and temporal extent, different-frequency waves are sensitive to different characteristics of the rupture process. Although introduced long ago, the construction of kinematic earthquake slip and rupture models has become routine. In this context, kinematic indicates that the models are constructed to allow slip at a particular location at a particular time, but the underlying physics of the slip evolution are not included in the modeling. Broadband source models are required to explain rupture over a frequency range of a few hertz to static offsets. Further, different wave types tend to have different dominant observable frequencies as a result of interference during rupture and propagation. Some inversions have spanned very broad bandwidths, combining coseismic GPS observations, long-period surface waves, and body

waves (e.g. Ammon et al., 2011). The U.S. Geological Survey routinely combines all the available seismic observations in a relatively rapid estimation of the kinematic rupture properties (Hayes, 2017).

Underlying these modeling efforts is the application of superposition of sources of limited spatial extent to represent a larger finite rupture (e.g. Hartzell and Heaton, 1983; Ji et al., 2002, 2004; Ide, 2015). Often a simple planar fault surface with a fixed geometry (strike, dip, and maximum dimensions along the strike and dip directions) is assumed and it is divided into a grid of subfaults (Fig. 19.7). Subfault dimensions are defined after consideration of the spatial resolution afforded by the data and then a relationship between the earthquake slip history and the observations is constructed. Consider a fault model with N subfaults and let $g_{s,i}$ and $g_{d,i}$ represent the response of the ith subfault to a unit slip in the strike and dip direction respectively. Define scale factors for slip in the strike and dip directions in the ith subfault as s_i and d_i respectively. Then the nth waveform can be represented by the sum

$$u_n(\mathbf{x}, t)$$
$$= \sum_{i=1}^{N} s_i \cdot g_{s,i}(\mathbf{x}, \mathbf{x}_i, t - \tau_i) + d_i \cdot g_{d,i}(\mathbf{x}, \mathbf{x}_i, t - \tau_i),$$
$$(19.40)$$

where τ_i is the time that slip initiates in the ith subfault. If the rupture is completely prescribed, then these equations can be transformed into a linear matrix equation

$$\boldsymbol{d} = \boldsymbol{G}\boldsymbol{m} \qquad (19.41)$$

where $\boldsymbol{m}$ is a vector of unknowns, the slip factors s_i and d_i, $\boldsymbol{d}$ is a column vector constructed by concatenating the observed seismograms, and the column of $\boldsymbol{G}$ are the Green's functions corresponding to each seismogram and each subfault (for both dip and strike slip). There is of course

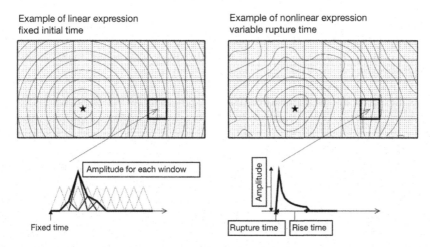

FIGURE 19.7 The underlying principle of finite-fault modeling is superposition of small, effective point sources, to produce the seismic wave radiation and geodetic deformation caused by a large earthquake. Typically, the fault is divided into rectangular subfaults with a fixed geometry but variable slip magnitudes. Two approaches are common, one in which the rupture initiation time is fixed prior to the inversion, the other starts with a specified average rupture pattern but allows the rupture from to deviate from a simple circular expansion. Each point source has a strike-slip and dip-slip component (that can be fixed or allowed to vary spatially and temporally) and a parameterized time function (from Ide, 2015).

nothing special about the strike and dip directions, and any two orthogonal directions of slip on fault surface will work. Often the directions are chosen so that both slip scale factors are positive and the equations are solved with positivity constraints – this prevents models with slip back and forth in the same direction, which is not physically realistic.

An example estimated source time function (computed for the direction of no directivity effects, orthogonal to the rupture) and slip model are shown in Fig. 19.8. The view of the rupture is from the same direction, perpendicular to the fault (in this case, looking downward at the fault). The model is a very simple parametrization that includes a fixed rupture initial time at each subfault, defined as $5\,km \times 5\,km$ squares. The figure includes an interpolated map of slip that shows two first-order asperities, a compact region of high slip near the hypocenter, and a 130 km long region of slip that includes several potential asperities. The rupture isochrons are circles centered on the hypocenter. The ampli-

tude of the moment-rate function at a particular time is proportional to an integral though the model along the corresponding isochron. Since the rupture is mostly unilateral, the temporal and spatial distributions of moment rate and slip can be roughly correlated in the figure (if you account for the finite rise time in each subfault).

The recovered slip distribution depends on the assumed rupture speed; faster ruptures expand the region of slip, slower ruptures compact the region of slip. The data provide some constraints on an optimal rupture speed, but a suite of values often fit comparably well. In contrast, the source time functions for six different speeds are shown and they are demonstrably insensitive to the assumed rupture speed. Seismic wave amplitudes are actually sensitive to the relative moment of each subfault, but most models are presented after converting the moment to slip using ($\Delta u_i = M_i/A_i\mu_i$) where Δu_i is the fault slip across the ith subfault, M_i is the subfault moment, A_i is the subfault area, and μ is the shear modulus in the model at the location of the

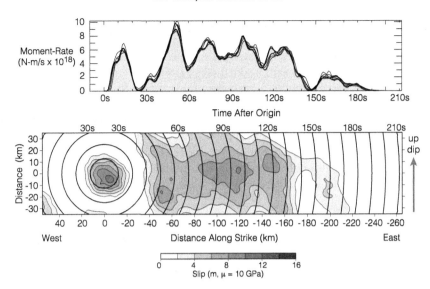

FIGURE 19.8 (Top) Moment rate functions estimated from six finite fault inversions with rupture speeds of 1.00, 1.25, and 1.50 km/s and with and without SH waveforms. The moment-rate function is perhaps the best-constrained quantity from this type of inversion. The thicker line and shading highlight the moment-rate function from the model above. (Bottom) Finite-fault inversion results for an assumed rupture speed of 1.25 km/s. The peak moment in the 5 km x 5 km cells is 0.34×10^{20} N m and the total moment is 7.0×10^{20} N m, (M_W = 7.8). The hypocenter location is (0, 0), the top of the fault is one km beneath the trench. Rupture front isochrons are shown at 10 s intervals (top axis); the left and bottom axes show distance from the focus in km. Slip is scaled from moment using $\mu = 10$ GPa – this value is lower than realistic for deeper events (not far off for tsunami earthquakes like this), it is used because it is easy to scale up to other values (modified from Ammon et al., 2006).

subfault. This earthquake was near the trench in regions of low shear modulus, but the very low value of 10 GPa is used more for convenience in scaling the values (for example, if $\mu \sim 20$ GPa double the slip values).

If the rupture initiation time is allowed to vary, the problem becomes nonlinear and we can use a standard linearized inversion (as described for earthquake location in Section 6.4) with the added initiation start times as an additional set of unknowns (the partial derivatives are easy to compute). Since the forward computations are straightforward sums, efficient algorithms can be developed to compute the predictions and nonlinear search-based inversion have proven to be effective in optimizing the fit to the observations. For large events, the slip in each subfault should be parameterized using a sequence

of basis functions (Fig. 19.3 and Fig. 19.7) to allow flexible recovery of the variable slip history (Ide, 2015). If a total of K time-function components is assumed for each subfault, then the number of additional unknowns is $K \times N \times 2$. The matrix dimensions can grow quickly, but usually the numerical inversions are relatively easy to handle. For the most complicated events, we can construct a kinematic model that includes multiple faults with changes in orientation. Such approaches are often necessary to model ruptures that propagate across several fault segments that may include a change in strike, or events that often begin with a subfault oriented different than the main rupture.

Finite-fault inversions can be performed using body-waves, surface-waves, tsunami waveforms (Box 19.2), relative source time-functions,

Box 19.2 Modeling tsunami waveforms for earthquake source parameters

In Chapter 9 we discussed tsunami, which are generated by a relatively sudden displacement of the ocean floor, often caused by an earthquake. Just as a seismic wave is a combination of source and propagation effects, recordings of a tsunami are sensitive to the source (slip distribution on a fault) and propagation (e.g. ocean bathymetry along the travel path) effects. Thus it is possible to invert tsunami observations (ocean height as a function of time) for fault slip (e.g. Satake (1989); Satake and Kanamori (1991)). Propagation effects are modeled using bathymetry information (and the properties of water) which are reasonably well known. Fig. B19.2.1 is a map of the slip distribution estimated for the 2011 Tohoku earthquake (Satake et al., 2013) and corresponding fits to a subset of tsunami waveforms used in the inversion.

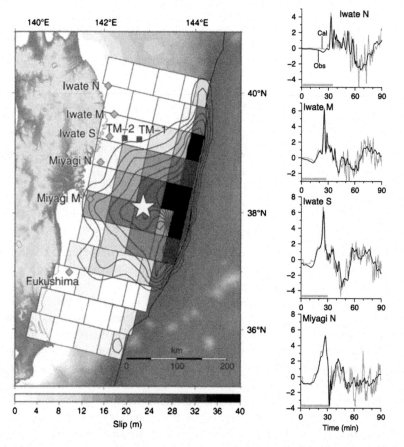

FIGURE B19.2.1 (left) Slip models (contours and shaded sub-faults) indicating the spatial distribution of slip of the M_W 9.0 2011 Tohoku Earthquake. Slip indicated in sub-faults was derived partly using tsunami observations; contours are the result of a separate strong-motion analysis. Contour interval is $4\,m$ and the white star indicates the epicenter. (right) Subset of fits to the tsunami signals at GPS wave gauges in shallow water along the Honshu coast (observation locations are shown on left) (modified from Satake et al., 2013).

The inversion of tsunami data is very useful for earthquakes that rupture near the trench, where excitation of the seismic waves is less sensitive. In addition, constraints on historic earthquakes benefit greatly from the tsunami information (e.g. Tanioka and Sataka 1996).

geodetic observations, or all of the above. The results from simultaneous inversions are generally better constrained than those based on individual data sets, but all the data have resolution issues, so some uncertainty always remains in the result. In fact, estimating and communicating accurate and meaningful uncertainty is one of the remaining challenges in the analysis of seismic sources. Accurate uncertainty computation in geophysics, where we must rely on band-limited data and relatively simple descriptive models of complex processes such earthquakes (and Earth's interior) is a difficult problem. The uncertainty includes not only a component from noise in the observations, but also a more difficult-to-quantify uncertainty associated with the calculation of model predictions,

the inclusion of a prior information, which is reasonable, but not necessarily easy to quantify. Seismologists often use the misfit to develop a representative estimate of the noise but also rely on experience and sensitivity analyses in which any major assumptions are modified, the analysis is repeated, and the differences in the results noted. Although not ideal, this is the standard practice that has led to significant advances in our understanding of earthquakes and other seismic sources.

In the next chapter, we explore how seismologists use the information in seismograms to illuminate Earth's interior and to construct models of the Earth appropriate at local, regional, and global scales.

Chapter goals

- Introduce classic seismological methods for imaging Earth's interior with a primary goal of highlighting the similarities of many seismic analysis approaches.
- Introduce seismic tomography for body-waves, surface-waves, and amplitude attenuation.
- Introduce seismological applications of linearized geophysical inversion to estimate earth structure.
- Introduce full-waveform modeling of seismograms to estimate earth structure.

Our knowledge of the Earth's seismic velocity structure is the result of *interpreting* seismograms. Many of the characteristics discovered over the last century or so were reviewed in Chapter 10. The more fully we quantify all of the ground motions in a seismogram, the more fully we understand the Earth's structure and its dynamic processes. Seismograms are a complicated mixture of source effects, such as radiation patterns and time history of energy release generated at the source, propagation phenomena such as multiple arrivals produced by reflection and transmission at seismic impedance boundaries or at the surface, and frequency band limitations of the recording instrument. Experience and a sound foundation in elastic-wave theory guide seismologists as they sort out coherent vibrations produced by reflections off deep layers

from background noise or from other arrivals scattered by the Earth's three-dimensional heterogeneity.

This chapter describes how we use seismograms to investigate large-scale Earth structure. Only a fraction of the methodologies that have been developed can be discussed, and new procedures are continually introduced in the scientific literature. Our approach is to present a spectrum of methods and the results of their application to various depth ranges and spatial scales in the Earth. Numerous texts are devoted to the techniques and interpretations of shallow crustal reflection seismology (where great effort is invested in imaging the shallow subsurface), so we do not discuss that field. Our discussions describe some basic methodologies common to all applications. Our goal is only an introduction for the student – more advanced texts and articles contain the details necessary for implementing these approaches.

20.1 Earth structure estimation using travel times

Earlier chapters in this text have provided a general characterization of seismic wave behavior in media that have smooth velocity variations or abrupt discontinuities in material properties. Mathematical expressions predict how the seismic wave amplitudes and travel times

537

should vary as a function of distance from the source. Comparison of observed travel-times or seismic waveforms with calculations for a suite of representations of the medium may yield a range of model parameter sets that satisfactorily match the observations. Exploration of potential model parameter sets is called *forward modeling*, a process in which we iteratively perturb model parameters in an effort to predict the observed behavior more accurately. For simple models, with only a few parameters, it is not unreasonable to perform forward modeling to get a "best" model, but even a realistic one-dimensional parameterization of Earth may include many parameters. The more complex the model and the model-data relationship, the more potential we have for model-parameter trade-off. To extract useful information from observed seismograms, we must define what we mean by an optimal model including criteria by which we judge the fit to the data, and model characteristics such as simplicity.

Although enormous increases in computer speed in recent years enable some "brute force" forward modeling optimization of model parameters by searching over vast suites of models, most current seismological methods employ a different strategy, involving geophysical inverse theory. Inverse problems were introduced in Chapter 6 in the context of earthquake location. Clearly, the location depends on the velocity structure, which can at best be an approximation of the actual Earth structure. How, then, can we determine the structure, at least well enough to enable a bootstrapping procedure of iteratively improving both the velocity model and the source locations? The key is to exploit the systematic relationship between seismic wave behavior as a function of distance and the velocity structure encountered along the path. We begin by considering a classic inverse procedure useful for determining a one-dimensional model of velocity variation as a function of depth.

20.1.1 Herglotz–Wiechert inversion

Consider the travel time–epicentral distance behavior for a spherical medium with smoothly varying velocity that is a function only of depth, $v(r)$. In Chapter 12 we developed parametric equations relating travel time and distance in a spherically symmetric earth model,

$$T(p, \xi) = p\,\Delta + 2 \int_{r_t}^{r_0} \frac{\sqrt{\xi^2 - p^2}}{r}\, dr \ ,$$

$$\Delta(p, \xi) = 2\,p \int_{r_t}^{r_0} \frac{dr}{r\sqrt{\xi^2 - p^2}} \ , \qquad (20.1)$$

where $\xi = r/v(r)$, r_t is the radius to the turning point of the ray, $p = r\sin i/v(r)$, and r_0 is the radius of the sphere. Our goal is to use the observed travel-time curve, $T(\Delta)$, to determine the velocity variation with radius. If the $T(\Delta)$ curve is well sampled, we can construct a smooth curve fit to the data and use that to determine $p(\Delta) = dT(\Delta)/d\Delta$, the instantaneous slope of the curve. In practice the data never form a perfectly smooth curve because of both measurement error and three-dimensional heterogeneity. The criteria used in fitting a curve to individual measurements $T_j(\Delta_i)$ affect the $p(\Delta)$ curve and ultimately the model values of $v(r)$, but the results are generally a useful approximation.

To use the determined $T(\Delta)$ and $p(\Delta)$ curves, we change the variables of integration in the Δ-integral,

$$\Delta = 2\,p \int_p^{\xi_0} \frac{1}{r\sqrt{\xi^2 - p^2}} \left(\frac{dr}{d\xi}\right) d\xi \ , \qquad (20.2)$$

where $\xi_t = r_t/v(r_t) = p$ and $\xi_0 = r_0/v_0$. Then apply the integral operator

$$\int_{p=p_1}^{p=\xi_0} \frac{dp}{\sqrt{p^2 - \xi_1^2}} \qquad (20.3)$$

to (20.2), where $\xi_1(r) = r_1/v(r_1) = p_1$ is the ray parameter for a ray bottoming at radius r_1. Thus

(20.3) is an integration over all rays from the ray at zero range ($p = \xi_0$, $\Delta = 0$) to a ray with turning point r_1 and ray parameter $\xi_1 = r_1/v(r_1)$:

$$\int_{\xi_1}^{\xi_0} \frac{\Delta \, dp}{\sqrt{p^2 - \xi_1^2}}$$

$$= \int_{\xi_1}^{\xi_0} dp \int_p^{\xi_0} \frac{2p}{r\sqrt{\xi^2 - p^2}\sqrt{p^2 - \xi_1^2}} \left(\frac{dr}{d\xi}\right) d\xi.$$
(20.4)

The left-hand side of (20.4) can be integrated by parts to give

$$\Delta \cdot \cosh^{-1}\left(\frac{p}{\xi_1}\right)\Big|_{p=\xi_1}^{p=\xi_0} - \int_{\xi_1}^{\xi_0} \frac{d\Delta}{dp} \cosh^{-1}\left(\frac{p}{\xi_1}\right) dp.$$
(20.5)

The first term vanishes because $\cosh^{-1}(1) = 0$, and $\Delta = 0$ at $p = \xi_0$ by definition. The second term in (20.5) can be written as an integral over Δ by

$$\int_0^{\Delta_1} \cosh^{-1}\left(\frac{p}{\xi_1}\right) d\Delta.$$
(20.6)

Turning now to the double integral in (20.4), we change the order of integration to obtain

$$\int_{\xi_1}^{\xi_0} \frac{d\xi}{r} \left(\frac{dr}{d\xi}\right) \int_{\xi_1}^{\xi} \frac{2p \, dp}{\sqrt{p^2 - \xi_1^2}\sqrt{\xi^2 - p^2}}.$$
(20.7)

The integral on p has a closed form, reducing to

$$\int_{\xi_1}^{\xi_0} \left(\frac{d\xi}{r}\right) \left(\frac{dr}{d\xi}\right)$$

$$\times \left\{ \sin^{-1}\left[\frac{2p^2 - (\xi^2 + \xi_1^2)}{\xi^2 - \xi_1^2}\right] \right\}_{p=\xi_1}^{p=\xi},$$
(20.8)

which reduces to

$$\pi \int_{\xi_1}^{\xi_0} \left(\frac{d\xi}{r}\right) \left(\frac{dr}{d\xi}\right) = \pi \int_{r_1}^{r_0} \frac{dr}{r}.$$
(20.9)

Integrating (20.9) and combining with (20.6), our Eq. (20.4) becomes

$$\ln\left(\frac{r_0}{r_1}\right) = \frac{1}{\pi} \int_0^{\Delta_1} \cosh^{-1}\left(\frac{p}{\xi_1}\right) d\Delta.$$
(20.10)

This is an expression for r_1 in terms of quantities measured from the $T(\Delta)$ plot. The quantity $\xi_1 = p_1 = (dT/d\Delta)_1$ is the slope of $T(\Delta)$ at distance Δ_1. The integral can be numerically evaluated with discrete values of $p(\Delta)$ for all Δ from 0 to Δ_1, to yield a value for r_1, $v(r_1)$ is obtained from $\xi_1 = r_1/v_1$.

Using the Herglotz–Wiechert formula we can "invert" a travel-time curve to construct a model of velocity as a function of depth. The approach has limitations, but it was used extensively in the development of the earliest P-wave and S-wave velocity models for deep Earth structure. The procedure is stable as long as $\Delta(p)$ is continuous, with $v(r)/r$ decreasing with r. Fig. 20.1 is an example of the procedure using a very limited data set. The example was chosen to show the power of travel time curves in imaging Earth's (or planetary) interiors for first order, approximate structure. Slight changes in the travel time curvature with distance provide valuable constraints on the speed as a function of radius. These results, from an introductory seismology laboratory exercise, benefit from the excellent data quality and event origin-time and location information that are available for modern earthquakes. If a low-velocity zone is present at depth, the formula cannot be used directly, although it is possible to "strip off" layers above the low-velocity zone and then use the contracted travel-time curve to construct smoothly increasing velocities at greater depth.

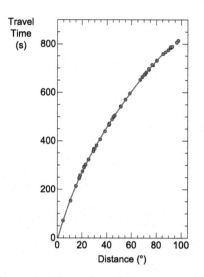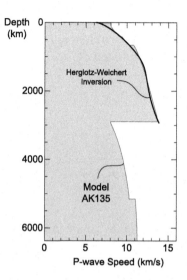

FIGURE 20.1 (left) Observed travel times from one seismic event (measured in an introductory computer laboratory exercise). The travel times were fit with a fourth-order polynomial in distance and that polynomial was used in a Herglotz-Wiechert inversion. (right) The resulting model is compared with reference Model AK135 (Kennett, 2005). The agreement is quite good – although the AK135 model includes important details on relatively sharp changes in the average mantle wave speed and has much lower uncertainties as a result of the large number of observations used in its construction.

Thus, one could build an Earth model for the mantle, then strip this off before determining the velocity structure of the low-velocity core.

20.1.2 Seismic traveltime tomography

To a fairly high level of approximation, the Earth can be viewed as a spherically layered planet, with primary chemical stratification between the crust, mantle, and core. A one-dimensional model such as PREM (Fig. 1.21) accurately predicts travel times for waves spreading throughout the planet with an error of less than 1% for teleseismic P waves. Thus, seismologists have exerted great effort to develop and refine a one-dimensional model for use in routine earthquake location and bulk geochemical/geological interpretations. The *iasp91* and *AK135* models (AK135 is used in Fig. 20.1) represent more recent attempts to develop radially-symmetric reference models with more empha-

sis on fitting body-wave travel times, which are used in earthquake location.

However, lateral heterogeneity exists throughout the Earth and is for example, directly indicated by the coda of seismic waves and by the scatter in seismic wave arrival times at all distances (Figs. 5.3 and 6.1). While the lateral variations below the crust are typically less than 10% for shear velocity (the most variable of elastic parameters), these fluctuations have great significance as markers of dynamic processes in the Earth's interior. For more than 50 years seismologists have been mapping gross velocity differences near the surface, associated with variations between continental and oceanic crust and upper mantle. More recently, but still spanning decades, a concerted effort has been made to image the three-dimensional structure everywhere inside Earth. The process evolved from localized one-dimensional characterizations of structure beneath specific regions to complete three-dimensional modeling using a method called **seismic tomography**. The central idea of tomographic imaging is that we can extract information from a large number of signals generated by waves that have traveled distinct but overlapping paths through the planet if we simultaneously include all the observations in a single modeling effort.

Seismic tomography was introduced in the mid-1960's in an earlier form called *regionalization*, in which surface waves traversing mixed oceanic and continental paths were analyzed to determine separate structures beneath each region. Regionalization involved calculating a travel-time anomaly relative to a reference symmetric Earth model for a surface-wave phase with a particular period and then partitioning the anomaly with respect to the percentage of path length in a specified tectonic regionalization of the surface. Regionalization was necessary because only a few paths occur for which the source, receiver, and entire path length are within a distinct tectonic region such as ocean basins less than 20 million years old. From these early beginnings seismology moved toward a smaller and smaller subdivision of the media, forgoing any regionalization based on surface geology and including both local and global two- and three-dimensional parameterizations of the media for which seismic-wave velocity or *slowness* (reciprocal velocity) perturbations would be estimated in each region.

The principle of travel-time tomography is that the particular seismic phase has a travel time, t_{obs}, is a path integral through the medium. In Chapter 6 we related the observed arrival time, t_{obs}, to the slowness (and event location, x_0, and origin time, t_0) using (Eq. (6.3))

$$t_{obs} = \underbrace{t_0 + \int_{\hat{\ell}} \hat{s}\,(\ell)\,d\ell}_{\text{Predicted Time}} + \underbrace{\int_{\hat{\ell}} \delta s(\ell)\,d\ell}_{\text{Slowness Correction}} + \underbrace{\delta t_0 + s_0\,\delta x_0}_{\text{Location Correction}}$$

$$+ \underbrace{\epsilon_{obs}}_{\text{Pick Error}} .$$

In that instance, our focus was on estimating the location corrections (δt_0 and δx_0) assuming that we had an adequate slowness model, $\hat{s}(x)$, related to the true slowness, $s(x)$, by $s(x) = \hat{s}(x) + \delta s(x)$. Here, we assume that the locations are accurate (so the location corrections are negligible), then

$$t_{obs} = t_0 + \int_{\hat{\ell}} \hat{s}\,(\ell)\,d\ell + \int_{\hat{\ell}} \delta s(\ell)\,d\ell + \epsilon_{obs} . \quad (20.11)$$

We do not have to assume perfect locations – the problems are often addressed together, but we do so here to simplify the problem. Still, the problem remains difficult because we also do not know the correct path along which to perform the integrations (it depends on the slowness). The expressions we have written assume that the reference model is sufficient enough to use the reference-model path to related the observed time and the slowness correction. This is a linear approximation to a nonlinear problem that can be iterated to provide an estimate of the slowness. Even today, most analyses assume that the paths through a one-dimensional reference model such as PREM are adequate to integrate to a reasonably reconstruction of the subsurface variations. Then,

$$t_{obs} - t_0 - \int_{\hat{\ell}} \hat{s}\,(\ell)\,d\ell = \int_{\hat{\ell}} \delta s(\ell)\,d\ell + \epsilon_{obs} ,$$

$$(20.12)$$

or, representing the predicted time at t_{pred},

$$\Delta T = t_{obs} - t_{pred} = \int_{\hat{\ell}} \delta s(\ell)\,d\ell + \epsilon_{obs} . \quad (20.13)$$

Now, we discretize the integral over the slowness correction. Almost all seismic tomography methods involve subdividing the slowness perturbation field into uniform blocks, cells, or functional spatial parameterizations such as spherical harmonics. The idea is that the path integral through the medium perturbations should equal the observed travel-time residual.

For example, if the medium is subdivided into uniform regions (such as cells), one can calculate the path length l_j in the jth block and discretize Eq. (20.13),

$$\Delta T = \sum_j l_j \, \Delta s_j. \qquad (20.14)$$

For a data set that includes observations from multiple event–station pairs, each with the ith raypath, we develop a system of i equations

$$\Delta T_i = \sum_j l_{ij} \, \Delta s_j . \qquad (20.15)$$

Eq. (20.15) is set of linear equations that relate the travel-time misfit to a slowness correction that can be solved to use the current travel-time misfit to construct slowness corrections that reduce the travel time misfit. In linear-algebraic form,

$$\mathbf{\Delta T} = \mathbf{L} \, \mathbf{\Delta s} , \qquad (20.16)$$

where $\mathbf{\Delta T}$ is a vector of travel time residuals (observed – predicted times), $\mathbf{L}$ is a matrix for which the ith row contains the ray-path segments for each cell of the ith ray path (most are zero since each ray samples only part of the model), and $\mathbf{\Delta s}$ is a vector for which the jth element is the slowness correction to be applied to the jth cell to reduce the travel time misfit. Note that if we make the approximation that the waves actually do follow the reference-model paths, then we can relate the travel times directly to the slowness values.

$$T_i = t_{obs} - t_0 = \sum_j l_{ij} \, s_j . \qquad (20.17)$$

This equation can be written as

$$\mathbf{T} = \mathbf{L} \, \mathbf{s} \qquad (20.18)$$

The difference now is that if we choose to apply constraints (such as smoothness) to our model, they apply directly to our slownesses, not slowness-corrections.

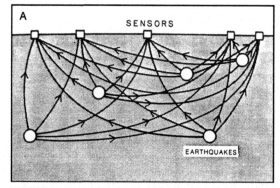

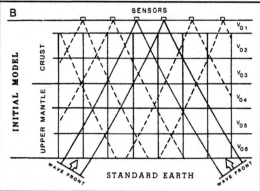

FIGURE 20.2 (A) Raypath coverage utilized in local tomography, where crisscrossing paths from many nearby earthquakes to seismic stations are used in joint inversions to determine both source locations and crustal structure. (B) Tomography that uses teleseismic signals recorded at a densely spaced seismic array. In this case the differential times between arrivals are mapped into velocity perturbations of three-dimensional blocks below the array (modified from Iyer, 1989).

When we solve the Eqs. (20.16) or (20.18) we use the information contained in crossing raypaths to reveal two- or three-dimensional variations in the medium. We usually have many more raypaths than model parameters, yielding an overdetermined system, but noise in the data and inadequacy of the model parameterization are sure to make the system require stabilization. We often stabilize the equations by appending equations that require a solution with minimum length or minimum roughness (smooth models),

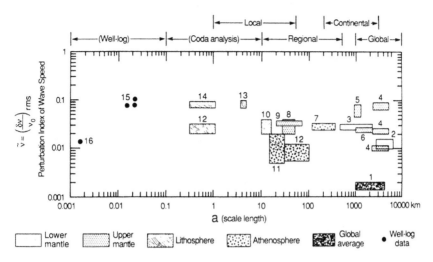

FIGURE 20.3 Summary of many investigations of velocity heterogeneity inside the Earth. The parameter a is the scale length, and $\bar{V} = (\delta V/V_0)$ rms is the velocity perturbation index of the heterogeneity. (1) Global average, from the analysis of mode splitting of free oscillations. (2) Lower mantle, from body-wave inversion. (3) Lower mantle, from body-wave tomography. (4) Upper mantle, from surface-wave waveform inversion. (5) Upper mantle (Pacific plate), from surface-wave full-wave inversion. (6) Upper mantle (United States), from travel-time inversion. (7) Asthenosphere (central United States, 125–225 km deep), from travel-time inversion. (8) Upper mantle (southern California), from body-wave tomography. (9) Upper mantle, a summary, from travel-time inversion. (10) Lithosphere, from transmission fluctuations at LASA. (11) Lithosphere, from transmission fluctuation at NORSAR. (12) Lithosphere, from transmission fluctuation at NORSAR. (13) Lithosphere, from coda wave analysis. (14) Lithosphere, from coda wave analysis. (15) Crust, from acoustic well log. (16) Crust, from acoustic well log (from Wu, 1989).

then use a least-squares approach based on or equivalent to the normal equations (Chapter 6),

$$\Delta s = \left[L^T L \right]^{-1} L^T \Delta T . \qquad (20.19)$$

Examples of seismic tomography analyses are shown in Figs. 1.22 and 1.24 and numerous figures in Chapter 10. Fig. 20.2A corresponds to the common application of local earthquakes recorded by an array of surface sensors. While it is possible to locate the events with a one-dimensional velocity model, the crossing ray coverage allows us to estimate variations in shallow crustal structure as well. Sometimes this involves holding the earthquake locations fixed, solving for the three-dimensional structure, and then iterating on both locations and structure, or the problem can be formulated for a simulta-

neous solution of velocity structure and source locations.

From seismic tomography applications, it has become well established that the Earth is heterogeneous at all depths at all scale lengths and that we can never achieve a complete deterministic understanding of the full internal spectrum of heterogeneity. However, great progress has already been made toward assessing the strength of heterogeneities in different regions (Fig. 20.3). Perturbations in seismic velocity from 1% to 10% appear to exist throughout the mantle and crust, with smaller perturbations possibly existing in the core. Seismic waves of different ranges and wavelengths detect this heterogeneity, and any given data set will be able to resolve only a limited portion of the length spectrum. Fortunately, the mantle heterogeneity spectrum appears to be "red," meaning that the longer-wavelength features have more variations. This is a favorable situation for seismic tomography, as much of the important internal structure can be imaged using models with large-scale parameterizations. The heterogeneity spectrum in the lithosphere, and possibly near the core–mantle boundary, appears to be "white," with a more uniform degree of heterogeneity occurring at all spatial scales. This greatly complicates attaining detailed seismic images at shallow depth. At some level there is simply not enough wavefield information to resolve the small-scale heterogeneity in detail, and *Statistical tomography* techniques are used to characterize parameters of a random medium representation of the interior. Both statistical and deterministic images of the interior are our main means for studying the dynamic processes presently occurring deep in the Earth, so there are many ongoing efforts to extend and improve both methods.

20.1.3 Amplitude attenuation tomography

Seismic-wave travel times are not the only observations that can be modeled with tomographic approaches. Intrinsic and scattering of seismic wave amplitudes can be modeled as a process that occurs in a manner distributed along the wave's path.

As described in Chapters 2 and 13, the Earth does not transmit seismic waves with perfect elasticity; small anelastic losses occur that progressively attenuate the wave energy. This anelasticity causes dispersion, changes pulse shapes, and affects amplitudes of the waves; therefore it can be modeled as well.

Assume a simple parameterization of the form

$$A_{obs}(f) = A_0(f) e^{-\pi f t^*(f)}, \qquad (20.20)$$

where A_{obs} is the observed amplitude at the seismic station, A_0 is the amplitude of the wave leaving the source and corrected for geometric

spreading, reflections, etc. The parameter *t-star*, t^*, in an integral over the wave's path of the arclength divided by the inverse of the product of the quality factor, $Q(f)$ and the wave speed, $c(\ell)$,

$$t^*(f) = \int_\ell \frac{d\ell}{c(\ell)\, Q(\ell, f)}, \qquad (20.21)$$

where ℓ represents the path. Taking the logarithm of both sides of Eq. (20.20) and using Eq. (20.21),

$$\ln A_{obs}(f) - \ln A_0(f)$$
$$= -\pi f t^*(f) = -\pi f \int_\ell \frac{d\ell}{c(\ell)\, Q(\ell, f)}. \qquad (20.22)$$

This equation has the same general form as a travel-time integral written in terms of slowness (e.g. Eq. (20.13)) and for a specific frequency (or range of frequencies) can be discretized and inverted for the inverse of the product of speed and Q. Similar as we described for travel times, a collection of narrow-band amplitudes observed for multiple events and multiple stations can be combined and inverted to estimate the attenuation parameters in a set of uniform regions (or parameterized model parameters such as spherical harmonic coefficients). For a straightforward implementation to estimate inverse-Q, much must be known about each source (the location and the factors included in $A_0(f)$) and a reasonably good velocity model is needed (for computation of the path as well as for providing the value of $c(\ell)$).

Seismic attenuation is caused by either *intrinsic anelasticity*, associated with small-scale crystal dislocations, friction, and movement of interstitial fluids, or *scattering attenuation*, an elastic process of redistributing wave energy by reflection, refraction, and conversion at irregularities in the medium. The latter process is not true anelasticity but has virtually indistinguishable effects that are not accounted for by simple Earth models. At frequencies with wavelengths much

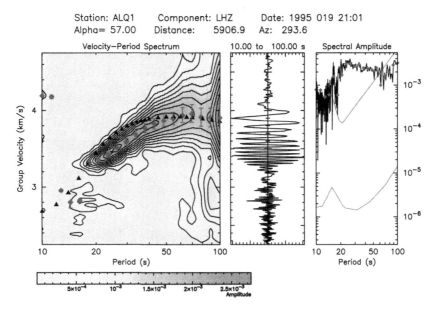

Station: ALQ1   Component: LHZ   Date: 1995 019 21:01
Alpha= 57.00   Distance: 5906.9 Az: 293.6

FIGURE 20.4 Surface wave dispersion measurements made using a multiple filter analysis. (Left) Contour diagram of the seismogram envelope showing the estimated group velocities (diamonds) and a reference curve corresponding to PREM (triangles). (Middle) The seismogram is displayed along the group-velocity axis. (Right) The amplitude spectrum and the sampled points (diamonds) are shown along with the USGS low- and high-noise reference curves. The shorter periods have low signal (review the spectrum), which makes the measurements for group velocity difficult. A typical method for dealing with this is to exclude group velocity measurements periods below 20 s in this example.

larger than the heterogeneities in the medium, intrinsic attenuation dominates. Unlike seismic velocities, the Earth does not have a simple layered attenuation structure. Instead, lateral variations in attenuation quality factor Q can involve many orders of magnitude at a given depth. In general, the upper mantle has lower Q values (higher attenuation) than the deep mantle, but there are paths through the upper mantle with little attenuation and others with strong attenuation. Thus, it has long been apparent that a three-dimensional model of Q is needed.

20.1.4 Surface-wave dispersion tomography

Surface-waves were described in Chapter 14 and perhaps the primary distinguishing trait of surface-wave signals is dispersion. Different fre-

quencies travel at different speeds, so as the wave travels, its energy is spread across larger distance and time ranges. The group delay is the time at which a particular frequency arrives at the station. The phase of the wave when it arrives can be calculated from the phase-delay. Both the group and phase delays can be written in terms of an integral along the wave's path.

For example, let the inverse of group velocity be represented by the group slowness, $s_g(\mathbf{x}, f)$. For a particular source-receiver pair, the group delay, $T_g(f)$ can be measured using the methods described in Chapter 14. The most common approach for group delay estimation is a multiple filter analysis in which the arrival of energy is estimated using a suite of narrow-band filters to isolate the arrival time of the energy in the narrow band (Fig. 20.4). The group delay accumulates as the wave travels across the surface, in

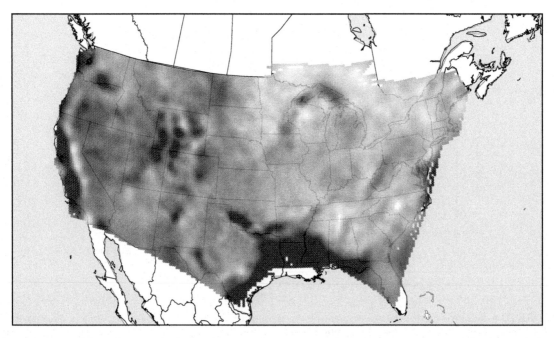

FIGURE 20.5 Variation in Rayleigh-wave phase velocity estimated using observations of waves traveling across the conterminous United States. The values shown correspond to a period of eight seconds. Plot clipping values of ±10% of the deviation from the mean speed (3.15 km/s) have been applied. The darker the shading the lower the group speed. Only regions with sufficient path coverage are included, leaving some gaps along the Gulf Coast, for example. The model values are a courtesy of Robert B. Herrmann, personal communication.

a manner analogous to body-wave travel time,

$$T_g(f) = \int_{\hat{\ell}} \delta s_g(\ell, f)\, d\ell + \epsilon_{obs} \; . \qquad (20.23)$$

Again, this is the same form we had for the body-wave travel times discussed above. So the equation can be discretized to form a set of linear equations for the group slowness. In practice, the paths are often assumed to be the great-circle arcs connecting the sources and stations. Provided enough observations from multiple sources and stations can be acquired, we can use the criss-crossing raypaths to localize spatially the variations in group velocity. An inversion must be performed for each frequency (or period), and smooth dispersion curves for each localized region can be enforced by inverting all the periods simultaneously with smooth-ness constraints between individual period images. An analogous expression for surface-wave phase velocities can be constructed, although the measurement of the phase delay must be performed differently and should include a correction for the initial seismic source phase that can affect the estimated speed.

The result of a surface-wave tomography is a set of spatially localized dispersion curves defined as function of location. Fig. 20.5 is an example tomographic inversion of eight-second period Rayleigh wave phase velocities for waves traversing the conterminous United States. The region has dense seismographic coverage and numerous high-resolution tomographic images have been produced (see examples in Chapter 10). The model shown is from Robert B. Herrmann (personal communication) and was constructed using earthquake and ambient noise measurements. Although values from only one period are shown in the figure, each circle corresponds to a phase-velocity curve (and each also has curves for Love and Rayleigh, group and phase speeds). Many of the known basins (California's central valley, the Gulf coast / Mississippi, the Mid-continent Rift, intermountain basins of the central Rocky Mts., etc.) across the region are illuminated as regions of relatively slow Rayleigh-wave speed. A dispersion model has some value when simply used to correct surface-wave phase in synthetic seismograms for laterally homogeneous models. Such a correction is employed in the GCMT analyses. However, most of the time, the tomography is followed by a discrete geophysical inversion of the dispersion curves to estimate shear-wave speeds (and occasionally density and P-wave speed as well). We discuss these approaches next.

20.2 Discrete geophysical inversion

Tomographic approaches are actually a special case of a broad class of problems often referred to as discrete geophysical inversion. When the ray paths are known, the relationship between the travel times and the slowness is linear. When the raypaths are considered unknown, the problem is nonlinear. Nonlinear problems introduce a number of challenges and a comprehensive discussion is beyond our scope (see, for example, Tarantola, 2005; Menke, 2018; Aster et al., 2013). We'll consider relatively well-behaved nonlinear problems that can be approached using an iterative solution of a sequence of linearized inversions for model-parameter corrections to an initial and sequentially updated models. This restriction is not too limiting, most models of earth structure are constructed as solutions of simultaneous equations relating perturbations of the model to the misfit

between observations and predictions. The approach was described within our discussion of earthquake location in Chapter 6. Problems and approaches including a finite number of model parameters are called *discrete inversions*. These approaches almost always involve solution of a system of equations of the form

$$\mathbf{d} = \mathbf{G}\,\mathbf{m}\,, \qquad (20.24)$$

where $\mathbf{d}$ is a vector of observations or differences between observations and model predictions, $\mathbf{G}$ is a matrix of partial derivatives, $G_{ij} = \partial d_i / \partial m_j$, and $\mathbf{m}$ is the vector of model parameters. In other words a column of $\mathbf{G}$ is a list of the sensitivity of each observation to an individual model parameter; a row of $\mathbf{G}$ corresponds to a list of the sensitivity of an individual observation to each model parameter. The partial derivatives are usually computed using perturbation theory (analytic formulas) or numerically (such as with a finite-difference formulation). The finite-difference approach is computed with repeated forward calculations and can be computationally challenging for large, three-dimensional model. Some full-waveform approaches rely on characteristics of the equations of motion and its adjoint differential equations to derived sensitivity of the data to the model parameters.

If we have n observations and m model unknowns, the solution is formally *overdetermined* if $n > m$ and $\mathbf{G}$ has rank m, formally *underdetermined* if $n < m$, and *exactly determined* if $m = n$. In practice, most seismological inversions are *mixed-determined*, the data may outnumber the model parameters, but the data are still not sufficient to tightly constrain each model parameter. At some level, all solutions and models are intrinsically non-unique (more than one significantly different model explains the data equally well). Although beyond our scope here, geophysicists (and most scientific disciplines) have developed powerful and widely-used tools for the analysis and solution of such inverse problems. Since observations have noise, the data

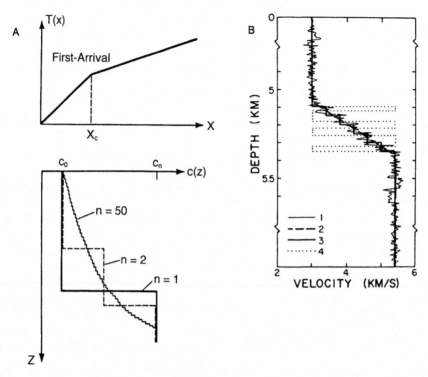

FIGURE 20.6 One of the primary challenges in Earth structure studies is determining the uniqueness of the model. (A) First-arrival travel-time curve that is exactly matched by the three velocity profiles shown below. (B) Four velocity profiles that are indistinguishable when examined using only 1-km-wavelength seismic waves (Part A is modified from Aki and Richards (2009); part B is modified from Spudich and Orcutt (1980)).

may not all be consistent, and one must define the criteria for matching them, such as choosing a least-squares fit. In some problems, the inversion can be cast in terms of continuous model functions $m(r)$ constrained by discrete, finite data sets that are often *noisy*. Inverse approaches provide powerful tools for exploring the resolution and uncertainty of inverse problems. For simple linear problems, the approaches are relatively clear and straightforward. Most geophysical problems of interest include nonlinear model dependences and often are based on an approximate solution to the forward problem (calculating the predictions). As described in the previous chapter on source-parameter estimation, uncertainty is a substantial challenge and requires time, effort, and computation.

The choice of inversion method is largely predicated on the nature of the data that are available and the extent of *a priori* constraints on the model. First, let us consider the nature of the data. The most reliable travel-time observations generally correspond to first arrivals because later arrivals overlap with multiple reverberations and scattered signal coda. Herglotz–Wiechert inversion is usually applied to first-arrival information alone. However, the information content of first-arrival times may be inadequate to constrain the structure. Fig. 20.6A shows an example of a first-arrival travel-time curve that is exactly compatible with an infinite set of structures with markedly different layering (three are shown). These different structures could be readily differentiated by the secondary travel-time branches in the full wavefield but not by the first arrivals alone. Another aspect of the limited resolution of seismic data is shown in Fig. 20.6B. Four structures are shown that represent different transition structures across a boundary. Each structure would have a different petrological interpretation, but the details of the transition cannot be resolved by signals with wavelengths greater than 1 km. The examples in Fig. 20.6 illustrate the need to use complete wavefield information to study Earth's interior, which has driven the development of many different approaches to imaging the planet at different scales and depths.

Although complete description of the methodologies used in each case must be deferred to more advanced seismology texts that develop the mathematical procedures for constructing seismograms (i.e., solving the forward problem), the seismology basics in Chapters 13–18 provide sufficient background for the diligent reader to appreciate the procedures. Before proceeding, we need to describe additional methodologies used for developing three-dimensional models and for studying attenuation structure in the Earth.

20.2.1 Surface-wave dispersion modeling

We discussed using surface-wave tomography to spatially localize observed dispersion within a region of Earth. We'll assume that the localization is complete and that we are now interested in recovering the average vertical variation in material properties in that region. A surface-wave dispersion curve (group or phase velocity) represents a suite of frequency-dependent averages of the material properties across overlapping depth ranges (controlled by the surface-wave eigenfunctions corresponding to each frequency). The maximum and mean depth of sensitivity increases with period, and separating the overlapping averages into a reasonably well-resolved profile of properties as a function of depth works best for observations

that span a larger bandwidth. When the bandwidth is limited, we must incorporate prior information or accept the fact that we can only constrain averages of the subsurface properties with limited vertical resolution.

For simplicity, we'll assume that the density and P-wave speed can be related to the S-wave speed, β, so that we can set up our problem to estimate only unknown shear-wave velocities. This approximation is not necessary, but it greatly simplifies the quantities that we have to include in our expressions. The relationship between the dispersion and the shear-wave speed is nonlinear but can be approximated with a first-order Taylor Series expansion that relates the observed dispersion to a set of model parameters. Let the current estimate of the shear-wave speed be represented by a model-parameter vector, β, composed of shear-wave speeds in M uniform-thickness layers or at equally-spaced depth nodes (individual layer or node parameters are represented by β_i). We'll represent the N period-dependent speeds using $v^{obs}(T_j)$, where v may represent phase or group velocity, and T_j represents one of the periods for which we have an estimate of the dispersion. Then using the Taylor Series approximation, for the jth observation (period T_j), we have

$$v^{obs}(T_j) - \hat{v}(T_j) \approx \sum_{i=1}^{M} \frac{\partial v(T_j)}{\partial \beta_i}\bigg|_{\hat{\beta}} \Delta\beta_i , \quad (20.25)$$

where $\hat{\beta}$ represents the current model parameter estimates and $\hat{v}(T_j)$ represents the current model's prediction. We call the $\Delta\beta_i$ model-parameter corrections, the partial derivatives are often referred to as sensitivity kernels. Example Rayleigh-wave kernels for an isotropic model are shown in Fig. 20.7.

The goal is to solve equations like (20.25) for quantities $\Delta\beta_i$ such that the misfit to the observations is reduced when the model corrections are applied. If we gather observations for a range of periods that are sensitive to the model

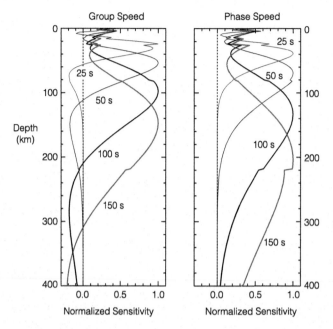

FIGURE 20.7 Sensitivity kernels for fundamental mode Rayleigh waves. (left) Group velocity sensitivity kernels at 25, 50, 100, and 150 s periods. (right) Phase velocity sensitivity kernels at the same periods. The kernels are amplitude normalized, have units of inverse length, and are defined as $\partial v(T_j)/\partial \beta_i$, where v is group or phase speed and β is shear-wave speed. The specific sensitivity depends on the model, so must be recalculated for each model. The strictly positive nature of the phase speed kernels makes their intuitive interpretation easier, but both indicate the group and phase speeds are a weighted average of the material properties (modified from Smith et al., 2004).

parameters, we can estimate model parameter corrections that reduce the misfit and provide new estimates of the model parameters, $\beta_i^{new} = \hat{\beta}_i + \Delta\beta_i$. To solve a set of equations of the form of Eq. (20.25) simultaneously, we gather the partial derivatives (measures of the sensitivity of each observation to each model parameter) into a matrix, G, the misfits, or residuals, into a vector, Δd, and the model parameter corrections into a vector Δm, then

$$\Delta d = G\, \Delta m \,. \tag{20.26}$$

This equation can be solved for Δm to minimize the difference between the left and right sides, producing a better fit to the model. We estimate Δm that when multiplied by the partial derivatives (linearized sensitivities) in G, produces a result similar to the differences between the observations and predictions. Then, when we add that correction to the current model-parameter estimates, the residuals are smaller and the fit to the observations is better. If the relationship between the dispersion and the shear-wave speeds (model parameters) was linear (the partial derivatives would be constant coefficients), we would only require one iteration. Since this is not the case, we must iterate to find the best fitting set of model parameters.

In general, the approach described is that used, but not all works so easily. However, a number of issues invariably arise that complicate the analysis. First, Earth's properties such as shear-wave speed are not simply a function of depth – the dispersion is an average of three dimensional structure that may influence differ-

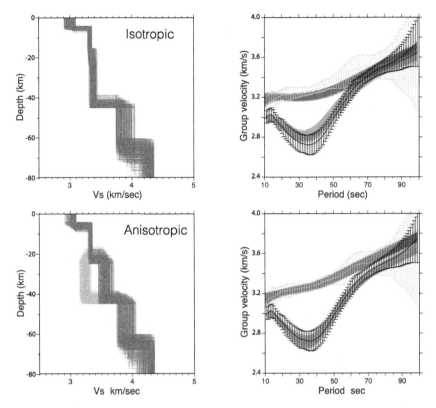

FIGURE 20.8 (Top) Models (left) and corresponding fits (right) for a suite of Rayleigh and Love dispersion curve inversions with a simple (few layers) isotropic parametrization. (Bottom) Models (left) and corresponding fits (right) for a suite of Rayleigh and Love dispersion curve inversions with a simple (few layers) and with radial anisotropy allowed in the middle crust (VSV and VSH are shown with light grey and dark grey lines, respectively). Rayleigh and Love wave observations are shown with black and light grey thin lines and error bars, respectively. Light and dark grey lines show Rayleigh and Love dispersion curves computed for models selected during the inversion (modified from Duret et al., 2010).

ent periods in different ways, which can lead to inconsistency within the observations and a one-dimensional model. Second, the band-limited nature of most observations and measurement error often allow more than one reasonable geological model to match the observations.

Results of analyses of Rayleigh and Love-wave dispersion measured in the 10-to-100 s period band for waves with paths within the Tibetan Plateau are shown in Fig. 20.8. The upper panels correspond to an isotropic analysis, the lower panels correspond to an analysis that allowed mid-crustal radial anisotropy (ver-

tical symmetry axis). Each analysis results in a suite of models that fit the data acceptably well (error bars are shown). Allowing mid-crustal anisotropy produces better fits in the band from about 25–60 s period. The examples illustrate that a range of models fits the data and this is generally true of surface-wave inversions. Allowing a more flexible model parameterization (more layers) would allow even more acceptable models than those shown. Such are the challenges seismologists face when imaging Earth's interior.

Overcoming the challenges that arise during geophysical inversions often means accepting a model close to some *a priori* model developed by others using independent observations or constructed based on analogy with geological characteristics of the region. Such prior models are often used as the initial model in the iterative process. To reduce non-uniqueness we often choose to favor simpler models (such as smooth models) or by acquiring and incorporating complementary data that provide the missing information. For example, the period-dependent polarization of a Rayleigh wave also provides information on the local material properties and can be included into the inversion with the dispersion values. Our introduction by no means illuminates the creativity with which seismologists have circumvented the limitations of observations to constrain the planet's interior on local, regional and global scales. But we must refer the student to more advanced books, reviews, and papers on the approaches.

20.3 Earth structure estimation using seismic amplitudes and waveforms

As an example of the improvement in structural sensitivity offered by waveform and secondary-arrival information, Fig. 20.9 shows synthetic seismogram profiles computed using a high-frequency wavelet in the crust for three models of the crust–mantle boundary. The *sharpness*, or depth distribution, of the velocity increase across the boundary clearly affects the amplitude of both reflected waves near vertical incidence (at close distances) and head waves (P_n) at long distances, which can potentially discriminate between various models. The overall waveform shape can constrain the complexity of the boundary as well.

When possible, the maximum amount of information can be extracted by fitting complete seismic waveforms – the broader the bandwidth

that can be modeled, the better the resolution. As with dispersion, the relationship between seismograms and simple earth-structure parameterizations leads to nonlinear problems. Again, we use a Taylor Series approach to relate the samples in a seismogram to the model parameters. We set up a linearized problem to minimize the sample-by-sample misfit between observed and predicted seismograms by adjusting the model parameters.

20.3.1 P-wave receiver-function modeling

Perhaps the simplest example of waveform modeling is the fit of a teleseismic P-wave receiver function – the response of the crust and upper mantle to a P or S wave from a distant earthquake well described as an incident plane wave from below (e.g., Langston, 1979; Owens et al., 1987; Ammon, 1991). To remove the effects of the source and near source structure, the vertical component of ground motion is deconvolved from the horizontal components, removing the common source and near source structure effects (Langston, 1979). We'll focus on an incident P wave and consider only the first 20–30 s following the direct-wave arrival. In that case the response is composed of the direct P arrival followed by P-to-S converted phases and reverberations within the lithosphere. (See Fig. 20.10.) These waves sample the crust and uppermost mantle in the vicinity (30–60 km) surrounding the stations. The amplitude (frequency dependent) of the conversions and reverberations are most sensitive to the nature (sharp, gradational, etc.) of the largest shear-velocity contrasts in the crust, generally the crust-mantle transition zone and the surface and near surface. The timing of the converted phases and reverberations constrains the vertical travel time between major velocity contrasts within the crust and/or shallow mantle.

The key to good receiver function resolution is a broad bandwidth and observations that span

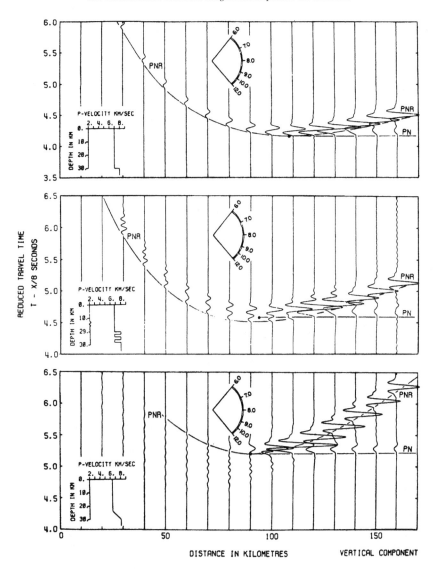

FIGURE 20.9 Seismic waveforms generally carry much more information than seismic travel times (similar to the extra information in waveforms compared with first-motion polarities for source investigations). Synthetic seismogram profiles for three different crust-mantle boundary velocity structures that indicate how waveform information can potentially distinguish between models of the transition. Reflected arrivals are designated *PNR*; head waves are designated *PN*. One can determine the apparent velocities of arrivals in km/s from the angles shown on the circular scales (modified from Braile and Smith, 1975).

a range of horizontal slowness, p (incidence angle) (e.g. Langston, 1979; Julia, 2007). The arrival time and amplitudes of the reverberations vary with p, so the broader the range of p the more in-

formation in the moveout of the phases. Zandt and Ammon (1995) used the relative moveout of the converted phase P_S and the multiples ($PpPms$, $PpSms$ and $PsSms$) to estimate bulk

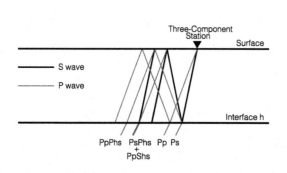
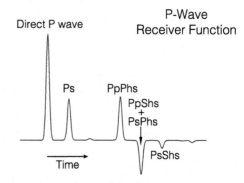

FIGURE 20.10 (left) Cartoon illustration of the principal arrivals observed in a *P*-wave receiver function generated by the wavefield interaction with a subsurface interface and Earth's surface. (right) The corresponding receiver function with the principal arrivals labeled. The most common interface of interest is the crust-mantle boundary (after Owens et al., 1987).

crustal Poisson's ratio for many regions and Zhu and Kanamori (2000) developed a robust approach using stacking to perform the calculation less reliant on travel time measurements. Their $H - \kappa$ stacking approach has become the first step in the analysis of typical, lithosphere-scale *P*-wave receiver functions.

Our focus is on waveform fitting – which requires that we construct a relationship between the observations and the model parameters. For a single station, resolving much detail on laterally varying structure is possible using the observed azimuthal receiver-function variations, but we consider the simpler case where we have a single receiver function to be modeled using a laterally uniform layered structure. As we did with the surface-wave dispersion, we assume that we can relate the density and the *P*-wave speed to the shear-wave speed (the *P*-wave receiver function is actually dominated by shear waves converted from the incident wave but larger on the radial component of motion). Let M be the number of layers included in the model and represent the N time samples in the receiver function using $R^{obs}(t_j)$, where t_j is the time of the jth sample in the waveform. Then using the Taylor Series approximation, for the jth obser-

vation, we have

$$R^{obs}(t_j) - \widehat{R}(t_j) \approx \sum_{i=1}^{M} \left. \frac{\partial R(t_j)}{\partial \beta_i} \right|_{\hat{\beta}} \Delta \beta_i , \quad (20.27)$$

where $\hat{\beta}$ represents the current model parameter estimates and $\widehat{R}(t_j)$ represents the current model's prediction. As before, we call the $\Delta \beta_i$ model-parameter corrections. Collect the sample-by-sample misfit (left-hand side of Eq. (20.27)) into a vector Δd and the model parameter corrections, $\Delta \beta_i$ into a vector Δm, and arrange the partial derivatives into a matrix G, for which each column corresponds to the change in each sample in the receiver function produced by a change in the model parameter. Considering all N data points (the whole sampled waveform) at once, we can combine equations into the familiar matrix form

$$\Delta d = G \, \Delta m . \quad (20.28)$$

As before, these equations can be used to iterate from an initial set of model parameters to a final set in a relatively straightforward and efficient manner.

Fig. 20.11 is an example inversion using data from a temporary deployment in Tibet (one of the first deployments of the PASSCAL program

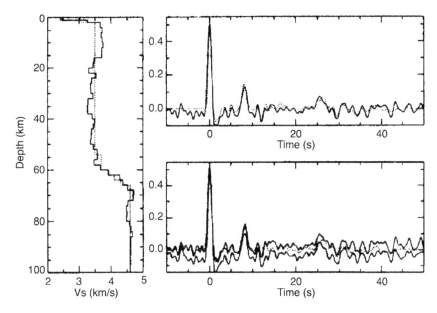

FIGURE 20.11 *P*-wave receiver function inversion results for observations from station WNDO, located in central Tibetan Plateau (the receiver function corresponds to the waves approaching the station from the southeast). The initial (dashed line) and final iteration (solid line) shear velocity models are shown on the left. The fit of final synthetic receiver function (dashed line) to both the mean receiver function (solid line) and the bounds (from Zhu et al., 1993).

of IRIS). Typically, the receiver functions used in an inversion are averages of all those available. Seismologists call the average of waveforms *stacking* and in this instance, stacking is done to reduce the noise in the individual observations and better isolate the signal arising from the Earth. The crust beneath Tibet is very thick, and the converted phase arrives at roughly eight seconds. The multiples are visible at a time of roughly 25–30 seconds. The inversion produces a model with relatively simple structure with a fast upper crust overling a slow mid-crust (which the surface-wave analysis described above suggested may be anisotropic). The crust-mantle boundary is gradational, starting at roughly 60 km depth and finishing at roughly 70 km depth.

As with most geophysical inversions, the approach described above is only a starting point. The observations may not provide unique solutions and more than one geologically reasonable

model may explain the observations. To handle numerical instabilities associated with the large number of layers in the model, smooth solutions are often favored, and that leads to questions about the sharpness of the crust-mantle boundary. Careful analysis often can provide insight into the most important and robust features of the model, but a simple error-bar view is often inadequate. Effort is necessary to explore the range of models that match the observations (e.g. Ammon et al., 1990).

Perhaps the best approach is to include complementary data that can provide the information needed to stabilize the result and reduce the number of reasonable models that can fit more than one data set. In fact, combining surface-wave dispersion with *P*-wave receiver functions (e.g. Julia et al., 2000) has proven quite successful in this regard (Box 20.1). Another approach is to fit waveforms that include multiple seismic arrivals including both body and surface waves.

Box 20.1 Joint receiver function and surface-wave-dispersion analysis

The similarity in the framework used to invert both dispersion observations and receiver functions enable a simultaneous inversion of both data sets for a single set of shear-wave speed model parameters. The receiver functions provide information on the relatively sharp velocity changes within the model (such as the crust-mantle boundary), the surface-wave dispersion provides information on the absolute averages of the wave speed. If one is careful to adjust the relative weights of the two data sets (e.g. Julia et al., 2000), then the inversion produces a more detailed and reliable model than either data set alone. Fig. B20.1.1 is an illustration of the idea.

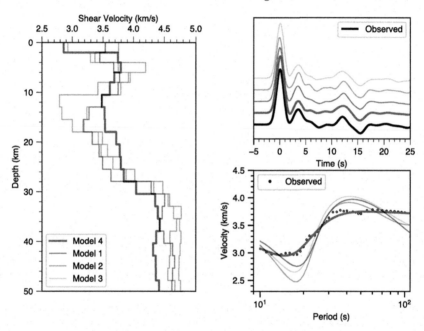

FIGURE B20.1.1 Joint receiver-function and surface-wave analysis of the crustal velocity structure under the US-Array TA station C09A (located near Davenport, Washington). The left panel shows four velocity structures obtained by inversions of the observed receiver function (Model 1–3). The fits to the observed receiver function is shown in the upper-right. The models differ because of different initial parameters in the inversion, but all produce reasonable fits to the receiver function. To better resolve the structure, short-period surface-wave dispersion observations (lower-right box) were inverted with the receiver functions to produce Model 4. The example is shared courtesy of Chengping Chai, and the data are from (Chai et al., 2015).

The joint-analysis model (thick gray line in the figure) fits both data sets well and is more geologically reasonable than the three receiver-function only inversion results. The simultaneous inversion of receiver functions and tomographically localized surface-wave dispersion is now routine and the data are in many ways ideally complementary. The method remains somewhat limited by the use of laterally homogeneous models, but such models are good average models and are essential starting points for three-dimensional analyses.

We refer to these inversions as full waveform inversions, and discuss them in the next section.

20.4 Full seismogram inversion

Seismologists have long sought approaches that allow complete waveforms to be modeled in a single analysis. The ideas have been under development for decades (often called *Full Waveform Inversion* in the exploration-seismic literature). Direct inversion of a complete waveform may not always be the best approach since the weighting of the information in an observed seismogram may favor particular frequency bands, but the idea of full seismogram fitting is very attractive because it has the potential to include all the information available in the seismogram, not simply measurements on a few specific waves or wave groups. More importantly, the approach generalizes to fully three-dimensional modeling, which is the obvious and ultimate goal of seismogram analysis. With the computational power available, at least for intermediate and long periods, effects such as anisotropy, topography of the surface and internal discontinuities that are important for global and/or regional seismology, rotational and gravitational effects, and attenuation can now be included in the forward computations.

Many 3D analyses start with a focus on reconstructing the broad wavenumber features and then iteratively refine the model to include more and more details (here we use the term wavenumber as the spatial analog of the temporal frequency). The fundamental equations do not separate clearly as a function of wavenumber, but experience suggests that the approach can lead to stable and relatively simple models of Earth's interior. Indeed, such an approach was natural for many tomographic models since the number of observations and the number of seismic stations was much smaller than available today (most of Earth's surface remains uninstrumented – the ocean-covered regions). For the investigation of structures spanning a broad range of wave-numbers, we can, and have used the perturbative approaches that we introduce earlier in the chapter.

Imaging broad spatial regions with high-resolution introduces an enormous number of model parameters and much larger dimensions of potential model parameter sets (called the *model space*). In theory, we could set up the problem for three dimensional, global waveform inversion using the perturbation-style approach that we outlined earlier for specific parts of measurements (times, amplitude, dispersion curves, etc.). However, computing sensitivity kernels for the large number of model parameters is prohibitive.

More fruitful approaches rely on the mathematical properties of the equations of motion and use related (adjoint) systems of equations to back-propagate the differences between observed and predicted waveforms into the model to identify regions where the discrepancies may have arisen (e.g. Liu and Gu, 2012). The sensitivity kernels for fully 3D models are computed by combining the results of the forward problem in which signals are propagated from the source throughout the model, and the adjoint problem in which signals are propagated backwards from the station throughout the model. The result is an elegant calculation of complete sensitivity kernels for the seismic wavefield (still based on the model through which the calculation is performed, but including the effects of all waves in the seismogram). Using these sensitivity kernels, an inversion can be performed for the 3D variations in earth structure (e.g. Modrak and Tromp, 2016).

Adjoint waveform tomography is computational demanding because of the 3D nature of the computations, but it is also relatively efficient compared to more traditional methods. The results can show substantial improvement in our ability to match the observed seismic ground motions. An example of the power of adjoint methods from a study of southern Cali-

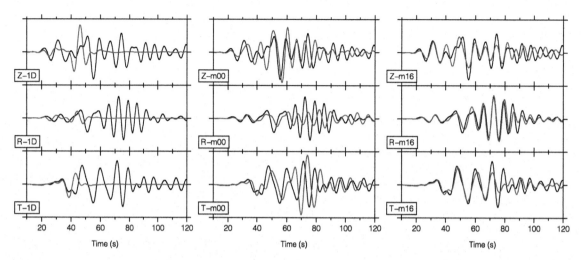

FIGURE 20.12 Observed (black) and predicted (gray) seismograms for the period range 6 to 30 s for an M_W 5.4 earthquake about 14 km beneath Chino Hills, east of the Los Angeles Basin, observed at seismic station STC.CI (Δ 137 km), located within the Ventura basin. Predicted seismograms are shown for a 1D layered model, an initial 3D model (m00), and the final 3D model (m16) for all three components, vertical (Z), radial (R), and transverse (T) (modified from Tape et al., 2009).

fornia by Tape et al. (2009) is shown in Fig. 20.12. Southern California has been investigated for many decades and includes a substantial, well-distributed number of high-quality seismic stations. The comparison shows an example of the dramatic improvement to short-to-intermediate period observations enabled by full-seismogram inversions. The observations were not included in the 3D inversion. The one-dimensional model phase is clearly difficult to match for the paths crossing the complex basin-structures of the Los Angeles area. The amplitudes are not too far off, and the fit to the longer-period components in the Rayleigh waves are not too bad. The Love wave dispersion is substantially underestimated. The initial, 3D model provides a reasonably fit to the amplitudes but still includes some important phase discrepancies. The predictions of the final 3D model are very good with

only minor misfit early in the vertical component Rayleigh waves.

Complete seismogram modeling approaches such as adjoint tomography demonstrate the power of seismic analysis to provide strong constraints on the subsurface geology. High-resolution will always require good seismic station coverage and bandwidth. Even by today's standards, adjoint methods are computationally intense, so they benefit from good starting models derived using more traditional analyses (traveltime analysis, body-wave fitting, surface-wave fitting, etc.). At present, seismology is a relatively data-rich field. The future is bright if seismologists continue to creatively extract the information available in seismograms, continue to invest in technological developments to improve data quality and to increase data quantity, and continue to value the hard work necessary to observe and study earthquakes and the Earth.

Bibliography

Abe, K., 1975. Reliable estimation of the seismic moment of large earthquakes. J. Phys. Earth 23 (4), 381–390. https://doi.org/10.4294/jpe1952.23.381.

Abercrombie, R.E., Ekström, G., 2001. Earthquake slip on oceanic transform faults. Nature 410 (6824), 74.

Akciz, S.O., Ludwig, L.G., Arrowsmith, J.R., Zielke, O., 2010. Century-long average time intervals between earthquake ruptures of the San Andreas fault in the carrizo plain, California. Geology 38 (9), 787–790. https://doi.org/10.1130/G30995.1.

Aki, K., 1968. Seismic displacements near a fault. J. Geophys. Res. 73 (16), 5359–5376. https://doi.org/10.1029/JB073i016p05359. WOS:A1968B642700032.

Aki, K., Richards, P.G., 2009. Quantitative Seismology, 2 ed. Univ. Science Books, Sausalito, Calif. OCLC: 845610339.

Algermissen, S., Perkins, D., Thenhaus, P., Hanson, S., Bender, B., 1990. Probabilistic earthquake acceleration and velocity maps for the United States and Puerto Rico. Report 2120, Reston, VA. https://doi.org/10.3133/mf2120.

Algermissen, S.T., Perkins, D.M., 1976. A probablistic estimate of maximum acceleration in rock in the contiguous United States. Open-File Report 76-416. United States Geological Survey. Series: Open-File Report.

Allen, R., Melgar, D., 2019. Earthquake early warning: advances, scientific challenges, and societal needs. Annu. Rev. Earth and Planet. Sci. 47, 361–388. https://doi.org/10.1146/annurev-earth-053018-060457.

Allen, T.I., Marano, K.D., Earle, P.S., Wald, D.J., 2009. PAGERCAT: a composite earthquake catalog for calibrating global fatality models. Seismol. Res. Lett. 80 (1), 57–62.

Allmann, B.P., Shearer, P.M., 2009. Global variations of stress drop for moderate to large earthquakes: global stress drop variations. J. Geophys. Res., Solid Earth 114 (B1). https://doi.org/10.1029/2008JB005821.

Ammon, C.J., 1991. The isolation of receiver effects from teleseismic p-waveforms. Bull. Seismol. Soc. Am. 81 (6), 2504–2510.

Ammon, C.J., Kanamori, H., Lay, T., 2008. A great earthquake doublet and seismic stress transfer cycle in the central Kuril Islands. Nature 451 (7178), 561–565. https://doi.org/10.1038/nature06521.

Ammon, C.J., Randall, G.E., Zandt, G., 1990. On the nonuniqueness of receiver function inversions. J. Geophys. Res. 95 (B10), 15,303–15,318. https://doi.org/10.1029/JB095iB10p15303.

Ammon, C.J., Kanamori, H., Lay, T., Velasco, A.A., 2006. The 17 July 2006 Java tsunami earthquake. Geophys. Res. Lett. 33 (24). https://doi.org/10.1029/2006GL028005.

Ammon, C.J., Lay, T., Kanamori, H., Cleveland, M., 2011. A rupture model of the 2011 off the Pacific coast of Tohoku earthquake. Earth Planets Space 63 (7), 693–696. https://doi.org/10.5047/eps.2011.05.015.

Anderson, D.L., Minster, B., Cole, D., 1974. The effect of oriented cracks on seismic velocities. J. Geophys. Res. 79 (26), 4011–4015.

Andronicos, C.L., Velasco, A.A., Hurtado, J.M., 2007. Large-scale deformation in the India-Asia collision constrained by earthquakes and topography. Terra Nova 19 (2), 105–119. https://doi.org/10.1111/j.1365-3121.2006.00714.x.

Argus, D.F., Gordon, R.G., DeMets, C., 2011. Geologically current motion of 56 plates relative to the no-net-rotation reference frame: NNR-MORVEL56. Geochem. Geophys. Geosyst. 12 (11). https://doi.org/10.1029/2011GC003751.

Aster, R.C., Borchers, B., Thurber, C.H., 2013. Parameter Estimation and Inverse Problems, 2 ed. Academic Press, Waltham, MA.

Astiz, L., Earle, P., Shearer, P., 1996. Global stacking of broadband seismograms. Seismol. Res. Lett. 67 (4), 8–18. https://doi.org/10.1785/gssrl.67.4.8.

Atwater, B.F., 2005. The Orphan Tsunami of 1700: Japanese Clues to a Parent Earthquake in North America, vol. 1707, 2 ed. U.S. Geological Survey, Seattle, Reston, Va.

Atwater, B.F., Griggs, G.B., 2012. Deep-Sea Turbidites as Guides to Holocene Earthquake History at the Cascadia Subduction Zone—Alternative Views for a Seismic-Hazard Workshop. Open-File Report 2012–1043. Series: Open-File Report.

Atwater, T., 1970. Implications of plate tectonics for the Cenozoic tectonic evolution of western North America. Bull. Geol. Soc. Am., 3513–3536.

Bakun, W.H., Lindh, A.G., 1985. The Parkfield, California earthquake prediction experiment. Science 229 (4714), 619–624.

Bakun, W.H., et al., 2005. Implications for prediction and hazard assessment from the 2004 Parkfield earthquake. Nature 437 (7061), 969–974. https://doi.org/10.1038/nature04067.

Ben-Menahem, A., Singh, S.J., 1981. Seismic Waves and Sources. Springer-Verlag, New York.

Bird, P., 2003. An updated digital model of plate boundaries: Updated model of plate boundaries. Geochem. Geophys. Geosyst. 4 (3). https://doi.org/10.1029/2001GC000252.

Blewitt, G., Kreemer, C., Hammond, W.C., Gazeaux, J., 2016. Midas robust trend estimator for accurate gps station velocities without step detection: Midas trend estimator for GPS velocities. J. Geophys. Res. 121 (3), 2054–2068.

Boettcher, M.S., Jordan, T.H., 2004. Earthquake scaling relations for mid-ocean ridge transform faults. J. Geophys. Res. 109 (B12), 302. https://doi.org/10.1029/2004JB003110.

Bolt, B.A., 1976. Nuclear Explosions and Earthquakes: The Parted Veil. A Series of Books in Geology. W.H. Freeman, San Francisco.

Bolt, B.A., 1982. Inside the Earth: Evidence From Earthquakes. W.H. Freeman, San Francisco.

Bolt, B.A., 2006. Earthquakes, 5 ed. W.H. Freeman, New York. OCLC: ocm63183027.

Bormann, P., Wielandt, E., 2013. Seismic signals and noise. In: Bormann, P. (Ed.), New Manual of Seismological Observatory Practice 2 (NMSOP2). In: Deutsches Geo-ForschungsZentrum GFZ, pp. 1–62.

Borrmann, P., Dewey, J.W., 2014. The new IASPEI standards for determining magnitudes from digital data and their relation to classical magnitudes. In: Bormann, P. (Ed.), New Manual of Seismological Observatory Practice (NMSOP-2). IASPEI, GFZ German Research Centre for Geosciences, Potsdam. http://nmsop.gfz-potsdam.de.

Brace, W., Byerlee, J., 1966. Stick-slip as a mechanism for earthquakes. Science 153 (3739), 990–992. https://doi.org/10.1126/science.153.3739.990. WOS:A19668140200024.

Braile, L.W., Smith, R., 1975. Guide to the interpretation of crustal refraction profiles. Geophys. J. R. Astron. Soc. 40, 145–176.

Bridgman, P.W., 1949. Polymorphic transitions and geologic phenomena. Am. J. Sci. 243A, 90–97.

Brocher, T.M., 2005. Empirical relations between elastic wavespeeds and density in the Earth's crust. Bull. Seismol. Soc. Am. 95 (6), 2081–2092. https://doi.org/10.1785/0120050077.

Brodsky, Emily E., Lay, Thorne, 2014. Recognizing Foreshocks from the 1 April 2014 Chile Earthquake. Science 344 (6185), 700–702. https://doi.org/10.1126/science.1255202.

Brudzinski, M.R., Thurber, C.H., Hacker, B.R., Engdahl, E.R., 2007. Global prevalence of double benioff zones. Science 316 (5830), 1472–1474. https://doi.org/10.1126/science.1139204.

Brune, J.N., 1970. Tectonic stress and spectra of seismic shear waves from earthquakes. J. Geophys. Res. 75 (26), 4997–5009. https://doi.org/10.1029/JB075i026p04997. WOS:A1970H283400009.

Buehler, J.S., Shearer, P.M., 2010. Pn tomography of the western United States using USARRAY. J. Geophys. Res. 115 (B9), 315. https://doi.org/10.1029/2009JB006874.

Bullen, K.E., Bolt, B.A., 1985. An Introduction to the Theory of Seismology, 4 ed. Cambridge University Press, New York.

Butler, R., 2006. Observations of polarized seismoacoustic T waves at and beneath the seafloor in the abyssal Pacific Ocean. J. Acoust. Soc. Am. 120 (6), 3599–3606. https://doi.org/10.1121/1.2354066.

Byerlee, J., 1978. Friction of rocks. Pure Appl. Geophys. 116 (4), 615–626. https://doi.org/10.1007/BF00876528. WOS:A1978FM31500003.

Båth, M., 1979. Introduction to Seismology, vol. bd. 27, 2 ed. Birkhäuser Verlag, Basel, Boston. oCLC: 913653751.

Calvert, A.J., Sawyer, E.W., Davis, W.J., Ludden, J.N., 1995. Archaean subduction inferred from seismic images of a mantle suture in the superior province. Nature 375 (6533), 670–674. https://doi.org/10.1038/375670a0.

Caputo, R., Helly, B., 2008. The use of distinct disciplines to investigate past earthquakes. Tectonophysics 453 (1–4), 7–19. https://doi.org/10.1016/j.tecto.2007.05.007.

Carvajal, M., Cisternas, M., Gubler, A., Catalán, P.A., Winckler, P., Wesson, R.L., 2017. Reexamination of the magnitudes for the 1906 and 1922 Chilean earthquakes using Japanese tsunami amplitudes: implications for source depth constraints. J. Geophys. Res. 122 (1), 4–17.

Chael, E.P., Anderson, D.L., 1982. Global Q estimates from antipodal Rayleigh waves. J. Geophys. Res., Solid Earth 87 (B4), 2840–2850.

Chai, C., Ammon, C.J., Maceira, M., Herrmann, R.B., 2015. Inverting interpolated receiver functions with surface wave dispersion and gravity: application to the western U.S and adjacent Canada and Mexico. Geophys. Res. Lett. 42, 4359–4366. https://doi.org/10.1002/2015GL063733.

Chapman, C.H., 2004. Fundamentals of Seismic Wave Propagation. Cambridge University Press, New York.

Christiansen, L., Hurwitz, S., Saar, M., Ingebritsen, S., Hsieh, P., 2005. Seasonal seismicity at western United States volcanic centers. Earth Planet. Sci. Lett. 240 (2), 307–321. https://doi.org/10.1016/j.epsl.2005.09.012.

Cicerone, R.D., Ebel, J.E., Britton, J., 2009. A systematic compilation of earthquake precursors. Tectonophysics 476 (3–4), 371–396. https://doi.org/10.1016/j.tecto.2009.06.008.

Civiero, C., et al., 2015. Multiple mantle upwellings in the transition zone beneath the northern East-African rift system from relative P-wave travel-time tomography: multiple upwellings beneath East-Africa. Geochem. Geophys. Geosyst. 16 (9), 2949–2968. https://doi.org/10.1002/2015GC005948.

Cleveland, K.M., Ammon, C.J., Kintner, J., 2018. Relocation of light and moderate-magnitude (M4-6) seismicity along the central mid-Atlantic. Geochem. Geophys. Geosyst. 19 (8), 2843–2856. https://doi.org/10.1029/2018GC007573.

Cochran, E.S., 2004. Earth tides can trigger shallow thrust fault earthquakes. Science 306 (5699), 1164–1166. https://doi.org/10.1126/science.1103961.

Collier, J.D., Helffrich, G.R., 2001. The thermal influence of the subducting slab beneath South America from 410 and 660 km discontinuity observations. Geophys. J. Int. 147 (2), 319–329. https://doi.org/10.1046/j.1365-246X.2001.00532.x.

Craig, T.J., Copley, A., Jackson, J., 2014. A reassessment of outer-rise seismicity and its implications for the mechanics of oceanic lithosphere. Geophys. J. Int. 197 (1), 63–89. https://doi.org/10.1093/gji/ggu013.

Crampin, S., 1981. A review of wave motion in anisotropic and cracked elastic-media. Wave Motion 3 (4), 343–391. https://doi.org/10.1016/0165-2125(81)90026-3.

Crotwell, H.P., Owens, T.J., Ritsema, J., 1999. The TauP toolkit: flexible seismic travel-time and ray-path utilities. Seismol. Res. Lett. 70 (2), 154–160. https://doi.org/10.1785/gssrl.70.2.154.

Cuéllar, A., Suárez, G., Espinosa-Aranda, J.M., 2017. Performance evaluation of the earthquake detection and classification algorithm of the seismic alert system of Mexico (SASMEX). Bull. Seismol. Soc. Am. 107 (3), 1451–1463.

Dahlen, F.A., Tromp, J., 1998. Theoretical Global Seismology. Princeton University Press, Princeton, N.J.

Dainty, A.M., Toksöz, M.N., Anderson, K.R., Pines, P.J., Nakamura, Y., Latham, G., 1974. Seismic scattering and shallow structure of the moon in oceanus procellarum. Moon 9, 11–29.

Dalton, C.A., Ekström, G., Dziewoński, A.M., 2008. The global attenuation structure of the upper mantle. J. Geophys. Res. B09, 303. https://doi.org/10.1029/2007JB005429.

Debayle, E., Kennett, B.L.N., 2000. The Australian continental upper mantle: structure and deformation inferred from surface waves. J. Geophys. Res. 105 (B11), 25,423–25,450. https://doi.org/10.1029/2000JB900212.

Dell, S., Gajewski, D., 2011. Common-reflection-surface-based workflow for diffraction imaging. Geophysics 76 (5), S187–S195. https://doi.org/10.1190/geo2010-0229.1.

DeMets, C., Gordon, R.G., Argus, D.F., 2010. Geologically current plate motions. Geophys. J. Int. 181 (1), 1–80. https://doi.org/10.1111/j.1365-246X.2009.04491.x.

Denolle, M.A., Shearer, P.M., 2016. New perspectives on self-similarity for shallow thrust earthquakes. J. Geophys. Res. 121 (9), 6533–6565.

Dewey, J.W., 1972. Seismicity and tectonics of western Venezuela. Bull. Seismol. Soc. Am. 62 (6), 1711.

Doornbos, D.J., 1989. Seismic diffraction. In: James, D.E. (Ed.), The Encyclopedia of Solid Earth Geophysics. Springer US, Boston, MA, pp. 1018–1024.

Douglas, A., 1967. Joint epicentre determination. Nature 215 (5096), 47–48. https://doi.org/10.1038/215047a0.

Douglas, A., 2013. Forensic Seismology and Nuclear Test Bans. Cambridge University Press, Cambridge.

Dragert, H., 2001. A silent slip event on the deeper Cascadia subduction interface. Science 292 (5521), 1525–1528. https://doi.org/10.1126/science.1060152.

Dragert, H., Wang, K., Rogers, G., 2004. Geodetic and seismic signatures of episodic tremor and slip in the northern Cascadia subduction zone. Earth Planets Space 56 (12), 1143–1150.

Duret, F., Shapiro, N.M., Cao, Z., Levin, V., Molnar, P., S, R., 2010. Surface wave dispersion across Tibet: direct evidence for radial anisotropy in the crust. Geophys. Res. Lett. 37 (16). https://doi.org/10.1029/2010GL043811.

Dziewonski, A., Chou, T.-A., Woodhouse, J., 1981. Determination of earthquake source parameters from waveform data for studies of global and regional seismicity. J. Geophys. Res. 86 (B4), 2825–2852. https://doi.org/10.1029/JB086iB04p02825. Cited by 1583.

Dziewonski, A.M., Anderson, D.L., 1981. Preliminary reference Earth model. Phys. Earth Planet. Inter. 25 (4), 297–356.

Eaton, D.W., Darbyshire, F., Evans, R.L., Grütter, H., Jones, A.G., Yuan, X., 2009. The elusive lithosphere–asthenosphere boundary (lab) beneath cratons. Lithos 109 (1–2), 1–22. https://doi.org/10.1016/j.lithos.2008.05.009.

Eissler, H., Kanamori, H., 1987. A single-force model for the 1975 Kalapana, Hawaii, earthquake. J. Geophys. Res. 92, 4827–4836.

Ekström, G., Dziewoński, A.M., 1998. The unique anisotropy of the Pacific upper mantle. Nature 394 (6689), 168–172. https://doi.org/10.1038/28148.

Ekström, G., Nettles, M., Abers, G.A., 2003. Glacial earthquakes. Science 302 (5645), 622–624.

Ekström, G., Nettles, M., Dziewoński, A., 2012. The global CMT project 2004–2010: centroid-moment tensors for 13,017 earthquakes. Phys. Earth Planet. Inter. 200–201, 1–9. https://doi.org/10.1016/j.pepi.2012.04.002.

Eliseevnin, V.A., 1965. Analysis of waves propagating in an inhomogeneous medium. Sov. Phys. Acoust. 10, 242.

Ellsworth, W.L., 2013. Injection-induced earthquakes. Science 341 (6142), 1225,942. https://doi.org/10.1126/science.1225942.

Evans, M.S., Kendall, J.-M., Willemann, R.J., 2006. Automated SKS splitting and upper-mantle anisotropy beneath Canadian seismic stations. Geophys. J. Int. 165 (3), 931–942.

Ewing, W.M., Jardetzky, W.S., Press, F., 1957. Elastic Waves in Layered Media. McGraw-Hill, New York.

Feng, G., Hetland, E.A., Ding, X., Li, Z., Zhang, L., 2010. Coseismic fault slip of the 2008 M_W 7.9 Wenchuan earthquake estimated from InSAR and GPS measurements:

Wenchuan coseismic fault slip. Geophys. Res. Lett. 37 (1). https://doi.org/10.1029/2009GL041213.

Field, E.H., et al., 2014. Uniform California earthquake rupture forecast, version 3 (UCERF3)–the time-independent model. Bull. Seismol. Soc. Am. 104 (3), 1122–1180. https://doi.org/10.1785/0120130164.

Frankel, A., 1995. Mapping seismic hazard in the central and eastern United States. Seismol. Res. Lett. 66 (4), 8–21.

Frankel, Arthur, Mueller, Charles, Barnhard, Theodore, Perkins, David, Leyendecker, E.V., 1996. National Seismic-Hazard Maps: Documentation. Open-File Report 96-532. United States Geological Survey.

Freed, A.M., 2005. Earthquake triggering by static, dynamic, and postseismic stress transfer. Annu. Rev. Earth Planet. Sci. 33 (1), 335–367. https://doi.org/10.1146/annurev.earth.33.092203.122505.

French, S., Lekic, V., Romanowicz, B., 2013. Waveform tomography reveals channeled flow at the base of the oceanic asthenosphere. Science 342 (6155), 227–230. https://doi.org/10.1126/science.1241514.

Fukao, Y., Obayashi, M., 2013. Subducted slabs stagnant above, penetrating through, and trapped below the 660 km discontinuity: subducted slabs in the transition zone. J. Geophys. Res. 118 (11), 5920–5938. https://doi.org/10.1002/2013JB010466.

Furlong, K.P., Lay, T., Ammon, C.J., 2009. A great earthquake rupture across a rapidly evolving three-plate boundary. Science 324 (5924), 226–229. https://doi.org/10.1126/science.1167476.

Garnero, E., Helmberger, D., Engen, G., 1988. Lateral variations near the core-mantle boundary. Geophys. Res. Lett. 15 (6), 609–612.

Gaherty, J.B., Jordan, T.H., 1995. Lehmann discontinuity as the base of an anisotropic layer beneath continents. Science 268 (5216), 1468–1471. https://doi.org/10.1126/science.268.5216.1468.

Gaherty, J.B., Lay, T., 1992. Investigation of laterally heterogeneous shear velocity structure in D″ beneath Eurasia. J. Geophys. Res., Solid Earth 97 (B1), 417–435.

Garth, T., Rietbrock, A., 2014. Order of magnitude increase in subducted H_2O due to hydrated normal faults within the Wadati-benioff zone. Geology 42 (3), 207–210. https://doi.org/10.1130/G34730.1.

Geiger, L., 1912. Probability method for the determination of earthquake epicenters from the arrival time only. Bull. St. Louis Univ. 8 (1), 56–71.

Gilbert, F., Dziewonski, A.M., 1975. An application of normal mode theory to the retrieval of structural parameters and source mechanisms from seismic spectra. Philos. Trans. R. Soc. Lond. Ser. A 278, 187–269.

Gubbins, D., 1990. Seismology and Plate Tectonics. Cambridge University Press, Cambridge.

Gung, Y., Romanowicz, B., 2004. Q tomography of the upper mantle using three-component long-period waveforms. Geophys. J. Int. 157 (2), 813–830. https://doi.org/10.1111/j.1365-246X.2004.02265.x.

Gutenberg, B., Richter, C.F., 1956. Magnitude and energy of earthquakes. Ann. Geophys. 9 (1), 1–15. https://doi.org/10.4401/ag-5590.

Hall, T.R., Nixon, C.W., Keir, D., Burton, P.W., Ayele, A., 2018. Earthquake clustering and energy release of the African–Arabian rift system. Bull. Seismol. Soc. Am. 108 (1), 155–162. https://doi.org/10.1785/0120160343.

Harjes, H.P., Henge, M., Stork, B., Seidi, C., Kind, R., 1980. Grf Array Documentation. Tech. Rep. Seismologiscles Zentrolobservatorium, Grafenberg.

Harmon, N., Forsyth, D.W., Weeraratne, D.S., Yang, Y., Webb, S.C., 2011. Mantle heterogeneity and off axis volcanism on young Pacific lithosphere. Earth Planet. Sci. Lett. 311 (3–4), 306–315. https://doi.org/10.1016/j.epsl.2011.09.038.

Hartzell, S.H., Heaton, T.H., 1983. Inversion of strong ground motion and teleseismic waveform data for the fault rupture history of the 1979 Imperial Valley, California, earthquake. Bull. Seismol. Soc. Am. 73 (6A), 1553–1583.

Hartzell, S.H., Heaton, T.H., 1985. Teleseismic time functions for large, shallow subduction zone earthquakes. Bull. Seismol. Soc. Am. 75, 965–1004.

Haskell, N.A., 1964. Total energy and energy spectra density of elastic waves from propagating faults. Bull. Seismol. Soc. Am. 54, 1811–1841.

Havskov, J., Alguacil, G., 2004. Instrumentation in Earthquake Seismology, vol. 358. Springer International Publishing. https://doi.org/10.1007/978-3-319-21314-9.

Hayes, G.P., 2017. The finite, kinematic rupture properties of great-sized earthquakes since 1990. Earth Planet. Sci. Lett. 468, 94–100. https://doi.org/10.1016/j.epsl.2017.04.003.

Heaton, T., 1990. Evidence for and implications of self-healing pulses of slip in earthquake rupture. Phys. Earth Planet. Inter. 64, 1–20.

Henry, C., Das, S., 2002. Aftershock zones of large shallow earthquakes: fault dimensions, aftershock area expansion and scaling relations. Geophys. J. Int. 147, 272–293. https://doi.org/10.1046/j.1365-246X.2001.00522.x.

Herrin, E., et al., 1968. 1968 seismological tables for P phases. Bull. Seismol. Soc. Am. 58, 1193–1241.

Herrmann, R., Park, S., Wang, C., 1981. The Denver earthquakes of 1967–1968. Bull. Seismol. Soc. Am. 71, 731–745.

Herrmann, R.B., 2013. Computer programs in seismology: an evolving tool for instruction and research. Seismol. Res. Lett. 84 (6), 1081–1088. https://doi.org/10.1785/0220110096.

Hetényi, G., Le Roux-Mallouf, R., Berthet, T., Cattin, R., Cauzzi, C., Phuntsho, K., Grolimund, R., 2016. Joint approach combining damage and paleoseismology observations constrains the 1714 A.D. Bhutan earthquake at magnitude 8 ± 0.5: Mind the Gap: The 1714 Bhutan Earthquake. Geophys. Res. Lett. 43 (20), 10,695–10,702. https://doi.org/10.1002/2016GL071033.

Hough, S.E., 2010. Predicting the Unpredictable: The Tumultuous Science of Earthquake Prediction. Princeton University Press, Princeton.

Hough, S.E., Armbruster, J.G., Seeber, L., Hough, J.F., 2000. On the modified Mercalli intensities and magnitudes of the 1811–1812 new Madrid earthquakes. J. Geophys. Res. 105 (B10), 23,839–23,864. https://doi.org/10.1029/2000JB900110.

Hubbard, J., Shaw, J.H., 2009. Uplift of the Longmen Shan and Tibetan Plateau and the 2008 wenchuan (M = 7.9) earthquake. Nature 458 (7235), 194–197. https://doi.org/10.1038/nature07837.

Hudson, J.A., 1980. The Excitation and Propagation of Elastic Waves. Cambridge Monographs on Mechanics and Applied Mathematics. Cambridge Univ. Pr, Cambridge. OCLC: 833040553.

Hussain, E., Wright, T.J., Walters, R.J., Bekaert, D.P.S., Lloyd, R., Hooper, A., 2018. Constant strain accumulation rate between major earthquakes on the North Anatolian fault. Nat. Commun. 9 (1), 1392. https://doi.org/10.1038/s41467-018-03739-2.

Ide, S., 2015. 4.09 – slip inversion. In: Schubert, G. (Ed.), Treatise on Geophysics (Second Edition), second. Elsevier, Oxford, pp. 215–241.

Imamura, A., 1937. Theoretical and Applied Seismology. Maruzen Co, Tokyo. 358 pp.

Isacks, B., Oliver, J., Sykes, L.R., 1968. Seismology and the new global tectonics. J. Geophys. Res. 73, 5855–5899.

Isacks, B.L., Molnar, P., 1971. Distribution of stresses in the descending lithosphere from a global survey of focal mechanism solutions of mantle earthquakes. Rev. Geophys. Space Phys. 9, 103–174.

ISC, 2016. International seismological centre on-line earthquake bulletin. https://doi.org/10.31905/D808B830.

Iyer, H., 1989. Seismic tomography. In: James, D.E. (Ed.), The Encyclopedia of Solid Earth Geophysics. Springer US, Boston, MA, pp. 1133–1151.

Jeffreys, H., Bullen, K.E., 1940a. Seismological Tables. British association for the advancement of science, Gray Milne Trust.

Jeffreys, H., Bullen, K.E., 1940b. Seismological Tables, the British Association for the Advancement of Science, OCLC Number: 559568850.

Ji, C., Wald, D., Helmberger, D., 2002. Source description of the 1999 Hector Mine, California, earthquake, part I: wavelet domain inversion theory and resolution analysis. Bull. Seismol. Soc. Am. 92 (4), 1192–1207. https://doi.org/10.1785/0120000916.

Ji, C., Larson, K.M., Tan, Y., Hudnut, K.W., Choi, K.H., 2004. Slip history of the 2003 San Simeon earthquake constrained by combining 1-Hz gps, strong motion, and teleseismic data. Geophys. Res. Lett. 31 (17). https://doi.org/10.1029/2004GL020448.

Julia, J., 2007. Constraining velocity and density contrasts across the crust-mantle boundary with receiver function amplitudes. Geophys. J. Int. 171 (1), 286–301. https://doi.org/10.1111/j.1365-2966.2007.03502.x.

Julia, J., Ammon, C.J., Herrmann, R.B., Correig, A.M., 2000. Joint inversion of receiver function and surface wave dispersion observations. Geophys. J. Int. 143 (1), 99–112. https://doi.org/10.1046/j.1365-246x.2000.00217.x.

Julian, B.R., 1970. Ray tracing in arbitrarily heterogeneous media. Tech. Rep. 1970-45. Lincoln Laboratory – MIT.

Kanamori, H., 1970. Synthesis of long-period surface waves and its application to earthquake source studies – Kurile Islands earthquake of October 13, 1963. J. Geophys. Res. 75, 5011–5027.

Kanamori, H., 1974. A new view of earthquakes. In: Kanamori, H., Boschi, E. (Eds.), Physics of the Earth (a Modern View of the Earth). Physical Society of Japan, Maruzen, Tokyo, pp. 261–282.

Kanamori, Hiroo, 1977. The energy release in great earthquakes. J. Geophys. Res. 82 (20), 2981–2987.

Kanamori, H., 1986. Rupture process of subduction-zone earthquakes. Annu. Rev. Earth Planet. Sci. 14, 293–322.

Kanamori, H., Given, J.W., 1983. Lamb pulse observed in nature. Geophys. Res. Lett. 10, 373–376.

Kanamori, H., Stewart, G.S., 1979. A slow earthquake. Phys. Earth Planet. Inter. 18, 167–175.

Kanamori, H., Mori, J., Hauksson, E., Heaton, T.H., Hutton, L.K., Jones, L.M., 1993. Determination of earthquake energy-release and m_l using TerraScope. Bull. Seismol. Soc. Am. 83 (2), 330–346.

Kasahara, K., 1981. Earthquake Mechanics. Cambridge Earth Science Series. Cambridge University Press, New York. Cambridge [Eng.].

Kato, A., Igarashi, T., 2012. Regional extent of the large coseismic slip zone of the 2011 M_W 9.0 tohoku-oki earthquake delineated by on-fault aftershocks. Geophys. Res. Lett. 39 (15). https://doi.org/10.1029/2012GL052220.

Kato, A., et al., 2012. Propagation of slow slip leading up to the 2011 M_W 9.0 tohoku-oki earthquake. Science 335 (6069), 705–708.

Kawasaki, I., 1989. Seismic anisotropy in the Earth. In: James, D.E. (Ed.), The Encyclopedia of Solid-Earth Geophysics. Van Nostrand-Reinhold, New York, pp. 994–1005.

Kennett, B., 2009. Seismic Wave Propagation in Stratified Media. Australian National University. OCLC: 1135541146.

Kennett, B.L.N., 2001. The Seismic Wavefield. Cambridge University Press, Cambridge, New York. OCLC: ocm47983842.

Kennett, B.L.N., 2005. Seismological Tables: ak135. Research School of Earth Sciences, the Australian National University, Canberra, AU.

Kennett, B.L.N., Engdahl, E.R., 1991. Traveltimes for global earthquake location and phase identification. Geophys. J. Int. 105 (2), 429–465. https://doi.org/10.1111/j.1365-246X.1991.tb06724.x.

Kikuchi, M., Kanamori, H., 1991. Inversion of complex body waves–iii. Bull. Seismol. Soc. Am. 81 (6), 2335–2350.

Kim, J., 1989. Complex seismic sources and time-dependent moment tensor inversion. Ph.D. thesis. University of Arizona.

Kind, R., Yuan, X., Mechie, J., Sodoudi, F., 2015. Structure of the upper mantle in the North-western and central United States from USARRAY s-receiver functions. Solid Earth 6 (3), 957–970. https://doi.org/10.5194/se-6-957-2015.

Kostrov, B., 1974. Seismic moment and energy of earthquakes, and seismic flow of rock. Izv. - Acad. Sci. USSR, Phys. Solid Earth, Engl. Transl. 1, 23–40.

Kostrov, B.V., 1966. Unsteady propagation of longitudinal shear cracks. J. Appl. Mech. 30, 1241–1248.

Kostrov, B.V., Das, S., 1988. Principles of Earthquake Source Mechanics, Cambridge Monographs on Mechanics and Applied Mathematics. Cambridge University Press, Cambridge, England.

Kuge, K., Kawakatsu, H., 1990. Analysis of a deep "non double couple" earthquake using very broadband data. Geophys. Res. Lett. 17 (3), 227–230. https://doi.org/10.1029/GL017i003p00227.

Kustowski, B., Ekström, G., Dziewoński, A.M., 2008. Anisotropic shear-wave velocity structure of the Earth's mantle: a global model. J. Geophys. Res. B06, 306. https://doi.org/10.1029/2007JB005169.

La Rocca, M., Galluzzo, D., Malone, S., McCausland, W., Saccorotti, G., Del Pezzo, E., 2008. Testing small-aperture array analysis on well-located earthquakes, and application to the location of deep tremor. Bull. Seismol. Soc. Am. 98 (2), 620–635.

Lamb, H., 1904. On the propagation of tremors over the surface of an elastic solid. Philos. Trans. R. Soc. Lond. Ser. A 203, 1–42.

Langston, C.A., 1979. Structure under Mount Ranier, Washington, inferred from teleseismic body waves. J. Geophys. Res. 84 (B9), 4749–4762. https://doi.org/10.1029/JB084iB09p04749.

Langston, C.A., 1981. Source inversion of seismic waveforms – the Koyna, India, earthquakes of 13 September 1967. Bull. Seismol. Soc. Am. 71 (1), 1–24.

Larson, K.M., 2013. A new way to detect volcanic plumes: GPS PLUMES. Geophys. Res. Lett. 40 (11), 2657–2660. https://doi.org/10.1002/grl.50556.

Laske, G., Widmer-Schnidrig, R., 2015. Theory and observations: normal mode and surface wave observations. In: Schubert, G. (Ed.), Treatise on Geophysics, 2 ed. Elsevier, Oxford, pp. 117–167.

Lawn, B.R., 1993. Fracture of Brittle Solids, 2nd ed. Cambridge University Press, Cambridge, New York.

Lay, T., 2015. Treatise on geophysics. In: Schubert, G. (Ed.), 1.22 - Deep Earth Structure: Lower Mantle and D″, Second Edition. Elsevier, pp. 683–723.

Lay, T., Helmberger, D.V., 1983. A lower mantle S-wave triplication and the shear velocity structure of D″. Geophys. J. R. Astron. Soc. 75 (3), 799–837.

Lay, T., Kanamori, H., 1981. An asperity model of large earthquake sequences. Maurice Ewing Ser. 4, 579–592.

Lay, T., Ammon, C.J., Kanamori, H., Yamazaki, Y., Cheung, K.F., Hutko, A.R., 2011. The 25 October 2010 Mentawi tsunami earthquake (M_W 7.8) and the tsunami hazard presented by shallow megathrust ruptures. Geophys. Res. Lett. 38. https://doi.org/10.1029/2010GL046552.

Lay, T., Ye, L., Kanamori, H., Satake, K., 2018. Constraining the dip of shallow, shallowly dipping thrust events using long-period Love wave radiation patterns: applications to the 25 October 2010 Mentawai Indonesia, and 4 May 2018 Hawaii Island Earthquakes. Geophys. Res. Lett. 45 (16). https://doi.org/10.1029/2018GL080042, 10,342–10,349.

LeFevre, L.V., Helmberger, D.V., 1989. Upper mantle P velocity structure of the Canadian shield. J. Geophys. Res. 94, 749–765.

Leonard, M., 2010. Earthquake fault scaling: self-consistent relating of rupture length, width, average displacement, and moment release. Bull. Seismol. Soc. Am. 100 (5), 1971–1988.

Lin, J., Stein, R.S., 2004. Stress triggering in thrust and subduction earthquakes and stress interaction between the southern San Andreas and nearby thrust and strike-slip faults: Stress triggering and fault interaction. J. Geophys. Res. 109 (B2). https://doi.org/10.1029/2003JB002607.

Liu, Q., Gu, Y.J., 2012. Seismic imaging: from classical to adjoint tomography. Tectonophysics 566–567 (C), 31–66.

Lockner, D.A., Morrow, C., Moore, D., Hickman, S., 2011. Low strength of deep San Andreas fault gouge from SAFOD core. Nature 472 (7341), 82–85. https://doi.org/10.1038/nature09927.

Long, M.D., Becker, T.W., 2010. Mantle dynamics and seismic anisotropy. Earth Planet. Sci. Lett. 297 (3–4), 341–354. https://doi.org/10.1016/j.epsl.2010.06.036.

Love, A.E.H., 1990. A Treatise on the Mathematical Theory of Elasticity, 4 ed. Dover Classics of Science and Mathematics. Dover Publications, New York. OCLC: 257990638.

Maceira, M., et al., 2017. Trends in nuclear explosion monitoring research & development – a physics perspective. U.S. Department of Energy (DOE) Report IA-UR-17–21274. https://doi.org/10.2172/1355758.

Madariaga, R., 1976. Dynamics of an expanding circular fault. Bull. Seismol. Soc. Am. 66 (3), 639–666.

Malagnini, L., Dreger, D.S., Bürgmann, R., Munafò, I., Sebastiani, G., 2019. Modulation of seismic attenuation at Parkfield, before and after the 2004 m 6 earthquake. J. Geophys. Res., Solid Earth 124 (6), 5836–5853.

Malvern, L.E., 1969. Introduction to the Mechanics of a Continuous Medium, Prentice-Hall Series in Engineering of the Physical Sciences. Prentice-Hall, Englewood Cliffs, NJ.

Marone, C., 1998. Laboratory-derived friction laws and their application to seismic faulting. Annu. Rev. Earth Planet. Sci. 26 (1), 643–696.

Massonnet, D., Feigl, K., Rossi, M., Adragna, F., 1994. Radar interferometric mapping of deformation in the year after the landers earthquake. Nature 369 (6477), 227–230. https://doi.org/10.1038/369227a0.

Masters, G., Ritzwoller, M., 1988. Low frequency seismology and three-dimensional structure – observational aspects. In: Vlaar, N.J., Nolet, G., Wortel, M.J.R., Cloetingh, S.A.P.L. (Eds.), Mathematical Geophysics: a Survey of Recent Developments in Seismology and Geodynamics. Reidel Publ., Dordrecht, the Netherlands, pp. 1–30.

Mayeda, K., 2003. Stable and transportable regional magnitudes based on coda-derived moment-rate spectra. Bull. Seismol. Soc. Am. 93 (1), 224–239. https://doi.org/10.1785/0120020020.

McCalpin, J.P., 2005. Late quaternary activity of the Pajarito fault, Rio Grande rift of northern New Mexico, USA. Tectonophysics 408 (1–4), 213–236. https://doi.org/10.1016/j.tecto.2005.05.038.

Meier, M.-A., Ampuero, J.P., Heaton, T.H., 2017. The hidden simplicity of subduction megathrust earthquakes. Science 357 (6357), 1277–1281. https://doi.org/10.1126/science.aan5643.

Meng, L., Zhang, A., Yagi, Y., 2016. Improving back projection imaging with a novel physics-based aftershock calibration approach: a case study of the 2015 Gorkha earthquake. Geophys. Res. Lett. 43 (2), 628–636. https://doi.org/10.1002/2015GL067034.

Menke, W., 2018. Geophysical Data Analysis: Discrete Inverse Theory, 4 ed. Elsevier/Academic Press, Amsterdam London Cambridge, MA, p. 1040698782. OCLC: 1040698782.

Merrifield, M.A., 2005. Tide gauge observations of the Indian Ocean tsunami, December 26, 2004. Geophys. Res. Lett. 32 (9), L09,603. https://doi.org/10.1029/2005GL022610.

Milne, J., 1903. Seismometry and geite. Nature 67, 538–539.

Modrak, R., Tromp, J., 2016. Seismic waveform inversion best practices: regional, global and exploration test cases. Geophys. J. Int. 206 (3), 1864–1889.

Montelli, R., 2004. Finite-frequency Tomography Reveals a variety of plumes in the mantle. Science 303 (5656), 338–343. https://doi.org/10.1126/science.1092485.

Montelli, R., Nolet, G., Dahlen, F.A., Masters, G., 2006. A catalogue of deep mantle plumes: New results from finite-frequency tomography. Geochem. Geophys. Geosyst. 7 (11). https://doi.org/10.1029/2006GC001248.

Mooney, W., 2015. Crust and lithospheric structure – global crustal structure. In: Treatise on Geophysics. Elsevier, pp. 339–390.

Moore, D.E., Rymer, M.J., 2007. Talc-bearing serpentinite and the creeping section of the San Andreas fault. Nature 448 (7155), 795–797. https://doi.org/10.1038/nature06064.

Mori, N., Takahashi, T., Yasuda, T., Yanagisawa, H., 2011. Survey of 2011 Tohoku earthquake tsunami inundation and run-up. Geophys. Res. Lett. 38 (7). https://doi.org/10.1029/2011GL049210.

Moulik, P., Ekström, G., 2014. An anisotropic shear velocity model of the Earth's mantle using normal modes, body waves, surface waves and long-period waveforms. Geophys. J. Int. 199 (3), 1713–1738.

Müller, G., Kind, R., 1976. Observed and computed seismogram sections for the whole world. Geophys. J. R. Astron. Soc. 44, 699–716.

Müller, R.D., Sdrolias, M., Gaina, C., Roest, W.R., 2008. Age, spreading rates, and spreading asymmetry of the world's ocean crust: digital models of the world's ocean crust. Geochem. Geophys. Geosyst. 9 (4). https://doi.org/10.1029/2007GC001743.

Nelson, P.L., Grand, S.P., 2018. Lower-mantle plume beneath the Yellowstone hotspot revealed by core waves. Nat. Geosci. 11 (4), 280–284.

Nettles, M., Wallace, T.C., Beck, S.L., 1999. The March 25, 1998 Antarctic plate earthquake. Geophys. Res. Lett. 26 (14), 2097–2100. https://doi.org/10.1029/1999GL900387.

National Geophysical Data Center, NOAA, 2018. Significant earthquake database. https://doi.org/10.7289/V5TD9V7K.

Obara, K., 2002. Nonvolcanic deep tremor associated with subduction in southwest Japan. Science 296 (5573), 1679–1681.

Officer, C.B., 1974. Introduction to Theoretical Geophysics. Springer-Verlag, New York Inc.

Ogata, Yosihiko, 1988. Statistical models for earthquake occurrences and residual analysis for point processes. J. Am. Stat. Assoc. 83 (401), 9–27.

Okada, A., Nagata, T., 1953. Land deformation of the neighborhood of muroto point after the Nankaido great earthquake in 1946. Bull. Earthq. Res. Inst. Univ. Tokyo 31, 169–177.

Okal, E.A., 1982. Mode-wave equivalence and other asymptotic problems in tsunami theory. Phys. Earth Planet. Inter. 30 (1), 1–11. https://doi.org/10.1016/0031-9201(82)90123-6.

Okal, E.A., 1988. Seismic parameters controlling far-field tsunami amplitudes: a review. Nat. Hazards 1 (1), 67–96.

Oldham, R.D., 1900. On the propagation of earthquake motion to great distances. Philos. Trans. R. Soc. Lond., Ser. A, Contain. Pap. Math. Phys. Character 194, 135–174.

Olive, J.-A., Behn, M.D., Ito, G., Buck, W.R., Escartin, J., Howell, S., 2015. Sensitivity of seafloor bathymetry to climate-driven fluctuations in mid-ocean ridge magma supply. Science 350 (6258), 310–313. https://doi.org/10.1126/science.aad0715.

Oliver, J., 1962. A summary of observed seismic surface wave dispersion. Bull. Seismol. Soc. Am. 52 (1), 81–86.

Onur, T., Muir-Wood, R., 2014. An extended global catalogue of 'giant' ($M_W \geq 8.8$) earthquakes. In: Proceedings of the 10th National Conference in Earthquake Engineering. Earthquake Engineering Research Institute, Anchorage, AK.

Owens, T.J., Taylor, S.R., Zandt, G., 1987. Crustal structure at regional seismic test network stations determined from inversion of broadband teleseismic P waveforms. Bull. Seismol. Soc. Am. 77 (2), 631–662.

Oxburgh, E.R., Turcotte, D.L., 1974. Membrane tectonics and the East African rift. Earth Planet. Sci. Lett. 22, 133–140.

Pacheco, J.F., Sykes, L.R., 1992. Seismic moment catalog of large shallow earthquakes, 1900 to 1989. Bull. Seismol. Soc. Am. 82 (3), 1306.

Park, J., 1988. Free-oscillation coupling theory. In: Vlaar, N.J., Nolet, G., Wortel, M.J.R., Cloetingh, S.A.P.L. (Eds.), Mathematical Geophysics: a Survey of Recent Developments in Seismology and Geodynamics. Reidel Publ., Dordrecht, the Netherlands, pp. 31–52.

Park, J., Yu, Y., 1992. Anisotropy and coupled free oscillations: simplified models and surface wave observations. Geophys. J. Int. 110 (3), 401–420. https://doi.org/10.1111/j.1365-246X.1992.tb02082.x.

Pattiaratchi, C.B., Wijeratne, E.M. Sarath, 2009. Tide gauge observations of 2004–2007 Indian Ocean tsunamis from Sri Lanka and western Australia. Pure Appl. Geophys. 166 (1–2), 233–258. https://doi.org/10.1007/s00024-008-0434-5.

Pavlis, G., Booker, J., 1983. Progressive multiple event location (pmel). Bull. Seismol. Soc. Am. 73 (6A), 1753–1777.

Peterson, J.R., 1993. Observations and modeling of seismic background noise. Tech. Rep. 2331-1258. US Geological Survey. https://doi.org/10.3133/ofr93322.

Press, W.H., Teukolsky, S.A., Vetterling, W.T., Flannery, B.P., 2007. Numerical Recipes: The Art of Scientific Computing, 3 ed. Cambridge University Press, USA.

Pujol, J., 1988. Comments on the joint determination of hypocenters and station corrections. Bull. Seismol. Soc. Am. 78 (3), 1179.

Reasenberg, Paul A., Jones, Lucile M., 1989. Earthquake hazard after a mainshock in California. Science 243 (4895), 1173–1176.

Richter, C.F., 1958. Elementary Seismology. W.H. Freeman, San Francisco. OCLC: 1117182733.

Ritsema, J., Deuss, A., van Heijst, H.J., Woodhouse, J.H., 2011. S40rts: a degree-40 shear-velocity model for the mantle from new Rayleigh wave dispersion, teleseismic traveltime and normal-mode splitting function measurements. Geophys. J. Int. 184 (3), 1223–1236. https://doi.org/10.1111/j.1365-246X.2010.04884.x.

Ritzwoller, M.H., Lin, F.-C., Shen, W., 2011. Ambient noise tomography with a large seismic array. C. R. Géosci. 343 (8–9), 558–570. https://doi.org/10.1016/j.crte.2011.03.007.

Romanowicz, B., Mitchell, B., 2015. Deep Earth structure: Q of the Earth from crust to core. In: Treatise on Geophysics. Elsevier, pp. 789–827.

Rost, S., 2002. Array seismology: methods and applications. Rev. Geophys. 40 (3), 305–327.

Rowley, D.B., 2019. Oceanic axial depth and age-depth distribution of oceanic lithosphere: comparison of magnetic anomaly picks versus age-grid models. Lithosphere 11 (1), 21–43. https://doi.org/10.1130/L1027.1.

Ruegg, J., Rudloff, A., Vigny, C., Madariaga, R., de Chabalier, J., Campos, J., Kausel, E., Barrientos, S., Dimitrov, D., 2009. Interseismic strain accumulation measured by gps in the seismic gap between Constitución and Concepción in Chile. Phys. Earth Planet. Inter. 175 (1), 78–85. https://doi.org/10.1016/j.pepi.2008.02.015.

Ruegg, J.C., 2002. Interseismic strain accumulation in South central Chile from gps measurements, 1996–1999. Geophys. Res. Lett. 29 (11), 1517. https://doi.org/10.1029/2001GL013438.

Ruiz-Constán, A., Galindo-Zaldívar, J., Pedrera, A., Célérier, B., Marín-Lechado, C., 2011. Stress distribution at the transition from subduction to continental collision (northwestern and central Betic cordillera): from subduction to continental collision. Geochem. Geophys. Geosyst. 12 (12). https://doi.org/10.1029/2011GC003824.

Sandvol, E., Hearn, T., 1994. Bootstrapping shear-wave splitting errors. Bull. Seismol. Soc. Am. 84 (6), 1971–1977.

Satake, K., 1989. Inversion of tsunami waveforms for the estimation of heterogeneous fault motion of large submarine earthquakes: the 1968 Tokachi-Oki and 1983 Japan Sea earthquakes. J. Geophys. Res. 94, 5627–5636.

Satake, K., 2003. Fault slip and seismic moment of the 1700 Cascadia earthquake inferred from Japanese tsunami descriptions. J. Geophys. Res. 108 (B11), 2535. https://doi.org/10.1029/2003JB002521.

Satake, K., 2015. Tsunamis. In: Treatise on Geophysics. Elsevier, pp. 477–504.

Satake, K., Kanamori, H., 1991. Use of tsunami waveforms for earthquake source study. Nat. Hazards 4 (2–3), 193–208.

Satake, K., Shimazaki, K., Tsuji, Y., Ueda, K., 1996. Time and size of a giant earthquake in Cascadia inferred from Japanese tsunami records of January 1700. Nature 379 (6562), 246–249. https://doi.org/10.1038/379246a0.

Satake, K., Fujii, Y., Harada, T., Namegaya, Y., 2013. Time and space distribution of coseismic slip of the 2011 Tohoku earthquake as inferred from tsunami waveform data. Bull. Seismol. Soc. Am. 103 (2B), 1473–1492. https://doi.org/10.1785/0120120122.

Savage, J.C., 1972. Relation of corner frequency to fault dimensions. J. Geophys. Res. 77 (20), 3788–3795. https://doi.org/10.1029/JB077i020p03788.

Scherbaum, F., 2001. Of Poles and Zeros: Fundamentals of Digital Seismology. Modern Approaches in Geophysics, vol. 15. Springer Science+Business Media, Dordrecht. OCLC: 845146557.

Scholz, C.H., 2019. The Mechanics of Earthquakes and Faulting, 3 ed. Cambridge University Press, Cambridge, New York, NY Port Melbourne. OCLC: 1042157802.

Scholz, C.H., Campos, J., 2012. The seismic coupling of subduction zones revisited: seismic coupling. J. Geophys. Res. 117 (B5). https://doi.org/10.1029/2011JB009003.

Schwartz, S., Lay, T., 1985. Comparison of long-period surface wave amplitude and phase anomalies for two models of global lateral heterogeneity. Geophys. Res. Lett. 12, 231–234.

Schwartz, S.Y., Dewey, J.W., Lay, T., 1989. Influence of fault plane heterogeneity on the seismic behavior in the southern Kurile Islands arc. J. Geophys. Res. 94 (B5), 5637–5649.

Segall, P., 2010. Earthquake and Volcano Deformation. Princeton University Press.

Selby, N.D., 2002. The Q structure of the upper mantle: constraints from Rayleigh wave amplitudes. J. Geophys. Res. 107 (B5), 2097. https://doi.org/10.1029/2001JB000257.

Services, I.D.M., 2012. SWS-DBs shear-wave splitting databases. https://doi.org/10.17611/DP/SWS.1.

Shearer, P.M., 1991. Constraints on upper mantle discontinuities from observations of long-period reflected and converted phases. J. Geophys. Res. 96, 147–182.

Shearer, P.M., 2019. Introduction to Seismology, 3 ed. Cambridge University Press, Cambridge, New York, NY.

Shimazaki, K., Nakata, T., 1980. Time-predictable recurrence model for large earthquakes. Geophys. Res. Lett. 7, 279–282.

Silver, P., Chan, W.W., 1991. Shear wave splitting and subcontinental mantle deformation. J. Geophys. Res. 96, 429–454.

Smith, D., Ritzwoller, M., Shapiro, N., 2004. Stratification of anisotropy in the Pacific upper mantle. J. Geophys. Res. 109, 11,309. https://doi.org/10.1029/2004JB003200.

Song, X., Helmberger, D.V., 1992. Velocity structure near the inner core boundary from waveform modeling. J. Geophys. Res. 97, 6573–6586.

Song, X., Richards, P.G., 1996. Seismological evidence for differential rotation of the Earth's inner core. Nature 382 (6588), 221–224. https://doi.org/10.1038/382221a0.

Spudich, P., Orcutt, J., 1980. A new look at the velocity structure of the crust. Rev. Geophys. Space Phys. 18, 627–645.

Stamps, D., Flesch, L., Calais, E., Ghosh, A., 2014. Current kinematics and dynamics of Africa and the East African rift system. J. Geophys. Res. 119 (6), 5161–5186. https://doi.org/10.1002/2013JB010717.

Stein, R.S., Barrientos, S.E., 1985. Planar high-angle faulting in the basin and range: geodetic analysis of the 1983 Borah Peak, Idaho, earthquake. J. Geophys. Res. 90 (B13), 1355–1366.

Stein, S., Wysession, M., 2003. An Introduction to Seismology, Earthquakes, and Earth Structure. Blackwell Publishers, Malden, MA.

Stixrude, L., Lithgow-Bertelloni, C., 2005. Thermodynamics of mantle minerals – I. physical properties. Geophys. J. Int. 162 (2), 610–632. https://doi.org/10.1111/j.1365-246X.2005.02642.x.

Storchak, D.A., Di Giacomo, D., Bondár, I., Engdahl, E.R., Harris, J., Lee, W.H.K., Villaseñor, A., Bormann, P., 2013. Public release of the ISC-GEM global instrumental earthquake catalogue (1900–2009). Seismol. Res. Lett. 84 (5), 810. https://doi.org/10.1785/0220130034.

Stump, B.W., 2002. Characterization of mining explosions at regional distances: implications with the international monitoring system. Rev. Geophys. 40 (4), 1011. https://doi.org/10.1029/1998RG000048.

Sykes, L.R., 1978. Intraplate seismicity, reactivation of pre-existing zones of weakness, alkaline magmatism and other tectonism postdating continental fragmentation. Rev. Geophys. Space Phys. 16, 621–688.

Takemura, S., Furumura, T., Maeda, T., 2015. Scattering of high-frequency seismic waves caused by irregular surface topography and small-scale velocity inhomogeneity. Geophys. J. Int. 201 (1), 459–474. https://doi.org/10.1093/gji/ggv038.

Tanioka, Y., Sataka, K., 1996. Fault parameters of the 1896 Sanriku tsunami earthquake estimated from tsunami numerical modeling. Geophys. Res. Lett. 23 (13), 1549–1552.

Tape, C., Liu, Q., Maggi, A., Tromp, J., 2009. Adjoint tomography of the southern California crust. Science 325 (5943), 988–992. https://doi.org/10.1126/science.1175298.

Tarantola, A., 2005. Inverse Problem Theory and Methods for Model Parameter Estimation. Society for Industrial and Applied Mathematics, Philadelphia, PA. oCLC: ocm56672375.

Taylor, S.R., Hartse, H.E., 1998. A procedure for estimation of source and propagation amplitude corrections for regional seismic discriminants. J. Geophys. Res., Solid Earth 103 (B2), 2781–2789. https://doi.org/10.1029/97JB03292.

Tkalčić, H., Young, M., Bodin, T., Ngo, S., Sambridge, M., 2013. The shuffling rotation of the Earth's inner core revealed by earthquake doublets. Nat. Geosci. 6 (6), 497–502. https://doi.org/10.1038/ngeo1813.

Toda, S., Stein, Ross S., Richard-Dinger, Keith, Bozkurt, Serkan B., 2005. Forecasting the evolution of seismicity in southern California: animations built on earthquake stress transfer. J. Geophys. Res. 110 (B5), B05S16. https://doi.org/10.1029/2004JB003415.

Tsuboi, C., 1959. Earthquake energy, earthquake volumes, aftershock area, and strength of the Earth's crust. J. Phys. Earth 4 (2), 63–66.

Udías, A., Buforn, E., 2018. Principles of Seismology, 2 ed. Cambridge University Press, Cambridge.

Udías, A., Madariaga, R., Buforn, E., 2014. Source Mechanisms of Earthquakes: Theory and Practice. Cambridge University Press, Cambridge.

Utsu, T., 2002. A list of deadly earthquakes in the world: 1500–2000. In: Lee, W.K., Kanamori, H., Jennings, P.C., Kisslinger, C. (Eds.), International Handbook of Earthquake Engineering and Seismology, vol. 81. Academic Press, Amsterdam and Boston, pp. 691–717.

Utsu, T., Seki, A., 1954. A relation between the area of aftershock region and the energy of main shock. J. Seism. Soc. Jpn. 7, 233–240.

van der Hilst, R.D., de Hoop, M.V., Wang, P., Shim, S.-H., Ma, P., Tenorio, L., 2007. Seismostratigraphy and thermal structure of Earth's core-mantle boundary region. Science 315 (5820), 1813–1817. https://doi.org/10.1126/science.1137867.

Veith, K.F., Clawson, G.E., 1972. Magnitude from short-period P-wave data. Bull. Seismol. Soc. Am. 62 (2), 435–452.

Velasco, A., Ammon, C.J., Farrell, J., Pankow, K., 2004. Rupture directivity of the 3 November 2002 Denali fault earthquake determined from surface waves. Bull. Seismol. Soc. Am. 94 (6B,S), S293–S299. https://doi.org/10.1785/0120040624.

Velasco, A.A., Ammon, C.J., Lay, T., 1994. Empirical green function deconvolution of broadband surface waves: rupture directivity of the 1992 Landers, California ($M_W = 7.3$), earthquake. Bull. Seismol. Soc. Am. 84 (3), 735–750.

Velasco, A.A., Hernandez, S., Parsons, T., Pankow, K., 2008. Global ubiquity of dynamic earthquake triggering. Nat. Geosci. 1 (6), 375–379. https://doi.org/10.1038/ngeo204.

Vigny, C., et al., 2011. The 2010 M_W 8.8 Maule megathrust earthquake of central Chile, monitored by GPS. Science 332 (6036), 1417–1421. https://doi.org/10.1126/science.1204132.

Wakita, H., 1981. Precursory changes in groundwater prior to the 1978 Izu-Oshima-Kinkai earthquake. Maurice Ewing Ser. 4, 527–532.

Waldhauser, F., 2000. A double-difference earthquake location algorithm: method and application to the Northern Hayward Fault, California. Bull. Seismol. Soc. Am. 90 (6), 1353–1368. https://doi.org/10.1785/0120000006.

Wallace, T.C., Helmberger, D.V., Engen, G.R., 1983. Evidence of tectonic release from underground nuclear explosions in long-period P waves. Bull. Seismol. Soc. Am. 73, 593–613.

Walter, W.R., Mayeda, K., Gok, R., Hofstetter, A., 2006. The scaling of seismic energy with moment: simple models compared with observations. In: Abercrombie, R., McGarr, A., Toro, G.D., Kanamori, H. (Eds.), Earthquakes: Radiated Energy and the Physics of Faulting. American Geophysical, Union, Washington D.C., pp. 25–41.

Wang, K., Chen, Q.-F., Sun, S., Wang, A., 2006. Predicting the 1975 Haicheng earthquake. Bull. Seismol. Soc. Am. 96 (3), 757–795. https://doi.org/10.1785/0120050191.

Ward, S., 1989. Tsunamis. In: James, D.E. (Ed.), The Encyclopedia of Solid Earth Geophysics. Springer US, Boston, MA, pp. 1279–1292.

Weber, M., Davis, J.P., 1990. Evidence of a laterally variable lower mantle structure from P- and S-waves. Geophys. J. Int. 102 (1), 231–255.

Wells, D.L., Coppersmith, K.J., 1994. New empirical relationships among magnitude, rupture length, rupture width, rupture area, and surface displacement. Bull. Seismol. Soc. Am. 84 (4), 974–1002.

West, J.D., Fouch, M.J., Roth, J.B., Elkins-Tanton, L.T., 2009. Vertical mantle flow associated with a lithospheric drip beneath the Great Basin. Nat. Geosci. 2 (6), 439–444. https://doi.org/10.1038/ngeo526.

Wilson, J.T., 1965. A new class of faults and their bearing on continental drift. Nature 207, 343–347.

Wu, R.S., 1989. Seismic wave scattering. In: James, D.E. (Ed.), The Encyclopedia of Solid Earth Geophysics. Springer US, Boston, MA, pp. 1166–1187.

Yamashita, T., 1976. On the dynamical process of fault motion in the presence of friction and inhomogeneous initial stress part I: rupture propagation. J. Phys. Earth 24 (4), 417–444. https://doi.org/10.4294/jpe1952.24.417.

Yang, B.B., Liu, K.H., Dahm, H.H., Gao, S.S., 2016. A uniform database of teleseismic shear-wave splitting measurements for the western and central United States: December 2014 update. Seismol. Res. Lett. 87 (2A), 295–300. https://doi.org/10.1785/0220150213.

Ye, L., Lay, T., Bai, Y., Cheung, K.F., Kanamori, H., 2017. The 2017 m_w 8.2 Chiapas, Mexico, earthquake: energetic slab detachment. Geophys. Res. Lett. 44 (23), 11–824.

Young, C.J., Lay, T., 1987a. Evidence for a shear velocity discontinuity in the lower mantle beneath India and the Indian Ocean. Phys. Earth Planet. Inter. 49 (1), 37–53.

Young, C.J., Lay, T., 1987b. The core-mantle boundary. Annu. Rev. Earth Planet. Sci. 15, 25–46.

Young, C.J., Lay, T., 1990. Multiple phase analysis of the shear velocity structure in the D″ region beneath Alaska. J. Geophys. Res. 95, 17385–17402.

Yue, H., Lay, T., Koper, K.D., 2012. En échelon and orthogonal fault ruptures of the 11 April 2012 great intraplate earthquakes. Nature 490 (7419), 245–249. https://doi.org/10.1038/nature11492.

Yue, H., et al., 2014. Rupture process of the 2010 M_W 7.8 Mentawai tsunami earthquake from joint inversion of near-field hr-GPS and teleseismic body wave recordings constrained by tsunami observations. J. Geophys. Res., Solid Earth 119 (7), 5574–5593. https://doi.org/10.1002/2014JB011082.

Zandt, G., Ammon, C.J., 1995. Continental-crust composition constrained by measurements of crustal poissons ratio. Nature 374 (6518), 152–154. https://doi.org/10.1038/374152a0.

Zhan, Z., Kanamori, H., Tsai, V.C., Helmberger, D.V., Wei, S., 2014. Rupture complexity of the 1994 Bolivia and 2013 sea of okhotsk deep earthquakes. Earth Planet. Sci. Lett. 385, 89–96. https://doi.org/10.1016/j.epsl.2013.10.028.

Zhu, L., Kanamori, H., 2000. Moho depth variation in southern California from teleseismic receiver functions. J. Geophys. Res. 105, 2969–2980.

Zhu, L., Rivera, L.A., 2002. A note on the dynamic and static displacements from a point source in multilayered media. Geophys. J. Int. 148 (3), 619–627. https://doi.org/10.1046/j.1365-246X.2002.01610.x.

Zhu, L.-P., Zeng, R.-S., Wu, F., Owens, T., Randall, G., 1993. Preliminary study of crust-upper mantle structure of the Tibetan Plateau by using broadband teleseismic body waveforms. Acta Seismol. Sin. 6, 305–316. https://doi.org/10.1007/BF02650943.

Zoback, m., Hickman, S., Ellsworth, W., 2010. Scientific drilling into the San Andreas fault zone. Eos, Trans. -Am. Geophys. Union 91 (22), 197–199. https://doi.org/10.1029/2010EO220001.

Zürn, W., Ferreira, A.M.G., Widmer-Schnidrig, R., Lentas, K., Rivera, L., Clévédé, E., 2015. High-quality lowest-frequency normal mode strain observations at the black forest observatory (SW-Germany) and comparison with horizontal broad-band seismometer data and synthetics. Geophys. J. Int. 203 (3), 1786–1803.

Index

Printed in the United States
By Bookmasters